METHODS IN MOLECULAR BIOLOGY

For further volumes:
http://www.springer.com/series/7651

DNA Methylation Protocols

Third Edition

Edited by

Jörg Tost

Laboratory for Epigenetics and Environment
Centre National de Recherche en Génomique Humaine,
CEA—Institut de Biologie Francois Jacob
Evry, France

Humana Press

Editor
Jörg Tost
Laboratory for Epigenetics and Environment
Centre National de Recherche en Génomique Humaine
CEA—Institut de Biologie Francois Jacob
Evry, France

ISSN 1064-3745 ISSN 1940-6029 (electronic)
Methods in Molecular Biology
ISBN 978-1-4939-7479-5 ISBN 978-1-4939-7481-8 (eBook)
https://doi.org/10.1007/978-1-4939-7481-8

Library of Congress Control Number: 2017960297

Cover caption: Merged immunostaining of a mouse brain for 5-methylcytosine, single-stranded DNA and total DNA – Image taken from Chapter 4 of this volume: "Antibody Based Detection of Global Nuclear DNA Methylation in Cells, Tissue Sections and Mammalian Embryos" by Sari Pennings and Nathalie Beaujean et al.

Printed on acid-free paper

This Humana Press imprint is published by Springer Nature
The registered company is Springer Science+Business Media, LLC
The registered company address is: 233 Spring Street, New York, NY 10013, U.S.A.

Preface

It is my great pleasure to introduce the third edition of *DNA Methylation: Methods and Protocols* to the scientific and medical community.

DNA methylation at cytosines forms one of the multiple layers of epigenetic mechanisms controlling and modulating gene expression through chromatin structure. It closely interacts with histone modifications and chromatin remodeling complexes to form the local genomic and higher-order chromatin landscape. DNA methylation is essential for proper mammalian development, crucial for imprinting, and plays a role in maintaining genomic stability. DNA methylation patterns are susceptible to change in response to environmental stimuli whereby the epigenome seems to be most vulnerable during early life. Changes of DNA methylation patterns have been widely studied in several diseases, especially cancer, where interest has focused on biomarkers for early detection of cancer development, accurate diagnosis, and response to treatment, but have also been shown to occur in many other complex diseases. Recent advances in epigenome engineering technologies allow now for the large-scale assessment of the functional relevance of DNA methylation. As a stable nucleic acid-based modification that is technically easy to handle and which can be analyzed with great reproducibility and accuracy by different laboratories, DNA methylation is a promising biomarker for many applications.

Although by no means complete, this edition of *DNA methylation: Methods and Protocols* gives a comprehensive overview of available technologies together with a detailed step-by-step protocol for all experimental procedures required to successfully perform DNA methylation analysis. As the degree of difficulty associated with performing these molecular assays is nowadays generally outweighed by the challenges associated with their analysis, many chapters have therefore also included detailed protocols for the analysis of data down to the details of the command lines to be used.

This is the third edition of the DNA methylation protocols, and the field has—again—dramatically changed since the second edition, which I edited six years ago. DNA methylation technologies and our knowledge of DNA methylation patterns have been advancing at a breathtaking pace over the past few years, and many important discoveries have been made due to technological advances. Most of the techniques described in the second edition have been further optimized and/or replaced by novel easier, refined, and/or more quantitative and resolutive technologies, many of which can now be (and have already been) performed on large cohorts. I have therefore again entirely remodeled and expanded the contents of this book, which in this third edition consists now of 35 chapters. Only three chapters have been retained from the last edition, and these have been completely rewritten by the authors to accommodate the changes and improvements made in the last years (Pyrosequencing, MethyLight, and the HELP assays). The selection of different technologies presented in this volume enables the analysis of the global DNA methylation content as well as precise quantification of DNA methylation levels on single CpG positions. Methods for the high-resolution analysis of CpG positions within a target region identified by one of the multiple available genome-wide technologies are presented, and many of the technologies included in a recent international study comparing the reliability and performance of locus-specific DNA methylation assays (Bock et al., Nat. Biotech. 2016) are included in this volume. While the second edition contained a large number of microarray-based analysis methods, they have been completely superseded by sequencing-based approaches. With sequencing costs further decreasing, whole genome bisulfite sequencing (WGBS) will soon become affordable in large cohorts and several protocols with different approaches such as MethylC-seq, PBAT, or tagmentation-based WGBS are presented in this volume. Special focus has been given to protocols that enable to start with low amounts of starting material allowing to analyze the cell-type or developmental-stage-specific DNA methylation patterns. Genome-wide sequencing approaches will lead to a large number of potential candidate genes, and therefore this volume now includes a novel section detailing methods analyzing a large number of target regions in parallel and the interested reader can choose from a variety of capture or amplification-based technologies.

For the first time, a chapter is dedicated to the experimental design of DNA methylation studies as well as the statistical considerations on their analysis, while a second chapter focuses on analysis strategies for the highly popular Illumina BeadArrays with the different parameters to be considered during data analysis and interpretation. Analysis of cell-free DNA at both the genetic and epigenetic level has attracted recently much interest for the diagnosis and clinical management of cancer and other complex diseases, and this volume provides a protocol for the DNA methylation analysis in body fluids. This collection also contains a protocol for DNA methylation analysis from archived Guthrie cards potentially enabling longitudinal studies and identifying DNA methylation changes prior to disease onset. With the increasing interest in analyzing several epigenetic modifications on the same samples, this volume contains also two chapters describing the integrated analysis of DNA methylation patterns in immunoprecipitated DNA and thus associated with specific histone modifications or transcription factor profiles as well as with nucleosome occupancy. As in the last edition, an introductory chapter summarizes briefly the different biological processes DNA methylation is involved in, the changes that have been described in diseases as well as potential clinical applications for which DNA methylation analysis might prove to be important allowing scientists to catch up with the current level of knowledge and learn about recent trends.

This volume of the Methods in Molecular Biology series contains widely used methods such as Pyrosequencing and methylation-specific PCR as well as protocols for special applications such as the hairpin bisulfite sequencing which allows for the assessment of the symmetry of DNA methylation in specific regions. Furthermore, research of the last years has shown that cytosine methylation is not the only DNA modification and that especially 5-hydroxymethylation is not only an intermediate in oxidative DNA demethylation, but constitutes a distinct layer in the complex process of epigenetic regulation with its own distribution and regulatory functions. As bisulfite-based technologies are not able to distinguish between these two modifications, special protocols have been developed and several of them are presented in the last section of this book.

This book is addressed to postdoctoral investigators and research scientists that are implicated in the different aspects of genetics and cellular and molecular biology as well as to clinicians involved in diagnostics or choice of treatment of diseases that have an epigenetic component, which nowadays means most biological and clinical questions. The presentation in this volume is equally suited for laboratories that already have a great deal of expertise in a certain technology to analyze DNA methylation, but might want to obtain other or complementary data using an orthogonal technique, and for genetics/genomics/biology groups that want to initiate research in this exciting area and want to identify the method best suited to answer their question. Notes and tips from experts of the different methods will enable a rapid implementation of the different protocols in the laboratory and avoid time-consuming and cost-intensive mistakes. With the tools and protocols available, our knowledge and understanding of DNA methylation will continue to increase rapidly with new groundbreaking and exciting discoveries to come, and this book will contribute to spreading of the "savoir faire" to analyze DNA methylation.

I am indebted to all the authors for their hard work and outstanding contributions to this third edition of *DNA Methylation Protocols*. It was a pleasure to work with them on this project. I hope that the protocols described in detail in this volume will help to accelerate the analysis and description of the "methylome" of different species, enhance our understanding of the molecular processes that determine the genomic DNA methylation landscape as well as provide robust technologies for the clinical implementation of DNA methylation-based biomarkers.

Evry, France ***Jörg Tost***

Contents

Contributors

KAROLINA A. ABERG • *Center for Biomarker Research and Precision Medicine, Virginia Commonwealth University, Richmond, VA, USA*

WILLIAM P. ACCOMANDO JR • *Department of Epidemiology, Harvard School of Public Health, Boston, MA, USA; Department of Obstetrics, Gynecology, and Reproductive Biology; Obstetrics and Gynecology Epidemiology Center, Brigham and Women's Hospital, Boston, MA, USA*

MARTYNA ADAMOWICZ • *MRC Clinical Sciences Centre, Faculty of Medicine, Imperial College London, London, UK*

ANDREW ADEY • *Department of Molecular and Medical Genetics, Oregon Health and Science University, Portland, OR, USA; The Knight Cardiovascular Institute, Oregon Health and Science University, Portland, OR, USA*

TIMOTHY J. AITMAN • *MRC Clinical Sciences Centre, Faculty of Medicine, Imperial College London, London, UK; Institute of Genetics and Molecular Medicine, University of Edinburgh, Edinburgh, UK*

GITTE BRINCH ANDERSEN • *Department of Biomedicine, Aarhus University, Aarhus, Denmark; Laboratory for Epigenetics and Environment, Centre National de Recherche en Génomique Humaine, CEA—Institut de Biologie Francois Jacob, Evry, France*

LILIT ATANESYAN • *MRC-Holland, Amsterdam, The Netherlands*

NATHALIE BEAUJEAN • *INRA, UMR1198 Biologie du Développement et Reproduction, Jouy-en-Josas, France; Univ Lyon, Université Claude Bernard Lyon 1, Inserm, INRA, Stem Cell and Brain Research Institute U1208, USC1361, Bron, France*

GABRIEL BEIKIRCHER • *Molecular Diagnostics Unit, AIT–Austrian Institute of Technology GmbH, Vienna, Austria*

ALEXANDRA M. BINDER • *Department of Epidemiology, Fielding School of Public Health, University of California, Los Angeles, CA, USA*

GLENN S. BROWN • *CSIRO Food and Nutrition Flagship, North Ryde, NSW, Australia*

DANIEL L. BURGESS • *Roche Sequencing Solutions, Madison, WI, USA*

FLORENCE BUSATO • *Laboratory for Epigenetics and Environment, Centre National de Recherche en Génomique Humaine, CEA–Institut de Biologie Francois Jacob, Evry, France*

MARIA JESÚS CAÑAL • *Area de Fisiología Vegetal, Dpto. Biología de Organismos y Sistemas, Universidad de Oviedo, Oviedo, Spain*

MIHAELA CAMPAN • *Clinical Sciences Building, University Hospital, Keck School of Medicine, University of Southern California, Los Angeles, CA, USA*

MATTEO CESARONI • *Fels Institute for Cancer Research and Molecular Biology, Temple University School of Medicine, Philadelphia, PA, USA*

ROBIN F. CHAN • *Center for Biomarker Research and Precision Medicine, Virginia Commonwealth University, Richmond, VA, USA*

TAMIR CHANDRA • *Epigenetics ISP, The Babraham Institute, Cambridge, UK; The Wellcome Trust Sanger Institute, Cambridge, UK; The Institute of Genetics and Molecular Medicine, University of Edinburgh, Edinburgh, UK*

SUSAN J. CLARK • *Epigenetics Research Laboratory, Genomics and Epigenetics Division, Garvan Institute of Medical Research, Sydney, NSW, Australia; St. Vincent's Clinical School, University of NSW, Sydney, NSW, Australia*

Christian Daviaud • *Laboratory for Epigenetics and Environment, Centre National de Recherche en Génomique Humaine, CEA-Institut de Biologie Francois Jacob, Evry, France*
Emelyne Dejeux • *Laboratory for Epigenetics and Environment, Centre National de Recherche en Génomique Humaine, CEA–Institut de Biologie Francois Jacob, Evry, France*
Jean-François Deleuze • *Laboratory for Epigenetics and Environment, Centre National de Recherche en Génomique Humaine, CEA-Institut de Biologie Francois Jacob, Evry, France; Laboratory for Bioinformatics, Fondation Jean Dausset – CEPH, Paris, France*
Dinh Diep • *Department of Bioengineering, University of California at San Diego, La Jolla, CA, USA*
Dimo Dietrich • *Institute of Pathology, University Hospital Bonn (UKB), Bonn, Germany*
Paul J. van Diest • *Department of Pathology, University Medical Centre Utrecht, Utrecht, The Netherlands*
Roland Eils • *Theoretical Bioinformatics, German Cancer Research Center (DKFZ), Heidelberg, Germany*
Hafida El abdalaoui • *Laboratory for Epigenetics and Environment, Centre National de Recherche en Génomique Humaine, CEA–Institut de Biologie Francois Jacob, Evry, France*
Mary Jo Fackler • *Breast and Ovarian Cancer Program, Sidney Kimmel Comprehensive Cancer Center, Johns Hopkins University School of Medicine, Baltimore, MD, USA*
Agustin F. Fernandez • *Cancer Epigenetics Laboratory, Institute of Oncology of Asturias (IUOPA), HUCA, Universidad de Oviedo, Oviedo, Spain; Fundación para la Investigación Biosanitaria de Asturias (FINBA), Grupo de Epigenética del Cáncer y Nanomedicina, Instituto de Investigación Sanitaria del Principado de Asturias (IISPA), Hospital Universitario Central de Asturias (HUCA), Oviedo, Spain*
Mario F. Fraga • *Cancer Epigenetics Laboratory, Institute of Oncology of Asturias (IUOPA), HUCA, Universidad de Oviedo, Oviedo, Spain; Department of Immunology and Oncology, National Center for Biotechnology, CNB-CSIC, Madrid, Spain; Fundación para la Investigación Biosanitaria de Asturias (FINBA), Grupo de Epigenética del Cáncer y Nanomedicina, Instituto de Investigación Sanitaria del Principado de Asturias (IISPA), Hospital Universitario Central de Asturias (HUCA), Oviedo, Spain*
Akram Ghantous • *Epigenetics Group, International Agency for Research on Cancer (IARC), Lyon, France*
Pascal Giehr • *Department of Biological Sciences, Genetics/Epigenetics, Saarland University, Saarbrücken, Saarland, Germany*
John M. Greally • *Department of Genetics, Albert Einstein College of Medicine of Yeshiva University, Bronx, NY, USA*
Anthony R. Green • *Cambridge Institute for Medical Research, University of Cambridge, Cambridge, UK; Department of Haematology, University of Cambridge, Cambridge, UK; Stem Cell Institute, University of Cambridge, Cambridge, UK; Department of Haematology, Addenbrooke's Hospital, Cambridge, UK*
Ivo Glynne Gut • *Biomedical Genomics Group, Centro Nacional de Analisis Genomico, CNAG-CRG, Center for Genomic Regulation, Barcelona Institute for Science and Technology, Barcelona, Spain*
Nur Annies Abd Hadi • *Centre for Cardiovascular Science, Queen's Medical Research Institute, University of Edinburgh, Edinburgh, UK*
Dali Han • *Department of Chemistry and Institute for Biophysical Dynamics, The University of Chicago, Chicago, IL, USA; Howard Hughes Medical Institute, The University of Chicago, Chicago, IL, USA*
Lise Lotte Hansen • *Institute of Biomedicine, Aarhus University, Aarhus C, Denmark*

TRAVER HART • *Donnelly Centre and Banting and Best Department of Medical Research, University of Toronto, Toronto, ON, Canada*
CHUAN HE • *Department of Chemistry and Institute for Biophysical Dynamics, The University of Chicago, Chicago, IL, USA; Howard Hughes Medical Institute, The University of Chicago, Chicago, IL, USA*
STEVEN R. HEAD • *Next Generation Sequencing Core, The Scripps Research Institute, La Jolla, CA, USA*
RUI HENRIQUE • *Cancer Biology and Epigenetics Group, Research Center (CI-IPOP), Portuguese Oncology Institute of Porto (IPO Porto), Porto, Portugal; Department of Pathology, Portuguese Oncology Institute of Porto (IPO Porto), Porto, Portugal; Department of Pathology and Molecular Immunology, Institute of Biomedical Sciences Abel Salazar, University of Porto (ICBAS-UP), Porto, Portugal*
ZDENKO HERCEG • *Epigenetics Group, International Agency for Research on Cancer (IARC), Lyon, France*
HECTOR HERNANDEZ-VARGAS • *Epigenetics Group, International Agency for Research on Cancer (IARC), Lyon, France*
THU HO • *CSIRO Food and Nutrition Flagship, North Ryde, NSW, Australia*
MANUELA HOFNER • *Molecular Diagnostics Unit, AIT–Austrian Institute of Technology GmbH, Vienna, Austria*
GARY C. HON • *University of Texas Southwestern Medical Center, Dallas, TX, USA*
DIANNA HUSSMANN • *Institute of Biomedicine, Aarhus University, Aarhus C, Denmark*
JEAN-PIERRE J. ISSA • *Fels Institute for Cancer Research and Molecular Biology, Temple University School of Medicine, Philadelphia, PA, USA*
TAKASHI ITO • *Department of Biochemistry, Kyushu University Graduate School of Medical Sciences, Fukuoka, Japan; Core Research for Evolutional Science and Technology (CREST), Japan Agency for Medical Research and Development (AMED), Fukuoka, Japan*
JEFFREY A. JEDDELOH • *Roche Sequencing Solutions, Madison, WI, USA*
JAROSLAV JELINEK • *Fels Institute for Cancer Research and Molecular Biology, Temple University School of Medicine, Philadelphia, PA, USA*
CARMEN JERÓNIMO • *Cancer Biology and Epigenetics Group; Research Center (CI-IPOP), Portuguese Oncology Institute of Porto (IPO Porto), Porto, Portugal; Department of Pathology and Molecular Immunology, Institute of Biomedical Sciences Abel Salazar, University of Porto (ICBAS-UP), Porto, Portugal; Portuguese Oncology Institute of Porto, Porto, Portugal*
PETER A. JONES • *Department of Biochemistry and Molecular Biology, Norris Comprehensive Cancer Center, Keck School of Medicine, University of Southern California, Los Angeles, CA, USA; Van Andel Institute, Grand Rapids, MI, USA*
MARIA JUNG • *Institute of Pathology, University Hospital Bonn (UKB), Bonn, Germany*
THERESA K. KELLY • *Active Motif, Carlsbad, CA, USA*
KRISTINA KIRSCHNER • *Cambridge Institute for Medical Research, University of Cambridge, Cambridge, UK; Department of Haematology, University of Cambridge, Cambridge, UK; Stem Cell Institute, University of Cambridge, Cambridge, UK*
H. KIYOMI KOMORI • *Department of Molecular and Experimental Medicine, The Scripps Research Institute, La Jolla, CA, USA*
GLEN KRISTIANSEN • *Institute of Pathology, University Hospital Bonn (UKB), Bonn, Germany*
FELIX KRUEGER • *Bioinformatics, The Babraham Institute, Cambridge, UK*

PEI-FEN KUAN • *Department of Applied Mathematics and Statistics, Stony Brook University, Stony Brook, NY, USA*
SONJA KUNZE • *Research Unit of Molecular Epidemiology/Institute of Epidemiology II, Helmholtz Zentrum München, Deutsches Forschungszentrum für Gesundheit und Umwelt (GmbH), Neuherberg, Germany*
PETER W. LAIRD • *Center for Epigenetics, Van Andel Research Institute, Grand Rapids, MI, USA*
SARAH A. LAMERE • *Department of Molecular and Experimental Medicine, The Scripps Research Institute, La Jolla, CA, USA*
FIDES D. LAY • *Department of Biochemistry and Molecular Biology, Norris Comprehensive Cancer Center, Keck School of Medicine, University of Southern California, Los Angeles, CA, USA; Program in Genetic, Molecular and Cellular Biology, Keck School of Medicine, University of Southern California, Los Angeles, CA, USA; Department of Molecular, Cell and Developmental Biology, University of California, Los Angeles, Los Angeles, CA, USA*
JUSTIN T. LEE • *Fels Institute for Cancer Research and Molecular Biology, Temple University School of Medicine, Philadelphia, PA, USA*
DAOFENG LI • *Department of Genetics, The Edison Family Center for Genome Sciences and Systems Biology, Washington University, St. Louis, MO, USA*
SHOUDAN LIANG • *Department of Bioinformatics and Computational Biology, The University of Texas MD Anderson Cancer Center, Houston, TX, USA*
YUE LU • *Department of Molecular Carcinogenesis, The University of Texas MD Anderson Cancer Center, Smithville, TX, USA*
JOZEF MADZO • *Fels Institute for Cancer Research and Molecular Biology, Temple University School of Medicine, Philadelphia, PA, USA*
KLIO MARATOU • *MRC Clinical Sciences Centre, Faculty of Medicine, Imperial College London, London, UK*
FLORENCE MAUGER • *Laboratory for Epigenetics and Environment, Centre National de Recherche en Génomique Humaine, CEA-Institut de Biologie Francois Jacob, Evry, France*
RICHARD R. MEEHAN • *MRC Human Genetics Unit, Institute of Genetics and Molecular Medicine, University of Edinburgh, Edinburgh, UK*
KARIN B. MICHELS • *Department of Epidemiology, Fielding School of Public Health, University of California, Los Angeles, CA, USA*
SUSAN M. MITCHELL • *CSIRO Food and Nutrition Flagship, North Ryde, NSW, Australia*
FUMIHITO MIURA • *Department of Biochemistry, Kyushu University Graduate School of Medical Sciences, Fukuoka, Japan; Core Research for Evolutional Science and Technology (CREST), Japan Agency for Medical Research and Development (AMED), Fukuoka, Japan; Precursory Research for Embryonic Science and Technology (PRESTO), Japan Science and Technology Agency (JST), Fukuoka, Japan*
CATHY B. MOELANS • *Department of Pathology, University Medical Centre Utrecht, Utrecht, The Netherlands*
PETER L. MOLLOY • *CSIRO Food and Nutrition Flagship, North Ryde, NSW, Australia*
CHRISTA NOEHAMMER • *Molecular Diagnostics Unit, AIT–Austrian Institute of Technology GmbH, Vienna, Austria*
EDWIN J.C.G. VAN DEN OORD • *Center for Biomarker Research and Precision Medicine, Virginia Commonwealth University, Richmond, VA, USA*
JULIAN R. PEAT • *Epigenetics Programme, Babraham Institute, Cambridge, UK*
SARI PENNINGS • *Centre for Cardiovascular Science, Queen's Medical Research Institute, University of Edinburgh, Edinburgh, United Kingdom*

Christoph Plass • *Division of Epigenomics and Cancer Risk Factors, German Cancer Research Center (DKFZ), Heidelberg, Germany*
Nongluk Plongthongkum • *Department of Bioengineering, University of California at San Diego, La Jolla, CA, USA*
Walter Pulverer • *Molecular Diagnostics Unit, AIT–Austrian Institute of Technology GmbH, Vienna, Austria*
João Ramalho-Carvalho • *Cancer Biology and Epigenetics Group; Research Center (CI-IPOP), Portuguese Oncology Institute of Porto (IPO Porto), Porto, Portugal; Biomedical Sciences Graduate Program, Institute of Biomedical Sciences Abel Salazar, University of Porto (ICBAS-UP), Porto, Portugal*
Keith N. Rand • *CSIRO Food and Nutrition Flagship, North Ryde, NSW, Australia*
Victor Renault • *Laboratory for Bioinformatics, Fondation Jean Dausset – CEPH, Paris, France*
Todd A. Richmond • *Roche Sequencing Solutions, Madison, WI, USA*
Heidi Rosenbaum • *Roche Sequencing Solutions, Madison, WI, USA*
Jason P. Ross • *CSIRO Food and Nutrition Flagship, North Ryde, NSW, Australia*
Daniel R. Salomon (Deceased) • *Department of Molecular and Experimental Medicine, The Scripps Research Institute, La Jolla, CA, USA*
Juliette Salvaing • *INRA, UMR1198 Biologie du Développement et Reproduction, Jouy-en-Josas, France; Univ. Grenoble Alpes, INRA, CEA, CNRS, BIG-LPCV, Grenoble, France*
Suvi P. Savola • *MRC-Holland, Amsterdam, The Netherlands*
Andrey A. Shabalin • *Center for Biomarker Research and Precision Medicine, Virginia Commonwealth University, Richmond, VA, USA*
Jay Shendure • *Department of Genome Sciences, University of Washington, Seattle, WA, USA*
Sébastien A. Smallwood • *Friedrich Miescher Institute for Biomedical Research, Basel, Switzerland; Epigenetics Programme, Babraham Institute, Cambridge, UK*
Jenny Z. Song • *Epigenetics Research Laboratory, Genomics and Epigenetics Division, Garvan Institute of Medical Research, Sydney, NSW, Australia*
Aaron L. Statham • *Epigenetics Research Laboratory, Genomics and Epigenetics Division, Garvan Institute of Medical Research, Sydney, NSW, Australia*
Clare Stirzaker • *Epigenetics Research Laboratory, Genomics and Epigenetics Division, Garvan Institute of Medical Research, Sydney, NSW, Australia; St. Vincent's Clinical School, University of NSW, Sydney, NSW, Australia*
Saraswati Sukumar • *Breast and Ovarian Cancer Program, Sidney Kimmel Comprehensive Cancer Center, Johns Hopkins University School of Medicine, Baltimore, MD, USA*
John P. Thomson • *MRC Human Genetics Unit, Institute of Genetics and Molecular Medicine, University of Edinburgh, Edinburgh, UK*
Ali Torkamani • *Department of Molecular and Experimental Medicine, The Scripps Research Institute, La Jolla, CA, USA*
Jörg Tost • *Laboratory for Epigenetics and Environment, Centre National de Recherche en Génomique Humaine, CEA—Institut de Biologie Francois Jacob, Evry, France*
Binh Trinh • *UCSF School of Medicine, University of California, San Francisco, CA, USA*
Luis Valledor • *Area de Fisiología Vegetal, Dpto. Biología de Organismos y Sistemas, Universidad de Oviedo, Oviedo, Spain*
Fernando Vallejo • *Centro de Edafología y Biología Aplicada del Segura (CEBAS-CSIC), Campus Universitario de Espinardo, Murcia, Spain*

JÖRN WALTER • *Department of Biological Sciences, Genetics/Epigenetics, Saarland University, Saarbrücken, Saarland, Germany*

QI WANG • *Applied Bioinformatics, German Cancer Research Center (DKFZ), Heidelberg, Germany; Division of Applied Bioinformatics, German Cancer Research Center (DKFZ), and German Cancer Consortium (DKTK), Heidelberg, Germany*

TING WANG • *Department of Genetics, The Edison Family Center for Genome Sciences and Systems Biology, Washington University, St. Louis, MO, USA*

DIETER WEICHENHAN • *Division of Epigenomics and Cancer Risk Factors, German Cancer Research Center (DKFZ), Heidelberg, Germany*

ANDREAS WEINHAEUSEL • *Molecular Diagnostics Unit, AIT–Austrian Institute of Technology GmbH, Vienna, Austria*

DANIEL J. WEISENBERGER • *Department of Biochemistry and Molecular Biology, USC Norris Comprehensive Cancer Center, Keck School of Medicine, University of Southern California, Los Angeles, CA, USA*

JENNIFER WENDT • *Roche Sequencing Solutions, Madison, WI, USA*

STEPHAN WOLF • *Core Facility High Throughput Sequencing, German Cancer Research Center (DKFZ), Heidelberg, Germany*

MICHAEL C. WU • *Public Health Sciences Division, Fred Hutchinson Cancer Research Center, Seattle, WA, USA*

LINYING XIE • *Center for Biomarker Research and Precision Medicine, Virginia Commonwealth University, Richmond, VA, USA*

XIAOYUN XING • *Department of Genetics, The Edison Family Center for Genome Sciences and Systems Biology, Washington University, St. Louis, MO, USA*

ZHENG-ZHOU XU • *CSIRO Food and Nutrition Flagship, North Ryde, NSW, Australia*

MIAO YU • *Ludwig Institute for Cancer Research, La Jolla, CA, USA*

BO ZHANG • *Department of Genetics, The Edison Family Center for Genome Sciences and Systems Biology, Washington University, St. Louis, MO, USA*

KUN ZHANG • *Department of Bioengineering, University of California at San Diego, La Jolla, CA, USA*

Part I

Introduction

Chapter 1

A Summary of the Biological Processes, Disease-Associated Changes, and Clinical Applications of DNA Methylation

Gitte Brinch Andersen and Jörg Tost

Abstract

DNA methylation at cytosines followed by guanines, CpGs, forms one of the multiple layers of epigenetic mechanisms controlling and modulating gene expression through chromatin structure. It closely interacts with histone modifications and chromatin remodeling complexes to form the local genomic and higher-order chromatin landscape. DNA methylation is essential for proper mammalian development, crucial for imprinting and plays a role in maintaining genomic stability. DNA methylation patterns are susceptible to change in response to environmental stimuli such as diet or toxins, whereby the epigenome seems to be most vulnerable during early life. Changes of DNA methylation levels and patterns have been widely studied in several diseases, especially cancer, where interest has focused on biomarkers for early detection of cancer development, accurate diagnosis, and response to treatment, but have also been shown to occur in many other complex diseases. Recent advances in epigenome engineering technologies allow now for the large-scale assessment of the functional relevance of DNA methylation. As a stable nucleic acid-based modification that is technically easy to handle and which can be analyzed with great reproducibility and accuracy by different laboratories, DNA methylation is a promising biomarker for many applications.

Key words DNA methylation, Nutrition, Environment, Complex disease, Epigenetics, Imprinting, Development, Cancer, Hydroxymethylation, Epidrugs, Epigenetic therapy, Epigenome engineering

1 Introduction

Almost all cells of an organism share the same genetic material encoded in the DNA sequence, but display a broad range of morphological and functional diversity. This heterogeneity is a result of differential gene expression patterns, governed by epigenetic changes. Epigenetics can be defined as the study of changes of a phenotype such as the gene expression patterns of a specific cell type not caused by underlying changes in the primary DNA sequence. These changes are mitotically and, maybe, in some cases meiotically heritable. Epigenetic regulation mediates genomic adaption to an environment thereby ultimately contributing toward the phenotype and "brings the phenotype into being" [1].

Jörg Tost (ed.), *DNA Methylation Protocols*, Methods in Molecular Biology, vol. 1708,
https://doi.org/10.1007/978-1-4939-7481-8_1, © Springer Science+Business Media, LLC 2018

Epigenetics consists of a variety of molecular mechanisms including post-transcriptional histone modifications, histone variants, ATP-dependent chromatin remodeling complexes, polycomb/trithorax protein complexes, small and other noncoding RNAs including siRNA and miRNAs, and DNA methylation [2, 3]. These diverse molecular mechanisms have all been found to be closely intertwined and stabilize each other to ensure the faithful propagation of an epigenetic state over time and especially through cell division. Nonetheless, epigenetic states are not definitive and changes occur with age in a stochastic manner as well as in response to environmental stimuli.

Chromatin modulations play a central role to shape the epigenome, and both open (euchromatin) and closed (heterochromatin) chromatin states are controlled by histone modifications and histone composition in close crosstalk with the binding of a myriad of non-histone proteins [4]. The basic building block of chromatin is the nucleosome consisting of an octamer of histone proteins. The protruding N-terminal tails of these histones are extensively modified by various modifications such as acetylation, methylation, phosphorylation, and ubiquitylation, which partly determine the transcriptional potential for a specific gene or a genomic region [5, 6]. DNA methylation is highly related to certain chromatin modifications; and enzymes that modify DNA and histones have been shown to directly interact and constitute links between local DNA methylation and regional chromatin structure [7].

This review gives an introduction to the multiple biological facets of the biology of DNA methylation and their potential use in clinical applications. A brief description of the DNA methylation landscape and the enzymes responsible for adding and removing methyl groups to the DNA will be given. The various biological processes, in which DNA methylation plays a key role, are described together with the recent possibilities to elucidate the functional relevance of DNA methylation patterns using epigenome engineering. The concept of Epigenome-wide Association Studies (EWAS) will be introduced and different pathologies for which changes in DNA methylation patterns have been investigated will be mentioned with a certain emphasis on cancer, as much of the DNA methylation literature on clinical applications of DNA methylation changes concerns so far changes in tumorigenesis. Due to space restrictions and the large field of research covered in this introduction, over-simplifications and omissions are inevitable and many of the references refer the interested reader to more detailed reviews.

2 The Biology of DNA Methylation

DNA methylation is a post-replication modification almost exclusively found on the 5 position of the pyrimidine ring of cytosines in the context of the dinucleotide sequence CpG, of which around 29 million are found in the human (haploid) genome [8, 9]. 5-Methylcytosine (5mC) accounts for ~1% of all bases, varying slightly in different tissue types and the majority (60–80%) of CpG dinucleotides throughout mammalian genomes are methylated. Other types of methylation such as methylation of cytosines in the context of CpNpG or CpA sequences have been detected in mouse embryonic stem cells, neurons, and plants, but are generally rare in somatic mammalian/human tissues. The sequence symmetry of CpG dinucleotides allows for the transmission of DNA methylation marks through cell division, having led to the hypothesis that DNA methylation marks are part of the cellular identity and memory. CpGs are underrepresented in the genome, as a result of their increased mutation potential with mutation rates at CpG sites to be about 10–50 times higher than other transitional mutations. As the deamination of methylated CpGs to TpGs yields a naturally occurring DNA base, it is less well corrected. Despite this general trend, relatively CpG-rich clusters, so called CpG islands, are found in the promoter region and first exons of ~65% of all genes containing about 7% of all CpGs [10]. Depending on the employed set of parameters, a CpG island is defined as having a C + G content of more than 50% (55%), an observed versus expected ratio for the occurrence of CpGs of more than 0.6 (0.65) and a minimum size of 200 (500) base pairs [11, 12]. They are mostly non-methylated in all tissues and throughout all developmental stages corresponding to an open chromatin structure and a potentially active state of transcription. There are around 30,000 CpG islands in the human genome. As CpG islands are mainly unmethylated in the germline, they are less susceptible to deamination and have therefore retained the expected frequency of CpGs. Binding of transcription factors, exclusion of nucleosomes, and presence of H3K4 methylation and the associated histone methyltransferases protect most CpG islands from DNA methylation. It should be noted that a number of CpG islands have been identified that are methylated in a tissue-specific manner in normal tissues, but concern mainly intragenic CpG islands [13–16]. CpG islands associated with genes not expressed in a specific cell type acquire the repressive histone modification H3K27Me3, but rarely DNA methylation. In contrast, regions located up- and downstream of CpG islands, termed CpG island shores, show variable tissue-specific DNA methylation patterns and these are often altered in tumorigenesis [17]. In contrast to CpG islands, gene bodies are commonly highly methylated, where DNA methylation has been associated with enhanced gene expression, maybe, by facilitating transcriptional elongation and preventing initiation of spurious transcription events [18]. Intragenic methylation has in addition been associated with the repression/use of alternative promoters or different splice variants [16, 19, 20].

2.1 DNA Methyltransferases

Both local and global epigenetic patterns are dictated by the composition of the genome depending on CpG spacing as well as sequence motifs and DNA structure [21, 22]. The transfer of a methyl-group from the universal methyl donor S-adenosyl-L-methionine (SAM) is carried out by DNA methyltransferases [23]. During the methylation reaction a methyl group is transferred from SAM to the cytosine, thereby leaving S-adenosylhomocysteine, which at high concentrations inhibits the action of DNA methyltransferases.

Four DNA methyltransferases have been identified (DNMT1, DNMT3A, DNMT3B, and DNMT3L) [24–26]. DNMT3L, however, lacks a catalytic domain, but is in complex with DNMT3A important for maternal genomic imprinting and male spermatogenesis.

Simplified, DNMT1 acts as a maintenance methyltransferase as it prefers hemi-methylated templates. It is located at the replication fork during the S phase of the cell cycle and methylates the newly synthesized DNA strand using the parent strand as a template with high fidelity [27]. The symmetric sequence of CpGs allows thereby passing the epigenetic information through cell generations. A number of proteins associated with the local chromatin structure such as LSD1 and URHF1 are required to ensure the specificity and stability of the DNA methylation reaction associated with DNA replication. However, DNMT3A and DNMT3B are also required for methylation maintenance [28]. De novo methylation is carried out by the methyltransferases DNMT3A and DNMT3B. These enzymes have certain preferences for specific targets (e.g., DNMT3A together with DNMT3L methylates maternal imprinted genes and DNMT3B localizes at minor satellite repeats as well as the gene bodies of actively transcribed genes), but also work cooperatively to methylate the genome [26, 29]. Possible trigger mechanisms to initiate de novo methylation include preferred target DNA sequences, RNA interference, but mostly chromatin structures induced by histone modifications and other protein-protein interactions [25, 30]. Histone modifications such as H3K9me3 are thought to initiate heterochromatin formation and DNA methylation comes in as a secondary molecular alteration to ensure the stable silencing of the repressed sequences.

2.2 5-Hydroxymethylation and Demethylation Processes

Mechanisms for DNA demethylation have long been searched for as active demethylation occurs at different stages of development and a global hypomethylation is associated with many cancers. DNA demethylation has been proposed to be either passive, where the 5mC is removed owing to a lack of maintaining the methylation during several cycles of replication, or as an active process, with direct removal of the methyl group independently of DNA replication [31]. The active process is conducted through the enzymatic oxidation of 5mC to 5-hydroxymethylcytosine

(5hmC) as described further below in this paragraph. However, 5hmC is now considered to be not only an intermediate in oxidative DNA demethylation, but constitutes a distinct layer in the complex process of epigenetic regulation with its own distribution and regulatory functions. 5hmC is catalyzed by the Ten-eleven translocation methylcytosine dioxygenase (TET) family of enzymes, consisting of three mammalian subtypes, TET1–3 [26, 32, 33]. 5hmC is most abundant in human brain tissue and embryonic stem cells, but at levels approximately tenfold lower than those of 5-methylcytosine [34]. TET enzymes are expressed in a tissue/cell-type and developmental stage-dependent manner with 5hmC decreasing during cell differentiation [35]. 5hmC levels do not correlate with 5mC levels of the respective tissue and 5hmC was found enriched at specific active functional elements of the genome, in particular enhancers, promoters, and gene bodies associating 5hmC with open chromatin and transcriptional activity [34, 36–38]. 5hmC levels are globally reduced in cancer and alterations of the TET enzymes have been reported for various cancers [33, 39–42]. This observation suggests that 5hmC alterations may have a distinct role in the development and progression of malignancies.

In addition to its regulatory function, 5hmC is an intermediate in an active demethylation process [43], where it is further oxidized to 5-formylcytosine (5fC) and 5-carboxylcytosine (5caC) again by the TET enzymes, with the latter two modifications being present at barely detectable levels in the human genome [44]. Both the carboxyl and the formyl groups can be removed enzymatically with or without base excision, generating an unmethylated cytosine. The discovery of 5hmC, 5fC, and 5caC has made the understanding of gene regulation through DNA methylation more complex, especially as the most widely used method for the analysis of DNA methylation, bisulfite conversion, cannot distinguish between the two cytosine modifications. Dedicated chemistries to distinguish 5-hydroxy cytosine methylation from cytosine methylation have since been developed and are described in chapters of this volume (*see* Chapters 33–35).

2.3 Transcription

Transcription does not occur on naked DNA but in the context of chromatin, which critically influences the accessibility of the DNA to transcription factors and the DNA polymerase complexes. DNA methylation, histone modifications, and chromatin remodeling are closely linked and constitute multiple layers of epigenetic modifications to control and modulate gene expression through chromatin structure [45]. DNMTs and histone deacetylases (HDACs) are found in the same multi-protein complexes and Methyl CpG-binding domain proteins (MBDs) interact with HDACs, histone methyltransferases as well as with the chromatin remodeling complexes. Furthermore, mutations or loss of members of the

SNF2 helicase/ATPase family of chromatin remodeling proteins such as *ATRX* or *LSH* lead to genome-wide perturbations of DNA methylation patterns and inappropriate gene expression programs.

Transcription may be affected by DNA methylation in several ways. First, the binding of transcriptional activators such as Sp1 and Myc may be inhibited directly by the methylated DNA through sterical hindrance, while other transcription factors especially homeodomain transcription factors are attracted by methylated target recognition sequences [46, 47]. Second, methylated DNA is bound by specific Methyl CpG-binding domain (MBD1, MBD2, and MBD4) proteins or methyl CpG-binding proteins (MeCP2) [24, 26, 48, 49]. They recruit transcriptional co-repressors such as histone deacetylating complexes, polycomb proteins, and chromatin remodeling complexes, thereby establishing a repressive chromatin configuration. Mbd3 binds specifically hydroxymethylated cytosines [50].

In many cases DNA methylation occurs subsequently to changes in the chromatin structure and is used as a molecular mechanism to permanently and thus heritably lock the gene in its inactive state [8]. It should be underlined that an unmethylated state of a CpG island or gene regulatory element does not necessarily correlate with the transcriptional activity of the gene, but rather that the gene can be potentially activated. It has, however, been shown that active histone marks such as H3K4Me3 at the transcription start site might permit transcription of a gene when stimulated even in the presence of a partly methylated CpG island immediately adjacent to the transcription start site [51]. The simple presence of methylation does therefore not necessarily induce silencing of nearby genes. Only when a specific core region of the promoter that is often—but not necessarily—spanning the transcription start site becomes hypermethylated, the expression of the associated gene is modified [52]. In CpG-poor intergenic gene regulatory regions, CpG dinucleotides are mostly highly methylated, but methylation is reduced when the region or the methylated CpG is bound by transcription factors [53, 54].

2.4 Genome Stability

DNA methylation plays an important role in the maintenance of genome integrity by transcriptional silencing of repetitive DNA sequences and endogenous transposons [55] and the absence of DNA methylation leads to a reactivation of retroviruses during embryonic development [56]. DNA methylation may prevent the potentially deleterious recombination events between non-allelic repeats caused by these mobile genetic elements. In addition, methylation increases the mutation rate leading to a faster divergence of identical sequences and disabling many retrotransposons [57]. The role of DNA methylation becomes evident during tumorigenesis where loss of DNA methylation both globally, but also at specific genes like *CHFR*, leads to chromosomal instability and aneuploidy [58] as discussed in more detail in Subheading 3.1.

2.5 Development

Cytosine methylation is essential for normal mammalian embryogenesis during which methylation levels change dynamically [59, 60]. During development and differentiation the mammalian organism creates a number of cell-type-specific differentially marked epigenomes, whose identity is partly defined by their respective DNA methylation patterns [61]. Thus, the human body with one genome contains several hundred different epigenomes. While embryonic stem cells can be maintained in the absence of DNMTs, their differentiation is inhibited and requires the presence of DNMTs [62]. Forced de-differentiation, as for example for the creation of induced pluripotent stem cells, does not completely reset the tissue-specific DNA methylation marks and cells might have a slight preference to re-differentiate into their cell-type of origin suggesting the DNA methylation marks function as a memory for the cell [63, 64].

Mammalian development is characterized by two waves of genome-wide epigenetic reprogramming, in the zygote and in the primordial germ cells [59, 65]. In mammals the genome becomes demethylated during pre-implantation, probably to initiate cellular differentiation. Most of the paternal genome is actively and rapidly demethylated by Tet3 with 5hmC accumulating in the paternal pro-nucleus leading *in fine* to the erasure of most paternal germline methylation marks, while the maternal genome remains methylated [65, 66]. Recent results do, however, also suggest the presence of 5hmC and some active demethylation in the maternal genome [67, 68]. Around implantation, where cells start to commit to different developmental lineages, DNA methylation levels are then restored by de novo methylation. Disruption of any of the DNA methyltransferases results in embryonic lethality and hypomorphic alleles of *DNMT1* result in genome-wide deregulation of gene expression. The second reprogramming event occurs also during embryogenesis, but only in the primordial germ cells where DNA methylation patterns are erased at most single-copy genes (including imprinted genes) and some repetitive elements in a clearly defined order [69, 70]. Depending on the sex of the newly formed germ line, imprints at paternally methylated loci are restored shortly after birth while maternally methylated loci occur only during the last stages of oogenesis. Incorrect establishment of DNA methylation patterns in the male germline has been associated with reduced male fertility [71, 72]. Modifications to the environment during early development including the embryo, the fetus, as well as the early post-natal period are the critical windows of exposure and associated with high epigenetic plasticity or reprogramming. Stimuli during these times can lead to permanent changes in the patterns of epigenetic modifications giving rise to the developmental origins of health and disease hypothesis setting the stage for disease or protection from disease during later life [73] (*see* also Subheading 2.9 below). In vitro fertilization and the associated cell culture are associated with changes of DNA methylation patterns, especially at imprinted genes in placental and cord blood samples in both human and mice [74–76]. There is a small but significant increase in the occurrence of imprinting disorders following assisted reproductive technologies [77, 78].

2.6 Aging

Epigenetic changes are also an integral part of aging and cellular senescence, whereby the overall content of DNA methylation in the mammalian and human genome decreases with age especially at repetitive elements [79–82]. Simultaneously, distinct genes acquire methylation at specific sites including their promoters, a situation that strikingly resembles the DNA methylation changes found in cancer [83, 84]. DNA methylation levels at a number of specific loci have been shown to correlate very well with lifetime (chronological) age and several DNA methylation signatures have been developed for the accurate prediction of the age through the analysis of DNA methylation patterns [81, 85, 86]. Accelerated epigenetic changing has been associated with earlier all-cause mortality in later life [87–89], while people with exceptional longevity show a decreased epigenetic age compared to their chronological age [90].

2.7 Imprinting

In mammals, the maternal and paternal genomes differ functionally and both are required for normal development. A subset of genes is asymmetrically expressed from only the maternal or the paternal allele in a parent-of-origin-specific manner in all somatic cells of the offspring [91]. These imprinted genes are generally located in clusters and the alleles are differentially marked by DNA methylation, histone acetylation/deacetylation, and histone methylation and often associated with antisense RNAs [92]. About 150 imprinted genes are known in mouse and man respectively, but some more imprinted genes have been computationally predicted [93, 94]. Imprints are established in the gametes by DNMT3A and—at least for maternally imprinted genes—the regulatory cofactor DNMT3L in a parent-of-origin specific manner. These epigenetic marks in imprinting control regions are not erased in the zygote. Imprinted genes are probably the most important buffering factors for regulating the day-to-day flux between mother and fetus in placental mammals. The *H19/Igf2* locus is one paradigm for imprinting and has been extensively studied, demonstrating that the physical contacts between differentially methylated regions, containing insulators, silencers, and activators, create a higher-order chromatin structure leading to transcriptional regulation of both *H19* and *Igf2* [95, 96].

2.8 X Inactivation

Random silencing of one of the two X chromosomes in embryonic tissues of female mammals to achieve dosage compensation is another paradigm for a stable and heritable epigenetic state in somatic cells [97]. DNA methylation occurs quite late during the inactivation process. Only after expression of the large noncoding *Xist* RNA, changes in the patterns of histone modifications and variants, and gene silencing, DNA methylation patterns are established on the inactive X chromosome, which contribute to maintaining it in the inactive state [98].

2.9 Environmental and Nutritional Epigenetics and the Possibility of Transgenerational Inheritance

Epigenetics holds the promise to explain at least a part of the influences the environment has on a phenotype. Studies in monozygotic twins have demonstrated that epigenetic differences in genetically identical humans accumulate with age and different environments create different patterns of epigenetic modifications [99, 100]. Differences are therefore largest in twin pairs of old age that have been raised separately. Transient nutritional or chemical stimuli occurring at specific developmental stages, especially early during development, may have long-lasting influences on gene expression by interacting with epigenetic mechanisms and altering chromatin compaction and thereby transcription factor accessibility. Developmental stages in multicellular organisms proceed according to a tightly regulated temporal and spatial pattern of gene expression accompanied by changes in DNA methylation patterns. Therefore, epigenetics provides a mechanism by which physiological homeostasis could be developmentally programmed and inherited.

The modulation of epigenetic patterns *in utero* has given rise to the developmental origin of disease hypothesis, which postulates that the *in utero* environment can cause permanent changes to metabolic processes that directly affect the postnatal phenotype, confers susceptibility to multifactorial disease at adult age, and may also be transmitted to subsequent generations [73]. Chemical toxins such as the fungicide vinclozolin or the plasticizer compound bisphenol A, life style factors such as diet, smoking and physical activity, but also stress and complex behaviors or the environmental temperature have shown to induce changes to DNA methylation patterns without altering the genetic sequence and leading to epimutation-associated phenotypes [101–105]. Imprinted genes might be particularly sensitive to such environmental stimuli [106]. Long after the stimulus is gone, "cellular memory" mechanisms enable cells to remember their chosen fate, thus perturbation at an early stage may have long-lasting consequences. Transgenerational epigenetic inheritance refers to the transfer of epigenetic information across generations, i.e., through meiosis requiring thus the transmission through epigenetically altered germ cells or molecules present in the oocyte [103, 107]. This mechanism would explain the inheritance of a phenotype in addition to the DNA sequence from the parents. There has been much attention regarding this mechanism, and there is some data supporting the transgenerational epigenetic inheritance theory [103, 105]. However, transgenerational epigenetic inheritance has not been clearly identified in humans despite some epidemiological evidence, many of the early animal models involve intergenerational (parental or grandparental exposure) rather than real transgenerational inheritance and many questions on the appropriate experimental design such as the time-point and the duration of the window of exposure, time-point of separation of the offspring from the parents, the transmission through the paternal or maternal line, the genetic background, the breeding strategies, the tissues analyzed, and many more parameters have to be further explored before conclusions can be made with certainty [102, 108, 109].

2.10 Elucidating the Function of DNA Methylation Using Epigenome Engineering

The quickly evolving field of genome engineering has now also embraced epigenomics and promises to revolutionize the possibilities to functionally investigate and validate the importance of epigenetic modifications at any locus in the genome. These approaches will allow for the first time to determine if epigenetic changes are causal to observed phenotypic changes and will substantially further our understanding of the multifaceted roles of epigenetic modifications and their combinations in chromatin, which is the key to the answer of many biomedical questions. Epigenome editing will help to identify and evaluate the role of intergenic gene regulatory elements such as enhancers in the regulation of endogenous gene expression in a cell-type specific context through targeted alteration of the local chromatin structure. Three major technological approaches have been pursued in the past to localize effector domains of epigenetic modifiers to a specific genomic context. They can all alter the epigenetic code of the target region by depositing or removing specific epigenetic marks close to their binding site and thereby provide evidence for the role of an epigenetic mark at a locus of interest. These include Transcription Activator-Like Effectors (TALEs), zinc finger-based artificial transcription factors (ATFs), and the Clustered Regularly Interspaced Short Palindromic Repeats (CRISPR)/CRISPR-associated protein (Cas) 9 system [110–113]. They can all be combined with a large number of effector domains, catalytically active protein fragments of epigenetic enzymes such as DNA methyltransferases, TETs, and histone (de)methylases, acetyltransferases and deacetylases as well as transcriptional repressors and activators that are able to recruit chromatin modifying and remodeling complexes [113]. Demethylation of specific CpGs in human cells has first been demonstrated combining TALEs (transcription activator-like effectors), which are peptides recognizing specific DNA sequences, and the TET1 hydroxylase catalytic domain [114]. Another approach for epigenetic editing is through the CRISPR/Cas9 system. This consists of a synthetic guide RNA (gRNA) containing the specific sequence of the DNA sequence one wish to target and bacterially derived protein (Cas9), which normally introduce double-stranded breaks [115]. The fusion of the core catalytic domain of a DNA methyltransferase (e.g., DNMT3A) or demethylase (e.g., TET1) to a modified nuclease-deficient Cas9 (dCas9) has been shown to induce specific epigenetic changes either locally if a promoter is targeted or more regionally if a distant gene regulatory element such as an enhancer is targeted [116–118]. Recently devised technologies to create guide RNAs at large-scale and low cost will allow these approaches to be performed at high-throughput in the future [119].

3 DNA Methylation and Disease

DNA methylation and chromatin structure are strikingly altered in many complex diseases. Although a number of genetic variations have been identified to confer susceptibility to a certain disease, in most cases even the worst combination of alleles of several disease susceptibility loci only explains a small percentage of disease occurrences. Consequently, environmental factors play undoubtedly a large role in the actual occurrence of disease. Epigenetic modifications constitute a memory of an organism to all the stimuli or insults it has ever been exposed to.

Disease-associated changes in epigenetic modifications can be classified into two categories: (1) genes that are epigenetically regulated; (2) genes that are part of the molecular machinery responsible for correct establishment and propagation of the epigenetic modifications through development and cell division. Aberrant DNA methylation patterns have been reported in various neurodevelopmental disorders including Fragile X, HSAN1 (hereditary sensory neuropathy with dementia and hearing loss), and ICF (immune deficiency, centromeric instability, and facial abnormalities) [120]. The latter is caused by mutations in the *DNA methyltransferase 3B*. Mutations in the methyl-binding protein *MeCP2* are found in Rett syndrome [121, 122], while mutations in *TET2* have been found in multiple hematological malignancies, where they are probably among the earliest genetic events of the disease [123]. Imprinting anomalies lead to disorders such as Prader-Willi, Angelman, and Beckwith-Wiedemann syndrome or transient neonatal diabetes [124, 125].

3.1 DNA Methylation Changes in Cancer

Cancer is by far the most studied disease with a strong epigenetic component [3, 58]. In tumors a global loss of DNA methylation (hypomethylation) of the genome is observed [126]. This hypomethylation has been suggested to initiate and propagate oncogenesis by inducing aneuploidy, genome instability, activation of retrotransposons, and transcriptional activation of oncogenes and pro-metastatic genes [127–129]. The overall decrease in DNA methylation is accompanied by a region- and gene-specific increase of methylation (hypermethylation) of multiple CpG islands [3, 130]. Hypermethylation of CpG islands in the promoter region of a tumor suppressor or otherwise cancer-related gene is frequently associated with transcriptional silencing of the associated gene although silencing occurs often prior to the onset of the hypermethylation suggesting that DNA methylation is not the causal mechanism. Genes of numerous pathways representing the different hallmarks of cancer are often inappropriately inactivated by DNA methylation [131]. These include genes involved in signal transduction (*APC*), self-sufficiency of growth signals (*RASSF1A*), DNA repair (*MGMT, MLH1, BRCA1*), detoxification (*GSTP1*), cell cycle regulation (*p15, p16, RB*), tissue invasion and metastasis (*CDH1*), genome integrity (*CHFR*), differentiation (*MYOD1*), angiogenesis (*THBS1, VHL*), and apoptosis (*Caspases, p14, DAPK*). In addition, loss of imprinting leading to biallelic activation of growth favoring genes such as *IGF2* is commonly observed in cancers [132]. However, it should be noted that so far no single gene has been identified that is always methylated in a certain type of cancer. Genes are not necessarily individually targeted, as entire chromosomal regions might be affected by altered profiles of epigenetic modifications, a phenomenon termed long-range epigenetic silencing (LRES) or activation (LREA) [133–135]. Large domains of hypomethylated DNA associate with the nuclear lamina leading to a reorganization of chromosomal domains within the nucleus of cancer cells [136]. The profound alteration of the DNA methylation landscape is therefore probably a prerequisite for the rewiring of the transcriptional and gene regulatory networks occurring in cancer cells. Furthermore, DNA methylation is—as described above—an efficient mechanism to guarantee and maintain cellular identity, which is lost during the cellular transformation of tumorigenesis.

Potential mechanisms include the genetic alteration of epigenetic enzymes, which have been frequently identified in large-scale sequencing projects and whose prevalence had previously been underestimated [137–139] or changes in the cellular metabolome such as mutations in *IDH1*, which will lead to the loss of activity of the TET enzymes through the competition for the metabolite for α-ketoglutarate and the establishment of a methylator phenotype [140, 141].

While the contribution of genetic factors to carcinogenesis such as the high-penetrance germ-line mutations in genes such as *BRCA1, TP53,* and *RB1* in familial cancers has long been recognized, it has become evident that epigenetic changes leading to transcriptional silencing of tumor suppressor genes constitute an at least equally contributing mechanism. For example, microarray expression profiles of breast tumors with *BRCA1* mutations are very similar to those of sporadic breast cancer cases with *BRCA1* promoter hypermethylation [142]. Further, recent large-scale sequencing projects showed that 91% of breast cancers with *BRCA1* mutations or promoter hypermethylation clustered in the same mutation signature group, and these were often basal-like, triple negative cancers demonstrating that disruption of *BRCA1* function by either genetic or epigenetic pathways leads to the same perturbations of cellular homeostasis [143]. With the exception of haploinsufficient genes, "two hits" are necessary to inactivate the two alleles of a gatekeeper tumor suppressor gene to enable oncogenic progression according to the two hit hypothesis [144]. DNA methylation can act as one hit having the same functional effect as a genetic mutation or deletion as proven by numerous experiments in which re-establishing expression of tumor suppressor genes could be achieved through drugs inducing demethylation. Epimutations can inactivate one of the two alleles while the other is lost through genetic mechanisms or silence both alleles [58, 145]. DNA methylation changes and genetic mutations co-occurring in the same tumors show different dynamics and evolve differently during tumor progression providing further support for the functional relevance of epigenetic changes on the phenotype of the tumor [146]. Epigenetic changes occur at higher frequency compared to genetic changes and may be especially important in early-stage human neoplasia [147, 148]. It has therefore been suggested that epigenetic lesions in the stem cell /progenitor population of a normal tissue set the stage for neoplasia [149]. DNA hypermethylation could for example not only be detected in dysplastic epithelium of patients with ulcerative colitis, a condition associated with an increased risk for the development of colon cancer, but already in histological normal epithelium [150]. Similarly, in breast cancer the genome-wide comparison of methylation profiles in normal tissue, ductal carcinoma in situ (pre-invasive precursor), and invasive breast cancer showed important differences in DNA methylation profiles between healthy tissue and the pre-invasive precursor [151]. Aberrant DNA methylation patterns are therefore probably not a consequence or by-product of malignancy, but contribute directly to the cellular transformation. It has been extrapolated that aberrant promoter methylation is initiated at ~1% of all CpG islands and as much as 10% become methylated during the multistep process of tumorigenesis [147]. This hypermethylation is not random and occurs primarily at gene promoters that are targets of the polycomb repressive complex 2 (PRC2) and marked by H3K27Me3 in (embryonic) stem cells [152, 153]. However, it is challenging to determine which of the DNA methylation changes are actually driving tumorigenesis and which ones are merely passengers. In contrast to 5mC, 5hmC is greatly reduced or absent in many cancers and TET enzymes are often downregulated [42, 154,

155] suggesting that DNA demethylation might have a tumor-suppressive function and potentially protect against cancer-associated hypermethylation [156, 157].

3.2 Application of DNA Methylation as a Biomarker in Cancer

The field of epigenetics is rapidly expanding as the technologies and our understanding of the complex interactions between different epigenetic and genetic changes improves. Biomarkers capable of distinguishing diseased or malignant cells from normal ones must be specific, sensitive, and detectable in specimens obtained through minimally invasive procedures to be clinically applicable. Many biomarkers on the protein, RNA or DNA level fulfilling these criteria have been discovered. In routine clinical practice, most tumor diagnostics is carried out by biochemical assays determining the presence and/or quantity of enzymes, receptors, growth factors, or hormones. Despite the widespread use of RNA detection techniques in research facilities, there are some potential pitfalls associated with its use in routine clinical diagnostics such as the required preservation of mRNA from the tissue, tissue heterogeneity, and the need for normalization. Attention to numerous details of sample extraction, storage, and handling has to be paid to ensure intra- and inter-laboratory reproducibility [158]. Methylgroups on cytosines are part of the covalent structure of the DNA and once methylation is acquired, it is in most cases chemically and biologically stable over time while expression of mRNA and/or proteins is more prone to vary. DNA-based molecular biomarkers can therefore be more easily transferred from a research laboratory setting into routine diagnostics in a clinic due to the amplifiable and stable nature of DNA. DNA methylation can be analyzed with a growing number of methods that are amenable to high throughput. Quantitative assays, many of which are described in this volume, eliminate the need for normalization. They are applicable to formalin-fixed paraffin-embedded (FFPE) clinical specimens and other archived material. Large-scale evaluation of locus-specific technologies has shown the high accuracy and reproducibility of results obtained by different technologies and in different laboratories for the same samples [159]. As most methods determine the ratio between methylated and unmethylated CpGs, DNA methylation analysis is independent of the total amount of starting material. It provides a binary and positive signal that can be detected independent of expression levels. However, it should be noted that so far no single gene has been identified that is always methylated in a certain type of cancer.

Detection of DNA methylation changes can be carried out in the tissue itself, but a high level of concordance of DNA methylation patterns has been demonstrated in tumor biopsies and matched DNA samples extracted from body fluids such as serum, plasma, urine, and sputum. DNA methylation thus fulfills one of the most important criteria for a clinically useful biomarker, which is enabling screening of individuals at potential risk and monitoring of therapy response or disease recurrence through the analysis of surrogate tissues, which can be obtained through minimal invasive procedures. The sensitive and specific detection of tumor-specific DNA methylation patterns at distal sites makes DNA methylation a biomarker of choice for the clinical management of cancer patients. DNA methylation-based markers are therefore promising tools for noninvasive detection of different tumor types. The most effective way to detect the aberrant methylation is to analyze fluids that have been in physical contact with the site of the respective cancer. A large number of novel sources have been successfully tested including nipple aspirate fluid, breast-fine needle washing, bronchial brush samples, buccal cells, needle biopsies, pancreatic juice, peritoneal fluid, prostate fluid or ejaculate, brochoalveolar lavages, saliva, exfoliated cells from bladder or cervix, urine, peritoneal fluid, or stool samples [160]. Tumors release a substantial amount of genomic DNA into the systemic circulation and this freely circulating DNA contains the same genetic and epigenetic alterations that are specific to the primary tumor [161]. As the analyzed gene-specific methylation patterns are in most cases absent in control patients, methylation analysis of DNA recovered from plasma and serum can be used as a biomarker for molecular diagnosis and prognosis in various types of malignancies [162, 163]. The commercial FDA-approved EpiPro Colon test analyzes the methylation profile of an intronic sequence of the *SEPT9* gene for the population-wide screening of colorectal cancer achieving a sensitivity of 50–80% depending on the stage of the cancer and a very high specificity (>95%) [164, 165]. A number of DNA methylation changes have been linked to prognosis of different cancers [166] and recent results suggest that overall increased heterogeneity of DNA methylation patterns correlates with a more aggressive disease and has a negative impact on prognosis [146, 167–169]. Besides early detection, the methylation status of CpG islands can be used to characterize and classify cancers as for example recently demonstrated for subtypes of Ewing sarcoma, which are all characterized by the same recurrent genetic alteration [169]. Furthermore, as the DNA methylation patterns of distant metastases will at least partly carry the tissue-specific DNA methylation signature of the primary tumor, analysis of the DNA methylation profile of the metastases has shown to permit the identification of the tissue of the primary tumor [170]. The predictive algorithm based on DNA methylation profiles of the metastases achieved very high sensitivity and specificity and outperformed other molecular profiling and imaging technologies predicting the tissue-of-origin in 87% of patients with an unknown primary cancer. Similarly, the DNA methylation profile of cell-free DNA from plasma/serum contains the information of the tissue it is released from and allows detecting the tissue-of-origin of a cancer [171–173].

DNA methylation changes detect tumor recurrence as well as predict and monitor patients' response and the effectiveness to a given anti-cancer therapy. Prediction of the response of patients with glioblastomas to the alkylating agent temozolomide based on the DNA methylation status of the DNA repair gene *MGMT* is probably the best-known example [174, 175]. Methylation of *SMAD1* as a predictor for the resistance to doxorubicin in patients with diffuse large B-cell lymphoma is another example [176]. As DNA methylation is a non-mutational and therefore—at least in principle—a reversible modification, it can be used as a point of departure for anti-neoplastic treatment by chemically induced demethylation as described in the next paragraph [162, 177].

3.3 Epigenetic Therapy

Epigenetic alterations are nowadays considered to be equally responsible for carcinogenesis as genetic mutations. However, epigenetic changes are, unlike genetic changes, reversible, which make them an ideal target for cancer therapy. Two types of epigenetic drugs have currently been approved by the US FDA and/or the European Medicines Agency (EMA) for cancer treatment: histone deacetylase inhibitors (HDACis) and DNA methyltransferase inhibitors (DNMTis) [178]. Two DNMTis (azacytidine (Vidaza)) and 5-aza-deoxycytidine (decitabine, Dacogen, Decitibine) have been approved for the treatment of several hematological malignancies [179]. Azacytidine is a ribonucleoside and is therefore predominantly incorporated into RNA, whereas decitabine, the deoxyribose analog, is only incorporated into DNA [180]. After the incorporation into the DNA, DNMTs are trapped through a covalent bound between the nitrogen instead of carbon in the 5-position of the modified pyrimidine. This results in passive loss of DNA methylation in the daughter cells after replication through depletion of DNMTs. At low doses azacytidine and decitabine induce their effect through demethylation of silenced genes associated with reduced apoptosis, cell differentiation, and proliferation, whereas at higher doses the main cytotoxic effect is due to DNA damage after incorporation [180]. Both azacytidine and decitabine have a poor chemical stability with in vivo half-lives of 41 and 20 min, respectively, and are most efficient at low concentration [181–183]. Second-generation DNMTis with improved pharmacology and lower toxicity such as the prodrug SGI-110 show high potential for the use in the treatment of several different malignancies [178, 184, 185]. The efficacy of the drugs is varying as only around 50% of the patients respond to treatment [186, 187] and the precise process by which DNMTis work in the patients is not fully understood. It is thought that tumor suppressor genes and genes involved in normal cell cycle control, which are epigenetically silenced, are reactivated [180]. However, evidence for the demethylation is scarce and other mechanisms such as effects on the immune response have repeatedly been reported [188]. The action of DNMTis in combination with standard chemotherapy is also a promising field for the treatment of patients with solid tumors to improve the efficacy of standard chemotherapy as well as immunotherapy and to—at least temporarily—overcome drug resistance [177, 189–192]. Today, no efficient DNMT is have been developed that do not require incorporation into DNA.

Epigenetic therapy is rapidly evolving with many combination therapies now under investigation, and even though the mechanisms of action still need to be uncovered and the lack of specificity of the currently used DNMTis remains an issue of major concern, the near future will probably see more results from the laboratory taken to the patients for a more widespread use of epigenetic drugs in solid cancers.

3.4 Epigenome-Wide Association Studies (EWAS)

In analogy to the quite successful genome-wide association studies (GWAS) analyzing the association of genetic variation with a specific phenotype such as a disease, a condition or a trait, EWAS investigate associations between the epigenetic variation—mainly DNA methylation changes—and these phenotypes. The study of epigenetic influences on complex diseases and traits has gained momentum after the introduction of technologies allowing large-scale studies of DNA methylation analyzing a considerable number of CpG sites throughout the human genome using microarray and sequencing-based profiling technologies, many of which are described in this volume [193, 194].

The design of an EWAS is important, especially with respect to sample selection, which is the main challenge for these studies [195] as DNA methylation patterns change with time and are tissue and developmental-stage specific (*see* also Chapter 2 for important issues on the design and analysis of EWAS). For most EWASs the most relevant tissue for a disease or phenotype under investigation is not available and often analyses are restricted to accessible tissue sources, e.g., saliva, urine, blood, and hair. Depending on the question asked, this might significantly complicate the analysis of, for example, brain-based diseases.

DNA methylation changes are the result of complex interactions between genetic and environmental factors [196, 197]. An example of an environmental factor strongly associated with DNA methylation changes is the exposure to cigarette smoke. Several EWAS have been conducted to examine the interaction between cigarette smoke and DNA methylation modifications. The first study used the Illumina 27K Methylation BeadChip, which identified a CpG site in the protease-activated receptor 4 gene (*F2RL3*) with a lower methylation degree in smokers than non-smokers [198]. This site was confirmed differentially methylated in two subsequent studies using the Illumina 450K Methylation BeadChip [199, 200]. DNA methylation changes associated with smoking are generally not influenced by leucocyte composition, suggesting that the association between DNA methylation changes and smoking is unlikely to be cell-type-specific [201]. Additional EWAS found a strong link between tobacco use and changes in the methylome in either blood cells or placental tissue with many CpGs associated as well as a direct association between smoking intensity, the level of DNA methylation at specific loci and serum or urine cotinine levels [202–204].

EWAS have been performed on many phenotypes, traits, and diseases with now more than 500 EWAS published and many more large-scale studies are likely to be conducted in the near future linking complex diseases and traits with changes in the epigenome. Very interestingly, in some cases EWAS explain much more of the phenotypic variability than the analysis of genetic variation as exemplified by the associations of DNA methylation with the levels of circulating IgE [205].

3.5 DNA Methylation and Complex Disease

Research on the epigenetic influence on complex disease etiology has increased rapidly in recent years, as DNA methylation is a modification that can mediate the effects of environmental exposures on gene expression and thereby disease-relevant phenotypes. Epigenetic mechanisms are consistent with various non-Mendelian features of multifactorial diseases such as the relatively high degree of discordance in monozygotic twins. Several neuropsychiatric disorders have been linked to epigenetic changes [206], with for example DNA methylation found to be altered at numerous genes in patients with schizophrenia [207]. Epigenetic dysregulation in cognitive disorders such as Alzheimer's and Huntigton's disease as well as age-related memory decline has also been reported [208–210]. DNA methylation patterns are also disturbed in inflammatory and autoimmune diseases such as the lupus erythematosus, rheumatoid arthritis, multiple sclerosis, and allergic diseases including asthma [211–214]. Epigenetic changes are also involved in the pathogenesis of atherosclerosis [215], diabetes [216], metabolic syndrome, and intermediated phenotypes, where disease susceptibility seems to be influenced by the maternal *in utero* environment and epidemiological evidence implicates also paternal behavior [217, 218]. It is interesting to note that in a number of studies the epigenetic changes are located in the same genomic region as genetic variation previously associated with the disease as, e.g., shown in autoimmune and inflammatory diseases [212, 219, 220]. It is difficult to infer causality from EWAS, but at least in some case DNA methylation seems to mediate the effects of genetic variation to yield the phenotype [221]. To further underline the scope of epigenetic alterations in disease, it is interesting to point out that studies have shown that even monogenetic diseases such as α-thalassemia that have previously been attributed solely to genetic alterations can also be caused by epigenetic alterations at the same locus [222]. Similar to the above-mentioned examples in cancer,

DNA methylation might also predict treatment response in other complex diseases exemplified by the level of DNA methylation of the *FMR1* promoter predicting the response to the mGluR5 antagonist AFQ056 in patients with fragile X syndrome [223]. The field of epigenetics of complex diseases is still relatively young, but epigenetics may provide the missing link between the genetic susceptibility and the phenotype by mediating and modulating environmental influences. While pioneered by the cancer field a number of recent studies have also shown the potential to detect DNA methylation changes of complex diseases in the plasma or other body fluids making use of tissue-specific DNA methylation signatures and constitute a proof-of-principle for a novel approach of detecting and monitoring complex diseases [224]. The absence of methylation at the insulin promoter is a specificity of insulin-producing pancreatic β-cells and the decrease of the DNA methylation degree at the *INS* promoter can therefore be used to monitor the proportion of β-cell-derived DNA molecules in ccfDNA, which was found to be significantly elevated in patients with type 1 diabetes [225–227]. Similarly, DNA methylation levels at specific CpGs of the *PPARγ* promoter, which has previously shown to stratify fibrosis severity in liver biopsies of patients with non-alcoholic fatty liver disease, have also been found to be elevated in circulating cell-free DNA in patients with non-alcoholic fatty liver disease or alcoholic liver disease compared to controls and the degree of hypermethylation correlated positively with the fibrosis score [228]. Common to these and other examples is the observation that DNA molecules of dying cells are shed into the blood stream, while DNA methylation changes are hardly detectable in circulating molecules during quiescent disease [224]. Detection of cell-free methylated (or unmethylated) fetal DNA can also be used to differentiate fetal DNA molecules, which are circulating within the maternal blood stream, from maternal DNA and allows for noninvasive prenatal diagnosis [229, 230].

Changes in 5-hydroxymethylation have been little studied in other diseases than cancer with the exception of neurodegenerative and other brain-related diseases [231, 232] as well as behavioral phenotypes [233]. However for the moment results have been conflicting with some studies reporting an increase [234], while others report a decrease in, e.g., Alzheimer's disease [235]. More resolutive analyses of 5-hydroxymethylation in larger sample cohorts are required to clarify current contradictions.

4 Conclusions

Although epigenetics in general and DNA methylation research in particular are advancing at a breathtaking speed, we are probably only at the tip of the iceberg and we will see a large increase in the number of large EWAS studies and the investigation of epigenetic changes in complex diseases. Our understanding of the epigenetic changes and their consequences in development and disease will improve as we gain more knowledge of the effect of environmental and chemical exposures. This knowledge together with the now available tools to modify the epigenome might ultimately improve existing treatments and create new options to prevent, slow down the progress, or eventually cure some diseases.

References

1. Waddington CH (1942) The epigenotype. Endeavour 1:18–20
2. Tost J (2008) Epigenetics. Horizon Scientific Press, Norwich, UK
3. Dawson MA, Kouzarides T (2012) Cancer epigenetics: from mechanism to therapy. Cell 150:12–27
4. Chen T, Dent SY (2014) Chromatin modifiers and remodellers: regulators of cellular differentiation. Nat Rev Genet 15:93–106
5. Zentner GE, Henikoff S (2013) Regulation of nucleosome dynamics by histone modifications. Nat Struct Mol Biol 20:259–266
6. Munshi A, Shafi G, Aliya N et al (2009) Histone modifications dictate specific biological readouts. J Genet Genomics 36:75–88
7. Cedar H, Bergman Y (2009) Linking DNA methylation and histone modification: patterns and paradigms. Nat Rev Genet 10:295–304
8. Bird A (2002) DNA methylation patterns and epigenetic memory. Genes Dev 16:6–21
9. Lister R, Pelizzola M, Dowen RH et al (2009) Human DNA methylomes at base resolution show widespread epigenomic differences. Nature 462:315–322
10. Illingworth RS, Bird AP (2009) CpG islands—'a rough guide'. FEBS Lett 583:1713–1720
11. Gardiner-Garden M, Frommer M (1987) CpG islands in vertebrate genomes. J Mol Biol 196:261–282
12. Takai D, Jones PA (2002) Comprehensive analysis of CpG islands in human chromosomes 21 and 22. Proc Natl Acad Sci U S A 99:3740–3745
13. Shen L, Kondo Y, Guo Y et al (2007) Genome-wide profiling of DNA methylation reveals a class of normally methylated CpG island promoters. PLoS Genet 3:2023–2036
14. Illingworth R, Kerr A, Desousa D et al (2008) A novel CpG island set identifies tissue-specific methylation at developmental gene loci. PLoS Biol 6:e22
15. Altun G, Loring JF, Laurent LC (2010) DNA methylation in embryonic stem cells. J Cell Biochem 109:1–6
16. Maunakea AK, Nagarajan RP, Bilenky M et al (2010) Conserved role of intragenic DNA methylation in regulating alternative promoters. Nature 466:253–257
17. Irizarry RA, Ladd-Acosta C, Wen B et al (2009) The human colon cancer methylome shows similar hypo- and hypermethylation at conserved tissue-specific CpG island shores. Nat Genet 41:178–186
18. Lee SM, Choi WY, Lee J et al (2015) The regulatory mechanisms of intragenic DNA methylation. Epigenomics 7:527–531
19. Kulis M, Queiros AC, Beekman R et al (2013) Intragenic DNA methylation in transcriptional regulation, normal differentiation and cancer. Biochim Biophys Acta 1829:1161–1174
20. Lev Maor G, Yearim A, Ast G (2015) The alternative role of DNA methylation in splicing regulation. Trends Genet 31:274–280
21. Bock C, Paulsen M, Tierling S et al (2006) CpG island methylation in human lymphocytes is highly correlated with DNA sequence, repeats, and predicted DNA structure. PLoS Genet 2:e26
22. Jia D, Jurkowska RZ, Zhang X et al (2007) Structure of Dnmt3a bound to Dnmt3L suggests a model for de novo DNA methylation. Nature 449:248–251
23. Ulrey CL, Liu L, Andrews LG et al (2005) The impact of metabolism on DNA methylation. Hum Mol Genet 14 Spec No 1: R139–147
24. Klose RJ, Bird AP (2006) Genomic DNA methylation: the mark and its mediators. Trends Biochem Sci 31:89–97
25. Cheng X, Blumenthal RM (2008) Mammalian DNA methyltransferases: a structural perspective. Structure 16:341–350
26. Ludwig AK, Zhang P, Cardoso MC (2016) Modifiers and readers of DNA modifications and their impact on genome structure,

expression, and stability in disease. Front Genet 7:115

27. Hermann A, Goyal R, Jeltsch A (2004) The Dnmt1 DNA-(cytosine-C5)-methyltransferase methylates DNA processively with high preference for hemimethylated target sites. J Biol Chem 279:48350–48359
28. Jones PA, Liang G (2009) Rethinking how DNA methylation patterns are maintained. Nat Rev Genet 10:805–811
29. Baubec T, Colombo DF, Wirbelauer C et al (2015) Genomic profiling of DNA methyltransferases reveals a role for DNMT3B in genic methylation. Nature 520:243–247
30. Chedin F (2011) The DNMT3 family of mammalian de novo DNA methyltransferases. Prog Mol Biol Transl Sci 101:255–285
31. Franchini DM, Schmitz KM, Petersen-Mahrt SK (2012) 5-Methylcytosine DNA demethylation: more than losing a methyl group. Annu Rev Genet 46:419–441
32. Tahiliani M, Koh KP, Shen Y et al (2009) Conversion of 5-methylcytosine to 5-hydroxymethylcytosine in mammalian DNA by MLL partner TET1. Science 324:930–935
33. Tan L, Shi YG (2012) Tet family proteins and 5-hydroxymethylcytosine in development and disease. Development 139:1895–1902
34. Nestor CE, Ottaviano R, Reddington J et al (2012) Tissue type is a major modifier of the 5-hydroxymethylcytosine content of human genes. Genome Res 22:467–477
35. Langemeijer SM, Aslanyan MG, Jansen JH (2009) TET proteins in malignant hematopoiesis. Cell Cycle 8:4044–4048
36. Pastor WA, Pape UJ, Huang Y et al (2011) Genome-wide mapping of 5-hydroxymethylcytosine in embryonic stem cells. Nature 473:394–397
37. Ficz G, Branco MR, Seisenberger S et al (2011) Dynamic regulation of 5-hydroxymethylcytosine in mouse ES cells and during differentiation. Nature 473:398–402
38. Jin SG, Wu X, Li AX et al (2011) Genomic mapping of 5-hydroxymethylcytosine in the human brain. Nucleic Acids Res 39:5015–5024
39. Ono R, Taki T, Taketani T et al (2002) LCX, leukemia-associated protein with a CXXC domain, is fused to MLL in acute myeloid leukemia with trilineage dysplasia having t(10;11)(q22;q23). Cancer Res 62:4075–4080
40. Mercher T, Quivoron C, Couronne L et al (2012) TET2, a tumor suppressor in hematological disorders. Biochim Biophys Acta 1825:173–177
41. Putiri EL, Tiedemann RL, Thompson JJ et al (2014) Distinct and overlapping control of 5-methylcytosine and 5-hydroxymethylcytosine by the TET proteins in human cancer cells. Genome Biol 15: R81
42. Jin SG, Jiang Y, Qiu R et al (2011) 5-Hydroxymethylcytosine is strongly depleted in human cancers but its levels do not correlate with IDH1 mutations. Cancer Res 71:7360–7365
43. Wu SC, Zhang Y (2010) Active DNA demethylation: many roads lead to Rome. Nat Rev Mol Cell Biol 11:607–620
44. Ito S, Shen L, Dai Q et al (2011) Tet proteins can convert 5-methylcytosine to 5-formylcytosine and 5-carboxylcytosine. Science 333:1300–1303
45. Geiman TM, Robertson KD (2002) Chromatin remodeling, histone modifications, and DNA methylation-how does it all fit together? J Cell Biochem 87:117–125
46. Yin Y, Morgunova E, Jolma A et al (2017) Impact of cytosine methylation on DNA binding specificities of human transcription factors. Science 356(6337). https://doi.org/10.1126/science.aaj2239
47. Hu S, Wan J, Su Y et al (2013) DNA methylation presents distinct binding sites for human transcription factors. eLife 2:e00726
48. Sasai N, Defossez PA (2009) Many paths to one goal? The proteins that recognize methylated DNA in eukaryotes. Int J Dev Biol 53:323–334
49. Baubec T, Ivanek R, Lienert F et al (2013) Methylation-dependent and -independent genomic targeting principles of the MBD protein family. Cell 153:480–492
50. Yildirim O, Li R, Hung JH et al (2011) Mbd3/NURD complex regulates expression of 5-hydroxymethylcytosine marked genes in embryonic stem cells. Cell 147:1498–1510
51. Brinkman AB, Pennings SW, Braliou GG et al (2007) DNA methylation immediately adjacent to active histone marking does not silence transcription. Nucleic Acids Res 35:801–811
52. Ushijima T (2005) Detection and interpretation of altered methylation patterns in cancer cells. Nat Rev Cancer 5:223–231
53. Stadler MB, Murr R, Burger L et al (2011) DNA-binding factors shape the mouse methylome at distal regulatory regions. Nature 480:490–495

54. Hodges E, Molaro A, Dos Santos CO et al (2011) Directional DNA methylation changes and complex intermediate states accompany lineage specificity in the adult hematopoietic compartment. Mol Cell 44:17–28
55. Lippman Z, Gendrel AV, Black M et al (2004) Role of transposable elements in heterochromatin and epigenetic control. Nature 430:471–476
56. Walsh CP, Chaillet JR, Bestor TH (1998) Transcription of IAP endogenous retroviruses is constrained by cytosine methylation. Nat Genet 20:116–117
57. Yoder JA, Walsh CP, Bestor TH (1997) Cytosine methylation and the ecology of intragenomic parasites. Trends Genet 13:335–340
58. Baylin SB, Jones PA (2016) Epigenetic determinants of cancer. Cold Spring Harb Perspect Biol 8. https://doi.org/10.1101/cshperspect.a019505
59. Reik W, Dean W, Walter J (2001) Epigenetic reprogramming in mammalian development. Science 293:1089–1093
60. Guibert S, Weber M (2013) Functions of DNA methylation and hydroxymethylation in mammalian development. Curr Top Dev Biol 104:47–83
61. Smith ZD, Meissner A (2013) DNA methylation: roles in mammalian development. Nat Rev Genet 14:204–220
62. Jackson M, Krassowska A, Gilbert N et al (2004) Severe global DNA hypomethylation blocks differentiation and induces histone hyperacetylation in embryonic stem cells. Mol Cell Biol 24:8862–8871
63. Kim K, Doi A, Wen B et al (2010) Epigenetic memory in induced pluripotent stem cells. Nature 467:285–290
64. Ohi Y, Qin H, Hong C et al (2011) Incomplete DNA methylation underlies a transcriptional memory of somatic cells in human iPS cells. Nat Cell Biol 13:541–549
65. Smith ZD, Chan MM, Humm KC et al (2014) DNA methylation dynamics of the human preimplantation embryo. Nature 511:611–615
66. Carrell DT (2012) Epigenetics of the male gamete. Fertil Steril 97:267–274
67. Nakatani T, Yamagata K, Kimura T et al (2015) Stella preserves maternal chromosome integrity by inhibiting 5hmC-induced gammaH2AX accumulation. EMBO Rep 16:582–589
68. Wang L, Zhang J, Duan J et al (2014) Programming and inheritance of parental DNA methylomes in mammals. Cell 157:979–991
69. Lees-Murdock DJ, Walsh CP (2008) DNA methylation reprogramming in the germ line. Epigenetics 3:5–13
70. Seisenberger S, Andrews S, Krueger F et al (2012) The dynamics of genome-wide DNA methylation reprogramming in mouse primordial germ cells. Mol Cell 48:849–862
71. Boissonnas CC, Abdalaoui HE, Haelewyn V et al (2010) Specific epigenetic alterations of IGF2-H19 locus in spermatozoa from infertile men. Eur J Hum Genet 18:73–80
72. Gunes S, Arslan MA, Hekim GNT et al (2016) The role of epigenetics in idiopathic male infertility. J Assist Reprod Genet 33:553–569
73. Hanson MA, Gluckman PD (2014) Early developmental conditioning of later health and disease: physiology or pathophysiology? Physiol Rev 94:1027–1076
74. Fauque P, Ripoche MA, Tost J et al (2010) Modulation of imprinted gene network in placenta results in normal development of in vitro manipulated mouse embryos. Hum Mol Genet 19:1779–1790
75. Nelissen EC, Dumoulin JC, Daunay A et al (2013) Placentas from pregnancies conceived by IVF/ICSI have a reduced DNA methylation level at the H19 and MEST differentially methylated regions. Hum Reprod 28:1117–1126
76. Castillo-Fernandez JE, Loke YJ, Bass-Stringer S et al (2017) DNA methylation changes at infertility genes in newborn twins conceived by in vitro fertilisation. Genome Med 9:28
77. Uyar A, Seli E (2014) The impact of assisted reproductive technologies on genomic imprinting and imprinting disorders. Curr Opin Obstet Gynecol 26:210–221
78. Chiba H, Hiura H, Okae H et al (2013) DNA methylation errors in imprinting disorders and assisted reproductive technology. Pediatr Int 55:542–549
79. Feil R, Fraga MF (2012) Epigenetics and the environment: emerging patterns and implications. Nat Rev Genet 13:97–109
80. Heyn H, Li N, Ferreira HJ et al (2012) Distinct DNA methylomes of newborns and centenarians. Proc Natl Acad Sci U S A 109:10522–10527
81. Hannum G, Guinney J, Zhao L et al (2013) Genome-wide methylation profiles reveal quantitative views of human aging rates. Mol Cell 49:359–367
82. Jones MJ, Goodman SJ, Kobor MS (2015) DNA methylation and healthy human aging. Aging Cell 14:924–932

83. Fraga MF, Esteller M (2007) Epigenetics and aging: the targets and the marks. Trends Genet 23:413–418
84. Issa JP (2014) Aging and epigenetic drift: a vicious cycle. J Clin Invest 124:24–29
85. Horvath S (2013) DNA methylation age of human tissues and cell types. Genome Biol 14:R115
86. Weidner CI, Lin Q, Koch CM et al (2014) Aging of blood can be tracked by DNA methylation changes at just three CpG sites. Genome Biol 15:R24
87. Marioni RE, Shah S, McRae AF et al (2015) DNA methylation age of blood predicts all-cause mortality in later life. Genome Biol 16:25
88. Zheng Y, Joyce BT, Colicino E et al (2016) Blood epigenetic age may predict cancer incidence and mortality. EBioMedicine 5:68–73
89. Perna L, Zhang Y, Mons U et al (2016) Epigenetic age acceleration predicts cancer, cardiovascular, and all-cause mortality in a German case cohort. Clin Epigenetics 8:64
90. Armstrong NJ, Mather KA, Thalamuthu A et al (2017) Aging, exceptional longevity and comparisons of the Hannum and Horvath epigenetic clocks. Epigenomics 9:689–700
91. Reik W, Walter J (2001) Genomic imprinting: parental influence on the genome. Nat Rev Genet 2:21–32
92. Skaar DA, Li Y, Bernal AJ et al (2012) The human imprintome: regulatory mechanisms, methods of ascertainment, and roles in disease susceptibility. ILAR J 53:341–358
93. Kelsey G, Bartolomei MS (2012) Imprinted genes ... and the number is? PLoS Genet 8: e1002601
94. Barbaux S, Gascoin-Lachambre G, Buffat C et al (2012) A genome-wide approach reveals novel imprinted genes expressed in the human placenta. Epigenetics 7:1079–1090
95. Bell AC, Felsenfeld G (2000) Methylation of a CTCF-dependent boundary controls imprinted expression of the Igf2 gene. Nature 405:482–485
96. Kurukuti S, Tiwari VK, Tavoosidana G et al (2006) CTCF binding at the H19 imprinting control region mediates maternally inherited higher-order chromatin conformation to restrict enhancer access to Igf2. Proc Natl Acad Sci U S A 103:10684–10689
97. Lee JT, Bartolomei MS (2013) X-inactivation, imprinting, and long noncoding RNAs in health and disease. Cell 152:1308–1323
98. Gendrel AV, Heard E (2014) Noncoding RNAs and epigenetic mechanisms during X-chromosome inactivation. Annu Rev Cell Dev Biol 30:561–580
99. Li C, Zhao S, Zhang N et al (2013) Differences of DNA methylation profiles between monozygotic twins' blood samples. Mol Biol Rep 40:5275–5280
100. Fraga MF, Ballestar E, Paz MF et al (2005) Epigenetic differences arise during the lifetime of monozygotic twins. Proc Natl Acad Sci U S A 102:10604–10609
101. Bollati V, Baccarelli A, Hou L et al (2007) Changes in DNA methylation patterns in subjects exposed to low-dose benzene. Cancer Res 67:876–880
102. Xin F, Susiarjo M, Bartolomei MS (2015) Multigenerational and transgenerational effects of endocrine disrupting chemicals: a role for altered epigenetic regulation? Semin Cell Dev Biol 43:66–75
103. Hanson MA, Skinner MK (2016) Developmental origins of epigenetic transgenerational inheritance. Environ Epigenet 2:dvw002
104. Ladd-Acosta C (2015) Epigenetic signatures as biomarkers of exposure. Curr Environ Health Rep 2:117–125
105. Bohacek J, Mansuy IM (2015) Molecular insights into transgenerational non-genetic inheritance of acquired behaviours. Nat Rev Genet 16:641–652
106. Kappil M, Lambertini L, Chen J (2015) Environmental influences on genomic imprinting. Curr Environ Health Rep 2:155–162
107. Cuzin F, Grandjean V, Rassoulzadegan M (2008) Inherited variation at the epigenetic level: paramutation from the plant to the mouse. Curr Opin Genet Dev 18:193–196
108. Heard E, Martienssen RA (2014) Transgenerational epigenetic inheritance: myths and mechanisms. Cell 157:95–109
109. Bohacek J, Mansuy IM (2017) A guide to designing germline-dependent epigenetic inheritance experiments in mammals. Nat Methods 14:243–249
110. de Groote ML, Verschure PJ, Rots MG (2012) Epigenetic editing: targeted rewriting of epigenetic marks to modulate expression of selected target genes. Nucleic Acids Res 40:10596–10613
111. Carroll D (2014) Genome engineering with targetable nucleases. Annu Rev Biochem 83:409–439
112. Jurkowski TP, Ravichandran M, Stepper P (2015) Synthetic epigenetics-towards intelligent control of epigenetic states and cell identity. Clin Epigenetics 7:18
113. Laufer BI, Singh SM (2015) Strategies for precision modulation of gene expression by

epigenome editing: an overview. Epigenetics Chromatin 8:34
114. Maeder ML, Angstman JF, Richardson ME et al (2013) Targeted DNA demethylation and activation of endogenous genes using programmable TALE-TET1 fusion proteins. Nat Biotechnol 31:1137–1142
115. Straubeta A, Lahaye T (2013) Zinc fingers, TAL effectors, or Cas9-based DNA binding proteins: what's best for targeting desired genome loci? Mol Plant 6:1384–1387
116. Choudhury SR, Cui Y, Lubecka K et al (2016) CRISPR-dCas9 mediated TET1 targeting for selective DNA demethylation at BRCA1 promoter. Oncotarget 7:46545–46556
117. Thakore PI, Black JB, Hilton IB et al (2016) Editing the epigenome: technologies for programmable transcription and epigenetic modulation. Nat Methods 13:127–137
118. Vojta A, Dobrinic P, Tadic V et al (2016) Repurposing the CRISPR-Cas9 system for targeted DNA methylation. Nucleic Acids Res 44:5615–5628
119. Koferle A, Worf K, Breunig C et al (2016) CORALINA: a universal method for the generation of gRNA libraries for CRISPR-based screening. BMC Genomics 17:917
120. Weng YL, An R, Shin J et al (2013) DNA modifications and neurological disorders. Neurotherapeutics 10:556–567
121. Delhommeau F, Dupont S, Della Valle V et al (2009) Mutation in TET2 in myeloid cancers. N Engl J Med 360:2289–2301
122. Leonard H, Cobb S, Downs J (2017) Clinical and biological progress over 50 years in Rett syndrome. Nat Rev Neurol 13:37–51
123. Chiba S (2017) Dysregulation of TET2 in hematologic malignancies. Int J Hematol 105:17–22
124. Ishida M, Moore GE (2013) The role of imprinted genes in humans. Mol Asp Med 34:826–840
125. Elhamamsy AR (2017) Role of DNA methylation in imprinting disorders: an updated review. J Assist Reprod Genet 34:549–562
126. Feinberg AP, Vogelstein B (1983) Hypomethylation distinguishes genes of some human cancers from their normal counterparts. Nature 301:89–92
127. Gaudet F, Hodgson JG, Eden A et al (2003) Induction of tumors in mice by genomic hypomethylation. Science 300:489–492
128. Ehrlich M, Lacey M (2013) DNA hypomethylation and hemimethylation in cancer. Adv Exp Med Biol 754:31–56
129. Hur K, Cejas P, Feliu J et al (2014) Hypomethylation of long interspersed nuclear element-1 (LINE-1) leads to activation of proto-oncogenes in human colorectal cancer metastasis. Gut 63:635–646
130. Virani S, Colacino JA, Kim JH et al (2012) Cancer epigenetics: a brief review. ILAR J 53:359–369
131. Hanahan D, Weinberg RA (2011) Hallmarks of cancer: the next generation. Cell 144:646–674
132. Kaneda A, Feinberg AP (2005) Loss of imprinting of IGF2: a common epigenetic modifier of intestinal tumor risk. Cancer Res 65:11236–11240
133. Clark SJ (2007) Action at a distance: epigenetic silencing of large chromosomal regions in carcinogenesis. Hum Mol Genet 16 Spec No 1:R88–R95
134. Frigola J, Song J, Stirzaker C et al (2006) Epigenetic remodeling in colorectal cancer results in coordinate gene suppression across an entire chromosome band. Nat Genet 38:540–549
135. Bert SA, Robinson MD, Strbenac D et al (2013) Regional activation of the cancer genome by long-range epigenetic remodeling. Cancer Cell 23:9–22
136. Berman BP, Weisenberger DJ, Aman JF et al (2011) Regions of focal DNA hypermethylation and long-range hypomethylation in colorectal cancer coincide with nuclear lamina-associated domains. Nat Genet 44:40–46
137. Simo-Riudalbas L, Esteller M (2014) Cancer genomics identifies disrupted epigenetic genes. Hum Genet 133:713–725
138. Shen H, Laird PW (2013) Interplay between the cancer genome and epigenome. Cell 153:38–55
139. Plass C, Pfister SM, Lindroth AM et al (2013) Mutations in regulators of the epigenome and their connections to global chromatin patterns in cancer. Nat Rev Genet 14:765–780
140. Turcan S, Rohle D, Goenka A et al (2012) IDH1 mutation is sufficient to establish the glioma hypermethylator phenotype. Nature 483:479–483
141. Figueroa ME, Abdel-Wahab O, Lu C et al (2010) Leukemic IDH1 and IDH2 mutations result in a hypermethylation phenotype, disrupt TET2 function, and impair hematopoietic differentiation. Cancer Cell 18:553–567
142. Hedenfalk I, Duggan D, Chen Y et al (2001) Gene-expression profiles in hereditary breast cancer. N Engl J Med 344:539–548

143. Nik-Zainal S, Davies H, Staaf J et al (2016) Landscape of somatic mutations in 560 breast cancer whole-genome sequences. Nature 534:47–54
144. Knudson AG (2001) Two genetic hits (more or less) to cancer. Nat Rev Cancer 1:157–162
145. Balmain A, Gray J, Ponder B (2003) The genetics and genomics of cancer. Nat Genet 33(Suppl):238–244
146. Li S, Garrett-Bakelman FE, Chung SS et al (2016) Distinct evolution and dynamics of epigenetic and genetic heterogeneity in acute myeloid leukemia. Nat Med 22:792–799
147. Costello JF, Fruhwald MC, Smiraglia DJ et al (2000) Aberrant CpG-island methylation has non-random and tumour-type-specific patterns. Nat Genet 24:132–138
148. Goelz SE, Vogelstein B, Hamilton SR et al (1985) Hypomethylation of DNA from benign and malignant human colon neoplasms. Science 228:187–190
149. Feinberg AP, Ohlsson R, Henikoff S (2006) The epigenetic progenitor origin of human cancer. Nat Rev Genet 7:21–33
150. Issa JP, Ahuja N, Toyota M et al (2001) Accelerated age-related CpG island methylation in ulcerative colitis. Cancer Res 61:3573–3577
151. Fleischer T, Frigessi A, Johnson KC et al (2014) Genome-wide DNA methylation profiles in progression to in situ and invasive carcinoma of the breast with impact on gene transcription and prognosis. Genome Biol 15:435
152. Ohm JE, McGarvey KM, Yu X et al (2007) A stem cell-like chromatin pattern may predispose tumor suppressor genes to DNA hypermethylation and heritable silencing. Nat Genet 39:237–242
153. Schlesinger Y, Straussman R, Keshet I et al (2007) Polycomb-mediated methylation on Lys27 of histone H3 pre-marks genes for de novo methylation in cancer. Nat Genet 39:232–236
154. Yang H, Liu Y, Bai F et al (2013) Tumor development is associated with decrease of TET gene expression and 5-methylcytosine hydroxylation. Oncogene 32:663–669
155. Lian CG, Xu Y, Ceol C et al (2012) Loss of 5-hydroxymethylcytosine is an epigenetic hallmark of melanoma. Cell 150:1135–1146
156. Neri F, Dettori D, Incarnato D et al (2015) TET1 is a tumour suppressor that inhibits colon cancer growth by derepressing inhibitors of the WNT pathway. Oncogene 34:4168–4176
157. Uribe-Lewis S, Stark R, Carroll T et al (2015) 5-Hydroxymethylcytosine marks promoters in colon that resist DNA hypermethylation in cancer. Genome Biol 16:69
158. Simon R (2005) Roadmap for developing and validating therapeutically relevant genomic classifiers. J Clin Oncol 23:7332–7341
159. BLUEPRINT consortium (2016) Quantitative comparison of DNA methylation assays for biomarker development and clinical applications. Nat Biotechnol 34:726–737
160. Laird PW (2003) Early detection: the power and the promise of DNA methylation markers. Nat Rev Cancer 3:253–266
161. Silva JM, Dominguez G, Garcia JM et al (1999) Presence of tumor DNA in plasma of breast cancer patients: clinicopathological correlations. Cancer Res 59:3251–3256
162. Akhavan-Niaki H, Samadani AA (2013) DNA methylation and cancer development: molecular mechanism. Cell Biochem Biophys 67:501–513
163. Warton K, Mahon KL, Samimi G (2016) Methylated circulating tumor DNA in blood: power in cancer prognosis and response. Endocr Relat Cancer 23: R157–R171
164. Lamb YN, Dhillon S (2017) Epi proColon (R) 2.0 CE: a blood-based screening test for colorectal cancer. Mol Diagn Ther 21:225–232
165. Church TR, Wandell M, Lofton-Day C et al (2014) Prospective evaluation of methylated SEPT9 in plasma for detection of asymptomatic colorectal cancer. Gut 63:317–325
166. How Kit A, Nielsen HM, Tost J (2012) DNA methylation based biomarkers: practical considerations and applications. Biochimie 94:2314–2337
167. Brocks D, Assenov Y, Minner S et al (2014) Intratumor DNA methylation heterogeneity reflects clonal evolution in aggressive prostate cancer. Cell Rep 8:798–806
168. Landau DA, Clement K, Ziller MJ et al (2014) Locally disordered methylation forms the basis of intratumor methylome variation in chronic lymphocytic leukemia. Cancer Cell 26:813–825
169. Sheffield NC, Pierron G, Klughammer J et al (2017) DNA methylation heterogeneity defines a disease spectrum in Ewing sarcoma. Nat Med 23:386–395
170. Moran S, Martinez-Cardus A, Sayols S et al (2016) Epigenetic profiling to classify cancer of unknown primary: a multicentre, retrospective analysis. Lancet Oncol 17:1386–1395
171. Guo S, Diep D, Plongthongkum N et al (2017) Identification of methylation

haplotype blocks aids in deconvolution of heterogeneous tissue samples and tumor tissue-of-origin mapping from plasma DNA. Nat Genet 49:635–642
172. Kang S, Li Q, Chen Q et al (2017) CancerLocator: non-invasive cancer diagnosis and tissue-of-origin prediction using methylation profiles of cell-free DNA. Genome Biol 18:53
173. Sun K, Jiang P, Chan KC et al (2015) Plasma DNA tissue mapping by genome-wide methylation sequencing for noninvasive prenatal, cancer, and transplantation assessments. Proc Natl Acad Sci U S A 112:E5503–E5512
174. Hegi ME, Diserens AC, Gorlia T et al (2005) MGMT gene silencing and benefit from temozolomide in glioblastoma. N Engl J Med 352:997–1003
175. Weller M, Stupp R, Reifenberger G et al (2010) MGMT promoter methylation in malignant gliomas: ready for personalized medicine? Nat Rev Neurol 6:39–51
176. Clozel T, Yang S, Elstrom RL et al (2013) Mechanism-based epigenetic chemosensitization therapy of diffuse large B-cell lymphoma. Cancer Discov 3:1002–1019
177. Ahuja N, Easwaran H, Baylin SB (2014) Harnessing the potential of epigenetic therapy to target solid tumors. J Clin Invest 124:56–63
178. Treppendahl MB, Kristensen LS, Gronbaek K (2014) Predicting response to epigenetic therapy. J Clin Invest 124:47–55
179. Gros C, Fahy J, Halby L et al (2012) DNA methylation inhibitors in cancer: recent and future approaches. Biochimie 94:2280–2296
180. Gnyszka A, Jastrzebski Z, Flis S (2013) DNA methyltransferase inhibitors and their emerging role in epigenetic therapy of cancer. Anticancer Res 33:2989–2996
181. Yang X, Lay F, Han H et al (2010) Targeting DNA methylation for epigenetic therapy. Trends Pharmacol Sci 31:536–546
182. Qin T, Jelinek J, Si J et al (2009) Mechanisms of resistance to 5-aza-2′-deoxycytidine in human cancer cell lines. Blood 113:659–667
183. Stresemann C, Brueckner B, Musch T et al (2006) Functional diversity of DNA methyltransferase inhibitors in human cancer cell lines. Cancer Res 66:2794–2800
184. Herranz M, Martin-Caballero J, Fraga MF et al (2006) The novel DNA methylation inhibitor zebularine is effective against the development of murine T-cell lymphoma. Blood 107:1174–1177
185. Fahy J, Jeltsch A, Arimondo PB (2012) DNA methyltransferase inhibitors in cancer: a chemical and therapeutic patent overview and selected clinical studies. Expert Opin Ther Pat 22:1427–1442
186. Schecter J, Galili N, Raza A (2012) MDS: refining existing therapy through improved biologic insights. Blood Rev 26:73–80
187. Lee YG, Kim I, Yoon SS et al (2013) Comparative analysis between azacitidine and decitabine for the treatment of myelodysplastic syndromes. Br J Haematol 161:339–347
188. Linnekamp JF, Butter R, Spijker R et al (2017) Clinical and biological effects of demethylating agents on solid tumours—a systematic review. Cancer Treat Rev 54:10–23
189. Cowan LA, Talwar S, Yang AS (2010) Will DNA methylation inhibitors work in solid tumors? A review of the clinical experience with azacitidine and decitabine in solid tumors. Epigenomics 2:71–86
190. Appleton K, Mackay HJ, Judson I et al (2007) Phase I and pharmacodynamic trial of the DNA methyltransferase inhibitor decitabine and carboplatin in solid tumors. J Clin Oncol 25:4603–4609
191. Fang F, Munck J, Tang J et al (2014) The novel, small-molecule DNA methylation inhibitor SGI-110 as an ovarian cancer chemosensitizer. Clin Cancer Res 20:6504–6516
192. Li H, Chiappinelli KB, Guzzetta AA et al (2014) Immune regulation by low doses of the DNA methyltransferase inhibitor 5-azacitidine in common human epithelial cancers. Oncotarget 5:587–598
193. Laird PW (2010) Principles and challenges of genomewide DNA methylation analysis. Nat Rev Genet 11:191–203
194. Tost J (2016) Current and emerging technologies for the analysis of the genome-wide and locus-specific DNA methylation patterns. Adv Exp Med Biol 945:343–430
195. Rakyan VK, Down TA, Balding DJ et al (2011) Epigenome-wide association studies for common human diseases. Nat Rev Genet 12:529–541
196. Kaminsky ZA, Tang T, Wang SC et al (2009) DNA methylation profiles in monozygotic and dizygotic twins. Nat Genet 41:240–245
197. Terry MB, Delgado-Cruzata L, Vin-Raviv N et al (2011) DNA methylation in white blood cells: association with risk factors in epidemiologic studies. Epigenetics 6:828–837
198. Breitling LP, Yang R, Korn B et al (2011) Tobacco-smoking-related differential DNA methylation: 27K discovery and replication. Am J Hum Genet 88:450–457
199. Shenker NS, Polidoro S, van Veldhoven K et al (2013) Epigenome-wide association study in the European Prospective

Investigation into Cancer and Nutrition (EPIC-Turin) identifies novel genetic loci associated with smoking. Hum Mol Genet 22:843–851

200. Monick MM, Beach SR, Plume J et al (2012) Coordinated changes in AHRR methylation in lymphoblasts and pulmonary macrophages from smokers. Am J Med Genet B Neuropsychiatr Genet 159B:141–151
201. Heiss JA, Brenner H (2017) Impact of confounding by leukocyte composition on associations of leukocyte DNA methylation with common risk factors. Epigenomics 9:659–668
202. Joubert BR, Felix JF, Yousefi P et al (2016) DNA methylation in newborns and maternal smoking in pregnancy: genome-wide consortium meta-analysis. Am J Hum Genet 98:680–696
203. Zhang Y, Florath I, Saum KU et al (2016) Self-reported smoking, serum cotinine, and blood DNA methylation. Environ Res 146:395–403
204. Lee DH, Hwang SH, Lim MK et al (2017) Performance of urine cotinine and hypomethylation of AHRR and F2RL3 as biomarkers for smoking exposure in a population-based cohort. PLoS One 12:e0176783
205. Liang L, Willis-Owen SA, Laprise C et al (2015) An epigenome-wide association study of total serum immunoglobulin E concentration. Nature 520:670–674
206. Ai S, Shen L, Guo J et al (2012) DNA methylation as a biomarker for neuropsychiatric diseases. Int J Neurosci 122:165–176
207. Hannon E, Dempster E, Viana J et al (2016) An integrated genetic-epigenetic analysis of schizophrenia: evidence for co-localization of genetic associations and differential DNA methylation. Genome Biol 17:176
208. Graff J, Mansuy IM (2009) Epigenetic dysregulation in cognitive disorders. Eur J Neurosci 30:1–8
209. Adwan L, Zawia NH (2013) Epigenetics: a novel therapeutic approach for the treatment of Alzheimer's disease. Pharmacol Ther 139:41–50
210. Day JJ, Sweatt JD (2011) Epigenetic mechanisms in cognition. Neuron 70:813–829
211. Nielsen HM, Tost J (2013) Epigenetic changes in inflammatory and autoimmune diseases. Subcell Biochem 61:455–478
212. Miceli-Richard C, Wang-Renault SF, Boudaoud S et al (2016) Overlap between differentially methylated DNA regions in blood B lymphocytes and genetic at-risk loci in primary Sjogren's syndrome. Ann Rheum Dis 75:933–940
213. Fogel O, Richard-Miceli C, Tost J (2017) Epigenetic changes in chronic inflammatory diseases. Adv Protein Chem Struct Biol 106:139–189
214. Potaczek DP, Harb H, Michel S et al (2017) Epigenetics and allergy: from basic mechanisms to clinical applications. Epigenomics 9:539–571
215. Valencia-Morales Mdel P, Zaina S, Heyn H et al (2015) The DNA methylation drift of the atherosclerotic aorta increases with lesion progression. BMC Med Genet 8:7
216. Ronn T, Ling C (2015) DNA methylation as a diagnostic and therapeutic target in the battle against Type 2 diabetes. Epigenomics 7:451–460
217. Martinez D, Pentinat T, Ribo S et al (2014) In utero undernutrition in male mice programs liver lipid metabolism in the second-generation offspring involving altered Lxra DNA methylation. Cell Metab 19:941–951
218. Jimenez-Chillaron JC, Ramon-Krauel M, Ribo S et al (2016) Transgenerational epigenetic inheritance of diabetes risk as a consequence of early nutritional imbalances. Proc Nutr Soc 75:78–89
219. Liu Y, Aryee MJ, Padyukov L et al (2013) Epigenome-wide association data implicate DNA methylation as an intermediary of genetic risk in rheumatoid arthritis. Nat Biotechnol 31:142–147
220. Low D, Mizoguchi A, Mizoguchi E (2013) DNA methylation in inflammatory bowel disease and beyond. World J Gastroenterol 19:5238–5249
221. Hong X, Hao K, Ladd-Acosta C et al (2015) Genome-wide association study identifies peanut allergy-specific loci and evidence of epigenetic mediation in US children. Nat Commun 6:6304
222. Tufarelli C, Stanley JA, Garrick D et al (2003) Transcription of antisense RNA leading to gene silencing and methylation as a novel cause of human genetic disease. Nat Genet 34:157–165
223. Jacquemont S, Curie A, des Portes V et al (2011) Epigenetic modification of the FMR1 gene in fragile X syndrome is associated with differential response to the mGluR5 antagonist AFQ056. Sci Transl Med 3:64ra61
224. Tost J (2016) Follow the trace of death: methylation analysis of cell-free DNA for clinical applications in non-cancerous diseases. Epigenomics 8:1169–1172

225. Lehmann-Werman R, Neiman D, Zemmour H et al (2016) Identification of tissue-specific cell death using methylation patterns of circulating DNA. Proc Natl Acad Sci U S A 113: E1826–E1834
226. Akirav EM, Lebastchi J, Galvan EM et al (2011) Detection of beta cell death in diabetes using differentially methylated circulating DNA. Proc Natl Acad Sci U S A 108:19018–19023
227. Zhang K, Lin G, Han Y et al (2017) Circulating unmethylated insulin DNA as a potential non-invasive biomarker of beta cell death in type 1 Diabetes: a review and future prospect. Clin Epigenetics 9:44
228. Hardy T, Zeybel M, Day CP et al (2017) Plasma DNA methylation: a potential biomarker for stratification of liver fibrosis in non-alcoholic fatty liver disease. Gut 66 (7):1321–1328
229. Wong AI, Lo YM (2015) Noninvasive fetal genomic, methylomic, and transcriptomic analyses using maternal plasma and clinical implications. Trends Mol Med 21:98–108
230. Wong FC, Lo YM (2016) Prenatal diagnosis innovation: genome sequencing of maternal plasma. Annu Rev Med 67:419–432
231. Al-Mahdawi S, Virmouni SA, Pook MA (2014) The emerging role of 5-hydroxymethylcytosine in neurodegenerative diseases. Front Neurosci 8:397
232. Sherwani SI, Khan HA (2015) Role of 5-hydroxymethylcytosine in neurodegeneration. Gene 570:17–24
233. McEwen BS, Bowles NP, Gray JD et al (2015) Mechanisms of stress in the brain. Nat Neurosci 18:1353–1363
234. Coppieters N, Dieriks BV, Lill C et al (2014) Global changes in DNA methylation and hydroxymethylation in Alzheimer's disease human brain. Neurobiol Aging 35:1334–1344
235. Condliffe D, Wong A, Troakes C et al (2014) Cross-region reduction in 5-hydroxymethylcytosine in Alzheimer's disease brain. Neurobiol Aging 35:1850–1854

Chapter 2

Considerations for Design and Analysis of DNA Methylation Studies

Karin B. Michels and Alexandra M. Binder

Abstract

The number of epigenetic studies is exponentially increasing. There is anticipation that DNA methylation may close gaps in our understanding of disease etiology, and how certain risk factors affect health and disease, but also that it has potential as a biomarker for disease. Human DNA methylation studies require careful considerations for design and analysis including population and tissue selection, population stratification, cell heterogeneity, confounding, temporality, sample size, appropriate statistical analysis, and validation of results. In this chapter, we discuss relevant aspects for the design of DNA methylation studies and delineate essential steps for their analysis. Specifically, we summarize methods used to extricate biologic signals from technical noise, and statistical approaches to capture meaningful variability based on the research hypothesis.

Key words Epigenetics, DNA methylation, Epidemiology, Cohort study, Case-control study, Cross-sectional study, Confounding, Cellular heterogeneity, Biomarkers, EWAS

1 Introduction: DNA Methylation Studies

The rapid advance in both high-throughput technologies and analytic tools to digest, sort, and align the resulting high dimensional data has paved the way for epigenome-wide explorations and moved DNA methylation center stage on the molecular map. Insights gained during the last decade or two have suggested a role for DNA methylation in many essential functions and stages of life [1, 2]. A more complete comprehension of the multifaceted role of DNA methylation and other epigenetic features in health and disease is warranted and is being enthusiastically pursued in many laboratories around the globe. The study of DNA methylation may have a multitude of goals: to understand cellular and developmental plasticity, to provide mechanistic explanations or gain etiologic insights, to find a new biomarker, to disentangle complex disease processes, to develop new therapeutic aids, to name just a few [3]. Depending on the aim, setting up the study,

Jörg Tost (ed.), *DNA Methylation Protocols*, Methods in Molecular Biology, vol. 1708,
https://doi.org/10.1007/978-1-4939-7481-8_2,

choosing the subjects and tissue(s) to study, and accounting for other factors that might affect interpretation of results may differ greatly. A number of challenges present themselves in designing and analyzing such a study [4]:

(a) Choosing the appropriate study design
(b) Selecting the study population
(c) Considering and accounting for confounding factors
(d) Settling on the tissue to study
(e) Cellular heterogeneity in DNA methylation
(f) Choosing the target: candidate genes vs. epigenome-wide DNA methylation
(g) Choosing the appropriate analytic method
(h) Validation and verification of results

In the following, we discuss the most important considerations in the design and analysis of a DNA methylation study.

2 Design of DNA Methylation Studies

2.1 Types of Studies

2.1.1 Biomarker Studies

1. Biomarkers for Disease

Trying to identify a biomarker for disease has mostly one of two goals: finding a biomarker of disease risk or finding a biomarker for early disease detection. Either way, our biomarker needs to be manifest in a tissue that can be readily harvested in order to be useful as a screening tool: e.g., blood, saliva, buccal cells, urine, stool, skin cells, hair, nails, subcutaneous adipose tissue. Note that these tissues may be different from the target tissue: even if we are interested in a biomarker for brain cancer we would not search for DNA methylation changes in brain tissue if we wanted to use the biomarker to screen populations. Given tissue and cell specificity of DNA methylation marks we can use two approaches: to identify DNA marks that are present in both brain tissue and an easily accessible tissue (e.g., buccal cells may lend themselves since they originate from the same developmental germ cell layer as brain cells) or to simply look for DNA methylation marks in the secondary tissue only irrespective of whether the mark is present in brain tissue. Consistency of DNA methylation marks across different populations is particularly important for the latter approach. Note that the yield will be substantially lower if the first option is selected as few DNA methylation marks track across multiple tissues.

Biomarker for Disease Risk: If the goal is to find a biomarker that indicates whether a person is at elevated risk for a disease before the onset of disease or before the disease is clinically manifest or detectable by screening, pre-diagnostic samples need to be available. A nested case-control study within a prospective cohort would be the most appropriate design for this scenario.

Biomarker for Early Disease Detection: A biomarker for the early detection of existing disease is in line with the concept of screening for disease with the goal of detecting preclinical disease early and preventing more severe disease and mortality by offering effective treatment. If this is the goal, discoveries of DNA methylation differences can be made among individuals with the disease compared to those free of the disease, but validation of the presence of the marks pre-diagnostically is essential to ensure that DNA methylation marks manifest early during the disease process and are not a result of the disease.

2. Biomarkers of Exposure

More recently, the value of DNA methylation as a marker of exposure has been appreciated. A number of lifestyle factors impact DNA methylation, e.g., smoking [5, 6] and alcohol consumption [7, 8], similarly environmental factors, e.g., exposure to endocrine-disrupting chemicals (EDCs) [9]. If a signature of DNA methylation patterns for smokers can be established that is consistent across several populations, this profile could be used in studies that have not (or poorly) assessed smoking behavior as an indicator for smoking.

The appropriate study design to capture the DNA methylation pattern associated with a particular exposure depends on the exposure. To determine the DNA methylation profile among smokers, a cross-sectional study is appropriate because smoking is a consistent and regular behavior pattern. The association with alcohol consumption is more difficult to establish because most individuals have varying alcohol intake over time; here, a prospective cohort design may be more appropriate.

Confounding by other environmental or lifestyle factors correlated with the exposure of interest and impacting DNA methylation is a considerable concern. For example, smokers generally consume more alcohol than non-smokers and alcohol intake affects DNA methylation. Such confounding may substantially distort findings and hampers interpretation about important DNA methylation differences associated with the primary exposure. While matching exposed and unexposed individuals on important potential confounding factors will increase efficiency, confounders still need to be accounted for in the analysis. Moreover, residual and unmeasured confounding may remain.

Furthermore, the exposure of interest may exert different effects on DNA methylation in distinct population strata, e.g., in women and men.

2.1.2 Mechanistic Studies

1. Etiologic Studies

The etiology of many diseases remains insufficiently understood. Epigenetic mechanisms may provide some of the missing links in our understanding of cancer, autoimmune diseases, allergies, autism, neurologic diseases, and many others. DNA methylation patterns are influenced by genetics and vice versa and may affect expression and translation levels. To elucidate the role of DNA methylation in disease etiology we would want to compare the DNA methylation profile of individuals with the disease of interest to individuals free of the disease similar to the study of biomarkers for disease (*see* Subheading 2.1.1). Since we are interested in mechanisms, we need to obtain the target tissue, i.e., brain tissue for the study of Alzheimer's disease and pancreatic tissue for the study of pancreatic cancer. It may be especially challenging to obtain comparison tissue from individuals free of the disease of interest, e.g., pancreatic tissue from healthy individuals. For this reason (and also to minimize confounding), adjacent histologically normal tissue has often been used in the past as a comparison; however, there is now substantial evidence that epigenetic changes may also manifest in adjacent tissue, producing false negative results. Another challenge is the uncertainty of the timing of any DNA methylation change: In a cross-sectional study any differences in DNA methylation between individuals with disease and individuals without disease may be a consequence of the disease, rather than a preceding event. Temporality, necessary to imply causality, may only be possible if pre-diagnostic biospecimens are available which for most tissues will be impossible: collecting breast tissue from thousands of healthy women and following them prospectively until a subset has developed breast cancer is not feasible from an ethical or practical perspective.

2. Understanding the Impact of Exposure

The impact of a particular exposure on the DNA methylation profile is best assessed in a prospective cohort study preserving temporality of the exposure event and subsequent DNA methylation assessment. Depending on the prevalence of exposure, exposed and unexposed may be selected in equal numbers and may be matched on potential confounding variables. If we are interested, for example, how gestational diabetes may impact the DNA methylation of the child, we might select pregnant women with gestational diabetes and match pregnant women free of gestational diabetes on a number of characteristics, e.g., maternal pre-pregnancy weight, weight gain during pregnancy, folic acid supplementation, smoking during pregnancy, method of conception, and others.

If a more persistent characteristic is studied, e.g., smoking or prolonged exposure to an occupational hazard, a cross-sectional study may be appropriate [10].

3. Connecting Exposure and Disease

Of particular interest are studies that explore the role of epigenetics as possible mechanistic underpinning of an established exposure-disease association. The Developmental Origins of Health and Disease (DOHaD) hypothesis is based on observations that the intrauterine environment shapes the future health and disease prospects of the unborn child throughout the life course. The mechanisms underlying these observations are not well understood, but epigenetic mechanisms are suspected to play a role [11]. First, the susceptible period for prenatal conditions to affect health coincides with the establishment of the epigenetic code; second, epigenetics may provide an explanation for fetal plasticity, i.e., how the fetus copes with various environmental stressors; third, the intrauterine experience is modifiable by the postnatal environment making epigenetic marks again a viable mechanistic candidate.

Mechanistic DOHaD studies are challenged by the distance between the intrauterine period and chronic disease occurrence later in life. Case-control studies nested within prospective cohorts are rarely feasible because biospecimens would have to be collected in close proximity to the exposure (i.e., at birth) and cohorts would have to be maintained for many decades. This makes intermediate endpoints a desirable compromise. Obtaining biospecimens in adulthood is less informative as it requires the strong assumption that epigenetic marks are maintained throughout life.

The study of epigenetic mechanisms linking an adult exposure to disease is best examined using a nested case-control approach. If exposure is persistent throughout a long time period, a cross-sectional approach may be feasible. For example, hypomethylation of the aryl hydrocarbon receptor repressor gene (AHRR) has been implicated in the association between smoking and cancer etiology [12].

2.1.3 EWAS: Epigenome-Wide Association Studies

Assessing DNA methylation of candidate genes may be sensible, especially for imprinted genes or genes with certain functions such as tumor suppressor genes or oncogenes. In many instances, whether we are interested in identifying biomarkers or trying to gain a better understanding of mechanisms, no immediate candidates may be apparent. An epigenome-wide scan provides the opportunity for discovery of important variation of loci-specific DNA methylation in a population. While whole genome bisulfite sequencing covers the entire methylome of an individual, this approach is still cost-prohibitive in population-based research.

The most popular methods currently used, the Illumina Infinium 450K Microarray, MeDIP, and Reduced Representation Bisulfite Sequencing (RRBS), cover parts of the methylome, each with their individual strengths and weaknesses. These methods are described in more detail in Chapters 5–10 and 16.

EWAS have recently gained popularity with a striking rise between 2010 and 2014 [4]. Most applications have addressed the role of DNA methylation in the etiology of cancer and other diseases.

2.1.4 Treatment Studies

Demethylating agents are exploited for cancer treatment; they work by demethylating hypermethylated and thus silenced tumor suppressor genes thus restoring expression. Cytidine analogs azacitidine and decitabine are currently used to treat Myelodysplastic syndrome [13]. Development of new applications of DNA methylation for the treatment of cancer or other diseases requires randomized clinical trials to evaluate the efficacy of the approach.

2.2 Challenges and Solutions in the Design of DNA Methylation Studies

1. Cell Specificity of Epigenetic Marks

Arguably, the biggest challenge in the design and analysis of human epigenetic studies is the cell specificity of the epigenetic marks. The cellular heterogeneity of blood and solid tissues deters easy interpretation. Any differences in DNA methylation between two individuals (or two population groups or within the same person over time) may reflect true differences in methylation or may be due to differences in the cellular composition of the biospecimens. Cell subpopulations may shift, e.g., due to infection or inflammation. While cell composition data are available for sorted blood cells, such information cannot be obtained for frozen blood and is not readily available for solid tissues. Thus, to disentangle diversity in DNA methylation from differences due to shifts in cell distribution it is imperative to understand the DNA methylation patterns of individual cell types within tissues. This will permit statistical adjustment for cellular heterogeneity described in more detail below (*see* also Subheading 3.2) [14]. Studies characterizing the cell-specific DNA methylation profile for various tissues are urgently needed.

2. DNA Methylation Differences Between Population Subgroups

Certain population subgroups differ in their DNA methylation profile. Ideally, stratification by ethnicity, sex, age group, and other factors is desirable. However, sample sizes rarely permit numerous stratifications. Restriction to a homogenous study population circumvents this problem, but precludes generalizability of the results to other populations.

3. Confounding

Another challenge in the design of epigenetic studies is the prevention of confounding. If the two groups we wish to compare (healthy and diseased or exposed and unexposed) with respect to their DNA methylation profile differ in any other factor, which affects DNA methylation confounding may arise and distort the results [15]. While matching on potential confounding variables may increase efficiency, confounding likely remains and needs to be accounted for in the analyses. Most essential in the design of the study is to measure potential confounding factors to permit statistical adjustment.

4. Sample Size

Whole genome-bisulfite sequencing, e.g., as part of the NIH Roadmap is mostly based on n-of-1. The challenge to the interpretation of an individual methylome is that it remains unknown how representative this individual is for the underlying population it is supposed to represent. Moreover, data on one individual does not permit capturing stochastic variation in DNA methylation between individuals of a certain population subgroup (e.g., healthy Asian females age 50–60 years). It is important, however, to assess this "uninteresting" variation and subtract it from the total variation observed between two population groups we wish to compare (e.g., healthy Asian females age 50–60 years and Asian females age 50–60 years with breast cancer) in order to obtain the "interesting" variation in DNA methylation that is informative about biologically relevant processes (i.e., breast cancer). The larger the sample size, the higher the precision with which we can capture the various components of variation. Many studies are "underpowered" to detect important differences in group comparisons. Sample size (power) calculations should be performed at the time the study is designed to ensure sufficient statistical power to detect differences of interest. Anticipated effect sizes for these calculations should be guided by previous publications. Additionally, the significance cutoff used to calculate power must take into account the impact of multiple testing on the identification of false positives. A conservative approach is to adjust the alpha-level using the Bonferroni correction.

5. Validation and Verification

Especially when employing an epigenome-wide interrogation method such as a DNA methylation microarray, it is imperative to validate the results. Validation entails using a different method to assess DNA methylation in a separate study population [4]. It is advisable to identify a validation population already during the design phase of the index study to allow efficient subsequent validation. The validation cohort should be comparable to the index cohort in characteristics that are associated with DNA methylation, e.g., ethnicity, sex, age distribution (*see* point 2 in this subsection) to avoid false negative results. In addition, verification should be performed, i.e., epigenome-wide methods should be confirmed by loci-specific DNA methylation techniques, e.g., pyrosequencing (*see* also Chapter 22) in the index cohort.

6. Functional Relevance

Considerable uncertainty remains about the functional relevance of DNA methylation differences even if they are reasonably large (e.g., 5–10%), statistically significant after adjustment for False Discovery Rate (FDR), and validated and/or verified. One approach to examining the functional relevance of variation in DNA methylation is to pair it with expression data, either on a locus/gene-specific or an epigenome-wide level.

3 Analysis of DNA Methylation Studies

3.1 Removing Technical Variation

Each technique for interrogating methylation can introduce unwanted technical variation that will obfuscate the biological signal of interest. The relative proportion of observed variation dependent on technical artifacts can be reduced at the study design stage and by normalizing the data prior to association testing. The ability to appropriately account for the potential bias introduced by technical variation is largely dependent on our ability to identify platform-specific sources of bias and estimate the proportion of variation that is explained by these artifacts.

1. Batch Effects

Batch effects may bias any biomarker study, and are an important consideration in the study design. Broadly defined, batch effects are differences between subgroups introduced during biomarker storage, handling, and/or measurement [16]. A few potential sources of batch effects include changes in personnel while processing the samples, processing the samples on different dates, and differences in the reagents or instruments used. While batch effects generally introduce extraneous variation into the measurements, the main concern is the potential confounding they can induce into the association of interest. If case and control samples are interrogated on separate plates, measurements may be perfectly confounded by batch, precluding the disentanglement of plate effects from biological signals. Even if samples are randomly allocated between plates, technical artifacts may introduce bias due to stochastic differences in case-status proportions at particular plate locations. If characterized only roughly, adjusting for surrogates for batch in downstream analysis may leave residual confounding in the estimated associations. Additionally, including indictors for batch in a mixed model will account for correlations in the error structure, but will not remove confounding in the main effect. The impact on subsequent interpretations is dependent on the proportion of the variation we assume is explained by the true batch effects. Among a range of high-throughput technologies, batch effects have been shown to contribute to a substantial, if not a majority of the observed variation (reviewed in [16]). Given the high proportion of variation assumed to be explained by batch, our power to detect biological signals is reduced by our need to control for these artifacts.

2. Reduction in Study Design

The first step to reducing the pervasive impact of batch is in the study design, the success of which can be evaluated by the judicious use of technical replicates. The three major sources of error are within-batch, between-batch, and within-individual. Intra-assay technical replicates can be used to identify the precision intrinsic to the technology and can be quantified by the coefficient of variation. Technical replicates between our perceived batches can help define the aspects of our measurements influenced by batch, such as the dynamic range, the measurement of central tendency, etc. The appropriate distribution of unique biospecimens across batches depends on the study design, the question of interest, and the ability to estimate the batch effects (reviewed in [17]). Assuming that adjusting for surrogates for batch does not completely control for their influence, samples should be randomly distributed across batches, with matched samples and biological replicates measured in the same batch [17]. Several methods have been proposed to better estimate batch effects from the error structure of genome-wide data, facilitating more complete control of batch effects if introduced. These data-driven estimates are generally not applicable to candidate gene approaches, making identification and reduction of batch effects in the study design of particular importance in these contexts.

3. Estimating Batch Effects from the Data

Assuming a major component of the variation in a genome-wide assay is due to batch, adjusting for latent structures in the data should help reduce associated bias. Beyond the contribution of batch, recent studies have suggested that a majority of published associations with methylation in heterogeneous tissues reflect shifts in cell populations. If the distribution of cell types confounds the association, adjusting for unknown components of the variation may reduce both sources of bias. However, if shifts in cell population are part of the effect of interest, methods that are not informed by our surrogates for batch may remove meaningful biological changes.

Surrogate variable analysis (SVA) is one approach for estimating unmeasured sources of bias (e.g., batch) without defining the confounding mechanism, detecting singular vectors in the methylation variation orthogonal to the primary variable [18]. After identifying the vectors that explain more variation in the residual matrix than expected due to chance, the surrogate variables are built on the full signature of the probes driving these significant singular vectors, allowing them to be correlated with the primary variable (i.e., a confounder). This method is appealing when there are many suspected unknown confounders, but assumes that these variables are not capturing measurement error/misclassification in the primary variable or interesting downstream cell shifts. Another method named "Remove Unwanted Variation, 2-step" (RUV-2) attempts to reduce the proportion of biological signal inadvertently removed by restricting this variance decomposition to negative control genes known *a priori* to be unassociated with the biological factor or interest [19].

Other data-driven approaches are guided by our surrogates for batch, and therefore assume that these have been adequately documented. These methods include ComBat, which uses a Hierarchical Empirical Bayesian regression model to remove known batch effects [20]. By pooling information across genes in each batch to shrink the batch effect parameter estimates toward the overall mean of the gene-specific estimates, ComBat can remove batch effects better than a simple linear model adjusting for our indicators of batch. However, ComBat may not be appropriate for complex sources of batch. Another method, independent surrogate variable analysis (ISVA), builds on the SVA framework, but restricts to independent components in the residual matrix significantly associated with our surrogate confounders, such as our measured indicators for batch [21].

4. Platform-Specific Technical Artifacts

Spurious associations or false negatives may result from technical artifacts associated with the method used to interrogate methylation levels. Reducing the influence of these factors requires a thorough understanding of the potential bias introduced by the measurement protocol. Accordingly, adjustment for these factors is often platform-specific. The "best practices" for preprocessing are constantly evolving as awareness of these sources of bias changes and technologies advance. More generally, the influence of technical artifacts can often be reduced by between-array normalization of signals. However, this approach makes the strong assumption that the global distribution of methylation is the same across samples.

This may not apply to certain research questions, such as comparing cancer and normal samples, or identifying tissue-specific patterns.

Several investigators have developed and compared preprocessing pipelines for methylation microarray data [22–33] (*see* also Chapter 16). This generally includes filtering nonspecific probes [34, 35], as well as removing those containing common SNPs and probes with low levels of detection across a majority of samples. Filtering is followed by within-array normalization, performing color bias adjustment and background correction, either using control probes or color channels opposite their designed base expression to estimate background [29]. To evaluate site-specific methylation levels, the Infinium HumanMethylation450 and EPIC BeadChips use two types of chemistries (type I and type II), which are designed to interrogate different genomic contexts. The less prevalent type I probes, found primarily in CpG islands, have a greater stability and dynamic range, increasing the power to detect differential methylation among these probes. To reduce this bias, several pipelines perform subset normalization of the methylation values estimated by each chemistry, accounting for the fact that these probes are enriched for different regions [25–27]. A detailed description of guidelines to analyze data and account for various biases from the Infinium HumanMethylation450 and EPIC BeadChips is given in Chapter 16.

Among sequencing-based methods, precision is dependent on sequencing depth. This may vary by genomic context if reads are enriched for certain regions, impacting relative power across the genome. Several methods to help stabilize estimates have been proposed, borrowing information across neighboring CpGs by performing local smoothing [36–38]. Additionally, methods have been developed to adjust for read coverage [38], as well as biological variation to reduce the frequency of false positives [39–41]. Instead of using read sequence to estimate methylation levels, enrichment-based methods characterize methylation levels based on the enrichment or depletion of reads mapped to a specific region. Analyses of these data require correcting for bias induced by differences in CpG density, which has been implemented in a number of analysis pipelines [42–46].

3.2 Overcoming Cellular Heterogeneity

As previously mentioned, cellular composition may explain a substantial proportion of observed variation in methylation. Adjustment for cell-type composition is a primary concern when cell mixture is suspected to confound the association with methylation or induce uninteresting variation downstream of the exposure. Ideally, estimated proportions of sorted cells are available for each individual, which can be included in the analytic models. Due to time or budget limitations, or use of stored frozen samples, this information is often not available. However, these proportions can be estimated by utilizing the methylation signatures of purified cells identified from array data [14]. Similar to regression calibration approaches used for measurement error correction, this method estimates cell proportions with relatively high accuracy when compared to gold standards [47]. Isolating purified cell data sets at the outset of a study is both time-consuming and expensive, but invaluable to all subsequent epigenetic studies based on the same tissue, assuming transportability. Application of these models requires attention to differences in the platforms and preprocessing used for the gold standard and target data, which can distort estimates. To date, models to estimate cell compositions based on the methylation profiles of purified cells have been derived for blood [14], cord blood [48], and the brain [49]. Additional models can be built using publically accessible data from previously published studies of cell-specific genome-wide methylation. The use of this reference-based approach to estimate cell proportions requires interrogating specific informative loci in the target dataset, which may not be captured by all methylation assays. In these circumstances, the extent to which shifts in cell population may explain the observed associations can sometimes be appraised by the exploration of cell-specific modifications within the large reference epigenome projects, such as the National Institutes of Health Roadmap Epigenomics Initiative [50], the International Human Epigenome Consortium [51], and BLUEPRINT [52]. Reference-free methods to adjust for cellular composition in array data have additionally been developed, which assume that a major component of the error structure is associated with cell mixture [53, 54]. Adjusting for unmeasured sources of bias will likely capture similar components of the variation, but without gold standards, it is impossible to gauge the degree to which any reference-free method is removing variation due to cell heterogeneity. Generally, adjustment for cell-proportion substantially attenuates associations with methylation across the genome [55, 56]. Whether cell proportions are estimated and adjusted for in the analysis or not, tissue composition is an important consideration for the interpretation of associations.

3.3 Capturing Biological Variation

After removing unwanted technical noise, the first step to identifying meaningful variation in methylation associated with the primary variable is to define important biological confounders and effect modifiers, a process guided by substantive knowledge. Differences in the distribution of unmeasured sources of bias and effect modifiers between cohorts will contribute to validation failures. The power to identify significant differences in methylation after adjusting for additional covariates can be increased in the study design by matching, which must be accounted for in the analysis of case-control studies. Disentangling confounders from characteristics on the causal pathway often requires temporality assumptions. Given that adjustment for intermediates can introduce bias into the estimated associations with methylation if there are unmeasured common causes of the intermediate and outcome of interest, these modeling assumptions should be explicitly stated.

The next step to identifying meaningful signals is to consider the patterns of methylation expected to be associated with the exposure of interest. Potential forms of variation include site-specific modifications, differentially methylated regions, shifts in global methylation content, and coordinated pathway-level changes. Several of these questions cannot be addressed by all methods used to interrogate methylation and therefore require consideration in the study design process. If possible, global patterns of methylation should be initially explored to investigate potential sources of bias or effect modification that should be accounted for in locus-specific models, and used to gauge the magnitude of changes associated with the primary variable. Methods to assess these changes include principle component analysis and unsupervised hierarchical clustering techniques, such as the recursive-partitioning mixture model developed for methylation data [57]. Most of the methods used to estimate methylation levels focus on identifying site-specific variation or differentially methylated regions. The models employed to estimate associations with methylation will depend on whether methylation is modeled as the exposure or the outcome, and the data generation mechanism. In addition to utilizing common regression models based on means, entropy may also be used to assess methylation pattern stability and diversity [58–61]. However, compared to more conventional models, this approach is not conducive to adjusting for bias in the analysis. When quantitative locus-level measurements of methylation are feasible, site-specific changes can be utilized to identify larger regional patterns. This may be performed by "bump hunting" [62] or using "probe lasso" [63], both of which require *a priori* definition of region boundaries, generally guided by probe proximity. For genome-wide platforms, higher order regulation of these regional changes can be assessed at the pathway level. Identification of pathways enriched for methylation differences often requires assuming that gene proximity is the major determinant of the correlation between methylation and expression, missing the influence of distal regulatory elements. Additional concerns for pathway analyses include how to deal with multiple probes per gene and adjusting the gene universe to account for the fact that not all genes are in proximity to measured CpG loci. A final consideration for association testing is adjustment for multiple comparisons, often by controlling for the family-wise error rate [64], or the less conservative false discovery rate [65–67].

Primarily these analysis considerations assume that the study goal is to capture all variation in methylation associated with the exposure or phenotype under study. However, if the goal is to identify the most predictive subset of regions, many features may be providing redundant information. In these circumstances, classification algorithms such as elastic net can be used to identify the features driving predictive precision [68, 69].

References

1. Feinberg AP (2007) Phenotypic plasticity and the epigenetics of human disease. Nature 447:433–440
2. Jaenisch R, Bird A (2003) Epigenetic regulation of gene expression: how the genome integrates intrinsic and environmental signals. Nat Genet 33(Suppl):245–254
3. Michels KB (2011) Epigenetic epidemiology. Springer, New York
4. Michels KB, Binder AM, Dedeurwaerder S et al (2013) Recommendations for the design and analysis of epigenome-wide association studies. Nat Methods 10:949–955
5. Breitling LP, Yang R, Korn B et al (2011) Tobacco-smoking-related differential DNA methylation: 27K discovery and replication. Am J Hum Genet 88:450–457
6. Zeilinger S, Kuhnel B, Klopp N et al (2013) Tobacco smoking leads to extensive genome-wide changes in DNA methylation. PLoS One 8:e63812
7. Philibert RA, Plume JM, Gibbons FX et al (2012) The impact of recent alcohol use on genome wide DNA methylation signatures. Front Genet 3:54
8. Zakhari S (2013) Alcohol metabolism and epigenetics changes. Alcohol Res 35:6–16
9. LaRocca J, Binder AM, McElrath T et al (2014) The impact of first trimester phthalate and phenol exposure on IGF2/H19 genomic imprinting and birth outcomes. Environ Res 133:396–406
10. Besingi W, Johansson A (2014) Smoke-related DNA methylation changes in the etiology of human disease. Hum Mol Genet 23:2290–2297
11. Waterland RA, Michels KB (2007) Epigenetic epidemiology of the developmental origins hypothesis. Annu Rev Nutr 27:363–388
12. Shenker NS, Polidoro S, van Veldhoven K et al (2013) Epigenome-wide association study in the European Prospective Investigation into Cancer and Nutrition (EPIC-Turin) identifies novel genetic loci associated with smoking. Hum Mol Genet 22:843–851
13. Jones PA, Baylin SB (2002) The fundamental role of epigenetic events in cancer. Nat Rev Genet 3:415–428
14. Houseman EA, Accomando WP, Koestler DC et al (2012) DNA methylation arrays as surrogate measures of cell mixture distribution. BMC Bioinformatics 13:86
15. Michels KB (2010) The promises and challenges of epigenetic epidemiology. Exp Gerontol 45:297–301
16. Leek JT, Scharpf RB, Bravo HC et al (2010) Tackling the widespread and critical impact of batch effects in high-throughput data. Nat Rev Genet 11:733–739
17. Tworoger SS, Hankinson SE (2006) Use of biomarkers in epidemiologic studies: minimizing the influence of measurement error in the study design and analysis. Cancer Causes Control 17:889–899
18. Leek JT, Storey JD (2007) Capturing heterogeneity in gene expression studies by surrogate variable analysis. PLoS Genet 3:1724–1735
19. Gagnon-Bartsch JA, Speed TP (2012) Using control genes to correct for unwanted variation in microarray data. Biostatistics 13:539–552
20. Johnson WE, Li C, Rabinovic A (2007) Adjusting batch effects in microarray expression data using empirical Bayes methods. Biostatistics 8:118–127
21. Teschendorff AE, Zhuang J, Widschwendter M (2011) Independent surrogate variable analysis to deconvolve confounding factors in large-scale microarray profiling studies. Bioinformatics 27:1496–1505
22. Dedeurwaerder S, Defrance M, Calonne E et al (2011) Evaluation of the Infinium Methylation 450K technology. Epigenomics 3:771–784
23. Du P, Kibbe WA, Lin SM (2008) lumi: a pipeline for processing Illumina microarray. Bioinformatics 24:1547–1548
24. Wang D, Yan L, Hu Q et al (2012) IMA: an R package for high-throughput analysis of Illumina's 450K Infinium methylation data. Bioinformatics 28:729–730
25. Maksimovic J, Gordon L, Oshlack A (2012) SWAN: subset-quantile within array normalization for illumina infinium HumanMethylation450 BeadChips. Genome Biol 13:R44
26. Touleimat N, Tost J (2012) Complete pipeline for Infinium((R)) Human Methylation 450K BeadChip data processing using subset quantile normalization for accurate DNA methylation estimation. Epigenomics 4:325–341
27. Teschendorff AE, Marabita F, Lechner M et al (2013) A beta-mixture quantile normalization

method for correcting probe design bias in Illumina Infinium 450 k DNA methylation data. Bioinformatics 29:189–196
28. Pidsley R, Y Wong CC, Volta M et al (2013) A data-driven approach to preprocessing Illumina 450K methylation array data. BMC Genomics 14:293
29. Triche TJ Jr, Weisenberger DJ, Van Den Berg D et al (2013) Low-level processing of Illumina Infinium DNA Methylation BeadArrays. Nucleic Acids Res 41:e90
30. Dedeurwaerder S, Defrance M, Bizet M et al (2013) A comprehensive overview of Infinium HumanMethylation450 data processing. Brief Bioinform 15:929–941
31. Marabita F, Almgren M, Lindholm ME et al (2013) An evaluation of analysis pipelines for DNA methylation profiling using the Illumina HumanMethylation450 BeadChip platform. Epigenetics 8:333–346
32. Yousefi P, Huen K, Schall RA et al (2013) Considerations for normalization of DNA methylation data by Illumina 450K BeadChip assay in population studies. Epigenetics 8:1141–1152
33. Wu MC, Joubert BR, Kuan PF et al (2014) A systematic assessment of normalization approaches for the Infinium 450K methylation platform. Epigenetics 9:318–329
34. Price ME, Cotton AM, Lam LL et al (2013) Additional annotation enhances potential for biologically-relevant analysis of the Illumina Infinium HumanMethylation450 BeadChip array. Epigenetics Chromatin 6:4
35. Zhang X, Mu W, Zhang W (2012) On the analysis of the illumina 450k array data: probes ambiguously mapped to the human genome. Front Genet 3:73
36. Hansen KD, Langmead B, Irizarry RA (2012) BSmooth: from whole genome bisulfite sequencing reads to differentially methylated regions. Genome Biol 13:R83
37. Hebestreit K, Dugas M, Klein HU (2013) Detection of significantly differentially methylated regions in targeted bisulfite sequencing data. Bioinformatics 29:1647–1653
38. Xu H, Podolsky RH, Ryu D et al (2013) A method to detect differentially methylated loci with next-generation sequencing. Genet Epidemiol 37:377–382
39. Park Y, Figueroa ME, Rozek LS et al (2014) methylSig: a whole genome DNA methylation analysis pipeline. Bioinformatics 30:2414–2422
40. Feng H, Conneely KN, Wu H (2014) A Bayesian hierarchical model to detect differentially methylated loci from single nucleotide resolution sequencing data. Nucleic Acids Res 42:e69
41. Sun D, Xi Y, Rodriguez B et al (2014) MOABS: model based analysis of bisulfite sequencing data. Genome Biol 15:R38
42. Down TA, Rakyan VK, Turner DJ et al (2008) A Bayesian deconvolution strategy for immunoprecipitation-based DNA methylome analysis. Nat Biotechnol 26:779–785
43. Pelizzola M, Koga Y, Urban AE et al (2008) MEDME: an experimental and analytical methodology for the estimation of DNA methylation levels based on microarray derived MeDIP-enrichment. Genome Res 18:1652–1659
44. Statham AL, Strbenac D, Coolen MW et al (2010) Repitools: an R package for the analysis of enrichment-based epigenomic data. Bioinformatics 26:1662–1663
45. Chavez L, Jozefczuk J, Grimm C et al (2010) Computational analysis of genome-wide DNA methylation during the differentiation of human embryonic stem cells along the endodermal lineage. Genome Res 20:1441–1450
46. Huang J, Renault V, Sengenes J et al (2012) MeQA: a pipeline for MeDIP-seq data quality assessment and analysis. Bioinformatics 28:587–588
47. Accomando WP, Wiencke JK, Houseman EA et al (2014) Quantitative reconstruction of leukocyte subsets using DNA methylation. Genome Biol 15:R50
48. Gervin K, Page CM, Aass HC et al (2016) Cell type specific DNA methylation in cord blood: a 450K-reference data set and cell count-based validation of estimated cell type composition. Epigenetics 11:690–698
49. Guintivano J, Aryee MJ, Kaminsky ZA (2013) A cell epigenotype specific model for the correction of brain cellular heterogeneity bias and its application to age, brain region and major depression. Epigenetics 8:290–302
50. Bernstein BE, Stamatoyannopoulos JA, Costello JF et al (2010) The NIH Roadmap Epigenomics Mapping Consortium. Nat Biotechnol 28:1045–1048
51. Satterlee JS, Schubeler D, Ng HH (2010) Tackling the epigenome: challenges and opportunities for collaboration. Nat Biotechnol 28:1039–1044
52. Adams D, Altucci L, Antonarakis SE et al (2012) BLUEPRINT to decode the epigenetic signature written in blood. Nat Biotechnol 30:224–226
53. Houseman EA, Molitor J, Marsit CJ (2014) Reference-free cell mixture adjustments in analysis of DNA methylation data. Bioinformatics 30:1431–1439
54. Zou J, Lippert C, Heckerman D et al (2014) Epigenome-wide association studies without

the need for cell-type composition. Nat Methods 11:309–311

55. Liu Y, Aryee MJ, Padyukov L et al (2013) Epigenome-wide association data implicate DNA methylation as an intermediary of genetic risk in rheumatoid arthritis. Nat Biotechnol 31:142–147
56. Jaffe AE, Irizarry RA (2014) Accounting for cellular heterogeneity is critical in epigenome-wide association studies. Genome Biol 15:R31
57. Houseman EA, Christensen BC, Yeh RF et al (2008) Model-based clustering of DNA methylation array data: a recursive-partitioning algorithm for high-dimensional data arising as a mixture of beta distributions. BMC Bioinformatics 9:365
58. Xie H, Wang M, de Andrade A et al (2011) Genome-wide quantitative assessment of variation in DNA methylation patterns. Nucleic Acids Res 39:4099–4108
59. Zhang Y, Liu H, Lv J et al (2011) QDMR: a quantitative method for identification of differentially methylated regions by entropy. Nucleic Acids Res 39:e58
60. He J, Sun X, Shao X et al (2013) DMEAS: DNA methylation entropy analysis software. Bioinformatics 29:2044–2045
61. Su J, Yan H, Wei Y et al (2013) CpG_MPs: identification of CpG methylation patterns of genomic regions from high-throughput bisulfite sequencing data. Nucleic Acids Res 41:e4
62. Jaffe AE, Murakami P, Lee H et al (2012) Bump hunting to identify differentially methylated regions in epigenetic epidemiology studies. Int J Epidemiol 41:200–209
63. Morris TJ, Butcher LM, Feber A et al (2014) ChAMP: 450k Chip Analysis Methylation Pipeline. Bioinformatics 30:428–430
64. Holm S (1979) A simple sequentially rejective multiple test procedure. Scand J Stat 6:65–70
65. Benjamini Y, Hochberg Y (1995) Controlling the false discovery rate - a practical and powerful approach. J R Stat Soc B Stat Methodol 57:289–300
66. Benjamini Y, Yekutieli D (2001) Controlling the false discovery rate - a practical and powerful approach. Ann Stat 29:1165–1188
67. Storey JD, Tibshirani R (2003) Statistical significance for genomewide studies. Proc Natl Acad Sci U S A 100:9440–9445
68. Friedman J, Hastie T, Tibshirani R (2010) Regularization paths for generalized linear models via coordinate descent. J Stat Softw 33:1–22
69. Zhuang J, Widschwendter M, Teschendorff AE (2012) A comparison of feature selection and classification methods in DNA methylation studies using the Illumina Infinium platform. BMC Bioinformatics 13:59

Part II

Global DNA Methylation Levels

Chapter 3

Quantification of Global DNA Methylation Levels by Mass Spectrometry

Agustin F. Fernandez, Luis Valledor, Fernando Vallejo, Maria Jesús Cañal, and Mario F. Fraga

Abstract

Global DNA methylation was classically considered the relative percentage of 5-methylcysine (5mC) with respect to total cytosine (C). Early approaches were based on the use of high-performance separation technologies and UV detection. However, the recent development of protocols using mass spectrometry for the detection has increased sensibility and permitted the precise identification of peak compounds based on their molecular masses. This allows work to be conducted with much less genomic DNA starting material and also to quantify 5-hydroxymethyl-cytosine (5hmC), a recently identified form of methylated cytosine that could play an important role in active DNA demethylation. Here, we describe the protocol that we currently use in our laboratory to analyze 5mC and 5hmC by mass spectrometry. The protocol, which is based on the method originally developed by Le and colleagues using Ultra Performance Liquid Chromatography (UPLC) and mass spectrometry (triple Quadrupole (QqQ)) detection, allows for the rapid and accurate quantification of relative global 5mC and 5hmC levels starting from just 1 μg of genomic DNA, which allows for the rapid and accurate quantification of relative global 5mC and 5hmC levels.

Key words 5-Methyl-cytosine, 5-Hydroxymethyl-cytosine, Mass spectrometry

1 Introduction

DNA methylation is the best-known epigenetic modification. In eukaryotic cells this term refers to the addition of a methyl group to the 5′ position of the pyrimidine ring of the cytosine to form 5-methyldeoxycytosine (5mdC) (Fig. 1), albeit that methylation of adenine has also been reported in some fungi and algae. This modification mainly occurs in cytosines that precede guanine and these dinucleotides are usually referred to as CpGs [1]. In mammalian genomes CpGs are asymmetrically distributed into CpG-poor and CpG-rich regions, the latter being called "CpG islands." In plants methylation can occur at virtually any base, but clonal

Agustin F. Fernandez and Luis Valledor contributed equally to this work.

Jörg Tost (ed.), *DNA Methylation Protocols*, Methods in Molecular Biology, vol. 1708, https://doi.org/10.1007/978-1-4939-7481-8_3, © Springer Science+Business Media, LLC 2018

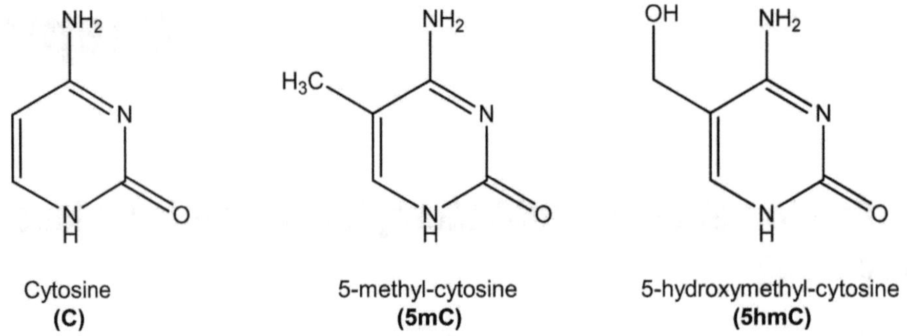

Fig. 1 Chemical structures of Cytosine (C), 5-methyl-cytosine (5mC) and 5-hydroxymethyl-cytosine (5hmC)

transmission of the methylation patterns only occurs at the symmetrical sequences CpG and CpNpG [2, 3]. CpG islands are often found in association with genes, most often in promoters and first exons, but also in almost half the genes, in regions more toward the 3′ end [4].

The methylation of cytosines is a dynamic process that takes place throughout the course of normal development and is essential for life. Methylation of CpG islands in promoter regions or first exons of genes alters the binding of transcriptional factors and other proteins, preventing RNA-Polymerase transcription of DNA and causing gene silencing. Thus, we can consider DNA methylation of these regions as a gene ON-OFF (unmethylated-methylated) regulatory mechanism [5], which recruits methyl-DNA-binding proteins and histone deacetylases that condense the chromatin around the gene-transcription start site. DNA methylation is also involved in genomic imprinting [6], X-chromosome inactivation in females [7], silencing of parasitic, foreign, and transposable elements [8, 9], and the loss-gain of morphogenetic competences in plant tissues [10], among other processes. Hypermethylation of CpG islands in promoter regions, and its associated silencing, contributes to the typical hallmarks of a cancer cell that result from tumor suppressor gene inactivation [11, 12].

In mammals, methylation of deoxycytosines is catalyzed by a group of enzymes called DNA methyltransferases (DNMTs), which transfer a methyl group from *S*-adenosylmethionine (SAM) to the 5′ position of the deoxycytosine. These methyltransferases fall into two groups: those that maintain or copy methylation marks after DNA replication, and those that initiate new methylation (de novo) of DNA. In plants, two more families have been described: chromomethylases (CMT), involved in the maintenance of methylation at CpNpG islands, and domains rearranged methylases (DRM), a family implicated in the establishment of new methylation patterns [3, 13]. The DNMT1 family (MET1 in plants) is the most abundant methyltransferase in somatic cells and is required not only for maintaining DNA methylation but also for correct embryo development, imprinting, and X-inactivation [14]. DNMT3a and b are responsible for de novo methylation [15], however it could be possible that all methyltransferases cooperate and can participate in the de novo methylation of DNA [16, 17].

DNA demethylation, required for cell adaptive-reprogramming processes, can occur passively across generations through the loss of methylation information during DNA replication. Alternatively, DNMT1 may not be able to recognize the oxidized form of 5-mdC, 5 hydroxy-methyl-deoxycytosine (5-hmdC) (Fig. 1) so another possible route starts with the oxidation of cytosine before DNA replication [18]. This oxidation is carried out by 10 or 11 translocation (TET) enzymes, members of the 2-oxoglutarate oxygenase family. In plants an active demethylation involving demethylases and not requiring DNA replication has been proposed [19, 20]. This demethylation is achieved through base excision repair after the recognition of 5-hmdC, beginning with the hydrolysis of the N-glycosidic bond of 5-mdC by a specific 5-mdC-DNA glycolsylase. This process has also recently been described in animals, where this excision is catalyzed by thymine-DNA glycosylases [18] belonging to the TET superfamily.

The importance of global DNA methylation is evidenced by its role in the principal cellular and biological processes of an organism. Loss of global DNA methylation during aging was described in fish early-on in this field of study [21] and then subsequently in mammals [22, 23]. Moreover, in mammals, aberrant global DNA methylation is a hallmark of cancer [24, 25]. Although the contributions of genetic and non-genetic factors in the establishment of proper DNA methylation patterns are still poorly defined, it has been shown that exposure to specific environmental factors is associated with alterations of normal global DNA methylation profiles, which in turn seems to mediate the appearance of developmental disorders and tumorigenesis [26–30]. In addition, in plants DNA methylation levels can be used for defining timing of floration, physiological status of the plant, and clonal capacity [31].

Global DNA methylation of specific samples can be achieved using various instrumentations although many rely on the extraction of DNA and the subsequent estimation of the relative amount of methyl-cytosine. Initial studies relied on the use of high-performance chromatography (HPLC) and High Performance Capillary Electrophoresis (HPCE) after the enzymatic digestion of DNA into single deoxynucleotides [32, 33]. These two instruments generally use UV-detectors to monitor global levels of mC and 5mC. The relative amount of 5mC is then calculated with the help of calibrating curves. Despite the frequent availability of separation instruments with UV-detection in most molecular biology laboratories, this type of technology lacks the power to distinguish between mdC and hmdC and, in some cases, it is also difficult to account for residual RNA and the co-migration of uracil with mdC. The increasing availability and affordability of mass spectrometers has enabled the design of new analytical procedures that can circumvent these drawbacks.

Currently, the most precise quantification of DNA is based on the determination of the absolute abundance of the different bases of the DNA and their modifications by using multiple reaction monitoring in a mass spectrometer (MS–MRM) coupled to nano-ultra HPLC (giving part per million resolution). This means that with a minimum amount of sample the absolute level of DNA methylation can be defined with unprecedented precision. The setting up of a mass spectrometry protocol requires specific sample-handling techniques and the fine-tuning of the instrument in order to get an exact calibration. The protocol described in this chapter is based on the method described by Le and collaborators [34] to measure not only global 5mC, but also global levels of 5hmC in the same sample.

2 Materials

2.1 Sample Preparation

1. DNA Degradase Plus™ kit (Zymo Research) for DNA hydrolysis.
2. DNA Degradase Plus™ nuclease mix.
3. 10× DNA Degradase™ Reaction Buffer.
4. ddH_2O.
5. 0.1% Formic acid.
6. Microcentrifuge Tubes 1.5 mL.
7. Heating device, for example Thermoblock.

All the reagents used in this step should be stored at −20 °C except for the formic acid (room temperature).

2.2 DNA Calibration Standards

1. 5-methylcytosine (5mC) and 5-hydroxymethylcytosine (5hmC) DNA standard set (e.g., Zymo Research) for the establishment of a calibration curve.
2. Cytosine DNA standard (e.g., Zymo Research).
3. 5mC DNA standard (e.g., Zymo Research).
4. 5hmC DNA standard (e.g., Zymo Research).
5. All the reagents required are listed under Subheading 2.1.

2.3 Analysis by Mass Spectrometry

1. Ultra Performance Liquid Chromatography (UPLC) 1290 Infinity system (Agilent Technologies) equipped with a triple quadrupole (QqQ) mass spectrometer (e.g., 6460 Jet Stream Series, Agilent Technologies).

2. C18 column (e.g., Agilent Technologies) (100 mm × 3.0 mm, 2.7 μm particle size) for chromatographic separation.
3. 0.1% Formic acid.
4. Methanol.

3 Methods

The protocol for global DNA methylation (5mC) and hydroxymethylation (5hmC) analyses using a mass spectrometry method can be divided into three main steps.

3.1 Sample Preparation

Test sample preparation is one of the most important steps for mass spectrometry-based global DNA methylation analysis. The DNA must be efficiently degraded to its individual nucleoside components before injection onto the mass spectrometry device (*see* **Note 1**).

1. Hydrolyze 1 μg of genomic DNA (extracted by standard procedures including treatment with proteinase K (*see* **Notes 2** and **3**)) by using DNA Degradase Plus™ kit following the manufacturer's instructions. Briefly, mix 2 μL of DNA (500 ng/μL) with 2.5 μL of 10× DNA Degradase™ Reaction Buffer, 1 μL of the DNA Degradase Plus™ nuclease mix, and 19.5 μL of ddH_2O. Incubate the reaction mix at 37 °C for 2 h.
2. Inactivate the reaction by adding 175 μL of 0.1% formic acid. At this point a concentration of 5 ng/μL of hydrolyzed DNA is obtained.
3. To ensure the efficiency of the degradation, 1 μg of control DNA can be treated and tested by gel electrophoresis.

3.2 DNA Calibration Standards

In this step, the standards are prepared for the construction of the calibration curves for calculating the actual percentage of 5mC and 5hmC in every test sample.

1. Hydrolyze 1 μg each of genomic DNA of the 897-bp DNA standard, cytosine, 5mC, and 5hmC, following the same instructions as described in Subheading 3.1.
2. Prepare calibration standards with a known degree of 5mC by mixing increasing amounts of hydrolyzed 5mC in the presence of the same amount of hydrolyzed cytosine (0, 0.5, 1, 5, and 10%). Results obtained from the measurement of these samples are used to make the calibration curve for 5mC.
3. Prepare calibration standards with a known degree of 5hmC by mixing increasing amounts of hydrolyzed 5hmC in the presence of the same amount of hydrolyzed cytosine (0, 0.1, 0.5, 1, and 2%). The results obtained from the measurement of these samples are used to make the calibration curve for 5hmC.

3.3 Analysis with Mass Spectrometry

In this step, the previously hydrolyzed DNA, of both test samples and standards, is analyzed by liquid chromatography electrospray ionization tandem mass spectrometry with multiple reaction monitoring (LC–ESI–MS/MS–MRM) (*see* **Note 4**). The 5mC and 5hmC are identified and quantified using a UPLC 1290 Infinity Series equipped with a triple quadrupole (QqQ) mass spectrometer. The optimal mass spectrometer parameters for the detection of both 5hmC and 5mC are optimized with pure standards, connecting the column directly to the "Jet Stream."

1. Inject 5 μL of the hydrolyzed DNA (25 ng, *see* **Note 5**) onto a UPLC-C18 column using water with 0.1% formic acid (A) and methanol with 0.1% formic acid (B) as the mobile phases, for chromatographic separation. An isocratic flow of 95% A and 5% B is maintained for 10 min. The column is operated at 30 °C and with a flow rate of 0.3 mL/min. The triple quadrupole parameters are optimized by selecting the positive ion polarity for the system. The operating conditions are as follows: gas temperature of 300 °C, drying nitrogen gas of 9 L/min, nebulizer pressure of 40 psig, sheath gas temperature of 350 °C, sheath gas flow of 10 L/min, capillary voltage of 3500 V, nozzle voltage of 1000 V. For nucleotide analysis, the fragmentor voltage, collision energy, and cell accelerator voltage are set to 90 V, 5 eV, and 7 V respectively, with a scan time of 100 ms, which gives a value of 3.2 cycles/s. The multiple reaction monitoring (MRM) method monitors three transitions for each analysis: that of 5hmC by monitoring the transitions of m/z 258.1 → 142.1, that of 5mC by monitoring the transitions of m/z 242.1 → 126.1, and that of deoxycytidine by monitoring the transitions of m/z 228.1 → 112.1 (Fig. 2).
2. Calculate the total amount of 5mC in test samples from the 5mC MRM peak area divided by the sum of the 5mC, 5hmC, and cytosine peak areas (5mC/5mC + 5hmC + C).
3. Calculate the total amount of 5hmC in the test samples from the 5hmC MRM peak area divided by the sum of the 5mC, 5hmC, and cytosine peak areas (5hmC/5mC + 5hmC + C).
4. The measurement of 5mC for the calibration standards (described in Subheading 3.2, **step 2**) is calculated from the 5mC MRM peak area divided by the total cytosine (5mC/5mC + C). The calibration curve for 5mC is constructed from these data plotted against known percentages (0, 0.5, 1, 5, and 10%).
5. The measurement of 5hmC for the calibration standards (described in Subheading 3.2, **step 2**) is calculated from the 5hmC MRM peak area divided by the total cytosine (5hmC/5hmC + C). The calibration curve for 5hmC is constructed from

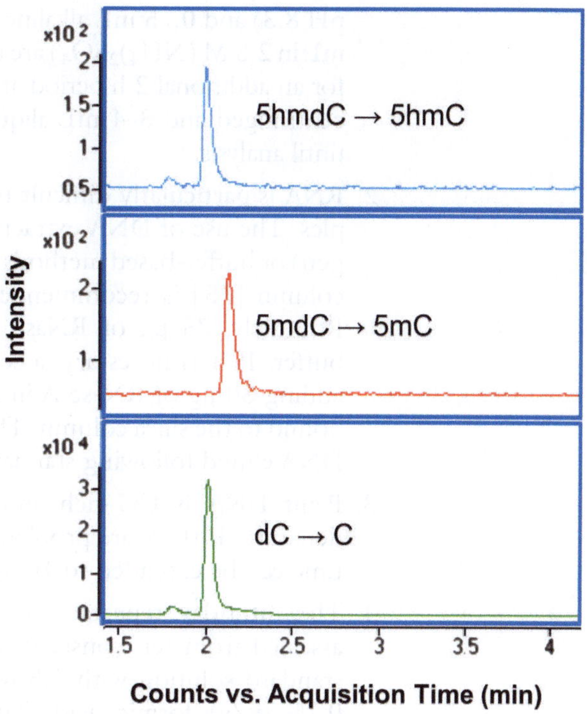

Fig. 2 Representative UHPLC-MS/MS chromatograms of methylated and hydroxymethylated cytosines at the optimized MRM ESI transitions in positive mode for quantification and confirmation of the target analytes

these data plotted against known percentages (0, 0.1, 0.5, 1, and 2%).

6. Finally, calculate the actual percentage of both 5mC and 5hmC in test samples by interpolation from the respective linear calibration curve (*see* **steps 4** and **5** of this paragraph), as previously described by Le and collaborators [34].

4 Notes

1. Specific conditions for plant DNA hydrolysis: Hydrolyze 1 μg of genomic DNA by using Nuclease S1 (e.g., Fermentas, 10 U/μL) or P1 (e.g., Sigma, 0.2 U/μL). Mix 2 μL of genomic DNA (500 ng/μL) with 6 μL of 6× Nuclease Reaction Buffer and 20 μL of ddH_2O in a 200 μL tube. Denature the DNA by heating at 99 °C for 5 min (3) and quickly cooling them down to 4 °C (the use of a thermocycler is recommended). After the sample has been cooled, add 2 μL of nuclease (10 U/μL) and incubate at 37 °C overnight. It is recommended that the digestion is tested by analyzing 2 μL of digestion mix on an electrophoresis gel. Then, 1.25 mL Tris–HCl (0.5 M,

pH 8.3) and 0.75 mL alkaline phosphatase (e.g., Sigma; 50 U/mL in 2.5 M $(NH_4)_2SO_4$) are added and the mixtures incubated for an additional 2 h period at 37 °C. Hydrolyzed solutions are centrifuged and 3–4 mL aliquots taken and stored at −20 °C until analysis.

2. RNA is particularly difficult to remove from plant DNA samples. The use of DNA extraction kits (i.e., plant DNeasy, Qiagen) or buffer-based methods integrating the use of a silica spin column [35] is recommended with long RNase treatments. Routinely, 75 μg of RNase A is added to DNA extraction buffer. If it is necessary, a second digestion is performed by adding 30 μg of RNase A in a volume of 20 μL after DNA is bound to the silica column. The column is then washed and the DNA eluted following standard protocols.
3. Plant DNA is CG rich and can be difficult to completely denature. If there are problems with digestion, the denaturing time can be extended to 10 min.
4. The intra-day repeatability of the UPLC–QqQ method is assessed from ten consecutive chromatographic runs using a standard solution with 2.5 μM of every standard in MeOH: 0.1% (v/v) formic acid. The interday repeatability of the method is assessed by analyzing the same standard solution on 2 consecutive days. The relative standard deviation (RSD) for peak area is in the range of 0.5–4.7% in the intra-day test and 1.3–3.5% in the case of the inter-day test. To avoid instrumental drift, an autotune recalibration employing reference parameters and masses is performed every day.
5. Although in the seminal work described by Le and collaborators [34], 50 ng of digested DNA is used, we successfully conducted the procedure with 25 ng, thereby reducing the amount of DNA required for injection in the UPLC.

Acknowledgments

We thank Ronnie Lendrum for editorial assistance. This work has been financially supported by the following: Fondo de Investigaciones Sanitarias FIS/FEDER (PI11/01728 to AF.F. and PI12/01080 to M.F.F.); the ISCIII-Subdirección General de Evaluación y Fomento de la Investigación (Miguel Servet contract: CP11/00131 to A.F.F.); the Spanish National Research Council (CSIC; 200820I172 to M.F.F.); Fundación Ramón Areces (to M.F.F). The IUOPA is supported by the Obra Social Cajastur, Spain.

References

1. Herman JG, Baylin SB (2003) Gene silencing in cancer in association with promoter hypermethylation. N Engl J Med 349:2042–2054
2. Cokus SJ, Feng S, Zhang X et al (2008) Shotgun bisulphite sequencing of the Arabidopsis genome reveals DNA methylation patterning. Nature 452:215–219
3. Finnegan EJ, Kovac KA (2000) Plant DNA methyltransferases. Plant Mol Biol 43:189–201
4. Zhang X, Yazaki J, Sundaresan A et al (2006) Genome-wide high-resolution mapping and functional analysis of DNA methylation in Arabidopsis. Cell 126:1189–1201
5. Bender J (2004) DNA methylation and epigenetics. Annu Rev Plant Biol 55:41–68
6. Feinberg AP, Cui H, Ohlsson R (2002) DNA methylation and genomic imprinting: insights from cancer into epigenetic mechanisms. Semin Cancer Biol 5:389–398
7. Payer B, Lee JT (2008) X chromosome dosage compensation: how mammals keep the balance. Annu Rev Genet 42:733–772
8. Doerfler W (1991) Patterns of DNA methylation—evolutionary vestiges of foreign DNA inactivation as a host defense mechanism. A proposal. Biol Chem 372:557–564
9. Lippman Z, Gendrel AV, Black M et al (2004) Role of transposable elements in heterochromatin and epigenetic control. Nature 430:471–476
10. Ronemus MJ, Galbiati M, Ticknor C et al (1996) Demethylation-induced developmental pleiotropy in Arabidopsis. Science 273:654–657
11. Esteller M (2008) Epigenetics in cancer. N Engl J Med 358:1148–1159
12. Jones PA, Baylin SB (2007) The epigenomics of cancer. Cell 128:683–692
13. Law JA, Jacobsen SE (2010) Establishing, maintaining and modifying DNA methylation patterns in plants and animals. Nat Rev Genet 11:204–220
14. Beard C, Li E, Jaenisch R (1995) Loss of methylation activates Xist in somatic but not in embryonic cells. Genes Dev 9:2325–2334
15. Okano M, Bell DW, Haber DA et al (1999) DNA methyltransferases Dnmt3a and Dnmt3b are essential for de novo methylation and mammalian development. Cell 99:247–257
16. Kim GD, Ni J, Kelesoglu N et al (2002) Co-operation and communication between the human maintenance and de novo DNA (cytosine-5) methyltransferases. EMBO J 21:4183–4195
17. Rhee I, Bachman KE, Park BH et al (2002) DNMT1 and DNMT3b cooperate to silence genes in human cancer cells. Nature 416:552–556
18. He YF, Li BZ, Li Z et al (2011) Tet-mediated formation of 5-carboxylcytosine and its excision by TDG in mammalian DNA. Science 333:1303–1307
19. Morales-Ruiz T, Ortega-Galisteo AP, Ponferrada-Marin MI et al (2006) DEMETER and REPRESSOR OF SILENCING 1 encode 5-methylcytosine DNA glycosylases. Proc Natl Acad Sci U S A 103:6853–6858
20. Wu SC, Zhang Y (2010) Active DNA demethylation: many roads lead to Rome. Nat Rev Mol Cell Biol 11:607–620
21. Berdyshev GD, Korotaev GK, Boiarskikh GV et al (1967) Nucleotide composition of DNA and RNA from somatic tissues of humpback and its changes during spawning. Biokhimiia 32:988–993
22. Bjornsson HT, Sigurdsson MI, Fallin MD et al (2008) Intra-individual change over time in DNA methylation with familial clustering. JAMA 299:2877–2883
23. Wilson VL, Smith RA, Ma S et al (1987) Genomic 5-methyldeoxycytidine decreases with age. J Biol Chem 262:9948–9951
24. Kulis M, Esteller M (2010) DNA methylation and cancer. Adv Genet 70:27–56
25. Moore LE, Pfeiffer RM, Poscablo C et al (2008) Genomic DNA hypomethylation as a biomarker for bladder cancer susceptibility in the Spanish Bladder Cancer Study: a case-control study. Lancet Oncol 9:359–366
26. Baccarelli A, Bollati V (2009) Epigenetics and environmental chemicals. Curr Opin Pediatr 21:243–251
27. Kim M, Bae M, Na H et al (2012) Environmental toxicants-induced epigenetic alterations and their reversers. J Environ Sci Health C Environ Carcinog Ecotoxicol Rev 30:323–367
28. Tajuddin SM, Amaral AF, Fernandez AF et al (2014) LINE-1 methylation in leukocyte DNA, interaction with phosphatidylethanolamine N-methyltransferase variants and bladder cancer risk. Br J Cancer 110:2123–2130
29. Woo HD, Kim J (2012) Global DNA hypomethylation in peripheral blood leukocytes as a biomarker for cancer risk: a meta-analysis. PLoS One 7:e34615
30. Feil R, Fraga MF (2012) Epigenetics and the environment: emerging patterns and implications. Nat Rev Genet 13:97–109
31. Valledor L, Hasbún R, Meijón M et al (2007) Involvement of DNA methylation in tree development and micropropagation. Plant Cell Tissue Organ Cult 91:75–86

32. Berdasco M, Fraga MF, Esteller M (2009) Quantification of global DNA methylation by capillary electrophoresis and mass spectrometry. Methods Mol Biol 507:23–34

33. Torano EG, Petrus S, Fernandez AF et al (2012) Global DNA hypomethylation in cancer: review of validated methods and clinical significance. Clin Chem Lab Med 50:1733–1742

34. Le T, Kim KP, Fan G et al (2011) A sensitive mass spectrometry method for simultaneous quantification of DNA methylation and hydroxymethylation levels in biological samples. Anal Biochem 412:203–209

35. Valledor L, Escandon M, Meijon M et al (2014) A universal protocol for the combined isolation of metabolites, DNA, long RNAs, small RNAs, and proteins from plants and microorganisms. Plant J 79:173–180

Chapter 4

Antibody-Based Detection of Global Nuclear DNA Methylation in Cells, Tissue Sections, and Mammalian Embryos

Nathalie Beaujean, Juliette Salvaing, Nur Annies Abd Hadi, and Sari Pennings

Abstract

Immunostaining is widely used in cell biology for the in situ detection of proteins in fixed cells. The method is based on the specificity of antibodies for recognizing and binding to a selected target, combined with immunolabeling techniques for microscopic imaging. Antibodies with high specificities for modified nucleotides have also been widely developed, and among those, antibodies that recognize modified cytosine: 5-methylcytosine (5mC), and more recently, its derivates 5-hydroxymethylcytosine (5hmC), 5-formylcytosine (5fC) and 5-carboxylcytosine (5caC). To allow for their detection, primary antibody signals can be amplified using secondary antibodies coupled to fluorophores for immunofluorescence, or other molecules for immunocytochemistry.

Immunostaining can be used to gain information on the spatial distribution and levels of DNA methylation states within the nucleus. Although the resolution remains quite low in genomic terms, advanced microscopy techniques and image analysis can obtain detailed spatial information content from immunostained sites. The technique complements genomic approaches that permit the assessment of DNA methylation on specific sequences, but that cannot provide global nuclear spatial context. Immunostaining is an accessible method of great benefit in several cases: when working with limited material (such as embryos or primary cells), to quickly assess at the level of individual cells the effect of siRNA, drugs, or biological processes that promote or inhibit DNA methylation or demethylation, or to study the 3D nuclear organization of regions with high DNA methylation, such as constitutive heterochromatin.

Here, we review and outline protocols for the fluorescent and enzymatic immunodetection of DNA methylation in the nuclei of cells, tissue sections, and mammalian embryos.

Key words DNA methylation, Epigenetics, Fluorescence immunocytochemistry, Immunohistochemistry, Microscopy

1 Introduction

DNA methylation in mammalian genomes occurs primarily at CpG dinucleotide sequences, most of which are found in a methylated meCpG state carrying 5-methylcytosine (5mC). This primary form of DNA methylation has been linked to silencing of repetitive

Jörg Tost (ed.), *DNA Methylation Protocols*, Methods in Molecular Biology, vol. 1708,
https://doi.org/10.1007/978-1-4939-7481-8_4, © Springer Science+Business Media, LLC 2018

1.1 Epigenetic Modifications of DNA, Localization, and Function

elements, genomic imprinting, and X chromosome inactivation. In addition, the propagation of stable gene repression patterns in cell types is commonly attributed to the epigenetic heritability of meCpG, which is maintained through DNA replication by DNA methyltransferases (DNMTs) [1, 2]. DNA methylation may coordinately modulate chromatin states with histone modifications via interactions between modifying enzymes and methyl binding domains [2, 3]. Silencing of genes or genomic regions by DNA methylation is associated with the repressive compartment of the nucleus [4]. Epigenetic states are set up in early embryos [5, 6]; however, highly condensed constitutive heterochromatin does not acquire a mature histone modification signature until late in development [7].

Other less abundant forms of DNA methylation, such as 5-hydroxymethylcytosine (5hmC), and the more recently identified rarer forms 5-formylcytosine (5fC) and 5-carboxylcytosine (5CaC) have been linked with DNA demethylation pathways during development [8]. 5hmC is thought to arise as an oxidation intermediate of 5mC in an iterative oxidation process mediated by the TET family of enzymes [8]. Its genomic distribution pattern is nevertheless different from 5mC, as 5hmC maps to the body of transcribed genes and enhancers and is a more dynamic epigenetic mark [9–11].

The current focus on genomic DNA methylation mapping has driven the development of new methodologies for detecting and localizing 5hmC by DNA sequencing [8, 12]. In addition, existing 5mC data sets have been reported to require re-evaluation after it emerged that some of the traditional 5mC mapping methods such as bisulfite sequencing and methylation-specific restriction analysis cannot differentiate between 5mC and 5hmC [13]. While newer chemical modification-assisted sequencing and subtraction methods offer single-base resolution of 5hmC and 5mC (*see* also Chapters 33 and 34), whole genome coverage is currently limiting due to the higher sequencing depth required for mapping 5hmC [8]. For this reason, affinity-based enrichment sequencing techniques based on methylated DNA immunoprecipitation (MeDIP and hMeDIP, *see* also Chapters 11 and 33) remain a standard methodology for genome mapping. These data sets have no retrospective issues, as they rely on immunoprecipitation grade antibodies with recognized specific affinity for either 5mC or 5hmC. The same antibodies can also be used for immunofluorescence microscopy detection of these epitopes in the nuclei of fixed cells, and were in several cases originally developed for this purpose.

1.2 Antibodies for Immunodetection of DNA Methylation

A mouse monoclonal antibody (clone 33D3) specifically recognizing 5mC was originally developed by Alain Niveleau for the detection of methylnucleosides in enzyme-linked immunoassays [14]. It was subsequently validated for microscopic visualization of methylcytosine-rich DNA on metaphase chromosomes and in thin sections from paraffin-embedded cancer tissue [15, 16]. Several other applications have predominantly relied on this antibody, including MeDIP [17] (*see* also Chapter 11).

The monoclonal mouse IgG1λ clone 33D3 is still the most cited antibody and is commercially available from several sources. More recently marketed monoclonal antibodies include clone 10G4 (mouse IgG1); clone MAb-006 (mouse IgG1); clone A1 (mouse IgG1); clone 87G31 (mouse IgG3κ); clone 5MC-CD (mouse IgM); clone EP4694 (rabbit IgG); and others. Alternative substitutes for mouse IgG are polyclonal antibodies raised in rabbit or sheep. It is worthwhile testing antibodies, which may have different affinities for the epitope or its context, having been raised against different hapten-carrier constructs.

The 5mC epitope may not be a strong immunogen due to its ubiquity in vertebrates. By contrast, the genome-wide abundance of 5hmC is an order of magnitude lower than that of 5mC [8]. The higher affinity 5hmC antibodies are a boon to immunostaining procedures for microscopy, as they can be combined with 5mC monoclonal antibodies to visualize both signals in similar experimental conditions. As a precaution, preliminary single and double immunostaining experiments should be performed to check that the antibodies are not physically cross-reacting when both marks are found in the vicinity of each other [18]. The market for 5hmC antibodies is still expanding; therefore, recent citations are a good guide. Polyclonal antibodies are frequently used to allow simultaneous anti-rabbit/anti-mouse secondary antibody immunofluorescence detection of 5hmC/5mC. Monoclonal antibodies are also available: clone HMC 31 (mouse IgG1κ); clone HMC-MA01 (mouse IgG1κ); and clone AB3/63.3 (rat IgG2aκ). While these antibodies have comparable specificities, rabbit polyclonal 5hmC antibody was reported to have a slightly higher immunoprecipitation efficiency, as well as higher dot blot efficiency, as compared with rat monoclonal antibody [19, 20]. Rabbit polyclonal 5-caC antibodies and 5-fC antibodies originally generated for immunostaining are now also available [21].

1.3 General Considerations in Antibody-Based Detection

Several steps are critical when performing immunostaining. First is the fixation: this step is meant to kill the cell while preserving it as much as possible in its original state. The main fixatives used are alcohols (e.g., methanol or a mixture of ethanol and acetic acid) or aldehydes (para/formaldehyde and glutaraldehyde). These have different effects on the cell; therefore, the choice of the fixative depends on the question addressed and on the antigen of interest. One of the major effects of alcohol-based fixatives is dehydration of the cell. As such, they affect the 3D organization of the cell/nucleus, which is why they are not a primary choice for our studies. On the other hand, paraformaldehyde is effective for the fixation of nucleic acids and allows better preservation of the 3D structure. Other parameters, such as length of fixation and concentration of the fixative, can also play an important role. Indeed, a weak fixation can result in protein degradation, whereas a strong fixation can alter the structure of protein antigens and their subsequent recognition by antibodies.

A second important step is the permeabilization of the cells, as antibodies are large molecules that cannot naturally enter the cells. Alcoholic fixatives extract the cell lipids and thus do not necessitate additional permeabilization treatment, but this is not the case with paraformaldehyde. Permeabilization is performed using detergents, most often Triton X-100, but also Tween 20, NP-40, or other. Insufficient permeabilization will result in poor signal quality due to the lack of penetration of the antibodies, whereas overtreatment can have dramatic effects on cell morphology.

1.4 DNA Methylation Immunocytochemistry on Cells

Antibodies raised against the 5-methylcytosine group and its derivatives have in most cases selective affinity for these epitopes on denatured DNA strands. Whereas genomic DNA immunoprecipitation or immunodetection is performed on isolated DNA, immunocytochemistry aims to visualize these epitopes in their cell nuclear context. DNA denaturation of samples can retain cell morphology, yet it is often incompatible with quantitative co-staining for protein components of the cell. The additional DNA denaturation step requires additional time and skills compared to standard immunostaining protocols. Incomplete denaturation may result in weak and uneven staining of nuclei, whereas excess denaturation will eventually damage the DNA, with similar outcomes. Before embarking on a new type of sample, whether cells, tissues, or whole embryo, an initial denaturation test series is strongly advisable in order to determine optimal conditions for DNA methylation staining.

A second point of consideration is the avidity of 5mC epitopes and their derivatives in genomic DNA, given their variable local frequency in the DNA sequence. The avidity stems from the multiplicity of affinities in sequences where a higher local concentration of single-binding sites exists. This results in different functional affinities of the antibody for sites of high density versus low density of DNA methylation. Binding saturation, as with overnight antibody incubation, is generally avoided in order to report on the dynamic range of DNA methylation densities and to visualize the densely 5mC methylated heterochromatin in particular. On the other hand, long incubation may provide a general overview of all the genomic regions bearing 5mC, allowing their localization in the nuclei.

1.5 DNA Methylation Immunohistochemistry on Tissue Sections

Standard histological analysis involves tissue fixation and dehydration followed by embedding in paraffin. Archived clinical samples, for example, are assayed in human tissue microarrays or whole tissue blocks, from which paraffin thin sections can be cut on a microtome as required. Immunohistochemistry on paraffin sections requires rehydration and antigen retrieval steps. While this methodology uniquely permits the analysis of archival samples, the organic solvent treatments affect cell morphology, which is best viewed at lower magnification. They may also reduce the epitope yield for immunostaining—although this is less an issue for DNA than for protein epitopes.

Immunofluorescence staining of animal tissues is often performed on frozen tissue cryosections. These tissues are fresh frozen and embedded in aqueous mounting medium to facilitate cryosectioning. This method uses paraformaldehyde (PFA) as the sole fixative and generally aims to preserve native protein epitopes in their original cell morphology. Identification of the tissue context is less straightforward than on paraffin sections, however; this may require counterstaining of alternate sections. Blocks can be stored frozen but cryosections are generally best processed without further freezing/defrosting.

DNA methylation antibodies have been shown to work in either of these fixative and preparative conditions. DNA methylation immunohistochemistry on tissue sections is nevertheless less commonly performed than immunocytochemistry on cultured cells or whole mount embryos. One issue is that thin sections are dense in extracellular as well as cellular material compared to cultured cells and thus produce more background signal. More importantly, a survey of the literature to date shows uneven results for DNA methylation staining of nuclei in tissue sections and immunofluorescence staining of cryosections in particular.

After testing several protocol variations from the literature, we have narrowed this problem down to a difficulty in achieving DNA denaturation in cryosections when relying on the HCl denaturation step of immunocytochemistry protocols. Of particular concern was that even when DNA denaturation did not take place, as evidenced by a lack of reduction in DAPI fluorescence, acid treatment often resulted in a high nonspecific nuclear background. This staining was similar with all DNA antibodies but dissimilar from the DAPI stained DNA pattern. At low magnification this could be mistaken (and was in a few instances reported) as a nuclear signal, but it was revealed at higher resolution to consist of aggregates. By contrast, other reports that employed the immunocytochemistry protocol for cryosections showed partial, mosaic nuclear staining for DNA methylation in some tissues. We observed similar mosaic results following longer HCl treatment and overnight antibody incubation but found that these apparently partial results could not be improved further.

As none of these issues are encountered with whole mount staining of tissues, we reasoned that DNA denaturation by boiling in sodium citrate buffer, as used in Southern hybridization and FISH methodology, might be more effective on substrate-bound dehydrated DNA, including DNA in sections adhered onto glass.

Furthermore, in immunohistochemistry this step serves as an antigen retrieval method by breaking fixative crosslinks, aiding the process. The original paraffin section 5mC staining protocol, also presented here, combines a sodium citrate boiling step with HCl denaturation [16]. Furthermore, the Pfeifer lab with experience in both clinical paraffin and cryosections employed a hybrid protocol for the latter in a rare demonstration of reliable DNA methylation staining [22]. For cryosections, we therefore do not recommend HCl treatment for DNA denaturation and present a sodium citrate boiling step protocol that we recently developed in our lab.

1.6 DNA Methylation Immunocytochemistry on Whole Mount Embryos

Immunofluorescence imaging techniques are widely used in the study of mammalian embryos, which due to their size, transparency, and low cell number are particularly suitable to microscopic methods. During the first stages of development, the genome of the embryos is silent, and development is sustained by maternally inherited RNA and proteins, which have accumulated during oogenesis. Thus, the zygote (1-cell embryo) is a huge cell with a very large cytoplasm, and even in the subsequent cleavage stages, the size of the cells remains large. This provides a challenge for immunostaining, as penetration of antibodies into the nucleus is more difficult in embryos than in most cell types. As a result, conditions for fixation and permeabilization may differ from what is typically used for cells, and overnight incubations with primary antibodies are usually preferable to short incubations.

Antibody concentrations are also generally slightly increased; in some cases, antibodies with a low affinity that still give a signal in regular cells do not perform in embryos, even though their target is present. It must be noted that in the first developmental stages, and in particular in the zygote, nuclei are very decondensed, which can hinder the detection of low density nuclear antigens.

As discussed previously, the 5mC epitopes in the DNA need to be exposed so the antibodies can bind to their target. This requires DNA denaturation, which in embryos is commonly performed using HCl. The conditions for HCl denaturation, concentration, incubation temperature, and time are again very important to test [18]. Tryptic digestion, which digests the proteins binding to 5mC, can also be used in addition to HCl to achieve valid measurements of immunolocalizations [23, 24].

2 Materials

2.1 DNA Methylation Immunocytochemistry on Cells

1. Tissue culture hood and CO_2 incubator; aspirator.
2. Cell culture plates: 24-well or 6-well.
3. Cell culture media; gelatine solution.

4. Phosphate-buffered saline "Dulbecco A" (PBS): 137 mM NaCl, 2.7 mM KCl, 8.1 mM Na_2HPO_4, 1.5 mM KH_2PO_4 (supplemented with 1 mM $CaCl_2$ for washing of cell cultures).
5. 4% Paraformaldehyde (PFA) in PBS.
6. 0.4% Triton X-100 in PBS.
7. PBST buffer: 0.2% Tween 20 in PBS.
8. 4 M Hydrochloric acid (HCl).
9. Blocking buffer: 2% BSA and (optional) 1% donkey serum in PBS.
10. Primary and secondary antibodies (*see* **Notes 1** and **2**).
11. DNA counterstain stock solution: 3 mM 4′,6-diamidino-2-phenylindole (DAPI).
12. Mounting medium (*see* **Note 3**).
13. Cover slips No. 1.5; sizes 13 mm round (24-well plate); 22 mm square (6-well plate) (*see* **Note 4**).
14. Microscope carrier slides.

2.2 DNA Methylation Immunohisto-chemistry on Paraffin-Embedded Tissues

1. Microtome.
2. Superfrost Plus adhesion microscope slides; cover slips.
3. Histology staining baths with glass slide holders.
4. Plastic Coplin jar.
5. Microwave oven.
6. Hydrophobic barrier PAP pen.
7. Humidity chamber.
8. Deionized water (H_2O).
9. Tris–HCl buffer: 50 mM Tris–HCl, 150 mM NaCl, 0.01% Triton X-100, pH 7.6.
10. Phosphate-buffered saline (PBS): 137 mM NaCl, 2.7 mM KCl, 8.1 mM Na_2HPO_4, 1.5 mM KH_2PO_4.
11. PBE buffer: 1% BSA (fraction V), 1 mM EDTA in PBS.
12. 0.01 M Citric acid, adjusted with 1 M sodium hydroxide to pH 6.0 (or made up as 0.01 M sodium citrate buffer, pH 6.0).
13. 3.5 M Hydrochloric acid (HCl).
14. 3% Hydrogen peroxide (H_2O_2).
15. 1% Goat serum (or other to match secondary antibody host).
16. Primary and secondary antibodies (*see* **Notes 1** and **2**).
17. Biotin-streptavidin peroxidase-based detection system.
18. Chromogen substrate (3,3′-diaminobenzidine tetrahydrochloride (DAB) or other).
19. Hematoxylin counterstain.

2.3 DNA Methylation Immunohisto-chemistry on Cryosections

1. Cryotome.
2. Superfrost Plus adhesion microscope slides; cover slips.
3. Schieferdecker jar (or heat resistant beaker with glass slide holder).
4. Microwave and microwave dish (or hotplate for heating beaker).
5. Hydrophobic barrier PAP pen.
6. Humidity chamber.
7. Phosphate-buffered saline (PBS): 137 mM NaCl, 2.7 mM KCl, 8.1 mM Na_2HPO_4, 1.5 mM KH_2PO_4.
8. 2% Paraformaldehyde (PFA) in PBS.
9. 0.5% Triton X-100 in PBS.
10. PBST buffer: 0.2% Tween 20 in PBS.
11. 0.01 M Citric acid adjusted with 1 M sodium hydroxide to pH 6.0 (or made up as 0.01 M sodium citrate buffer, pH 6.0).
12. Blocking buffer: 2% BSA and (optional) 1% donkey serum in PBS.
13. Primary and secondary antibodies (*see* **Notes 1** and **2**).
14. DNA counterstain stock solution: 3 mM 4′,6-diamidino-2-phenylindole (DAPI).
15. Mounting medium (*see* **Note 3**).

2.4 For DNA Methylation Immunocytochemistry on Whole Mount Embryos

Prepare all the solutions using ultrapure water (18.2 MΩ cm resistivity at 25 °C).

1. Fixative: 20% paraformaldehyde (PFA) solution. Upon opening of the stock solution, adjust pH to ~7.0 with 0.2 M NaOH if necessary and store maximum for 1 month at 4 °C with light protection (aluminum foil for example). Before each experiment, pH should be checked with a pH indicator strip; discard if below pH 7.0 or above pH 7.4.
2. Phosphate-Buffered Saline (PBS): 137 mM NaCl, 2.7 mM KCl, 8.1 mM Na_2HPO_4, 1.5 mM KH_2PO_4. We usually prepare PBS by dissolving commercially available tablets in deionized water according to the manufacturer's instructions, autoclave and store at 4 °C. Filter through a 0.22 μm sterile cellulose acetate membrane before use.
3. Permeabilization stock solution: prepare 10% Triton X-100 stock solution with ultrapure water (w/v) and mix slowly to avoid bubbles. We recommend storage at 4 °C for maximum 1 month.
4. PBS-BSA 2% solution: dissolve BSA powder slowly in PBS (w/v) and filter through 0.22 μm sterile cellulose acetate

membranes. Aliquots of 1 or 2 mL can be kept at −20 °C for several months and thawed just before use. After thawing, filter through a 0.22 μm sterile cellulose acetate membrane, store at 4 °C.

5. Concentrated Hydrochloric acid (HCl 37%).
6. Primary and secondary antibodies (*see* **Notes 1** and **2**).
7. Mineral oil, embryo tested: keep at room temperature and protect from direct light.
8. Nucleic acid stain (*see* **Note 5**).
9. Mounting antifading medium and clear nail polish.
10. Dissecting microscope with appropriate working distance and heating block set at 27 °C (*see* **Note 6**).
11. Aspirator tube (~40 cm length) and thin glass pipettes for embryo manipulation (*see* **Note** 7).
12. Embryo glass dishes with glass lids and parafilm to secure the lids (*see* **Note 8**).
13. Petri dishes (35 or 60 mm diameter); micro test tubes (0.5 or 1.5 mL).
14. 37 °C oven or incubator.
15. SuperFrost Plus slides and glass cover slips (*see* **Note 9**).

3 Methods

3.1 DNA Methylation Immunocytochemistry on Cells

The cell immunostaining protocol outlined below is widely used in the field, with minor variations (*see* **Note 10**).

1. Grow the cells on gelatine-coated No. 1.5 glass cover slips inside a multi-well culture plate.
2. Remove the medium by washing cells twice in PBS containing 1 mM $CaCl_2$ (5 min each). Remove solutions by aspiration. Do not let the cells dry during the following steps performed at room temperature.
3. Add 200 μL of 4% PFA in PBS to each cover slip (13 mm diameter) for 10 min to fix cells. Remove the fixation solution and rinse cells twice in 200 μL of PBS (5 min each).
4. Add 200 μL of 0.4% Triton X-100 in PBS to each cover slip for 15 min to permeabilize cells. Remove the permeabilization solution and rinse the cells twice in PBS (10 min each).
5. Add 200 μL of 4 M HCl to each cover slip for 10 min to denature the DNA. Neutralize with 100 mM Tris–HCl pH 8.5 (this optional neutralization step ensures reproducible conditions when multiple slides are processed). Remove the solution and rinse the cells three times in 200 μL of PBS (5 min each).

6. Add 200 μL of 2% BSA in PBS containing (optional) 1% donkey serum and incubate for 1 h to prevent nonspecific staining. Remove the blocking solution (except from those cells that will serve as the "no antibody" controls, *see* **Note 11**).
7. Add the primary antibody diluted 1:300–1:500 in the blocking solution. Incubate the cells under closed cover for 2 h at room temperature. Remove the antibody solution and rinse the cells three times in 200 μL of PBST (15 min each), removing the solutions between each step.
8. Add the fluorescent secondary antibody diluted 1:250 in blocking solution and incubate for 1 h as above, in the dark. Remove the antibody solution and rinse the cells in 200 μL of PBST (15 min).
9. Add 0.3 μM of DAPI solution in PBST for 15 min to counterstain for DNA (denatured DNA has a reduced DAPI fluorescence; for alternative DNA staining *see* **Note 5**). Remove the counterstaining solution and rinse once with PBS (15 min).
10. Mount the cover slips onto standard microscope slides using mounting medium. Leave the slides to cure overnight at room temperature.

3.2 DNA Methylation Immunohisto-chemistry on Paraffin-Embedded Tissues

This protocol is based on standard immunohistochemistry methodology starting from formalin-fixed tissue embedded in a paraffin block. It is adapted from a protocol developed by Niveleau and colleagues for the original 5mC antibody [16].

3.2.1 Preparation of Paraffin Sections and Rehydration

1. Using a microtome, cut thin 5 μm paraffin sections and adhere 1–3 sections per coated microscope slide.
2. To remove paraffin from the sections, heat the slides in an oven for 1 h at 58 °C. Transfer to a xylene bath with three changes of xylene (2 min each).
3. Rehydrate tissues by putting the slides successively through 100%, 95%, 75% ethanol (3 min each), and finally in Tris–HCl buffer (3–30 min).
4. Drain excess buffer and carefully dry an area around each section using a piece of tissue, then draw a hydrophobic barrier around the sections using a PAP pen. Do not allow sections to dry from this point onward.
5. If not proceeding right away, cover the sections with 50–200 μL (depending on size) of Tris–HCl buffer. Place in a humidity chamber, which can keep specimens stable for a few hours.

3.2.2 Antigen Retrieval and DNA Denaturation

1. Drain the buffer from the slides before placing slides in a plastic Coplin jar filled with H_2O (1–3 min).
2. Stand the jar in a wide microwave dish filled 2.5 cm high with H_2O and heat in the microwave for 8 min until the water is near boiling point.
3. Taking great care, remove the very hot H_2O from the Coplin jar and fill the jar to the top with antigen retrieval solution (0.01 M citric acid, pH 6.0).
4. Place the Coplin jar in the microwave dish containing the hot H_2O. Heat in the microwave oven on high power to boiling point; then continue heating to allow a further 5 min of boiling (*see* **Note 12**). Continue heating in the microwave for a further 5 min of boiling.
5. Carefully remove the hot Coplin jar with slides from the microwave oven and allow cooling for 10–15 min.
6. Rinse the slides twice with H_2O and transfer to 3.5 M HCl for 15 min to denature the DNA. Rinse again with H_2O and place the slides in Tris–HCl buffer.
7. Drain and circumscribe sections on the slides with a hydrophobic barrier pen, as described above. Proceed without allowing the sections to dry.

3.2.3 Immunostaining Using Peroxidase Conjugates

1. In the preparation for peroxidase staining, quench the endogenous peroxidase activity in the sections by adding 50–200 μL of 3% H_2O_2 for 3 min. Rinse in Tris–HCl buffer for 3 min, draining the solutions from the slides between each step.
2. Arrange the slides in a humidity chamber. Add 50–200 μL of 1% goat serum in PBE onto each section to prevent nonspecific staining, and incubate for 10 min. Drain the blocking solution from the slide by gently tapping the side onto paper tissue (except from those sections that will serve as the "no antibody" control).
3. Add the primary antibody diluted in PBE buffer to the concentration recommended by the manufacturer. Incubate the slides in the humidity chamber for 1 h at room temperature. Rinse the slides twice in Tris–HCl (5 min each), draining the solutions between steps.
4. Add the biotinylated secondary antibody diluted in PBE buffer. Incubate the slides in the humidity chamber for 30 min at room temperature. Rinse twice in Tris–HCl (5 min each) as before.
5. Add streptavidin-peroxidase diluted in PBE buffer to the slides and incubate for 30 min (or as indicated by the manufacturer). Rinse the slides twice in Tris–HCl (5 min each) as before.
6. Add the DAB mixture (or other substrates) to the slides and incubate until the desired intensity of staining is achieved (2–15 min). Rinse the slides with H_2O.

3.2.4 Hematoxylin Counterstaining

1. Place the slides with stained sections in Mayers hematoxylin for 1 min (less in a fresh solution; avoid overstaining). Rinse with tap water for 1 min.
2. Depending on the enzyme substrate used: either mount the sections with cover slips using aqueous mounting medium such as glycerol gelatine; alternatively, dehydrate the slides through graded alcohols 70; 95 and 100% alcohol (3 min each), followed by three changes of xylene; then mount cover slips using Permount or similar.

3.3 DNA Methylation Immunohisto -chemistry on Cryosections

Following comparative testing of published methods, we do not recommend using cell-based immunocytochemistry protocols for tissue cryosections (*see* Subheading 1 and **Note 13**). We outline below an alternative protocol for cryosections recently developed in our lab, in which the DNA is denatured by boiling in sodium citrate buffer. The representative results are given in Fig. 1.

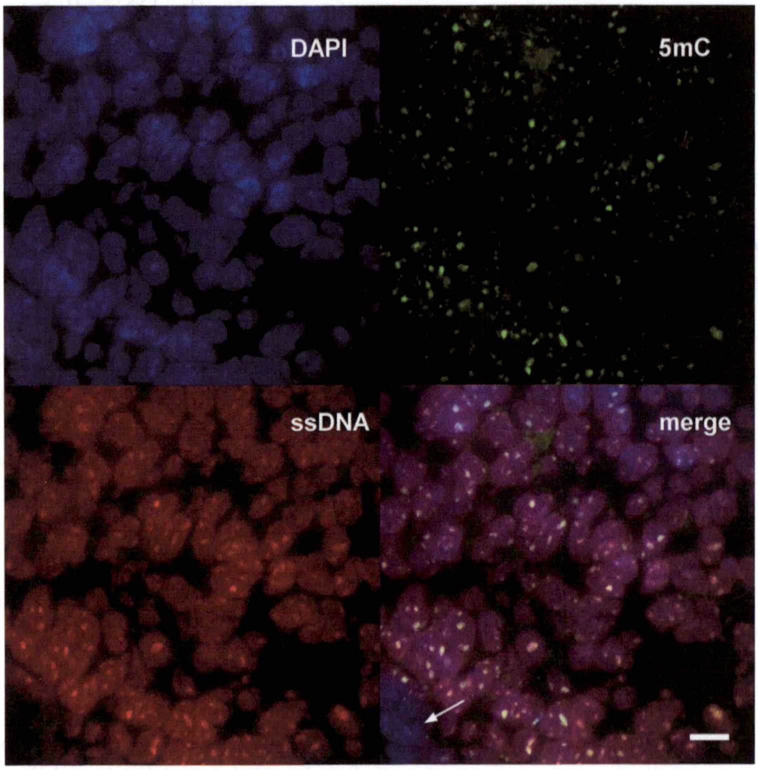

Fig. 1 Immunodetection of 5mC in cryosections. Cryosection of mouse E13.5 embryonic brain illustrating immunostaining protocol results for 5mC (*upper right panel*) and ssDNA (*lower left panel*), compared with DAPI counterstaining of DNA (*upper left panel*) and merged channels (*lower right panel*). The 5mC signal overlaps with high density DAPI nuclear foci, whereas the ssDNA signal shows a nuclear DNA pattern similar to DAPI. Antibodies (1:500 dilution) and conditions were as described in the protocol. *Arrow*, small unstained area for comparison; scale bar, 10 μm

1. Using a cryotome, cut 5–7 μm cryosections from fresh frozen embedded tissues and adhere 1–3 sections per coated microscope slide. Trace a hydrophobic barrier around each section using a PAP pen. Proceed with the staining procedure in a humidity chamber at room temperature. Do not allow sections to dry between steps.
2. Cover the sections with 200 μL of 2% PFA in PBS for 50 min to fix the tissues. Drain the fixation solution from the slide by gently tapping the side onto paper tissue. Rinse twice with 200 μL of PBS (5 min each).
3. Add 200 μL of 0.5% Triton X-100 in PBS for 30 min to permeabilize the tissues. Drain the permeabilization solution and rinse twice with PBS (10 min each).
4. Preheat a wide microwavable dish filled with 2.5 cm of H_2O in a microwave oven for 8 min at full power (alternatively, preheat a heat resistant beaker containing 0.01 M sodium citrate pH 6.0 on a hotplate to near boiling point).
5. Fill a Schieferdecker jar with 0.01 M sodium citrate pH 6.0 and put the slides in. Place jar (without the lid) in the previously preheated dish, which has hot H_2O in the bottom (or transfer the slides in a holder to the beaker containing preheated 0.01 M sodium citrate pH 6.0).
6. Heat the jar containing the slides in the microwave oven at full power for 3 min (*see* **Note 12**). Heat the slides for an additional 3 min (or bring the beaker to a boil on the hotplate, and then continue boiling for a further 10 min to complete DNA denaturation and antigen retrieval).
7. Carefully remove the Schieferdecker jar from the microwave oven and allow the slides to cool for 10 min (or remove the beaker from the hotplate and allow the slides to cool while submerged in the solution).
8. Rinse the slides in PBS in a rocking jar for 15 min to remove any loose material.
9. Add 200 μL of 2% BSA in PBS containing (optional) 1% donkey serum and incubate for 1 h to prevent nonspecific staining. Drain the blocking solution (except from those sections that will serve as the "no antibody" control; *see* **Note 11**).
10. Add the primary antibody diluted 1:300–1:500 in the blocking solution. Incubate the slides in the humidity chamber for 2 h at room temperature. Remove the antibody solution and rinse the slides three times with 200 μL of PBST (15 min each), draining the solutions between each step.
11. Add the secondary antibody diluted 1:250 in blocking solution and incubate for 1 h as above. Remove the antibody solution and rinse the slides with 200 μL of PBST (15 min).

12. Add 0.3 μM of DAPI solution in PBST for 20 min to counterstain for DNA (denatured DNA has a reduced DAPI fluorescence; for alternative DNA staining *see* **Note 5**). Remove the counterstaining solution and rinse once with PBS (15 min).
13. Mount the sections with cover slips using high refractive index mounting medium. Leave the slides to cure overnight at room temperature.

3.4 DNA Methylation Immunocytochemistry on Whole Mount Embryos

All the steps are performed at room temperature (22–25 °C or 27 °C on the heating plate) with a total volume of 500 μL per glass dish, unless otherwise mentioned. Between each step diligently follow hazardous waste recycling regulations. Carefully wash each glass dish with distilled water and dry it with a non-fluffy rag. The representative results are given in Fig. 2.

1. Label the glass dishes and lids with appropriate signs in order to identify the different groups (control, NT, etc.), date and sample size ($n = \ldots$).
2. Choose between these two modes of fixation:

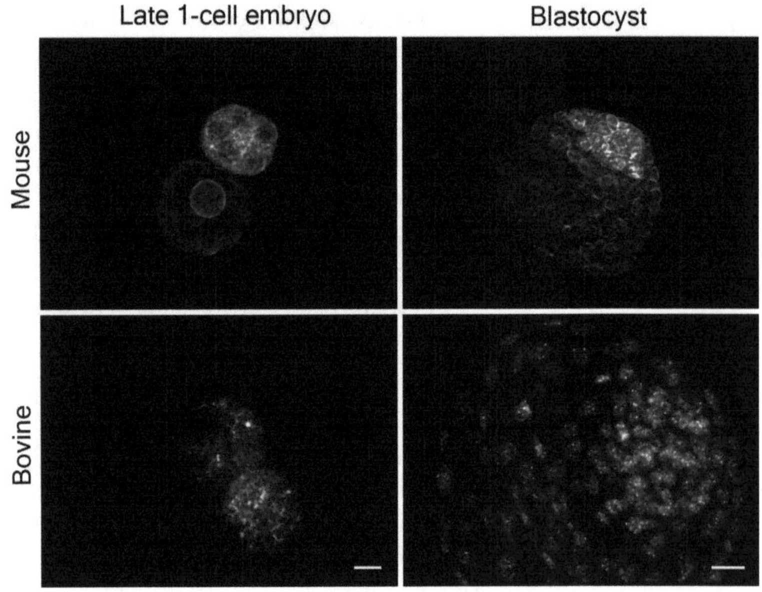

Fig. 2 Immunodetection of 5mC in embryos. Mouse (*upper panels*) and bovine (*lower panels*) 1-cell embryos (*left column*) and blastocysts (*right column*) after immunostaining for 5mC using the 33D3 clone (1:500 dilution). Permeabilization was performed for 30 min (mouse) or 1 h (bovine) in 0.5% Triton X100; HCl denaturation was performed with 2 M HCl for 1 h either at 37 °C (mouse) or at room temperature (bovine); and incubation with the primary antibody was performed overnight at 4 °C. Scale bar: 10 μm for 1-cell embryos and 20 μm for blastocysts

- Short fixation: 20 min at room temperature (22–25 °C or 27 °C on the heating plate) with 2% PFA freshly diluted in PBS, if you plan to perform immunostaining immediately after.
- Long fixation: overnight at +4 °C with 4% PFA, freshly diluted in PBS, if you plan to perform immunostaining the following day. If necessary keep the embryos in PFA for up to 1 week—in that case, change the 4% PFA every 2 days (*see* **Note 14**).

3. Prepare in each glass dish a volume of ~500 μL PFA, diluted in PBS.
4. With a thin glass manipulation pipette transfer the embryos group by group into the corresponding glass dish, directly from the collection/culture medium into the fixative solution, without rinse step (embryos are known to be very sensitive to environmental changes).
5. Cover the glass dishes with the glass lids; in the case of a long fixation, wrap each dish with parafilm to avoid drying out of the fixative at 4 °C.
6. Incubate in PFA, either 20 min at room temperature (or 27 °C on the heating plate) or overnight at +4 °C.
7. Transfer the embryos with the glass embryo manipulation pipettes from the fixative into PBS:
   (a) If PFA incubation was ≤30 min, two times 5 min PBS rinse is enough.
   (b) If PFA incubation was longer, wash for at least 30 min (three times 10 min).

   In both the cases, and at each further step, embryos have to be aspirated in the pipette with a minimum amount of solution, breathed out (i.e., expelled gently) into the next solution and moved within the glass dish several times (*see* **Note 15**).
8. For permeabilization, transfer the embryos and incubate for 30 min in 0.5% Triton X-100, freshly diluted in PBS (mix gently to avoid bubbles) (*see* **Note 16**).
9. Prepare 2 M HCl from the stock solution (37% HCl corresponds to 12 M) diluted with ultrapure water. Glass dishes containing drops of 2 M HCl (typically 120 μL drops, *see* **Note 17**) must be wrapped with parafilm or placed in a humid chamber to prevent drying out of the HCl, and preheated at 37 °C in the incubator (or oven).
10. For DNA denaturation, transfer the embryos in the preheated HCl dilution and incubate at 37 °C for 1 h (*see* **Note 18**).
11. Transfer the embryos into PBS and rinse two times 10 min (*see* **Note 19**).

12. Saturation of nonspecific sites is performed by 1 h incubation of the embryos in PBS containing 2% BSA (PBS-BSA 2%).
13. Dilute the primary antibody directed against 5mC with PBS containing 2% BSA in a micro test tube (*see* **Notes 20** and **21**). You will need a total of 20 μL per group +20 μL for equilibration. In a Petri dish prepare 20 μL drops of the primary antibody for each group and add one drop for equilibration. Cover the drops with mineral oil.
14. Embryos are transferred group by group in the primary antibody: first in the equilibration drop and then in their respective antibody incubation drops.
15. Perform incubation in the first antibody overnight at 4 °C.
16. Transfer the embryos in glass dishes with PBS. Wash three times 10 min to remove antibody excess (*see* **Note 22**).
17. Perform second antibody incubation: dilute in PBS-BSA 2% (as above), prepare 20 μL drops under mineral oil in a Petri dish and transfer the embryos group by group as above (first in the equilibration drop and then in their respective antibody incubation drops). Incubate for 1 h at room temperature (22–25 °C or 27 °C on the heating plate). From this step to the end, always carefully protect dishes from light, with aluminum foil for example.
18. Transfer the embryos in glass dishes with PBS. Wash three times 10 min to remove excess antibody (*see* **Note 23**).
19. Embryos can be transferred in 2% PFA (diluted in PBS) for 30 min + washed once with PBS (2 min) (*see* **Note 24**).
20. Transfer the embryos in glass dishes containing a nucleic acid stain diluted in PBS and incubate for 20 min at 37 °C (*see* **Note 25**).
21. Prepare SuperFrost Plus slides with the identification of the experiment on the side (date/species/embryonic stage/number of embryos and antibody used) and as many areas as groups drawn on the back on the slide with a permanent pen (*see* **Note 26**).
22. Load a 100 μL pipetman with 40 μL of the antifading agent.
23. Aspirate the DNA staining solution to halfway up the glass manipulation pipette and then aspirate the embryos of the first group in a minimum volume of solution (*see* **Note 27**).
24. Breath out the embryos in the delimited area of the slide.
25. With a very thin glass pipette aspirate as much solution around the embryos as possible; be particularly careful not to dry the embryos.

26. With the pipetman containing the antifading agent immediately put a little drop of antifading agent on embryos after the removal of the solution.
27. Repeat all the steps for the other groups of embryos.
28. Put a little drop of clear nail polish on each corner of the slide, next to the groups, in order to preserve the 3D shape of the embryos.
29. Carefully put the coverslip on the slide and secure with clear nail polish around the coverslip; do not push on the coverslip in order to avoid squashing.
30. Keep the slides at 4 °C until observation.

4 Notes

1. Primary antibodies recognizing specific antigens can be raised in various species (mouse, rabbit, rat, human autoantibodies) as mentioned in Subheading 1. The dilution of the primary antibody should be determined in preliminary experiments to find the lowest concentration that still provides an immunostaining signal with a good signal-to-noise ratio. We recommend the use of mouse monoclonal 5mC antibody clone 33D3 and rabbit polyclonal 5hmC antibody.
2. Primary antibodies can be directly coupled to fluorophores, but in order to amplify the signal, in most cases fluorescently tagged secondary antibodies are used (indirect immunostaining). The range of available fluorophores (such as FITC, Rhodamine, Alexa Fluor) allows for flexibility. If you want to use several antibodies simultaneously, make sure that (1) primary antibodies were produced in different species, (2) each corresponding secondary antibody is coupled to a different fluorophore, and (3) each chosen fluorophore can be distinguished by the microscope that will be used. We recommend "green" Alexa Fluor conjugated IgG that is easier to observe with red or blue DNA counterstaining. Secondary antibodies are diluted according to the manufacturer's recommendations; we commonly use 1/200–1/300 dilutions with antibodies from Jackson ImmunoResearch or Molecular Probes.
3. We recommend the use of mountant with a high refractive index and antifade properties, such as Prolong Gold from Molecular Probes.
4. The thickness of the cover slip is crucially important for high-resolution microscopy. Most microscope objectives are designed for use with No. 1 cover slips (0.13–0.17 mm thickness). Use of cover slips that deviate from 0.17 mm thickness (including mountant) will result in spherical aberration and a

reduction in resolution and image intensity. If the cells are grown directly on cover slips, No. 1.5 (0.16–0.19 mm thickness) provides the optimal distance and ease of handling.

5. An extensive assortment of nucleic acid stains is commercially available for DNA counterstaining of the embryos. Their fluorescence absorption and emission spectra span the visible-light spectrum from blue to near-infrared, making them compatible with many different types of secondary antibodies. We often use Propidium Iodide (red fluorescence) or Ethidium Homodimer 2 (red fluorescence), which bind to DNA even after HCl denaturation in embryos. Alternatively, we use a single-strand DNA antibody, which requires strong protocol adjustments (*see* **Note 25**).
6. Ideally room temperature should be stable, between 22 and 25 °C, otherwise we recommend using a heating block set to 27 °C.
7. Aspirator tubes can be hand-made with clear HPLC tubing (diameter 2 mm), but are also commercially available (e.g., from Sigma-Aldrich). To this system you can add a filter tip and aspirate by mouth, or add a 1 or 2 mL syringe to control the valve manually. To manufacture glass pipettes for the manipulation of embryos, stretch a glass capillary (or tip of a Pasteur pipette) in the flame in order to produce very thin pipettes with diameters to fit the embryo. It is easier to work with quite short pipettes (not more than 8 cm in total), which are a little oblique.
8. Immunostaining is usually performed in commercially available glass dishes with one or several wells. Plastic 4-cell plates can also be used but are usually less convenient for small preimplantation embryos. The capacity of each well has to be $\geq$500 μL.
9. Glass cover slips are commercially available in a variety of widths, lengths, and thicknesses (*see* also **Note 4**). We recommend square or rectangular cover slips that are more easily placed on the embryos, measuring 22 × 22 mm up to 22 × 40 mm.
10. Cells for microscopy are best grown on collagen-coated cover slips to reduce the distance to the objective for high-resolution imaging. Chamber slides are a more expensive option and most are designed to let the cells grow on the carrier slide, resulting in a longer optical path through mountant.
11. Standard controls for the immunostaining procedure may include: a “no primary antibody” control to assess secondary antibody nonspecific binding; a “no antibody” control to assess tissue autofluorescence; a “normal serum” control to assess IgG nonspecific binding; single label controls for double

staining experiments to assess spectral bleed-through and antibody interference; ssDNA primary antibody control and DAPI counterstaining to assess DNA denaturation.

12. Check that the solution has not evaporated to a level that might expose the sections to drying; add more solution if necessary.
13. The HCl denaturation step is ineffective on cryosections that are adhered to microscope slides. This is not an issue for whole mount tissues or thick sections that are kept submerged in a hydrated state, which should be processed using the embryo protocol. Cryosections on microscope slides are denatured in a Schieferdecker type jar holding the slides submerged sideways in the microwave. Metal slide holders require a heat resistant beaker heated on a hotplate.
14. A comparative experiment with a short and a long fixation in parallel, using the same antibodies, should be performed to confirm that a long fixation step does not alter the immunostaining.
15. The overall success of the immunostaining, especially the quality of the signal, depends on the perfect contact of the embryos with each incubation solution. It is thus necessary at each step to correctly transfer the embryos and move them regularly in the glass dish.
16. In some cases, especially in other species than mouse: should the immunodetection not produce a good signal, permeabilization may be either (1) extended to 1 h, (2) performed with 1% Triton X-100, (3) performed at higher temperature, e.g., 37 °C or performed, (4) after the removal of the zona pellucida by HCl or pronase treatments.
17. Transfer of embryos in 2 M HCl results in an osmotic shock. As a result, the embryos tend to float in the HCl. The use of small drops makes it easier to gather the embryos and bring them back to the bottom of the dish. We therefore recommend the use of smaller glass dishes, or glass plates with several wells, and drops of around 100 μL. A second consequence of the osmotic shock is the reduction of embryo size, which may necessitate a change of the glass pipette used to transfer them.
18. This denaturation step is required to increase DNA accessibility to the primary 5mC antibodies. Two concentrations of HCl are found in the literature: 2 and 4 M. Similarly, different incubation times (from 10 min to 1 h) and temperatures (room temperature or 37 °C) can be found. 4 M HCl is more often used at room temperature with shorter incubation times (10–20 min), while 2 M HCl is used with longer incubation times and/or higher temperature. In our experience, temperature is the most sensitive parameter, which is why when using room temperature, we recommend using the heating block for more reproducibility. It is also for this reason that we prefer

using the conditions outlined in the main article (2 M HCl, 1 h, 37 °C). In addition, those conditions work best for the combination with the DNA staining using the single-stranded DNA antibody (*see* **Note 25**).

19. An additional step to neutralize the HCl is often added here. Different neutralization solutions can be used: 0.1 M Tris–HCl, pH 8 or 0.1 M Sodium Borate ($Na_2B_4O_7$), pH 8.5 for 10 min at room temperature. This step is essential when working with cells, as not all the HCl can be efficiently and rapidly removed, but is optional with embryos.
20. You may combine several primary antibodies if they are not derived from the same species (to avoid cross-reactivity upon addition of the secondary antibodies). However, the combined use of a 5mC antibody and an antibody detecting another modified form of cytosine (5hmC, 5fC, 5caC) should be very carefully tested, especially if the aim of the experiments is to quantify the signals. Indeed, in our experiments on mouse embryos, using both anti-5mC and anti-5hmC resulted in a weaker signal for both antibodies [18].
21. Antibodies have to be prepared in sterile conditions, either under an appropriate hood or next to a flame. For the 5mC antibody clone 33D3 from Eurogentec we recommend 1:500 dilution. If the immunostaining results are not satisfactory, showing too much background or large fluorescent antibody aggregates, we recommend centrifuging the antibody (10 min at 11500 × *g* with a microcentrifuge) after the dilution, before taking the supernatant to prepare the drops.
22. Tween 20 (0.05%) may be added in these washing steps in order to reduce the background signal if necessary.
23. The "post-fixation" is used to preserve the immunostaining when immediate observation on the microscope is not possible.
24. DNA stain concentrations: 0.002 mM Ethidium Homodimer-2; 1 μg/mL Propidium Iodide.
25. While Ethidium Homodimer-2 and Propidium Iodide can be used for counterstaining in most cases, we could not obtain a reproducible staining with those dyes in mouse 1-cell embryos. We therefore used a single-stranded DNA antibody, which requires additional steps in the protocol. After **step 8** of Subheading 3.4, (1) post-fix in 4% PFA for 25 min at RT; (2) block the nonspecific sites with PBS/0.05% Tween/4% low fat milk for 4 h at RT; (3) incubate with the anti-single-stranded DNA antibody diluted at 1:10 in PBS/0.05% Tween/4% low fat milk for around 40 h (overnight is not sufficient as it results in great variability of the signal obtained, at least two nights yield good results); (4) repeat **step 8** of Subheading 3.4 and continue with

the rest of the protocol. As the 5mC antibody we recommend is mouse IgG and the anti-single-stranded DNA we used is mouse IgM, secondary antibodies need to be chosen carefully to avoid cross-reactions.

26. Do not use the extremity of the slide: when using an inverted microscope for observations the objective could be damaged.
27. Each step has to be performed carefully in order to preserve the 3D-shape of the samples. If the group contains many embryos, we recommend taking five embryos at a time.

Acknowledgments

We thank Tatiana Chebotareva, Cristina Aguilar, and other members of the S.P lab for their help in developing some of the protocols presented, and we thank Richard Meehan for helpful suggestions. All present and past members from N. B's lab are acknowledged for their work and contributions, especially Claire Boulesteix. We wish to acknowledge the Niveleau lab for their development of the original 5mC antibody and some of the protocols outlined in this chapter. N.A.A.H. is a recipient of a MARA (Malaysia) scholarship; work in S.P.'s lab is supported by the BBSRC and the BHF. Work in N.B.'s lab is supported by the REVIVE Labex (Investissement d'Avenir, ANR-10-LABX-73).

References

1. Bird A (2002) DNA methylation patterns and epigenetic memory. Genes Dev 16:6–21
2. Reddington JP, Pennings S, Meehan RR (2013) Non-canonical functions of the DNA methylome in gene regulation. Biochem J 451:13–23
3. Rothbart SB, Strahl BD (2014) Interpreting the language of histone and DNA modifications. Biochim Biophys Acta 1839:627–643
4. Politz JC, Scalzo D, Groudine M (2013) Something silent this way forms: the functional organization of the repressive nuclear compartment. Annu Rev Cell Dev Biol 29:241–270
5. Meehan RR, Dunican DS, Ruzov A et al (2005) Epigenetic silencing in embryogenesis. Exp Cell Res 309:241–249
6. Beaujean N (2014) Histone post-translational modifications in preimplantation mouse embryos and their role in nuclear architecture. Mol Reprod Dev 81:100–112
7. Wongtawan T, Taylor JE, Lawson KA et al (2011) Histone H4K20me3 and HP1α are late heterochromatin markers in development, but present in undifferentiated embryonic stem cells. J Cell Sci 124:1878–1890
8. Wu H, Zhang Y (2014) Reversing DNA methylation: mechanisms, genomics, and biological functions. Cell 156:45–68
9. Ficz G, Branco MR, Seisenberger S et al (2011) Dynamic regulation of 5-hydroxymethylcytosine in mouse ES cells and during differentiation. Nature 473:398–402
10. Shen L, Zhang Y (2013) 5-Hydroxymethylcytosine: generation, fate, and genomic distribution. Curr Opin Cell Biol 25:289–296
11. Nestor CE, Ottaviano R, Reinhardt D et al (2015) Rapid reprogramming of epigenetic and transcriptional profiles in mammalian culture systems. Genome Biol 16:11
12. Booth MJ, Ost TW, Beraldi D et al (2013) Oxidative bisulfite sequencing of 5-methylcytosine and 5-hydroxymethylcytosine. Nat Protoc 8:1841–1851

13. Nestor C, Ruzov A, Meehan R et al (2010) Enzymatic approaches and bisulfite sequencing cannot distinguish between 5-methylcytosine and 5-hydroxymethylcytosine in DNA. BioTechniques 48:317–319
14. Reynaud C, Bruno C, Boullanger P et al (1992) Monitoring of urinary excretion of modified nucleosides in cancer patients using a set of six monoclonal antibodies. Cancer Lett 61:255–262
15. Barbin A, Montpellier C, Kokalj-Vokac N et al (1994) New sites of methylcytosine-rich DNA detected on metaphase chromosomes. Hum Genet 94:684–692
16. Piyathilake C, Niveleau A, Grizzle W (2004) Role of global methylation of DNA in lung carcinoma (Ch. 10). In: Hayat MA (ed) Immunohistochemistry and in situ hybridization of human carcinomas. Elsevier Acad. Press, Burlington, MA, pp 181–187
17. Keshet I, Schlesinger Y, Farkash S et al (2006) Evidence for an instructive mechanism of de novo methylation in cancer cells. Nat Genet 38:149–153
18. Salvaing J, Aguirre-Lavin T, Boulesteix C et al (2012) 5-Methylcytosine and 5-hydroxymethylcytosine spatiotemporal profiles in the mouse zygote. PLoS One 7:e38156
19. Wu H, D'Alessio AC, Ito S et al (2011) Genome-wide analysis of 5-hydroxymethylcytosine distribution reveals its dual function in transcriptional regulation in mouse embryonic stem cells. Genes Dev 25:679–684
20. Thomson JP, Lempiäinen H, Hackett JA et al (2012) Non-genotoxic carcinogen exposure induces defined changes in the 5-hydroxymethylome. Genome Biol 13:R93
21. Inoue A, Shen L, Dai Q et al (2011) Generation and replication-dependent dilution of 5fC and 5caC during mouse preimplantation development. Cell Res 21:1670–1676
22. Jin SG, Jiang Y, Qiu R et al (2011) 5-Hydroxymethylcytosine is strongly depleted in human cancers but its levels do not correlate with IDH1 mutations. Cancer Res 71:7360–7365
23. Li Y, O'Neill C (2013) 5′-Methylcytosine and 5′-hydroxymethylcytosine each provide epigenetic information to the mouse zygote. PLoS One 8:e63689
24. Salvaing J, Li Y, Beaujean N, O'Neill C (2014) Determinants of valid measurements of global changes in 5′-methylcytosine and 5-′-hydroxymethylcytosine by immunolocalisation in the early embryo. Reprod Fertil Dev 27:755–764

Part III

Genome-Wide DNA Methylation Analysis

Chapter 5

Whole-Genome Bisulfite Sequencing Using the Ovation® Ultralow Methyl-Seq Protocol

Christian Daviaud, Victor Renault, Florence Mauger, Jean-François Deleuze, and Jörg Tost

Abstract

The analysis of genome-wide epigenomic alterations including DNA methylation has become a subject of intensive research for many complex diseases. Whole-genome bisulfite sequencing (WGBS) using next-generation sequencing technologies can be considered the gold standard for a comprehensive and quantitative analysis of cytosine methylation throughout the genome. Several approaches including tagmentation- and post bisulfite adaptor tagging (PBAT)-based WGBS have been devised. Here, we provide a detailed protocol based on a commercial kit for the preparation of libraries for WGBS from limited amounts of input DNA (50–100 ng) using the classical approach of WGBS by ligation of methylated adaptors to the fragmented DNA prior to bisulfite conversion. The converted library is then amplified with an optimal number of PCR cycles to ensure high sequence diversity and low duplicate rates. Spike-in of unmethylated DNA allows for the precise estimation of bisulfite conversion rates. We also provide a step-by-step description of the data analysis using publicly available bioinformatic tools. The described protocol has been successfully applied to different human samples as well as DNA extracted from plant tissues and yields robust and reproducible results.

Key words Whole-genome bisulfite sequencing, DNA methylation, Data analysis, Bisulfite conversion, Low-input, Spike-in

1 Introduction

Epigenetic modifications add additional layers of information on top of the bare genomic sequence, thereby dramatically extending the information potential of the genetic code. The field has gained great momentum in recent years as it has become clear that epigenetics plays a key role in normal development as well as in disease. The technical revolution of massively parallel sequencing, which allows the interrogation of multiple epigenetic modifications using the same analytical technologies, as well as the development of advanced high-throughput epigenotyping technologies has enormously spurred our knowledge on gene regulatory mechanisms.

Jörg Tost (ed.), *DNA Methylation Protocols*, Methods in Molecular Biology, vol. 1708,
https://doi.org/10.1007/978-1-4939-7481-8_5, © Springer Science+Business Media, LLC 2018

DNA methylation occurring at CpG dinucleotides is probably the best-studied epigenetic modification due to the extensive mapping of DNA methylation patterns in different diseases. DNA methylation-based biomarkers bear the promise to contain valuable information for early diagnosis, prognosis, disease classification and might assist in the prediction of response to therapy [1]. DNA methylation patterns add thereby valuable information to the picture of the disease-underlying processes that cannot be obtained solely through genetic or transcriptomic analysis. While initially epigenetic research has focused on the analysis of epigenetic alterations in cancer [2], research in recent years has demonstrated that alterations are also present in nearly all complex diseases including autoimmune and inflammatory diseases, allergic disorders, metabolic as well as neurodegenerative and psychological disorders [3–9] (*see* also Chapter 1). Whole-genome bisulfite sequencing or MethylC-seq can be considered the current gold standard for the genome-wide identification of differentially methylated CpGs (DMCs) and regions (DMRs) at single-nucleotide resolution. It overcomes the limitations of cloning and Sanger sequencing, a low-throughput method limited to a small number of loci of interest, in which the quantitative resolution was limited by the number of clones analyzed (in most studies <20). Furthermore, whole-genome bisulfite sequencing avoids the problem with the primer design that often introduces multiple biases [10, 11]. However, this unprecedented quantitative and spatial resolution that is currently transforming DNA methylation analysis comes still at a high cost and requires substantial sequencing to obtain a proper and even coverage and requires some bioinformatic expertise and resources. The most widely used protocol consists of the fragmentation of genomic DNA, adapter ligation, bisulfite conversion, and limited amplification using adapter-specific PCR primers. While initially several micrograms of DNA were required to perform whole-genome bisulfite sequencing, the replacement of electrophoretic steps and gel extraction for purification by magnetic beads has enabled creating libraries suitable for sequencing from ~100 ng of input material [12]. Libraries have been reported to be constructed from even lower input material, but require in most cases a high number of PCR cycles (up to 25 cycles [13]) inducing potentially a large bias for the estimation of the DNA methylation levels and a much more substantial sequencing effort to obtain a homogeneous and sufficient coverage. To assess the bisulfite conversion efficiency, DNA of the bacteriophage λ is spiked in the reaction (*see* also Chapter 20). Mapping the reads against the bisulfite-converted genome of the phage and counting any remaining cytosines allows estimation of the conversion rates and identification of problems during bisulfite conversion. A large number of programs have been developed to perform the quality control, preprocessing steps (such as adaptor, barcode, and quality score trimming), mapping of the reads to a bisulfite-converted reference genome, and scoring of DNA methylation levels (count statistics) and identification of DMCs and DMRs [14]. WGBS has been widely used for the methylome-wide analysis of a large number of organisms and plants as well as human tissues and ~90–95% of the cytosines present in the genome are covered [15–21].

While most commonly performed on the Illumina sequencing platform, which allows for a much higher coverage, protocols and analytical pipelines have also been devised for the SOLiD platform [22, 23] and have been applied in some studies [24, 25].

Commonly used alternative approaches such as tagmentation-based whole genome bisulfite sequencing [26] and post bisulfite adaptor tagging (PBAT)-based WGBS [27] are described in detail in Chapters 6, 7, and 9 of this book. It should be pointed out that bisulfite sequencing is not able to distinguish 5-methylcytosine and 5-hydroxymethylcytosine. The measured DNA methylation degree is thus a mixture of both marks and specialized approaches such as OxBS-seq [28] (*see* also Chapter 34) or TAB-seq [29] (*see* also Chapter 33) are required to distinguish methylcytosine from its oxidative derivative. In the present protocol, we describe the protocol used in routine in our laboratory, which is based on the commercial Ovation® UltraLow Methyl-seq DR Multiplex System (NuGen, San Carlos, CA). The workflow of the library preparation protocol is depicted in Fig. 1 and follows the classical library preparation protocol, in which methylated adaptors are ligated to the fragmented DNA prior to bisulfite conversion. 50–100 ng of genomic DNA is fragmented to a size of approximately 200 base pairs (bp), purified and methylated adaptors compatible with sequencing on an Illumina HiSeq instrument are ligated. The resulting DNA library is purified and bisulfite converted. A qPCR assay determines the optimal number of PCR amplification cycles required to yield a high diversity library with minimal duplicates.

After sequencing using a dedicated sequencing primer for read 1, the data is quality controlled and mapped to the genome to obtain the percentage of DNA methylation at all cytosines in the genome using Bismark [30]. Further downstream processing such as calling of DMRs can be performed using a variety of recently reviewed tools [31].

With the protocol described in this chapter, we obtain per Hiseq 2000 lane routinely 100M reads, of which 85–90% can be mapped to a bisulfite-treated human genome resulting in a genome-wide coverage of 7–8×. Duplicate rates are below 2% (Picard) and a conversion efficiency of ~99% is routinely achieved as measured by using either unmethylated lambda phage DNA or methylation outside CpG context in the human genome. Coverage was found to be comparable to a standard bisulfite sequencing protocol starting with 4 μg of DNA and quantitative values correlated well with data obtained from Illumina Infinium Human

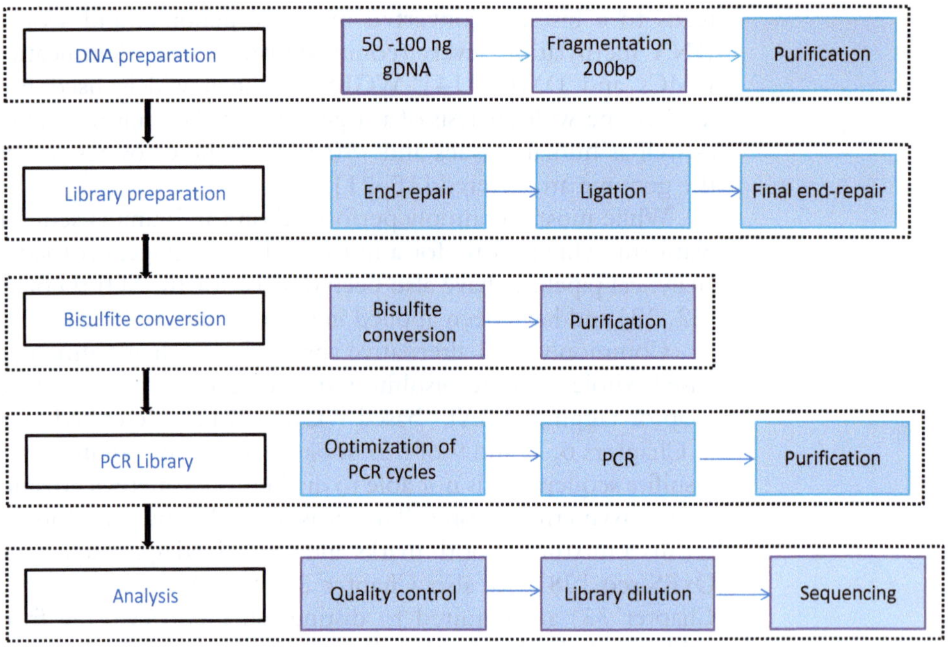

Fig. 1 Workflow of the whole-genome bisulfite sequencing protocol using the Ovation® Ultralow Methyl-seq Kit. See text for a detailed description

Methylation 450K BeadArrays (Renault et al., submitted). The presented protocol has been found to be very robust and applicable without modifications to DNA extracted from various human, mammalian, and plant tissues. With the advent of new sequencing instruments with higher output and lower price per sequenced base, WGBS will become more and more affordable and will eventually become a valuable alternative to the currently widely used 450K and EPIC microarrays [32, 33] (*see* also Chapter 16) as it allows for the comprehensive and quantitative analysis of all cytosines in a genome.

2 Materials

2.1 Experimental Workflow

1. 50–100 ng genomic DNA (*see* **Note 1**).
2. Ovation® Ultralow Methyl-Seq DR Multiplex System 1–8 32rx (NuGEN, cat#0335-32) containing: End Repair Buffer Mix (ER1 ver 8), End Repair Enzyme Mix (ER2 ver 4), End Repair Enhancer (ER3 ver 2), Ligation Buffer Mix (L1 ver 4), Ligation Adaptor Mix (L2V11DR-BC1) to Ligation Adaptor Mix (L2V11DR-BC8), Ligation Enzyme Mix (L3 ver 4), Final Repair Buffer Mix (FR1 ver 4), Final Repair Enzyme (FR2), Amplification Primer Mix (P2 ver 8), Amplification Enzyme Mix (P3 ver 3), 25 μM Sequencing Primer (MetSeq Primer 1), nuclease-free water (D1), and Agencourt RNAClean XP Beads.

3. Unmethylated cl857 Sam7 Lambda DNA (e.g., Promega).
4. EpiTect FAST DNA Bisulfite kit 50rx (Qiagen) containing: Bisulfite Solution, DNA Protect Buffer, RNase-Free Water, MinElute® DNA Spin Columns, collection Tubes (2 mL), BL Buffer, BW Buffer, BD Buffer, EB Buffer, and Carrier RNA.
5. DNA 1000 kit (Agilent).
6. 100 and 70% Ethanol.
7. 1× TE Buffer: 10 mM Tris–HCl, 1 mM EDTA, pH 8.0.
8. EvaGreen® 20×.
9. Covaris tube: 6 × 16 mm Round-bottom glass, AFA fiber, and cap system (100 μL) (e.g., Kbioscience).
10. 0.2 mL reaction tube "PCR clean" with lid.
11. 1.5 mL reaction tube "PCR clean" with lid.
12. LightCycler® 480 Multiwell Plate 384.
13. LightCycler ®480 Sealing Foil.
14. LightCycler480® (Roche) or similar qPCR instrument.
15. DiaMag02 magnetic rack (Diagenode) or similar.
16. ThermoCycler ProS (Eppendorf) or similar.
17. Qiacube (Qiagen) (optional).
18. Ultrasonicator E210 (Covaris).
19. Bioanalyzer 2100 instrument (Agilent).
20. Vortex.
21. Centrifuge.
22. Laminar Flow Cabinets (optional, *see* **Note 2**).

2.2 Data Analysis

1. One linux/unix multi-core machine with at least 16GB of RAM and 5 cores (see Bismark user guide referenced in http://www.bioinformatics.babraham.ac.uk/projects/download.html#bismark).
2. FastQC: http://www.bioinformatics.babraham.ac.uk/projects/fastqc/
3. Cutadapt for read trimming + Illumina adapter removal: https://cutadapt.readthedocs.org/en/stable/. Requires python and module python-pip for installation.
4. Bismark: http://www.bioinformatics.babraham.ac.uk/projects/bismark/ [30].
5. bowtie2 (available from http://bowtie-bio.sourceforge.net/bowtie2/index.shtml) (*see* **Note 3**).
6. samtools: http://samtools.sourceforge.net/ [34].
7. perl, zcat.
8. Picard: http://broadinstitute.github.io/picard/

3 Methods

3.1 DNA Fragmentation

1. Dilute 50–100 ng of gDNA into 50 µL of 1× TE buffer.
2. Add 0.25% w/w of unmethylated Lambda DNA (e.g., 0.25 ng of Lambda DNA for 100 ng DNA).
3. Transfer the sample to a Covaris tube.
4. Fragment using the settings of the Covaris focused ultrasonicator listed in Table 1 for a 200 bp target length.
5. After fragmentation, briefly spin the Covaris tube and transfer the entire sample to a new 0.2 mL tube.

3.2 DNA Purification

1. Thirty minutes before proceeding take the Agencourt RNA Clean XP beads and nuclease-free water out of the fridge and leave to warm up at room temperature.
2. Resuspend the RNAClean XP beads by vortexing the tube. Ensure beads are fully resuspended before adding to the sample.
3. Prepare a fresh 70% ethanol wash solution (1.5 mL per sample).
4. Add 90 µL (1.8 volumes) of the bead suspension to the 50 µL of fragmented DNA and mix by pipetting ten times.
5. Incubate at room temperature for 10 min.
6. Place the tube into the magnetic rack and incubate for 5 min at room temperature to completely clear the solution.
7. Carefully remove 130 µL of the binding buffer and discard it (*see* **Note 4**).
8. Add 200 µL of freshly prepared 70% ethanol, wash and leave to stand for 30 s.
9. Remove and discard the wash solution using a pipette.

Table 1
Settings of the Covaris E210 for the fragmentation of genomic DNA

Parameters	
DNA volume	50 µL
Duty cycle	10%
Intensity	5.0
Cycles per burst	200
Duration	60 s (×2)[a]
Power	36 W
Temperature	8 °C

[a]Spin between two runs of fragmentation

10. Repeat **steps 8** and **9** one more time, for a total of two 70% ethanol washes (*see* **Note 5**).
11. Air-dry the bead pellets on the magnetic rack for 5–10 min at room temperature. Inspect each tube carefully to ensure that all the ethanol has evaporated.
12. Add 14 μL room temperature nuclease-free water (provided in the kit) to the dried beads. Mix thoroughly to ensure all the beads are resuspended.
13. Place the tube into the magnetic separation rack and incubate for 5 min at room temperature to completely clear the solution.
14. Carefully transfer 13 μL of the eluate (ensuring as few beads as possible are carried over) into a new 0.2 mL tube and place on ice.

3.3 End Repair

1. Thaw the End Repair Buffer (ER1) of the Ovation Ultralow Methyl-Seq DR Multiplex System.
2. Make an End-Repair master mix by combining in a 0.2 mL tube 2 μL of End Repair Buffer (ER1), 0.5 μL of End Repair Enzyme mix (ER2 ver 4) and 0.5 μL of End Repair Enhancer (ER3 ver 2).
3. Add 3 μL of the End Repair master mix to the fragmented DNA sample (from Subheading 3.2). Mix by pipetting, cap and briefly spin down the tubes and place on ice.
4. Place the tubes in a preheated thermal cycler using program 1 described below (*see* **Note 6**):

Step	Temperature (°C)	Time	Number of cycles	Comment
1	25	30 min	1	Open the lid
2	70	10 min	1	
3	4	Hold		

5. Remove the tubes from the thermal cycler, briefly centrifuge and place on ice.

3.4 Ligation

1. Choose and thaw one adaptor per sample from the Ligation Adaptor Mixes (L2V11DR-BC1 through L2V11DR-BC8). For multiplex sequencing refer to Table 2 for possible combinations. Thaw the Ligation Buffer Mix (L1 ver4) on ice. Thaw the nuclease-free water (D1) at room temperature.
2. Add 3 μL of L2 Ligation Adaptor Mix to each sample (from Subheading 3.3). Mix thoroughly by pipetting.
3. Prepare a Ligation master mix by combining in a 0.2 mL tube 4.5 μL of nuclease-free water (D1), 6 μL of Ligation Buffer Mix, and 1.5 μL of Ligation Enzyme Mix (L3 ver 4). Mix by

Table 2
Barcode sequences for dedicated read (DR) adaptors used in ovation ultralow methyl-seq DR multiplex system 1–8

Ligation adaptor mix	6 nt barcode sequence as read by the sequencer	Barcode pairing (2-plex)	Barcode pairing (>2-plex)
L2V11DR-BC1	AACCAG	Duplex set 1	One of the duplex sets from the column to the left should be used in combination with any of the other remaining six individual barcodes
L2V11DR-BC2	TGGTGA		
L2V11DR-BC3	AGTGAG	Duplex set 2	
L2V11DR-BC4	GCACTA		
L2V11DR-BC5	ACCTCA	Duplex set 3	
L2V11DR-BC6	GTGCTT		
L2V11DR-BC7	AAGCCT	Duplex set 4	
L2V11DR-BC8	GTCGTA		

pipetting slowly, without introducing bubbles, briefly spin down and place on ice (*see* **Note 7**).

4. Add 12 μL of Ligation Master Mix to each reaction tube. Mix thoroughly by pipetting slowly, spin and place on ice. Proceed immediately to the next step.
5. Place the reaction in a preheated thermal cycler using program 2 (*see* **Note 6**):

Number	Temperature (°C)	Time	Number of cycles	Comment
1	25	30 min	1	Open the lid
2	70	10 min	1	
3	4	Hold		

6. Remove the tubes from the thermal cycler, briefly spin down, and place on ice.

3.5 Post-ligation Purification

1. Thirty minutes before proceeding take the Agencourt RNA Clean XP beads and nuclease-free water out of the fridge and leave to warm up at room temperature.
2. Resuspend beads by vortexing the tube. Ensure beads are fully resuspended before adding to the sample.
3. Prepare a fresh 70% ethanol wash solution (1.5 mL per sample).
4. Add 45 μL (1.5 volumes) of the bead suspension to the 30 μL of sample (from Subheading 3.4) and mix by pipetting ten times.
5. Incubate at room temperature for 10 min.
6. Place the tube into the magnetic separation rack and incubate for 5 min at room temperature to completely clear the solution.

7. Carefully remove 70 μL of the binding buffer and discard it (*see* **Note 4**).
8. Add 200 μL of freshly prepared 70% ethanol and leave to stand for 30 s at room temperature.
9. Remove and discard the wash solution using a pipette.
10. Repeat **steps 8** and **9** for a total of two 70% Ethanol washes (*see* **Note 5**).
11. Air-dry the beads pellets on the magnetic rack for 5–10 min at room temperature. Inspect each tube carefully to ensure that all the ethanol has evaporated.
12. Add 16 μL of the room temperature nuclease-free water (green: D1) to the dried beads. Mix thoroughly to ensure all the beads are resuspended.
13. Transfer the tube into the magnetic separation racks and incubate for 5 min at room temperature to completely clear the solution.
14. Carefully transfer 15 μL of the eluate (ensuring that as few beads as possible are carried over) into a new PCR tube and place on ice.

3.6 Final Repair

1. Thaw the Final Repair Buffer Mix (Purple: FR1 VER 4) on ice.
2. Make a Final Repair master mix by combining 4.5 μL of Final Repair Buffer Mix (FR1) and 0.5 μL of Final Repair Enzyme Mix (FR2) in a 0.2 mL capped tube. Mix by pipetting, spin down briefly, and place the tube on ice.
3. Add 5 μL of Final Repair master mix to each sample (from Subheading 3.5).
4. Mix thoroughly by pipetting slowly, spin and place on ice.
5. Place the reaction in a preheated thermal cycler using program 3 below:

Step	Temperature (°C)	Time	Number of cycles
1	60	10 min	1
2	4	Hold	

6. Remove the tubes from the thermal cycler, spin down briefly, and place on ice.

3.7 Bisulfite Conversion (See Note 8)

1. Thaw the DNA to be used in the bisulfite reactions. Make sure the Bisulfite Solution is completely dissolved (*see* **Note 9**).
2. At room temperature (15–25 °C), add 85 μL of Bisulfite Solution and 35 μL of DNA Protect Buffer to each sample (20 μL from Subheading 3.6). Mix the bisulfite reaction thoroughly and gently spin down (*see* **Note 10**).

3. Place the reaction in a preheated thermal cycler using program 4 below:

Step	Temperature (°C)	Time (min)	Number of cycles
1	95	5	1
2	60	20	1
3	95	5	1
4	60	20	1
5	20		Hold

3.8 Cleanup of Bisulfite-Converted DNA

1. Briefly centrifuge the tubes containing the bisulfite reactions, and then transfer the totality of the bisulfite reactions into new 1.5 mL microcentrifuge tubes.
2. Add 310 μL of freshly prepared BL Buffer containing 10 μg/mL carrier RNA to each sample. Mix the solutions by vortexing and then centrifuge briefly.
3. Add 250 μL ethanol (96–100%) to each sample. Mix the solutions by pulse vortexing for 15 s, and centrifuge briefly.
4. Place a MinElute DNA spin column and a collection tube into a suitable rack and transfer the entire mixture from each tube into the corresponding MinElute DNA spin column.
5. Centrifuge the spin column at 15,000 × *g* for 1 min at room temperature. Discard the flow-through, and place the spin column back into the collection tube.
6. Add 500 μL of BW Buffer (wash buffer) to each spin column, and centrifuge at maximum speed for 1 min at room temperature. Discard the flow-through, and place the spin column back into the collection tube.
7. Add 500 μL of BD Buffer (desulfonation buffer) to each spin column (*see* **Notes 11** and **12**), and incubate for 15 min at room temperature.
8. Centrifuge the spin columns at 15,000 × *g* for 1 min at room temperature. Discard the flow-through, and place the spin column back into the collection tube.
9. Add 500 μL of BW Buffer to each spin column and centrifuge at 15,000 × *g* for 1 min at room temperature. Discard the flow-through and place the spin column back into the collection tube.
10. Add 500 μL of BW Buffer to each spin column and centrifuge at 15,000 × *g* for 1 min at room temperature. Discard the flow-through and place the spin column back into the collection tube.

11. Add 250 μL of ethanol (96–100%) to each spin column and centrifuge at 15,000 × *g* for 1 min at room temperature.
12. Place the spin column into a new 2 mL collection tube and centrifuge at 15,000 × *g* for 1 min to remove any residual liquid.
13. Place the spin column into a new 1.5 mL tube. Add 22 μL of EB Buffer (elution buffer) directly onto the center of each spin-column membrane and close the lid gently.
14. Incubate the spin column at room temperature for 1 min.
15. Centrifuge for 1 min at 15,000 × *g* at room temperature to elute the DNA.
16. Transfer 20 μL of the eluate into a new 0.2 mL tube and place on ice (*see* **Note 13**).

3.9 Determination of Optimal Amplification Cycles by qPCR

1. Thaw the Amplification Primer Mix (P2 ver 5) and Amplification Enzyme Mix (P3) on ice.
2. Transfer 4 μL of Converted DNA (from Subheading 3.8) into a LightCycler® 480 Multiwell Plate 384.
3. Prepare a qPCR master mix by combining 1 μL of Amplification Primer Mix (P2 ver 5), 0.5 μL of EvaGreen® 20×, and 4.5 μL of Amplification Enzyme Mix (P3) in a 0.5 mL capped tube. Mix by pipetting, spin down briefly, and place on ice.
4. Add 6 μL of qPCR master mix to each 4 μL of the sample. Seal the LightCycler® Plate 384 with LightCycler® 480 Sealing Foil.
5. Centrifuge at 4000 × *g* for 1 min at room temperature and place the tube on ice.
6. Place the reaction into a LightCycler® 480 using program 5 below:

Number	Temperature (°C)	Time	Number of cycles
1	95	2 min	1
2	95	15 s	35
3	60	1 min	
4	72[a]	30 s[a]	
5	4 °C	Hold	

[a]Optical reading (SyBRgreen channel)

7. Determine the optimal number of cycles needed (*N*): plot linear Rn (*see* **Note 14**) versus cycle and determine the cycle number that corresponds to 1/2 of the maximum fluorescent intensity (Fig. 2).

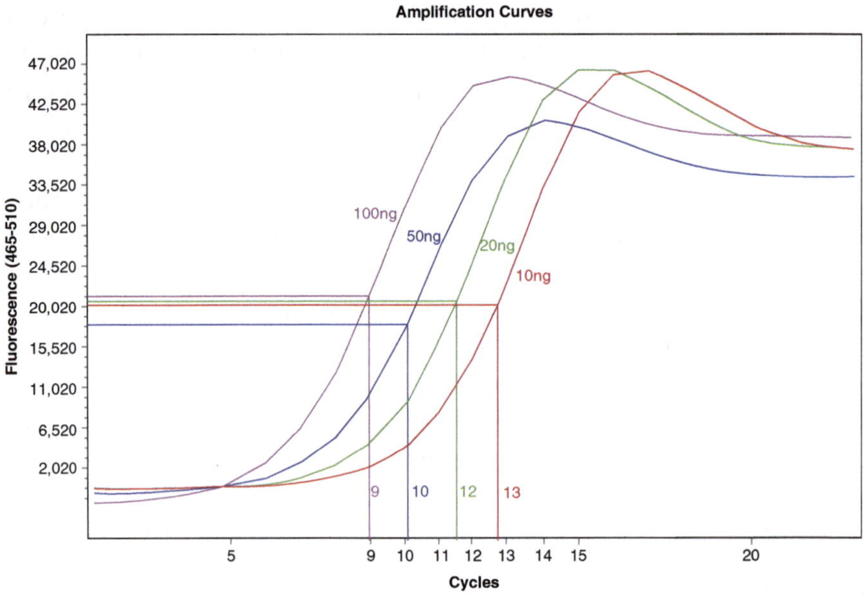

Fig. 2 Determination of the optimal number of PCR cycles. Relative fluorescence is plotted against the number of PCR cycles. For 10, 20, 50, and 100 ng of input DNA, the optimal number of cycles is respectively, 13, 12, 10, and 9 cycles. 50 and 100 ng input DNA do allow for a high quality and diversity library, with low duplicate reads

3.10 PCR Amplification

1. Thaw the Amplification Primer Mix (P2 ver 5) and Amplification Enzyme Mix (P3) on ice.
2. Make a PCR master mix by combining 4 μL of Amplification Primer Mix (P2 ver 5) and 20 μL of Amplification Enzyme Mix (P3) in a 1.5 mL capped tube. Mix by pipetting, spin down the tube, and place on ice.
3. Add 24 μL of PCR master mix to each sample (16 μL from Subheading 3.8).
4. Place the reaction in a preheated thermal cycler using program 6 below:

Step	Temperature (°C)	Time	Number of cycles	Comment
1	95	2 min	1	
2	95	15 s		
3	60	1 min		*N* = *see* Subheading 3.9
4	72	30 s		
5	4	Hold	*N*	

5. Remove the tube from the thermal cycler, spin down briefly, and place on ice.

3.11 Post-PCR Library Purification

1. Thirty minutes before proceeding, take the Agencourt RNA-Clean XP beads and nuclease-free water out of the fridge and leave to warm up to room temperature.
2. Resuspend beads by vortexing the tube. Ensure beads are fully resuspended before adding to the sample.
3. Prepare a fresh 70% ethanol wash solution (1.5 mL for each sample).
4. Add 50 μL (1 volume) of the room temperature bead suspension to the 50 μL of sample (from Subheading 3.10) and mix by pipetting ten times.
5. Incubate at room temperature for 10 min.
6. Place the sample on the magnetic separation rack and incubate for 5 min at room temperature to completely clear the solution.
7. Carefully remove 100 μL of the binding buffer and discard it (*see* **Note 4**).
8. Add 200 μL of freshly prepared 70% ethanol and incubate for 30 s at room temperature.
9. Remove and discard the 70% ethanol wash solution using a pipette.
10. Repeat **steps 8** and **9**, for a total of two 70% ethanol washes (*see* **Note 5**).
11. Air-dry the bead pellets on the magnetic rack for 5–10 min. Inspect each tube carefully to ensure that all the ethanol has evaporated.
12. Add 20 μL of the nuclease-free water (green: D1) to the dried beads. Mix thoroughly to ensure all the beads are resuspended.
13. Place the tube on a magnetic separation rack and incubate for 5 min at room temperature to completely clear the solution.
14. Carefully, transfer 18 μL of the eluate (ensuring as few beads as possible are carried over) into a new PCR tube and place on ice.

3.12 Quantitative and Qualitative Assessment of the Library

1. Thirty minutes before proceeding, take the DNA 1000 kit reagents out of the fridge and leave to warm up to room temperature.
2. Run 1 μL of the libraries on a Bioanalyzer DNA 1000 Chip.
3. Profiles should display a peak at approximately 300 bp (270 and 420 bp) as shown in Fig. 3 corresponding to 150–200 base pair inserts (*see* **Note 15**).

3.13 Sequencing

Sequencing is performed using standard Illumina sequencing protocols in 2 × 101 or 2 × 150 base pairs following the instructions of Illumina. However, the design of the Ovation Ultralow Methyl-Seq DR Multiplex System requires the use of a custom Read

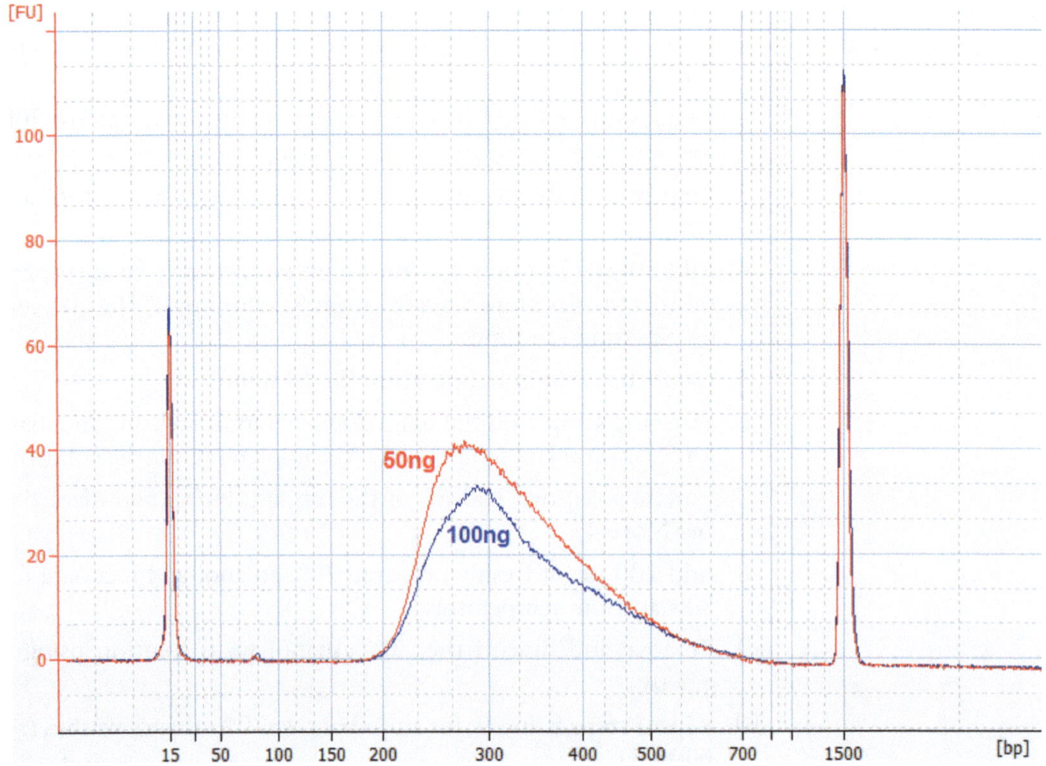

Fig. 3 A representative Bioanalyzer profile using the DNA 1000 Chips. The *x* axis shows the size distribution of the amplification products in base pairs, and the *y* axis shows units of fluorescence. The *red* and *blue curves* represent libraries made from 50 to 100 ng DNA, respectively. The average library size is approximately 290 base pairs

1 sequencing primer, called MetSeq Primer 1, that is included in this kit at a concentration of 25 μM. 6 μL of this primer are added manually during the Cluster Generation Process on a cBOT for sequencing on a HiSeq 2000. The standard primers provided in the Illumina sequencing kit are appropriate for Read 2 and for sequencing the DR barcodes.

3.14 Data Analysis

The following paragraph is based on the assumption that the requirements defined in the Materials section are met. Each required software should have its associated bin folder in the PATH environment variable (for example for Bismark, export PATH = $PATH:PATH_TO_BISMARK_BIN). As current sequencers tend to generate longer reads (≥ 50 bp), bowtie2 will be used in the protocol below with Bismark (*see* Bismark user guide at http://www.bioinformatics.babraham.ac.uk/projects/download.html#bismark for more details). **Steps 5–9** should be performed successively for both the lambda genome and the sequenced genome (e.g., human, mouse, plant, etc.). **Step 8** could be omitted for the lambda genome to save time and computational resources. Representative results are shown in Fig. 4.

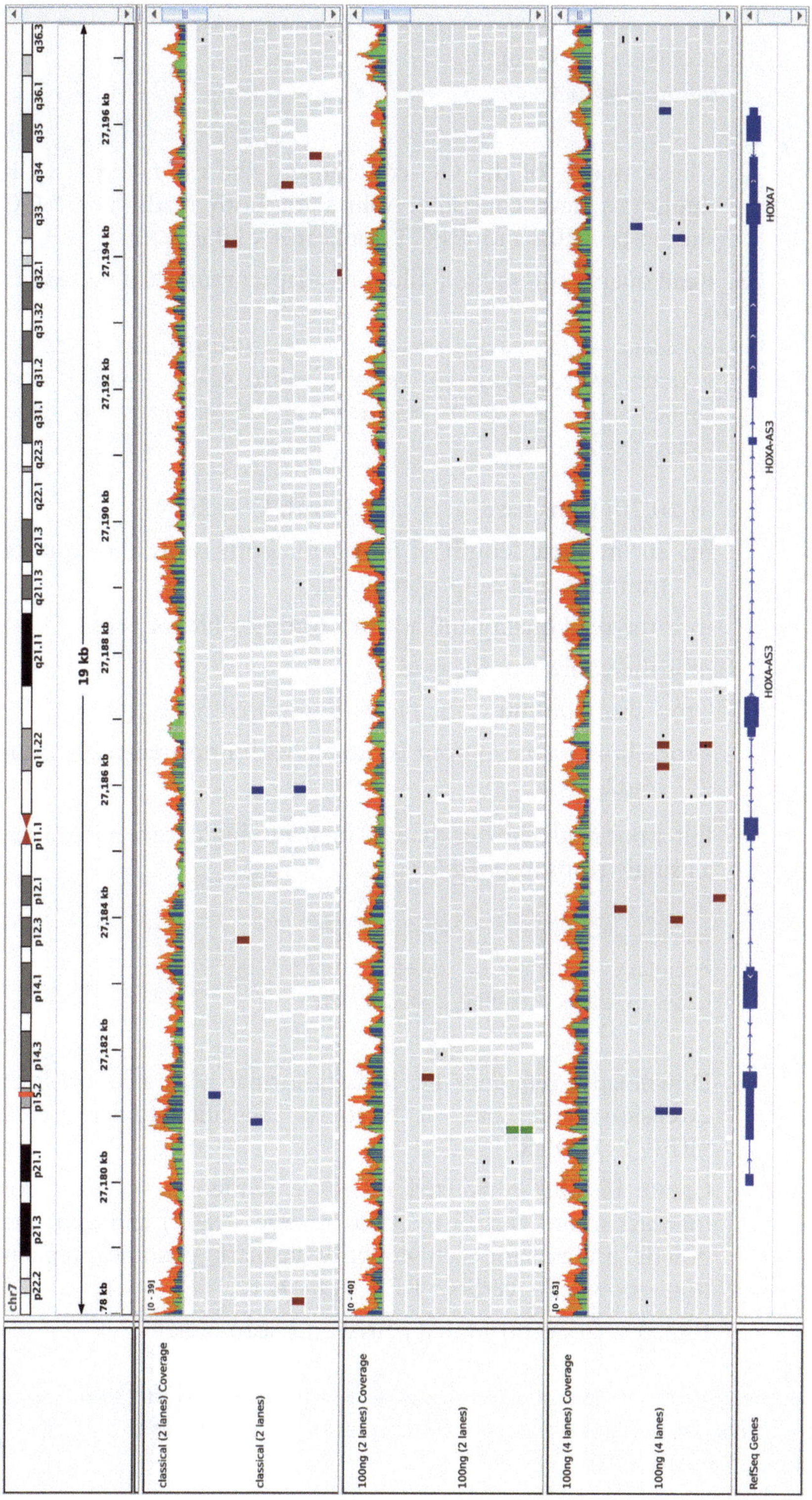

Fig. 4 Comparative analysis of the coverage of a locus in the *Hox* gene cluster by WGBS. *Upper panel* sequence coverage obtained by sequencing two lanes on a HiSeq2000 starting from 4 μg of DNA (average coverage 15×); *middle panel*: 2 lanes sequenced of a library prepared from 100 ng of input DNA using the protocol described in this chapter (average coverage 16×); *lower panel*: 4 lanes sequenced of a library prepared from 100 ng of input DNA using the protocol described in this chapter (average coverage 33×). Libraries prepared from 4 μg to 100 ng of DNA yield nearly indistinguishable results. At the recommended coverage of 30×, a sufficient coverage for accurate estimation of the DNA methylation degree is obtained

1. Run fastQC on read 1 and 2:

```
fastqc FASTQ_READ1
fastqc FASTQ_READ2.
```

2. Check that % of the GC content and the estimated % PCR duplicates have consistent values for both reads, i.e., %GC should be ~20% and % PCR duplicates <20%.
3. Read quality trimming + Illumina adapter removal if necessary:

```
cutadapt -q 20,20 -a ADAPTER_FWD -A ADAPTER_REV -m 50 -o
TRIMMED_FASTQ_READ1 -p TRIMMED_FASTQ_READ2 --too-short-
output SHORT1 --too-short-paired-output SHORT2 FASTQ_READ1
FASTQ_READ2
```

-q 20,20: trims bases at 5′ and 3′ with Q < 20

-m 50: minimal length of 50 bp for both reads after quality trimming and adapter removal

-o is the read quality and adapter trimmed file for read1

-p is the equivalent for read2

--too-short-output is the file with unpaired reads from read1

--too-short-paired-output is the file with unpaired reads from read2

4. Filter out reads shorter than 50 bp from single reads and merge all single reads in one file:

```
cutadapt -m 50 -o S1 SHORT1
cutadapt -m 50 -o S2 SHORT2
zcat S1 S2 > TRIMMED_SINGLE_FASTQ
```

5. Download a reference file in fasta format with a .fa or .fasta extension and put it into a specific folder (e.g., "REF_FOLDER").
6. Create bisulfite-converted reference sequences (C to T for forward strand and G to A for reverse strand) and generate bowtie2 indexes for each sequence: bismark_genome_preparation --bowtie2 --verbose REF_FOLDER
7. Run Bismark on trimmed paired end fastq files:

```
cd TARGET_DIR; bismark --ambiguous --un --bowtie2 --gzip --fastq -p
NB_CORES REF_FOLDER -X 1000 -1 TRIMMED_FASTQ_READ1 -2
TRIMMED_FASTQ_READ2
```

--ambiguous writes reads that fail to map uniquely to files with suffix
"_ambiguous_reads_N.fq.gz"

--un writes unmapped reads to files with suffix
"_unmapped_reads_N.fq.gz"

--gzip writes temporary bisulfite conversion files in GZIP format

--fastq to tell input files are in FASTQ format

-p gives the number of cores that will be passed to bowtie2

-X is the maximum allowed insert size to consider for a paired-end alignment

"cd TARGET_DIR" is necessary in order to have all Bismark output files written in TARGET_DIR.

NB_CORES, i.e., the total number of CPU cores available on the linux/unix server that will be used for the analyses, should be set to TOTAL_CORES/2 as by default Bismark assumes reads come from directional sequencing (current Illumina protocol) and thus read1 has gone through the bisulfite conversion C to T and read2 the reverse complement of this conversion (G to A).

8. Run Bismark on single reads:

```
cd TARGET_DIR; bismark --ambiguous --un --bowtie2 --gzip --
fastq -p NB_CORES --non_directional REF_FOLDER TRIMMED_-
SINGLE_FASTQ
```

--non_directional indicates all four possible alignments will be run as reads come from both read1 and read2

NB_CORES thus needs to be TOTAL_CORES/4

9. Check the Bismark report file to make sure that the mapping efficiency is good enough (i.e., >70%) and the methylation percentage in the different contexts are consistent with what is expected (~60% for cytosines in the CpG context for human genome and close to 0 for the other sequence contexts, close to 0 for all sequence contexts for the lambda phage) (*see* **Note 16**).
10. Sort bam files by chromosome position:

```
java -Xmx13g -jar PICARD_DIR/SortSam.jar INPUT=BAM_FILE
OUTPUT=SORTED_BAM SORT_ORDER=coordinate
MAX_RECORDS_IN_RAM=3000000 CREATE_INDEX=true
TMP_DIR=FAST_DIR VALIDATION_STRINGENCY=LENIENT
```

-Xmx gives how much RAM is available for java to run the command

MAX_RECORDS_IN_RAM: 250,000 reads per each GB given to the –Xmx

TMP_DIR: the location for all temporary folder. This parameter should be set if a location efficient in writing is available and different from the current working directory

VALIDATION_STRINGENCY: set to LENIENT to have Picard warning messages displayed

11. Merge all bam files related to the same sample if necessary:

```
java -Xmx14G -jar PICARD_DIR/MergeSamFiles.jar I=BAM_FILE1
I=BAM_FILE2 O=OUT_BAM CREATE_INDEX=true USE_THREADING=true
ASSUME_SORTED=true MAX_RECORDS_IN_RAM=3250000
TMP_DIR=FAST_DIR
```

USE_THREADING=true: uses 20% more CPU and decreases running time by ~20%

I=BAM_FILE1 I=BAM_FILE2 … I=BAM_FILEN: lists all bam files to merge

12. Mark PCR duplicates:

```
java -Xmx7g -jar PICARD_DIR/MarkDuplicates.jar
ASSUME_SORTED=true REMOVE_DUPLICATES=false
VALIDATION_STRINGENCY=LENIENT INPUT=BAM_FILE
OUTPUT=OUT_BAM METRICS_FILE=OUT_METRICS
MAX_FILE_HANDLES_FOR_READ_ENDS_MAP=1000
MAX_RECORDS_IN_RAM=1500000 TMP_DIR=FAST_DIR
CREATE_INDEX=true
```

ASSUME_SORTED=true: assumes that the bam file is coordinate sorted

METRICS_FILE: gives the location of Picard's output file

MAX_FILE_HANDLES_FOR_READ_ENDS_MAP: maximum number of simultaneous opened file handles for temporary files. This value should be lower than 'ulimit -n'

13. Extract methylation values from paired-end reads in bam (the .cov.gz file will contain the methylation information at CpG sites) (*see* **Note 17**):

```
bismark_methylation_extractor -p --counts --gzip --genome_folder
REF_FOLDER BAM_FILE --report --bedGraph --comprehensive --
multicore NB_CORES
```

-p indicates BAM_FILE is paired-end

--bedGraph --counts creates CpG context bedGraph (chr positions + methylation values) and coverage file (same as bedGraph but with #Cs and #Ts). For other methylation contexts, add option "--CX" to the above command

--gzip makes Bismark output files in GZIP format

--comprehensive creates one methylation file per methylation context

NB_CORES needs to be set to TOTAL_CORES/3

14. Optional: For the analysis of DMS or DMRs, a number of freely available tools can be used including MethylKit [35].

4 Notes

1. Between 50 and 100 ng of input DNA allow obtaining high diversity libraries with comprehensive genomic coverage. We have successfully sequenced samples with input material as limited as 30 ng. However, below this amount the duplicate rates increase significantly and for sufficient coverage extensive sequencing is required. We therefore recommend a minimum input of 30 ng and ideally 50 ng for routine use.
2. For Subheadings 3.1–3.8, it is recommended to work in Laminar Flow Cabinets.
3. Bowtie2 is recommended if the read length exceeds ≥50 base pairs. Use bowtie1 if the read length is below 50 bp.
4. The beads should not disperse; otherwise, they will stay on the walls of the tubes and be lost during the wash process. Significant loss of beads at this stage will impact the amount of purified DNA, so ensure beads are not removed with the binding buffer or during the wash steps.
5. With the final wash, it is critical to remove as much of the ethanol as possible. If necessary, use extra pipetting steps and allow excess ethanol to collect at the bottom of the tubes after removing most of the ethanol in the first pipetting step.
6. Leave the thermal cycler lid open during the incubation at 25 °C (**step 1**).
7. The L1 Ligation Buffer Mix is very viscous. Please be sure to pipet this reagent slowly.
8. We have been evaluating several commercial bisulfite conversion kits and all kits tested have been found to be compatible with the library preparation protocol. However, the kit cited within this protocol allowed us to obtain the most consistent

and the best yields after bisulfite conversion with excellent conversion rates.

9. If necessary, heat the Bisulfite Solution to 60 °C and vortex until all precipitates are dissolved again. Do not place dissolved bisulfite solution on ice.
10. DNA Protect Buffer should turn from green to blue after the addition to the DNA–Bisulfite Solution mixture, indicating sufficient mixing and correct pH for the bisulfite conversion reaction.
11. If there are precipitates in the BD Buffer, avoid transferring them to the spin column.
12. To avoid acidification from carbon dioxide in the air, it is important to close the bottle containing the BD Buffer immediately after use and close the lids of the spin columns before incubation.
13. Proceed immediately or store purified DNA at −20 °C.
14. Rn is the fluorescence of the reporter dye divided by the fluorescence of a passive reference dye.
15. Slightly bigger inserts have not been found to negatively impact the sequencing of the library. For significantly larger inserts, this might however have an influence on library diversity and as inserts are penalized during cluster generation on the cBot.
16. Plants might have in some species high methylation levels for cytosines in all sequence contexts. In the same line, cellular systems such as iPS cells might show cytosine methylation in other sequence context. In this case, only the spike-in DNA can be used to calculate reliably the bisulfite conversion efficiency.
17. Methylation values can also be extracted using MethylKit [35] from the whole bam, i.e., paired + single reads. For further details see the read.bismark function in the methylKit R package.

Acknowledgments

We thank Benjamin G Schroeder (NuGEN, San Carlos, CA) for help with setting up the technology in the laboratory and Doug Amorese (NuGEN, San Carlos, CA) and Steven McGinn (CNRGH) for the critical reading of the manuscript. Work in the laboratory of Jörg Tost is supported by grants from the ANR (ANR-13-EPIG-0003-05 and ANR-13-CESA-0011-05), Aviesan/INSERM (EPIGI2014-18 and EPIG2014-01), INCa (PRT-K14-049) and the joint CEA-EDF-IRSN program (CP-PHE-102).

References

1. How Kit A, Nielsen HM, Tost J (2012) DNA methylation based biomarkers: practical considerations and applications. Biochimie 94:2314–2337
2. Baylin SB, Jones PA (2011) A decade of exploring the cancer epigenome - biological and translational implications. Nat Rev Cancer 11:726–734
3. Lardenoije R, Iatrou A, Kenis G et al (2015) The epigenetics of aging and neurodegeneration. Prog Neurobiol 131:21–64
4. Nielsen HM, Tost J (2012) Epigenetic changes in inflammatory and autoimmune diseases. Subcell Biochem 61:455–478
5. Harb H, Renz H (2015) Update on epigenetics in allergic disease. J Allergy Clin Immunol 135:15–24
6. Zhang Z, Zhang R (2015) Epigenetics in autoimmune diseases: pathogenesis and prospects for therapy. Autoimmun Rev 14:854–863
7. Shorter KR, Miller BH (2015) Epigenetic mechanisms in schizophrenia. Prog Biophys Mol Biol 118:1–7
8. Abdolmaleky HM, Zhou JR, Thiagalingam S (2015) An update on the epigenetics of psychotic diseases and autism. Epigenomics 7:427–449
9. Ronn T, Ling C (2015) DNA methylation as a diagnostic and therapeutic target in the battle against Type 2 diabetes. Epigenomics 7:451–460
10. Grunau C, Clark SJ, Rosenthal A (2001) Bisulfite genomic sequencing: systematic investigation of critical experimental parameters. Nucleic Acids Res 29:e65
11. Warnecke PM, Stirzaker C, Song J et al (2002) Identification and resolution of artifacts in bisulfite sequencing. Methods 27:101–107
12. Urich MA, Nery JR, Lister R et al (2015) MethylC-seq library preparation for base-resolution whole-genome bisulfite sequencing. Nat Protoc 10:475–483
13. Kobayashi H, Sakurai T, Imai M et al (2012) Contribution of intragenic DNA methylation in mouse gametic DNA methylomes to establish oocyte-specific heritable marks. PLoS Genet 8:e1002440
14. Adusumalli S, Mohd Omar MF, Soong R et al (2015) Methodological aspects of whole-genome bisulfite sequencing analysis. Brief Bioinform 16:369–379
15. Lister R, Pelizzola M, Dowen RH et al (2009) Human DNA methylomes at base resolution show widespread epigenomic differences. Nature 462:315–322
16. Lister R, O'Malley RC, Tonti-Filippini J et al (2008) Highly integrated single-base resolution maps of the epigenome in Arabidopsis. Cell 133:523–536
17. Lister R, Pelizzola M, Kida YS et al (2011) Hotspots of aberrant epigenomic reprogramming in human induced pluripotent stem cells. Nature 471:68–73
18. Li Y, Zhu J, Tian G et al (2010) The DNA methylome of human peripheral blood mononuclear cells. PLoS Biol 8:e1000533
19. Chalhoub B, Denoeud F, Liu S et al (2014) Plant genetics. Early allopolyploid evolution in the post-Neolithic Brassica napus oilseed genome. Science 345:950–953
20. Lyko F, Foret S, Kucharski R et al (2010) The honey bee epigenomes: differential methylation of brain DNA in queens and workers. PLoS Biol 8:e1000506
21. Guo JU, Su Y, Shin JH et al (2014) Distribution, recognition and regulation of non-CpG methylation in the adult mammalian brain. Nat Neurosci 17:215–222
22. Kreck B, Marnellos G, Richter J et al (2012) B-SOLANA: an approach for the analysis of two-base encoding bisulfite sequencing data. Bioinformatics 28:428–429
23. Bormann Chung CA, Boyd VL, McKernan KJ et al (2010) Whole methylome analysis by ultra-deep sequencing using two-base encoding. PLoS One 5:e9320
24. Kreck B, Richter J, Ammerpohl O et al (2013) Base-pair resolution DNA methylome of the EBV-positive endemic Burkitt lymphoma cell line DAUDI determined by SOLiD bisulfite-sequencing. Leukemia 27:1751–1753
25. Hansen KD, Timp W, Bravo HC et al (2011) Increased methylation variation in epigenetic domains across cancer types. Nat Genet 43:768–775
26. Adey A, Shendure J (2012) Ultra-low-input, tagmentation-based whole-genome bisulfite sequencing. Genome Res 22:1139–1143
27. Miura F, Enomoto Y, Dairiki R et al (2012) Amplification-free whole-genome bisulfite sequencing by post-bisulfite adaptor tagging. Nucleic Acids Res 40:e136
28. Booth MJ, Branco MR, Ficz G et al (2012) Quantitative sequencing of 5-methylcytosine and 5-hydroxymethylcytosine at single-base resolution. Science 336:934–937

29. Yu M, Hon GC, Szulwach KE et al (2012) Base-resolution analysis of 5-hydroxymethylcytosine in the mammalian genome. Cell 149:1368–1380
30. Krueger F, Andrews SR (2011) Bismark: a flexible aligner and methylation caller for Bisulfite-Seq applications. Bioinformatics 27:1571–1572
31. Klein HU, Hebestreit K (2015) An evaluation of methods to test predefined genomic regions for differential methylation in bisulfite sequencing data. Brief Bioinform 17:796–807
32. Moran S, Arribas C, Esteller M (2015) Validation of a DNA methylation microarray for 850,000 CpG sites of the human genome enriched in enhancer sequences. Epigenomics 8:389–399
33. Sandoval J, Heyn H, Moran S et al (2011) Validation of a DNA methylation microarray for 450,000 CpG sites in the human genome. Epigenetics 6:692–702
34. Li H, Handsaker B, Wysoker A et al (2009) The Sequence Alignment/Map format and SAMtools. Bioinformatics 25:2078–2079
35. Akalin A, Kormaksson M, Li S et al (2012) MethylKit: a comprehensive R package for the analysis of genome-wide DNA methylation profiles. Genome Biol 13:R87

Chapter 6

Tagmentation-Based Library Preparation for Low DNA Input Whole Genome Bisulfite Sequencing

Dieter Weichenhan, Qi Wang, Andrew Adey, Stephan Wolf, Jay Shendure, Roland Eils, and Christoph Plass

Abstract

Aberrations of the DNA methylome contribute to onset and progression of diseases. Whole genome bisulfite sequencing (WGBS) is the only analytical method covering the complete methylome. Alternative methods requiring less DNA than WGBS analyze only a minor portion of the methylome and do not cover important regulatory features like enhancers and noncoding RNAs. In tagmentation-based WGBS (TWGBS), several DNA and time-consuming steps of the conventional WGBS library preparation are circumvented by the use of a hyperactive transposase, which simultaneously fragments DNA and appends sequencing adapters. TWGBS requires only nanogram amounts of DNA and, thus, is well suited to study precious biological specimens such as sorted cells or micro-dissected tissue samples.

Key words Whole genome bisulfite sequencing, 5-Methyl cytosine, Methylome, Tagmentation, Transposase

1 Introduction

1.1 Methods to Analyze the Methylome

DNA methylation, the covalent attachment of a methyl group to the carbon-5 atom of a cytosine in a CpG context, is a stable, yet reversible epigenetic modification of the mammalian genome. A large majority of the roughly 28 million CpGs in the genome of humans and other mammals is methylated [1, 2]. Shaping of the DNA methylome, the entirety of all methylated CpGs in a genome, is a critical process in early development, X-chromosome inactivation, differentiation, and gene regulation. Deviations from normal methylome patterns in a cell type contribute to developmental malformations and human disorders, particularly cancer [3].

The methylome can be profiled by a variety of methods, which usually target only a minor portion of all CpGs in a genome. Methods such as reduced representation bisulfite sequencing (RRBS) [4, 5] (*see* also Chapter 8) or *Hpa*II tiny fragment-enrichment by ligation-mediated PCR (HELP) [6, 7]

Jörg Tost (ed.), *DNA Methylation Protocols*, Methods in Molecular Biology, vol. 1708, https://doi.org/10.1007/978-1-4939-7481-8_6, (*see* also Chapter 11) are based on the use of methylation sensitive or insensitive restriction enzymes and require only low input DNA amounts; these approaches analyze single CpGs, mostly in CpG islands which are CpG-rich genomic regions often overlapping with gene promoters. A further, widespread technology is the HumanMethylation450 BeadChip which interrogates more than 485,000 CpG dinucleotides of the genome (recently updated as MethylationEPIC BeadChip interrogating more than 850,000 CpGs) including promoters, gene bodies, micro RNAs, and CpG islands [8] (*see* also Chapter 16). In contrast to the aforementioned methods, which directly determine the methylation status of single CpGs, methods enriching for methylated CpGs like MeDIP using antibodies [9–11] (*see* also Chapter 12) or like MethylCap [12] (*see* also Chapter 10) or MCIp [13, 14] using methylated DNA binding proteins followed by array analysis or next-generation sequencing interrogate only indirectly the methylation status of CpGs. Such enrichment methods require additional targeted validation of methylation states. A major drawback of all these methods is the incomplete analysis of the DNA methylome, sparing important regulatory genomic regions of lower CpG density like enhancers, CpG-poor promoters, and CpG island adjacent sequences as well as noncoding RNA genes.

Whole-genome bisulfite sequencing (WGBS) analyzes the methylation status of all CpGs in a genome. Bisulfite treatment of the DNA converts unmethylated cytosines to uracil while methylated cytosines remain unchanged, by this means enabling discriminating between methylated and unmethylated CpGs. The conventional protocol for the WGBS library preparation includes a number of enzymatic and purification steps and requires μg quantities of DNA [2] (Fig. 1a). Publications describing conventional WGBS with less input DNA in the ng range are either lacking methodological details on how library preparation has been performed or present data of low genomic coverage and only limited value to assess the methylation status at the single-base level [15–17]. Alternative protocols for WGBS library preparation which require only ng quantities of input DNA have been published recently such as post-bisulfite adaptor tagging (PBAT) [18] (*see* also Chapter 7) or are available as commercial kits [19, 20].

Tagmentation-based WGBS (TWGBS) also requires only ng quantities of input DNA by avoiding several enzymatic and purification steps of the conventional method for the preparation of the sequencing library [21, 22] (*see* Fig. 1a). In the conventional protocol, the DNA is fragmented, usually by sonication, followed by end polishing, A-tailing, and ligation of the sequencing adapter. Subsequently, adapter ligated fragments are size selected by the separation on an electrophoresis gel and elution from the gel. In TWGBS, a hyperactive Tn5 transposase fragments the DNA and appends sequencing adapters in a single short reaction [23], which

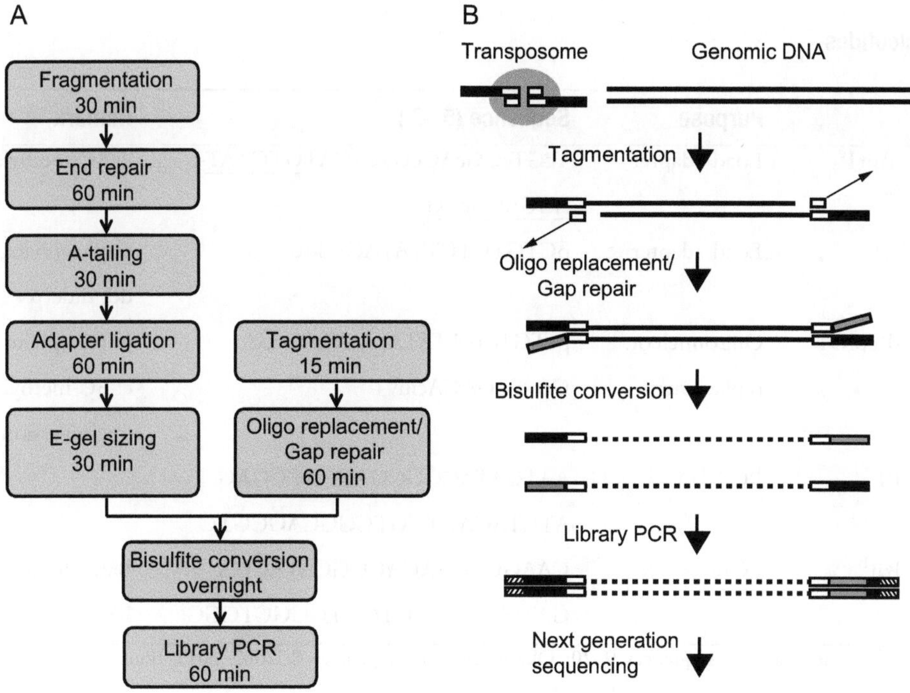

Fig. 1 Principle and timing of TWGBS library preparation. (**a**) Flowchart comparison between conventional WGBS and TWGBS library preparation. (**b**) Components and steps of TWGBS library preparation. Each step includes solid phase reversible immobilization (SPRI) bead or column purification, respectively. The transposome is formed from the Tn*5* transposase (ellipse) and two partially methylated oligonucleotide adapters containing the sequencing primer 1 (*black bar*) and the transposase recognition sequence (*white bar*; *see* also Table 1). Tagmentation fragments the genomic DNA and attaches the adapters at the fragment ends. The unmethylated, shorter oligonucleotide is then replaced by a methylated oligonucleotide containing the sequencing primer 2 sequence (*gray bar*); the gap at each end is filled and covalently closed. The tagmented DNA fragments are now flanked by identical methylated adapters at both the ends. Bisulfite treatment leads to conversion of unmethylated cytosines to uracil in the genomic sequence (*interrupted lines*) of the single-stranded fragments. Sequencing libraries are finally generated by PCR with low cycle numbers using a primer bearing the flowcell primer 1 (*left striated bar*) and the sequencing primer 1 sequences and a second primer bearing the flowcell primer 2 (*right striated bar*), the barcode (*right black bar*) and the sequencing primer 2 sequences

is followed by the replacement of an unmethylated with a methylated oligonucleotide and gap closure between oligonucleotide and genomic DNA. Gel sizing is not required. In both the methods, conventional WGBS and TWGBS, final library PCR is preceded by bisulfite treatment of the DNA fragments. Though starting with a much lower input DNA amount, TWGBS proved as comprehensive as conventional WGBS with respect to the determination of methylation levels and coverage of the human methylome [21, 22].

1.2 Principle and Application of TWGBS

Tagmentation has been initially described as an efficient method to generate next-generation sequencing libraries from low input DNA amounts, making use of a transposome assembled from a

Table 1
Oligonucleotides

Name	Purpose	Sequence (5′–3′)	Remark
Tn5mC-Apt1[a]	Load adapter	TcGTcGGcAGcGTcAGATGTGTAT-AAGAGAcAG	c: 5C-methylated
Tn5mC1.1-A1block	Load adapter	pCTGTCTCTTATACAddC	p: phosphate, dd: dideoxy
Tn5mC-ReplO1	Oligonucleot. replacement	pcTGTcTcTTATAcAcATcTccGA-GcccAcGAGAcinvT	p: phosphate c: 5C-methylated inv: inversion
Tn5mCP1	PCR	AATGATACGGCGACCACCGAG-ATCTACACTCGTCGGCAGCGTC	
Tn5mCBar[b]	PCR	CAAGCAGAAGACGGCATACGA-GAT(8–9N)GTCTCGTGGGCTCGG	Barcode of 8 or 9 Bases

[a]The transposon end recognition sequence is underlined

[b]Barcode sequences: (1) GGATGTTCT, (2) CTTATCCAG, (3) GTAAGTCAC, (4) TTCAGTGAG, (5) CTCGTAATG, (6) CATGTCTCA, (7) AATCGTGGA, (8) GTATCAGTC, (9) TCGCCTTA, (10) CTAGTACG, (11) TTCTGCCT, (12) GCTCAGGA, (13) AGGAGTCC, (14) CATGCCTA, (15) GTAGAGAG, (16) CCTCTCTG, (17) AGCGTAGC, (18) CAGCCTCG, (19) TGCCTCTT, (20) TCCTCTAC

hyperactive Tn5 transposase and a sequencing adapter [23]. The concept of tagmentation was then extended to TWGBS using a transposome assembled from the transposase and an adapter with methylated cytosine bases [21] (Fig. 1b and Table 1). Upon tagmentation of genomic DNA, random genomic fragments with adapters appended at both ends are generated (*see* Fig. 1b). A gap of nine bases remains at both the ends between the genomic DNA and the shorter adapter oligonucleotide (bottom adapter, *see* Table 1), which has to be replaced by a methylated oligonucleotide; at the stage of oligonucleotide replacement, the gap is filled by DNA polymerase and closed by DNA ligase. The resulting genomic fragments are then flanked by identical methylated adapters. The DNA is then bisulfite treated to enable distinction between methylated and unmethylated cytosines, the former remaining cytosines, the latter being converted to uracil. It should be noted, however, that bisulfite treatment does not enable distinction between different cytosine modifications like 5-methyl, 5-hydroxymethyl, 5-carboxyl, or 5-formyl which are present as well in the mammalian genome, though usually to a very low percentage [24]. The final step is the sequencing library generation by PCR using primers, which match the adapter oligonucleotides and are compatible with sequencing oligonucleotides from Illumina (TruSeq Dual Index Sequencing Primer Box, Paired End). Monitoring library generation by real-time PCR prevents over-amplification to avoid high duplication frequencies in subsequent sequencing. Libraries are analyzed by 101 bp paired-end sequencing on a next-generation sequencing platform, e.g., Illumina HiSeq 2000. Quality controls have been implemented to monitor fragmentation efficiency, DNA loss through the bisulfite treatment, and the conversion frequency of unmethylated cytosine to uracil bases.

TWGBS enables the determination of the methylome of rare cell types, collected, for example, by flow-cytometry or by laser-capture microdissection. Starting with only 30 ng human or mouse genomic DNA in a single-tagmentation reaction, an amount equivalent to about 15,000 cells assuming a DNA recovery of 30%, generation of four sequencing libraries by independent PCR amplifications is feasible. Although the Tn*5* transposase used to generate the transposome has a slightly higher sequence bias than the fragmentation by sonication [23], this bias did neither compromise genomic coverage nor base composition of sequencing reads in TWGBS [21, 22].

2 Materials (*See* Note 1)

2.1 Preparation of the Load Adapter and Assembly of the Transposome

1. Oligonucleotides Tn5mC-Apt1 and Tn5mC1.1-a1block (*see* Table 1 and **Note 2**).
2. Sterile deionized H_2O and sterile 100% glycerol.
3. 1.5 mL reaction tube and eight-well strip with lid.
4. 96-well gradient thermocycler.
5. Heater block.
6. EZ-Tn5^{TM}Transposase (Lucigen, cat. no. TNP92110).

2.2 Tagmentation

1. 10–30 ng genomic DNA (*see* **Note 3**).
2. 2× buffer for tagmentation (2×TMA-DMF; composition *see* **Note 4**).
3. Unmethylated phage λ DNA.
4. 1.5 mL reaction tube and eight-well strip with lid.
5. 96-well thermocycler.

2.3 Post Tagmentation DNA Purification and Determination of the Fragment Size Range

1. Agencourt AMPure XP beads.
2. 5 M Guanidinium thiocyanate; dissolved in H_2O, filter sterilize and store at −20 °C.
3. AMPure buffer (*see* **Note 5**).
4. Sterile deionized H_2O and 80% ethanol.

5. Eight-well strip with lid.
6. Magnetic separator for eight-well strips.
7. Quant-iT dsDNA HS Assay Kit (Agilent).
8. Bioanalyzer 2100 with Electrophoresis Set (Agilent).

2.4 Oligonucleotide Replacement and Gap Repair

1. Oligonucleotide Tn5mC-ReplO1 (*see* Table 1 and **Note 2**).
2. Deoxynucleotide mix (100 mM of each dNTP). Dilute 1:40 in H_2O to have a final concentration of 2.5 mM of each dNTP.
3. Ampligase, thermostable DNA ligase, with 10× buffer.
4. T4 DNA polymerase.
5. 96-well gradient thermocycler.

2.5 DNA Purification After Oligo Replacement/Gap Repair

1. Refer to Subheading 2.3 for material.

2.6 Bisulfite Treatment and Purification of the DNA

1. EZ DNA Methylation™ Kit (ZYMO Research; *see* **Note 6**).
2. DNA LoBind reaction tubes (e.g., Eppendorf) and eight-well strip with lid.
3. 96-well thermocycler.
4. Benchtop centrifuge.

2.7 TWGBS Library Preparation by Real-Time PCR

1. Oligonucleotide primers Tn5mCP1 and Tn5mCBar (*see* Table 1 and **Note 2**).
2. KAPA 2G Robust Hot Start Ready Mix.
3. SYBR® Green I nucleic acid gel stain, 10,000×. 100× SYBR-Green is 1 μL 10,000× SYBR® Green I nucleic acid gel stain dissolved with 99 μL H_2O. This solution can be stored protected from light at −20 °C for at least 6 months.
4. 96-well real-time PCR plate for Lightcycler 480.
5. Real-time thermocycler, e.g., Lightcycler 480 with 96-well block.

2.8 TWGBS Library Purification and Determination of the Library Size Range and DNA Concentration

1. Refer to Subheading 2.3 for material.
2. EB elution buffer (Qiagen).
3. Qubit dsDNA HS assay kit.
4. Qubit 2.0 fluorometer.
5. Qubit assay tubes.
6. Quant-iT dsDNA HS Assay Kit.
7. Bioanalyzer 2100.

2.9 Next-Generation Sequencing of the TWGBS Libraries

1. TruSeq Dual Index Sequencing Primer Box, Paired End (Illumina, cat. no. PE-121-1003).
2. Illumina HiSeq2000 Sequencer.

3 Methods

3.1 Preparation of the Load Adapter and Assembly of the Transposome

1. Prepare a homogeneous mix of each 20 μL of oligonucleotide Tn5mC-Apt1 and Tn5mC1.1-A1block (100 μM each; *see* Table 1) and 160 μL H_2O. The final concentration of each oligonucleotide is 10 μM.
2. Distribute the mix in 4 × 50 μL aliquots to an eight-well strip and carry out adapter assembly in a thermocycler with the following conditions:

Cycle number	Denature	Anneal	Ramp to 26 °C	Hold
1	95 °C, 3 min	70 °C, 3 min		
2–46		70 °C, 30 s	−1 °C/cycle, 30 s	
47				25 °C, infinite

3. Heat about 300 μL 100% glycerol to 90 °C in a heater block to enable exact pipetting of the hot glycerol.
4. Transfer 200 μL hot glycerol to a 1.5 mL reaction tube and cool down to room temperature (RT).
5. Add the 4 × 50 μL annealed adapter from **step 2** to the 200 μL glycerol and mix by repeated pipetting. This adapter-glycerol mix is designated load adapter (stable at −20 °C for at least 6 months).
6. To assemble the transposome, transfer 10 μL load adapter to 10 μL Ez-Tn*5* transposase and mix by repeated pipetting; avoid air bubbles.
7. Maintain the mixture at RT for 30 min during which the transposome forms. The transposome mixture is sufficient for 8 tagmentation reactions and stable at −20 °C for at least 6 months.

3.2 Tagmentation

1. For a single-tagmentation reaction, mix on ice in an eight-well strip 10 μL 2× TMA-DMF buffer and 7.5 μL DNA solution containing 10–30 ng genomic DNA and 5–15 pg unmethylated phage λ DNA (5 pg λ DNA/10 ng genomic DNA). Add 2.5 μL transposome (from **step 7** of Subheading 3.1, and mix by repeated pipetting; avoid air bubbles (*see* **Note 7**).

2. Perform the tagmentation reaction in a thermocycler as follows:

Cycle number	Tagmentation	Cool
1	55 °C, 8 min	4 °C, infinite

3.3 Post Tagmentation DNA Purification

1. Add to each tagmentation reaction mix in an 8-well strip 15 μL 5 M guanidinium thiocyanate (DNA solution has then a total volume of 35 μL), 10 μL AMPure beads, and 36 μL AMPure buffer (total volume of the bead solution is then 46 μL) and homogenize by repeated pipetting.
2. Keep the eight-well strip at RT for 10 min.
3. Transfer the strip to the magnetic separator and wait until the bead suspensions have been completely cleared (about 1 min). Remove the clear supernatants thoroughly without disturbing the bead pellets and discard the supernatants.
4. Still keeping the strip in the magnetic separator, add 50 μL of 80% ethanol to each bead pellet. Wash the beads with the 80% ethanol by repeated (5–10 times) pipetting without touching the beads with the pipette tips, then remove and discard the wash solution.
5. Keep the strip without lid in the magnetic separator for about 10 min at RT until the beads are completely dry.
6. Remove the strip from the magnet and resuspend each bead pellet thoroughly in 12 μL sterile deionized H_2O.
7. Transfer the strip back to the magnetic separator and wait until the bead suspensions have been completely cleared (about 1 min).
8. Transfer the clear supernatants containing the purified tagmented DNA carefully to a new eight-well strip; avoid bead carryover.
9. For quality control of the tagmentation reaction, analyze 1 μL of each supernatant using the Bioanalyzer 2100.
10. If the DNA has been efficiently fragmented (Fig. 2a and *see* **Note 8**), proceed to the next step or freeze the DNA at −20 °C.

3.4 Oligonucleotide Replacement and Gap Repair

1. To each well containing the 11 μL supernatant with the tagmented DNA, add 2 μL of dNTP mix (each dNTP 2.5 mM), 2 μL of 10× Ampligase buffer and 2 μL of the oligonucleotide Tn5mC-ReplO1 (10 μM; *see* Table 1); mix by repeated pipetting and make sure, the complete mixtures reside at the bottom of the wells (*see* **Note 9**).

2. Start the incubation for oligonucleotide replacement in a thermocycler:

Cycle number	Denature	Anneal	Ramp to 37 °C	Hold
1	50 °C, 1 min	45 °C, 10 min		
2		45 °C, 0.1 s	−0.1 °C/s	37 °C, infinite

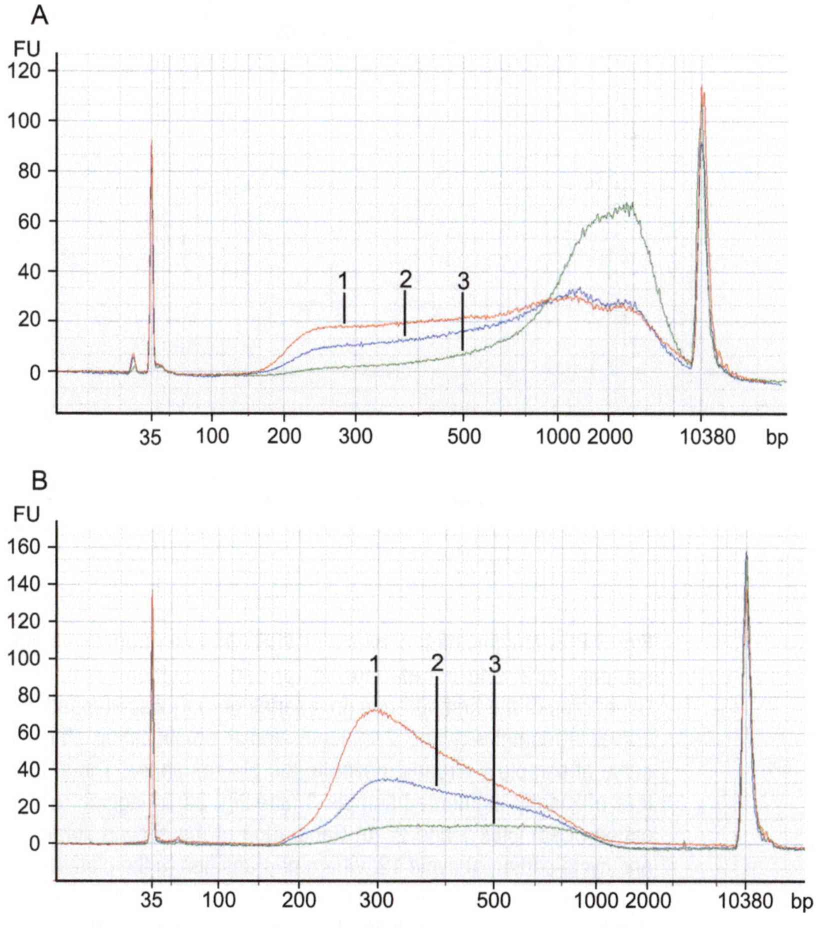

Fig. 2 Quality control after tagmentation and library preparation. (**a**) Fragment size ranges of three identical human DNA samples, 30 ng input each, after tagmentation. Samples 1 and 2 indicate successful tagmentation with broad-shouldered size ranges from about 200 bp to less than 10 kb. Sample 3, in contrast, shows a broad peak around 2000 bp indicating insufficient fragmentation, which was caused by a less active transposome. Four TWGBS libraries were subsequently generated per tagmented sample. (**b**) Fragment size ranges of each one of the four libraries generated from the three tagmented samples shown in Fig. 2a. Library fragments of samples 1 and 2 both range in size from about 200 to 1000 bp with a peak around 300 bp. A size average of 300–350 bp enables efficient 101 bp paired-end sequencing on an Illumina HiSeq 2000. The library of sample 3 has a size range similar to that of samples 1 and 2 but shows no peak and less efficient fragment amplification, indicated by the smaller area under the curve and a lower DNA concentration (*see* also Fig. 3b and Table 2). *FU* fluorescence units

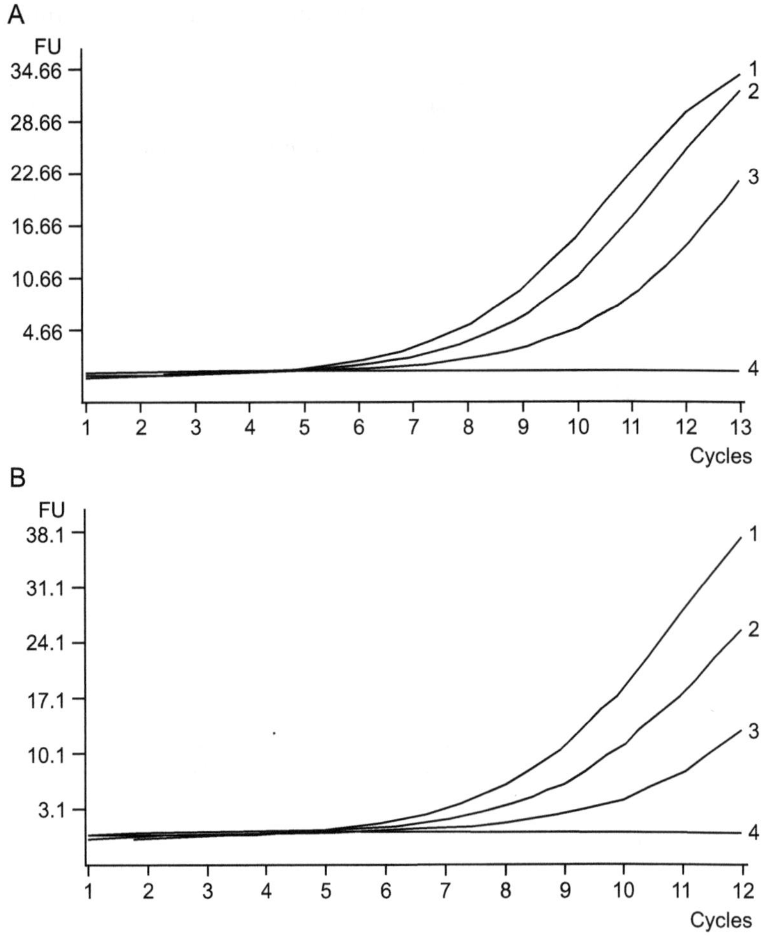

Fig. 3 PCR amplification curves of TWGBS libraries from the tagmented human samples 1, 2, and 3 (refer to Fig. 2) and a tagmented mock sample 4. (**a**) Generation of the first libraries (in a series of four) was stopped at cycle 13. The curves of samples 1 and 2 indicate similar amplification efficiencies, and the curve of sample 1 already moves to the plateau phase. The amplification of the less efficiently tagmented sample 3 (*see* Fig. 2a) is lagging behind the amplification of samples 1 and 2. (**b**) Generation of the second libraries was stopped one cycle earlier, at cycle 12, where no transition to the plateau phase is visible. The difference in the amplification efficiencies between samples 1 and 2 is likely due to experimental variability. For the size ranges and DNA concentrations of the corresponding TWGBS libraries refer to Fig. 2b and Table 2). *FU* fluorescence units

3. While the strip remains in the thermocycler at 37 °C, add for enzymatic gap repair 1 μL T4 DNA Polymerase and 2.5 μL Ampligase to each reaction mixture and mix by repeated pipetting (*see* **Note 9**).

Table 2
Tagmentation PCR library characteristics

Sample	ng/μL	ng total	Range (bp)	Peak (bp)
1	1.6	32	~200–1000	300
2	1.1	22	~200–1000	300
3	0.4	8	~250–1000	No peak

4. Continue the incubation:

Cycle number	Gap repair	Hold
1	37 °C, 30 min	4 °C, infinite

3.5 DNA Purification After Oligo Replacement/Gap Repair

1. Add to each reaction mix 10 μL AMPure beads and 26 μL AMPure buffer (volumes are 20.5 μL of reaction mix and 36 μl of bead solution) and homogenize by repeated pipetting.
2. Repeat **steps 2–8** from Subheading 3.3, but resuspend the beads in 50 μL (instead of 12 μL) of sterile deionized H_2O.
3. Transfer 5 μL of each 50 μL supernatant to a second, new eight-well strip and keep at −20 °C for later control in the final PCR (*see* **Note 10**).
4. Either proceed to the bisulfite treatment or store the supernatants at −20 °C.

3.6 Bisulfite Treatment and Purification of the DNA

1. Add to each supernatant (45 μL) 5 μL M-Dilution Buffer and mix by repeated pipetting.
2. Incubate in a thermocycler:

Cycle number	Gap repair	Hold
1	37 °C, 15 min	25 °C, infinite

3. Add to each reaction mixture 100 μL CT Conversion Reagent and mix by repeated pipetting; distribute each mixture in 75 μL aliquots to two wells of an eight-well strip and incubate overnight in a thermocycler:

Cycle number	Denature	Conversion	Hold
1–16	95 °C, 15 s	50 °C, 1 h	
17			4 °C, Infinite

4. For each sample, transfer 400 μL M-Binding Buffer to a Zymo-Spin™ IC Column assembled in a collection tube and then load the 2 × 75 μL of each sample into the Zymo-Spin™ IC Column containing the M-Binding Buffer. Close the column lids and invert the columns several times. Centrifuge the column-tube assemblies with 11,000 × *g* for 30 s and discard the flow-through.
5. Add 100 μL M-Wash Buffer to each column and centrifuge with 16,000 × *g* for 30 s; discard the flow-through.
6. Add 200 μL M-Desulphonation Buffer to each column, keep the assemblies at RT for 15–20 min, and then centrifuge with 16,000 × *g* for 30 s; discard the flow-through.
7. Add 200 μL M-Wash Buffer to each column, centrifuge with 16,000 × *g* for 30 s, and discard the flow-through. Repeat washing once and finally centrifuge with 16,000 × *g* for 3 min.
8. Transfer the columns to LoBind reaction tubes and pipette 11.5 μL M-Elution Buffer onto each column matrix. Centrifuge with 16,000 × *g* for 30 s to elute the converted DNA. Repeat this elution step (*see* **Note 11**).
9. Proceed to the real-time PCR or store the eluates at −80 °C.

3.7 TWGBS Library Preparation by Real-Time PCR

1. For each tagmented, bisulfite-treated or untreated sample, prepare in a 96-well real-time PCR plate the following mixture on ice (*see* **Note 12**):

Component	Amount (μL)	Final
Kapa 2G Robust HotStart ReadyMix (2×)	12.5	1×
Primer Tn5mCP1n (10 μM; *see* Table 1)	0.75	0.3 μM
Primer Tn5mCBar (10 μM; *see* Table 1)	0.75	0.3 μM
SYBRGreen (100×)	0.25	1×
DNA	10.75	
Total	25	

2. Incubate in a real-time thermocycler using the following conditions:

Cycle number	Denature	Annealing	Read	Extension
1	95 °C, 3 min			
1–20	95 °C, 20 s	62 °C, 15 s	72, 30 s	72, 10 s

3. Stop the PCR at the latest, when the amplification curves move from the exponential to the plateau phase; this transition should occur between cycles 10 and 15 (Fig. 3).
4. Prepare up to three additional TWGBS libraries, depending on the input DNA amount (*see* **Notes 11** and **12**).

3.8 TWGBS Library Purification and Determination of the Library Size Range and DNA Concentration (See Note 13)

1. Transfer the PCR reaction mixes to an eight-well strip and add 45 μL (1.8-fold volume) AMPure beads to each; homogenize by repeated pipetting.
2. Repeat **steps 2–7** from Subheading 3.3, but wash the beads with 200 μL 80% ethanol (instead of 50 μL) and resuspend the beads in 22 μL EB elution buffer.
3. Transfer the clear supernatants corresponding to the TWGBS libraries carefully to 1.5 mL reaction tubes; avoid bead carryover.
4. Transfer 1 μL of each library to a LoBind reaction tube and dissolve with 4 μL H_2O (1:5) for the subsequent determination of the fragment size distributions with the Bioanalyzer and of the DNA concentrations by Qubit fluorimetry (Fig. 2b and Table 2).

3.9 Next-Generation Sequencing of the TWGBS Libraries and Bioinformatics Analysis

1. Analyze the up to four differently barcoded libraries of a tagmented sample on a single lane of a 101 bp paired-end Illumina HiSeq2000 sequencing run, using the appropriate sequencing primers (TruSeq Dual Index Sequencing Primer Box, Paired End) compatible with the TWGBS library primers Tn5mCP1 and Tn5mCBar (*see* Table 1 and **Note 14**).
2. For read alignments of human or mouse DNA, use the most recent reference genomes hg38 (December 2013) and mm10 (December 2011), respectively, according to [25] with the adjustment to bisulfite converted genomes as follows. Transform the top strand (C to T) and the bottom strand (G to A) of the respective genome in silico. Remove the adapter sequences of all reads using SeqPrep (https://github.com/jstjohn/SeqPrep). Convert in each read pair C to T in the first read and G to A in the second read. Align the converted reads to the transformed top and bottom strands of the reference genome using BWA [26] (bwa-0.6.1-tpx) with default parameters. Disable the quality threshold for read trimming (-q) of 20 and the Smith-Waterman for the unmapped mate (-s). Re-convert the reads after the alignment to their original sequences and remove reads mapped to the complementary strand of the top or bottom strand, respectively. Remove duplicate reads within a TWGBS library and determine the complexity using (Picard MarkDuplicates—http://picard.sourceforge.net/). Remove reads with alignment scores less than 1. Calculate the genomic

coverage of a library using the total number of bases aligned from uniquely mapped reads relative to the total number of mappable bases in the genome.

3. For the determination of the methylation state within reads, consider an identified cytosine as methylated and a cytosine converted to thymine as unmethylated. Take only bases with a Phred-scaled quality score of ≥ 20 into account. Exclude the first nine bases of the second read and the last nine bases before the adapter of the first read (*see* **Note 15**).

4 Notes

1. Use only filter pipetting tips throughout the whole procedure of TWGBS library generation. Moreover, to control for a possible DNA contamination, particularly through previously generated tagmentation libraries, use a DNA-free H_2O control in parallel to the DNA samples throughout the whole procedure. For convenience, use multichannel pipettes, if several samples are processed in parallel.
2. All oligonucleotides (*see* Table 1) were custom-made. Some oligonucleotides contain methylated carbon 5 atoms in the cytosine bases to prevent cytosine to uracil conversion through the bisulfite treatment. The replacement oligonucleotide Tn5mC-ReplO1 bears a 5′ phosphate to enable final gap closure; the oligonucleotide's inverted 3′ deoxythymidylate leads to a 3′–3′ linkage which prevents degradation by 3′ exonucleases and extension by DNA polymerases. The 5′ phosphate of the load adapter oligonucleotide Tn5mC1.1-A1block may not be required; its 3′ dideoxycytidylate prevents unwanted extension during the PCR if it has not been completely removed before. The TWGBS library primers Tn5mCP1 and Tn5mCBar are compatible with the sequencing primers of the TruSeq Dual Index Sequencing Primer Box, Paired End from Illumina.
3. Recommended is the use of a commercial kit, e.g., the QIAmp DNA mini or micro kit (Qiagen) for the isolation of genomic DNA. Standard phenol/chloroform extraction might also be suitable for genomic DNA isolation, yet, requires particular care to prevent carryover of phenol or chloroform in the final DNA solution.
4. 2× tagmentation buffer (2× TMA-DMF) is 20 mM Tris (hydroxymethyl)aminomethane, 10 mM $MgCl_2$, 20% (vol/vol) dimethylformamide [27]. Adjust the pH to 7.6 with 100% acetic acid before the addition of DMF. The buffer is filter sterilized and can be stored at −20 °C for at least 6 months.

5. AMPure buffer is 2.5 M NaCl, 20% Polyethylene glycol 8000. The solution is filter sterilized and can be stored at 4 °C for at least 6 months.
6. The EZ DNA Methylation™ Kit (ZYMO Research) contains the CT conversion reagent as a solid substance. Immediately before use of the reagent, add 750 μL of H_2O and 210 μL of M-dilution buffer, also a component of the EZ DNA methylation kit, and mix by frequent vortexing for 10 min. The dissolved CT reagent can be stored protected from light at −20 °C for up to 1 month. As a further component of the EZ DNA methylation kit, the M-wash buffer is supplied without ethanol and must be completed by the addition of 96 mL of 100% ethanol to 24 mL of buffer. The buffer mixture can be stored at RT for at least 1 month.
7. The composition of the tagmentation reaction mix described in Subheading 3.2 works best with an input amount of 10–30 ng of human or mouse DNA. A DNA amount as low as 5 ng may be compensated by a lower amount of the transposome to maintain a suitable fragment size for final sequencing. However, DNA amounts less than 10 ng require more PCR cycles for library generation and lead, as a consequence, to higher duplication frequencies of sequencing reads. Unmethylated λ DNA is used to enable calculation of the conversion frequency of unmethylated cytosines to uracils in the bioinformatics analysis. Using multichannel pipettes in subsequent purification steps, up to eight samples (seven DNA samples and a H_2O control) can readily be processed in parallel.
8. Quality control of the tagmentation reaction is recommended, since insufficient fragmentation of the genomic DNA results in inefficient final PCR amplification (*see* Fig. 2, Fig. 3 and Table 2). Efficient fragmentation is indicated by a relatively even distribution of fragments ranging from about 200 bp to less than 10 kb. Insufficiently fragmented DNA shows a less even distribution of fragment sizes and peaks at a size of 2 kb or larger (*see* Fig. 2a). Inefficient tagmentation may be caused by DNA contaminations, a too high amount of the input DNA and an inappropriately prepared buffer or transposome. Even batch differences in Tn5 transposase activity may be observed. Contaminated DNA may be cleaned by mini dialysis, precipitation or gel purification, provided a sufficient amount is available. To avoid loss of precious input DNA by failing tagmentation, initial testing of the activity of each transposome with a tested control DNA is highly recommended.
9. If multiple samples are processed in parallel, the oligonucleotide replacement/gap repair step can be simplified by the preparation of reaction master mixes. For example, if eight samples

are processed at a time, a mastermix of each 17 μL of dNTPs, 10× Ampligase buffer and Tn5mC-ReplO1 can be prepared and distributed in 6 μL aliquots to each sample. Similarly, an enzyme mastermix of 9 μL T4 DNA polymerase and 22.5 μL Ampligase may be prepared and then distributed in 3.5 μL aliquots to each reaction mixture.

10. The set-aside 5 μL aliquots of the 50 μL supernatants serve as a positive control in the final library PCR to monitor, if the bisulfite treatment of the remaining 45 μL may cause severe DNA loss and, hence, failure in PCR amplification. In addition, the control aliquots may also be used to generate sequencing libraries for genomic profiling such as screening for copy number and single-nucleotide variants [28].
11. The total volume used for elution may range from 23 to 46 μL depending on the initial amount of input DNA for tagmentation. A 30 ng DNA input allows for the generation of four independent PCR libraries requiring 44 μL eluate in total, while an input of 10 ng DNA is good for two libraries requiring 22 μL eluate.
12. Preparation of a PCR master mix without DNA templates and barcode primers Tn5mCBar is recommended. Different barcode primers may be used for different template DNAs and for additional libraries from the same template DNA. Non-tagmented genomic DNA and the mock sample without DNA processed in parallel to the DNA samples serve as negative control, the tagmented but not bisulfite-treated aliquot serves as positive control (*see* **Note 10**). As long as the bisulfite-treated and corresponding untreated aliquots exhibit amplification curves as shown exemplarily in Fig. 3, the loss of DNA by the bisulfite treatment should not compromise successful TWGBS library generation. However, if no or substantially delayed amplification of the bisulfite-treated sample is observed, the time of treatment may be shortened.
13. To avoid contamination of any equipment and solutions used for tagmentation and subsequent steps by the final PCR amplification products, post PCR purification and subsequent quality and quantity control should be performed at a different working place with extra pipettes, AmPure beads, magnetic separator, etc.
14. If four TWGBS libraries are analyzed per lane of a HiSeq2000 paired-end run, barcode combinations 6, 16, 19, and 20 or 8, 11, 17, and 18 (*see* Table 1) are recommended. Many other barcode combinations are possible with the barcode primers presented in Table 1, depending on the number of differently barcoded samples to be analyzed on a single lane. For more

information on barcode combinations, refer to [29]. To enable an in-depth statistical methylation analysis at the single-base level, a 20-fold genomic coverage (tenfold per each Watson and Crick strand) should be achieved [30]. To acquire such a sequencing depth, analysis of the libraries on one or two more HiSeq2000 lanes is required. High frequencies of sequencing read duplicates (>20%) within the libraries of a tagmented sample may be prohibitive to reach this goal and, hence, may require generation of additional TWGBS libraries. Too high duplication frequencies can be avoided by efficient tagmentation (*see* Fig. 2a), proper bisulfite treatment, and a moderate number of PCR cycles (*see* Fig. 3). Input DNA amounts of less than 30 ng human or mouse DNA for tagmentation may also be not sufficient to achieve a 20-fold genomic coverage.

15. Unmethylated bases are incorporated in vitro during the gap repair step into the nine bases gaps which remain after tagmentation [21] (*see* Fig. 1b). As a consequence, the first nine bases of the second read and the last 9 bases before the adapter of the first read must be excluded from the methylation analysis of the genomic DNA but, on the other hand, can be used to calculate the conversion frequency instead of or together with the spiked in unmethylated phage λ DNA. Appropriate bisulfite treatment is reflected by a conversion frequency of ≥99.5%.

Acknowledgments

We gratefully acknowledge excellent technical support by Marion Bähr, Monika Helf and helpful discussions with Daniel Lipka. We also acknowledge the excellent support from the sequencing core facility at the DKFZ, particularly from André Götze. Work in the Plass laboratory was supported by the Helmholtz Foundation and the German Federal Ministry of Education and Science in the program for medical genome research (FKZ: 01KU1001A). Q.W. obtained support by the Humboldt Research Fellowship for Postdoctoral Researchers. A.A. is funded by an NSF Graduate Research Fellowship.

Author Contributions: D.W., Q.W., A. A. and J.S. conceived the study. S.W., R.E. and C.P. contributed materials. D.W. did the experiments and analyzed data. D.W. and C.P. wrote the manuscript.

Competing financial interests: The authors declare no competing financial interests. A provisional patent application has been deposited for aspects of these methods (A.A., J.S.).

References

1. Fazzari MJ, Greally JM (2004) Epigenomics: beyond CpG islands. Nat Rev Genet 5:446–455
2. Lister R, Pelizzola M, Dowen RH et al (2009) Human DNA methylomes at base resolution show widespread epigenomic differences. Nature 462:315–322
3. Weichenhan D, Plass C (2013) The evolving epigenome. Hum Mol Genet 22:R1–R6
4. Gu H, Smith ZD, Bock C et al (2011) Preparation of reduced representation bisulfite sequencing libraries for genome-scale DNA methylation profiling. Nat Protoc 6:468–481
5. Meissner A, Gnirke A, Bell GW et al (2005) Reduced representation bisulfite sequencing for comparative high-resolution DNA methylation analysis. Nucleic Acids Res 33:5868–5877
6. Khulan B, Thompson RF, Ye K et al (2006) Comparative isoschizomer profiling of cytosine methylation: the HELP assay. Genome Res 16:1046–1055
7. Oda M, Glass JL, Thompson RF et al (2009) High-resolution genome-wide cytosine methylation profiling with simultaneous copy number analysis and optimization for limited cell numbers. Nucleic Acids Res 37:3829–3839
8. http://www.illumina.com/products/methylation_450_beadchip_kits.ilmn
9. Mohn F, Weber M, Schubeler D et al (2009) Methylated DNA immunoprecipitation (MeDIP). Methods Mol Biol 507:55–64
10. Taiwo O, Wilson GA, Morris T et al (2012) Methylome analysis using MeDIP-seq with low DNA concentrations. Nat Protoc 7:617–636
11. Weber M, Davies JJ, Wittig D et al (2005) Chromosome-wide and promoter-specific analyses identify sites of differential DNA methylation in normal and transformed human cells. Nat Genet 37:853–862
12. Brinkman AB, Simmer F, Ma K et al (2010) Whole-genome DNA methylation profiling using MethylCap-seq. Methods 52:232–236
13. Gebhard C, Schwarzfischer L, Pham TH et al (2006) Genome-wide profiling of CpG methylation identifies novel targets of aberrant hypermethylation in myeloid leukemia. Cancer Res 66:6118–6128
14. Sonnet M, Baer C, Rehli M et al (2013) Enrichment of methylated DNA by methyl-CpG immunoprecipitation. Methods Mol Biol 971:201–212
15. Seisenberger S, Andrews S, Krueger F et al (2012) The dynamics of genome-wide DNA methylation reprogramming in mouse primordial germ cells. Mol Cell 48:849–862
16. Popp C, Dean W, Feng S et al (2010) Genome-wide erasure of DNA methylation in mouse primordial germ cells is affected by AID deficiency. Nature 463:1101–1105
17. Kobayashi H, Sakurai T, Imai M et al (2012) Contribution of intragenic DNA methylation in mouse gametic DNA methylomes to establish oocyte-specific heritable marks. PLoS Genet 8:e1002440
18. Miura F, Enomoto Y, Dairiki R et al (2012) Amplification-free whole-genome bisulfite sequencing by post-bisulfite adaptor tagging. Nucleic Acids Res 40:e136
19. http://www.nugeninc.com/nugen/index.cfm/products/cs/ngs/methyl-seq/
20. http://www.zymoresearch.com/downloads/dl/file/id/628/d5455i.pdf
21. Adey A, Shendure J (2012) Ultra-low-input, tagmentation-based whole-genome bisulfite sequencing. Genome Res 22:1139–1143
22. Wang Q, Gu L, Adey A et al (2013) Tagmentation-based whole-genome bisulfite sequencing. Nat Protoc 8:2022–2032
23. Adey A, Morrison HG, Asan et al (2010) Rapid, low-input, low-bias construction of shotgun fragment libraries by high-density in vitro transposition. Genome Biol 11:R119
24. Ito S, Shen L, Dai Q et al (2011) Tet proteins can convert 5-methylcytosine to 5-formylcytosine and 5-carboxylcytosine. Science 333:1300–1303
25. Johnson MD, Mueller M, Game L et al (2012) Single nucleotide analysis of cytosine methylation by whole-genome shotgun bisulfite sequencing. Curr Protoc Mol Biol 99:21.23
26. Li H, Durbin R (2009) Fast and accurate short read alignment with Burrows-Wheeler transform. Bioinformatics 25:1754–1760
27. Grunenwald HL, Caruccio N, Jendrisak J et al (2010) Transposon end compositions and methods for modifying nucleic acids. USA Patent US20100120098A1
28. Jager N, Schlesner M, Jones DT et al (2013) Hypermutation of the inactive X chromosome is a frequent event in cancer. Cell 155:567–581
29. http://res.illumina.com/documents/products%5Ctechnotes%5Ctechnote_nextera_low_plex_pooling_guidelines.pdf
30. Hansen KD, Langmead B, Irizarry RA (2012) BSmooth: from whole genome bisulfite sequencing reads to differentially methylated regions. Genome Biol 13:R83

Chapter 7

Post-Bisulfite Adaptor Tagging for PCR-Free Whole-Genome Bisulfite Sequencing

Fumihito Miura and Takashi Ito

Abstract

Post-bisulfite adaptor tagging (PBAT) is a highly efficient procedure to construct libraries for whole-genome bisulfite sequencing (WGBS). PBAT attaches adaptors to bisulfite-converted genomic DNA to circumvent bisulfite-induced degradation of library DNA inherent to conventional WGBS protocols. Consequently, it enables PCR-free WGBS from nanogram quantities of mammalian DNA, thereby serving as an invaluable tool for methylomics.

Key words Methylome, Next-generation sequencing, PCR-free, Single-nucleotide resolution

1 Introduction

The current gold standard for single-base resolution methylome analysis is whole-genome bisulfite sequencing (WGBS). WGBS was first applied to the *Arabidopsis* methylome [1, 2] and then applied to a variety of organisms. While its power is remarkable, it has a practical drawback in terms of the amount of input DNA: it typically requires a few micrograms of DNA (i.e., approximately a million mammalian diploid cells). This number of cells is difficult or sometimes even prohibitive for various biologically interesting samples, such as mammalian oocytes, early embryonic tissues, and tissue stem cells. Extensive PCR amplification to compensate limitation of input DNA not only exacerbates biased genomic representation, but makes the estimate of methylation level inaccurate. This is due to the bisulfite conversion that induces sequences differences in the methylated and unmethylated alleles of the same locus leading to differential amplification.

To expand the range of samples suitable for WGBS, a novel protocol that requires a much smaller amount of input DNA than the conventional ones is required. Although bisulfite treatment induces DNA fragmentation, the conventional (*see* also

Jörg Tost (ed.), *DNA Methylation Protocols*, Methods in Molecular Biology, vol. 1708,
https://doi.org/10.1007/978-1-4939-7481-8_7, © Springer Science+Business Media, LLC 2018

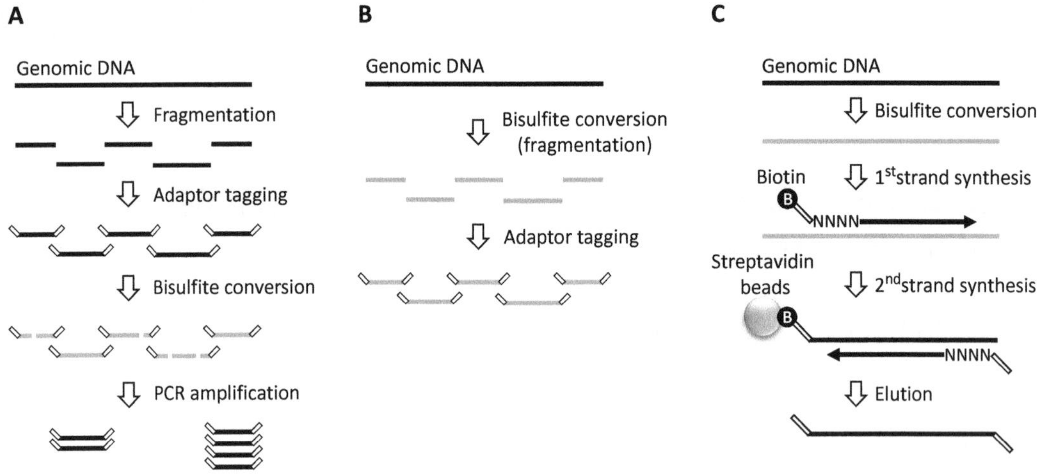

Fig. 1 Principle of PBAT. (**a**) Conventional WGBS protocols. Bisulfite treatment follows adaptor-tagging and degrades adaptor-tagged library DNAs. (**b**) PBAT strategy. Adaptor-tagging follows bisulfite treatment to circumvent bisulfite-induced fragmentation of adaptor-tagged library DNAs. (**c**) Random priming-mediated PBAT. Two rounds of random primer extension on bisulfite-converted genomic DNA generate directionally adaptor-tagged library DNAs

Chapter 5) as well as tagmentation-based protocols [3] (*see* also Chapter 6) include the step for bisulfite-treatment of adaptor-tagged DNAs that results in the degradation and low yield of DNA for the generation of the library (Fig. 1a). To circumvent this adverse effect, we conceived a novel principle termed Post-Bisulfite Adaptor Tagging (PBAT), in which adaptor tagging follows bisulfite treatment, in contrast to the other protocols [4] (Fig. 1b). Since it is difficult to efficiently ligate adaptors to bisulfite-treated, denatured DNAs, we developed a simple adaptor-tagging protocol using two rounds of random primer extension (Fig. 1c).

The random priming-based PBAT protocol can generate a PCR-free library from as little as 125 pg of DNA. It typically allows generating a WGBS library of sufficient quality and diversity to achieve ~30-fold, PCR-free coverage of the mammalian genomes from ~30 ng of DNA. We and others have successfully applied PBAT to various samples ranging from plants, fungi, and animals, especially those with limited amounts of DNA (*see* also Chapter 9). For example, PBAT has been used to perform PCR-free WGBS from only 1000 oocytes mouse and a few thousand flow-sorted primordial germ cells [5, 6]. We have also applied PBAT to target-enriched genomic DNA, thereby achieving highly efficient targeted methylome sequencing [7]. In contrast to conventional PCR-assisted WGBS, PBAT can preferentially cover GC-rich genic regions and CpG islands. We expect that the PBAT protocol described below will help readers conduct various novel WGBS applications.

2 Materials

2.1 Reagents

1. Qubit dsDNA BR Assay Kit.
2. Qubit dsDNA HS Assay Kit.
3. Qubit ssDNA Assay Kit.
4. Agencourt AMPure XP beads (Beckman Coulter).
5. 10× PCR buffer: 100 mM Tris–HCl, 500 mM KCl, 15 mM $MgCl_2$. pH 8.3.
6. RNA 6000 Pico Kit (Agilent).
7. High Sensitivity DNA Kit (Agilent).
8. 10 mM Tris-HAcetate, pH 8.0.
9. 10 mM Tris–HCl, pH 7.5.
10. Klenow fragment (3′ → 5′ exo^-): high concentration (50 U/μ L) (New England Biolabs, NEB, *see* **Note 1**).
11. Bst DNA polymerase large fragment (80 U/μL) (NEB).
12. Exonuclease I (20 U/μL).
13. Phusion Hot Start II High-Fidelity DNA Polymerase (2 U/μ L).
14. EZ DNA Methylation-Gold Kit (ZYMO Research).
15. Dynabeads M-280 Streptavidin.
16. 2× BW buffer: dissolve 6.3 g LiCl in 40 mL of double-distilled water (ddH_2O), after LiCl has completely dissolved, add 0.5 mL of 1 M Tris–HCl, pH 8.0 and 0.1 mL of 0.5 M EDTA, and adjust the volume to 50 mL with ddH_2O (*see* **Note 2**).
17. 0.1 M NaOH (*see* **Note 3**).
18. KAPA Library Quantification Kit for Illumina (KAPA Biosystems).
19. Hybridization Buffer A: combine 9 mL of 5 M NaCl and 9 mL of 1 M Tris–HCl, pH 7.4, and bring to a final volume of 50 mL with ddH_2O.

2.2 Oligonucleotides

1. Bio-PEA2-N4: 100 μM

   5′-biotin-ACA CTC TTT CCC TAC ACG ACG CTC TTC CGA TCT NNN N-3′.

2. PE-reverse-N4: 100 μM

   5′-CAA GCA GAA GAC GGC ATA CGA GAT NNN N-3′.

3. PBAT-PE-iX-N4: 100 μM each

5′-CAA GCA GAA GAC GGC ATA CGA GAT XXX XXX GTA AAA CGA CGG CCA GCA GGA AAC AGC TAT GAC NNN N-3′.

Replace the underlined hexamer with one of the following index sequences, which are complementary to those used in the Illumina's TruSeq DNA LT Sample Prep Kit, so that the same index numbers as those in the kit can be used. The index numbers 17, 24, and 26 are reserved by Illumina for unknown reasons.

Index #	Sequence	Index #	Sequence	Index #	Sequence
1	CGTGAT	9	CTGATC	18	GCGGAC
2	ACATCG	10	AAGCTA	19	TTTCAC
3	GCCTAA	11	GTAGCC	20	GGCCAC
4	TGGTCA	12	TACAAG	21	CGAAAC
5	CACTGT	13	TTGACT	22	CGTACG
6	ATTGGC	14	GGAACT	23	CCACTC
7	GATCTG	15	TGACAT	25	ATCAGT
8	TCAAGT	16	GGACGG	27	AGGAAT

4. Primer 3: 100 μM:

   5′-AAT GAT ACG GCG ACC ACC GAG ATC TAC ACT CTT TCC CTA CAC GAC GCT CTT CCG ATC T-3′.

5. PBAT-PE-Seq:

   5′-GTA AAA CGA CGG CCA GCA GGA AAC AGC TAT GAC-3′.

6. PBAT-PE-Idx:

   5′-GTC ATA GCT GTT TCC TGC TGG CCG TCG TTT TAC-3′.

2.3 Plastic Disposables

1. Microcentrifuge and PCR tubes: In all the steps, use of low-retention 1.5 and 0.2 mL tubes is recommended.
2. Pippette tips: use of low-retention tips is recommended for dispensing streptavidin-coated magnetic beads.

2.4 Equipment

1. DynaMag-2 Magnet (Thermo Fisher Scientific, or equivalent).
2. SPRIPlate 96R Magnet Plate (Beckman Coulter, or equivalent).
3. High-speed refrigerated microcentrifuge.
4. Agilent Bioanalyzer 2100.
5. Qubit Fluorometer or Qubit 2.0 Fluorometer.

6. TOMY PMC-060 Capsulefuge (or equivalent).
7. Thermal cycler.
8. StepOnePlus Real-Time PCR System (Thermo Fisher Scientific or equivalent).

3 Methods

3.1 Bisulfite Treatment (Day 1)

1. Measure the concentration of the DNA sample with the Qubit dsDNA BR Assay Kit and Qubit Fluorometer according to the manufacturer's instructions (*see* **Note 4**).
2. Add 900 μL of ddH_2O, 50 μL of M-Dissolving Buffer, and 300 μL of M-Dilution Buffer to one tube of CT Conversion Reagent from the EZ DNA Methylation-Gold Kit. Dissolve the material by rotating the tube of CT Conversion Reagent for 10 min at room temperature.
3. Mix the following components well: 130 μL of CT conversion reagent, $(20 - x)$ μL ddH_2O, and x μL of sample DNA (*see* **Notes 5** and **6**).
4. Divide the solution into three 50 μL aliquots in 0.2 mL tubes.
5. Place the tubes on a thermal cycler, and start the following program: 98 °C for 10 min, 64 °C for 150 min, followed by a 4 °C hold.
6. Place a Zymo-Spin IC Column in a Collection Tube and add 600 μL of M-Binding Buffer to the column.
7. Add the sample from **step 5** to the M-Binding Buffer in the column. Close the cap and mix by inverting several times.
8. Centrifuge at full speed ($\geq$10,000 × g) for 30 s.
9. Reload the flow-through onto the same column again (*see* **Note 7**).
10. Centrifuge at full speed ($\geq$10,000 × g) for 30 s. Discard the flow-through.
11. Add 100 μL of M-Wash Buffer prepared with ethanol to the column, and centrifuge at full speed for 30 s. Discard the flow-through.
12. Add 200 μL of M-Desulphonation Buffer to the column and let the column stand at room temperature for 15 min.
13. Centrifuge at full speed for 30 s. Discard the flow-through.
14. Add 200 μL of M-Wash Buffer with ethanol to the column and centrifuge at full speed for 30 s. Discard the flow-through.
15. Repeat the wash in **step 14** one more time and then transfer the spin column to a new clean 1.5 mL tube.

16. Add 22 μL of M-Elution Buffer directly to the column matrix and let the column stand at room temperature for 2 min. Centrifuge at full speed for 30 s to elute the DNA (*see* **Notes 8** and **9**).

3.2 First-Strand Synthesis (Day 1)

1. Prepare the first-strand synthesis reaction mix as follows: Add 5 μL of 10× NEB Buffer 2, 5 μL of 2.5 mM dNTPs, 16 μL of ddH$_2$O, 4 μL of 100 μM primer Bio-PEA2-N4 and 20 μL of bisulfite-treated sample DNA from Subheading 3.1.
2. Place the tube on a thermal cycler and start the following program: 94 °C for 5 min, 4 °C for 20 min, gradual increase from 4 to 37 °C at a rate of +1 °C/min, 37 °C for 90 min, 70 °C for 10 min, followed by a hold at 4 °C (*see* **Note 10**).
3. After 5 min of incubation at 4 °C (i.e., the second step of the program), pause the program and remove the tube from the thermal cycler. Add 1.5 μL of Klenow fragment (3′ → 5′ exo$^-$) to the first-strand synthesis mix and mix well.
4. Place the tube on the thermal cycler again and resume the program to complete the first-strand synthesis reaction (*see* **Note 11**).

3.3 Removal of Excess Primers (Day 2)

1. Transfer the solution of the first-strand reaction (~50 μL) into a new 1.5 mL tube, add 50 μL of AMPure XP beads, mix well, and spin the tube briefly (*see* **Note 12**).
2. Let the tube stand at room temperature for 10 min.
3. Place the tube on a magnetic stand and wait for the beads to be collected. Then, remove the supernatant carefully and make sure not to aspirate the beads.
4. Add 200 μL of 75% ethanol to wash the beads and then remove the supernatant.
5. Add 45 μL of 10 mM Tris-HAcetate buffer and vortex the tube well to suspend the beads. After a brief centrifugation, place the tube on the magnetic stand and wait for the beads to be collected.
6. Transfer the supernatant into a new 1.5 mL tube. Add 5 μL of 10× PCR Buffer and 50 μL of AMPure XP beads to the supernatant. Then mix well and spin briefly (*see* **Note 13**).
7. Let the tube stand at room temperature for 10 min.
8. Place the tube on the magnetic stand and wait for the beads to be collected. Then remove the supernatant carefully not to aspirate the beads.
9. Add 200 μL of 75% ethanol to wash the beads and then remove the supernatant.

10. Add 51 μL of 10 mM Tris-HAcetate buffer and vortex the tube well to suspend the beads. After a brief centrifugation, place the tube on the magnetic stand and wait for the beads to be collected.
11. Transfer the supernatant in a new 1.5 mL tube.
12. Use 1 μL of the supernatant to measure the DNA concentration using the Qubit dsDNA HS kit (*see* **Note 14**).

3.4 Capturing Biotinylated DNA on Streptavidin Beads (Day 2)

1. Take 20 μL of a well-dispersed suspension of Dynabeads M280 streptavidin beads into a 1.5 mL tube. Place the tube on a magnet stand to collect the beads.
2. Remove the supernatant, and add 50 μL of 2× BW buffer to suspend the beads.
3. Add the suspension of beads to the product obtained in Subheading 3.3.
4. Incubate the tube at room temperature for 30 min while gently rotating the tube.
5. Place the tube on the magnet stand to collect the beads and then remove the supernatant.
6. Add 180 μL of 2× BW buffer to the beads, vortex well, and spin the tube briefly.
7. Place the tube on the magnet stand to collect the beads and then remove the supernatant.
8. Suspend the beads in 180 μL of 0.1 M NaOH, vortex well, incubate at room temperature for 2 min, and spin briefly.
9. Place the tube on the magnet stand to collect the beads and then remove the supernatant.
10. Repeat **steps 8** and **9**.
11. Add 180 μL of 2× BW buffer to the beads, vortex well, and spin the tube briefly.
12. Place the tube on the magnet stand to collect the beads and then remove the supernatant.
13. Add 180 μL of 10 mM Tris–HCl to the beads, vortex well, and spin the tube briefly.

3.5 Second-Strand Synthesis (Day 2)

1. Place the tube on a magnet stand to collect the beads and remove the supernatant.
2. Prepare the second-strand synthesis reaction mix as follows and add to the beads: 5 μL of 10× NEB Buffer 2, 5 μL of 2.5 mM dNTPs, 36 μL of ddH_2O, 4 μL of 100 μM PE-reverse-N4, for single-end sequencing, or PBAT-PE-iX-N4 for paired-end and index sequencing.

3. Suspend the beads by vortexing, and transfer the beads suspension into a new 0.2 mL tube.
4. Place the tube on a thermal cycler and start the following program: 94 °C for 5 min, 4 °C for 20 min, gradual increase from 4 to 37 °C at a rate of +1 °C/min, 37 °C for 30 min, 70 °C for 10 min, followed by a 4 °C hold (*see* **Note 10**).
5. After 5 min of the incubation at 4 °C (i.e., the second step of the program), pause the program and remove the tube from the thermal cycler. Add 1.5 μL of Klenow Fragment (3′ → 5′ exo$^-$) to the second-strand synthesis solution and mix well.
6. Place the tube on the thermal cycler again and resume the program to complete the second-strand synthesis reaction.

3.6 Chase Reaction (Day 2)

1. Place the tube on a magnet stand to collect the beads and remove the supernatant.
2. Prepare the chase reaction mix as follows and add to the beads: 5 μL of 10× ThermoPol Buffer (provided with the Bst polymerase large fragment), 5 μL of 2.5 mM dNTPs, 40 μL of ddH_2O and 1 μL of Bst DNA polymerase large fragment.
3. Incubate the reaction mix at 65 °C for 30 min.

3.7 Elution/Extension of Template DNA (Day 2) (See Note 15)

1. Place the tube on a magnet stand to collect the beads and remove the supernatant.
2. Prepare the elution/extension reaction mix as follows and add to the beads: 10 μL 5× Phusion HS buffer, 5 μL of 2.5 mM dNTPs, 35 μL of ddH_2O, 0.4 μL of 100 μM Primer 3 and 1 μL of Phusion Hot Start High-fidelity DNA polymerase.
3. Start the following program: 94 °C for 5 min, 55 °C for 15 min, 68 °C for 30 min, followed by a 4 °C hold.
4. Place the tube on the magnet stand to collect the beads, and transfer the supernatant into a new 1.5 mL tube.
5. Add 1 μL of exonuclease I to the supernatant, mix well, and incubate the tube at 37 °C for 30 min followed by heat inactivation at 70 °C for 10 min.
6. Use 1 μL of the eluted DNA with the Qubit dsDNA HS Kit to measure the concentration of DNA.

3.8 Size Fractionation (Day 2)

1. Add 50 μL of AMPure XP beads to the eluted DNA (50 μL), mix well, and spin briefly.
2. Place the tube on a magnetic stand and wait for the beads to be separated. Then, remove the supernatant carefully not to aspirate the beads.
3. Add 200 μL of 75% ethanol to wash the beads and then remove the supernatant.

4. Add 45 μL of 10 mM Tris-HAcetate buffer and vortex the tube well to suspend the beads. After a brief centrifugation, place the tube on the magnetic stand and wait for the beads to be collected.
5. Transfer the supernatant to a new 1.5 mL tube.
6. Add 5 μL of 10× PCR Buffer and 50 μL of AMPure XP beads to the supernatant. Mix well and spin briefly.
7. Let the tube stand at room temperature for 10 min.
8. Place the tube on the magnetic stand and wait for the beads to be collected. Then, remove the supernatant carefully not to aspirate the beads.
9. Add 200 μL of 75% ethanol to wash the beads, and then remove the supernatant.
10. Add 22 μL of 10 mM Tris-HAcetate buffer and vortex well to re-suspend the beads. After a brief centrifugation, place the tube on the magnetic stand and wait for the beads to be collected.
11. Transfer the supernatant to a new 1.5 mL tube.
12. Use 1 μL of the supernatant to measure the concentration of DNA using the Qubit dsDNA HS kit (*see* **Note 16**).
13. Subsequently, determine the exact molar concentration of template DNA using an appropriate qPCR assay (*see* **Note 17**).

3.9 Illumina Sequencing

Here, we provide guidance about sequencing PBAT libraries on Illumina HiSeq2000, HiSeq2500 and MiSeq instruments.

3.9.1 Calculation of Template Volume Required for Sequencing

1. Calculate the volume of template required in cluster generation using the following equation and parameters (*see* **Note 18**).

Platform	HiSeq 2000	HiSeq 2500 in rapid mode with cBot	HiSeq 2500 in rapid mode without cBot or MiSeq
Target concentration of denatured template (pM)	10[a]	10[a]	10[a]
Target volume of template (μL)	120	70	480
Molar concentration of template (pM)	y	y	y
Volume of template required = x (μL)	$\frac{120\times10}{y}$	$\frac{70\times10}{y}$	$\frac{480\times10}{y}$

[a]For target concentration, 10 pM is a good point to start optimization

3.9.2 Running a Single Lane of Illumina HiSeq2000

1. Dispense 110 μL of hybridization buffer A to a new 1.5 mL tube and place it on ice.
2. Prepare a 2 M NaOH by diluting a 10 M NaOH solution.
3. Denature sequencing templates by combining x μL of template DNA solution, (19 − *x*) μL of ddH_2O and 1 μL of 2 M NaOH.
4. Let the tube stand at room temperature for 5 min.
5. Add 100 μL of the ice-cooled hybridization buffer A to the denatured template, mix well, and place the tube on ice.
6. Start cluster generation according to the manufacturer's instructions (*see* **Note 19**).

3.9.3 Running a Single Lane of Illumina HiSeq2500 in Rapid Mode with cBot

1. Dispense 70 μL of hybridization buffer A to a new 1.5 mL tube and place it on ice.
2. Prepare a 2 M NaOH by diluting a 10 M NaOH solution.
3. Denature sequencing templates by combining x μL of template DNA solution, (11 − *x*) μL of ddH_2O and 0.6 μL of 2 M NaOH.
4. Let the tube stand at room temperature for 5 min.
5. Add 58 μL of the ice-cooled hybridization buffer A to the denatured template, mix well, and place the tube on ice.
6. Add 70 μL of ice-cold 8 pM denatured phiX control to the tube and mix well (*see* **Note 20**).
7. Start cluster generation according to the manufacturer's instruction.

3.9.4 Running a Single Lane of Illumina HiSeq2500 in Rapid Mode without cBot or Illumina MiSeq

1. Dispense 440 μL of hybridization buffer A to a new 1.5 mL tube and put it on ice.
2. Prepare a 2 M NaOH by diluting a 10 M NaOH solution.
3. Denature sequencing templates by combining *x* μL of template DNA solution, (76 − *x*) μL of ddH_2O and 4 μL of 2 M NaOH.
4. Let the tube stand at room temperature for 5 min.
5. Add 400 μL of the ice-cold hybridization buffer A, mix well, and place the tube on ice.
6. Add 120 μL of ice-cold 8 pM denatured phiX control to the tube and mix well (*see* **Note 20**).
7. Start run according to the manufacturer's instruction (*see* **Note 21**).

4 Notes

1. Be sure to use high concentration enzymes (i.e., 50,000 U/mL).

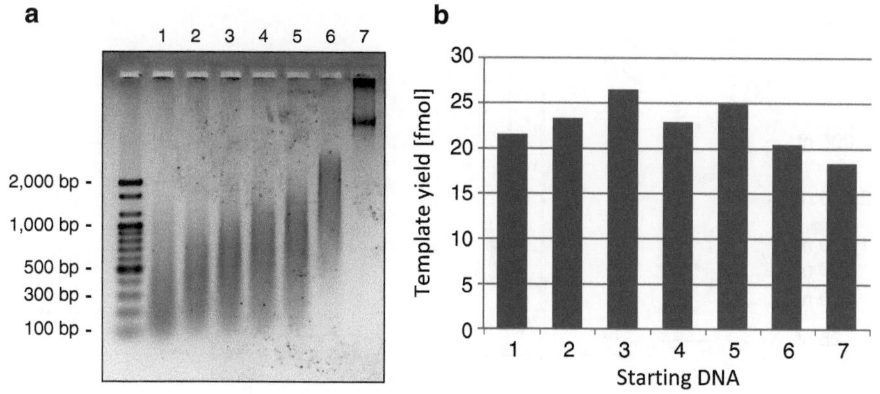

Fig. 2 Effect of input DNA size. (**a**) Agarose gel electrophoresis of DNAs variously fragmented using the Covaris S220 acoustic sonicator. (**b**) Yields of libraries generated from 100 ng of the variously fragmented DNA shown in (**a**)

2. Dissolving LiCl is an exothermic process. To avoid bumping of the solution, add LiCl to 40 mL of ddH_2O. Then, add Tris and EDTA, and adjust the volume to 50 mL with ddH_2O.
3. Dilute from 10 M NaOH stock before use.
4. Accurate estimation of DNA concentration is critical. We routinely use the Qubit dsDNA BR Assay Kit for the purpose. Avoid measuring at an optical density of 260 nm, because various materials other than DNA absorb light at 260 nm and often result in an overestimation of DNA concentration. The size of input DNA seems to be less critical, as it does not affect the yield of library (Fig. 2).
5. We routinely start with 100 ng of DNA, because this amount is easy to handle. However, note that the maximum efficiency of template preparation is achieved with ~1 ng of DNA as a starting material (*see* also Chapter 9). Thus, the bisulfite-treated DNA may be divided into several aliquots before first-strand synthesis to further increase the efficiency of template preparation. All the reagents used in this step are provided in EZ DNA methylation kit.
6. Use freshly prepared CT conversion reagent to ensure high yield and efficient bisulfite conversion.
7. Because we occasionally encountered "shunts" in the column through which the solution flows with minimal contact with the resin, the column should be carefully inspected before use. Reloading of the flow-through increases the contact of the solution with the resin.
8. The elution volume (22 μL) includes 1 μL for the determination of the yield using the Qubit ssDNA Assay kit and 1 μL for the QC with the Agilent Bioanalyzer using the RNA 6000 Pico

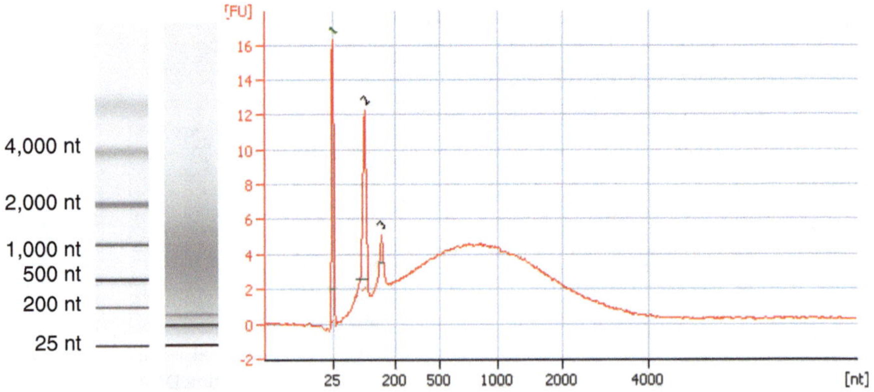

Fig. 3 Typical size distribution of bisulfite-treated DNA. Bisulfite-treated genomic DNA was separated on an Agilent Bioanalyzer 2100 using the RNA 6000 pico kit

Kit. Typically, the yield of DNA is between 30 and 70% of the input. The typical size range of denatured DNA is 100–1000 nt with a peak around 600 nt (Fig. 3). When the starting amount of DNA is <30 ng, both the platforms will fail to detect the eluted DNA. Thus, omit these QC steps and reduce the elution volume in **step 17** to 20 μL.

9. *Do not stop here.* Proceed immediately to the first-strand synthesis step, because the bisulfite-treated DNA is labile.
10. If your thermal cycler cannot generate a temperature ramp of +1 °C/min, you may use a two-step PCR cycling program with an increment of temperature by 1 °C for every step. If the first cycle of the program is set as 4.0 °C for 30 s followed by 4.5 °C for 30 s, then the temperature will reach to 37 °C after 33 cycles with the intended rate of +1 °C/min.
11. You can stop here either by leaving the tube at 4 °C or storing it at −20 °C until use. This is presumably because the bisulfite-treated DNAs are now double-stranded and excessive primers in the solution serve as a carrier DNA to prevent the adsorption of template DNA to the tube wall.
12. At this mixing ratio (i.e., DNA solution:AMPure XP of 1:1), DNA fragments less than 200 bp are effectively removed in the supernatant. While primers and primer dimers are less than 100 nt, the products of the first-strand synthesis are larger than 200 bp.
13. Addition of 10× PCR Buffer at this step increases the reproducibility and yield of AMPure XP-based purification of DNA.
14. Typical yields at this step are between 40 and 60% of the input. When the starting amount of DNA is <30 ng, the kit will fail to quantify the DNA. Thus, omit this QC step and reduce the elution volume in **step 10** to 50 μL.

15. This step not only enables the precise selection of double-stranded DNA by SPRI beads, but also synthesizes the sequence required for bridge PCR.
16. Typical yield at this step is between 20 and 40% of the input DNA. When the starting amount of DNA is <30 ng, the kit will fail to quantify DNA. Thus, omit this QC step and reduce the elution volume in **step 9** to 21 μL.
17. Note that the product obtained in Subheading 3.8 contains not only intact sequencing templates but also several-fold greater amounts of by-products. Thus, it is essential to determine the correct concentration of the template DNA by qPCR but not by fluorometry. We perform quantification using Library Quantification Kits for Illumina (KAPA biosystems) according to the manufacturer's instruction, because it is easy to use and highly reproducible. Typical mass yield at this step is calculated to be 2–8% relative to the starting amount of DNA. If the starting DNA is 100 ng, this number corresponds to ~20 fmol of template, which is sufficient for 20 lanes or 4 runs of sequencing on a Illumina HiSeq 2000 or MiSeq, respectively. In addition, the by-products make it impossible to examine the size of template DNA directly by electrophoresis. Accordingly, we analyze the size of qPCR product. Because the size distribution of PCR-amplified templates becomes unreliable after the PCR reaches the plateau, we run the amplified product on a 6% polyacrylamide/7 M urea gel in TBE buffer (89 mM Tris, 89 mM boric acid, 2 mM EDTA). As shown in Fig. 4, the typical size is between 200 and 500 bp with a peak around 300 bp.

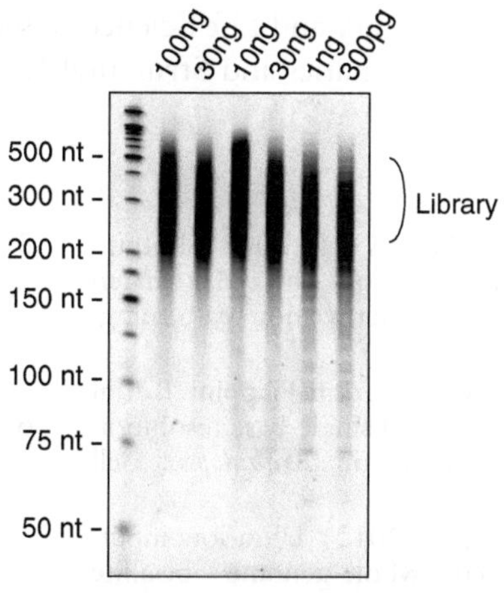

Fig. 4 Typical size distribution of sequencing templates. Amplification products of qPCR to quantify libraries generated from the indicated amount of input DNA were separated on 6% Novex TBE-Urea gel and stained with SYBR Gold

18. Be sure to add PBAT-PE-Seq and PBAT-PE-Idx to the Illumina primer mix at the final concentration of 0.5 μM each for paired-end and index sequencing.
19. Since bisulfite-converted DNA has a very biased base composition, each flow cell must include a control lane for the phiX control template to enable correct normalization of fluorescent signals.
20. For the same reason as described in **Note 19**, the phiX control template must be added to the sample. Follow the instruction provided by Illumina for the preparation of denatured phiX control template.
21. We usually use 101 cycles for sequencing. The reads are adequately processed and used for mapping. While a number of bisulfite mappers have been made available, it often happens that unique characteristics of PBAT reads not only necessitate appropriate preprocessing of the data but hamper their efficient mapping. We recommend a popular bisulfite mapper, bismark, which has an option "pbat" (http://www.bioinformatics.babraham.ac.uk/projects/bismark/). We also provide BMap that uses an algorithm termed adaptive seeds for highly efficient mapping of PBAT reads (http://itolab.med.kyushu-u.ac.jp/BMap/).

Acknowledgments

This work was supported by Research Program of Innovative Cell Biology by Innovative Technology (Cell Innovation) of the Ministry of Education, Culture, Sports, Science and Technology (MEXT) and the Platform Project for Supporting in Drug Discovery and Life Science Research (Platform for Drug Discovery, Informatics and Structural Life Science) of AMED.

References

1. Cokus SJ, Feng S, Zhang X et al (2008) Shotgun bisulphite sequencing of the *Arabidopsis* genome reveals DNA methylation patterning. Nature 452:215–219
2. Lister R, O'Malley RC, Tonti-Filippini J et al (2008) Highly integrated single-base resolution maps of the epigenome in *Arabidopsis*. Cell 133:523–536
3. Adey A, Shendure J (2012) Ultra-low-input, tagmentation-based whole-genome bisulfite sequencing. Genome Res 22:1139–1143
4. Miura F, Enomoto Y, Dairiki R et al (2012) Amplification-free whole-genome bisulfite sequencing by post-bisulfite adaptor tagging. Nucleic Acids Res 40:e136
5. Shirane K, Toh H, Kobayashi H et al (2013) Mouse oocyte methylomes at base resolution reveal genome-wide accumulation of non-CpG methylation and role of DNA methyltransferases. PLoS Genet 9:e1003439
6. Kobayashi H, Sakurai T, Miura F et al (2013) High-resolution DNA methylome analysis of primordial germ cells identifies gender-specific reprogramming in mice. Genome Res 23:616–627
7. Miura T, Ito F (2015) Highly sensitive targeted methylome analysis by post-bisulfite adaptor tagging. DNA Res 22:13–18

Chapter 8

Multiplexed Reduced Representation Bisulfite Sequencing with Magnetic Bead Fragment Size Selection

William P. Accomando Jr. and Karin B. Michels

Abstract

Reduced representation bisulfite sequencing (RRBS) is a technique for assessing genome-wide DNA methylation in an organism whose genome has been fully sequenced. It allows researchers to target gene regions with particular CpG densities, thereby selecting the desired genomic contexts. Here, we describe an approach that uses magnetic beads to accomplish this selection. In addition, the use of indexed, methylated adapters enables up to 12 samples to be pooled, and subjected to multiplexed RRBS in a single-sequencing lane. First, genomic DNA is fragmented via restriction endonuclease digestion that ensures at least two CpG loci per fragment. The fragmented DNA is then end-repaired and A-tailed. Indexed, methylated adapters are ligated to the A-tailed DNA fragments to create a DNA library. A combination of negative and positive selections, using magnetic beads that preferentially bind to larger DNA fragments, ensures that only the desired sizes of adapter-ligated DNA fragments are included in a library. This allows researchers to dictate what types of genomic regions will be sequenced, since fragment size depends on the proximity of restriction sites. The DNA libraries are then quantified, and up to 12 libraries are pooled in order to be sequenced on a single lane of an Illumina HiSeq2500. The pools are next treated with sodium bisulfite, and then PCR amplified. A final bead cleanup removes any residual contaminants prior to sequencing, which is followed by base calling and alignment to a sequenced genome.

Key words Reduced representation bisulfite sequencing (RRBS), mRRBS, Epigenetics, DNA methylation

1 Introduction

Epigenetics refers to the study of stable, mitotically heritable regulation of gene expression that is independent of the primary sequence of DNA nucleotides [1]. The archetypical epigenetic mark is DNA methylation occurring on cytosine residues in the context of cytosine-guanine dinucleotides, or CpG loci [2, 3]. Though underrepresented in the genome, CpG loci tend to be concentrated in regions referred to as CpG islands, which are found in 60% of gene promoters [4]. It is widely believed that CpG island methylation in gene promoters indicates transcriptional silencing, which may be due to chromatin condensation

Jörg Tost (ed.), *DNA Methylation Protocols*, Methods in Molecular Biology, vol. 1708,
https://doi.org/10.1007/978-1-4939-7481-8_8, © Springer Science+Business Media, LLC 2018

[5, 6]. For this reason, many DNA methylation microarrays target CpG loci that are located in gene promoters. However, DNA methylation is meaningful in other genomic contexts as well, and the function of DNA methylation is likely to vary depending on context [7]. Therefore, a technique that allows for the assessment of DNA methylation in a variety of user-definable genomic contexts throughout the genome is highly useful, since it allows researchers to tailor the approach to address their particular research questions.

Reduced representation bisulfite sequencing (RRBS) is a focused approach to measuring DNA methylation across the genome of an organism that offers a number of advantages over other approaches to genome-wide DNA methylation assessment [8–12] (*see* also Chapters 5–16). RRBS can assess DNA methylation throughout a genome that has been fully sequenced. Moreover, RRBS allows for the selection of particular genomic contexts by choosing the density of CpG loci in the DNA regions that are examined. The original RRBS procedure achieved this selection using a gel-based approach, which was labor intensive and not always precise. Here, we describe a method for achieving the same selection using magnetic beads that is less labor intensive and more precise. In addition, RRBS was originally carried out by analyzing only one sample per sequencing lane, but here we describe a multiplexed approach that allows up to 12 different samples to be analyzed in a single-sequencing lane through the use of indexed adapters.

An overview of the required steps in this protocol is shown in Fig. 1. To carry out multiplexed RRBS, highly pure genomic DNA (gDNA) is needed as a starting material [11]. While it is possible to perform multiplexed RRBS using anywhere from 10 to 300 ng of gDNA [10], we have found that 100 ng of starting DNA at a concentration of 20–50 ng/μL provides the best balance of high-quality results and sample conservation. The isolated gDNA is fragmented by digestion with a restriction endonuclease that cleaves at a restriction site containing a CpG locus. This results in a variety of different sized DNA fragments that will each contain at least two CpG loci. The two restriction enzymes that are most commonly used for this purpose are MspI (C↓CGG) and TaqI (T↓CGA), though in this protocol we will focus on MspI. The resulting fragments have overhangs at each end, therefore the next step in the protocol is to perform an end-repairing reaction that removes the 3′ overhangs and fills in the 5′ gaps, generating blunt ends. Then, a single-adenosine residue is added to each terminus of the end-repaired fragments. These A-tails facilitate the ligation of methylated adapters, which are oligonucleotides containing universal sequences that enable PCR amplification of the fragments. Each of the adapters used in multiplexed RRBS also contains a unique six-base index sequence, sometimes referred to as a barcode, which are used to tag different samples. This allows multiple samples to be pooled, run in a single-sequencing lane,

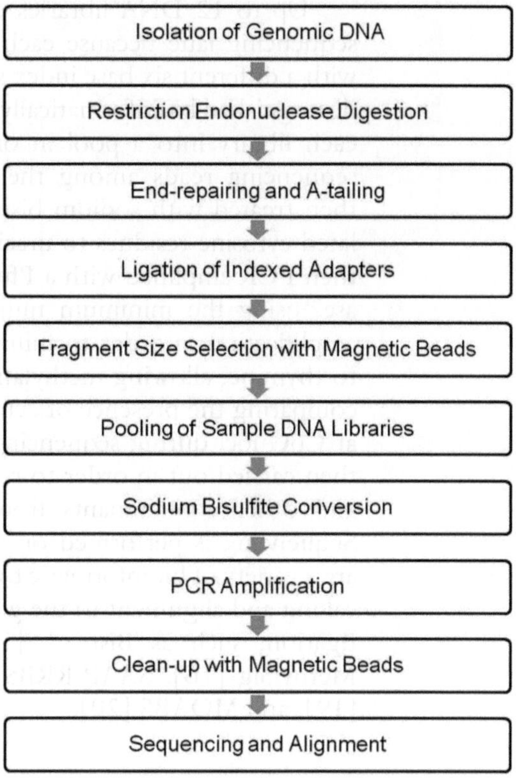

Fig. 1 Overview of steps in multiplexed reduced representation bisulfite sequencing with magnetic bead fragment size selection

and then distinguished bioinformatically. The Illumina HiSeq 2500 technology allows for a maximum of 12 multiplexed samples per lane in order to yield meaningful results, but more recent sequencers such as the Hiseq3000/4000 may allow for more than 12 samples to be pooled and run in a single lane.

Next, a particular size range of adapter-ligated DNA fragments is obtained using a combination of negative and positive selections with magnetic beads. The magnetic beads preferentially bind to larger DNA fragments, and the volume of beads relative to the volume of fragmented DNA in each selection step determines the fragment sizes that are included in the final library. Depending on the research question, investigators can adjust this selection process in order to customize their library to include fragment sizes that correspond to their desired genomic features. This step also cleans up the DNA libraries by removing any un-ligated adapters and residual dNTPs. To verify that the adapters are successfully ligated to the fragmented DNA and that the proper adapter-ligated fragment sizes have been selected, a small amount of each RRBS library can be amplified by PCR using universal primers, which anneal to a sequence common to all of the indexed methylated adapters, and then visualized on an agarose gel.

Up to 12 DNA libraries can be pooled and run in a single-sequencing lane because each is ligated to a methylated adapter with a different six base index sequence allowing the samples to be distinguished bioinformatically. It is best to add an equal amount of each library into a pool in order to improve the uniformity of sequencing reads among the libraries. The pooled libraries are then treated with sodium bisulfite in order to convert unmethylated cytosine residues to uracil. The bisulfite converted pools are then PCR amplified with a PfuTurbo Cx Hotstart DNA polymerase, using the minimum number of cycles required for robust amplification in order to minimize bias. This also converts uracil to thymine, allowing methylation to be quantified at CpG loci by comparing the presence of cytosine and thymine that are detected at CpG loci during sequencing. A final magnetic bead cleanup is then carried out in order to remove any residual dNTPs, primers, and other contaminants from the pools prior to sequencing. Sequencing is performed on an Illumina HiSeq machine. There are a variety of bioinformatic tools that can be used to perform base calling and alignment to the genome of the organism under investigation, such as Bismark [13], MAQ [14], RRBSMap [15], MethylSig [16], SAAP-RRBS [17], RRBS-analyzer [18], PASH [19], and MOABS [20].

2 Materials

1. DNA extraction kit, such as the DNeasy Blood and Tissue Kit (Qiagen) or equivalent.
2. Microcentrifuge.
3. (Optional) Column-based genomic DNA cleanup and concentration kit, such as DNA Clean & Concentrator (Zymo Research) or equivalent.
4. TE buffer: 10 mM Tris–HCl, 0.1 mM EDTA, pH 8.0.
5. NanoDrop spectrophotometer.
6. 1.5 mL nuclease-free microcentrifuge tubes.
7. Nuclease-free 96-well plates.
8. Centrifuge suitable for 96-well plates.
9. MspI restriction endonuclease [20 U/μL] and associated restriction digest buffer e.g., from New England Biolabs.
10. Nuclease-free Water.
11. Thermal cycler.
12. NEBNext® Ultra™ DNA Library Prep Kit for Illumina® (New England Biolabs).

13. Agencourt AMPure XP magnetic beads (Beckman Coulter, Inc.), or equivalent.
14. Magnetic rack for 96-well plate, such as DynaMag 96-side magnet (Thermo Fisher Scientific) or equivalent.
15. 80% Ethanol.
16. 10 mM Tris–HCl, pH 8.0, EDTA free.
17. Indexed methylated adapters: TruSeq adapters sets A and/or B can be purchased in the TruSeq Nano DNA LT Sample Prep Kits (Illumina) or the TruSeq PCR-free DNA LT Sample Prep Kits (Illumina), and can be diluted 1:20 in nuclease-free water prior to use in this protocol. Alternatively, one can obtain the TruSeq adapter sequences from Illumina and have the oligonucleotides synthesized, with all cytosines methylated, by a commercial supplier such as Integrated DNA Technologies (Coralville, IA, USA).
18. DNA gel materials including: DNA ladder, agarose, ethidium bromide, trays, combs, running apparatus, and UV visualizer.
19. Qubit fluorometer.
20. Qubit Assay Tubes.
21. PCR Mastermix: 50 units/mL of *Taq* DNA polymerase, proprietary reaction buffer (pH 8.5), 400 μM dATP, 400 μM dGTP, 400 μM dCTP, 400 μM dTTP, and 3 mM $MgCl_2$ (Promega, or equivalent).
22. Universal PCR primers: forward primer, 5′-AATGATACGGCGACCACCGAGAT-3′; reverse primer, 5′-CAAGCAGAAGACGGCATACGA-3′, included in the TruSeq Nano DNA LT Sample Prep Kits (Illumina), or can be purchased as custom oligonucleotides by companies such as Integrated DNA Technologies (Coralville, IA, USA).
23. Sodium bisulfite treatment kit, such as EZ DNA Methylation-Gold Kit (Zymo Research), or equivalent.
24. PfuTurbo Cx hotstart DNA polymerase (Agilent Technologies).
25. dATP, dTTP, dCTP, and dGTP: Separate dNTPs at concentrations of 100 mM or higher, or a mixture of all four dNTPs at a concentration of 25 mM each.
26. Magnetic rack for 1.5 mL microcentrifuge tubes, such as DynaMag-2 magnet (Thermo Fisher Scientific) or equivalent.
27. Illumina HiSeq2500 sequencer and corresponding reagents.

3 Methods

3.1 Sample Preparation

Highly pure genomic DNA (gDNA) is needed to carry out reduced representation bisulfite sequencing. Between 10 and 300 ng of gDNA at a concentration of 20–50 ng/μL in TE buffer is needed for each sample. However, using less than 60 ng of gDNA does not always yield reliable results. We recommend using 100 ng of gDNA per sample, and standardizing all samples to a concentration of 20 ng/μL in TE prior to endonuclease digestion. All the experiments using human and/or animal samples must comply with the pertinent institutional and government guidelines.

1. Isolate gDNA using a standard protocol or commercial kit, including steps to remove protein and RNA contamination. The presence of cellular proteins, particularly proteins that bind DNA, leads to problems with enzymatic reactions as well as sodium bisulfite treatment. In addition, RNA contamination can lead to inaccurate quantification of the DNA.
2. Quantify the concentration of gDNA using a NanoDrop spectrophotometer, or using a Qubit fluorometer. Ideally, DNA should be of high quality as indicated by a 260/280 ratio of 1.8 or higher.
3. (Optional) If a gDNA sample is of poor quality, it can be cleaned up using the Genomic DNA Clean and Concentrator kit in order to remove residual contaminants.
4. Dilute gDNA to between 20 and 50 ng/μL using TE buffer. Standardizing all DNA samples to one concentration will simplify subsequent steps.
5. Store DNA at 4 °C until needed.

3.2 Restriction Endonuclease Digestion

There are many restriction enzymes that will fragment genomic DNA in order to generate reduced representations. In this protocol we focus on MspI (C↓CGG), which cleaves genomic DNA into fragments that contain at least two CpG loci, one at each terminus, regardless of fragment length or CpG methylation status. Typically, 1 unit of MspI per μg of DNA is recommended for digestion. However, altered specificity (i.e., "star activity") has not been observed when genomic DNA is overdigested with up to 1000 units of MspI per μg of DNA [11]. Alternative protocols include digesting with TaqI (T↓CGA), or co-digestion with both MspI and TaqI.

1. In a nuclease-free 96-well plate, combine MspI enzyme (and/or other restriction endonucleases), NEB buffer 2, -nuclease-free water, and 10–300 ng of gDNA, according to Table 1. The total volume of each reaction should be 30 μL. In our experience, 100 ng of DNA input is optimal to generate a

Table 1
MspI Digestion

Component	Volume or quantity	Final concentration or amount
H_2O	Up to 21.5 μL	
NEB buffer 2 (10×)	3 μL	1×
MspI (20 U/μL)	0.5 μL	10 U
Genomic DNA	10–300 ng (in 5–26.5 μL)	0.33–10 ng/μL
Total	30 μL	

high-quality library for sequencing while conserving the sample. To perform multiple digests at once, a mastermix containing enzyme, buffer, and water can be prepared first, and then combined with each DNA sample in separate reaction wells.

2. Mix well by pipetting, and then seal the plate.
3. Centrifuge briefly to ensure liquid is at the bottom of the tube or well.
4. Incubate overnight (16–18 h) at 37 °C in a thermal cycler with the heated lid turned off (*see* **Note 1**).

3.3 End-Repairing and A-Tailing

Enzymatic digestion of genomic DNA yields fragments with overhangs at each end. In this step, an end repair reaction will remove the 3′ overhangs and fill in the 5′ gaps, thereby generating blunt ends. Then, a single-adenosine residue will be enzymatically added to each terminus of the end-repaired fragments. These A-tails will facilitate the ligation of methylated adapters in a subsequent step. It is not necessary to deactivate MspI or clean up the digested DNA prior to end-repairing and A-tailing.

1. Briefly centrifuge the plate containing MspI digested DNA to transfer all liquid to the bottom of the reaction wells.
2. Combine MspI digested DNA, NEBNext Ultra End Prep Enzyme Mix, NEBNext Ultra End Repair Reaction Buffer (10×), and nuclease-free water as shown in Table 2. The total volume of each reaction should be 65 μL. All the components can be combined in the well that already contains the 30 μL of digested DNA. If preparing multiple reactions, you can create a mastermix containing 110% of all the components except the digested DNA in a nuclease-free 1.5 mL microcentrifuge tube, and then add 35 μL of the mix to each 30 μL MspI digested DNA sample.
3. Mix well by pipetting.

Table 2
End repairing and A-tailing

Component	Volume per reaction (μL)
NEBNext ultra end prep enzyme mix	3
NEBNext ultra end repair reaction buffer (10×)	6.5
Nuclease-free H_2O	25.5
Digested DNA	30
Total	65

4. Centrifuge briefly to ensure the liquid is at the bottom of the wells, and to remove bubbles.
5. In a thermal cycler with the heated lid turned off (*see* **Note 1**) incubate at 20 °C for 30 min followed by 65 °C for 30 min.
6. Store at 4 °C until ready to begin adapter ligation.

3.4 Adapter Ligation

Indexed, methylated adapters are oligonucleotides that contain 5-methylcytosine in place of cytosine residues in order to maintain integrity during bisulfite sequencing. All of these adapters include universal sequences that enable PCR amplification of adapter-ligated DNA fragments. In addition, each of the different adapters contains a unique six-base index sequence, sometimes referred to as a barcode. In this step, a different adapter is ligated to A-tailed DNA fragments from each of the samples that will later be pooled and run in a single-sequencing lane. The unique index sequences will tag each of the different samples in a pool, allowing them to be distinguished bioinformatically after sequencing. Since there will only be a finite number of reads per sequencing lane, e.g., the HiSeq2500 allows for a maximum of 12 samples to be pooled (*see* **Note 2**). Different adapters and/or batches of adapters may perform differently (*see* **Note 3**).

1. Briefly centrifuge the plate containing MspI digested, end-repaired, A-tailed DNA to ensure all liquid is at the bottom of the wells.
2. Combine indexed methylated adapter, NEBNext Ultra Enhancer, and NEBNext Ultra T4 Ligase Master Mix to the 65 μL of MspI digested, end-repaired, A-tailed DNA according to Table 3. Use an adapter with a different index sequence for each of the samples that will be pooled and run in a single lane. These reagents can be combined in the well already containing DNA.
3. Mix well by pipetting.
4. Incubate at 20 °C for 15 min in an incubator or in a thermal cycler with the heated lid turned off (*see* **Note 1**).

Table 3
Adapter ligation

Component	Volume per reaction (μL)
Digested, end repaired, A-tailed DNA	65
Indexed methylated adapter	2.5
NEBNext ultra enhancer	1
NEBNext ultra T4 ligase master mix	15
Total	83.5

5. Add 3 μL of NEBNext Ultra USER enzyme to each reaction. Total reaction volume should now be 86.5 μL (*see* **Note 4**).
6. Incubate at 37 °C for 15 min in an incubator or a thermal cycler with the heated lid turned off (*see* **Note 1**).
7. Store at 4 °C until ready to perform fragment size selection.

3.5 Size Selection Using Magnetic Beads

A particular size range of adapter-ligated DNA fragments is obtained using a combination of negative and positive selections with magnetic beads. Depending on the research question, researchers will want to adjust this selection process to acquire adapter-ligated fragment sizes that correspond to their desired genomic context(s). The MspI restriction site contains a CpG locus; therefore, smaller fragments will be from DNA regions that are dense in CpG loci whereas larger fragments will be from DNA regions that are CpG sparse. The volume of beads relative to the volume of the DNA library determines the adapter-ligated fragment sizes that will bind to the beads. The magnetic beads preferentially bind to larger fragments before smaller fragments, therefore adding a larger volume of beads will result in the beads binding to smaller sized DNA fragments. Table 4 shows the recommended volumes for first and second bead selections to obtain specific size ranges from a 100 μL library volume [21]. However, we advocate first using a DNA ladder to experimentally test magnetic bead-based size selections in order to ensure that the desired sizes are obtained (*see* **Note 5**). In addition to (or in place of) size selection, this step will also remove any un-ligated adapters and residual dNTPs from the sample libraries. It is possible to carry out this step in multiple samples simultaneously using a multi-channel pipette.

1. Briefly centrifuge the plate containing the DNA libraries to ensure the liquid is at the bottom of the tube or well.
2. Resuspend the Agencourt AMPure XP magnetic beads by vortexing.

Table 4
Recommended magnetic bead volumes for DNA fragment size selection

Insert size, approximate	150 bp	200 bp	250 bp	300–400 bp	400–500 bp	500–700 bp
Library size (insert + adapter)	270 bp	320 bp	400 bp	400–500 bp	500–600 bp	600–800 bp
First bead selection	65 μL	55 μL	45 μL	40 μL	35 μL	30 μL
Second bead selection	25 μL	25 μL	25 μL	20 μL	15 μL	15 μL

3. Add 13.5 μL of nuclease-free water to each DNA library sample to bring the total volume up to 100 μL. If removal of large fragments is not desired, skip to **step 10**.
4. Add the first bead selection volume of resuspended Agencourt AMPure XP magnetic beads to each library as specified in Table 4, as or determined experimentally using a DNA ladder (*see* **Note 5**). These beads will bind to unwanted large DNA fragments.
5. Mix well by pipetting up and down at least 10 times. The solution should become homogeneous.
6. Incubate at room temperature for 5 min.
7. Centrifuge the plate briefly.
8. Place the plate into the DynaMag-96 side magnet.
9. Once the beads have separated from the supernatant, and the solution becomes clear (about 5–10 min), carefully remove the supernatant containing the desired smaller DNA fragments and transfer it to a new reaction well. Tilt the plate and the magnetic rack to avoid collecting any beads with the supernatant. Discard the beads containing unwanted larger DNA fragments.
10. Add the second bead selection volume of resuspended Agencourt AMPure XP magnetic beads to each DNA library as specified in Table 4, or as determined experimentally using a DNA ladder (*see* **Note 5**). These beads will bind to the desired DNA fragments. For clean-up only, without fragment size selection, add a volume of Agencourt AMPure XP magnetic beads equal to 1.8 times the volume of fragmented DNA (e.g., 180 μL of beads to 100 μL of DNA library).
11. Mix well by pipetting up and down at least ten times to create a homogeneous mixture.
12. Incubate at room temperature for 5 min.
13. Centrifuge the plate briefly.
14. Place the tube or plate into the DynaMag-96 side magnet.
15. Once the beads have separated from the supernatant and the solution becomes clear (about 5–10 min), remove and discard

the supernatant containing unwanted DNA fragments in addition to un-ligated adapters and residual dNTPs. Avoid disturbing the beads that now contain the desired DNA fragments. Tilting the plate and the magnetic rack will help avoid disturbing the beads.

16. Leave the tube or plate in the magnetic rack, and add 200 μL of 80% ethanol to wash the beads that are bound to desired DNA fragments. Incubate for at least 30 s at room temperature, then carefully remove and discard the supernatant without disturbing the beads. Repeat this step two more times for a total of three washes.
17. Allow the beads to air-dry with the plate uncovered and still in the magnetic rack for at least 10 min, until all moisture has evaporated. Cracks should develop along the surface of the beads, indicating that they are fully dry.
18. Remove the plate from the magnetic rack, and then add 35 μL of 10 mM Tris–HCl in order to elute the target DNA from the beads. Do not use TE, despite what is stated in the manufacturer's instructions, because the EDTA can interfere with subsequent PCR reactions by chelating magnesium. If a prognostic gel to determine the quality of a library is desired, use 40 μL for the elution so that an extra volume is available for PCR amplification prior to gel visualization.
19. Mix well by pipetting to create a homogeneous solution.
20. Place the plate into the DynaMag-96 side magnet in order to separate the beads from the liquid supernatant, which now contains the target DNA fragments that have been eluted from the beads.
21. Once the liquid becomes clear (about 5–10 min), carefully transfer the liquid supernatant to a fresh well. Be careful to avoid pipetting the beads along with the liquid. Tilting the plate and magnetic rack will help avoid disturbing the beads. The beads may now be discarded. Approximately 5 μL of liquid will be retained by the beads.
22. Store at 4 °C until needed.

3.6 (Optional) Diagnostic Gel to Assess Library Quality

To verify that the adapter is successfully ligated to the fragmented DNA and that the proper sized adapter-ligated fragments have been selected, a small amount (3 μL) of each RRBS library can be amplified by PCR using universal primers, which anneal to a sequence common to all of the indexed methylated adapters. The PCR product is then viewed on a 2% agarose gel with Ethidium Bromide using UV transillumination. Any PCR product that is visible on the gel will indicate that the adapters have ligated to the DNA fragments. Ideally, the product should appear as a smear that spans the desired adapter-ligated DNA fragment size range.

Table 5
Post ligation diagnostic PCR

Component	Volume per reaction (μL)	Final concentration or amount
Promega PCR MasterMix, 2×	12.5	1×
Primer mix, 2.5 μM each, 5 μM	5.5	1.1 μM
Nuclease-free water	4	
Total	22	
Combine 22 μL of the mixture with 3 μL library DNA		

1. Combine a mixture of forward and reverse universal PCR primers so that they are each at a concentration of 2.5 mM in nuclease-free water, creating a total primer concentration of 5 mM.
2. Create a PCR mixture containing Promega PCR Mastermix, the 5 mM primer mix from **step 1**, and nuclease-free water, according to Table 5.
3. In a nuclease-free 96-well plate, combine 22 μL of the mixture from **step 2** with 3 μL of library DNA.
4. Mix well by pipetting, then centrifuge briefly to eliminate bubbles, and ensure that the reaction is at the bottom of the well.
5. In a thermal cycler with the heated lid on, carry out PCR as follows. Step 1: 95 °C for 2 min. Step 2: 16 cycles of 95 °C for 30 s, 65 °C for 30 s, and 72 °C for 45 s. Step 3: 72 °C for 7 min. Step 4: hold at 4 °C until ready to run the gel.
6. Run the PCR products on a 2% agarose gel containing ethidium bromide at 200 V for 60 min and visualize the gel under UV transillumination (Fig. 2).

3.7 Pooling of DNA Libraries

Up to 12 DNA libraries can be pooled and run in a single-sequencing lane, provided that each is ligated to a methylated adapter containing a different six base index sequence. It is possible to simply combine the entire volume of the libraries into a pool, but differing concentrations can result in discrepancies in the number of sequencing reads between the libraries. Therefore, we recommend first quantifying the concentration of each library and then adjusting the volumes that are added to a pool accordingly. The goal is to add an equal amount of each library into the pool in order to improve the uniformity of sequencing reads among the libraries that are run in a single lane (*see* **Note 2**). However, it is not advisable to add less than 5 μL of any one library into a pool, even if it is significantly more concentrated than the other libraries in that pool. Likewise, we advise to set a minimum concentration

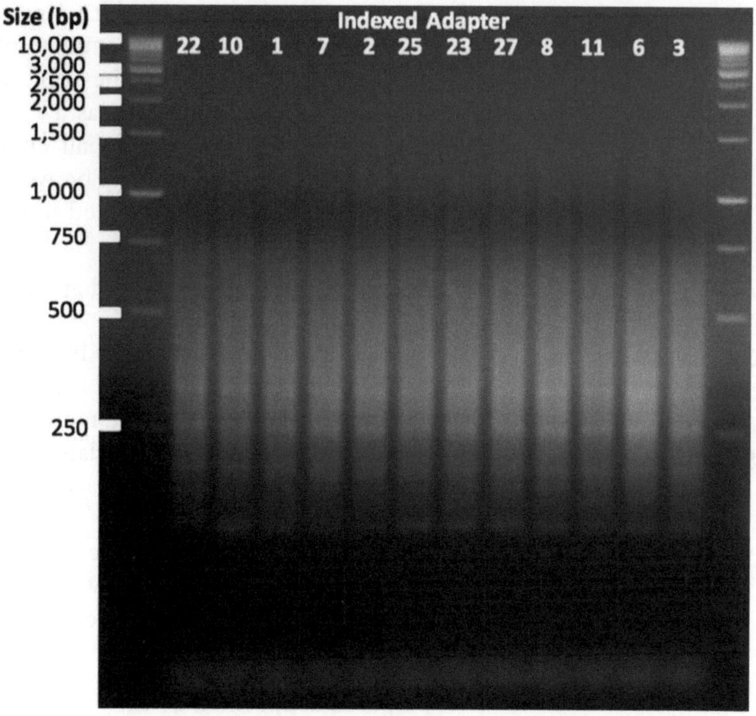

Fig. 2 Example of a diagnostic gel showing high-quality RRBS libraries. Twelve RRBS libraries were prepared starting with 200 ng of genomic DNA input from the same person. Each library was ligated to a different indexed adapter (indicated across the *top* of the figure), and then subjected to magnetic bead selection to eliminate DNA fragments smaller than 100 bp in size. Three microliters of the library was PCR amplified and then visualized via transillumination on a 2% agarose gel with ethidium bromide

below which the entire 30 μL volume of the library is added to the pool.

1. Quantify each library using a Qubit flourometer. It is possible to instead use a NanoDrop for this quantification, but Qubit will provide superior accuracy.
2. Calculate the volume of each library (up to 30 μL) that should be added such that an equal quantity of each library will be present in the final pool. However, for libraries that are significantly more concentrated than the others, a minimum volume of 5 μL should be added to the pool. Likewise, for libraries that are significantly less concentrated than the others, the entire 30 μL volume of the library should be used.
3. In a nuclease-free 1.5 μL microcentrifuge tube, combine the volumes of each library that were calculated in **step 2** of this section to create a pool containing up to 12 libraries.
4. Mix the pool well by pipetting.

5. To conserve sodium bisulfite treatment reagents, you may concentrate the pool by vacuum centrifugation, or by passive evaporation (i.e., leaving the tube open, but covered with a sterile, dust-free cloth such as a kimwipe at 4 °C for several days) until the volume is small enough for a single-bisulfite treatment reaction (generally between 20 and 50 μL). Alternatively, each pool can be divided into multiple sodium bisulfite treatments.
6. Pooled libraries can be stored at 4 °C until ready to proceed.

3.8 Treatment of Pooled DNA Libraries with Sodium Bisulfite

The pooled libraries should next be treated with sodium bisulfite using the DNA Methylation Gold Kit, or another preferred bisulfite conversion kit, according to the manufacturer's instructions. This treatment will convert unmethylated cytosine residues to uracil (which in turn become thymine during PCR) while methylated cytosine residues are protected and remain cytosine. Use the instructions from the manufacturer for specific details (e.g., reagent volumes), which may vary between different kits. In general, sodium bisulfite treatment involves the following steps:

1. Combine the pooled library DNA with a conversion reagent prepared as specified by the manufacturer.
2. Mix well and incubate as indicated by the manufacturer's instructions. Some kits have special indications for this incubation that pertain to Illumina sequencing.
3. After incubation, add the mixture to a column containing binding buffer, placed in a collection tube. Incubate at room temperature as indicated by the manufacturer's instructions.
4. Centrifuge as indicated by the manufacturer's instructions.
5. Wash the column with wash buffer containing ethanol, then centrifuge as indicated by the manufacturer's instructions.
6. Add desulfonation buffer to the column and incubate at room temperature for 15 min. Do not let this incubation go longer than the time recommended by the manufacturer. In our experience, it is best to minimize the length of this desulfonation step (*see* **Note 6**).
7. Centrifuge as indicated by the manufacturer's instructions.
8. Wash two more times with wash buffer containing ethanol, centrifuging after each wash. Add extra time to the last centrifugation to ensure that the entire volume of wash buffer has moved out of the column, and to aid in drying the column.
9. Dry the column by leaving the top open, covered with a sterile, dust-free cloth (e.g., a kimwipe) at room temperature for 5 or more minutes.

10. Elute the bisulfite-converted DNA using a total of 40 μL of elution buffer for each pool. It is best to add the elution buffer directly onto the column matrix. If pools were first concentrated and treated in a single column, perform two successive elutions using 20 μL of elution buffer. If a pool was divided into multiple conversions, add an equal amount of elution buffer for each elution such that the total volume used for the pool totals 40 μL, then combine these into a single tube.
11. Bisulfite-converted library pools should be stored at −20 °C temporarily or at −80 °C for up to 6 months. Prolonged storage (more than 6 months) is not recommended as bisulfite-converted DNA is prone to degradation.

3.9 Amplification of Bisulfite-Converted Library Pools

Prior to sequencing, the bisulfite-converted library pools should be PCR amplified using PfuTurbo Cx hotstart DNA polymerase. However, to reduce the effects of amplification bias (*see* **Note 7**) PCR should be performed using the fewest number of cycles that results in robust amplification. Thus, a small quantity of pooled library DNA should first be used to test whether amplification occurs after 13, 16, or 19 cycles. Three separate 10 μL PCRs, each using 1 μL of a pool, are run using the three different reaction conditions. Successful amplification is determined by visualizing the products under UV transillumination on a 2% agarose gel containing ethidium bromide. Then, the remainder of the pool is amplified using the minimum number of cycles, as determined from the small-scale tests.

1. Create a mix dATP, dTTP, dCTP, and dGTP such that each dNTP is at a concentration of 25 mM in nuclease-free water. Thus, the total concentration is 100 mM. This will be used in all subsequent PCR amplifications.
2. Combine forward and reverse universal PCR primers for Illumina sequencing so that they are each at a concentration of 2.5 mM in nuclease-free water. Thus, the total concentration is 5 mM. Mix well by pipetting. This will be used in all subsequent PCR amplifications.
3. To test the minimum number of PCR cycles required for the final amplification, create a Pfu PCR mastermix by combining the dNTP mix from **step 1**, the universal primer mix from **step 2**, 10× PCR buffer, PfuTurbo Cx hotstart DNA polymerase, and nuclease-free water according to Table 6 under the column for 10 μL reactions. Mix well by pipetting.
4. In three separate PCR plates or tubes, add 9 μL of the mastermix to 1 μL of the pool. Mix well by pipetting, then centrifuge briefly to eliminate bubbles, and ensure that the reactions are at the bottom of the wells or tubes.

Table 6
PfuTurbo PCR mastermix recipe

Component	Volume per 10 μL reaction (μL)	Volume per 200 μL reaction (μL)	Final concentration or amount
10× PCR buffer	1	20	1×
dNTP mix, 25 mM each, 100 mM total	0.1	2	0.25 mM each, 1 mM
Primer mix, 2.5 μM each, 5 μM total	1.6	32	0.4 μM each, 0.8 μM
Pfu turbo cx Hotstart DNA polymerase, 2.5 U/μL	0.2	4	2.5 U
Nuclease-free water	6.1	112	
Total volume per reaction	9	170	

For 10 μL reaction: combine 9 μL of mastermix with 1 μL of pooled library DNA
For 200 μL reaction: combine 170 μL of mastermix with 30 μL of pooled library DNA

5. In a thermal cycler with the heated lid on, carry out PCR as follows. Step 1: 95 °C for 2 min. Step 2: 13, 16 or 19 cycles of 95 °C for 30 s, 65 °C for 30 s, and 72 °C for 60 s. Step 3: 72 °C for 7 min. Step 4: hold at 4 °C until ready to run the gel.
6. Run the PCR products on a 2% agarose gel containing ethidium bromide at 200 V for 60 min and visualize the gel under UV transillumination.
7. Determine the least number of cycles required for amplification of the pooled libraries, as indicated by presence or absence of PCR product under each of the different reaction conditions. If need be, run additional PCR reactions using 1 μL of the pool to fine tune the correct minimum number of cycles (Fig. 3).
8. Create enough Pfu PCR mastermix for all pools according to Table 6 using the column for 200 μL reactions.
9. In a nuclease-free 1.5 mL microcentrifuge tube, combine 170 μL of the Pfu PCR mastermix with 30 μL of the pool to create a 200 μL mixture and mix well by pipetting.
10. Divide each 200 μL mixture containing a pool into four 50 μL (or eight 25 μL) PCR reactions in separate wells of a nuclease-free 96-well plate.
11. Seal the plate and then centrifuge briefly to ensure that liquid is at the bottom of the reaction wells, and to eliminate bubbles in the mixture.
12. In a thermocycler, amplify the pooled libraries using the same Pfu conditions as the test PCR runs, with the number of cycles set to the minimum that was determined from the tests. Thus, the reaction conditions are as follows. Step 1: 95 °C for 2 min.

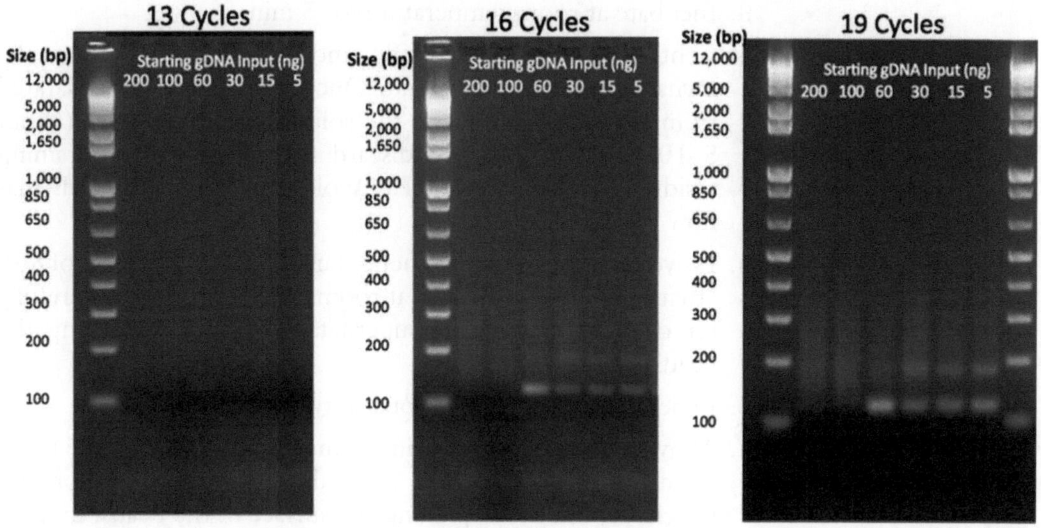

Fig. 3 Example of final PCR amplification cycle number optimization. One microliter of pooled, sodium bisulfite-treated RRBS libraries generated from different amounts of starting DNA input (indicated across the *top* of each gel) were amplified in a total volume of 10 μL using 13, 16, or 19 cycles to determine the minimum number of cycles required for sequencing. Products were visualized via transillumination on a 2% agarose gel with ethidium bromide. In this experiment, PCR product appeared at 16 cycles, indicating that the minimum (ideal) number of cycles to use for final amplification of the bisulfite-converted RRBS pools is 16 cycles. Satellite bands are evident for PCR products for pools of RRBS libraries generated from 60 ng of DNA or less, and become more pronounced after 19 cycles

Step 2: 13, 16 or 19 cycles of 95 °C for 30 s, 65 °C for 30 s, and 72 °C for 60 s. Step 3: 72 °C for 7 min. Step 4: hold at 4 °C.

13. Centrifuge the plate briefly to ensure that all liquid is at the bottom of the wells.
14. In a nuclease-free 1.5 mL microcentrifuge tube, combine the four (or eight) PCR products for the pool back into one 200 μL volume.
15. Store at −20 °C until ready to proceed.

3.10 Bead Cleanup of Amplified Library Pool

A final AMPure XP magnetic bead cleanup needs to be carried out in order to remove any residual dNTPs, primers, and other contaminants from the bisulfite converted, PCR amplified pools prior to sequencing. To avoid size selection, use a volume of beads that is 1.8 times the volume of the PCR product.

1. Vortex AMPure XP beads to resuspend.
2. Add at 360 μL of AMPure XP beads to the 200 μL volume of bisulfite converted, PCR amplified library pool, which should already be in a 1.5 mL microcentrifuge tube. Mix well by pipetting up and down at least 10 times until the mixture becomes homogeneous.

3. Incubate at room temperature for 5 min.
4. Centrifuge the tube briefly and then place it onto the DynaMag-2 magnetic stand. Once the beads have separated from the supernatant and the solution becomes clear (about 5–10 min), remove and discard the supernatant containing residual primers and dNTPs. Avoid disturbing the beads that now contain the library.
5. Leave the tube in the magnetic stand, and add 1000 μL of 80% ethanol. Incubate for 30 s at room temperature, then carefully remove and discard the supernatant without disturbing the beads.
6. Repeat **step 5** once for a total of two washes.
7. Allow the beads to air-dry in the magnetic stand with the tubes open for at least 10 min, until all moisture has evaporated. Cracks should develop along the surface of the beads, indicating that they are fully dry.
8. Remove the tube from the magnetic stand and add 40 μL of 10 mM Tris–HCl in order to elute the library from the beads. Do not use TE, despite what is stated in the manufacturer's instructions, because the EDTA can interfere with subsequent reactions by chelating magnesium.
9. Mix well by pipetting up and down at least 10 times to create a homogeneous solution. Be sure to run the liquid down the side of the tube over the beads to resuspend them into the solution.
10. Centrifuge briefly, then place the tube back into the DynaMag-2 magnetic stand in order to separate the beads from the liquid supernatant. The supernatant will contain the eluted library.
11. Once the liquid becomes clear (about 5–10 min), carefully transfer the liquid supernatant to a fresh nuclease-free tube or well in a 96-well plate. Be careful to avoid pipetting the beads along with the liquid; it is imperative that no beads end up in the final library. The beads may now be discarded. Some of the liquid will be retained by the beads and some may remain in the tube, but the final pooled library volume should be at least 30 μL.
12. Store at −20 °C until ready to sequence.

3.11 Sequencing, Base Calling, and Alignment

Sequencing should be performed by experienced technicians (generally at a fee-for-service core facility) on an Illumina HiSeq using Illumina reagents according to the manufacturer's instructions. We generally employ 75-base single-end sequencing reads, though it is possible to use shorter reads, or to perform pair-end sequencing, depending on the research questions and the desired genomic context. Some sequencing facilities may be able to perform dark sequencing, which delays imaging and cluster localization until the fourth sequencing cycle, past the bias introduced by the MspI cut site [10]. However, we have had difficulty accomplishing dark sequencing using the HiSeq 2500, and instead account for this bias bioinformatically. To calculate the dye matrix for base calling, a standard (non-RRBS) library generated from the bacteriophage PhiX 174 should be run in a lane on the flow cell and/or spiked into each RRBS flow cell lane. RRBS sequencing requires special alignment and data processing, the specifics of which will depend on the research questions, though it is always necessary to remove the adapters and barcodes during the alignment step [10, 11]. We recommend working with a bioinformatics core team with powerful computing resources to perform base calling and alignment. There are a variety of tools that can be used to perform base calling and alignment to the genome of the organism under investigation, such as Bismark [13], MAQ [14], RRBSMap [15], MethylSig [16], SAAP-RRBS [17], RRBS-analyzer [18], PASH [19], and MOABS [20]. Our group uses Bismark to perform base calling and alignment of multiplexed RRBS sequencing results for human blood and tissue samples, but other tools may be preferable for different interests.

4 Notes

1. When using a thermal cycler for certain enzymatic reactions, it is critical that the heated lid is turned off in order to avoid raising the temperature and prematurely deactivating the enzymatic activity. In this protocol, we have found that turning off the heated lid is necessary for MspI digestion reactions, end-repairing and A-tailing reactions, and adapter ligation reactions. In some instances, the manufacturer's instructions may suggest keeping the heated lid on. However, we have found that these recommendations are incorrect and should be disregarded. Instead, it is best to turn off the heated lid to ensure a uniform temperature distribution throughout the reaction, and to avoid deactivating the enzymes that have relatively low (e.g., 65 °C) kill temperatures.
2. The HiSeq 2500 sequencing platform from Illumina uses flowcells that are each capable of performing two billion single-end reads. These reads are divided equally between the eight lanes in the flowcell, resulting in 250 million reads per sequencing lane. These reads are, in turn, divided roughly evenly among the adapter-ligated DNA fragments from the libraries in each lane. If the same amount of each library is added into a pool, the reads will be allocated fairly equally among the libraries. Thus, adding more libraries to a pool will reduce the number of reads per library. In general, ten million reads per library is

desired. In addition to the RRBS libraries, a standard (non-RRBS) library generated from the bacteriophage PhiX 174 is often spiked into each sequencing lane to facilitate the calculation of the dye matrix for base calling. Theoretically, adding 12 libraries per lane without Phi-X spiked in will yield 20 million reads per library, but in reality it is usually less due to undetermined reads and other technical issues. The Phi-X spike in will further reduce the number of reads per library, resulting in 12 libraries per lane yielding the desired ten million reads per library.

3. There seem to be differences in the ligation efficiencies of the 24 different indexed, methylated TruSeq adapters from Illumina, which are the adapters that we have used in this protocol. We have empirically tested all 24 different indexed, methylated TruSeq adapters from Illumina and found that the following 12 adapters worked best in concert for multiplexed RRBS sequencing: 1, 2, 3, 6, 7, 8, 10, 11, 22, 23, 25, and 27. Adapter number 9 tends to be the most problematic, possibly due to dimerization, and its use should be avoided. In addition, adapter number 12 resulted in downstream bioinformatic problems and should be avoided. We found that adapter number 4 significantly outperformed the other adapters, and we therefore excluded it to avoid disproportionate sequencing reads for the sample to which it is ligated. It is also possible that there are batch-specific differences in the adapters. Testing efficacy of adapter ligation with each new batch of indexed methylated adapters prior to using them with valuable samples may be beneficial. To do this, obtain at least 2.4 μg of the same non-valuable gDNA sample and divide it into 24 aliquots containing 100 ng each. Carry out MspI digestion, and then end repairing and A-tailing reactions as if these were 24 different samples. Next, ligate a different indexed methylated adapter to each of the 24 samples. Finally, amplify by PCR using universal primers and run a diagnostic gel to determine the ligation efficiency. A diagnostic qPCR with SYBR green, designed to amplify a specific size of adapter-ligated DNA fragments, will also aid in determining differences in the quality of libraries that result from different adapters.

4. Adding the USER enzyme may be unnecessary for standard Illumina TruSeq adapters, as the USER enzyme is meant to remove the hairpin structure from New England Biolab's methylated adapter. The USER enzyme is very viscous and tends to run out before other reagents. Using indexed methylated TruSeq adapters from Illumina, we tested the protocol with and without USER enzyme (compensating for water instead, and still including the 15 min 37 °C incubation) and did not see any difference in library quality.

5. There can be differences between different batches of magnetic beads, and lab specific differences in the sizes of fragments that are obtained using magnetic bead-based selection. It is best to verify that the proper sized DNA fragments are selected before performing the final selection using a valuable sample library. To do this, use a DNA ladder (such as the Thermo Scientific GeneRuler 1 kb Plus DNA Ladder 75–20,000 bp) as your DNA input. Perform multiple size selections in parallel using the protocol described in Subheading 3.5, but test different relative bead to sample volumes for both the negative and positive selections. Then, run the size selected DNA ladders, as well as an unaltered ladder, on a 2% agarose gel containing ethidium bromide at 200 V for 60 min (or longer) and visualize under UV transillumination. The genomic regions that are obtained using this protocol will depend on the enzyme that is used in the digest, as the size of a fragment will be determined by the proximity of the restriction sites. In addition, the adapters add length to the fragments. In the case of the indexed, methylated TruSeq adapters for which this protocol was designed, the adapters will add 63 bp to each end of the fragment, totaling 126 bp of additional length. It is possible to perform in silico experiments to determine what genomic features are covered in a library, for example most promoter sequences and CpG islands will be represented in MspI digested fragments of 40–220 bp in size (approximately 160–380 bp including the adapters) [11, 22, 23].
6. Sodium bisulfite deaminates unmethylated cytosines, and subsequent desulfonation at alkaline pH converts these deaminated cytosine residues into uracil by removing HSO_3^- [24]. However, the harsh desulfonation reagent employed in sodium bisulfite treatment can potentially damage DNA during a prolonged exposure, degrading the quality of a DNA library. Thus, it is best to minimize the length of the desulfonation step.
7. When we PCR amplify RRBS libraries generated from human samples, we typically see discrete satellite bands that appear around 150, 175, and 225 bp on an agarose gel. These satellite bands become more pronounced with increasing number of PCR cycles. It is likely that they are a result of sequences that occur frequently in the human genome, such as repetitive elements. This banding pattern has been observed by others, as have characteristic satellite bands that are distinct for other organisms [11]. Such sequences have the potential to overshadow other sequences by using disproportionately more sequencing reads if too many PCR cycles are employed in the final amplification.

Acknowledgments

The authors would like to acknowledge Alexander Meissner, Juan Carmona, Benedetta Izzi, and Alexandra Binder for their significant contributions to the development and optimization of this protocol. Dr. Accomando was supported by Training Grant T32CA09001 in Cancer Epidemiology from the National Cancer Institute, National Institutes of Health. Dr. Michels was supported in part by research grant R01CA158313 from the National Cancer Institute, National Institutes of Health.

References

1. Berger SL, Kouzarides T, Shiekhattar R et al (2009) An operational definition of epigenetics. Genes Dev 23:781–783
2. Zaidi SK, Young DW, Montecino M et al (2011) Bookmarking the genome: maintenance of epigenetic information. J Biol Chem 286:18355–18361
3. Khavari DA, Sen GL, Rinn JL (2010) DNA methylation and epigenetic control of cellular differentiation. Cell Cycle 9:3880–3883
4. Shi H, Wang MX, Caldwell CW (2007) CpG islands: their potential as biomarkers for cancer. Expert Rev Mol Diagn 7:519–531
5. Lande-Diner L, Cedar H (2005) Silence of the genes--mechanisms of long-term repression. Nat Rev Genet 6:648–654
6. Miranda TB, Jones PA (2007) DNA methylation: the nuts and bolts of repression. J Cell Physiol 213:384–390
7. Jones PA (2012) Functions of DNA methylation: islands, start sites, gene bodies and beyond. Nat Rev Genet 13:484–492
8. Bock C, Tomazou EM, Brinkman AB et al (2010) Quantitative comparison of genome-wide DNA methylation mapping technologies. Nat Biotechnol 28:1106–1114
9. Smallwood SA, Kelsey G (2012) Genome-wide analysis of DNA methylation in low cell numbers by reduced representation bisulfite sequencing. Methods Mol Biol 925:187–197
10. Boyle P, Clement K, Gu H et al (2012) Gel-free multiplexed reduced representation bisulfite sequencing for large-scale DNA methylation profiling. Genome Biol 13:R92
11. Gu H, Smith ZD, Bock C et al (2011) Preparation of reduced representation bisulfite sequencing libraries for genome-scale DNA methylation profiling. Nat Protoc 6:468–481
12. Meissner A, Gnirke A, Bell GW et al (2005) Reduced representation bisulfite sequencing for comparative high-resolution DNA methylation analysis. Nucleic Acids Res 33:5868–5877
13. Krueger F, Andrews SR (2011) Bismark: a flexible aligner and methylation caller for Bisulfite-Seq applications. Bioinformatics 27:1571–1572
14. Li H, Ruan J, Durbin R (2008) Mapping short DNA sequencing reads and calling variants using mapping quality scores. Genome Res 18:1851–1858
15. Xi Y, Bock C, Muller F et al (2012) RRBSMAP: a fast, accurate and user-friendly alignment tool for reduced representation bisulfite sequencing. Bioinformatics 28:430–432
16. Park Y, Figueroa ME, Rozek LS et al (2014) MethylSig: a whole genome DNA methylation analysis pipeline. Bioinformatics 30:2414–2422
17. Sun Z, Baheti S, Middha S et al (2012) SAAP-RRBS: streamlined analysis and annotation pipeline for reduced representation bisulfite sequencing. Bioinformatics 28:2180–2181
18. Wang T, Liu Q, Li X et al (2013) RRBS-analyser: a comprehensive web server for reduced representation bisulfite sequencing data analysis. Hum Mutat 34:1606–1610
19. Coarfa C, Yu F, Miller CA et al (2010) Pash 3.0: a versatile software package for read mapping and integrative analysis of genomic and epigenomic variation using massively parallel DNA sequencing. BMC Bioinformatics 11:572
20. Sun D, Xi Y, Rodriguez B et al (2014) MOABS: model based analysis of bisulfite sequencing data. Genome Biol 15:R38

21. NEB Next Ultra DNA Library Prep Kit for Illumina: Instruction Manual (2013). New England Biolabs, Inc., Ipswich, MA
22. Meissner A, Mikkelsen TS, Gu H et al (2008) Genome-scale DNA methylation maps of pluripotent and differentiated cells. Nature 454:766–770
23. Gu H, Bock C, Mikkelsen TS et al (2010) Genome-scale DNA methylation mapping of clinical samples at single-nucleotide resolution. Nat Methods 7:133–136
24. Darst RP, Pardo CE, Ai L et al (2010) Bisulfite sequencing of DNA. Curr Protoc Mol Biol Chapter 7:Unit 7.9.1–Unit 7.917

Chapter 9

Low Input Whole-Genome Bisulfite Sequencing Using a Post-Bisulfite Adapter Tagging Approach

Julian R. Peat and Sébastien A. Smallwood

Abstract

The epigenetic mark 5-methylcytosine confers heritable regulation of gene expression that can be dynamically modulated during transitions in cell fate. With the development of high-throughput sequencing technologies, it is now possible to obtain comprehensive genome-wide maps of the mammalian DNA methylation landscape, but the application of these techniques to limited material remains challenging. Here, we present an optimized protocol to perform whole-genome bisulfite sequencing on low inputs (100–5000 somatic cells) using a post-bisulfite adapter tagging approach. In this strategy, bisulfite treatment is performed prior to library generation in order to both convert unmethylated cytosines and fragment DNA to an appropriate size. Then sequencing adapters are added by complementary strand synthesis using random tetramer priming, and libraries are subsequently amplified by PCR.

Key words DNA methylation, High-throughput sequencing, Bisulfite sequencing, Low input, Epigenetics

1 Introduction

Methylation of cytosine bases is a critical epigenetic modification that directs gene expression in diverse biological contexts, including cell fate specification and reprogramming, parental imprinting, repression of repetitive elements, and X-chromosome inactivation [1–5]. In light of this regulatory function, a range of techniques has been developed to detect and quantify the presence of 5mC in DNA. In combination with high-throughput sequencing, approaches such as methylated DNA immunoprecipitation (MeDIP-Seq, *see* also Chapter 12) and reduced representation bisulfite sequencing (RRBS, *see* also Chapter 8) can generate global methylation profiles [6–9]; however, these techniques provide data that is either low resolution and relative rather than absolute (MeDIP-Seq), or limited to certain genomic features (RRBS, i.e., CpG-rich regions). Whole-genome bisulfite sequencing (WGBS, *see* also Chapter 5) is currently the only method able to interrogate

Jörg Tost (ed.), *DNA Methylation Protocols*, Methods in Molecular Biology, vol. 1708,
https://doi.org/10.1007/978-1-4939-7481-8_9, © Springer Science+Business Media, LLC 2018

absolute methylation status at base resolution across the entire genome.

In order to generate a suitable length for Illumina high-throughput sequencing, DNA for WGBS is subjected to fragmentation, usually by sonication. The ends of these fragments must then be repaired, and adaptors required for PCR amplification and binding to the sequencer's flow cell are then ligated. Unmethylated cytosines are then converted to uracil by bisulfite treatment, and the library amplified by PCR to a concentration adequate for sequencing. However, the bisulfite treatment used for cytosine conversion is a potent inducer of DNA degradation, so a large fraction of informative adaptor-tagged fragments will be cleaved. This limitation has hindered the application of WGBS to nonabundant cell-types, which include many systems with biologically important methylation dynamics such as the early embryo and gametes.

A recently described strategy, post-bisulfite adaptor tagging (PBAT, *see* also Chapter 7), circumvents this issue with a simple innovation: bisulfite treatment is used to both convert cytosines and fragment DNA at the start of the protocol [10]. This combination eliminates the need for fragmentation by sonication, and avoids degrading adaptor-tagged sequences during bisulfite treatment. Since bisulfite-treated DNA is single-stranded, adaptors cannot be ligated in the same manner as standard protocols; instead, they are added by two rounds of complementary strand synthesis using random tetramer priming. This strategy has permitted WGBS libraries to be prepared from biologically interesting rare cell-types including mouse oocytes [11, 12].

Here, we describe an enhanced PBAT-based protocol that allows the generation of high-quality WGBS libraries from as few as 100 somatic cells (600 pg DNA equivalent). We significantly optimized the original protocol by modifying the conditions of bisulfite treatment and complementary strand synthesis to improve DNA recovery. In addition, we introduced an Exonuclease I step that prevents build-up of adapter artifacts. It also improves accuracy of methylation calls by limiting bias generated by the same bisulfite-treated fragment being tagged at multiple positions. Finally, our protocol includes library amplification by PCR in order to generate a sufficiently concentrated library from extremely low inputs, allowing visualization for quality control and the use of a classic Illumina pipeline for sequencing. Using this protocol, we recently generated DNA methylome from a limited number of mouse oocytes and zygotes [13, 14], allowing us for example to investigate the contribution of TET3 in the epigenetic reprogramming triggered by fertilization.

2 Materials

1. 0.5 mL DNA LoBind tubes (e.g., Eppendorf).
2. Low-binding pipette tips (RPT, Starlab).
3. PCR tubes with individual caps.
4. Magnets for 0.2 mL PCR tubes and 0.5 mL tubes.
5. PCR thermocycler.
6. Fluorometer.
7. Bioanalyzer (Agilent Technologies).
8. 1 M Tris–HCl, pH 8.0 (UltraPure).
9. 10% SDS.
10. 10 mM Tris–HCl, pH 8.0 (equivalent to "EB" buffer).
11. H_2O (molecular biology grade).
12. PCR-Grade dNTPs.
13. Quant-iT PicoGreen dsDNA Assay Kit (Invitrogen).
14. Proteinase K (0.8 U/μL).
15. Lambda DNA.
16. Imprint DNA Modification Kit (Sigma-Aldrich).
17. Purelink PCR Micro Kit (Life Technologies).
18. BD buffer from the EpiTect Bisulfite Kit (Qiagen).
19. NEB Buffer 2 (New England Biolabs).
20. Klenow Fragment (3′ → 5′ exo-) (50 U/μL).
21. Exonuclease I (20 U/μL).
22. AMPure XP (Agencourt).
23. M-280 streptavidin Dynabeads (Life Technologies).
24. Binding and washing (B&W) Buffer (2×): 10 mM Tris–HCl, 1 mM EDTA, 2 M NaCl, pH 7.5.
25. HiFi HotStart DNA Polymerase (KAPA Biosystems).
26. Library Quantification Kit—Illumina/Universal (KAPA Biosystems).
27. PBAT-F oligo: 5′biotin-ACACTCTTTCCCTACACGACGCTCTTCCGATCTNNNN (where N are random nucleotides).
28. PBAT-R oligo: 5′CGGTCTCGGCATTCCTGCTGAACCGCTCTTCCGATCTNNNN (where N are random nucleotides).
29. Illumina Library PCR primer PE1.0: 5′AATGATACGGCGACCACCGAGATCTACACTCTTTCCCTACACGACGCTCTTCCGATC*T.

30. iPCR-tagged PE2.0: 5′CAAGCAGAAGACGGCATACGAGATXXXXXXXXGAGATCGGTCTCGGCATTCCTGCTGAACCGCTCTTCCGAT*C (where XXXXXXXX are unique sequences (tags); for more details *see* ref. 15).

3 Methods

This protocol describes the preparation of WGBS libraries using a PBAT approach from 100 to 5000 somatic cells (0.6–30 ng DNA; *see* **Note 1**). It is recommended to perform a negative control (no DNA) in parallel to detect potential contaminations or other artifacts. For library generation from a few hundred cells, a positive control (10–50 ng of genomic DNA) can also be performed. Bench, pipettes, and racks used should be thoroughly cleaned with DNA decontamination solution. It is recommended to use low-DNA-binding plastic ware.

3.1 Cell Lysis and DNA Purification

This approach results in no loss of material associated with DNA purification, but in a lower conversion efficiency, albeit high enough for most analysis (>95%).

3.1.1 Option 1: Perform Bisulfite Conversion Directly on Cell Lysate (See Note 2)

1. Lyse cells in 10 mM Tris–HCl, pH 8.0, 0.6% SDS, 0.5 mM EDTA buffer, with 1 μL proteinase K to give a final volume of 24 μL.
2. Incubate at 55 °C for 5 min, followed by 55 min at 37 °C.
3. Unmethylated lambda DNA (1/5000 final) may be spiked in to assess bisulfite conversion efficiency (*see* **Note 3**).
4. Proceed to Subheading 3.2.

3.1.2 Option 2: DNA Purification Prior to Bisulfite Conversion

This option can be used when over 1000 cells are being processed.

1. Use the QIAamp DNA microkit according to the manufacturer's protocol for tissues. Elute the DNA using 25 μL of 10 mM Tris–HCl, pH 8.0.
2. Quantify DNA using the Quant-iT kit according to the manufacturer's instructions (*see* **Note 4**).
3. Unmethylated lambda DNA (1/5000 final) may be spiked in to assess bisulfite conversion efficiency (*see* **Note 3**).
4. Proceed to Subheading 3.2.

3.2 Bisulfite Conversion

1. Perform bisulfite treatment on 24 μL of lysate or purified DNA from Subheading 3.1 using the Imprint DNA Modification Kit "two-step" protocol with the following modification: after the addition of the DNA Modification solution, incubate at 65 °C for 90 min, 95 °C for 3 min, and then 65 °C for 30 min. Pause at 4 °C (*see* **Note 5**).

2. Purify DNA with the PureLink PCR Micro column according to the manufacturer's protocol (*see* **Note 6**). After binding, desulfonate DNA on-column by incubation with the BD Buffer from the EpiTect Bisulfite Kit for 8 min at RT. Elute DNA in 20 μL of 10 mM Tris–HCl, pH 8.0.

3.3 Oligo-I Tagging

1. To 19.5 μL of bisulfite-converted DNA (Subheading 3.2), add 1 μL dNTPs (10 mM stock), 1 μL of PBAT-F (10 μM stock), and 2.5 μL of 10× NEB2 buffer.
2. Incubate at 65 °C in a PCR cycler for precisely 3 min and place the samples on ice immediately.
3. Add 1 μL of Klenow Fragment 3′ → 5′ exo- and mix by pipetting.
4. Incubate for 5 min at 4 °C then raise to 37 °C at a rate of 1 °C every 15 s, and incubate at 37 °C for 90 min. Pause at 4 °C.

3.4 Exonuclease Treatment and Purification

1. Add 1 μL of Exonuclease I directly into the reaction mix from Subheading 3.3.
2. Incubate at 37 °C for 60 min. Pause at 4 °C.
3. Dilute the reaction by adding 75 μL of 10 mM Tris–HCl, pH 8.0 and transfer into new 0.5 mL tubes (*see* **Note 7**).
4. Purify using AMPure XP beads according to the manufacturer's instructions with a bead ratio of 0.8× (i.e., add 80 μL of beads).
5. Elute from beads in 50 μL of 10 mM Tris–HCl, pH 8.0 and transfer to new 0.5 mL tubes.

3.5 Capture of the Tagged DNA Strands

1. Prepare 20 μL of M-280 Streptavidin Dynabeads per sample by washing them twice with 2× B&W buffer according to the supplier's protocol. Resuspend beads in 50 μL of 2× B&W buffer per sample.
2. Add beads to the sample from Subheading 3.4 and incubate for 30 min at RT on a rotating wheel.
3. Place the tubes on a magnet and wash the beads containing the tagged DNA strands twice with 100 μL of freshly prepared 0.1 M NaOH, followed by two washes with 100 μL of 10 mM Tris–HCl, pH 8.0.

3.6 Oligo-II Tagging

1. Remove all liquid from the beads and resuspend them with the following enzymatic reaction mix (per sample): 2 μL dNTPs (10 mM stock), 2 μL of PBAT-R oligos (10 μM stock), 5 μL of 10× NEB2 buffer, and 38 μL of H_2O.
2. Transfer the beads and reaction mix into new PCR tubes.
3. Incubate at 95 °C in a PCR cycler for 45 s exactly and place samples on ice immediately.

4. Add 2 μL of Klenow Fragment 3′ → 5′ exo- and mix by pipetting.
5. Incubate for 5 min at 4 °C and then raise to 37 °C at a rate of 1 °C every 15 s, and incubate at 37 °C for 90 min. Pause at 4 °C.

3.7 Amplification of the Tagged Fragments

1. Place the tubes on a PCR magnet.
2. Remove the reaction mix and wash the beads twice with 50 μL of 10 mM Tris–HCl, pH 8.0.
3. Remove the 10 mM Tris–HCl, pH 8.0 buffer and resuspend in the following the PCR reaction mix (per sample): 1 μL dNTPs (10 mM stock), 10 μL of 5× Fidelity buffer, 1 μL of PE1.0 oligo (10 μM stock), 1 μL of iPCR-tagged PE2.0 (10 μM stock), 1 μL of KAPA HiFi Hotstart DNA polymerase, and 36 μL of H_2O.
4. Place the reaction in a PCR cycler and incubate at 95 °C for 2 min, followed by *N* cycles of 94 °C for 80 s, 65 °C for 30 s, 72 °C for 30 s, and a final elongation at 72 °C for 3 min. Pause at 4 °C. *See* **Note 8** for guidance on determining the number of PCR cycles.

3.8 Library Purification

1. Place the samples from Subheading 3.7 on a magnet.
2. Transfer the supernatant into 0.5 mL tubes.
3. Dilute the reaction by adding 50 μL of 10 mM Tris–HCl, pH 8.0.
4. Purify using AMPure XP beads according to the manufacturer's instructions with a bead ratio of 0.8× (i.e., add 80 μL of beads).
5. Elute the library in 15 μL of 10 mM Tris–HCl, pH 8.0.

3.9 Quality Control

Prior to sequencing, perform quality control using the Bioanalyzer platform according to the manufacturer's instructions (1 μL required). PBAT libraries are generally characterized by a unique wide peak (~300–800 bp) (*see* Fig. 1). In addition, use 1 μL for quantification by qPCR using the Library Quantification Kit (KAPA Biosystems) according to the manufacturer's protocol.

3.10 Sequencing and Methylation Calls

Libraries can be sequenced on the HiSeq 1000/2000/2500 platforms (Illumina). We recommend using 100 bp single-read sequencing kits. Using paired-end sequencing is also an option, but some hybrid molecules can be formed during the PBAT protocol, resulting in a relatively lower mapping efficiency. In addition, variability exists regarding the fragments size distribution (Fig. 1) as assessed on the Bioanalyzer and therefore 100 bp paired-end sequencing can be in some cases an "overkill." For subsequent

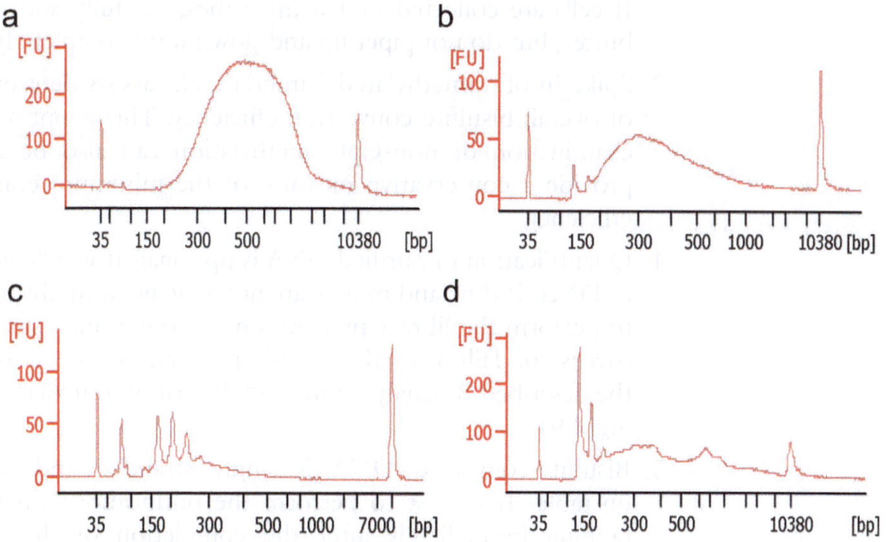

Fig. 1 Quality assessment of PBAT libraries using Bioanalyzer. (**a**) Typical Bioanalyzer profile of good quality and successfully sequenced PBAT libraries generated with approx. 1000 somatic cells. (**b**) Typical Bioanalyzer profile of good quality and successfully sequenced PBAT libraries generated with approx. 100 somatic cells. Note the size distribution shift toward shorter fragments compared to (**a**) due to higher DNA fragmentation associated with bisulfite treatment for lower cell input. (**c**, **d**) Examples of Bioanalyzer profiles of poor quality PBAT libraries, with peaks corresponding to adapters artifacts, suggesting that the ratio adapters/DNA is inaccurate due to a loss of DNA at the cell collection or cell lysis or bisulfite purification steps. It is not recommended to sequence this library

bioinformatics analysis, we recommend that sequencing quality is first assessed using "FastQC" and poor quality sequences be trimmed using "Trim Galore!". In addition, the first 4 bp corresponding to the Oligo random nucleotides can be trimmed. Subsequently, mapping and methylation calls can be performed using "Bismark" [16], which has a specific PBAT option. These three tools can be freely downloaded at the Babraham Bioinformatics website (http://www.bioinformatics.babraham.ac.uk/projects/).

4 Notes

1. The main limitation of our protocol arises from the use of a library amplification step, which increases the PCR duplication level of the sequenced libraries. A useful strategy to circumvent this is to generate replicate libraries in parallel, which are pooled together before sequencing. For example, if 300 cells are being processed, better results will be obtained if three libraries of 100 cells are generated in parallel and pooled.
2. To avoid losing material during the lysis step, cells should be collected in PCR tubes and lysis performed in a PCR cycler.

If cells are collected in 1.5 mL tubes, carefully add the lysis buffer, but do not pipet up and down until complete lysis.

3. Spike-in of unmethylated Lambda DNA assists determination of overall bisulfite conversion efficiency. This is optional, and examination of non-CpG methylation can also be used to provide a conservative measure of the minimum conversion efficiency.
4. Quantification of purified DNA is optional. It will "cost" 1 μL of DNA. If time and money are not an issue, it might be better to perform the library preparation regardless and estimate the success or failure of the sample preparation afterward using the described Bioanalyzer and qPCR quality control (Subheading 3.9).
5. Bisulfite-converted DNA is single stranded and relatively unstable. It is best to perform the purification and Oligo I tagging immediately after the completion of the bisulfite treatment.
6. Better results are obtained using columns from the PureLink PCR Micro Kit rather than the Imprint DNA Modification Kit. It is currently unknown whether the differences are due to the column itself or the different desulfonation buffer. Purification with the EZ DNA Methylation Kit (Zymo Research) also appears to work well.
7. Diluting the enzymatic reaction improves purification with AMPure XP beads by reducing the viscosity of the solution. It also improves the efficacy of ethanol washes.
8. It is important to avoid over-amplification of the PBAT libraries at the risk of increasing the duplication level. With this protocol, 14 cycles of PCR are normally sufficient for 100 cells and 4–5 cycles for 5000 cells. To accurately determine the optimal number of PCR cycles, one strategy is to perform preliminary cycles (10 cycles for 100 cells, 4 cycles for 5000 cells, other inputs tailored accordingly), then pause the reaction at 4 °C and use 1 μL to perform qPCR quantification. It can then be calculated how many extra cycles are required to achieve a concentration of ~10 nM based on a twofold increase at each cycle. Simply continue the paused reaction.

References

1. Smith ZD, Meissner A (2013) DNA methylation: roles inmammalian development. Nat Rev Genet 14:204–220
2. Ferguson-Smith AC (2011) Genomic imprinting: the emergence of an epigenetic paradigm. Nat Rev Genet 12:565–575
3. Smallwood SA, Kelsey G (2012) De novo DNA methylation: a germ cell perspective. Trends Genet 28:33–42
4. Seisenberger S, Peat JR, Hore TA et al (2012) Reprogramming DNA methylation in the mammalian life cycle: building and breaking

epigenetic barriers. Philos Trans R Soc Lond Ser B Biol Sci 368:20110330

5. Cedar H, Bergman Y (2012) Programming of DNA methylation patterns. Annu Rev Biochem 81:97–117
6. Weber M, Davies JJ, Wittig D et al (2005) Chromosome-wide and promoter-specific analyses identify sites of differential DNA methylation in normal and transformed human cells. Nat Genet 37:853–862
7. Harris RA, Wang T, Coarfa C et al (2010) Comparison of sequencing-based methods to profile DNA methylation and identification of monoallelic epigenetic modifications. Nat Biotechnol 28:1097–1105
8. Smallwood SA, Tomizawa S-I, Krueger F et al (2011) Dynamic CpG island methylation landscape in oocytes and preimplantation embryos. Nat Genet 43:811–814
9. Smallwood SA, Kelsey G (2012) Genome-wide analysis of DNA methylation in low cell numbers by reduced representation bisulfite sequencing. In: Genomic imprinting. Humana, Totowa, NJ, pp 187–197
10. Miura F, Enomoto Y, Dairiki R et al (2012) Amplification-free whole-genome bisulfite sequencing by post-bisulfite adaptor tagging. Nucleic Acids Res 40:e136
11. Shirane K, Toh H, Kobayashi H et al (2013) Mouse oocyte methylomes at base resolution reveal genome-wide accumulation of non-CpG methylation and role of DNA methyltransferases. PLoS Genet 9:e1003439
12. Kobayashi H, Sakurai T, Imai M et al (2012) Contribution of intragenic DNA methylation in mouse gametic DNA methylomes to establish oocyte-specific heritable marks. PLoS Genet 8:e1002440
13. Peat JR, Dean W, Clark SJ et al (2014) Genome-wide bisulfite sequencing in zygotes identifies demethylation targets and maps the contribution of TET3 oxidation. Cell Rep 9 (6):1990–2000
14. Stewart KR, Veselovska L, Kim J et al (2015) Dynamic changes in histone modifications precede de novo DNA methylation in oocytes. Genes Dev 29(23):2449–2462
15. Quail MA, Otto TD, Gu Y et al (2012) Optimal enzymes for amplifying sequencing libraries. Nat Methods 9:10–11
16. Krueger F, Andrews SR (2011) Bismark: a flexible aligner and methylation caller for Bisulfite-Seq applications. Bioinformatics 27:1571–1572

Chapter 10

Methyl-CpG-Binding Domain Sequencing: MBD-seq

Karolina A. Aberg, Robin F. Chan, Linying Xie, Andrey A. Shabalin, and Edwin J.C.G. van den Oord

Abstract

Detailed biological knowledge about the potential importance of the methylome is typically lacking for common diseases. Therefore, methylome-wide association studies (MWAS) are critical to detect disease relevant methylation sites. Methyl-CpG-binding domain sequencing (MBD-seq) offers potential advantages compared to antibody-based enrichment, but performance depends critically on using an optimal protocol. Using an optimized protocol, MBD-seq can approximate the sensitivity/specificity obtained with whole-genome bisulfite sequencing, but at a fraction of the costs and time to complete the project. Thus, MBD-seq offers a comprehensive first pass at the CpG methylome and is economically feasible with the samples sizes required for MWAS.

Key words Methyl-CpG-binding domain, Sequencing, Affinity-based capture, MBD-seq, Methylome-wide association studies, MWAS, CpG, MethylMiner, Blood spots, Differentially methylated regions, High-dimensional data analysis, RaMWAS

1 Introduction

Detailed biological knowledge about the potential importance of the methylome is typically lacking for common diseases. Therefore, genome-wide approaches, which also proved fruitful in the context of finding sequence variants [1], will likely be critical to detect disease relevant methylation sites [2–4]. Massively parallel sequencing is an appealing technology as the availability of fast semi-automated sample preparation, the increase in the amount of data generated per run, and the decrease in reagent costs allow a comprehensive interrogation of the methylome in a cost-effective way.

The most comprehensive method for ascertaining methylation ($5^{me}C$) status at each nucleotide position is whole-genome sequencing of bisulfite-converted DNA [5] (WGBS, *see* also Chapter 5). However, due to the combination of high costs of sequencing entire genomes and the large number of samples required for methylome-wide association studies MWAS [2], this

Jörg Tost (ed.), *DNA Methylation Protocols*, Methods in Molecular Biology, vol. 1708,
https://doi.org/10.1007/978-1-4939-7481-8_10,

approach is not yet economically feasible. Affinity-based capture approaches present a cost-effective alternative as they aim to sequence only the methylated part of the genome. In these approaches, genomic DNA is fragmented and the methylated fragments are bound to antibodies [6] (*see* also Chapter 12) or other proteins [7] with high affinity for methylated DNA. The unmethylated fraction is removed, and the methylation-enriched portion of the genome is eluted and sequenced [7–10].

Among affinity-based capture methods, there are differences in assay properties depending on the DNA-binding protein or antibody used. As opposed to antibody-based approaches that may capture DNA fragments containing any methylated cytosine, proteins with a methyl-CpG-binding domain (MBD) strictly bind methylated CpGs. This is generally not a limitation as, with some exceptions [11], in most mammalian somatic tissues DNA methylation occurs almost exclusively at CpG dinucleotides [12–14]. Furthermore, taking association studies of sequence variants as an example, the different polymorphisms (common variants, rare variants, copy-number variations, etc.) are typically not examined in a single study as different study designs and assays are more suitable for some polymorphisms than for others. Thus, we view MBD-seq as providing a first pass at the methylome by only considering sites where methylation is likely to occur. It is perhaps this more focused approach that explains why MBD-seq has demonstrated to be sensitive, capable of identifying differentially methylated regions [7, 9, 10, 15–17], able to detect previously reported robust associations [18] and produce MWAS findings that replicate when using "gold standard" technologies such as targeted pyrosequencing of bisulfite-converted DNA [19] (*see* also Chapter 22).

MBD enrichment offers potential advantages compared to antibody-based enrichment, but we have found that performance depends critically on using an optimal protocol [20]. First, different (combinations of) proteins can be used, after working with a number of kits, we have obtained the best results with the MBD2 protein. Second, fragments with few methylated sites are more difficult to capture [10]. Through using a low salt concentration for the elution buffer, instead of a single or double elution with a higher salt concentration, MBD enrichment is made sensitive also to fragments with few methylated sites. Sensitivity can be further enhanced by using the shortest possible DNA fragments that still can be sequenced properly.

To evaluate our optimized MBD-seq protocol, we compared it with WGBS. Assays were performed in duplicate using DNA from the same subject. For MBD-seq we obtained an average of 81 million 75 base pair (bp) single-end reads/sample, and for WGBS 783 million 150 bp (paired-end) reads/sample. For MBD-seq 99.5% of the reads mapped successfully, and for WGBS 67.7% of the read pairs aligned.

MBD enrichment extract fragments with one or more methylated sites. The number of fragments covering a CpG is then used to estimate the amount of methylation. Thus, rather than providing methylation information at single-base resolution, MBD-seq assays the total amount of methylation for a locus that is about the size of the extracted fragments. To study how well it covers the methylome, we calculated the sensitivity, or proportion of methylated loci detected by enrichment methods. The sensitivity needs to be balanced against the specificity, the proportion of times that it correctly identifies unmethylated loci as unmethylated. For example, by simply calling all loci methylated, sensitivity would be one but specificity zero. To obtain an overall agreement index that reflects both sensitivity and specificity, we also calculated the proportion of times MBD-seq arrived at the same conclusion regarding the methylation status as WGBS. To avoid unreliable estimates, for WGBS we only used methylation calls that were based on five or more reads. This results in conservative estimates of the relative performance of MBD-seq as WGBS was not penalized for sites dropping out because of having too few reads. The calculation of sensitivity/specificity requires distinguishing methylated versus unmethylated loci. Methylation levels showed a bimodal pattern (Fig. 1a). The first *modus* was located at zero suggesting that a proportion of fragments did not contain methylated sites. The vertical line is drawn at the minimum value between the two *modi* (i.e., estimated inflection point). Taking this minimum as the threshold to distinguish methylated from unmethylated loci, 92.4% all loci contained mCG.

Result showed that the sensitivity of MBD-seq approximates the sensitivity of WGBS, and that specificity was generally better for enrichment compared WGBS (Fig. 1b). A reasonable balance is obtained using an MBD-seq fragment coverage cutoff of 0.5. Relative to the sensitivity/specificity between WGBS duplicates, at this cut-off MBD-seq achieved 0.91 of the sensitivity and 1.09 of the specificity of WGBS. Because there are more methylated versus unmethylated loci, the overall agreement more closely tracks the sensitivity and equaled 0.92.

MBD-seq cannot pinpoint the exact CpG that causes the MWAS signal at a locus. However, this limitation is mitigated by three factors. First, our top findings will be followed up using a targeted bisulfite-based approach. This means that for the sites of main interest we will obtain single-base resolution. Second, very often the CpGs in a region have similar methylation statuses, thereby mitigating the possible disadvantage, as even WGBS would detect a set of CpGs rather than a single site. To quantify this phenomenon, we used the WGBS data to calculate the variance of methylation sites in MBD-seq assayed loci. Figure 1c shows that over 90% of the regions have a variance <0.05, meaning that for the majority of loci the methylation differences between CpGs located close to each other are very small. Third, even if there would be

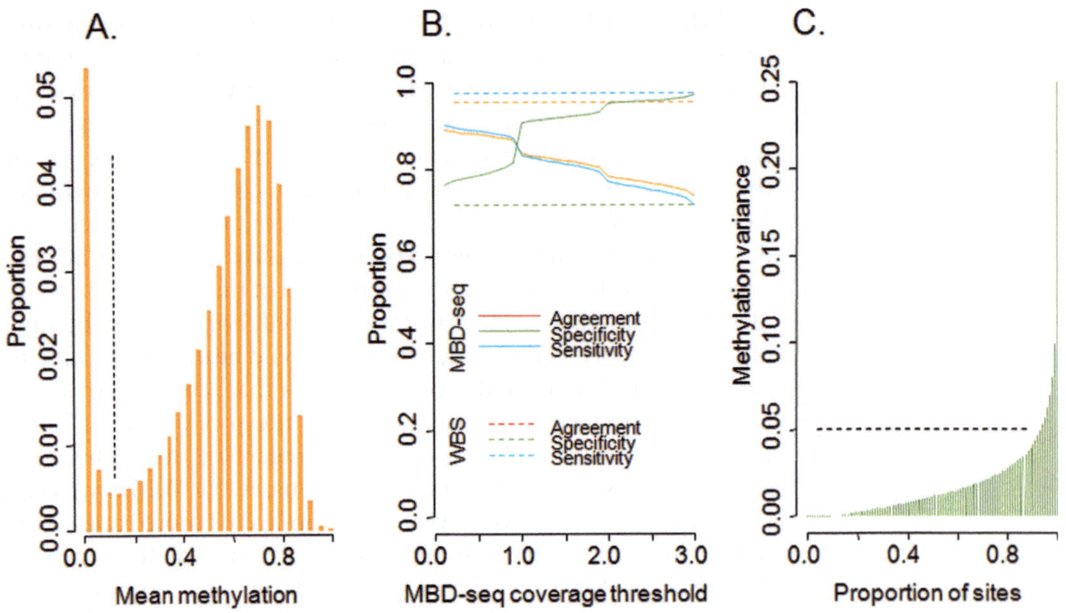

Fig. 1 Comparison MBD-seq and whole genome bisulfite sequencing. (**a**) Histogram of mean WGBS methylation for fragment-sized loci. (**b**) Sensitivity, specificity, agreement for MBD-seq coverage, and WGBS duplicates, (**c**) Percentiles (*X*-axis) for the variance of WGBS methylation density (*Y*-axis)

local variability in methylation status, the resolution of MBD-seq is very reasonable as it is approximately equal to the fragment size (typically 140–150 bp).

In summary, MBD-seq approximates the sensitivity/specificity obtained with WGBS. However, in terms of reagents, costs will be less than 5%. In addition, because WGBS uses longer and paired-end reads (in our case 300 vs. 75 cycles) and requires more total runs (in our case 1 sample per run vs. 10 samples per run), it incurs further expenses (lab techs, overheads, sequencer depreciation, etc.) and increases the time to complete a project.

Collecting MBD-seq data on a large number of samples to perform MWAS imposes demands on computer hardware and software. In addition to storing and backing up the raw data, hardware will be needed to store the aligned reads that will be too time-consuming to re-produce for each analysis. In addition, it may often not be possible to use standard statistical packages because of the ultra-high dimension of MBD-seq data and specific nature of some steps in the analyses. In this chapter, we will therefore also discuss hardware requirements, a data analysis pipeline, and RaMWAS an analysis tool specifically designed for MWAS.

2 Materials

2.1 Isolation of Genomic DNA

1. Gentra Puregene Blood/Tissue Kit (Qiagen) or QIAamp DNA Mini Kit (Qiagen) with Buffer AW1 and AW2 prepared according to the manufacturer's instructions.
2. GenSolve Spin Pack Tubes and Baskets (GenTegra).
3. Proteinase K (100 μg/mL).
4. RNase A (100 mg/mL).
5. 3 M Sodium Acetate buffer, pH 5.2 (adjusted with glacial acetic acid).
6. Glycogen (20 μg/μL).
7. 10% Bleach (0.6% w/v NaOCl): 1 volume 6% w/v sodium hypochlorite and 9 volumes water.
8. Stainless steel forceps and small surgical scissors.
9. Centrifuge capable of performing 8000 × *g*.
10. Water bath or heat block for incubation at 56, 70, and 85 °C.
11. Qubit 2.0 Fluorometer (Life Technologies).
12. Qubit dsDNA HS Assay Kit (Life Technologies).

2.2 DNA Fragmentation

1. Covaris™ System (Covaris).
2. Snap-Cap microTUBE (Covaris).
3. S-Series Holder microTUBE (Covaris).
4. 2100 Bioanalyzer (Agilent Technologies).
5. High Sensitivity DNA Kit (Agilent Technologies).

2.3 MethylMiner

1. MethylMiner Methylated DNA Enrichment Kit (Thermo-Fisher Scientific).
2. 3 M Sodium Acetate buffer, pH 5.2 (adjusted with glacial acetic acid).
3. 70 and 100% Ethanol.
4. Magnetic rack for microcentrifuge tubes.
5. Tube rotator (rotisserie) capable of end-over-end mixing (~20 rpm) at 4 °C.
6. Centrifuge capable of performing 12,000 × *g* at 4 °C.
7. 1× Low TE Buffer: 10 mM Tris–HCl, 0.1 mM EDTA, pH 8.0.
8. Qubit 2.0 Fluorometer (Life Technologies).
9. Qubit dsDNA HS Assay Kit (Life Technologies).

2.4 Library Construction and Sequencing

1. Single-end library construction kit for short reads (typically 50–75 bp) compatible with barcoding for the sequencing platform to be used.

2. Library Barcode Adaptors specific for the library type and sequencing platform to be used.
3. Agencourt AMPure XP DNA Purification Beads (Beckman Coulter).
4. 2100 Bioanalyzer (Agilent Technologies).
5. High Sensitivity DNA Kit (Agilent Technologies).
6. Sequencing platform and required reagents and consumables for sequencing barcoded single-end libraries (*see* **Note 1**).

2.5 Software

1. Alignment software: Good and fast aligners include Bowtie2 [21] and BWA [22].
2. MBD-seq software: Although packages are available to analyze MBD-seq data [23, 24], they are not specifically tailored for large-scale MWAS studies. The ultra-high dimensional data collected for many samples creates specific demands in terms of software's speed and memory use. For example, it is common for statistical packages to load the full matrix of methylation measurements in computer memory in order to perform principal component analysis or statistical tests. This becomes increasingly impractical as the full set of methylation measurements for, say, 1000 samples across 27 million CpGs comprises a 216 GB data matrix. This greatly exceeds memory capacity of most desktop computers and typical nodes of computational clusters. Therefore, to analyze MWAS data we developed RaMWAS that integrates new methods and software we have published previously in a single package [25–27]. RaMWAS, for example, can perform tasks such as principal component analysis, needed to control for unmeasured confounders, without loading the matrix of methylation measurements into computer memory.

2.6 Computer Hardware

1. It is important to plan data storage for MDB-seq analysis. In our projects the raw sequencing data and alignment files use about 10 GB per sample. The full run of the analysis pipeline (RaMWAS) requires less than 0.5 GB per sample.
2. We strongly recommend having a reliable (off site) backup copy of the raw sequencing data as it may be expensive to reproduce in case of a storage failure. The alignment files can be recreated from the raw sequencing data and thus do not necessitate equally reliable backup.
3. The major computational burden lies in aligning raw sequencing data to the reference genome. The alignment process requires multiple CPU-hours for each sample. For quality control and to spread the computations over time we recommend aligning the raw sequencing data as soon as it is produced and using RaMWAS to calculate multiple quality scores.

4. With RaMWAS, methylation wide association study across thousands of samples can be performed on a regular desktop computer in a matter of hours. While 8 GB of RAM would suffice, greater amount of RAM improves the performance. RaMWAS can also be run on a computing cluster.

3 Methods

3.1 Genomic DNA Isolation from Fresh/Frozen Tissue OR from Dry Blood Spots

Genomic DNA can be isolated from fresh or frozen tissue using any standard extraction protocol. We have successfully used the Gentra Puregene Blood/Tissue Kit to isolate DNA from blood and brain tissue. To isolate genomic DNA from dry blood spots (*see* also Chapter 31), we have successfully used a modified version of the QIAamp DNA Mini Kit.

1. Cut a dry blood spot into small pieces, and place into a 2 mL Lo-Bind tube (*see* **Note 2**). Soak used scissors in 10% bleach (v/v) for 5 min, and wash twice with ddH_2O before re-using.
2. Add 720 μL of Buffer ATL (provided in the QIAamp DNA Mini Kit) to each tube and vortex to saturate paper.
3. Incubate on heat block/bath at 85 °C for 10 min. Cool for a few min at room temperature and then briefly centrifuge to collect condensation.
4. Add 80 μL proteinase K solution, vortex, and incubate on heat block/bath at 56 °C for 1 h. Vortex every 15 min while incubating. Following incubation briefly centrifuge to collect condensation.
5. Place a GenSolve spin basket into a matching collection tube. Transfer the lysate (~800 μL) and paper fragments into the spin basket. Centrifuge at ≥16,000 × *g* for 2 min. Remove and discard the spin basket (*see* **Note 3**).
6. Add 16 μL RNase A solution to the filtered lysate, vortex, and incubate 5 min at room temperature.
7. Add 800 μL Buffer AL, vortex immediately and vigorously, then incubate on heat block/batch at 70 °C for 10 min. Following incubation briefly centrifuge to collect condensation.
8. Add 400 μL of 100% ethanol to each tube, mix thoroughly by vortexing, and pulse spin to bring the liquid down from the cap. Total volume should be ~1.2 mL.
9. Transfer 600 μL of the mixture from **step 8** to a new QIAamp Mini spin column and centrifuge at 8000 × *g* for 1 min. Discard flow-through and place the column back into the collection tube. Repeat until all the lysate for the sample has been passed through the column (*see* **Note 3**).

10. Add 600 μL Buffer AW1 to the column without wetting the rim and centrifuge 8000 × *g* for 1 min. Discard the flow-through and place the column in a new 2 mL collection tube.
11. Add 600 μL Buffer AW2 to the column and centrifuge at full speed (20,000 × *g*) for 30 s. Discard the flow-through.
12. Add 600 μL Buffer AW2 to the column and centrifuge at full speed (20,000 × *g*) for 2 min. Carefully remove the column and place into a new 1.5 mL Lo-Bind tube. Discard the collection tube and flow-through.
13. Add 100 μL Buffer AE (***do not use heated buffer***; *see* **Note 4**). Incubate at room temperature for 3 min and then centrifuge at 8000 × *g* for 1 min. Reload the 100 μL of flow-through to the column and centrifuge at 8000 × *g* for 1 min. Eluate contains isolated gDNA. Evaluate the concentration of the genomic DNA extracted by Qubit dsDNA HS Assay. A yield of 500–1000 ng of gDNA can be expected for a single ~12.5 mm dia. Blood spot.

3.2 DNA Fragmentation

Using ultrasonication the genomic DNA is sheered into fragments with a median fragment length that is optimal for the downstream sequencing approach. The following shearing conditions are optimized for the Covaris™ System. To ensure that the expected fragment size is obtained, we recommend evaluating the fragment length distribution using the High Sensitivity DNA kit with the Bioanalyzer.

3.2.1 DNA Fragmentation Generating Fragments with a Mean Length of ~150 bp

This fragment size is suitable for, for example, 75 bp single-end fragment libraries for Illumina sequencers such as the NextSeq 500.

1. Dilute 1–3 μg of genomic DNA in Low TE to a final volume of 100 μL and load it into a Snap-Cap microTUBE.
2. Place the tube in the microTUBE holder and perform the fragmentation in frequency sweeping mode for 6 cycles using the following conditions: duty cycle 10%, intensity 5, 100 cycles per burst, and duration 60 s. Set the water temperature limit to 15 °C and maintain the water temperature at 5 °C.
3. Confirm the concentration of the fragmented DNA using a Qubit dsDNA HS Assay.

3.3 Methylation Enrichment with MethylMiner

3.3.1 Preparation of MBD-Beads

1. Combine 1 part 5× Bind/Wash Buffer with 4 parts DNase-free water to prepare 1× Bind/Wash Buffer. 2 mL 1× Bind/Wash Buffer is needed for each 1–2 μg capture reaction.
2. Resuspend Dynabeads by gentle pipetting until homogenized. *Do not vortex—always mix by pipetting!*
3. Transfer 10 μL of beads for each microgram of total input DNA to a 1.5 mL LoBind tube. Beads for multiple reactions can be pooled together (*see* **Note 5**).

4. Bring the volume of the beads up to 250 μL with 1× Bind/Wash Buffer and mix by pipetting.
5. Place the tube on a magnetic rack for at least 1 min to collect the beads against the tube wall.
6. Keeping the tube on the magnetic rack, remove and discard the clear supernatant from the bead pellet.
7. Remove the tube from the magnetic rack and resuspend in 250 μL of 1× Bind/Wash Buffer.
8. Wash the Dynabeads once more by repeating **steps 5**–7,
9. Remove MBD-Biotin Protein from freezer and thaw on ice.
10. To a new 1.5 mL LoBind Tube add 7 μL of MBD-Biotin Protein for each microgram of total input DNA. Keep on ice.
11. Dilute aliquoted protein with 1× Bind/Wash Buffer to a final volume of 200 μL (500 μL if diluting >70 μL of protein).
12. Transfer diluted MBD-Biotin protein to washed Dynabeads. Make sure beads are fully mixed into a homogenous suspension.
13. Incubate mixture on a rotating mixer at room temperature for at least 1 h.
14. After incubation, *lightly* pulse spin tube to collect liquid and place in a magnetic rack until the solution becomes clear and a tight bead pellet forms.
15. Discard the supernatant and resuspend MBD-beads in 250 μL of 1× Bind/Wash Buffer.
16. Incubate beads on a rotating mixing at room temperature for 5 min to wash the beads.
17. Discard the supernatant and repeat wash with 250 μL of 1× Bind/Wash Buffer twice more.
18. Resuspend the washed MBD-beads in the same volume of 1× Bind/Wash Buffer as originally transferred from the Dynabeads stock (10 μL for each microgram of total input DNA) (*see* **Note 6**).

3.3.2 Capture of Methylated DNA Fragments

1. Bring the volume of each fragmented DNA sample (1–2 μg) up to 100 μL with DNase-free water.
2. Add 25 μL of 5× Bind/Wash Buffer to each DNA sample.
3. To each sample tube, add 10 μL of MBD-beads for every microgram of fragmented DNA present. Continually mix the MBD-beads while aliquoting to keep them suspended.
4. Bring the final volume of each reaction to 200 μL with 1× Bind/Wash Buffer (e.g., add 65 μL for a 1 μg reaction).
5. Close each capture reaction tube tightly (optionally seal with parafilm). Mix by inverting and incubate on a rotating mixer for 1 h at room temperature or at 4 °C overnight.

3.3.3 Elution of Methylated DNA Fragments

1. Prepare 0.5 M Elution Buffer by combining three parts Low-Salt Elution Buffer (no NaCl) with one part High-Salt Elution Buffer (2.0 M NaCl). 400 μL of 0.5 M Elution Buffer is needed for each μg of DNA input in the capture reaction.
2. After overnight incubation, remove the capture reaction tubes from the rotating mixer. Pulse spin to pull down the mixture and place in a magnetic rack for at least 1 min.
3. Transfer the clear supernatant from each capture tube to a new 1.5 mL LoBind tube. This fraction contains the non-captured DNA (*NC*).
4. Add 200 μL of 1× Bind/Wash Buffer to each tube containing the beads and wash on a rotating mixer for 3 min at room temperature (*see* **Note 7**).
5. Pulse spin and return tubes to the magnetic rack to collect the beads, and transfer the clear supernatant to a new 1.5 mL LoBind tube. This is the first wash fraction (*W1*).
6. Repeat **steps 4** and **5** once more to collect a second wash fraction (*W2*). If greater than 1 μg of DNA was used as input repeat for a total of four washes.
7. Next, add 200 μL (400 μL when input >1 μg) of 0.5 M Elution Buffer to each tube containing beads and place on a rotating mixer for 3 min at room temperature.
8. Return tubes to the magnetic rack to collect the beads, and transfer the clear supernatant to a new 1.5 mL LoBind tube. This is the first 0.5 M elution (*E1a*).
9. Repeat **steps 7** and **8** once more to collect a second 0.5 M elution (*E1b*). Perform a third 0.5 M elution if greater than 1 μg was used as input.
10. Repeat the elution protocol in **steps 7–9** with High-Salt Elution Buffer to obtain a first (*E2a*) and second (*E2b*) 2.0 M elution. Perform a third 2.0 M elution if greater than 1 μg was used as input.
11. To each of the fractions (*NC, W1, W2, E1a, E1b,* etc.) add 1 μL glycogen, 0.1× sample volumes of 3 M sodium acetate, and 2.2× sample volumes of ice-cold 100% ethanol.
12. Mix each tube well by vortexing and cool on ice or at ≤−20 °C for at least 30 min to precipitate the DNA in each fraction.
13. Centrifuge each tube at 12,000 × *g* for 15 min at ≥4 °C to pellet the precipitated DNA. Carefully discard the supernatant (*see* **Note 8**).
14. Wash each pellet with 800 μL of ice-cold 70% ethanol. Remove the ethanol wash by pipetting. Stop and spin tubes at ≥12,000 × *g* for 3 min at ≥4 °C if the pellet becomes dislodged at any time. Repeat to wash the pellet at least twice.

15. Open each tube and allow the pellet to air-dry for 3–5 min or until the pellet becomes translucent.
16. Resuspend the pellet in 31 μL of Low TE.
17. Determine DNA concentration of each fraction with a Qubit dsDNA HS assay (*see* **Note 9**).
18. Combine the 500 mM fractions. Together they will form the starting material for library construction in Subheading 3.4 (*see* **Note 10**).

3.4 Library Construction and Sequencing

The specific protocol for library construction will depend on, for example, the sequencing platform, library construction kit, and read length. Below we give general directions for the library construction. However, for details we refer to the standard protocol for each approach.

3.4.1 Library Construction

1. In contrast to some standard protocols we prefer to perform size-selection first so that only fragments that are suitable for sequencing are used in library construction. Therefore, prior to fragment end repairing, perform size selection to obtain fragments of suitable size for sequencing. For maximal DNA recovery, we prefer the use of SPRI beads (Agencourt AMPure XP) over gel extraction methods for fragment size selection (*see* **Notes 11** and **12**).
2. Confirm successful size-selection and output concentration of the DNA fragments on a 2100 Bioanalyzer with a High Sensitivity DNA assay.
3. Perform end repair and adaptor ligation steps for library construction, on the size-selected DNA, following standard protocols for the relevant sequencing platform.
4. Prior to any amplification of the library, check for the presence of low molecular weight (<100 bp) species by a High Sensitivity DNA assay. If present, perform a round of purification with a 1.8× volume of AMPure XP Beads.
5. Complete the amplification for the library construction following the manufacturer's instructions.
6. After the completion of library construction perform quantitation and measure the size distribution of the library by a High Sensitivity DNA assay. Check for the expected increase in fragment length as a result of adaptor ligation. Repeat quantitation of library concentration to increase the accuracy of library pooling.

3.4.2 Pooling Libraries in Equal Molarities

Multiplexed sequencing of barcoded/indexed libraries requires careful pooling of libraries in order to obtain the desired number of reads for each sample. To obtain equal numbers of reads for each sample, libraries must be pooled so that their final concentrations are equimolar. If a different number of reads for separately barcoded/indexed libraries is required, the libraries need to be pooled in a weighted approach. As the correct pooling of libraries is critical for successful sequencing, we recommend developing a spreadsheet that can aid in the calculations using the formulas in the steps explained below.

1. Determine the representation of each library based on the desired number of reads and the maximum output of the sequencing platform with:

   Representation = Desired reads/Maximum output
2. Using the library concentration measured (average of technical replicates) in **step 6** of subheading 3.4.1, the volume to be loaded into the pool, from each library, can be calculated with:

   Volume = Representation × Molar amount × [Library], where the molar amount pertains to the final library pool (i.e., the molarity of the final pool multiplied by the total volume of the final pool to be used for sequencing). For advice on how to optimize the molarity of the final pool *see* **Note 13**.
3. After mixing libraries according to the volumes calculated above add buffer to bring up the final pool volume if necessary.

3.4.3 Sequencing

Perform the sequencing following the standard procedure for the specific sequencing platform to be used (*see* **Note 1**).

3.5 Data Analysis

Except for the alignment, all other steps described in this section can be performed with RaMWAS.

3.5.1 Align Reads

To align the reads we use a seed-and-extend approach combined with local alignment while allowing for gaps [28]. Specifically, for 50–75 bp reads we typically use a 25 bp seed and allow for a 1-base mismatch in the seed. Rather than considering the entire extension, local alignment is used to improve sensitivity by finding the maximum similarity score between the reference sequence and a substring of the extension that may be might "trimmed" or "clipped." Gaps are allowed to account for small indels.

3.5.2 Read Quality Control

We quality control multi- and duplicate-reads. Reads often map to multiple genomic locations but in most cases a single alignment can be selected because it is clearly better than the others. In the case of multi-reads, multiple alignments are about equally good. Duplicate-reads are reads that start at the same nucleotide positions. When sequencing a whole genome duplicate-reads often arise from template preparation or amplification artifacts. In our context of sequencing an enriched genomic fraction, duplicate-reads are increasingly likely to occur by chance because reads are expected to align to a much smaller fraction of the genome. In instances where >3 (duplicate) reads start at the same position, we typically reset the read count to 1 implicitly assuming that these reads all tagged a single fragment.

3.5.3 Obtain Locations of Target Sites

To identify individual methylation biomarkers, we will perform methylome-wide association studies (MWAS) on the ~28 million common CpGs (defined as a site that is a CpG in at least 95% of the subjects).

Four types of target sites are calculated that serve different purposes in the data analyses. CpGs can be created/destroyed by single-nucleotide polymorphisms (SNPs). By considering SNP allele frequencies we distinguish "regular" CpG dinucleotides (e.g., a site that is a CpG in at least 95% of the subjects) versus common CpG-SNPs (e.g., a site that is a CpG in 5–95% of the subjects). Whereas regular CpG can be analyzed using traditional multiple regression analyses, the analysis of CpG-SNPs requires some adaptations [29]. In addition, we calculate isolated CpGs defined as a site C that is a CpG for which the interval [C − d, C + d] contains no other CpGs but C and where d is larger than the longest possible fragment size. These isolated CpGs are used to estimate the fragment size distribution (see next section). Finally, non-CpGs defined as a site C that is not a CpG and for which the interval [C − d, C + d] contains no CpGs and where d is larger than the longest possible fragment size. Because these sites cannot be methylated they provide a QC index for the quality of the MBD assay.

3.5.4 Estimating the Fragment Size Distribution and Calculate Coverage

A natural way to quantify enrichment is to count the number of fragments covering each CpG. However, with single-end libraries the fragment sizes are not observed. Instead, the number of reads covering the CpG may be counted. This seriously underestimates coverage as the fragment that generated the read is longer than the read and the CpG site might not be in the read itself. To remedy this, read length is sometimes extended to the expected fragment length. However, this is still imprecise as one cannot assume an equal size for all fragments and expected fragment size may be mis-specified if the sequenced fragment pool differs from the size obtained directly after fragmentation. A more precise method is to estimate the sample-specific fragment size distributions from the empirical sequencing data using isolated CpGs [26] (*see* **Note 14**). These sample-specific estimated fragment size distributions were used to calculate the probability for each read that the fragment it is tagging covers the CpG under consideration. Coverage for each CpG can then be calculated by taking the sum of the probabilities that all fragments in its neighborhood cover the CpG. For example, this probability equals 1.0 for fragments with reads starting within a read length to the CpG, but will be ≤ 1.0 for fragments with reads starting further away.

A CpG on the forward strand is a GpC on the reverse strand. Because the use of the reference genome will yield coordinates on the forward strand, the constant 1 needs to be added to the coordinates when calculating coverage using reads tagging fragments from the reversed strand. Coverage is affected by the total number of used reads per sample that is a function of the lab protocol (e.g., sequencer output and degree of multiplexing) rather than methylation. The estimates therefore need to be standardized using the total number of reads that remained after quality control (*see* **Note 15**).

3.5.5 Site Quality Control

Our site quality control carefully evaluates three types of sites that, if not handled correctly, increase the risk of false positive findings. The two types are: (1) Sites with very low coverage (e.g., coverage less than the 99th percentile of the coverage estimates at the non-CpG sites) could be excluded to avoid statistical artifact due to handling sparse data and because sites that are unmethylated in all subjects cannot show case-control differences. (2) Repetitive elements constitute about 45% of the human genome. Reads may be difficult to align to these loci because of their high sequence similarity. Because methylation markers in repeats can be associated with diseases [30–32], rather than eliminating all repeats we propose to eliminate only those causing alignment problems according to an in silico alignment experiment described elsewhere [25]. Depending on the platform and read length we have observed that this may eliminate 2–10% of the sites.

3.5.6 Principle Component Analysis

Controlling for confounders to avoid false positive findings presents a major challenge in MWAS. In addition to technical factors associated with processing samples and the risk of false positives due to population structure, there are many possible differences between cases and controls that may affect the methylome and consequently produce significant association results. The list of potential confounders is long, only a subset will have been measured, and some confounders may simply be unknown. In addition to regressing out measured covariates, we perform principal components analysis to capture the major sources of remaining unmeasured variation in the methylation data.

3.5.7 Association

We use the data obtained after the data reduction stage as input for multiple regression analyses to test for association with case-control status while regressing out covariates. Covariates include possible assay related artifacts such as the quantity of genomic DNA starting material for MethylMiner, quantity of methylation-enriched DNA captured, and sample batch. In addition, we typically regress out demographic and life style variables such as substance use that can leave methylation signatures but do not reflect disease processes.

Principal components (PCs) may be included to account for unmeasured confounders. The number of components is determined by a "scree" test. This method consists of plotting the components in descending order on the *X* axis and their corresponding eigenvalues on the *Y*-axis. As one moves to the higher components, the eigenvalues will decrease. The screen test says to drop all further components after this decreases tapers off.

Most tissues consist of multiple cell types. This heterogeneity can create significant results for many CpGs when two conditions hold (1) the relative abundance of specific cell types is affected by alcohol, (2) there are differences in methylation patterns across these cell types. As these significant results are related to shifts in the general cell type population, findings for individual CpGs may not provide any specificity. To handle possible cell type heterogeneity we can create reference methylomes by performing MBD-seq in isolated cells. These reference methylomes can then be used to estimate those proportions using the methylation data in the study sample [33, 34] and regress out those estimates in the MWAS. In a study of common blood cell types in 337 subjects, we have found the precision of the MBD-seq cell type proportions to be comparable to that of the automated counts obtained with the Abbott Sapphire system that uses optical scatter and impedance.

3.5.8 Annotation

To interpret the MWAS findings, they need to be annotated with features such as CpG island/shores, histone binding sites, potential gene promoter, transcription factor binding sites, information about the location of exons, introns, untranslated regions. For this purpose, a wealth of biological data is available in public data repositories. RaMWAS results are conveniently annotated using the biomaRt [35] package in R that provides easy access to an increasing number of databases such as Ensembl, Uniprot, and dbSNP to retrieve the desired information.

4 Notes

1. We have successfully performed MBD-seq using the SOLiD platform (ThermoFisher Scientific), the Ion Proton™ System (ThermoFisher Scientific) as well as both the MiSeq (Illumina) and NextSeq 500 (Illumina). Other well-performing platforms suitable for MBD-seq are, for example, HiSeq 2500 (Illumina) and HiSeq 3000/4000 (Illumina). Although a large number of sequencing platforms, suitable for MBD-seq, are available, it is important to note that the sequence output, the run time, and the read quality vary between platforms and therefore some platforms may be more cost efficient than others.

2. We have found that greatly increasing the volume of lysis buffers allows full saturation and submersion of the filter paper and results in increased yields of gDNA from dry blood spots. If extracting from a spot larger than 12.7 mm in diameter, divide the material between two 2 mL microcentrifuge tubes and process in parallel.
3. Use of the spin baskets improves yields of gDNA from dry blood spots by removing additional lysate from the filter paper. If unavailable, the use of two spin columns during Subheading 3.1, **step 8** may help prevent clogging by paper fibers.
4. We have found that use of heated elution buffers (70 °C) leads to significant protein and salt contamination of the eluted gDNA, leading to failure of subsequent methylation enrichment.
5. MBD-beads can be prepared on a per-sample basis in multiple tubes, but it is far less cumbersome to prepare all the beads needed for all capture reactions in a single tube.
6. Reducing the volume of the prepared MBD-bead suspension aids in keeping the beads suspended during aliquoting when using a pooled batch.
7. While loading tubes onto the rotating mixer make sure the buffer runs over the bead pellet during rotation or else the beads will not be properly washed/eluted.
8. When using Lo-Bind tubes the pellet is particularly easy to dislodge. Removing the supernatant with progressively smaller pipette tips is usually necessary.
9. The DNA yield obtained from the wash and elution fractions from 1 to 2 μg of input are usually too low to be quantified by spectrophotometer. We recommend the use of a fluorometric assay (Qubit or Bioanalyzer) to measure DNA concentrations from MBD enrichment. For a 1 μg reaction, 50–100 ng of DNA is typically recovered from the combined wash and elution fractions. The majority of the input DNA is collected in the non-captured fraction.
10. The high-salt elutions (with 2 M Elution Buffer), as well as the non-captured (*NC*) and wash (*W1, W2,* etc.) fractions, may have potential use for other applications studying, e.g., CpG rich regions or non-CpG methylation, respectively. We recommend storing these fractions at −80 °C.
11. While using Agencourt AMPure XP beads for size selection we have found that 0.9× and 0.6× the sample volume of XP beads to DNA are best at binding >250 and >350 bp fragments respectively. These ratios of beads differ greatly from many standard protocols.

12. Important: when purifying sub-microgram amounts of DNA with Agencourt AMPure XP beads it is critical to prevent over-drying of the beads during ethanol washes. If the bead pellet is allowed to dry to the point in which it appears cracked, the DNA will become permanently bound and cannot be eluted.
13. Due to differences in individual techniques and equipment, adjustments to the standard manufacturer sequencing protocols may help maximize sequencing output. If the expected theoretical maximum output (expected) is not achieved, it may be possible to increase the observed output (observed) for an upcoming run by increasing the molarity of the library pool to be sequenced. However, it should be noted that overloading the sequencing pool may result in failed sequencing or low quality data. To adjust the observed output compare the expected number of reads versus the observed number of reads obtained for a given concentration of input library:

    Adjustment factor = Theoretical max output/Observed output

    For example, with our NextSeq 500, we have found that increasing the molarity of the final library pool from 1.8 (standard) to 2.45 pM boosts instrument output to 150% of the theoretical maximum.
14. Samples may have been split over multiple sequencing lanes, rerun to supplement reads, or sequenced in duplicate. Our rule of thumb is that if the sequencing is done on the same libraries, we add the raw (read starts) data prior to analyses. If different libraries were created for the same samples we use a weighted sum (weights summing to one and being proportional to the total number of reads) of the coverage estimates. This latter approach can account differences between libraries that may affect the coverage (e.g., fragment size distributions) estimates.
15. The enrichment is more challenging for fragments that contain a single versus multiple methylated CpGs. Because the estimation procedure uses data from these isolated CpGs, the quality of the library or actual methylation patterns can affect the precision of the estimated fragment size distribution. Using the "mean fragment size distribution" of the batch may be an alternative in situations where the individual distribution(s) are considered unstable. This can be diagnosed by inspecting the decay of number of reads starting near the isolated CpG [26].

References

1. Altshuler D, Daly MJ, Lander ES (2008) Genetic mapping in human disease. Science 322:881–888
2. Rakyan VK, Down TA, Balding DJ et al (2011) Epigenome-wide association studies for common human diseases. Nat Rev Genet 12:529–541
3. Laird PW (2010) Principles and challenges of genomewide DNA methylation analysis. Nat Rev Genet 11:191–203

4. Beck S, Rakyan VK (2008) The methylome: approaches for global DNA methylation profiling. Trends Genet 24:231–237
5. Li Y, Zhu J, Tian G et al (2010) The DNA methylome of human peripheral blood mononuclear cells. PLoS Biol 8:e533
6. Mohn F, Weber M, Schubeler D et al (2009) Methylated DNA immunoprecipitation (MeDIP). Methods Mol Biol 507:55–64
7. Serre D, Lee BH, Ting AH (2010) MBD-isolated genome sequencing provides a high-throughput and comprehensive survey of DNA methylation in the human genome. Nucleic Acids Res 38:391–399
8. Brinkman AB, Simmer F, Ma K et al (2010) Whole-genome DNA methylation profiling using MethylCap-seq. Methods 52:232–236
9. Li N, Ye M, Li Y et al (2010) Whole genome DNA methylation analysis based on high throughput sequencing technology. Methods 52:203–212
10. Harris RA, Wang T, Coarfa C et al (2010) Comparison of sequencing-based methods to profile DNA methylation and identification of monoallelic epigenetic modifications. Nat Biotechnol 28:1097–1105
11. Lister R, Mukamel EA, Nery JR et al (2013) Global epigenomic reconfiguration during mammalian brain development. Science 341:1237905
12. Bernstein BE, Meissner A, Lander ES (2007) The mammalian epigenome. Cell 128:669–681
13. Bird AP (1986) CpG-rich islands and the function of DNA methylation. Nature 321:209–213
14. Lister R, Pelizzola M, Dowen RH et al (2009) Human DNA methylomes at base resolution show widespread epigenomic differences. Nature 462:315–322
15. Hogart A, Lichtenberg J, Ajay SS et al (2012) Genome-wide DNA methylation profiles in hematopoietic stem and progenitor cells reveal over-representation of ETS transcription factor binding sites. Genome Res 22:1407–1418
16. Lan X, Adams C, Landers M et al (2011) High resolution detection and analysis of CpG dinucleotides methylation using MBD-Seq technology. PLoS One 6:e22226
17. Nair SS, Coolen MW, Stirzaker C et al (2011) Comparison of methyl-DNA immunoprecipitation (MeDIP) and methyl-CpG binding domain (MBD) protein capture for genome-wide DNA methylation analysis reveal CpG sequence coverage bias. Epigenetics 6:34–44
18. McClay JL, Aberg KA, Clark SL et al (2014) A methylome-wide study of aging using massively parallel sequencing of the methyl-CpG-enriched genomic fraction from blood in over 700 subjects. Hum Mol Genet 23:1175–1185
19. Aberg KA, McClay JL, Nerella S et al (2014) Methylome-wide association study of schizophrenia: identifying blood biomarker signatures of environmental insults. JAMA Psychiat 71:255–264
20. Aberg KA, Xie L, Chan RF et al (2015) Evaluation of methyl-binding domain based enrichment approaches revisited. PLoS One 10: e132205
21. Langmead B, Salzberg SL (2012) Fast gapped-read alignment with bowtie 2. Nat Methods 9:357–359
22. Li H, Durbin R (2009) Fast and accurate short read alignment with burrows-wheeler transform. Bioinformatics 25:1754–1760
23. Lienhard M, Grimm C, Morkel M et al (2014) MEDIPS: genome-wide differential coverage analysis of sequencing data derived from DNA enrichment experiments. Bioinformatics 30:284–286
24. Bock C (2012) Analysing and interpreting DNA methylation data. Nat Rev Genet 13:705–719
25. Aberg KA, McClay JL, Nerella S et al (2012) MBD-seq as a cost-effective approach for methylome-wide association studies: demonstration in 1500 case-control samples. Epigenomics 4:605–621
26. van den Oord EJ, Bukszar J, Rudolf G et al (2013) Estimation of CpG coverage in whole methylome next-generation sequencing studies. BMC Bioinformatics 14:50
27. Chen W, Gao G, Nerella S et al (2013) methylPCA: a toolkit for principal component analysis in methylome-wide association studies. BMC Bioinformatics 14:74
28. Li H, Homer N (2010) A survey of sequence alignment algorithms for next-generation sequencing. Brief Bioinform 11:473–483
29. van den Oord EJ, Clark SL, Xie LY et al (2016) A whole methylome CpG-SNP association study of psychosis in blood and brain tissue. Schizophr Bull 42:1018–1026
30. Ushida H, Kawakami T, Minami K et al (2012) Methylation profile of DNA repetitive elements in human testicular germ cell tumor. Mol Carcinog 51:711–722
31. Bollati V, Galimberti D, Pergoli L et al (2011) DNA methylation in repetitive elements and Alzheimer disease. Brain Behav Immun 25:1078–1083
32. Bollati V, Fabris S, Pegoraro V et al (2009) Differential repetitive DNA methylation in

multiple myeloma molecular subgroups. Carcinogenesis 30:1330–1335

33. Houseman EA, Accomando WP, Koestler DC et al (2012) DNA methylation arrays as surrogate measures of cell mixture distribution. BMC Bioinformatics 13:86
34. Koestler DC, Christensen B, Karagas MR et al (2013) Blood-based profiles of DNA methylation predict the underlying distribution of cell types: a validation analysis. Epigenetics 8:816–826
35. Durinck S, Spellman PT, Birney E et al (2009) Mapping identifiers for the integration of genomic datasets with the R/Bioconductor package biomaRt. Nat Protoc 4:1184–1191

Chapter 11

The HELP-Based DNA Methylation Assays

John M. Greally

Abstract

Restriction enzymes have been valuable tools for representing the genome for DNA methylation assays, whether by using methylation-dependent enzymes or by sampling a reduced representation of the genome using a methylation-insensitive enzyme. These survey assays have remained mainstays of genome-wide approaches even with the development of more comprehensive shotgun genome-wide bisulphite sequencing-based assays, as they are significantly more affordable. DNA methylation survey assays are numerous and include reduced representation bisulphite sequencing (RRBS), the Illumina HumanMethylation450K and EPIC microarray system, and our evolving series of HELP-based assays. The HELP (*H*paII tiny fragment *E*nrichment by *L*igation-mediated *P*CR) assays initially involved microarray-based reporting of DNA methylation, but have now migrated to the use of massively parallel sequencing. In this chapter, we describe the latest HELP-tagging assay that uses Illumina Tru-Seq adapters, and mention the extension of the HELP-tagging assay to quantify 5-hydroxymethylation using the HELP-GT assay.

Key words Cytosine methylation, CpG dinucleotide, Massively parallel sequencing, Epigenetic, Epigenome

1 Introduction

In general, assays testing cytosine methylation fall into three categories, those based on selective sensitivity of certain restriction enzymes to methylation, those based on selective affinity of certain proteins for methylated DNA, and those based on selective conversion of unmethylated cytosines to uracil following bisulphite exposure. All three categories are described in this volume. The HELP-based assays represent examples of the use of methylation-sensitive restriction enzymes. The advantages offered by comparative isoschizomer profiling is the use of the methylation-insensitive representation for comparison, which allows much better accuracy for both microarray [1–3] and massively parallel sequencing-based [4] versions of the assay. The signal at a given locus from a HpaII representation can be influenced not only by the methylation status of that locus, but also by the size of the fragment, its base composition (both variables influencing PCR amplification), and whether

Jörg Tost (ed.), *DNA Methylation Protocols*, Methods in Molecular Biology, vol. 1708,
https://doi.org/10.1007/978-1-4939-7481-8_11, © Springer Science+Business Media, LLC 2018

the locus is mutated in any way (copy number, mutations of the CG-containing and therefore highly mutable restriction enzyme target site). As the MspI representation will be influenced to the same extent by these variables, expressing the HpaII signal as a function of the MspI representation at each locus allows more robust comparison of different loci from the same DNA sample, an "intragenomic" comparison. In common with other assays, HELP also allows two different samples to be compared ("intergenomic"), looking for differences in methylation between cell types.

Base composition affects all cytosine methylation assays in different ways. Affinity-based assays are influenced by the density of CG dinucleotides in a given sequence, while restriction enzyme-based assays will interrogate more or fewer loci depending on the number of restriction sites locally (*see* also Chapters 12 and 13). Reduced representation sequencing-based assays based on the use of restriction enzymes are designed to take advantage of this tendency to interrogate different genomic loci, with RRBS enriching CG dinucleotide-dense regions preferentially [5] (*see* also Chapter 8).

Finally, it should be stressed that the degree of difficulty associated with performing these molecular assays is generally outweighed significantly by the challenges associated with their analysis. We have published the analytical pipeline supporting HELP-tagging [6], which allows DNA methylation to be quantified as a continuous variable including intermediate levels. The assay has now been performed on thousands of samples in many countries, testing multiple organisms [7] and resulting in a number of publications based on its use. Here, we describe the updated protocol that takes advantage of the Illumina Tru-Seq adapters in library preparation.

2 Materials

2.1 Genomic DNA Extraction

1. 1× PBS: 137 mM NaCl, 2.7 mM KCl, 4.3 mM Na_2HPO_4-$7H_2O$, 1.4 mM KH_sPO_4.
2. 1.0 M Tris–HCl, pH 8.0: Dissolve 121.14 g Tris into approximately 800 mL of distilled water and adjust pH to 8.0 with HCl. Fill up the solution up to 1.0 L and autoclave.
3. 0.5 M EDTA pH 8.0: Dissolve 93.06 g EDTA into distilled water; adjust pH, make up to 1.0 L and autoclave.
4. 20% SDS: Dissolve 20 g of SDS in autoclaved water. Bring the solution to a final volume of 100 mL.
5. RNAse A: Resuspend in water to a final concentration of 10 mg/mL. Aliquot and store at −20 °C.

6. Extraction buffer: 10 mM Tris–HCl pH 8.0, 0.1 M EDTA pH 8.0, 0.5% SDS, 20 μg/mL RNAse A. Make up fresh each time.
7. Proteinase K 20 mg/mL.
8. TE: 10 mM Tris–HCl pH 8.0, 1.0 mM EDTA pH 8.0.
9. Tris-EDTA (TE)-saturated phenol.
10. Chloroform.
11. Isoamyl alcohol.
12. Phenol-Chloroform-Isoamyl alcohol (PCI): Phenol: Chloroform: Isoamyl Alcohol 25:24:1 by volumes. Make up fresh each time.
13. Spectra/Por dialysis tubing (Spectrum Laboratories, Rancho Dominguez CA) MWCO 12–14K kDa, pretreated according to the manufacturer's recommendations and washed in double-distilled water prior to use.
14. 20× SSC: 3.0 M NaCl, 0.3 M Sodium Citrate, pH 7.0. Dissolve 175.3 g NaCl and 88.2 g Sodium Citrate in 800 mL water. Adjust pH to 7.0 with NaOH or HCl. Make up to 1.0 L with water and autoclave.
15. Polyethylene glycol (PEG) molecular weight 20,000.

2.2 Protocol: TrueSeq-HELP Tagging

2.2.1 Reagents

1. Agarose for molecular biology and low range agarose for PCR product.
2. 50× TAE buffer for electrophoresis.
3. Saturated Phenol.
4. Chloroform.
5. Glycogen.
6. Isopropanol.
7. Ethachinmate (carrier solution for DNA precipitation).
8. 3.0 M Sodium Acetate: Dissolve 123.04 g of Sodium Acetate into distilled water, adjust pH to 5.2, and fill up to 500 mL.
9. TE Buffer: 10 mM Tris–HCl, pH 8.0, 1 mM EDTA.
10. 4 and 10 mM dNTP nucleotides: Add up 40 or 100 μL of each nucleotides (100 mM) and 840 or 600 μL of sterile water to make up 1 mL mix.
11. ATP (10 mM).
12. BSA.
13. HpaII: 50,000 U/mL (high concentrate/recommended) or 10,000 U/mL (e.g., New England Biolabs).
14. MspI: 50,000 U/mL (e.g., New England Biolabs).
15. EcoP15I: 10,000 U/mL (e.g., New England Biolabs).

16. Phusion High-Fidelity DNA Polymerase 2000 U/mL.
17. Klenow Fragment (3′ → 5′ exo-) 5000 U/mL.
18. Quick Ligation Kit (New England Biolabs).
19. T4 DNA Polymerase 3000 U/mL and associated buffer.
20. T4 DNA Polynucleotide kinase 10,000 U/mL.
21. MEGAshortscript kit (Thermo Fisher Scientific).
22. DNase Turbo 2 U/mL (Thermo Fisher Scientific).
23. SuperScriptIII reverse transcription kit including DTT (Thermo Fisher Scientific).
24. RNase out (Thermo Fisher Scientific).
25. QIAquick PCR purification kit QIAGEN) or similar.
26. MinElute PCR purification kit (QIAGEN) or similar.
27. MinElute Gel extraction kit (QIAGEN) or similar.
28. RNeasy clean up kit MinElute (QIAGEN) or similar.
29. Agencourt AMPure XP 5 mL Kit (Beckman Coulter Inc.).
30. Qubit dsDNA HS Assay Kit (Thermo Fisher Scientific).
31. GelGreen (Biotium).

2.2.2 Materials and Equipment

1. PCR thermal cycler.
2. Gel system and power source.
3. Water bath at 37 °C.
4. 1.5 mL Eppendorf tubes.
5. PCR tubes.
6. Benchtop microcentrifuge.
7. Vortex.
8. SPRIPlate® 96R-Ring Magnet Plate (Beckman Coulter Inc.).
9. Dark Reader transilluminators.
10. Qubit® 2.0 Fluorometer (Thermo Fisher Scientific).

2.2.3 Primer Sequences

All primer sequences are given in Table 1 (*see* **Note 1**).

3 Methods

3.1 Genomic DNA Extraction

The DNA preparation technique we recommend involves standard cell lysis, proteinase K digestion, and extraction in organic solvents, but proceeds to a dialysis step to purify and concentrate the material prior to use (*see* **Note 2**). If the sample is limited in amount, it may be more appropriate to use ethanol precipitation following the extraction step, which we have also found to be reliable.

Table 1
Adapter sequences

Adapter TS_AS	
TS_HT_AS adapter top	5′-ATTGATACGGCGACCACCGAGATCTACACTCTTTCCCTACACGACGCTCTTCCGATC[a]T
TS_HT_AS adapter bot	5′-Phos-GATCGGAAGAGCGTCGTGTAGGGAAAGAGTGTAGATCTCGGTGGTCGCCGTATCAAT
Adapters TS_AE multiplex	
TS_HT_AE *Index 1* (ATCACG) adapter sequence	
TS_ AE1 top	5′-ACAGTAATACGACTCACTATAGGGAGAACAAGCAGAAGACGGCATACGAGATCGTGATGTGACTGGAGTTCAGACGTGTGCTCTTCCGATCTCAGCAG
TS_AE1 bot	5′-Phos-CGCTGCTGAGATCGGAAGAGCACACGTCTGAACTCCAGTCACATCACGATCTCGTATGCCGTCTTCTGCTTGTTCTCCCTATAGTGAGTCGTATTACTG[b]T
S_HT_AE *Index 2* (CGATGT) adapter sequence	
TS_AE2 top	5′-ACAGTAATACGACTCACTATAGGGAGAACAAGCAGAAGACGGCATACGAGATACATCGGTGACTGGAGTTCAGACGTGTGCTCTTCCGATCTCAGCAG
TS_AE2 bot	5′-Phos-CGCTGCTGAGATCGGAAGAGCACACGTCTGAACTCCAGTCACCGATGTATCTCGTATGCCGTCTTCTGCTTGTTCTCCCT ATAGTGAGTCGTATTACTG[b]T
TS_HT_AE *Index 3* (TTAGGC) adapter sequence	
TS_AE3 top	5′-ACAGTAATACGACTCACTATAGGGAGAACAAGCAGAAGACGGCATACGAGATGCCTAAGTGACTGGAGTTCAGACGTGTGCTCTTCCGATCTCAGCAG
TS_AE3 bot[a]	5′-Phos-CGCTGCTGAGATCGGAAGAGCACACGTCTGAACTCCAGTCACTTAGGCATCTCGTATGCCGTCTTCTGCTTGTTCTCCCT ATAGTGAGTCGTATTACTG[b]T
TS_HT_AE *Index 4* (TGACCA) adapter sequence	
TS_AE4 top	5′-ACAGTAATACGACTCACTATAGGGAGAACAAGCAGAAGACGGCATACGAGATTGGTCAGTGACTGGAGTTCAGACGTGTGCTCTTCCGATCTCAGCAG
TS_AE4 bot	5′-Phos-CGCTGCTGAGATCGGAAGAGCACACGTCTGAACTCCAGTCACTGACCAATCTCGTATGCCGTCTTCTGCTTGTTCTCCCT ATAGTGAGTCGTATTACTG[b]T
TS_HT_AE *Index 5* (ACAGTG) adapter sequence	
TS_AE5 top	5′-ACAGTAATACGACTCACTATAGGGAGAACAAGCAGAAGACGGCATACGAGATCACTGTGTGACTGGAGTTCAGACGTGTGCTCTTCCGATCTCAGCAG
TS_AE5 bot	5′-Phos-CGCTGCTGAGATCGGAAGAGCACACGTCTGAACTCCAGTCACACAGTGATCTCGTATGCCGTCTTCTGCTTGTTCTCCCTATAGTGAGTCGTATTACTG[b]T
TS_HT_AE *Index 6* (GCCAAT) adapter sequence	
TS_AE6 top	5′-ACAGTAATACGACTCACTATAGGGAGAACAAGCAGAAGACGGCATACGAGATATTGGCGTGACTGGAGTTCAGACGTGTGCTCTTCCGATCTCAGCAG
TS_AE6 bot	5′-Phos-CGCTGCTGAGATCGGAAGAGCACACGTCTGAACTCCAGTCACGCCAATATCTCGTATGCCGTCTTCTGCTTGTTCTCCCTATAGTGAGTCGTATTACTG[b]T

(continued)

Table 1 (continued)

TS_HT_AE *Index 7* (CAGATC) adapter sequence	
TS_AE7 top	5′-ACAGTAATACGACTCACTATAGGGAGAACAAGCAGAAGACGGCATACGAGATGATCTGG TGACTGGAGTTCAGACGTGTGCTCTTCCGATCTCAGCAG
TS_AE7 bot	5′-Phos-CGCTGCTGAGATCGGAAGAGCACACGTCTGAACTCCAGTCACCAGATCATCTCGTA TGCCGTCTTCTGCTTGTTCTCCCTATAGTGAGTCGTATTACTG[b]T
TS_HT_AE *Index 8* (ACTTGA) adapter sequence	
TS_AE8 top	5′-ACAGTAATACGACTCACTATAGGGAGAACAAGCAGAAGACGGCATACGAGATTCAAGTGTG ACTGGAGTTCAGACGTGTGCTCTTCCGATCTCAGCAG
TS_AE8 bot	5′-Phos-CGCTGCTGAGATCGGAAGAGCACACGTCTGAACTCCAGTCACACTTGAATCTCGTAT GCCGTCTTCTGCTTGTTCTCCCTATAGTGAGTCGTATTACTG[b]T
TS_HT_AE *Index 9* (GATCAG) adapter sequence	
TS_AE9 top	5′-ACAGTAATACGACTCACTATAGGGAGAACAAGCAGAAGACGGCATACGAGATCTGATCGTG ACTGGAGTTCAGACGTGTGCTCTTC CGATCTCAGCAG
TS_AE9 bot	5′-Phos-CGCTGCTGAGATCGGAAGAGCACACGTCTGAACTCCAGTCACGATCAGATCTCGTAT GCCGTCTTCTGCTTGTTCTCCCTATAGTGAGTCGTATTACTG[b]T
TS_HT_AE *Index 10* (TAGCTT) adapter sequence	
TS_AE10 top	5′-ACAGTAATACGACTCACTATAGGGAGAACAAGCAGAAGACGGCATACGAGATAAGCTAGTG ACTGGAGTTCAGACGTGTGCTCTTCCGATCTCAGCAG
TS_AE10 bot[a]	5′-Phos-CGCTGCTGAGATCGGAAGAGCACACGTCTGAACTCCAGTCACTAGCTTATCTCGTAT GCCGTCTTCTGCTTGTTCTCCCTATAGTGAGTCGTATTACTG[b]T
TS_HT_AE *Index 11* (GGCTAC) adapter sequence	
TS_AE11 top	5′-ACAGTAATACGACTCACTATAGGGAGAACAAGCAGAAGACGGCATACGAGATGTAGCCGTG ACTGGAGTTCAGACGTGTGCTCTTCCGATCTCAGCAG
TS_AE11 bot	5′-Phos-CGCTGCTGAGATCGGAAGAGCACACGTCTGAACTCCAGTCACGGCTACATCTCGTA TGCCGTCTTCTGCTTGTTCTCCCTATAGTGAGTCGTATTACTG[b]T
TS_HT_AE *Index 12* (CTTGTA) adapter sequence	
TS_AE12 top	5′-ACAGTAATACGACTCACTATAGGGAGAACAAGCAGAAGACGGCATACGAGATTACAAGGT GACTGGAGTTCAGACGTGTGCTCTTCCGATCTCAGCAG
TS_AE12 bot	5′-Phos-CGCTGCTGAGATCGGAAGAGCACACGTCTGAACTCCAGTCACCTTGTAATCTCGTA TGCCGTCTTCTGCTTGTTCTCCCTATAGTGAGTCGTATTACTG[b]T
TS_HT_AE *Index 13* (AGTCAA) adapter sequence	
TS_AE13 top	5′-ACAGTAATACGACTCACTATAGGGAGAACAAGCAGAAGACGGCATACGAGATTTGACTGT GACTGGAGTTCAGACGTGTGCTCTTCCGATCTCAGCAG
TS_AE13 bot	5′-Phos-CGCTGCTGAGATCGGAAGAGCACACGTCTGAACTCCAGTCACAGTCAAATCTCGTA TGCCGTCTTCTGCTTGTTCTCCCTATAGTGAGTCGTATTACTG[b]T

TS_HT_AE *Index 14* (AGTTCC) adapter sequence	
TS_AE14 top	5′-ACAGTAATACGACTCACTATAGGGAGAACAAGCAGAAGACGGCATACGAGATGGAACTGTGACTGGAGTTCAGACGTGTGCTCTTCCGATCTCAGCAG
TS_AE14 bot	5′-Phos-CGCTGCTGAGATCGGAAGAGCACACGTCTGAACTCCAGTCACAGTTCCATCTCGTATGCCGTCTTCTGCTTGTTCTCCCTATAGTGAGTCGTATTACTG[b]T
TS_HT_AE *Index 15* (ATGTCA) adapter sequence	
TS_AE15 top	5′-ACAGTAATACGACTCACTATAGGGAGAACAAGCAGAAGACGGCATACGAGATTGACATGTGACTGGAGTTCAGACGTGTGCTCTTCCGATCTCAGCAG
TS_AE15 bot	5′-Phos-CGCTGCTGAGATCGGAAGAGCACACGTCTGAACTCCAGTCACATGTCAATCTCGTATGCCGTCTTCTGCTTGTTCTCCCTATAGTGAGTCGTATTACTG[b]T
TS_HT_AE *Index 16* (CCGTCC) adapter sequence	
TS_AE16 top	5′-ACAGTAATACGACTCACTATAGGGAGAACAAGCAGAAGACGGCATACGAGATGGACGGGTGACTGGAGTTCAGACGTGTGCTCTTCCGATCTCAGCAG
TS_AE16 bot	5′-Phos-CGCTGCTGAGATCGGAAGAGCACACGTCTGAACTCCAGTCACCCGTCCATCTCGTATGCCGTCTTCTGCTTGTTCTCCCTATAGTGAGTCGTATTACTG[b]T
TS_HT_AE *Index 17* (GTAGAG) adapter sequence	
TS_AE17 top	5′-ACAGTAATACGACTCACTATAGGGAGAACAAGCAGAAGACGGCATACGAGATCTCTACGTGACTGGAGTTCAGACGTGTGCTCTTCCGATCTCAGCAG
TS_AE17 bot	5′-Phos-CGCTGCTGAGATCGGAAGAGCACACGTCTGAACTCCAGTCACGTAGAGATCTCGTATGCCGTCTTCTGCTTGTTCTCCCTATAGTGAGTCGTATTACTG[b]T
TS_HT_AE *Index 18* (GTCCGC) adapter sequence	
TS_AE18 top	5′-ACAGTAATACGACTCACTATAGGGAGAACAAGCAGAAGACGGCATACGAGATGCGGACGTGACTGGAGTTCAGACGTGTGCTCTTCCGATCTCAGCAG
TS_AE18 bot	5′-Phos-CGCTGCTGAGATCGGAAGAGCACACGTCTGAACTCCAGTCACGTCCGCATCTCGTATGCCGTCTTCTGCTTGTTCTCCCTATAGTGAGTCGTATTACTG[b]T
TS_HT_AE *Index 19* (GTGAAA) adapter sequence	
TS_AE19 top	5′-ACAGTAATACGACTCACTATAGGGAGAACAAGCAGAAGACGGCATACGAGATTTTCACGTGACTGGAGTTCAGACGTGTGCTCTTCCGATCTCAGCAG
TS_AE19 bot	5′-Phos-CGCTGCTGAGATCGGAAGAGCACACGTCTGAACTCCAGTCACGTGAAAATCTCGTATGCCGTCTTCTGCTTGTTCTCCCTATAGTGAGTCGTATTACTG[b]T
TS_HT_AE *Index 20* (GTGGCC) adapter sequence	
TS_AE20 top	5′-ACAGTAATACGACTCACTATAGGGAGAACAAGCAGAAGACGGCATACGAGATGGCCACGTGACTGGAGTTCAGACGTGTGCTCTTCCGATCTCAGCAG
TS_AE20 bot	5′-Phos-CGCTGCTGAGATCGGAAGAGCACACGTCTGAACTCCAGTCACGTGGCCATCTCGTATGCCGTCTTCTGCTTGTTCTCCCTATAGTGAGTCGTATTACTG[b]T

(continued)

Table 1
(continued)

TS_HT_AE *Index 21* (GTTTCG) adapter sequence	
TS_AE21 top	5′-ACAGTAATACGACTCACTATAGGGAGAACAAGCAGAAGACGGCATACGAGATCGAAACGT GACTGGAGTTCAGACGTGTGCTCTTCCGATCTCAGCAG
TS_AE21 bot	5′-Phos-CGCTGCTGAGATCGGAAGAGCACACGTCTGAACTCCAGTCACGTTTCGATCTCGTA TGCCGTCTTCTGCTTGTTCTCCCTATAGTGAGTCGTATTACTG[b]T
TS_HT_AE *Index 22* (CGTACG) adapter sequence	
TS_AE22 top	5′-ACAGTAATACGACTCACTATAGGGAGAACAAGCAGAAGACGGCATACGAGATCGTACGGT GACTGGAGTTCAGACGTGTGCTCTTCCGATCTCAGCAG
TS_AE22 bot	5′-Phos-CGCTGCTGAGATCGGAAGAGCACACGTCTGAACTCCAGTCACCGTACGATCTCGTA TGCCGTCTTCTGCTTGTTCTCCCTATAGTGAGTCGTATTACTG[b]T
TS_HT_AE *Index 23* (GAGTGG) adapter sequence	
TS_AE23 top	5′-ACAGTAATACGACTCACTATAGGGAGAACAAGCAGAAGACGGCATACGAGATCCACTCGT GACTGGAGTTCAGACGTGTGCTCTTCCGATCTCAGCAG
TS_AE23 bot	5′-Phos-CGCTGCTGAGATCGGAAGAGCACACGTCTGAACTCCAGTCACGAGTGGATCTCGTA TGCCGTCTTCTGCTTGTTCTCCCTATAGTGAGTCGTATTACTG[b]T
TS_HT_AE *Index 24* (GGTAGC) adapter sequence	
TS_AE24 top	5′-ACAGTAATACGACTCACTATAGGGAGAACAAGCAGAAGACGGCATACGAGATGCTACCGT GACTGGAGTTCAGACGTGTGCTCTTCCGATCTCAGCAG
TS_AE24 bot	5′-Phos-CGCTGCTGAGATCGGAAGAGCACACGTCTGAACTCCAGTCACGGTAGCATCTCGT ATGCCGTCTTCTGCTTGTTCTCCCTATAGTGAGTCGTATTACTG[b]T
Reverse transcription-Tru-Seq primer	
TS_RT	5′-ATTGATACGGCGACCACCGAGATCTACACTCTTTCCCTACACGACGCTCTTCCGATC
PCR primers	
Primer_TS_P7_MS	5′-CAAGCAGAAGACGGCATACGAGAT
Primer_TS_P5_MS	5′-ATTGATACGGCGACCACCGA
Sequencing primer	
Illumina's Tru-Seq standard sequencing primer	

Top top strand, *bot* bottom strand
[a]phosphorothiate bond, *Phos* 5′ Phosphorylation
[b]5′ Phosphorylation 3′ Inverted dT

1. Pellet down 2–3 million cells at room temperature for 5 min at 200 × *g*. Remove the supernatant, resuspend the cell pellet in 1.0 mL of 1× PBS and wash once by spinning 5 min at 200 × *g*. Discard the supernatant.
2. Resuspend the cell pellet in 50 μL of 1× PBS. Pipette gently up and down until no cell clumps are visible. Add 500 μL of extraction buffer and incubate at 37 °C for 1 h in a water bath.
3. Add 2.75 μL of Proteinase K (20 mg/mL) to a final concentration of 100 μg/mL and incubate overnight at 50 °C in a water bath.
4. Add 1 volume (~550 μL) of TE-saturated phenol and mix completely but gently by inversion (10 min on a rocking platform is best); centrifuge for 5 min at room temperature at top speed in microcentrifuge (16,000 × *g*). Do not mix by vortexing since this may shear the DNA.
5. Transfer the supernatant (aqueous phase) into a new tube, leaving behind any impurities and being careful not to disturb the interface. Add an equal volume of PCI (25:24:1) and mix well by rocking for 10 min at room temperature, then centrifuge for 5 min at room temperature at 16,000 × *g*.
6. Transfer the supernatant into a new tube and measure the volume. If the supernatant is not clear, then repeat **step 5** until it becomes completely clear.
7. Transfer to cleared supernatant to pretreated dialysis tubing, clamp open end, and dialyze against 2.0 L of 2× SSC overnight at 4 °C, stirring gently.
8. When dialysis is complete, remove from 2× SSC, dry outside of tubing gently, and dredge with PEG 20,000 to cause water to exit by osmosis. Do not over-extract water, reduce volume to generate an expected DNA concentration in excess of 200 ng/μL, then wipe off PEG gently, unclamp the end of the tubing, and gently transfer DNA solution to an Eppendorf tube. Store at −20 °C (*see* **Note 3**).
9. Quantify the DNA using a spectrophotometer and run 1 μL on a 1% agarose gel.
10. Do not proceed with digestion or amplification if the DNA does not appear to be intact (*see* **Note 4**).

3.2 Library Preparation

Day 1

1. Set up a restriction digestion of 500 ng of genomic DNA with 2 μL (100 U) of either HpaII or MspI in separate 50 μL reactions, using NEB buffer #1 for HpaII and buffer #4 for MspI as recommended by the manufacturer (*see* **Note 5**).
2. Incubate overnight at 37 °C.

Day 2

3. Run 2 μL of the digested DNA on a 1% agarose gel. The two digests should appear different: for the HpaII digest, most of the DNA will remain high molecular weight, whereas with MspI there should appear an almost even smear with no remnant of high molecular weight DNA (*see* **Note 6**).
4. Add 450 μL of TE pH 8.0 to the digested DNA and 500 μL of saturated phenol:chloroform mix (1:1) and mix well. Centrifuge at 16,000 × *g* for 10 min at room temperature.
5. For isopropanol precipitation, remove the upper (aqueous) phase (about 500 μL) from the last step and transfer into a clean tube. Add 1 μL of Ethachinmate and 50 μL of 3 M sodium acetate and mix well.
6. Add 800 μL of isopropanol, incubate −20 °C for at least 30 min, and centrifuge at 16,000 × *g* for 20 min.
7. Remove the supernatant and wash the pellet with 70% ethanol and air dry the pellet. Resuspend the pellet in 5 μL of 10 mM Tris pH 8.0. Set up the linker ligation on the same day, since the digested DNA will have single-stranded overhangs that may degrade.
8. Set up a 10 μL reaction for ligating the Index Adapter TS_AE in a PCR tube as follows:

2× Quick ligase buffer	6.5 μL
DNA	5 μL
Index adapter TS_AE (0.1 μM)[a]	0.5 μL
Quick ligase	1 μL
	10 μL

[a]1:100 dilution of the 10 μM stock adapter

9. Incubate room temperature for 15 min.
10. Purify ligation products using a QiaQuick PCR purification kit according to the manufacturer's instructions. Elute in 30 μL elution buffer (EB).
11. Purify and size select ligation products using 28 μL AMPure XL beads according to the manufacturer's instructions. Elute in 11 μL water.
12. Set up the EcoP15I digestion reaction in a PCR tube with a total volume of 25 μL for each sample as follows:

DNA	11 μL
NEB buffer 3	2.5 μL
ATP (10 mM)	2.5 μL

(continued)

BSA	0.25 μL
EcoP15I	1.5 μL
Water	7.25 μL
	25 μL

13. Incubate at 37 °C overnight.

Day 3

14. Purify digestion products using a QiaQuick PCR purification kit according to the manufacturer's instructions. Elute in 21 μL elution buffer (EB) (*see* **Note 7**).
15. Set up the end repair reaction in a PCR tube with a total volume of 24.5 μL as follows:

DNA	19 μL
T4 DNA ligase buffer	2.5 μL
dNTP (10 mM)	0.5 μL
T4 DNA polymerase	1.25 μL
T4 DNA polynucleotide kinase	1.25 μL
	24.5 μL

16. Incubate 37 °C for 30 min.
17. Purify end repair products using a QiaQuick PCR purification kit according to the manufacturer's instructions. Elute in 13 μL elution buffer (EB) (*see* **Note 7**).
18. Set up the dA tailing reaction in a PCR tube with a total volume of 25 μL as follows:

DNA	20.5 μL
NEB buffer 2	2.5 μL
Klenow (3′ → 5′ exo)	1.5 μL
dATP (10 mM)	0.5 μL
	25 μL

19. Incubate 37 °C for 30 min.
20. Purify dA tailing products using a QiaQuick PCR purification kit according to the manufacturer's instructions. Elute in 13 μL elution buffer (EB) (*see* **Note 7**).
21. Set up a 25 μL reaction for ligating the TS_AS Adapter ligation in a PCR tube as follows:

2× Quick ligase buffer	12.5 μL
DNA	10 μL
Adapter TS_AS (1 μM)[a]	1.25 μL
Quick ligase	1.25 μL
	25 μL

[a]1:10 dilution of the 10 μM stock adapter

22. Incubate at room temperature for 15 min.
23. Purify ligation products using a QiaQuick PCR purification kit according to the manufacturer's instructions. Elute in 10 μL elution buffer (EB) (*see* **Note 7**).
24. Set up a 20 μL reaction for the In vitro transcription with the MEGAshortscript kit (RNA STEP) in a PCR tube as follows:

T7 10× reaction buffer	2 μL
T7 ATP solution (75 mM)	2 μL
T7 CTP solution (75 mM)	2 μL
T7 GTP solution (75 mM)	2 μL
T7 UTP solution (75 mM)	2 μL
DNA	8 μL
T7 Enzyme mix	2 μL
	20 μL

25. Incubate at 37 °C overnight.

Day 4

26. Add 1 μL of DNase Turbo and incubate at 37 °C for 15 min.
27. Purify ligation products using a RNeasy cleanup according to the manufacturer's instructions. Elute in 10 μL RNase-free water (*see* **Note 7**).
28. Set up a 10 μL reaction for the reverse transcription step using SuperScript III in a PCR tube as follows:

RNA	8 μL
Primer TS_RT (2 μM)[a]	1 μL
dNTPs	1 μL
	10 μL

[a]1:5 dilution of the 10 μM stock primer

29. Incubate at 65 °C for 5 min and then chill on ice.

30. Add the following reagents to a final volume of 20 μL:

10× RT buffer	2 μL
0.1 M DTT	2 μL
$MgCl_2$	4 μL
RNase out	1 μL
SuperScriptIII	1 μL
	20 μL final volume

31. Incubate at 25 °C for 5 min, 50 °C for 1 h, and 80 °C for 10 min.
32. PCR amplify the RT products using Phusion polymerase. Do single reaction first, then redo in triplicate. Set up an amplification mix as follows in a total volume of 50 μL in a thermocycler using the below described protocol:

5× Phusion buffer	10 μL
RT product	5 μL
Primer TS_P5_MS (10 μM)	0.5 μL
Primer TS_P7_MS (10 μM)	0.5 μL
dNTPs	1 μL
Water	32.5 μL
Phusion polymerase	0.5 μL
	50 μL

PCR cycles:
98 °C for 2 min
18 cycles: 98 °C for 15 s 60 °C for 15 s 72 °C for 15 s
Followed by
72 °C for 5 min

33. Verify the size of the PCR product (~160–170 bp) on a 3% agarose gel.
34. Combine all four PCR products using the MinElute PCR purification kit. Elute in 10 μL of EB buffer.
35. Size select products for sequencing on a 3–4% agarose gel (*see* **Note 8**). Purify products using the MinElute Gel Extraction Kit. Do not heat any buffers! Elute in 12 μL of EB buffer.

36. Confirm desired product size on a 3% agarose gel.
37. QC the library by Qubit and bioanalyzer, The library size should be around 160 bp.

3.3 DNA Sequencing

Multiplexed libraries are loaded onto the Illumina HiSeq 2500 or higher aiming to generate ~20 million clusters per library, usually eight samples per lane. Single-end 50 bp sequencing is performed.

3.4 Data Analysis and Interpretation

The data analysis is also described in Jing et al. [6].

3.4.1 Library Quality Assessment and Pre-alignment Tag Processing

Library quality is assessed by determining the proportion and position of the 3′ adapter sequence within the reads. A high percentage of reads containing adapter sequence starting at around position 27 is indicative of a high-quality library. Reads with excessive tracts of unknown bases and those failing to contain 5′ adapter sequence near the 3′ end are to be discarded.

3.4.2 Alignment

The trimmed and masked sequence tags are mapped back to the appropriate reference genome. During the alignment process, a maximum of two mismatches are allowed and indels are ignored. Statistics generated from the alignment included the number of sequence tags that are rejected because of there being too many matches to the reference genome or for which there was no match at all.

3.4.3 HpaII Site Tag Counting and Angle Calculation

Counts of sequence tags aligned adjacent to every annotated HpaII/MspI site across the genome are assessed. Tags that aligned to more than one locus are given a proportional count relative to the number of mapped loci, e.g., a tag mapping to two loci is counted as 0.5 and one mapping to 10 loci counted as 0.1. Cumulative counts at each annotated HpaII site are then normalized to represent a fraction of the total number of sequence tags aligned to annotated HpaII sites genome-wide.

In order to quantify the level of methylation at each HpaII/MspI site, the normalized accumulative proportional (NAP) count for the HpaII digested sample can be compared to the NAP count for the MspI digested sample. If the site is hypermethylated, the HpaII NAP count of this site should be less than the MspI NAP count since HpaII restriction enzyme is unable to cut DNA when the internal cytosine is methylated, whereas MspI enzyme is able to cut regardless of the cytosine methylation state. More specifically, the more the HpaII NAP count approaches the MspI NAP count, the less methylated the CCGG site is. To quantify the methylation level of each CCGG locus, the NAP counts for HpaII and MspI digested samples can be represented as Cartesian coordinates for vectors projected in two-dimensional space where the Y-axis represents the HpaII NAP count and the *X*-axis represents the MspI NAP count. Represented as a vector, the direction (angle relative to the origin) corresponds to a quantification of hypomethylation and the magnitude represents a measure of the tag counts, i.e., information content. Thus, the magnitude of each vector provides a level of confidence in the quantification (the greater the magnitude, the more tag counts contributed to the quantification and therefore the more confident we are in the result). The direction of each vector is calculated as the arc tangent of the ratio of the HpaII NAP count and the MspI NAP count. The final step in data processing is a linear scaling of these angular values to a range from 0 to 100, representing the proportional methylation at that locus. These data are stored in a WIG format file for easy viewing of the data as a histogram within the UCSC genome browser. As fully methylated CCGG sites have an angle value of 0, which would not appear on a genome browser view, the site is marked with a short tag underneath the *X*-axis (negative value) when the track is displayed within the genome browser.

An algorithm was developed to provide a level of confidence in methylation quantification based on vector magnitude (see explanation above) and it was decided to classify confidence as either high, medium, or low based on the distribution of the logarithm of magnitudes. A typical density plot of the logarithm of the vector magnitudes for data derived from an arbitrary HpaII digested sample showed that the data approximate a normal distribution. The three categories of confidence level are defined by using the mean value and standard deviation of this distribution. If the magnitude of a vector representing a particular CCGG site falls in the range of the mean value plus/minus the standard deviation, it would be categorized as medium confidence; if the magnitude is below or above this range, it would be categorized as being of low or high confidence respectively.

3.5 HELP-GT Modification

While the assay was originally developed to measure 5-methylcytosine, it is possible to discriminate the signal from 5-hydroxymethylcytosine by adding a step to the above protocol. We describe this protocol in more detail in our publication [8]. The assay is based on the ability to glycosylate 5-hmC residues using beta-glucosyl transferase (β-GT), protecting these sites from MspI digestion. By performing parallel assays in which the sample is divided and one pretreated with β-GT, followed by MspI digestion, a comparison of the β-GT+ and β-GT− MspI representations shows a reduction of sequence tags at 5-hydroxymethylated sites. As the same CCGG site is tested for both HELP-tagging and HELP-GT assays, the results are readily integrated.

4 Notes

1. The primers are pre-annealed in advance for convenience. If preferred, they can be pre-annealed immediately prior to use. Mix equal volumes of the pairs of linkers (6 OD/mL and 12 OD/mL, respectively) in a screw-top Eppendorf. Boil for 5 min and then allow the reaction to cool down to room temperature. The annealed linkers can then be stored at −20 °C.
2. Dialysis is not absolutely required, we have had numerous instances of successful HELP assays with ethanol-precipitated DNA. We recommend dialysis because of the more consistent quality obtained.
3. Osmosis will cause increased salt concentration in the DNA solution, which would be a problem if TE were used at this step, and may be a problem even with SSC if excessive water removal and salt concentration occurs. If the volume will need to be reduced substantially (e.g., tenfold), it is probably worth adding a further dialysis step against a more dilute buffer prior to osmotic water removal.
4. A small amount of DNA shearing is probably OK, as the later adapter ligation step involves compatible cohesive ends, so nonspecifically broken DNA molecules will not be amplified.
5. We routinely get <300 ng of DNA to work for HELP-tagging.
6. Some DNA samples (e.g., from DNA methyltransferase-deficient or 5-aza-2-deoxycytidine-treated cells) may have so little DNA methylation that the *Hpa*II digestions resemble those for *Msp*I.
7. Before loading EB, use a fine tip to remove any remaining wash from the column.
8. Occasionally two distinct bands can be seen between 160 and 180 bp. Both bands may be extracted for sequencing. The primary band is the lower band.

Acknowledgments

This work was originally supported by a grant from the National Institutes of Health (NCI) R03 CA111577 and subsequently by NICHD grant R01 HD044078 to J.M.G. The author also acknowledges the contributions of the Center for Epigenomics at the Albert Einstein College of Medicine.

References

1. Khulan B, Thompson RF, Ye K et al (2006) Comparative isoschizomer profiling of cytosine methylation: the HELP assay. Genome Res 16:1046–1055
2. Oda M, Glass JL, Thompson RF et al (2009) High-resolution genome-wide cytosine methylation profiling with simultaneous copy number analysis and optimization for limited cell numbers. Nucleic Acids Res 37:3829–3839
3. Suzuki M, Greally JM (2010) DNA methylation profiling using HpaII tiny fragment enrichment by ligation-mediated PCR (HELP). Methods 52:218–222
4. Suzuki M, Jing Q, Lia D et al (2010) Optimized design and data analysis of tag-based cytosine methylation assays. Genome Biol 11:R36
5. Smith ZD, Gu H, Bock C et al (2009) High-throughput bisulfite sequencing in mammalian genomes. Methods 48:226–232
6. Jing Q, McLellan A, Greally JM et al (2012) Automated computational analysis of genome-wide DNA methylation profiling data from HELP-tagging assays. Methods Mol Biol 815:79–87
7. Gissot M, Choi SW, Thompson RF et al (2008) Toxoplasma gondii and Cryptosporidium parvum lack detectable DNA cytosine methylation. Eukaryot Cell 7:537–540
8. Bhattacharyya S, Yu Y, Suzuki M et al (2013) Genome-wide hydroxymethylation tested using the HELP-GT assay shows redistribution in cancer. Nucleic Acids Res 41:e157

Chapter 12

Comprehensive Whole DNA Methylome Analysis by Integrating MeDIP-seq and MRE-seq

Xiaoyun Xing, Bo Zhang, Daofeng Li, and Ting Wang

Abstract

Understanding the role of DNA methylation often requires accurate assessment and comparison of these modifications in a genome-wide fashion. Sequencing-based DNA methylation profiling provides an unprecedented opportunity to map and compare complete DNA CpG methylomes. These include whole genome bisulfite sequencing (WGBS), Reduced-Representation Bisulfite-Sequencing (RRBS), and enrichment-based methods such as MeDIP-seq, MBD-seq, and MRE-seq. An investigator needs a method that is flexible with the quantity of input DNA, provides the appropriate balance among genomic CpG coverage, resolution, quantitative accuracy, and cost, and comes with robust bioinformatics software for analyzing the data. In this chapter, we describe four protocols that combine state-of-the-art experimental strategies with state-of-the-art computational algorithms to achieve this goal. We first introduce two experimental methods that are complementary to each other. MeDIP-seq, or methylation-dependent immunoprecipitation followed by sequencing, uses an anti-methylcytidine antibody to enrich for methylated DNA fragments, and uses massively parallel sequencing to reveal identity of enriched DNA. MRE-seq, or methylation-sensitive restriction enzyme digestion followed by sequencing, relies on a collection of restriction enzymes that recognize CpG containing sequence motifs, but only cut when the CpG is unmethylated. Digested DNA fragments enrich for unmethylated CpGs at their ends, and these CpGs are revealed by massively parallel sequencing. The two computational methods both implement advanced statistical algorithms that integrate MeDIP-seq and MRE-seq data. M&M is a statistical framework to detect differentially methylated regions between two samples. methylCRF is a machine learning framework that predicts CpG methylation levels at single CpG resolution, thus raising the resolution and coverage of MeDIP-seq and MRE-seq to a comparable level of WGBS, but only incurring a cost of less than 5% of WGBS. Together these methods form an effective, robust, and affordable platform for the investigation of genome-wide DNA methylation.

Key words MeDIP-seq, MRE-seq, M&M, methylCRF

1 Introduction

DNA methylation typically refers to the methylation of the 5′ position of cytosine (mC) by DNA methyltransferases (DNMT). It is a major epigenetic modification in human and many other species [1]. In mammals, most DNA methylations occur within the context of CpG dinucleotides [2, 3], although some non-CpG (i.e.,

Jörg Tost (ed.), *DNA Methylation Protocols*, Methods in Molecular Biology, vol. 1708,
https://doi.org/10.1007/978-1-4939-7481-8_12, © Springer Science+Business Media, LLC 2018

CHG, CHH) methylation is also observed in embryonic stem cell [4, 5]. DNA methylation is thought to be a repressive chromatin modification. The abnormal loss of DNA methylation in the genome, especially in repetitive elements, may result in genome destabilization [1, 6–8]. DNA methylation plays a crucial role in normal development [1, 8, 9]. Aberrant methylation can lead to many diseases including cancers [10, 11].

Analyses of DNA methylation have been traditionally focused on CpG dense regions including gene promoters and CpG islands, and more recently on CpG island shores [12], partly due to limitation in technology. These regions only account for a small fraction of the genome-wide CpGs (7.4% of the 28 M CpGs in human genome are in CpG islands, and an additional 4.9% are in island shores). Just in the past several years, the rapid development of next-generation sequencing-based technology allowed much more comprehensive views of genome-wide DNA methylation patterns. Perhaps the most exciting findings are related to the dynamic nature of DNA methylation changes on regulatory elements, especially on distal enhancers, which usually locate in intergenic and intronic regions [13–16]. For example, Stadler et al. discovered that DNA binding factors could lead to demethylation of distal regulatory regions in the mouse. By investigating intergenic hypomethylated regions in various human cell types, Schlesinger et al. suggested that de novo DNA demethylation defines distal regulatory elements [17]. Xie et al. discovered that thousands of transposable elements undergo DNA hypomethylation in a tissue-specific manner, and could serve as tissue-specific enhancers [18]. Hon et al. pointed out that identifying tissue-specific DMRs (tsDMRs) can be an alternative strategy for finding putative regulatory elements [13]. By profiling 30 different tissues and cell types, Ziller et al. estimated that 21.8% of the DNA methylome is dynamic [16]. These recent studies highlight the importance of measuring CpG methylation in a genome-wide, unbiased fashion.

Sequencing-based DNA methylation profiling methods provide an opportunity to map complete DNA methylomes. These technologies include whole genome bisulfite sequencing (WGBS, MethylC-seq [19, 20] or BS-seq [21] *see also* Chapters 5–7 and 9), Reduced-Representation Bisulfite-Sequencing (RRBS, *see also* Chapter 8) [22, 23], enrichment-based methods (MeDIP-seq [24, 25], MBD-seq [26] *see also* Chapter 10, mTAG-seq [27], and methylation-sensitive restriction enzyme-based methods (HELP [27] *see also* Chapter 11, MRE-seq [25]). These methods yield largely concordant results, but differ significantly in the extent of genomic CpG coverage, resolution, quantitative accuracy, and cost [28, 29]. For example, WGBS-based methods produce the most comprehensive and high-resolution DNA methylome maps, but typically require sequencing to 30× coverage which is still expensive for the routine analysis of many samples, particularly those with a large methylome (e.g., human). Additionally, bisulfite-based methods including WGBS and RRBS conflate methylcytosine (mC) and hydroxymethylcytosine (hmC) [30] unless combined with additional experiments [31, 32] (*see also* Chapters 34 and 35).

Both MeDIP-seq and MRE-seq are technologies for measuring DNA methylation at the genome-wide level. They represent complementary ways to enrich for either the methylated portion of the genome, or the unmethylated portion of the genome. Each technology has its own advantages and disadvantages, and can be applied independently or jointly. We will first describe each technology, and then discuss how integrating the two complementary technologies significantly increase their values (Fig. 1).

MeDIP-seq stands for methylation-dependent immunoprecipitation followed by sequencing (Fig. 1). In a typical MeDIP-seq protocol, a monoclonal antibody against 5-methylcytidine is used to enrich methylated DNA fragments of select sizes (typically 150–300 bp), then these fragments are sequenced and analyzed. MeDIP-seq was first reported by Weber et al. [24]. A protocol working on 50 ng of input DNA was published in 2012 [33]. It has been estimated that MeDIP-seq saturates at 5 Gb of sequencing [34].

MRE-seq stands for methylation-sensitive restriction enzyme digestion followed by sequencing (Fig. 1). In a typical MRE-seq protocol, genomic DNA is digested in parallel with 3–5 methylation-sensitive restriction endonucleases. These methylation-sensitive enzymes cut only restriction sites with unmethylated CpGs. Sequencing of digested DNA fragment ends allows identification of cut sites, which provides information of the methylation status of these sites. While methylated CpGs could be inferred by the absence of reads at cutting sites, this would require assuming perfect digestion which is not typically achieved in practice. MRE-seq was first reported in Manuakea et al. [25]. Utilizing multiple cut-sites, MRE-seq can cover close to 30% of the genome and saturates at ~3 Gb of sequencing [34].

Both MeDIP-seq and MRE-seq detect 5-methylcytosine exclusively. MRE-seq provides DNA methylation estimates at single CpG resolution, but is considered low coverage due to the limit of CpG containing recognition sites. An important advantage of MeDIP over enzymatic digestion-based methods is the lack of bias for a specific nucleotide sequence, other than CpGs. However, the relationship of enrichment to absolute methylation levels is confounded by variables such as CpG density [35]. Another inherent limitation of MeDIP-seq is its lower resolution (~150 bp) compared to MRE-seq or bisulfite-based methods in that one or more of the CpGs in the immunoprecipitated DNA fragment could be responsible for the antibody binding.

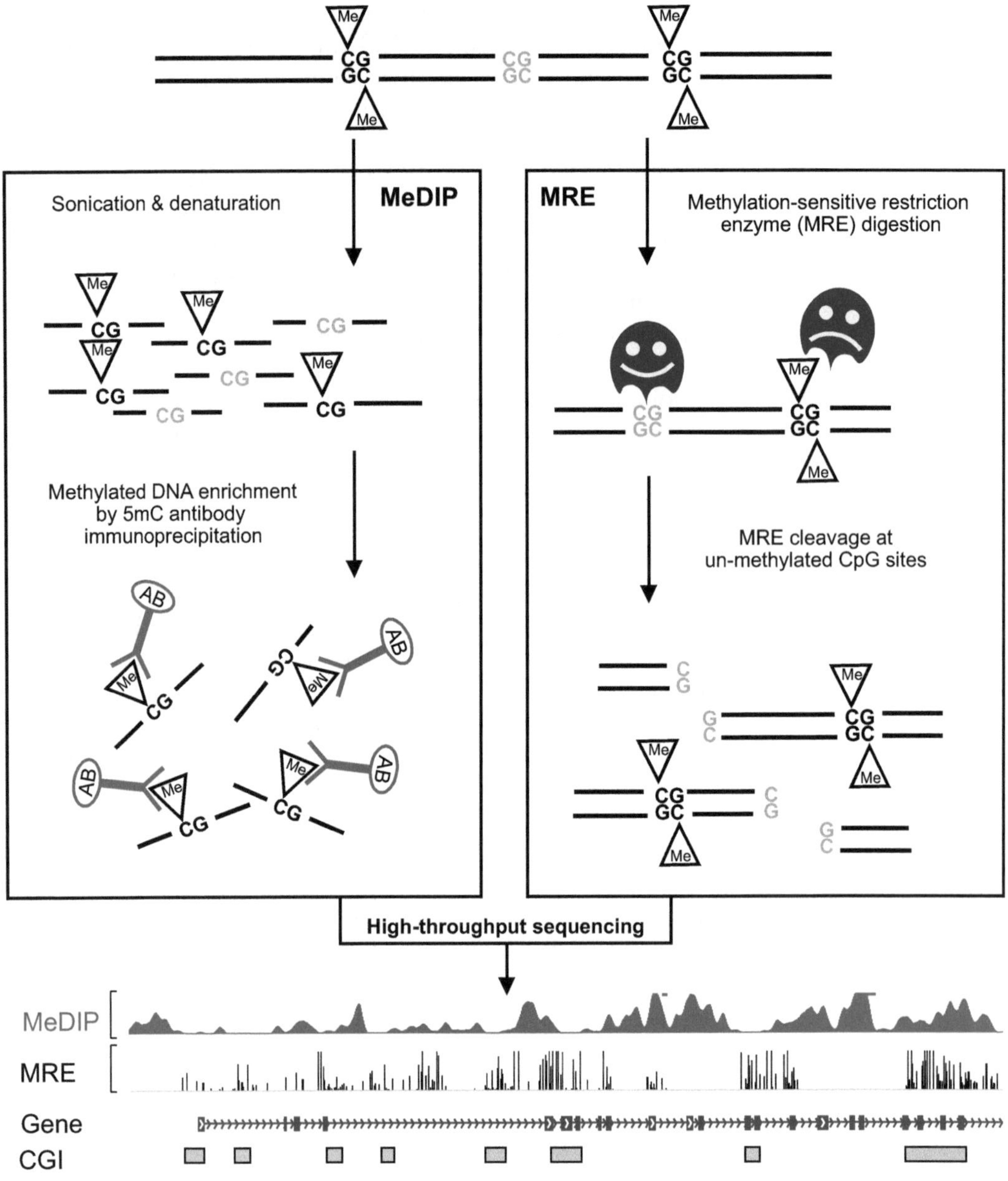

Fig. 1 Integration of MeDIP-seq and MRE-seq. Workflow of MeDIP-seq and MRE-seq. Genomic DNA is isolated and purified. For MeDIP-seq (*left panel*), genomic DNA is sonicated to a specific size range, and a monoclonal anti-5′methylcytidine antibody is used to enrich for methylated DNA fragments. Immunoprecipitated DNA fragments are then sequenced and mapped back to the reference genome assembly to review methylated regions. For the MRE-seq approach (*right panel*), several methylation-sensitive restriction endonucleases are used to digest intact genomic DNA. The resulting DNA fragments are then size-selected and sequenced. When mapped back to the reference genome assembly, these sequencing reads can reveal locations of unmethylated CpG sites, which are located within recognition sites of specific restriction enzymes. Both MeDIP-seq and MRE-seq results can be easily visualized in a Genome Browser

The MeDIP-seq and MRE-seq technologies each have their own advantages and disadvantages compared to other genome-wide DNA methylation profiling technologies, and both are widely and independently used by investigators. Because MeDIP-seq and MRE-seq are independent and complementary, combining the two assays provides additional, synergistic advantages, as we have demonstrated previously [25, 28]. Their combined value can be significantly increased with the application of the two algorithms we developed: M&M [15] and methylCRF [36].

M&M is a new statistical framework that identifies differentially methylated regions (DMRs) by jointly modeling MeDIP-seq and MRE-seq data. Experimental measurements can be modeled as a function of the underlying methylation state and genomic context. Thus, independent measurements (i.e., MeDIP-seq and MRE-seq) of the same sample can be integrated on the same underlying methylation state. Detecting DMRs between two samples can then be modeled as a test of the null hypothesis, namely that the two samples have the same methylation state, given the observed measurements and genomic context. We formulated the statistical test and provided a numerical solution to compute a probability that a genomic region is differentially methylated given observed MeDIP-seq and MRE-seq measurements [15].

methylCRF is a Conditional Random Fields-based [37, 38] algorithm that integrates MeDIP-seq and MRE-seq data to predict genome-wide DNA methylation levels at single CpG resolution. Because MeDIP-seq is enrichment-based, its resolution is limited by the size of the DNA fragments from immunoprecipitation. MRE-seq is a single CpG resolution method but is limited by the availability of restriction enzyme recognition sites in the genome. Based on the same principle that experimental measurements can be modeled as a function of the underlying methylation state and genomic context, Conditional Random Fields provide a machine learning framework to infer the absolute values of the underlying methylation state at single CpG level. We demonstrated that methylCRF transforms MeDIP-seq and MRE-seq data to the equivalent of a whole genome bisulfite sequencing (WGBS) experiment, which typically costs 20 times more to produce [36].

The main advantage of these two algorithms is the integration of independent, heterogeneous experiments on the same biological state that they measure—in this case, DNA methylation. All current genome-wide technologies for measuring DNA methylation have their inherent biases and limitations, but our confidence in inferring methylation state increases when results from two independent methods are integrated. For example, a decrease of MeDIP-seq signal could reflect a biological event (we infer that this region is demethylated) or could be a methodological artifact; but if it is corroborated by an increase in MRE-seq signal, then the inference of demethylation can be much more accurate. Taken together, the integrated protocols we describe in this chapter provide a streamlined platform for investigators to explore DNA methylation with high coverage, high resolution, and low cost.

2 Materials

2.1 MeDIP-seq

2.1.1 Genomic DNA Extraction

1. Extraction buffer: 50 mM Tris–HCl (pH 8.0), 1 mM EDTA (pH 8.0), 0.5% SDS and 1 mg/mL proteinase K (*see* **Note 1**).
2. Phase lock gel light 2 mL.
3. Phenol/chloroform/isoamyl alcohol.
4. RNase A (DNase and protease-free).
5. Chloroform.
6. 3 M Sodium Acetate, pH 5.5.
7. Ethanol.
8. Elution buffer (EB) (Qiagen).
9. NanoVue spectrophotometer (GE Healthcare) or similar.
10. Qubit fluorometer for DNA quantification (Thermo Fisher Scientific).
11. Qubit dsDNA HS Assay Kit (Thermo Fisher Scientific).
12. Qubit assay tubes (Thermo Fisher Scientific).
13. Agarose, Molecular Biology Grade.
14. Ethidium bromide.
15. TAE 50× buffer.
16. Orange DNA loading dye.
17. 2-Log DNA ladder.
18. Gel electrophoresis system.
19. Gel Imaging System (e.g., Gel Doc imaging system, Bio-Rad).

2.1.2 DNA Fragmentation

1. Bioruptor Pico system for DNA sonication (Diagenode).
2. Gel electrophoresis system for fragments size validation.
3. Gel Imaging System (e.g., Gel Doc imaging system, Bio-Rad).

2.1.3 Library Preparation I

1. Reagents for DNA end repair.
   - (a) DNA Polymerase I, Large (Klenow) Fragment (e.g., New England Biolabs).
   - (b) T4 DNA Polymerase (e.g., New England Biolabs).
   - (c) T4 Polynucleotide Kinase (e.g., New England Biolabs).
   - (d) Deoxynucleotide Solution Mix (e.g., New England Biolabs).

(e) Adenosine 5′-Triphosphate (ATP) (e.g., New England Biolabs).

2. Reagents for dA-tailing.

(a) Klenow Fragment (3′ → 5′ exo-) (e.g., New England Biolabs).

(b) dATP solution (e.g., New England Biolabs).

3. Reagents for adapter ligation.

(a) T4 ligase reaction buffer for adapter annealing (e.g., New England Biolabs).

(b) Adapter oligonucleotides (HPLC purification; *see* Table 1) (*see* **Note 2**).

Spin down lyophilized adapter oligos and resuspend each oligo in 1× T4 ligase reaction to 20 μM. Add equimolar quantities of each adapter into a 1.5-mL microcentrifuge tube. Divide into small aliquots (50 μL) and incubate at 95 °C for 10 min, and then leave to cool down to room temperature. Store aliquots at—20 °C until required. Final concentration is 10 μM.

(c) Quick Ligation Kit (New England Biolabs).

Table 1
Oligonulceotides used in the protocol

Name	Sequence (5′-3′)	Length (bp)	Amplicon size (bp)
Adapter: PE 1.0	ACACTCTTTCCCTACA CGACGCTCTTCCGATC*T	33	NA
Adapter: PE 2.0	**P**-GATCGGAAGAGCACACG TCTGAACTCCAGTCAC	33	NA
PCR primer: PE 1.0	AATGATACGGCGACCACCGAG ATCTACACTCTTTCCCTACACG ACGCTCTTCCGATCT	58	NA
PCR primer: PE 2.0	CAAGCAGAAGACGGCATACGAGA T**NNNNNNN**GTGACTGGAGTT CAGACGTGTGCTCTTCCGA	62	NA
SNRPN-F	CGCTCAACACCCCCTAAATA	20	156
SNRPN-R	GGTGGAGGTGGGTACATCAG	20	
GABRB3-F	CCTGCAACTTTACTGAATTTAGC	23	207
GABRB3-R	GGAATCTCACTTTCACCACTGG	22	
qPCR primer 1.0	AATGATACGGCGACCACCGAGAT	23	NA
qPCR primer 2.0	CAAGCAGAAGACGGCATACGA	21	NA

*phosphorothiate linkage, *P* phosphate group, *N* index code, *NA* not applicable

4. Agencourt AMPure XP beads for DNA purification (Beckman Coulter).
5. Thermal cycler for incubation.

2.1.4 MeDIP

1. 2 M NaCl.
2. 0.2 M Na_2HPO_4.
3. 0.2 M NaH_2PO_4.
4. 0.1 M Na_2HPO_4/NaH_2PO_4 (pH 7.0), sterile-filter and store at room temperature.
5. 10% Triton X-100.
6. Monoclonal Antibody against 5-Methylcytidine (Eurogentec; cat no. BI-MECY-0100, also available from other vendors).
7. AffiniPure Rabbit Anti-Mouse IgG, Fcγ Fragment Specific.
8. Protein A/G agarose beads.
9. MeDIP wash buffer: 10 mM Na_2HPO_4/NaH_2PO_4, 140 mM NaCl and 0.05% Triton X-100. Prepare fresh on the day of use.
10. MeDIP elution buffer: 0.25 mg/ml proteinase K and 0.25% SDS in 1× TE (0.1 M Tris–HCl, 0.01 M EDTA, pH 8.0). Prepare fresh on the day of use.
11. MinElute PCR Purification Kit for DNA purification (Qiagen).
12. Rotating wheel.

2.1.5 Library Preparation II

1. Phusion High-Fidelity DNA Polymerase (New England Biolabs).
2. Deoxynucleotide Solution Mix.
3. Indexed PCR primers used for adapter-modified PCR enrichment (HPLC purified; *see* Table 1).
   - Spin down primer oligos and resuspend each oligo in ultrapure water to create 100 μM stocks. Dilute an aliquot tenfold in ultrapure water to create 10 μM dilutions. Store dilutions at −20 °C until required.
4. Agencourt AMPure XP beads for DNA purification (Beckman Coulter).

2.1.6 Library Size Selection

1. MinElute Gel extraction kit (Qiagen).
2. Isopropanol.
3. 2-Log DNA ladder.
4. 2% agarose TAE gel (wt/vol).
5. Gel electrophoresis system for size selection.
6. Gel Imaging System (e.g., Gel Doc imaging system, Bio-Rad).

2.1.7 Library Quality Control

1. Primers (SNRPN, *GABRB3*) for validation of human MeDIP-seq libraries (*see* Table 1).
2. qPCR primers for identification of libraries (*see* Table 1).
3. 0.1% Tween-20.
4. iTaq universal SYBR Green supermix (2×) (Bio-Rad).
5. Thermal cycler for qPCR (e.g., CC007277 CFX from Bio-Rad or similar).
6. Qubit fluorometer for quantification (Thermo Fisher Scientific).
7. Bioanalyzer (Agilent).

2.1.8 Next-Generation Sequencing

1. HiSeq2500 or similar (Illumina).

2.2 MRE-seq

2.2.1 Genomic DNA Extraction

Same as Subheading 2.1.1.

2.2.2 DNA Fragmentation

1. Restriction enzymes for DNA digestion.
   (a) HpaII (e.g., New England Biolabs).
   (b) HinP1I (e.g., New England Biolabs).
   (c) AciI (e.g., New England Biolabs).
   (d) HpyCH4IV (e.g., New England Biolabs).
   (e) Bsh1236I (e.g., Thermo Fisher Scientific).
2. Gel electrophoresis system for size selection.
3. Gel Imaging System (e.g., Gel Doc imaging system, Bio-Rad).

2.2.3 Library Preparation I

Same as Subheading 2.1.3, except that only Klenow fragment is used for end repair.

2.2.4 Library Preparation II

Same as Subheading 2.1.5.

2.2.5 Library Size Selection

Same as Subheading 2.1.6.

2.2.6 Library QC

Same as Subheading 2.1.7, except that MeDIP-seq-specific primers are not used.

2.2.7 Next-Generation Sequencing

Same as Subheading 2.1.8.

2.3 Computational Environment for Running M&M and methylCRF

Program executions are demonstrated in a Linux/Unix environment. Typical Linux system should have Python and Perl installed already. In the *software* section, we recommend adding directory or file to the PATH environment. Setting up this variable makes access to the computational programs easier. For more information on the PATH variable in a Linux system, please consult this page: http://www.linfo.org/path_env_var.html. Please also notice software version might change. Newest version is always recommended.

Command lines are numbered if they are broken into more than one line to avoid confusion.

1. Cutadapt (Optional)

   Description: adapter trimming and quality filtering of raw reads.
   Website: https://code.google.com/p/cutadapt/.
   Documentation: https://code.google.com/p/cutadapt/wiki/documentation.
   Installation instructions:
   Using following command to install cutadpt;

```
easy_install cutadapt
```

   Or using pip command instead:

```
pip install cutadapt
```

2. BWA
   Description: Align raw or filtered reads to corresponding genome assembly.
   Website: http://bio-bwa.sourceforge.net/.
   Documentation: http://bio-bwa.sourceforge.net/bwa.shtml.
   Installation instructions:
   (a) Download BWA from http://sourceforge.net/projects/bio-bwa/files/, the latest version is recommended (here we used 0.7.10 as an example).
   (b) Uncompress the downloaded file:

```
tar -jxvf bwa-0.7.10.tar.bz2
```

   (c) Go to the newly generated folder, type make command to generate bwa binary file:

```
make
```

   (d) (optional) move the bwa binary file to your PATH (requires administrator privileges):

```
sudo mv bwa /usr/local/bin/
```

3. Samtools

   Description: Operating aligned result files.
   Website: http://samtools.sourceforge.net/.
   Documentation: http://samtools.sourceforge.net/SAM1.pdf.
   Installation instructions:

   (a) Download samtools from http://sourceforge.net/projects/samtools/files/samtools/, the latest version is recommended (here we used version 0.7.19 as example).

   (b) Decompress the downloaded file:

```
tar -jxvf samtools-0.1.19.tar.bz2
```

   (c) Go to the newly created folder, type make command to generate samtools binary file:

```
make
```

   (d) (optional) Move the samtools binary file to your PATH (requires administrator privileges), for example:

```
sudo mv samtools /usr/local/bin/
```

4. R

   Description: Statistical computing environment.
   Website: http://www.r-project.org/.
   Documentation: http://cran.r-project.org/manuals.html.
   Installation instructions:
   R installation steps may vary between different Linux systems; users should follow documentation listed at http://www.r-project.org/.

5. methylQA

   Description: Parsing aligned results, generating bed files and bedGraph files.
   Website: http://methylqa.sourceforge.net/.
   Documentation: http://methylqa.sourceforge.net/tutorial.php.
   Installation instructions:

   (a) Download methylQA from http://sourceforge.net/projects/methylqa/files/, the latest version is recommended (here we used 0.1.5 as an example).

   (b) Decompress the downloaded file:

```
tar -zxvf methylQA-0.1.5.tar.gz
```

(c) Go to the newly created folder, type make command to generate the methylQA binary file:

```
make
```

(d) (optional) move the methylQA binary file to your PATH (requires administrator privileges):

```
sudo mv methylQA /usr/local/bin/
```

6. methylM&M

Description: Calculating differentially methylated region (DMR) between two samples by integrating MeDIP-seq and MRE-seq data.

Website: http://epigenome.wustl.edu/MnM/.

Documentation: http://epigenome.wustl.edu/MnM/methylMnM.pdf.

Installation instructions:

(a) Go to the R environment by typing R command. Enter an R interactive interface. Install methylMnM package from Bioconductor using the biocLite command:

```
> source("http://bioconductor.org/biocLite.R")
> biocLite("edgeR")
> biocLite("statmod")
> biocLite("methylMnM")
```

(b) Test loading the methylMnM package using the ibrary command; Make sure there is no error message.

```
> library(methylMnM)
```

7. methylCRF

Description: Predicting methylation level at single CpG resolution by combining MeDIP-seq and MRE-seq data.

Website: http://methylcrf.wustl.edu/.

Documentation: http://methylcrf.wustl.edu/.

Installation instructions:

(a) Download the methylCRF package and decompress the downloaded file:

```
wget http://methylcrf.wustl.edu/methylCRF.tgz
tar -zxf methylCRF.tgz
```

(b) Go to the newly created folder and type make command:

```
cd methylCRF
make
```

(c) Add the methylCRF folder to your PATH environment variable.

(d) Download and decompress methylCRF model and data files. These files are needed for methylCRF and should be stored in the methylCRF folder:

```
wget http://methylcrf.wustl.edu/h1es_mdl.tgz
tar -zxf h1es_mdl.tgz
wget http://methylcrf.wustl.edu/hg19/hg19_gdat.tgz
tar -zxf hg19_gdat.tgz
```

(e) Download MRE fragment files, 3enz for three enzymes and 5enz for five enzymes (*see* **Note 23**):

```
1. wget http://methylcrf.wustl.edu/hg19/MRE_frags/
   MRE_3enz_4_6000.bed
2. wget http://methylcrf.wustl.edu/hg19/MRE_frags/
   MRE_5enz_4_6000.bed
```

3 Methods

3.1 MeDIP-seq

3.1.1 Genomic DNA Extraction

High-quality, high molecular weight genomic DNA without signs of degradation and denaturation and without protein and RNA contaminations is an important criteria for the success of the library construction (*see* **Note 3**).

1. Resuspend cell pellets or finely sliced tissues in a 1.5 mL centrifuge tube containing 100–600 μL of extraction buffer (*see* **Note 4**).
2. Pipette up and down immediately to ensure that sample suspends well (*see* **Note 5**).

3. Incubate tubes at 55 °C for 1 h or overnight if needed (*see* **Note 6**).
4. Centrifuge at maximum speed at 4 °C for 10 min.
5. Quickly transfer the supernatant to a phase lock gel (pre-pellet phase lock gel (PLG) at 16,000 × *g* for 30 s).
6. Add an equal volume of phenol/chloroform/isoamyl alcohol directly to the PLG tube.
7. Thoroughly mix the organic and aqueous phases to form a transiently homogeneous suspension. Do not vortex.
8. Centrifuge at 16,000 × *g* for 5 min to separate the phases.
9. Carefully pipet upper phase to a fresh tube.
10. Add 1 μL of DNase-free RNaseA (10 mg/mL) and incubate for 1 h at 37 °C.
11. Transfer the solution from **step 10** to a new pre-pelleted phase lock gel and repeat **steps 6–8** one more time.
12. Carefully transfer the upper phase to a new pre-pelleted phase lock gel and repeat **steps 6–9** one more time using chloroform instead of phenol/chloroform/isoamyl alcohol.
13. Precipitate with 1/10 volume of 3 M sodium acetate (pH 5.5) and 2.5 volumes of 100% ethanol. Invert tubes several times to mix.
14. A "ball" of DNA should appear in the tube.
15. Incubate for 15 min at room temperature. If there is no "ball" in **step 14**, incubate the tube at −20 °C overnight.
16. Spin tubes at maximum speed for 30 min at 4 °C.
17. Wash DNA pellet with 70% ethanol and spin 5 min.
18. Resuspend DNA in a desired volume of elution buffer (EB). Purified genomic DNA can be placed at −20 °C for long-term storage.
19. Quantitate DNA on a NanoVue Spectrophotometer (*see* **Note 7**).
20. Validate DNA quality by running it on 1% TAE agarose gel (*see* **Note 8**).

3.1.2 DNA Fragmentation

1. Dilute 500–1000 ng of intact genomic DNA obtained in **step 18** of Subheading 3.1.1 in 30 μL of EB in a 1.5 mL sterile DNase/RNase-free tube and seal the tube with parafilm.
2. Set the temperature of the Bioruptor Pico to 4 °C and prechill the sonicator water bath.
3. Sonicate for 10–15 min with 30-s on/off. Fragmented DNA can be stored at −20 °C until needed.
4. Run 3 μL of the sonicated DNA on 1% TAE agarose gel to validate DNA fragment size range (*see* **Note 9**).

3.1.3 Library Preparation I

DNA purification is necessary after each reaction (*see* **Note 10**).

1. DNA end repair is performed to get blunt 5′-phosphorylated ends. The reaction is set up on ice in a 50 μL volume with 1× NEB T4 Polynucleotide Kinase buffer, fragmented DNA (obtained in Subheading 3.1.2), 5 μL of 10 mM ATP, 2 μL of 10 mM dNTPs, 3 U of T4 DNA polymerase, 5 U of Klenow DNA polymerase, and 10 U of T4 Polynucleotide Kinase. Incubate the mixture in a thermocycler at 20 °C for 30 min.
2. Spin briefly. Purify the DNA sample with AMPure XP beads. Elute in 34 μL of EB. Purified DNA can be stored at −20 °C until needed.
3. dA-tailing reaction is set up in a 50 μL volume with 1× NEB Klenow buffer 2, end-repaired DNA (**step 2** of this subsection), 10 μL of 10 mM dATPs and 5 U of Klenow 3′-5′ exo minus. Incubate the mixture in a heat block for 30 min at 37 °C.
4. Spin briefly. Purify the DNA sample with AMPure XP beads. Elute in 23 μL of EB. Purified DNA can be stored at −20 °C until needed.
5. Adapter ligation is set up in a 50 μL volume with 1× Quick ligation reaction buffer, end-repaired, dA-tailed DNA (**step 4** of this subsection), 1 μL of 10 μM pre-annealed adapter oligo mix (*see* Subheading 2.1.3) and 2000 U of T4 DNA ligase. Incubate the mixture at room temperature for 15 min.
6. Purify the DNA sample with AMPure XP purification beads. Elute in 30 μL of EB. Purified DNA can be stored at −20 °C until needed.

3.1.4 MeDIP

1. Denature the adapter-ligated DNA obtained in the previous subsection at 95 °C for 10 min, and then transfer immediately to ice to prevent re-annealing. Keep on ice for 10 min (*see* **Note 11**).
2. A premix is set up on ice in a 500 μL volume with 50 μL of 0.1 M Na_2HPO_4/NaH_2PO_4, 35 μL of 2 M NaCl, 2.5 μL of 10% Triton X-100, and 1 μL of anti-methylcytidine antibody (1 mg/mL).
3. Keep the mix on ice. Add the mix to the tube, which contains the denatured adapter-ligated DNA. Incubate the reaction on a rotating wheel at 4 °C overnight.
4. After overnight incubation, add 1 μL of secondary antibody of rabbit anti-mouse IgG (2.5 μg/μL) and 80 μL of protein A/G agarose beads. Incubate at 4 °C for 2 h on a rotating wheel.
5. Centrifuge the sample at 2000 × *g* for 1 min in 4 °C microfuge. Carefully discard the supernatant using a p1000 tip.

6. Resuspend the pellet of the beads in 1000 μL precooled MeDIP wash buffer (freshly prepared). Make sure that the pellet is fully resuspended.
7. Repeat **steps 5** and **6** 7–10 times.
8. For the final wash, centrifuge at 5000 × *g* for 2 min, remove and discard the supernatant. Spin briefly and remove all remaining liquid with a p10 tip.
9. Add 200 μL of MeDIP elution buffer to the beads pellet. Resuspend thoroughly.
10. Incubate at 55 °C for 2 h and mix occasionally.
11. Let the tube cool down to room temperature. Purify DNA with Qiagen MinElute PCR Purification Kit. Elute DNA in 30 μL of EB (*see* **Note 12**). Purified DNA can be stored at −20 °C until needed.

3.1.5 Library Preparation II

1. An adapter-modified PCR is carried out on ice in a 50 μL volume with 1× Phusion HF buffer, 15 μL of MeDIP DNA (obtained in the previous subsection), 1 μL of 10 μM dNTP, 2.5 μL of PCR primer PE 1.0, 2.5 μL of PCR primer PE 2.0, and 0.5 U of Phusion DNA polymerase.
2. Perform PCR in a preheated thermocycler under the following conditions: denaturation for 30 s at 98 °C, followed by 12 cycles of 10 s at 98 °C, 30 s at 64 °C and 30 s at 72 °C, with a final extension of 5 min at 72 °C (*see* **Note 13**).
3. Purify the DNA sample with AMPure XP purification beads. Elute in 15 μL EB. Purified DNA can be stored at −20 °C until needed.

3.1.6 Library Size Selection

A regular agarose gel extraction is performed for the final library size selection. Inserts of 100–400 bp are selected, which are around 220–520 bp after the adapter ligation and PCR amplification. Gel selection excludes all the free adapters, primers, and dimerized oligonucleotides.

1. Prepare a 100 mL 2% TAE agarose gel containing ethidium bromide.
2. Mix the PCR amplified DNA with 3 μL of 6× loading dye and load in the gel, leaving space between the ladder and sample wells.
3. Carry out gel electrophoresis in 1× TAE buffer at 120 V until the orange dye has run to the two-thirds of the gel.
4. After gel electrophoresis, transfer the gel tray onto the gel doc imaging system under UV light, cut out the 220–520 bp size range with a clean blade.

5. Purify each gel slice with MinElute Gel DNA Extraction kit and elute in 15 μL EB buffer (*see* **Note 14**).
6. Measure the concentration of each library by analyzing 2 μL of size-selected DNA on Qubit Fluorometer using the Qubit dsDNA High-Sensitivity Assay kit according to the manufacturer's instructions. Purified libraries can be stored at −20 °C until needed.

3.1.7 Library QC

To ensure library quality before sequencing, two independent qPCR reactions including one positive control (*SNRPN* promoter) and one negative control (a CpG-less sequence in the intron of *GABRB3*) are performed for MeDIP-seq libraries to confirm enrichment of methylated DNA and depletion of unmethylated DNA. Another qPCR quantification step is applied using the primers, which match sequences within the linkers flanking library inserts. This step measures templates that have linker sequences on both the ends. They will subsequently form clusters on a flowcell.

1. Two sets of primers are designed to test for MeDIP enrichment for human samples (Table 1).
2. qPCRs are set up in triplicates in a 20 μL volume with 1× iTaq universal SYBR Green supermix, 2 μL of MeDIP library (prediluted to 10 nM in EB) (*see* **Note 15**) and 0.8 μL of 10 μM primer mix (*SNPRN* or *GABRB3*).
3. Perform qPCR in a thermocycler such as the CFX under the following conditions: denaturation for 30 s at 95 °C, followed by 50 cycles of 5 s at 95 °C and 5 s at 60 °C.
4. Calculate the fold enrichment for methylated sequences as follows:

   $$\text{Fold Enrichment} = 2\text{E}(\text{Ct}_{\text{unmeth}} - \text{Ct}_{\text{meth}})$$

   where E is primers efficiency which can be calculated from a serial dilution standard curve of a genomic DNA ($E = 10^{(-1/\text{slope})} - 1$) and Ct is the cycle threshold (*see* **Note 16**).
5. Primers for the second qPCR quantification are designed according to linker sequences, which match sequences at both the ends of the adapter-modified PCR primers (Table 1).
6. A control MeDIP-seq library that has been successfully sequenced should be selected for this assay (*see* **Note 17**).
7. Before starting the qPCR, the control template and libraries for quantification are diluted to 10 nM in EB buffer.
8. Add 2 μL of the control template (10 nM) to 198 μL of 0.1% Tween-20 solution to make a 100-fold dilution. Mix the dilution thoroughly.

9. Add 100 μL of the diluted template to 100 μL of 0.1% Tween-20 solution to make a titration curve of six 2× serial dilutions. This will give seven control template dilutions in the range of 100–1.6 pM.
10. Mix the dilution thoroughly.
11. Repeat **steps 7–10** to produce three independent serial dilutions of the control template (omit if the control template is limited).
12. Add 2 μL of each unknown library (10 nM) to 998 μL of 0.1% Tween-20 solution to make a 500-fold dilution. This will give an approximate concentration of 20 pM.
13. Mix the dilutions thoroughly.
14. Repeat **steps 12–13** to produce three independent serial dilutions of each newly constructed library (omit if the newly constructed library is limited).
15. Set up qPCR reactions in triplicates in 20 μL volume with 1× iTaq universal SYBR Green supermix, 2 μL of diluted control templates (**step 9**) or unknown libraries (**step 12**), 0.4 μL of 10 μM qPCR primer 1.0 and 0.4 μL of 10 μM qPCR primer 2.0.
16. Perform qPCR in a thermocycler such as the CFX under the following conditions: denaturation for 30 s at 95 °C, followed by 40 cycles of 5 s at 95 °C and 5 s at 60 °C.
17. Generate a standard curve from the control template dilutions by plotting the Ct values (*Y*-axis) against the log initial concentration (*X*-axis).
18. Ensure that the efficiency of the standard curve in **step 17** is 90–110% and that the $R^2 > 0.9$.
19. Lock the threshold fluorescence based on the standard curve, calculate the initial concentration of the newly constructed libraries.
20. The concentration of the newly constructed libraries should be close to that of the control library. Determine the loading concentration of the samples based on the loading concentration and cluster density on the control template.

3.1.8 Next-Generation Sequencing

Indexed MeDIP-seq libraries quantified by a Qubit fluorometer are pooled after QC and are quantified again on an Agilent 2100 Bioanalyzer. 101-bp paired-end sequencing is run for MeDIP-seq libraries following Illumina's standard protocol. On an Illumina Hi-Seq 2500 platform, typically 150–200 million raw reads per lane can be obtained. Three MeDIP-seq libraries can be typically pooled per lane.

3.2 MRE-Seq

3.2.1 Genomic DNA Extraction

Same as Subheading 3.1.1.

3.2.2 DNA Fragmentation

Five methylation-sensitive restriction endonucleases are used in DNA digestion step for fragmentation (*see* **Note 18**). All five enzymes are sensitive to CpG methylation. Sequencing of the 5′ ends of these digested fragments therefore provides information on unmethylated CpG sites.

1. Transfer 500 ng of genomic DNA to a 1.5 mL sterile DNase/RNase-free tube for each reaction mix. A premix is set up in a 20 μL volume with 1× restriction enzyme buffer and 2.5 U of restriction enzyme (HpaII, HinP1I, AciI, HpyCH4IV, or Bsh1236I). Keep the mixture on ice. Incubate the reaction in a heat block at 37 °C for 3 h. Five digests for each sample can be set up in parallel.
2. Add additional 2.5 U of enzymes to each reaction after 3 h of incubation. Mix and incubate for another 3 h at 37 °C.
3. After total 6 h digestion, incubate the reaction at 65 °C for 20 min to deactivate all enzymes except HpaII, which is deactivated at 80 °C.
4. Combine the five reactions. Each sample should have a total volume of 100 μL.
5. Purify the DNA sample with AMPure XP purification beads (*see* **Note 19**). Elute in 15 μL EB. Purified DNA can be stored at −20 °C until needed.
6. Prepare a 100 mL 1% TAE agarose gel containing ethidium bromide.
7. Mix the DNA (**step 5**) with 3 μL of 6× loading dye and load in the gel, leaving space between the ladder and sample wells.
8. Carry out gel electrophoresis in 1× TAE buffer at 120 V until the orange dye runs to the two third of the gel.
9. After gel electrophoresis, transfer the gel tray onto the gel doc imaging system under UV light, cut out the 100–500 bp size range with a clean blade.
10. Purify each gel slice with the MinElute Gel DNA Extraction kit and elute in 30 μL EB buffer (*see* **Note 14**).

3.2.3 Library Preparation I

Same as Subheading 3.1.3, except that only Klenow fragment is used for end repair (*see* **Note 20**).

3.2.4 Library Preparation II

Same as Subheading 3.1.5.

3.2.5 Library Size Selection

Same as Subheading 3.1.6, except gel electrophoresis should run until the orange dye runs until the bottom of the gel (*see* **Note 21**).

3.2.6 Library QC

Same as Subheading 3.1, starting from **step 3** (*see* **Note 22**).

3.2.7 Next-Generation Sequencing

Indexed MRE-seq libraries quantified on a Qubit fluorometer are pooled after QC and are quantified again on a 2100 Bioanalyzer. 50-bp single-end sequencing is run for MRE-seq libraries following Illumina's standard protocol. On the Hi-Seq 2500 platform, typically 150–200 million raw reads can be obtained per lane. Six MRE-seq libraries can be pooled per lane.

3.3 Integrative Analysis of MeDIP-seq and MRE-seq

We provide sample data to illustrate how to use M&M and methylCRF.

3.3.1 Obtaining Sample Data

1. Working directory.
   Make a new directory as your working folder, for example, /workbench/example, using the mkdir command:

```
mkdir /workbench/example
```

   All the following commands are run under this directory:

```
cd /workbench/example
```

   In this protocol, we start from aligning raw reads to the reference genome. For illustration purposes, we included MeDIP-seq and MRE-seq from two human samples—human embryonic stem cell H1 (H1ES), and human Brain. We included 1 million raw reads from each dataset. These sample reads were mostly derived from chromosome 6. Therefore, in the alignment step, we used human chromosome 6 as our reference sequence.

2. Genomic annotation data.
   Genomic annotation files (CpG coordinates, restriction enzyme recognition sites, and chromosome sizes) are needed for data processing. In this example, create a directory named "ann" for downloading and storing the annotation files:

```
mkdir ann
cd ann
```

3. Getting genomic annotation files for methylQA (*see* **Note 23**).

```
wget http://wang.wustl.edu/MeDIP-MRE/ann/hg19.size
wget http://wang.wustl.edu/MeDIP-MRE/ann/CpG.bed.gz
wget http://wang.wustl.edu/MeDIP-MRE/ann/TriMRE_frags.bed
wget http://wang.wustl.edu/MeDIP-MRE/ann/FiveMRE_frags.bed
```

File description:

- hg19.size is a simple 2-column file that indicates the size of human chromosomes;
- CpG.bed.gz contains all CpG sites in the human genome (decompress after download);
- TriMRE_frags.bed contains all MRE fragments in the human genome based on three MRE enzymes (*see* **Note 23**).
- FiveMRE_frags.bed file contains all MRE fragments in the human genome based on five MRE enzymes (*see* **Note 23**).

4. Getting files genomic annotation files for M&M.

```
1. wget http://wang.wustl.edu/MeDIP-MRE/ann/nu-
   m500_allcpg_hg19.bed
2. wget http://wang.wustl.edu/MeDIP-MRE/ann/num500_Fi-
   ve_mre_cpg_hg19.bed
3. wget http://wang.wustl.edu/MeDIP-MRE/ann/num500_-
   Three_mre_cpg_hg19.bed
```

File description:

- num500_allcpg_hg19.bed contains coordinates of all CpG sites in 500 bp windows of the hg19 genome.
- num500_Five_mre_cpg_hg19.bed contains five MRE enzyme cut sites in 500 bp windows genome-wide.
- num500_Three_mre_cpg_hg19.bed contains three MRE enzyme cut sites in 500 bp windows genome-wide.

Download of the annotation files for methylCRF is described in software section of methylCRF (**item** 7 of Subheading 2.3).

After downloading annotation files, go back to the working directory:

```
cd /workbench/example
```

5. Getting sample MeDIP- and MRE-seq data (raw reads)
   Download sample data from the table below (Please note the sample data has been created by the Roadmap Epigenomics Project: http://roadmapepigenomics.org/):

Sample	GEO#	Assay	URL
H1ES	GSM543016	MeDIP-seq	http://wang.wustl.edu/MeDIP-MRE/H1Es_MeDIP.fq.gz
	GSM428286	MRE-seq	http://wang.wustl.edu/MeDIP-MRE/H1Es_MRE.fq.gz
Brain	GSM669614	MeDIP-seq	http://wang.wustl.edu/MeDIP-MRE/Brain_MeDIP.fq.gz
	GSM669604	MRE-seq	http://wang.wustl.edu/MeDIP-MRE/Brain_MRE.fq.gz

Use the wget command to download the sample reads in the table above:

```
wget URL
```

6. Getting the genome assembly.
   Build a genome index for BWA alignment (for illustration purposes only. Users should choose their favorite short-read aligner):
   (a) Download the compressed human genome sequence:

```
wget http://wang.wustl.edu/MeDIP-MRE/hg19.fa.gz
```

(b) Decompress the downloaded file:

```
gunzip hg19.fa.gz
```

(c) Build the BWA index:

```
bwa index -a bwtsw -p hg19 hg19.fa
```

3.3.2 MeDIP-seq Data Processing

1. Reads alignment.

Use the following command lines to generate the alignment result for H1ES (*see* **Note 24**):

```
1.bwaaln-t4-f Brain_MeDIP.fq.gz.sai hg19 Brain_MeDIP.fq.gz
2. bwa samse -n 10 -f Brain_MeDIP.sam hg19 Brain_MeDIP.fq.
   gz.sai Brain_MeDIP.fq.gz
3. samtoolsview-bSBrain_MeDIP.sam-oBrain_MeDIP_unsort.bam
4. rm -f Brain_MeDIP.sam
5. samtools sort Brain_MeDIP_unsort.bam Brain_MeDIP
6. rm -f Brain_MeDIP_unsort.bam
```

These command line operations will perform the following functions:

(a) Generated suffix array file for sequence reads.
(b) Generate alignment result in SAM format.
(c) Convert SAM formatted result to BAM format, which is binary and smaller in size.
(d) Remove the SAM file.
(e) Sorti the BAM file.
(f) Removed the unsorted BAM file.

Parameter description:

- -t: the number of threads used for the alignment process
- -n 10: maximum edit distance for alignment is 10. More details of edit distance parameters can be found in: http://en.wikipedia.org/wiki/Edit_distance.

Similarly, Brain data can be aligned using the following commands (*see* **Note 24**):

```
1. bwa aln -t 4 -f H1Es_MeDIP.fq.gz.sai hg19 H1Es_MeDIP.fq.gz
2. bwa samse -n 10 -f H1Es_MeDIP.sam hg19 H1Es_MeDIP.fq.gz.sai
   H1Es_MeDIP.fq.gz
3. samtools view -bS H1Es_MeDIP.sam -o H1Es_MeDIP_unsort.bam
4. rm -f H1Es_MeDIP.sam
5. samtools sort H1Es_MeDIP_unsort.bam H1Es_MeDIP
6. rm -f H1Es_MeDIP_unsort.bam
```

2. Process MeDIP-seq alignment file with methylQA (*see* **Note 25**)

```
1. methylQA medip -m ann/CpG.bed ann/hg19.size H1Es_MeDIP.bam
2. methylQA medip -m ann/CpG.bed ann/hg19.size Brain_MeDIP.bam
```

The methylQA package processes the MeDIP alignment result file (.bam) to generate alignment quality report and files formatted for various downstream needs. Resulting files include .bed file (aligned read location), .bedGraph file (aligned read density), and .bigWig file (aligned read density, for display on a Genome Browser). The option -m specifies the annotation file for CpG genomic locations.

Figure 2 displays processed MeDIP-seq data in the Wash U EpiGenome Browser [12, 39–42].

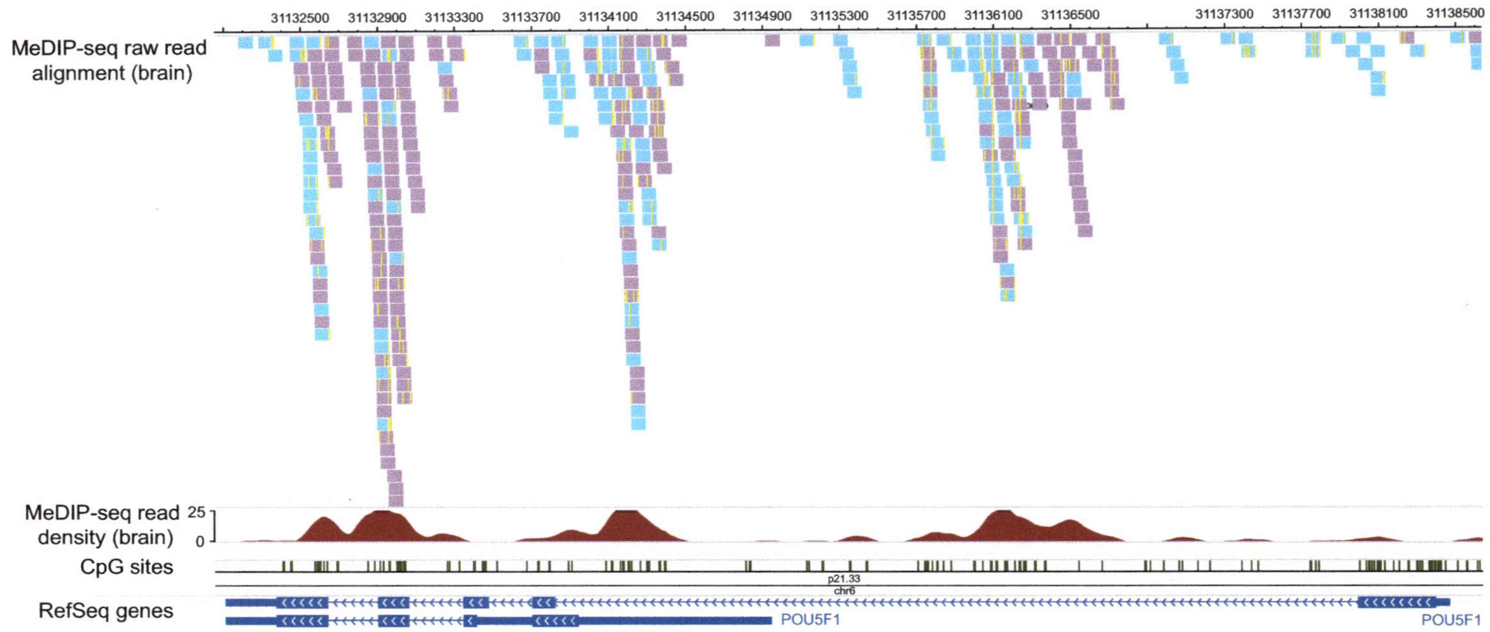

Fig. 2 Visualizing MeDIP-seq data in a Genome Browser. Processed MeDIP-seq data are displayed using the Wash U EpiGenome Browser. The genomic region encompassing *POU5F1* (a.k.a. Oct4, an important stem cell-specific gene) is shown. Four tracks are included. The first track displays MeDIP-seq reads that are uniquely aligned to this region, with redundant reads removed. Note that reads mapped to the forward or the reverse strand are colored differently. Mismatches between each read and the genome assembly are visible as *yellow ticks* within the reads. The second track shows MeDIP-seq read density. The third track shows locations of CpG sites, and the fourth track displays RefSeq genes, in this case the two isoforms of *POU5F1*

3.3.3 MRE-seq Data Processing

1. Align MRE-seq reads (*see* **Note 24**).

```
1. bwa aln -t 4 -f H1Es_MRE.fq.gz.sai hg19 H1Es_MRE.fq.gz
2. bwa samse -n 10 -f H1Es_MRE.sam hg19 H1Es_MRE.fq.gz.sai
   H1Es_MRE.fq.gz
3. samtools view -bS H1Es_MRE.sam -o H1Es_MRE_unsort.bam
4. rm -f H1Es_MRE.sam
5. samtools sort H1Es_MRE_unsort.bam H1Es_MRE
6. rm -f H1Es_MRE_unsort.bam
```

```
1. bwa aln -t 4 -f Brain_MRE.fq.gz.sai hg19 Brain_MRE.fq.gz
2. bwa samse -n 10 -f Brain_MRE.sam hg19 Brain_MRE.fq.gz.
   sai Brain_MRE.fq.gz
3. samtools view -bS Brain_MRE.sam -o Brain_MRE_unsort.bam
4. rm -f Brain_MRE.sam
5. samtools sort Brain_MRE_unsort.bam Brain_MRE
6. rm -f Brain_MRE_unsort.bam
```

2. Process MRE-seq alignment file with methylQA (*see* **Note 25**).

```
1. methylQA mre -m ann/CpG.bed ann/hg19.size ann/ TriMRE_-
   frags.bed H1Es_MRE.bam
2. methylQA mre -m ann/CpG.bed ann/hg19.size ann/ TriMRE_-
   frags.bed Brain_MRE.bam
```

The methylQA package processes the MRE alignment result file (.bam) to generate alignment quality report and files formatted for various downstream needs. Resulting files include .bed file (aligned read location), .bedGraph file (MRE score), and .bigWig file (MRE score, for display on a Genome Browser). The option -m specifies the annotation file for CpG genomic locations. Please note the MRE data used in this protocol were based on three enzymes, so all MRE fragment files and MRE window file were from the 3-enzyme version. For experiments with different numbers or combinations of enzymes, please use the corresponding annotation files.

Figure 3 displays processed MRE-seq data in a Genome Browser.

3.3.4 Run Methyl&M to Predict DMRs Between H1ES and Brain

1. In this example, we partition the genome into genomic bins of 500 bp in size. M&M is then performed for each genomic bin to generate a statistical assessment of the probability that the methylation levels of the two samples within each bin are different.
2. Use your favorite text editor to create an R script code "MnM. r" following the text box (*see* **Notes 26–29**):

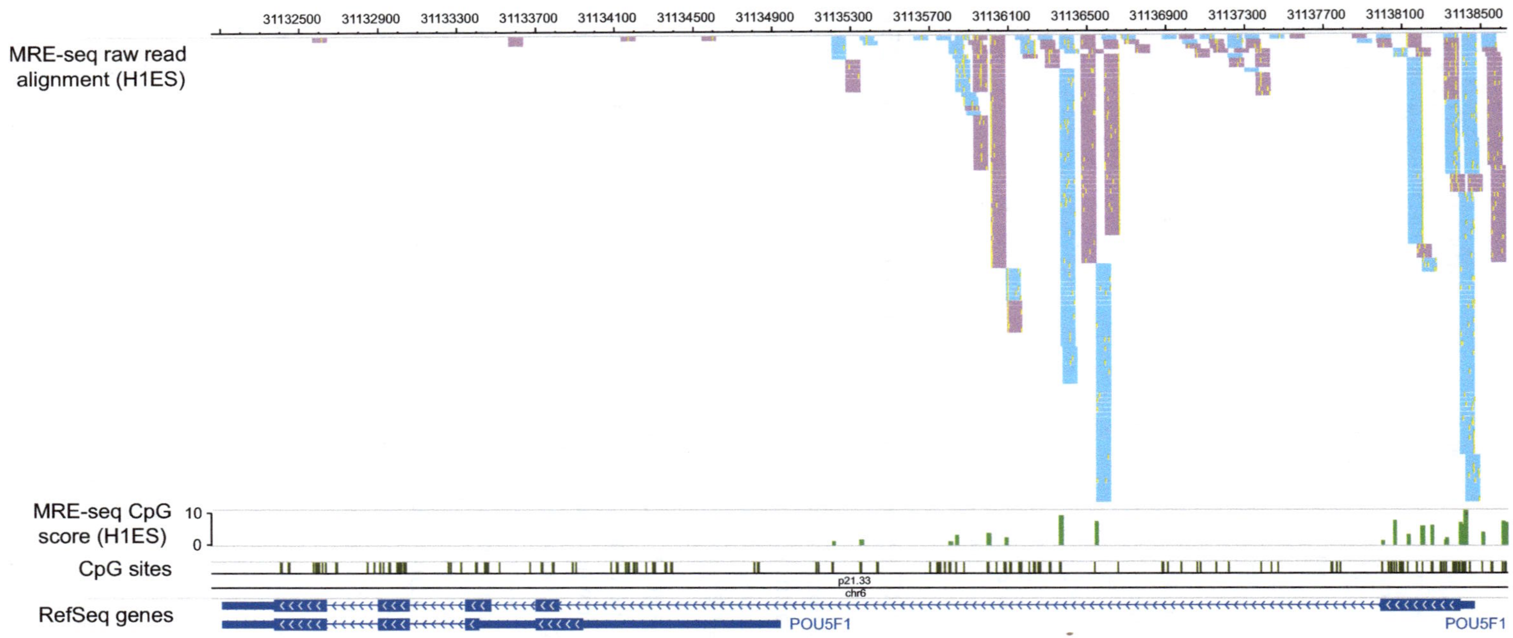

Fig. 3 Visualizing MRE-seq data on a Genome Browser. Processed MRE-seq data are displayed in the Wash U EpiGenome Browser, across the same genomic region as in Fig. 2. Four tracks are included. The first track displays MRE-seq reads that are uniquely aligned to this region, and filtered by restriction enzyme recognition sites. Because independent enzymatic cleavage of the same site will result in identical sequencing reads, all reads are kept for analysis. Note that reads mapped to the forward or the reverse strand are colored differently. Mismatches between each read and the genome assembly are visible as *yellow ticks* within the reads. The second track shows the MRE-seq score, which is a normalized read count per enzyme at a given CpG site. The third and fourth tracks are identical to those shown in Fig. 2

```
# Rscript MnM.r MeDIP_1 MeDIP_2 MRE_1 MRE_2 output_file
ts = proc.time()
args = commandArgs(TRUE)
(MeDIP_1 = args[1])
(MeDIP_2 = args[2])
(MRE_1 = args[3])
(MRE_2 = args[4])
(Out_file = args[5])
library(methylMnM)
cpg_bin = 'ann/num500_allcpg_hg19.bed'
mre_cpg_bin = 'ann/num500_Three_mre_cpg_hg19.bed'
Num500_MeDIP_1 = paste('num500_', MeDIP_1,sep = '')
Num500_MeDIP_2 = paste('num500_', MeDIP_2,sep = '')
Num500_MRE_1 = paste('num500_', MRE_1,sep = '')
Num500_MRE_2 = paste('num500_', MRE_2,sep = '')
countMeDIPbin(file.Medipsite = MeDIP_1,file.bin =
cpg_bin,writefile = Num500_MeDIP_1, binlength = 500)
countMeDIPbin(file.Medipsite = MeDIP_2,file.bin =
cpg_bin,writefile = Num500_MeDIP_2, binlength = 500)
 countMREbin(file.MREsite = MRE_1,file.bin = cpg_bin,wri-
tefile = Num500_MRE_1, binlength = 500)
countMREbin(file.MREsite = MRE_2,file.bin = cpg_bin,wri-
tefile = Num500_MRE_2, binlength = 500)
# calculate p-value of each bin.
datafile = c(Num500_MeDIP_1,Num500_MeDIP_2,Num500_MRE_1,
Num500_MRE_2)
pv_output = paste('pv_',Out_file,sep = '')
MnM.test(file.dataset = datafile,chrstring ="chr6", file.
cpgbin=cpg_bin,file.mrecpgbin=mre_cpg_bin,writefi-
le=pv_output,mreratio=3/7,method="XXYY", psd=2,
mkadded=1,a=1e-10,cut=100,top=500)
# calculate q-value of each bin
qv_output = paste('qv_',Out_file,sep = '')
MnM.qvalue(pv_output,qv_output)
qv = read.table(qv_output,header = T)
DMR_outfile = paste('DMRs_',qv_output,sep = '')
DMR_bg = paste(DMR_outfile,'.bedGraph',sep='')
dmr = MnM.selectDMR(qv,up = 1.45,down = 1/1.45,q.value =
1e-5,cutoff = 'q-value',quant = 0.9)
write.table(dmr,file=DMR_outfile,sep='\t',quote=F,row.
names=F)
Dbg = dmr[,1:4]
Dbg[,4]= log10(dmr[,12]+1e-40)* (dmr[,11]/abs(dmr[,11]))
write.table(Dbg,file=DMR_bg,sep='\t',quote=F,row.name-
s=F,col.names=F)
te = proc.time()
print('costing seconds:');te-ts
```

The main commands are described in the following:

(a) CountMeDIPbin(): computes the total MeDIP-seq read count in each bin.
   - Parameter *file.Medipsite* denotes the MeDIP-seq data.
   - Parameter *file.bin* denotes CpG sites information in the 500 bp windows.
   - Parameter *writefile* denotes output file name of MeDIP-seq read counts in 500 bp windows.
   - Parameter *binlength* denotes the sliding window length. In this example, 500 bp is used. Users can choose to use genomic bins of any arbitrary size, in which case the CpG sites information should be calculated by the countcpgbin function. Details can be found in the manual of methylMnM.

(b) CountMREbin(): computes the total MRE-seq read counts of each bin.
   - Parameter file.MREsite denotes the MRE-seq data.
   - Parameter file.bin denotes CpG site information in 500 bp windows.
   - Parameter writefile denotes output file name of MRE-seq read counts in 500 bp windows.
   - Parameter binlength denotes the sliding window length. In this example, 500 bp is used.

(c) MnM.test(): computes a *p*-value for each bin between two input samples.
   - Parameter file.dataset denotes a vector, which contains the names of MeDIP-seq 500 bp information and MRE-seq 500 bp information.
   - Parameter chrstring denotes the chromosome used in calculation. When using parameter chrstring = NULL, calculation will be performed for all chromosomes. In this example, we set chrstring = "chr6" to limit computation to chromosome 6.
   - Parameter filecpg.bin denotes CpG sites information in 500 bp windows.
   - Parameter filemrecpg.bin denotes MRE CpG sites information in 500 bp windows.
   - Parameter writefile denotes output file name of *p*-value in 500 bp windows.

(d) MnM.qvalue(): estimate the *q*-values for a given set of *p*-values.

(e) MnM.selectDMR(): select significant DMRs based on given parameters.
   - Parameter *q*-value denotes *q*-value cutoff, with default value being $1e-5$.
   - Parameter cutoff denotes *q*-value or *p*-value cutoff, with default being *p*-value.

3. Select DMRs.

```
1. Rscript MnM.r H1Es_MeDIP.extended.bed H1Es_MRE.filter.bed
   Brain_MeDIP.extended.bed Brain_MRE.filter.bed H1Es_vs_-
   Brain
```

The input files of methylMnM are methylQA processed alignment results in BED format.

This command generates several output result files:

- qv_H1Es_vs_Brain: all *q*-values from the run.
- DMRs_qv_H1Es_vs_Brain: statistically significant DMRs between H1ES and Brain based on *q*-values.
- DMRs_qv_H1Es_vs_Brain.bedGraph: this file can be directly visualized in genome browsers. It can also be transformed into .bigWig format for visualization.

The file DMR_qv_H1Es_vs_Brain looks like this:

```
1. chr chrSt chrEnd Medip1 Medip2 MRE1 MRE2 cg mrecg pvalue
   Ts qvalue
2. chr6 26756000 26756500 0.33604930970431
      0.0521568192129032 0.505034914010546
      4.35538343254228 15 8 1.54198003077371e-09
      6.15213657535548 3.89125898336746e-07
3. chr6 27146500 27147000 0.0775498407009947
      0.48679697932043 6.81797133914237
      1.61491745251568 22 11 5.32321414472478e-09
      -5.93993646148064 1.22580113946516e-06
4. chr6 27181000 27181500 0.310199362803979
      0.712809862576343 7.51239434590687
      0.734053387507126 25 10 2.19209717973469e-12
      -7.17996054783314 8.76472716079456e-10
5. chr6 27181500 27182000 0.0775498407009947
      0.556339404937634 22.4740536734693
      4.15963586254038 23 5 4.25659507641285e-13
      -7.24711688926306 1.89419388550094e-10
6. chr6 27250500 27251000 0.853048247710942 0 0 0 15
      0 4.01774138370403e-15 7.55689511636996
      4.77027539691851e-11
```

The 1st line is a header line, and the following lines represent identified DMRs, one DMR per line. Each column in the file has the following meaning:

1st column: chromosome of this DMR.

2nd column: start position of this DMR.

3rd column: end position of this DMR.

4th column: normalized MeDIP signal (RPKM) of this DMR from sample 1 (H1ES here).

5th column: normalized MeDIP signal (RPKM) of this DMR from sample 2 (Brain here).

6th column: normalized MRE signal (RPKM) of this DMR from sample 1 (H1ES here).

7th column: normalized MRE signal (RPKM) of this DMR from sample 2 (Brain here).

8th column: CpG site counts.

9th column: MRE CpG site counts.

10th column: *p*-value.

11th column: Ts value. Negative Ts represent hypermethylated DMRs in brain, and positive Ts represent hypomethylated DMRs in brain.

12th column: *q*-value.

The file DMR_qv_H1Es_vs_Brain.bedGraph looks like this:

```
1. chr6 26756000 26756500 -6.4099098637
2. chr6 27146500 27147000 5.91157997928
3. chr6 27181000 27181500 9.05726159861
4. chr6 27181500 27182000 9.72257556963
5. chr6 27250500 27251000 -10.3214565476
```

Each column in the file has the following meaning:

1st column: chromosome of this DMR.

2nd column: start position of this DMR.

3rd column: end position of this DMR.

4th column: negative log10 transformed *p*-value. If the value is negative, it represents hypomethylation in sample 1 and hypermethylation in sample 2; if the value is positive, it represents hypermethylation in sample 1 and hypomethylation in sample 2.

Figure 4 shows a DMR between H1ES and brain, detected by M&M.

3.3.5 Predict Single CpG Methylation Level for H1ES and Brain by methylCRF

Note: If your methylCRF folder is not in your working directory, please make a symbolic link as (*see* **Notes 28–30**):

```
ln -s /path/to/your/methylCRF/folder/workbench/example
```

1. Run methylCRF for H1ES

The input files of methylCRF are methylQA processed MeDIP-seq alignment files in BED format, and original MRE-seq alignment result in BAM format. Run the following commands:

```
1. awk '{OFS="\t"; print $1,$2,$3,".",1,$6}' H1Es_MeDIP.
   extended.bed > H1Es_MeDIP.extended.bed.crfdip
2. (samtools view H1Es_MRE.bam | methylCRF/sam2bed.pl -r -q
  10 - ) | methylCRF/MRE_norm.pl - methylCRF/MRE_3enz_4_6000.
  bed H1Es_MRE.bam.crfmre
3. methylCRF.pl methylCRF/h1es_mdl methylCRF/hg19_gdat H1Es_-
   MeDIP.extended.bed.crfdip H1Es_MRE.bam.crfmre_MRE.bed
   methylCRF/MRE_3enz_4_6000_cpg.bin 750 H1Es
```

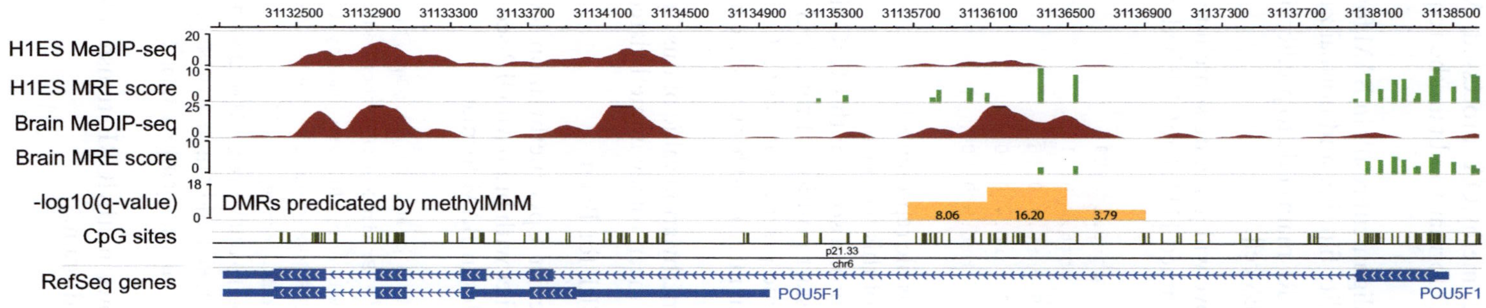

Fig. 4 Visualizing M&M results. Differentially methylated regions between H1ES cells and brain tissue are detected using the M&M program. The genomic region encompassing the *POU5F1* gene is shown in the Wash U EpiGenome Browser. MeDIP-seq and MRE-seq data of H1ES and of the brain are displayed as the first four Browser tracks. The fifth track displays the minus log10 transformed *q*-values, which result from the M&M comparison between the two samples at a 500 bp resolution. Three genomic regions are detected as DMRs, with the middle region being the most statistically significant at a *q*-value of 1*e*−16.2. The difference between H1ES and brain, and the complementary difference between MeDIP-seq and MRE-seq, can be visually appreciated. This DMR is biologically significant, and is directly linked to the regulation of *POU5F1*, which is expressed in embryonic stem cells but repressed in differentiated neuronal cells

These commands will perform the following functions:

(a) Convert the methylQA processed MeDIP-seq alignment result in BED format to obtain the required input format for methylCRF.

(b) Filter MRE alignment result in BAM format by MRE cut sites. Perform quality filtering and normalization.

(c) Run methylCRF main code to predict single CpG methylation level for H1ES cell.

The code above will generate some intermediate files, and the final result file is named H1ES_mCRF.bed, which looks like this:

```
chr6 60469 60471 chr6.1 0.89
chr6 60657 60659 chr6.2 0.91
chr6 60797 60799 chr6.3 0.95
chr6 60857 60859 chr6.4 0.90
chr6 60909 60911 chr6.5 0.90
```

Each line of this file represents methylation status of 1 CpG site, and each column of this file means:

1st column: chromosome of this CpG site.

2nd column: start position of this CpG site.

3rd column: end position of this CpG site.

4th column: an internal id for CpG site in methylCRF, could be ignored.

5th column: methylation value of this CpG site, 1 means totally methylated and 0 mean totally unmethylated.

2. Run methylCRF for Brain

Similar to predict CpG methylation level for H1ES cell, just replace the input files with brain MeDIP-seq file in BED format and brain MRE-seq files in BAM format.

```
1. awk '{OFS="\t"; print $1,$2,$3,".",1,$6}' Brain_MeDIP.
   extended.bed > Brain_MeDIP.extended.bed.crfdip
2. (samtools view Brain_MRE.bam | methylCRF/sam2bed.pl -r -q
   10 - ) | methylCRF/MRE_norm.pl - methylCRF/MRE_5enz_4_6000.
   bed Brain_MRE.bam.crfmre
3. methylCRF.pl methylCRF/h1es_mdl methylCRF/hg19_gdat Brain_-
   MeDIP.extended.bed.crfdip Brain_MRE.bam.crfmre_MRE.bed
   methylCRF/MRE_5enz_4_6000_cpg.bin 750 Brain
```

Figure 5 shows methylCRF transformed DNA methylation data for H1ES and brain.

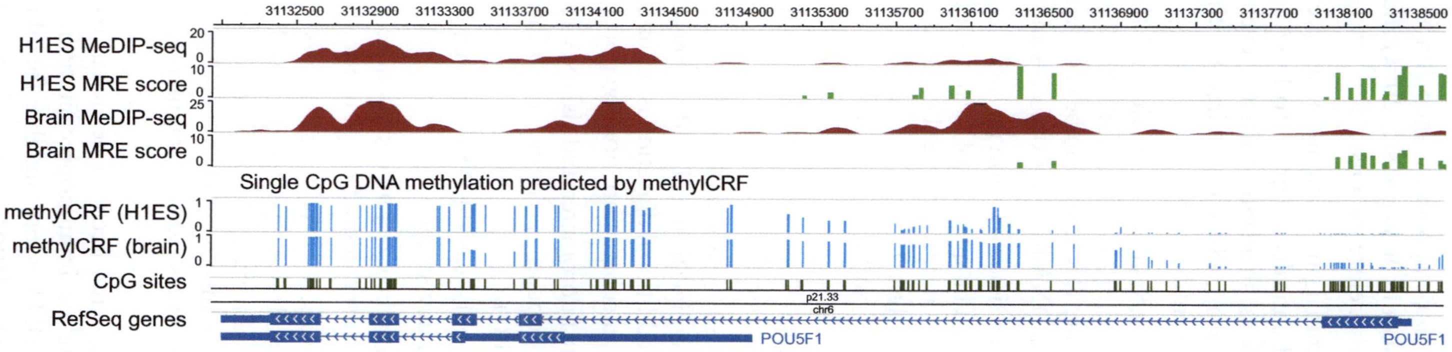

Fig. 5 Visualizing methylCRF results. The same genomic region as in previous figures is represented in a Browser view. MeDIP-seq and MRE-seq data from H1ES and brain are displayed as in Fig. 4. In addition, two methylCRF tracks are included, displaying DNA methylation levels at single CpG resolution. CpGs in the DMR shown in Fig. 4 have low methylation levels in H1ES, and high methylation levels in brain

4 Notes

1. Usually each component is prepared as a stock solution. Proteinase K should be stored at −20 °C and should be added only immediately before each use.
2. The 5-terminal phosphate in the adapter PE 2.0 facilitates the ligation of annealed adapters to DNA inserts. The phosphorothiate linkage in adapter PE 1.0 prevents nuclease cleavage of the T overhang, which is required for the adapter and A-tailed DNA fragment ligation.
3. Compared with commercial kits and the original DNA isolation protocol [25], this optimized protocol can generate higher quality DNA, and can be easily adapted for the isolation of genomic DNA from various cells and tissue types.
4. Allow frozen tissue to warm up slightly (5–10 s), and chop it in a petri dish. The quantity of extraction buffer is dependent on cell numbers and tissue size. For instance, 600 μL of buffer is sufficient to lyse 1–5 million cells or up to 20 mg of tissue. Alternatively, small samples can be resuspended in 100 μL extraction buffer.
5. Pipet gently! The DNA can be easily sheared by vigorous mixing or pipetting.
6. The solution should look homogenous following digestion. Inadequate digestion will lead to lower yield and lower DNA quality. If additional digestion is required, add more extraction buffer and incubate until homogenous.
7. The Qubit fluorometer and dsDNA HS assay kit can also be used to obtain accurate quantification of the DNA concentration.
8. High molecular weight DNA without random shearing should not appear as a smear.
9. DNA fragment size range should be 100–500 bp. Continue sonication if the desired range is not achieved.
10. A broad variety of commercial kits can be used for DNA purification. It was reported that AMPure XP beads can reduce DNA loss during purification when compared to Qiagen kit [33]. The beads selectively bind DNA fragments of 100 bp and larger, thus excess adapter fragments and primer dimers can be removed.
11. DNA fragments need to be denatured to single strands for methylcytidine immunoprecipitation. The denatured single-stranded DNA is not suitable for library construction. The immunoprecipitation step is therefore sandwiched between the two library preparation steps.

12. We choose not to use AMPure XP beads in this step because the elution buffer used in the immunoprecipitation step affects binding of DNA to the beads, especially when the DNA fragments are shorter than 300 bp.
13. Twelve cycles are sufficient in this step. Higher cycle numbers will result in higher PCR bias.
14. We modified the Qiagen gel extraction protocol to melt agarose gel slices at 37 °C instead of at 50 °C to reduce the G + C bias [43]. Use two Qiagen columns if a gel slice is over 400 mg and elute each column in 10 μL EB. Combine the two eluates.
15. A non-MeDIPed sonicated input control should be included if sufficient material is available. The ng/μL concentration measured by Qubit fluorometer can be converted to nanomolar with the following formula:

    nM DNA concentration = (DNA concentration in ng/μL/ Median selected DNA size in bp × 650) × 1000,000.

16. It is recommended that the methylated fragments are at least 25-fold enriched compared to the unmethylated fragments [36]. In practice, our libraries show enrichment scores higher than 2500. Non-MeDIPed sonicated input control should not exhibit any enrichment.
17. The control library should be as closely related as possible to the newly constructed libraries for quantification. For example, if you quantitate human MeDIP-seq libraries, you should use a human MeDIP-seq library as a control (from the same cell or tissue type if possible).
18. Three restriction endonucleases of HpaII (C↓CGG), HinP1I (G↓CGC), and AciI (C↓CGC) were used in the original protocol [25]. Five enzymes can be used to increase genome coverage.
19. AMPure XP beads are used in this step instead of phenol chloroform extraction, which was used in the original protocol [25].
20. Restriction endonuclease digested fragments contain 5-′-phosphorylated and 3′-terminal recessive ends. Therefore, a Klenow DNA polymerase is used to finish filling-in.
21. In MeDIP-seq, the unmethylated adapter cannot be precipitated; therefore, there are no adapter dimers in the PCR product. In contrast, in MRE-seq, the dimerized oligonucleotides (~120 bp in size) should be removed by running a gel as long as possible.
22. The control library should be as similar as possible to the newly constructed libraries for quantification. For example, if you quantitate human MRE-seq libraries, you should choose a

human MRE-seq library as control (from the same cell or tissue type if possible).

23. In both M&M and methylCRF, the MRE CpG information should match experimental design, which may use either three, four or five enzymes [15, 36].
24. In this chapter, both MeDIP-seq and MRE-seq data are single-end reads. For paired-end sequencing data, the alignment should follow instructions of alignment software that users choose.
25. For paired-end sequencing data, methylQA has different parameters to optimize the processing results.
26. When comparing two samples with different MeDIP sequencing methods (i.e., paired-end vs. single-end) by M&M, the paired-end data should be processed as single-end data for fair comparison [15].
27. When comparing two samples with different MRE usage (i.e., 3MRE vs. 5MRE) by M&M, MRE data should be processed as if the same set of enzymes were used for both the samples, with the limiting factor being the fewer of the two. For example, if one experiment used five MRE enzymes, then its data should be reduced to three MREs (i.e., data generated by the two additional enzymes should be filtered) before compared to the experiment that used three MRE enzymes [15].
28. Both methylM&M and methylCRF work with either MeDIP-seq data or MRE-seq data alone. However this will result in increased false positives. Thus, we do not recommend using methylM&M and methylCRF with only MeDIP-seq data or only MRE-seq data [15, 36].
29. Depending on the total read counts from MeDIP-seq and MRE-seq data, methylM&M and methylCRF will need up to 20 CPU hours to finish calculation for a typical human experiment. We recommend using "nohup" to run the processes in the background [15, 36].
30. methylCRF requires that MRE-seq data is provided in BAM format for proper normalization [36].

Acknowledgments

We thank Joseph F. Costello, Ravi Nagarajan, Chibo Hong for developing the experimental protocols described in this chapter. We thank Michael Stevens for developing methylCRF. We thank Nan Lin, Yan Zhou, Boxue Zhang for developing M&M. We thank members of the Wang laboratory for testing and improving various parts of the methods. This work was supported by NIH grant

U01ES017154 (T.W.), R01HG007354 (T.W.), NIDA's R25 program DA027995 (B.Z.), and American Cancer Society grant RSG-14-049-01-DMC (T.W.).

References

1. Bird A (2002) DNA methylation patterns and epigenetic memory. Genes Dev 16:6–21
2. Ziller MJ, Muller F, Liao J et al (2011) Genomic distribution and inter-sample variation of non-CpG methylation across human cell types. PLoS Genet 7:e1002389
3. Lister R, Pelizzola M, Kida YS et al (2011) Hotspots of aberrant epigenomic reprogramming in human induced pluripotent stem cells. Nature 471:68–73
4. Ramsahoye BH, Biniszkiewicz D, Lyko F et al (2000) Non-CpG methylation is prevalent in embryonic stem cells and may be mediated by DNA methyltransferase 3a. Proc Natl Acad Sci U S A 97:5237–5242
5. Yan J, Zierath JR, Barres R (2011) Evidence for non-CpG methylation in mammals. Exp Cell Res 317:2555–2561
6. Aran D, Sabato S, Hellman A (2013) DNA methylation of distal regulatory sites characterizes dysregulation of cancer genes. Genome Biol 14:R21
7. Suzuki MM, Bird A (2008) DNA methylation landscapes: provocative insights from epigenomics. Nat Rev Genet 9:465–476
8. Jones PA (2012) Functions of DNA methylation: islands, start sites, gene bodies and beyond. Nat Rev Genet 13:484–492
9. Smith ZD, Meissner A (2013) DNA methylation: roles in mammalian development. Nat Rev Genet 14:204–220
10. Robertson KD (2005) DNA methylation and human disease. Nat Rev Genet 6:597–610
11. Bergman Y, Cedar H (2013) DNA methylation dynamics in health and disease. Nat Struct Mol Biol 20:274–281
12. Irizarry RA, Ladd-Acosta C, Wen B et al (2009) The human colon cancer methylome shows similar hypo- and hypermethylation at conserved tissue-specific CpG island shores. Nat Genet 41:178–186
13. Hon GC, Rajagopal N, Shen Y et al (2013) Epigenetic memory at embryonic enhancers identified in DNA methylation maps from adult mouse tissues. Nat Genet 45:1198–1206
14. Stadler MB, Murr R, Burger L et al (2011) DNA-binding factors shape the mouse methylome at distal regulatory regions. Nature 480:490–495
15. Zhang B, Zhou Y, Lin N et al (2013) Functional DNA methylation differences between tissues, cell types, and across individuals discovered using the M&M algorithm. Genome Res 23:1522–1540
16. Ziller MJ, Gu H, Muller F et al (2013) Charting a dynamic DNA methylation landscape of the human genome. Nature 500:477–481
17. Schlesinger F, Smith AD, Gingeras TR et al (2013) De novo DNA demethylation and noncoding transcription define active intergenic regulatory elements. Genome Res 23:1601–1614
18. Xie M, Hong C, Zhang B et al (2013) DNA hypomethylation within specific transposable element families associates with tissue-specific enhancer landscape. Nat Genet 45:836–841
19. Cokus SJ, Feng S, Zhang X et al (2008) Shotgun bisulphite sequencing of the Arabidopsis genome reveals DNA methylation patterning. Nature 452:215–219
20. Lister R, Pelizzola M, Dowen RH et al (2009) Human DNA methylomes at base resolution show widespread epigenomic differences. Nature 462:315–322
21. Laurent L, Wong E, Li G et al (2010) Dynamic changes in the human methylome during differentiation. Genome Res 20:320–331
22. Meissner A, Gnirke A, Bell GW et al (2005) Reduced representation bisulfite sequencing for comparative high-resolution DNA methylation analysis. Nucleic Acids Res 33:5868–5877
23. Meissner A, Mikkelsen TS, Gu H et al (2008) Genome-scale DNA methylation maps of pluripotent and differentiated cells. Nature 454:766–770
24. Weber M, Davies JJ, Wittig D et al (2005) Chromosome-wide and promoter-specific analyses identify sites of differential DNA methylation in normal and transformed human cells. Nat Genet 37:853–862
25. Maunakea AK, Nagarajan RP, Bilenky M et al (2010) Conserved role of intragenic DNA methylation in regulating alternative promoters. Nature 466:253–257
26. Serre D, Lee BH, Ting AH (2010) MBD-isolated genome sequencing provides a high-throughput and comprehensive survey of

DNA methylation in the human genome. Nucleic Acids Res 38:391–399

27. Kriukiene E, Labrie V, Khare T et al (2013) DNA unmethylome profiling by covalent capture of CpG sites. Nat Commun 4:2190
28. Harris RA, Wang T, Coarfa C et al (2010) Comparison of sequencing-based methods to profile DNA methylation and identification of monoallelic epigenetic modifications. Nat Biotechnol 28:1097–1105
29. Bock C, Tomazou EM, Brinkman AB et al (2010) Quantitative comparison of genome-wide DNA methylation mapping technologies. Nat Biotechnol 28:1106–1114
30. Huang Y, Pastor WA, Shen Y et al (2010) The behaviour of 5-hydroxymethylcytosine in bisulfite sequencing. PLoS One 5:e8888
31. Booth MJ, Branco MR, Ficz G et al (2012) Quantitative sequencing of 5-methylcytosine and 5-hydroxymethylcytosine at single-base resolution. Science 336:934–937
32. Yu M, Hon GC, Szulwach KE et al (2012) Tet-assisted bisulfite sequencing of 5-hydroxymethylcytosine. Nat Protoc 7:2159–2170
33. Taiwo O, Wilson GA, Morris T et al (2012) Methylome analysis using MeDIP-seq with low DNA concentrations. Nat Protoc 7:617–636
34. Nair SS, Coolen MW, Stirzaker C et al (2011) Comparison of methyl-DNA immunoprecipitation (MeDIP) and methyl-CpG binding domain (MBD) protein capture for genome-wide DNA methylation analysis reveal CpG sequence coverage bias. Epigenetics 6:34–44
35. Pelizzola M, Koga Y, Urban AE et al (2008) MEDME: an experimental and analytical methodology for the estimation of DNA methylation levels based on microarray derived MeDIP-enrichment. Genome Res 18:1652–1659
36. Stevens M, Cheng JB, Li D et al (2013) Estimating absolute methylation levels at single-CpG resolution from methylation enrichment and restriction enzyme sequencing methods. Genome Res 23:1541–1553
37. Lafferty J, McCallum A, Pereira F (2001) Conditional random fields: probabilistic models for segmenting and labeling sequence data. Departmental Papers CIS-159
38. Wallach H (2004) Conditional random fields: an introduction. Technical report MS-CIS-04-21 Department of computer and information science, University of Pennsylvania
39. Zhou X, Li D, Lowdon RF et al (2014) methylC track: visual integration of single-base resolution DNA methylation data on the WashU EpiGenome browser. Bioinformatics 30:2206–2207
40. Zhou X, Lowdon RF, Li D et al (2013) Exploring long-range genome interactions using the WashU Epigenome browser. Nat Methods 10:375–376
41. Zhou X, Wang T (2012) Using the Wash U Epigenome Browser to examine genome-wide sequencing data. Curr Protoc Bioinformatics Chapter 10:Unit10 10
42. Zhou X, Maricque B, Xie M et al (2011) The human epigenome browser at Washington university. Nat Methods 8:989–990
43. Quail MA, Kozarewa I, Smith F et al (2008) A large genome center's improvements to the Illumina sequencing system. Nat Methods 5:1005–1010

Chapter 13

Digital Restriction Enzyme Analysis of Methylation (DREAM)

Jaroslav Jelinek, Justin T. Lee, Matteo Cesaroni, Jozef Madzo, Shoudan Liang, Yue Lu, and Jean-Pierre J. Issa

Abstract

Digital Restriction Enzyme Analysis of Methylation (DREAM) is a method for quantitative mapping of DNA methylation across genomes using next-generation sequencing (NGS) technology. The method is based on sequential cuts of genomic DNA with a pair of restriction enzymes (*Sma*I and *Xma*I) at CCCGGG target sites. Unmethylated sites are first digested with *Sma*I. This enzyme cuts the sites in the middle at CCC^GGG, leaving behind blunt ended fragments. CpG methylation completely blocks *Sma*I; therefore, only unmethylated sites are cleaved. The remaining methylated sites are digested with *Xma*I in the next step. This enzyme is not blocked by CpG methylation. It cuts the recognition site sideways at C^CCGGG forming 5′-CCGG overhangs. The sequential cuts thus create distinct methylation-specific signatures at the ends of restriction fragments: 5′-GGG for unmethylated CpG sites and 5′-CCGGG for methylated sites. The DNA fragments resulting from the digestions are ligated to NGS adapters. Sequencing libraries are prepared using hexanucleotide barcodes for sample identification. Individual libraries with distinct barcodes are pooled and sequenced using a paired ends protocol. The sequencing reads are aligned to the genome and mapped to unique CCCGGG target sites. Methylation at individual CpG sites is calculated as the ratio of sequencing reads with the methylated signature to the total number of reads mapping to the site. Sequencing of 25 million reads per sample typically yields 50,000 unique CpG sites covered with hundreds of reads enabling accurate determination of DNA methylation levels. DREAM does not require bisulfite conversion, has a very low background, and has high sensitivity to low levels of methylation. The method is simple, cost-effective, quantitative, highly reproducible, and can be applied to any species.

Key words DNA methylation, Next-generation sequencing, Restriction endonuclease, *Sma*I, *Xma*I

1 Introduction

Bacterial restriction endonucleases can precisely distinguish specific patterns of DNA methylation. This feature has been utilized for locus-specific and genome-wide methylation analyses [1]. We have focused on a unique pair of neoschizomeric enzymes, *Sma*I and *Xma*I, which recognize the same target site, CCCGGG. The enzymes differ in their sensitivity to CpG methylation. Additionally, they also cut the target site at different positions, creating distinct fragment ends. *Sma*I cleaves the site in the middle (at CCC^GGG)

Jörg Tost (ed.), *DNA Methylation Protocols*, Methods in Molecular Biology, vol. 1708,
https://doi.org/10.1007/978-1-4939-7481-8_13, © Springer Science+Business Media, LLC 2018

creating blunt-ended fragments starting with 5′-GGG and ending with CCC-3′. CpG methylation of the site completely blocks the cleavage. The *Xma*I enzyme cuts the site after the first C (at C^CCGGG) and it is not blocked by methylation at the CpG site. The *Xma*I fragments start with the 5′-CCGGG sequence and have a 5′-CCGG overhang. This *Sma*I/*Xma*I approach was first utilized for cloning of methylated CpG islands [2] and later adapted for microarrays [3, 4]. The main drawback of these methods was that only the methylated *Sma*I/*Xma*I sites were included in the analyses while the unmethylated sites were removed. Massively parallel next-generation sequencing (NGS) paved the way for including both the unmethylated and methylated sites in the same sequencing library. The signatures at the start of sequencing reads indicate the CpG methylation status. Mapping the reads to the genome shows their location. Digital counting of methylated and unmethylated reads in the single library ensures precise quantification of DNA methylation levels. Additionally, because DNA is not converted by bisulfite, the background is below 1% and mapping of sequencing reads is computationally straightforward. The accuracy and reproducibility of the method is further increased by the addition of standards with defined methylation levels at CCCGGG sites. These calibrators are spiked in the genomic DNA before enzymatic processing and used for fine adjustment of raw methylation values [5].

The outline of the procedure is as follows: one to two micrograms of high molecular weight genomic DNA are spiked with a set of calibrators with known levels of methylation ranging from 0% to 100%. The DNA with spiked in calibrators is first digested with *Sma*I restriction endonuclease, cleaving all unmethylated CCCGGG target sites. After complete *Sma*I digestion, *Xma*I restriction endonuclease is added to cut the remaining CC^{me}CGGG sites with methylated CpGs. Restriction fragments are filled in, A-tailed, ligated to NGS adapters and barcodes unique for each sample are added by PCR amplification. Pools of sequencing libraries with distinct barcodes are sequenced using a paired end protocol, and the resulting sequences are mapped back to the genome. Reads with methylated and unmethylated signatures mapping to individual *Sma*I/*Xma*I sites are counted and methylation values calculated as the percentage of reads with the methylated signature (Fig. 1).

2 Materials

2.1 Calibration Standards

1. *E. coli* genomic DNA, unsheared.
2. Lambda bacteriophage DNA.
3. Calibrator LA248:

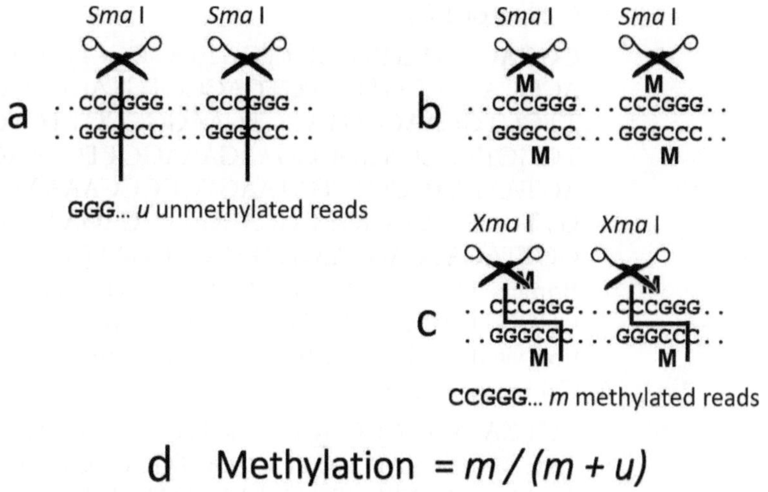

Fig. 1 Principle of DREAM. Genomic DNA is digested with *Sma*I restriction endonuclease. This enzyme cleaves all CCCGGG sites provided the cytosine at the CpG site is not methylated. The restriction fragments have blunt ends starting with 5′-GGG, the unmethylated signature *u* (**a**). CCCGGG sites with methylated CpGs are left intact, since CpG methylation blocks SmaI (**b**). After the completion of the SmaI digestion, *Xma*I restriction endonuclease is added to the reaction. This enzyme cleaves the remaining CCCGGG sites with methylated CpGs. The restriction fragments have 5′-CCGG overhangs and start with the methylated signature *m* (**c**). The restriction fragments are filled, ligated to sequencing adapters and the resulting libraries are sequenced on Illumina HiSeq instrument. The sequencing reads are mapped to the genome. Methylation level at individual CCCGGG sites is calculated as the proportion of reads with the methylated signature (**d**)

TCGAAAAAGAGCAGCACAGTGATGCCCGGGGAGGAT
ACGTTTCACTATGAGAGCCTGCGTGGACGTTATGTG
AGCGTGATGGCCGGACCGGTTTTACAAATCAGTAAG
CAGGTCAGTGCGTACGCCATGGCCGGAGTGGCTCAC
AGTCGGTGGTCCGGCAGTACAATGGATTACCGTAAG
ACGGAAATCACTCCCGGGTATATGAAAGAGACGACCA
CTGCCAGGGACGAAAGTGCAATGCGGCATAC

Primers TCGAAAAAGAGCAGCACAGTGATGCCC and GTATGCCGCATTGCACTTT. The final product is 248 bp. *Sma*I digestion yields 168 bp, 53 bp and 27 bp fragments.

4. Calibrator EC293:

TTCGTGATGCTGCAACTCTGCTACGTCTGGGGCTGG
CTTACGGCCCCGGGGGGATGTCATTACGTGAAGTCA
CTGCATGGGCTCAGCTCCATGACGTTGCAACATTATC
TGACGTGGCTCTCCTGAAGCGGCTGCGGAATGCCGC
CGACTGGTTTGGCATACTTGCCGCACAAACACTTGC
TGTACGCGCCGCAGTTACGGGTTGTACAAGCGGAAA
GAGATTGCGTCTTGTCGATGGAACAGCAATCAGTGC
GCCCGGGGGCGGCAGCGCTGAATGGCGACTACATAT
GGGA

Primers TTCGTGATGCTGCAACTCTG and TCCCATATG TAGTCGCCATTC. The final product is 293 bp. *Sma*I digestion yields 210 bp, 47 bp and 36 bp fragments.

5. Calibrator EC247:

   CGGACGGTGTTGACTTCGTTATCCCGGGTAACGACG
   ACGCAATCCGTGCTGTGACCCTGTACCTGGGCGCTG
   TTGCTGCAACCGTACGTGAAGGCCGTTCTCAGGATC
   TGGCTTCCCAGGCGGAAGAAAGCTTCGTAGAAGCTG
   AGTAATAAGGCTTGATAACTCCCCCAAAATAGTTCGA
   GTTGCAGAAAGGCGGCAAGCTCGAGAATTCCCGGGA
   GCTTACATCAGTAAGTGACCGGGATGAGCG

   Primers CGGACGGTGTTGACTTCGTT and CGCTCATC CCGGTCACTTAC. The final product is 247 bp. *Sma*I digestion yields 188 bp, 34 bp and 25 bp fragments.

6. Calibrator EC466:

   ACTGAGCTACCCGGGACTGTGTAGCCAGATCTGCGC
   TACATTCTTTACCCTCGCGGTGCGTGAGTGGATTGC
   ACAGGTTAATACTGAGAAAAACATTCTCAGTTTGCTT
   CTGCATCCACGTCTTGGTGCGGTAATACAGCAAATGC
   TGGAAATGCCAGGACACGCCTGGACCGTCGAATCGC
   TGGCCAGCATCGCTCACATGTCCCGGGCAAGTTTTG
   CCCAGCTTTTCCGTGATGTTTCCGGAACCACGCCGC
   TGGCTGTATTAACAAAGTTGCGTCTACAAATAGCGGC
   CCAGATGTTTTCCCGGGAAACGCTCCCTGTTGTGGT
   GATCGCTGAGTCAGTAGGCTATGCCAGTGAATCATCT
   TTTCACAAGGCGTTTGTCCGCGAGTTTGGTTGTAC
   CCCGGGAGAATATCGGGAAAGGGTCAGACAGCTTGC
   ACCCTGAATAAAACCGCCCGGGTAGCTCAGT

7. Primers ACTGAGCTACCCGGGACTGTGTAGCCAGATC TGCG and ACTGAGCTACCCGGGCGGTTTTATTCAGG GTGC. The final product is 466 bp. *Sma*I digestion yields 194, 99, 97 and 51 bp fragments.

8. Calibrator LA371:

   ACTGAGCTACCCGGGCAGGTAGCGCAGATCATCAAC
   GGTGTGTTCAGCCAGTTACTGGCAACTTTCCCGGCG
   AGCCTGGCTAACCGTGACCAGAACGAAGTGAACGAA
   ATCCGTCGCCAGTGGGTTCTGGCTTTTCGGGAAAAC
   GGGATCACCACGATGGAACAGGTTAACGCAGGAATG
   CGCGTAGCCCGTCGGCAGAATCGACCATTTCTGCCA
   TCACCCGGGCAGTTTGTTGCATGGTGCCGGGAAGAA
   GCATCCGTTACCGCCGGACTGCCAAACGTCAGCGAG
   CTGGTTGATATGGTTTACGAGTATTGCCGGAAGCGA
   GGCCTGTATCCGGATGCGGAGTCTTATCCGTG
   CCCGGGTAGCTCAGT

   Primers ACTGAGCTACCCGGGTGCCTGCTTGCCGAA TATCA and ACTGAGCTACCCGGGAAGCACGAGGAAG CGGTCAG. The final product is 371 bp. *Sma*I digestion yields 210 and 137 bp fragments.

9. Calibrator LA404:
   ACTGAGCTACCCGGGAAACCAATTTCAGCCAGTGCC
   TCGTCCATTTTTTCGATGAACTCCGGCACGATCTCGT
   CAAAACTCGCCATGTACTTTTCATCCCGCTCAATCAC
   GACATAATGCAGGCCTTCACGCTTCATACGCGGGTC
   ATAGTTGGCAAAGTACCAGGCATTTTTTCGCGTCACC
   CACATGCTGTACTGCACCTGGGCCATGTAAGCTGAC
   TTTATGGCCTCGAAACCACCGAGCCGGAACTTCATGA
   AATCCCGGGAGGTAAACGGGCATTTCAGTTCAAGGC
   CGTTGCCGTCACTGCATAAACCATCGGGAGAGCAGG
   CGGTACGCATACTTTCGTCGCGATAGATGATCGGGG
   ATTCAGTAACATTCACGCCGGAAGTCCCGGGTAGCT
   CAGT
   Primers ACT GAGCTACCCGGGAAACCAATTTCAGCC AGTGC and ACTGAGCTACCCGGGACTTCCGGCGTGA ATGTTAC. The final product is 404 bp. *Sma*I digestion yields 250 and 130 bp fragments.
10. *Sma*I restriction endonuclease (New England Biolabs).
11. *M.Sss*I CpG Methyltransferase (New England Biolabs).
12. Oligonucleotide primers.
13. *Taq* DNA Polymerase with ThermoPol® Buffer (New England Biolabs) or similar.
14. 10 mM Tris–HCl pH 8.0.
15. Triton X 100, molecular biology grade.
16. PCR thermal cycler.
17. Agarose gel electrophoresis supplies.
18. QIAquick PCR Purification Kit (Qiagen).

2.2 Construction of Sequencing Libraries

1. *Sma*I restriction endonuclease (New England Biolabs).
2. *Xma*I restriction endonuclease (New England Biolabs).
3. Klenow Fragment (3′ → 5′ exo-) (New England Biolabs).
4. T4 DNA Ligase (New England Biolabs).
5. 10 mg/mL BSA.
6. dNTP Set (100 mM each).
7. NEBNext® Multiplex Oligos for Illumina® (Index Primers Set 1 and 2) (New England Biolabs).
8. KAPA HiFi HotStart ReadyMix PCR Kit (Kapa Biosystems).
9. PCR plates 96-well, not skirted. Adhesive seals for PCR plates.
10. PCR thermal cycler.

2.3 Cleaning, Separation, and Quantitation of Sequencing Libraries

1. Agencourt AMPure XP magnetic beads (Beckman Coulter).
2. DynaMag 96 Side Magnetic Particle Concentrator (Invitrogen Cat. No. 123.31D).
3. Ethanol.
4. Molecular biology grade water.
5. PCR plates 96-well, not skirted.
6. Qubit 2.0 fluorometer, ds DNA HS Assay kit (ThermoFisher Scientific).
7. NanoDrop UV-Vis Spectrophotometer (ThermoFisher Scientific).
8. 2100 Bioanalyzer, DNA 1000 Kit (Agilent Technologies).

2.4 Next-Generation Sequencing

1. Illumina HiSeq 2500 or a similar instrument for next-generation sequencing.
2. Linux server and/or bioinformatics support for processing of the sequencing data.
3. Hard disk storage for the sequencing data.
4. bowtie2 [6].
5. SAMtools [7].
6. Scripts to compute, adjust, and annotate methylation levels https://github.com/jaroslavj/DREAM_tools/.

3 Methods

3.1 Methylation Standards for Calibration

Methylation standards should be spiked into genomic DNA before the enzymatic processing and construction of sequencing libraries in order to provide internal controls for the calibration of the raw methylation data. The standards consist of PCR products distinct from the genomic DNA to be analyzed. Each standard has 2–3 CCCGGG sites with a distance suitable for the sequencing procedure. Sequences of the calibrators and primers for making PCR products containing desired CCCGGG sites are given in Subheading 2.1. In the following paragraph the PCR products are created using *Escherichia coli* DNA as the template, CpG sites are methylated in part of the PCR products using the *M.Sss*I CpG methyltransferase and the methylated and unmethylated PCR products are mixed in the ratios that result in 0, 25, 50, 75, and 100% methylation.

1. PCR conditions for making all calibrators are as follows: plan for 100 μL reactions and mix reagents on ice. Each reaction will consist of 200 ng *Escherichia coli* DNA in 85 μL water, 10 μL of PCR buffer, 1 μL of 10 μM forward and reverse primer (final 100 nM), 1 μL of 25 mM dNTP mix (final 250 μM) and 2 μL

(10 U) of *Taq* polymerase. Aliquot the reaction mix to 4 wells in a PCR plate.

2. Set up the PCR program as follows: Step 1-initial denaturation at 94 °C for 3 min, Step 2-denaturation at 94 °C for 15 s, Step 3-primer annealing at 60 °C for 30 s, Step 4-elongation at 72 °C for 30 s, Step 5-return to Step 2 for 32 more cycles, Step 6- final extension at 72 °C for 5 min. This program totals 33 cycles. Pool the PCR products from the 4 aliquots into a single tube.
3. Verify the presence of CCCGGG sites by *Sma*I digestion of 5 μL of the PCR product. Compare the digested and undigested PCR products using 2% agarose gel electrophoresis. If you obtain a single band from the undigested PCR product and fragments of expected sizes after *Sma*I digestion (*see* Subheading 2.1), make 500 μL more of the PCR product for each calibrator. Use 2 μL of the first PCR product as the template instead of *E. coli* DNA. Purify the PCR products using the PCR purification kit following the manufacturer's protocol. Measure DNA concentration using NanoDrop spectrophotometer and/or Qubit fluorometer. Set aside 25% of the purified PCR products as unmethylated DNA.
4. Methylate the remaining 75% of amplified calibrators with the *M.Sss*I CpG methyltransferase. For each calibrator, take 10 μg of purified unmethylated PCR product. Add water to adjust the volume to 440 μL. Add 50 μL NEB Buffer 2, 2.5 μL of 32 mM S-adenosylmethionine (SAM), and 5 μL (20 U) of *M. Sss*I CpG methyltransferase. Incubate the solution at 37 °C. After 3 h, add 2.5 μL of SAM and 5 μL (20 U) of *M.Sss*I and continue the incubation overnight. Inactivate the enzyme at 95 °C for 5 min. Check the methylation level by *Sma*I digestion and gel electrophoresis, comparing 20 ng of methylated and unmethylated DNA. Estimate the methylation level as the proportion of DNA resistant to *Sma*I digestion. If a significant proportion of the *M.Sss*I-treated DNA still shows cleavage, continue with the *M.Sss*I treatment by adding SAM and *M. Sss*I once more and incubating at 37 °C for 3–6 h. Inactivate the enzyme at 95 °C for 5 min. Purify the methylated DNA using the PCR purification kit following the manufacturer's protocol. Measure DNA concentration using NanoDrop spectrophotometer and/or Qubit fluorometer.
5. For each calibrator, calculate the amount of unmethylated and methylated DNA to achieve a mix with a desired level of DNA methylation to make standards with 0, 25, 50, 75, and 100% methylation. Select the calibrators with 3–4 CCCGGG sites (EC466, LA404) to make the 50% methylated standard. Mix the calculated amount of unmethylated and methylated DNA

in a new tube. Estimate the level of methylation in each calibrator as the proportion of DNA resistant to cleavage by *Sma*I digestion of 20 ng DNA using gel electrophoresis. Once you have calibrators with desired levels of DNA methylation, pool equal amounts of them in a new tube. Measure the DNA concentration of this stock pool using NanoDrop spectrophotometer and/or Qubit fluorometer. Make a working dilution of the pooled calibrators at 10 pg/μL in 10 mM Tris containing 1% Triton X 100 as a detergent and lambda phage DNA at 10 ng/μL as a carrier. Store small aliquots of the working dilution and the stock at −20 °C.

3.2 Preparation of DNA Samples

The DREAM method relies on the signatures of the unmethylated and methylated CpG sites at the ends of the DNA fragments created by the restriction enzymes. It is essential that the genomic DNA is of high molecular weight and clean from impurities that would interfere with the enzyme actions.

1. Check the integrity of DNA by electrophoresis of 200 ng of DNA in 1% agarose gel. Good quality DNA should migrate as a single band above 10 kb. Any visible smear below 1 kb indicates fragmented DNA. While these random fragments will not affect the readings of DNA methylation at the bona fide (CCCGGG) target sites, their potential incorporation in the sequencing library will diminish the proportion of informative reads.
2. Assess the DNA purity with a NanoDrop spectrophotometer. The 260/230 nm ratio should be close to 2.0 and the 260/280 nm ratio above 1.8. A low 260/230 ratio is a sign of impurities that may adversely affect the enzymatic reactions. DNA containing low molecular weight fragments or impurities that have a low 260/230 nm ratio can be purified with AMPure XP beads (*see* **Note 1**).
3. Finally, measure the concentration of double-stranded DNA using a Qubit fluorometer. Ideally, the DNA concentration should be 100–300 ng/μL. If the DNA concentration is below 100 ng/μL, adjust the reaction volumes of restriction digests accordingly.

3.3 Construction of DREAM Libraries

It is convenient to process up to 12 samples in parallel and make master mixes for enzymatic reactions. The minimum recommended DNA amount for making complex sequencing libraries is around 500 ng. This represents approximately 100,000 copies of the diploid genome. It is important to consider potential losses of DNA during library preparation and also the fact that only a small portion of the final library is used for sequencing. Ideally, use 1–2 μg DNA for library preparation. Keep the tubes with samples and reagents on ice, unless specified otherwise. First, spike the DNA samples with calibration standards. Spiking genomic DNA with a set of calibration standards with defined methylation levels provides a tool for adjustments for small sample-to-sample variation and potential batch effects. Digest genomic DNA with spiked in calibrators with neoschizomeric restriction endonucleases *SmaI* and *XmaI*, in that specific order.

1. Pipette 2 μg of the DNA sample in a 1.5 mL tube and add 2 μL (4 pg) of pooled calibration standards. Adjust the volume to 20 μL with water.
2. First, set up the *Sma*I digest of unmethylated CCCGGG sites. Prepare 70 μL of the *Sma*I master mix for 12 reactions: 35 μL NEB Buffer #4, 3.5 μL BSA 10 mg/mL, 14 μL molecular grade water, 17.5 μL *Sma*I 20 U/μL. Add 5 μL of the *Sma*I master mix to each tube with spiked in sample DNA, tap the tube to mix gently, and spin briefly. Incubate the samples at **25 °C** for 3 h.
3. Digest the remaining methylated CC^{me}CGGG sites with *Xma*I. Make the *Xma*I master mix: 35 μL NEB Buffer #4, 3.5 μL BSA 10 mg/mL, 294 μL molecular grade water, 17.5 μL *Xma*I 10 U/μL. Add 25 μL of the *Xma*I master mix to each sample tube, tap the tube to mix gently, and spin briefly. Incubate the sample tubes overnight (for 16 h) at 37 °C.
4. Proceed with 3′-end filling and A-tailing. Briefly spin the samples from overnight digestion and put them on ice. Make 70 μL of a Klenow master mix: 42 μL of the CGA deoxynucleotide triphosphate mix (dCTP, dGTP, dATP, 10 mM each) and 28 μL Klenow Fragment (3′ → 5′ exo–). Add 5 μL of the Klenow master mix to each tube with the 50 μL of digested DNA, mix by gentle tapping, and spin briefly. Incubate for 30 min at 37 °C (*see* **Note 2**).
5. Purify the DNA fragments with AMPure XP beads. Use the 2.0× volume ratio of beads to DNA so that small DNA fragments are retained. Take an aliquot of thoroughly mixed AMPure XP beads and allow it to equilibrate to room temperature.
6. For each sample, pipette 110 μL of beads into a 96-well plate filling into as many wells as samples to be processed ($n = 12$). Be sure to leave space in between samples to prevent crossover contamination and note the location of the samples in the wells. Then, transfer each DNA sample into its own, individual well of AMPure XP beads. Mix by pipetting up and down ten times. Incubate the plate for 15 min at room temperature.
7. Transfer this plate to the 96-well magnetic particle concentrator and wait 5 min for beads to clear. Once the beads have cleared, carefully remove and discard 150 μL (15 μL less than

total volume) of the supernatant ensuring not to disturb the beads. Keep the 96-well plate on the magnet throughout the following washes.

8. Add 200 μL of freshly prepared 70% ethanol and wait 30 s. Remove and discard the 200 μL of 70% ethanol by pipette, then again add 200 μL of 70% ethanol and wait 30 s. After this second wash, carefully pipette out all the remaining liquid. It is critical to remove all ethanol because of interference with enzyme activity in the following steps. Allow the beads to air dry for 5–10 min, but do not over dry the beads as this will decrease yield (*see* **Note 3**).
9. Remove the 96-well plate from the magnet and resuspend the dried beads in 30 μL molecular grade water, pipetting up and down 10 times or until homogenous, then incubate at room temperature for 5 min. If the sample is not resuspended properly, wait 15 min and then repeat pipetting up and down 10 times until homogenous, followed by another 5 min incubation at room temperature. Place the plate back onto the magnet for 2 min once the beads are in suspension.
10. Once the beads have cleared from the solution, carefully pipette out 25 μL with as few beads as possible and transfer these individual samples to a new set of wells inside the 96-well plate (*see* **Note 4**). Proceed to ligation of NEBNext sequencing adaptors.
11. Prepare adaptors for ligation. The NEBNext adaptors are hairpin oligonucleotides with 3′-dT overhangs for the TA ligation to 3′-dA-tailed DNA fragments. Dilute the NEBnext 15 μM stock adapters to 3 μM in water and freeze in small aliquots.
12. Digested 3′-dA-tailed DNA is in 25 μL in a new set of wells for the ligation reaction. Set up the ligation by adding 3 μL 10× ligation buffer, 2 μL T4 DNA ligase, *X* μL 3 μM NEBNext adapter (using 1 μL per 1 μg starting DNA). Mix carefully by pipetting up and down. Seal the plate and incubate the ligation reactions at 16 °C overnight.
13. Once overnight ligation is completed, cleave the hairpin loop by adding 3 μL USER enzyme from the NEBNext adaptor kit and pipetting up and down to mix. Incubate at 37 °C for 15 min and continue to the clean-up step.
14. Post-ligation cleanup is a purification by AMPure XP beads (1.2× volume). Take an aliquot of thoroughly mixed AMPure XP beads and allow it to equilibrate to room temperature. Pipette 41 μL of beads into each single DNA sample, pipetting up and down 10 times and then incubating for 15 min at room temperature.

15. Transfer this plate to the 96-well magnetic particle concentrator and wait 5 min for beads to clear. Once beads have cleared, carefully remove and discard 60 μL (15 μL less than total volume) of supernatant ensuring not to disturb the beads. The 96-well plate remains on the magnet throughout the washes.
16. Add 200 μL of freshly prepared 70% ethanol and wait 30 s. Remove and discard the 200 μL of 70% ethanol, then again add 200 μL of 70% ethanol and wait 30 s. At this second wash, pipette out more than once to ensure removal of all liquid. It is critical to remove all ethanol because of interference with enzyme activity in the following steps. Allow the mixture to air dry for 5–10 min, but do not over-dry beads (*see* **Note 3**).
17. Remove the 96-well plate from the magnet and resuspend the dried beads in 100 μL molecular grade water, pipetting up and down 10 times or until homogenous.
18. Perform dual-SPRI size selection (*see* **Note 5**). Add 60 μL of AMPure XP beads (0.6× volume) into the wells containing the 100 μL of resuspended DNA with the beads, mix well by pipetting up and down 10 times. Incubate at room temperature for 15 min.
19. Transfer the plate to the magnet and let it stand for 5 min to completely clear the solution of beads. Carefully remove 155 μL of the supernatant (containing DNA fragments smaller than 450 bp) and transfer it to a new set of wells.
20. Add 20 μL of AMPure XP beads to the 155 μL and mix well by pipetting up and down 10 times. Allow it to incubate at room temperature for 15 min.
21. Transfer the plate to the magnetic stand for 5 min to completely clear the solution of beads. Carefully remove only 160 μL of the supernatant and discard it. Leaving some of the volume behind minimizes bead loss at this step. The beads will now bind DNA > 250 bp (*see* **Note 6**).
22. While the plate is still on the magnet, add 200 μL of freshly prepared 70% ethanol and wait 30 s. Remove and discard the 200 μL of 70% ethanol with a pipette, and again add 200 μL of 70% ethanol and wait 30 s. At this second wash, pipette out more than once to ensure removal of all liquid. It is critical to remove all ethanol because of interference with enzyme activity in the following steps. Allow this to air dry for 5–10 min, but do not over dry the beads (*see* **Note 3**).
23. Remove the 96-well plate from the magnet and resuspend the dried beads in 25 μL molecular grade water, pipetting up and down 10 times or until homogenous, and incubate at room temperature for 5 min. If the sample is not resuspended

properly, wait 15 min and repeat pipetting up and down 10 times, followed by another 5 min incubation at room temperature. Place the plate back onto the magnet for 2 min once the beads are in the solution.

24. Upon clearing of the beads in the solution, carefully pipette out 20 μL without any beads. Transfer these individual samples to a new set of wells inside the 96-well plate keeping space in between samples to prevent crossover contamination.
25. Proceed to the ligation-mediated PCR. At this step, samples will be assigned barcodes using the NEBNext barcoded primers. Two NEBNext multiplex oligonucleotide sets provide 24 different barcode-specific primers and only one is necessary for use with each sample in PCR. These primers come as 25 μM stocks. Dilute the primers to 10 μM in water and store at −20 °C in small aliquots.
26. Take note of which sample will be assigned which individual barcode. Plan for 50 μL reactions and mix reagents on ice. Each reaction will consist of 20 μL AMPurified DNA with ligated NEBNext adaptors, 25 μL KAPA Hifi HotStart ReadyMix 2×, 2.5 μL molecular grade water, 1.25 μL NEBNext universal primer 10 μM (250 nM final concentration), 1.25 μL assigned NEBNext barcoded primer 10 μM (250 nM final concentration).
27. Set up the PCR program as follows: Step 1-initial denaturation at 98 °C for 45 s, Step 2-denaturation at 98 °C for 15 s, Step 3-primer annealing at 60 °C for 30 s, Step 4-elongation at 72 °C for 30 s, Step 5-return to Step 2 for 10 more cycles, Step 6-final extension at 72 °C for 1 min. This program totals 11 cycles (*see* **Note 7**).
28. Upon completion of PCR, the final cleanup will be AMPure XP bead purification (1.2× volume). Take an aliquot of thoroughly mixed AMPure XP beads and allow it to equilibrate to room temperature. Pipette 60 μL of beads into each sample, pipetting up and down 10 times, and then incubate it for 15 min at room temperature.
29. Transfer this plate to the 96-well magnetic particle concentrator and wait 5 min for beads to clear. Once beads have cleared, carefully remove and discard 95 μL (15 μL less than total volume) of supernatant ensuring not to disturb the beads. The 96-well plate remains on the magnet throughout the following washes.
30. Add 200 μL of freshly prepared 70% ethanol and wait 30 s. Remove and discard the 200 μL of 70% ethanol by pipetting, then again add 200 μL of 70% ethanol and wait 30 s. At this second wash, pipette out the ethanol more than once to ensure removal of all liquid. It is critical to remove all ethanol because

of interference with enzyme activity in the following steps. Allow this to air-dry for 5–10 min, but do not over-dry beads (*see* **Note 3**).

31. Remove the 96-well plate from the magnet and resuspend dried beads in 30 μL molecular grade water, pipetting up and down 10 times or until homogenous, and incubate at room temperature for 5 min. If the sample is not resuspended properly, wait 15 min and repeat pipetting up and down 10 times, followed by another 5 min incubation at room temperature. Place the plate back onto the magnet for 2 min once the beads are in the solution.
32. Upon clearing of beads in the solution, carefully pipette out 25 μL without any beads, and transfer these individual samples to a new set of 1.5 mL tubes. This completes the procedure of making the DREAM libraries. These will be stored at 4 °C until samples have undergone DNA quantification by HS Qubit, Agilent electrophoresis, and pooling for sequencing.

3.4 DREAM Library Quantification, Quality Control, and Pooling

The concentration of the finished DREAM libraries should be in the 20–40 ng/μL range, however, the acceptable minimum is approximately 3 ng/μL.

1. Use Qubit fluorometry with high sensitivity reagents to measure low concentrations of ds DNA accurately. Use microelectrophoresis with Agilent 2100 Bioanalyzer instrument and DNA 1000 reagents to assess the distribution of fragment sizes and estimate the molarity of the library.
2. Validate the molarity calculated by Agilent Bioanalyzer software using the calculation based on the DNA concentration measured by Qubit. The formula is as follows:

   HS Qubit reading (ng/μL) $\times 10^6 = X$ ng/L

   Median base pair length from Agilent electrophoresis $\times 2 \times 292.7 = Y$ molecular weight

   X ng/μL/Y molecular weight $= Z$ nM (molarity in nanomoles per liter)
3. Based upon the molarity in nmol/L (fmol/μL), determine the amount of femtomoles of each sample to be included in the pool (*see* **Note 8**).
4. Pool 6-12 DREAM libraries with distinct barcodes using equal number of femtomoles from each library.
5. Measure the DNA concentration in the pooled DREAM libraries with the Qubit fluorometer and the high sensitivity protocol.
6. Estimate the size of DNA, molarity and DNA concentration of the pool with the Agilent 2100 Bioanalyzer and the DNA 1000 reagents.

3.5 Sequencing the Pooled DREAM Libraries

Sequence the pool of DREAM libraries using a paired end 2 × 40 bases protocol on Illumina HiSeq 2500 instrument and a Rapid Run flow cell. Accurate estimate of the effective molarity of the DREAM libraries is critical for optimal loading of the sequencing flow cell. Effective molarity is based on the molarity of fragments between 100 and 1000 bp. The molarity of the pool for sequencing should be above 10 nM. The expected amount of reads from an optimally loaded Rapid Run flow cell is 300 million. This will give 25 million reads per sample in a 12-sample pool. Approximately 15 million reads should map their ends to unique CCCGGG sites across the genome. Due to the nature of library preparation and the next-generation sequencing technique, DREAM has a preference for CCCGGG sites spaced within 500 bp from each other. This condition is met by approximately 100,000 sites in the human genome. The sites with a neighbor within 500 bp are typically covered by 90% of the sequencing reads (Fig. 2).

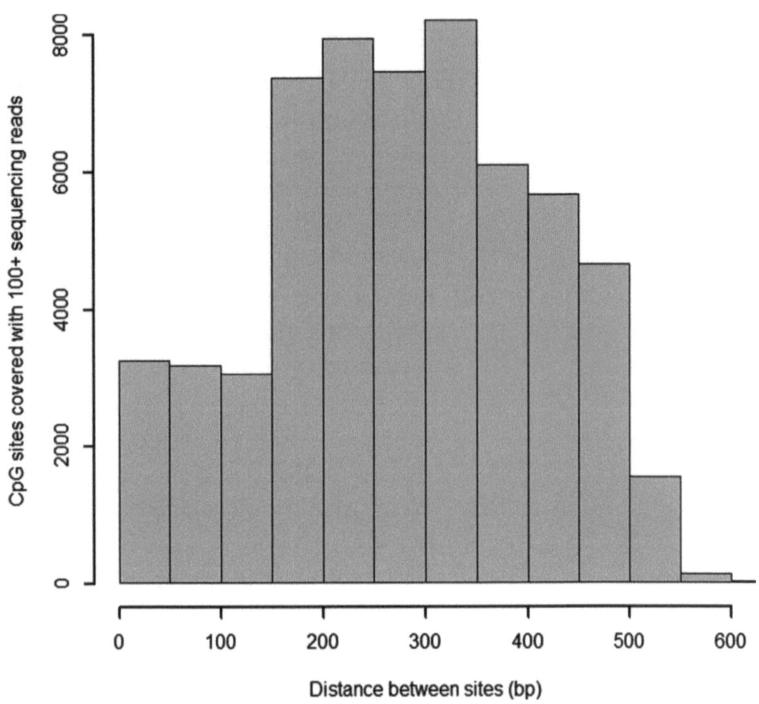

Fig. 2 Closely spaced target sites are covered with high frequency. CCCGGG sites with the distance less than 500 bp from each other (*x* axis) have consistent high coverage with the sequencing reads (*y* axis). The histogram shows frequency distribution of 50,000 CpG sites with the minimum coverage of 100 sequencing reads per site

3.6 Mapping the Reads to the Genome

1. First, align sequencing files in the fastq.gz format to the genome. Use bowtie2 [6] with –sensitive (-D 15 -R 2 -L 22 -i S,1,1.15) option for mapping the sequencing reads. Bowtie2 generates files in a generic SAM (Sequence Alignment/Map) format (*see* **Note 9**).
2. Next, use SAMtools [7] for the conversion of SAM files into sorted indexed binary (BAM) files.
3. Count methylation signatures and compute methylation levels. Use the sorted BAM files as input for the Python script, which assigns the reads to individual CCCGGG sites positioned in the genome. The script counts the reads with methylated and unmethylated signatures and calculates raw methylation values as the percentage of reads with the methylated signature in the total number of reads. The Python script is available at https://github.com/jaroslavj/DREAM_tools/.
4. Restriction enzymes are less than 100% efficient, which can distort the estimate of methylation levels. In order to correct for systematic distortions of methylation values, use the spiked in calibrators. Compute the ratios of methylated to unmethylated reads of the spikes and compare the measured and the expected values. Calculate the correction coefficient and use it for computing calibrator-adjusted methylation values at all CCCGGG sites (for details *see* **Note 10**). Finally, assemble tables containing numbers of reads, raw and adjusted methylation values for all annotated CCCGGG sites and import the tables into a Microsoft Access database for data queries. The pipeline for data processing, R scripts and annotation tables are available at https://github.com/jaroslavj/DREAM_tools/.

4 Notes

1. The ratio of 0.4 volume of AMPure XP beads to 1 volume of DNA allows for capture of high molecular weight DNA while leaving behind small fragments and impurities. Ensure that AMPure XP beads have reached room temperature before proceeding and resuspend them well. The maximum volume of DNA sample per well in a 96-well PCR plate is 140 μL. This would be in accordance with using 56 μL of AMPure beads, totaling 196 μL (for 96-well magnetic plate separation, 200 μL is recommended maximum volume per well). If the sample is at a higher volume, split into multiple wells. Add 56 μL of beads (**0.4× volume**) into the wells of a 96-well PCR plate. Add 140 μL of DNA containing low molecular weight fragments into respective wells with AMPure beads. Mix well by pipetting up and down 10 times, then incubate at room temperature for 15 min. Transfer the plate to the magnetic separator and let it

stand for 5 min to completely clear the solution of beads. Carefully remove and discard only 180 μL of the supernatant; leaving some of the volume behind minimizes bead loss at this step. The beads should not disperse; instead, they will stay on the walls of the wells. Significant loss of beads at this stage will impact the yield, so ensure beads are not removed with the supernatant or the wash. With the plate still on the magnet, add 200 μL of freshly prepared 70% ethanol and wait 30 s. Remove the 70% ethanol wash using a pipette. Repeat the 70% ethanol wash one more time, for a total of two washes. With the final wash, it is critical to remove as much of the ethanol as possible. Use at least two pipetting steps and allow excess ethanol to collect at the bottom of the tubes after removing most of the ethanol in the first pipetting step. Air-dry the beads on the magnet for a minimum of 10 min. Inspect each tube carefully to ensure that all the ethanol has evaporated. It is critical that all residual ethanol be removed prior to continuing. Remove the tubes from the magnet and add 30 μL room temperature nuclease-free water to the dried beads. Mix thoroughly to ensure all the beads are resuspended. Incubate at room temperature for 5 min. Transfer the plate to the magnetic separator and let it stand for 2 min. Carefully remove 25 μL of the eluate, ensuring as few beads as possible are carried over and transfer to a fresh set of wells. When pipetting any portion of this eluted library downstream, be sure to keep the plate on the magnetic separator to minimize bead carryover into any ensuing reactions.

2. Genomic DNA has now been digested at the CCCGGG target sites creating blunt-ended fragments starting with 5′-GGG at unmethylated sites and fragments with 5′-CCGG overhangs at methylated sites. The 3′ → 5′ exonuclease deficient Klenow Fragment of *E. coli* DNA polymerase fills in the 3′ recesses and adds 3′-dA tails to all fragments for the subsequent TA ligation of sequencing adapters. dTTP is intentionally omitted from the triphosphate mix to suppress the repair of uninformative DNA fragments. It is not necessary to purify DNA between the restriction digests and end filling/A-tailing. Klenow Fragment DNA polymerase works well in NEB Buffer #4.
3. Do not over-dry the AMPure beads. Beads are overly dried if they appear flaky with cracks showing. If this happens, allow ample time for suspension in molecular grade water. You can even attempt to manually disturb the beads from tube wall with a pipette tip if necessary.
4. When pipetting any portion of this eluted library downstream, be sure to use a magnet stand to minimize bead carryover into any subsequent reactions.

5. Dual-SPRI size selection permits the stepwise removal of fragments larger than 450 bp and also those smaller than 250 bp. First, the 0.6× AMPure step removes unwanted large DNA fragments because fragments larger than 450 bp are bound to beads. The supernatant with DNA fragments smaller than 450 bp is collected and transferred to a new well. Addition of AMPure XP beads to the collected supernatant raises the concentration of salts and polyethylene glycol. This results in binding of DNA fragments larger than 250 bp to the beads. The smaller fragments are washed away and then the fraction containing fragments between 250 and 450 bp is recovered.
6. The beads should not disperse; instead, they will stay on the walls of the tubes. Significant loss of beads at this stage will impact the yield, so ensure beads are not removed with the supernatant or the wash.
7. If the double-stranded DNA concentration of the library after purification is below 3 ng/μL, amplify the product with four more cycles of PCR following the setup in **step 27** of Subheading 3.3.
8. The most concentrated and the most diluted sample are good leads for the optimum amount of femtomoles to pool. For example, 2 μL of a library with the molarity of 100 nmol/L will yield 200 femtomoles. Pooling volumes under 2 μL may lead to significant pipetting errors and result in imbalanced representation of libraries in the pool. It is important that all barcoded libraries pooled for sequencing have well-balanced representation. Therefore, it is optimal that the libraries to be combined in a pool have approximately the same size and molarity to minimize potential errors.
9. Build the bowtie2 index that includes the sequences of the calibrators.
10. Use spiked in calibrators to compensate for potential distortions of raw methylation values expressed as the percentage of methylated reads. Compare the expected and observed ratios of methylated to unmethylated reads mapped to the calibrators and calculate the correction coefficient. The calculation is based on the assumption that the efficiency of the *Sma*I enzyme for cutting unmethylated CCCGGG sites is 100%. Verify this by checking the background methylation level of the completely unmethylated calibrator. This value is typically lower than 0.5%.

    Let s be the efficiency e_x of the *Xma*I enzyme for methylated DNA

$$s = e_x$$

where $0 < e_x \leq 1$. Let m be the number of molecules with methylated cytosines and u the number of molecules with unmethylated cytosines at a particular *Sma*I site. After digestion with two restriction enzymes, there are u molecules with unmethylated signature and sm molecules with methylated signature. Therefore the *measured* methylation ratio is

$$p_m = \frac{\mathrm{sm}}{\mathrm{sm} + u} \tag{1}$$

The *true* methylation ratio is

$$p_t = \frac{m}{m + u} \tag{2}$$

which is related to the measured methylation (1) by

$$p_t = \frac{1}{1 + s\left(\frac{1}{p_m} - 1\right)} \tag{3}$$

To adjust methylation levels measured by DREAM based on the values obtained from the spiked in standards, calculate log ratios $\ln(m/u)$ and $\ln(sm/u)$ for each standard, where m/u is the expected ratio of methylated and unmethylated reads, and sm and u are observed numbers of methylated and unmethylated reads. Obtain differences in the expected minus observed log ratios

$$\ln(m/u) - \ln(\mathrm{sm}/u) = \ln(1/s) \tag{4}$$

for each standard. Correction factor $c = 1/s$ is expressed as an antilog of the average log difference (expected – observed) from all the standards. Compute adjusted methylation values for each CpG site as

$$[100\%{}^*\mathrm{c}^*(\mathrm{sm} + 0.5)]/[\mathrm{c}^*(\mathrm{sm} + 0.5) + u + 0.5] \tag{5}$$

Acknowledgments

We thank Amy B. Hart for editorial help. Research in the authors' laboratories is supported by grants CA100632, CA158112, and CA049639 from the NIH.

References

1. Laird PW (2010) Principles and challenges of genomewide DNA methylation analysis. Nat Rev Genet 11:191–203
2. Toyota M, Ahuja N, Ohe-Toyota M et al (1999) CpG island methylator phenotype in colorectal cancer. Proc Natl Acad Sci U S A 96:8681–8686

3. Estécio MR, Yan PS, Ibrahim AE et al (2007) High-throughput methylation profiling by MCA coupled to CpG island microarray. Genome Res 17:1529–1536
4. Shen L, Kondo Y, Guo Y et al (2007) Genome-wide profiling of DNA methylation reveals a class of normally methylated CpG island promoters. PLoS Genet 3:2023–2036
5. Jelinek J, Liang S, Lu Y et al (2012) Conserved DNA methylation patterns in healthy blood cells and extensive changes in leukemia measured by a new quantitative technique. Epigenetics 7:1368–1378
6. Langmead B, Salzberg SL (2012) Fast gapped-read alignment with Bowtie 2. Nat Methods 9:357–359
7. Li H, Handsaker B, Wysoker A et al (2009) The sequence alignment/map format and SAMtools. Bioinformatics 25:2078–2079

Chapter 14

Nucleosome Occupancy and Methylome Sequencing (NOMe-seq)

Fides D. Lay, Theresa K. Kelly, and Peter A. Jones

Abstract

Various methodologies are available to interrogate specific components of epigenetic mechanisms such as DNA methylation or nucleosome occupancy at both the locus-specific and the genome-wide level. It has become increasingly clear, however, that comprehension of the functional interactions between epigenetic mechanisms is critical for understanding how cellular transcription programs are regulated or deregulated during normal and disease development. The *N*ucleosome *O*ccupancy and *Me*thylome *seq*uencing (NOMe-seq) assay allows us to directly measure the relationship between DNA methylation and nucleosome occupancy by taking advantage of the methyltransferase M.CviPI, which methylates unprotected GpC dinucleotides to create a footprint of chromatin accessibility. This assay generates dual nucleosome occupancy and DNA methylation information at a single-DNA molecule resolution using as little as 200,000 cells and in as short as 15 min reaction time. DNA methylation levels and nucleosome occupancy status of genomic regions of interest can be subsequently interrogated by cloning PCR-amplified bisulfite DNA and sequencing individual clones. Alternatively, NOMe-seq can be combined with next-generation sequencing in order to generate an integrated global map of DNA methylation and nucleosome occupancy, which allows for comprehensive examination as to how these epigenetic components correlate with each other.

Key words DNA methylation, Nucleosome occupancy, M.CviPI, Chromatin accessibility, Bisulfite sequencing, Genome-wide

1 Introduction

Traditionally, nucleosome occupancy is determined by MNase-seq, DNAse-seq, or FAIRE-seq. However, these methods measure nucleosome occupancy on a population level and interrogate only one component of the layered regulatory process of gene expression. To demonstrate the interplay between DNA methylationand nucleosome positioning, we use a footprinting approach in which a commercially available DNA methyltransferase M.CviPI, first described by Michael Kladde and colleagues, is used to methylate cytosines that are present in a GpC context without altering endogenous DNA methylation patterns present in CpG dinucleotides [1]. In this assay, GpC sites that are not protected by nucleosomes

Jörg Tost (ed.), *DNA Methylation Protocols*, Methods in Molecular Biology, vol. 1708,
https://doi.org/10.1007/978-1-4939-7481-8_14,

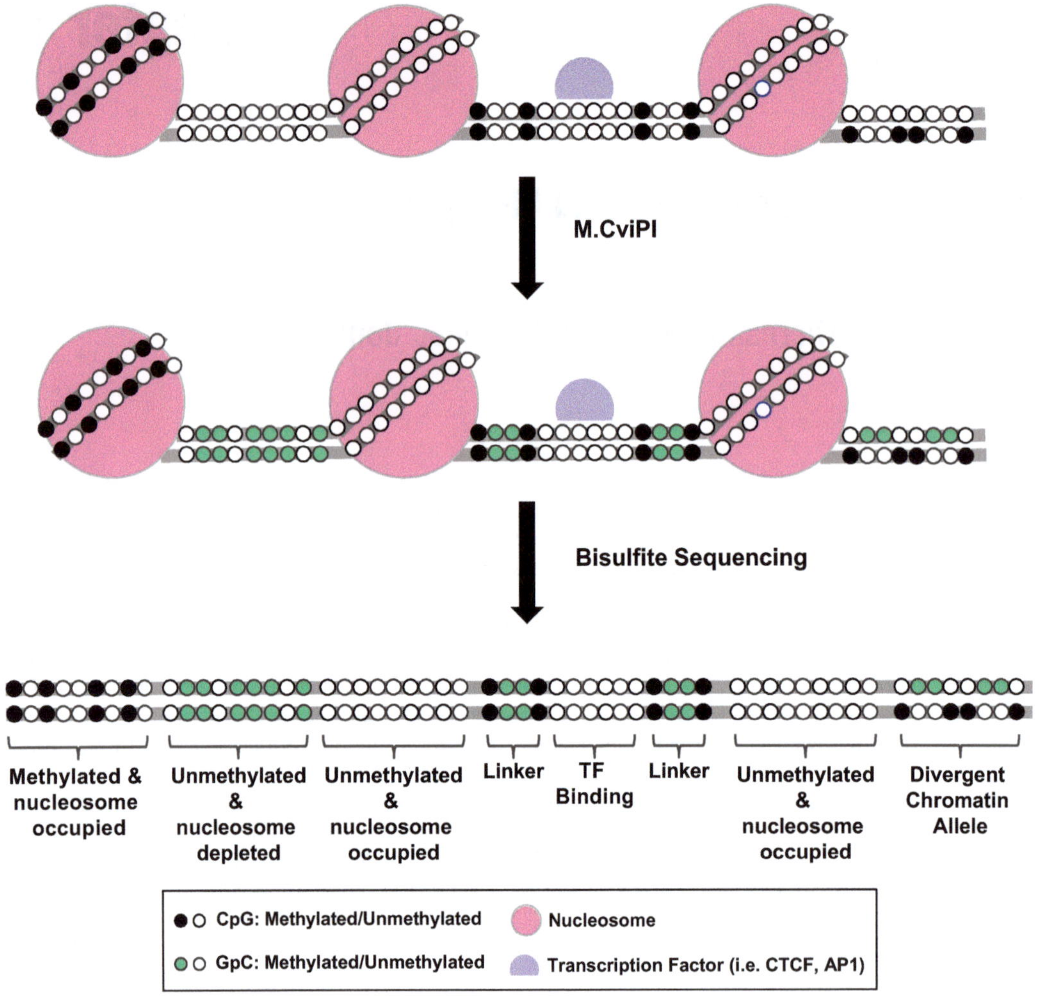

Fig. 1 Overview of the NOMe-seq assay. In this assay, the treatment of nuclei or chromatin with M.CviPI results in the methylation of cytosines present in the GpC dinucleotide context that are unprotected by nucleosomes or transcription factors. Various combinations of endogenous DNA methylation status and nucleosome/transcription factor occupancy pattern can be detected by using bisulfite sequencing. The single-molecule resolution of the assay can furthermore resolve loci that are present in multiple chromatin configurations, here termed divergent chromatin alleles

and other tight DNA binding factors are accessible and can be methylated by the enzyme, creating a footprint of nucleosome depletion across the genome. Unlike CpG dinucleotides, GpC dinucleotides are more widely distributed throughout the genome and endogenous GpC methylation does not occur in mammalian cells. The methylation status of both the CpG and GpC sites can thus be simultaneously detected at single-DNA molecule resolution by bisulfite sequencing, providing dual epigenetic information of both nucleosome occupancy and DNA methylation (Fig. 1). At the locus-specific level, this method has been used to investigate nucleosome positioning during the cell-cycle [2], the presence of nucleosome depleted regions (NDRs) in enhancers [3], the maintenance of NDR by OCT4 [4], and the epigenetic cross-talk between enhancers and promoters [5]. Genome-wide NOMe-seq has also been used to demonstrate the correlation between DNA methylation and nucleosome depletion in the linkers of CTCF regions as well as to identify regions present in multiple chromatin configurations such as imprinted or X-inactivated loci [6].

This chapter covers the procedures for performing locus-specific and genome-wide NOMe-seq [4–6]. Both the protocols begin by incubating freshly isolated nuclei from cell culture or primary tissue with the M.CviPI enzyme and its cofactor, S-adenosylhomocysteine (SAM) for two 7.5 min incubation times. Once the reaction is stopped, proteins are degraded and enzyme-treated DNA is purified. DNA methylation and nucleosome occupancy patterns in specific genomic loci are determined by cloning the products of PCR-amplified bisulfite converted DNA and sequencing individual colonies (Fig. 2, left panel). For global analysis, a whole-genome bisulfite sequencing library is generated by ligating methylated adapters to the end-repaired and A-tailed DNA fragments before bisulfite conversion (Fig. 2, right panel). Following whole-genome amplification of bisulfite-treated DNA, gel purification is performed to remove residual and self-ligated adapters and select a size-range of templates to be sequenced [7, 8].

Here, we also delineate an alternative method for treating formaldehyde-fixed chromatin isolated from fresh or frozen primary human tissues. The fixing of chromatin reduces the effects of dynamic nucleosome movement that may occur during the mechanical isolation of nuclei and inadvertently results in the methylation of GpC sites otherwise inaccessible. This approach, however, involves a much longer enzyme incubation time, which is necessary in order to ensure the efficient methylation of all accessible GpC sites while reducing the bias that may arise from enzyme overtreatment. Active Motif has adapted this formaldehyde-fixed version of the assay as the basis for a commercially available kit for performing NOMe-seq (cat. 54000). Similar to our standard protocol, the isolated DNA can be subsequently used for locus-specific and genome-wide analyses to examine the chromatin landscape of human tissues.

2 Materials

2.1 Isolation of Nuclei

1. 1× PBS (sterile): 8 mM Sodium Phosphate dibasic (Na_2HPO_4-$7H_2O$), 1.5 mM Potassium Phosphate monobasic (KH_2PO_4), 2.67 mM Potassium Chloride (KCl), 138 mM Sodium Chloride (NaCl).

2. Trypsin.
3. Trypan blue and hemocytometer.
4. Lysis Buffer: 10 mM Tris–HCl, pH 7.4, 10 mM NaCl, 3 mM $MgCl_2$, 0.1 mM EDTA 0.5% NP-40.

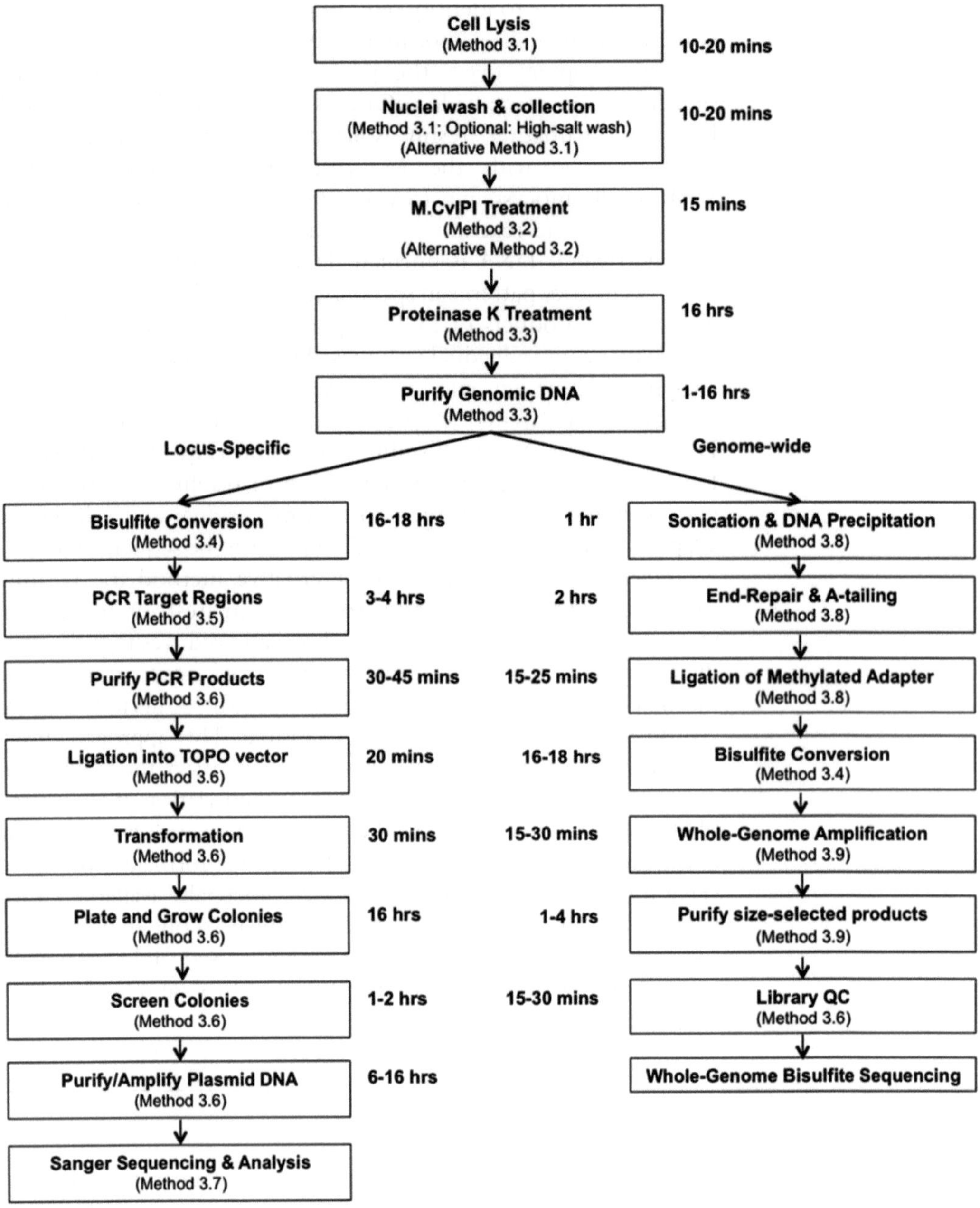

Fig. 2 Overview of the NOMe-seq protocol. The assay starts by treating freshly isolated nuclei with M.CviPI in a total reaction time of 15 min. Enzyme-treated DNA can be analyzed in a locus-specific or genome-wide manner, the steps of which are outlined including the approximate time required for each step

5. Wash Buffer: 10 mM Tris–HCl, pH 7.4, 10 mM NaCl, 3 mM $MgCl_2$, 0.1 mM EDTA.
6. EDTA-free proteinase-inhibitor cocktail (PIC) (Roche).
7. Optional: High-Salt Wash Buffer: 10 mM Tris–HCl, pH 7.4, 3 mM $MgCl_2$, 0.3 M Sucrose, 400 mM NaCl.
8. Optional: Dounce Homogenizer.
9. Optional: Dulbecco's Modified Eagle's Medium
10. Optional: 37% Formaldehyde to crosslink cells.
11. Optional: 1.25 M glycine.
12. Optional: Bioruptor (Diagenode), EpiShear Probe Sonicator (Active Motif) or other commercially available sonicators.

2.2 GpC Methyltransferase Reaction

1. 10× GpC Buffer (New England Biolabs).
2. 32 mM S-adenosylhomocysteine (SAM) (New England Biolabs).
3. 4 Units/μL M.CviPI (New England Biolabs).
4. 1 M Sucrose.
5. Nuclease-free water.
6. Stop Buffer: 20 mM Tris–HCl, pH 7.4, 600 mM NaCl, 1% SDS, 10 mM EDTA.
7. EDTA-free proteinase-inhibitor cocktail (Roche).

2.3 DNA Extraction

1. TE buffer: 10 mM Tris–HCl, 1 mM EDTA, pH 8.0.
2. Proteinase K.
3. 1:1 Phenol: Chloroform.
4. 100% Ethanol.
5. Nanodrop Spectrophotometer.
6. Optional: Glycogen.
7. Optional: Phase-lock gel tubes.
8. Optional: 5 M NaCl to reverse crosslink cells if using the alternative method.

2.4 Bisulfite Conversion

1. EZ DNA Methylation kit (Zymo Research) or other commercially available kits.
2. Nuclease-free water.

2.5 PCR

1. Target Primers.
2. PCR Master Mix containing Taq polymerase that is able to tolerate the presence of Uracil (e.g., MyTaq Red Mix from Bioline).
3. Gel Extraction Kit (e.g., Qiagen or Promega).

2.6 TA Cloning

1. TOPO TA Cloning Kit for Sequencing with OneShot TOP10 Competent Cells (Thermo Fisher Scientific).
2. LB agar plate supplemented with ampicillin.
3. 20 mg/mL X-gal (5-bromo-4-chloro-3-indolyl-β-D-galactopyranoside).
4. Insert Primers or M13 Primers (included in TOPO TA cloning kit).
5. TempliPhi (GE Healthcare Life Sciences).

2.7 Locus-Specific Sequence Analysis

1. BiQ Analyzer (free software available through Max Planck Institute which can be found at http://biq-analyzer.bioinf.mpi-inf.mpg.de/, *see* also Chapter 29) or similar software to align Sanger sequencing.
2. Optional: Methylviewer (free software available through University of Leeds, download from http://dna.leeds.ac.uk/methylviewer/).

2.8 Genome-Wide Bisulfite Sequencing Library Construction

1. End-Repair DNA kit (Epicentre).
2. Klenow and 10× Klenow Buffer or 10× NEB Buffer2 (New England Biolabs).
3. 1 mM dATP.
4. Minelute Reaction Cleanup Kit (Qiagen).
5. Rapid T4 DNA Ligase (Enzymatics).
6. 2× Rapid Ligation Buffer (Enzymatics).
7. 10 mM dNTP Mix.
8. DNA Methyl Adapter or DNA Adapter Index (Illumina).
9. Covaris S2 system.
10. Optional: Agencourt AMPure XP magnetic beads (Beckman Coulter).
11. Optional: Pair-End Sample Prep Kit (Illumina).
12. Optional: Truseq DNA Sample Prep Kit (Illumina).

2.9 Whole-Genome Amplification

1. KAPA HiFi HotStart Uracil + ReadyMix PCR Kit (KAPA Biosystems).
2. Gel Purification Kit (e.g., Qiagen or Promega).
3. Optional: Agencourt AMPure XP magnetic beads (Beckman Coulter).
4. Bioanalyzer (Agilent).

3 Method

3.1 Isolation of Nuclei

1. Trypsinize exponentially growing adherent cells or collect suspension cells and centrifuge at 250 × *g* for 5 min at 4 °C.
2. Remove the media and wash cells with 10 mL sterile PBS. Take care to keep cells on ice or at 4 °C at all times. Centrifuge at 250 × *g* at 4 °C for 5 min and remove PBS wash.
3. Resuspend 1 million cells in 1 mL ice-cold lysis buffer and let it sit undisturbed on ice for 5–10 min (*see* **Note 1**). We use 250,000 cells per reaction, but recommend performing each reaction in duplicate and including no enzyme control. Check a small aliquot of cells under the microscope using trypan blue and hemocytometer for intact nuclei and to ensure that the lysis is complete before proceeding to the next step (*see* **Note 2**).
4. Following the incubation, centrifuge for 5 min at 750 × *g* at 4 °C and discard the supernatant, taking care not to disturb the nuclear pellet.
5. Gently resuspend nuclei in 1 mL ice-cold wash buffer (*see* **Note 3**). Centrifuge for 5 min at 750 × *g* at 4 °C, discard the supernatant, and immediately proceed to M.CviPI treatment.

3.2 Treating Nuclei with M.CviPI

1. Resuspend nuclei in 1× GpC buffer to obtain a final concentration of 250,000 cells per 282 μL volume (*see* **Note 4**).
2. In a microcentrifuge tube, prepare the reaction mixture in the following order (*see* **Note 5**):

1 M sucrose	150 μL
10× GpC buffer	17 μL
Nuclei (250,000)	282 μL
32 mM SAM	1.5 μL
4 U/μL M.CviPI	50 μL
Total	500 μL

   We recommend using fresh GpC buffer and SAM each time. SAM in particular is unstable and thus proper storage based on the manufacturer's instruction is crucial. Freeze-thawing of reagents should also be avoided.

3. Incubate the reaction for 7.5 min at 37 °C.
4. Add a boost of enzyme and substrate in the following amount (*see* **Note 6**):

32 mM SAM	1.5 μL
4 U/μL M.CviPI	25 μL
Total	26.5 μL

5. Incubate for additional 7.5 min at 37 °C. Do not incubate for longer as over-treatment may occur and result in methylation of inaccessible regions.
6. Stop the reaction by adding an equal volume of stop buffer (526.5 μL), which was pre-warmed at 37°C to dissolve precipitate (*see* **Note 7**).

3.3 Alternative Method: Isolation of Nuclei

The following alternative method is adapted for fresh or fresh frozen primary tissue (*see* **Note 8**).

1. Using clean scalpel, dissect tissue into 1–3 mm pieces and resuspend them in 10 mL of DMEM per gram of tissue (*see* **Note 9**).
2. Add formaldehyde to a final concentration of 1% and gently rotate for 10–15 min at room temperature.
3. Stop the crosslinking by adding glycine to a final concentration of 0.125 M and rotate for 5 min.
4. Wash the crosslinked tissue twice with ice-cold PBS supplemented with proteinase inhibitors (PIC). Centrifuge at 300 × *g* for 5 min at 4 °C and remove the supernatant.
5. Resuspend tissue in 1 mL of ice-cold lysis buffer per 100 mg of tissue and incubate on ice for 5–10 min. For subsequent steps, scale up the reaction if using more than 100 mg of tissue, and separate reactions into multiple tubes if necessary.
6. Transfer the mixture into a chilled dounce homogenizer. Using the glass pestle, grind the tissue on ice for 10–20 strokes to release the nuclei. Depending on the tissue type, extra strokes may be required. We recommend monitoring the lysis process using a microscope.
7. Transfer nuclei into a 1.5 mL microcentrifuge tube and centrifuge at 750 × *g* for 10 min at 4 °C. Remove the supernatant.
8. Resuspend tissue in ice-cold wash buffer and centrifuge at 750 × *g* for 10 min at 4 °C. Remove the supernatant.

3.4 Alternative Method: Treating Nuclei with M.CviPI

1. Resuspend pellet in 282 μL of ice-cold 1× GpC buffer supplemented with PIC.
2. Sonicate the chromatin to a fragment of 1-2 kb at 4 °C. We regularly perform 3 cycles of 30 s on and 30 s off on the Bioruptor at 4 °C.
3. To the microcentrifuge tube, add the following reaction mixture

Chromatin	282 μL
1 M sucrose	150 μL
10× GpC buffer	17 μL
32 mM SAM	1 μL
4 U/μL M.CviPI	25 μL
	475 μL

4. Incubate at 37 °C for 1 h.
5. Boost with 1 μL of SAM and incubate for an additional hour.
6. Repeat **step 5**.
7. After a total of 3 h incubation time, add an equal volume (477 μL) of stop buffer (*see* **Note 10**).

3.5 Purification of Enzyme-Treated DNA

Perform **step 1** (and skip **step 2**) if using the standard protocol and start from **step 2** if using the alternative method.

1. Add 200 μg/mL of Proteinase K to each reaction tube and incubate for 16 h at 55 °C to remove excess enzyme.
2. Reverse cross-link chromatin by adding 5 M NaCl to a final concentration of 0.2 M and incubating at 65 °C for 4–6 h. Add 200 μg/mL of Proteinase K and incubate for 16 h at 55 °C.
3. Purify DNA using a standard phenol/chloroform extraction method followed by ethanol precipitation or DNA purification columns. (Optional: during phenol/chloroform extraction, phased-lock gel can be used to assist in the separation of the aqueous and the organic phase). Ethanol precipitation can be carried out at −20 °C overnight or at −80 °C for 1–2 h. During ethanol precipitation, 20 μg/mL of glycogen can be added as a carrier.
4. Resuspend DNA pellet in 20 μL of nuclease-free water or TE buffer. Quantify DNA and store long term in −20 °C.

3.6 Bisulfite Conversion

1. Bisulfite conversion can be performed using a commercially available kit according to the manufacturer's instructions. We regularly use the EZ DNA Methylation kit to convert 500 ng-1 μg of DNA per reaction although other kits, such as EZ-DNA Methylation Lightning kit (Zymo) or Epitect Bisulfite kit (Qiagen), are also available.
2. Elute DNA in 20 μL of nuclease-free water. Bisulfite converted DNA is stable in −20 °C for up to a year.

3.7 PCR of Bisulfite-Converted DNA

1. Design primers for regions of interest that are specific for bisulfite-converted DNA (Fig. 3, *see* **Note 11**).
2. Perform PCR for 40–45 cycles using 2 μL of bisulfite converted DNA and Taq polymerase that adds 3′–A overhangs to the end

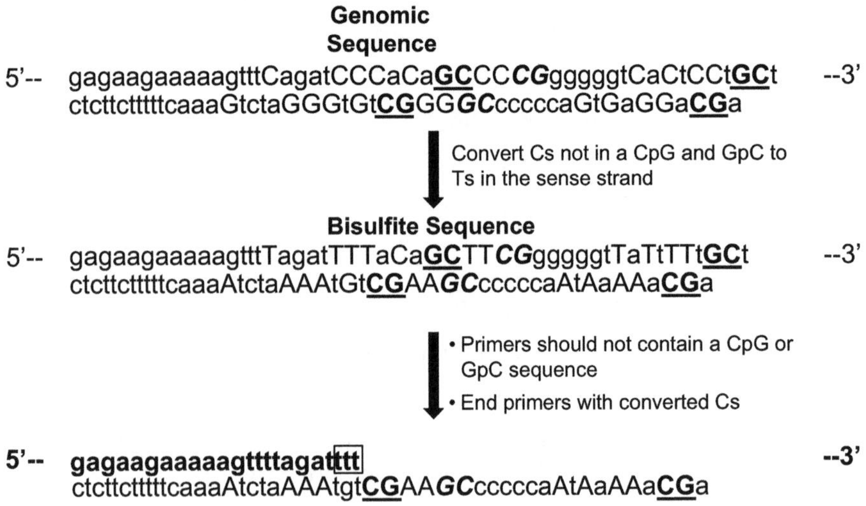

Fig. 3 NOMe-seq primer design. In the sense strand, all C's that are not part of CpG (*bold, italicized*) or GpC (*bold, underlined*) dinucleotides are converted to T (*upper case*). Design a forward primer that is complementary to the antisense strand. The primer should be 20–30 bp long with an annealing temperature of 50–60 °C, should not contain a CpG or GpC sites, and if possible, end with converted Cs. The reverse primer should follow the same criteria, but designed to complement the sense strand

of PCR products which is necessary for subsequent TA cloning. Make sure that the Taq polymerase is able to tolerate the presence of uracils in the DNA template. A final extension time of up to 10 min, depending on the robustness of the chosen Taq, should be included to ensure that the polymerase appends the end of the PCR products with an –A overhang. The following is an example of PCR condition using Bioline's MyTaq Red Mix:

2× MyTaq Red Mix	12.5 μL
10 μM forward primer	1 μL
10 μM reverse primer	1 μL
Nuclease-free water	8.5 μL
Bisulfite-treated DNA	2 μL
	25 μL

Cycling condition:	
(a) 95 °C	1 min
(b) 95 °C	30 s
(c) Ta (52–62 °C)	30 s
(d) 72 °C	30 s
Repeat (b–d) for 40–45 cycles	
(e) 72 °C	5 min

Annealing temperature (Ta) should be optimized for each primer set so that it is specific to the bisulfite converted DNA and does not amplify genomic DNA.

3. It is recommended to always include no template and genomic DNA control to ensure purity and efficiency (*see* **Note 12**).

3.8 TA-Cloning

1. PCR products that are specific to target region can be used directly for TA cloning although it is highly recommended to perform gel extraction to remove excess primers and polymerase, which might interfere with the subsequent ligation step.
2. Clone freshly purified PCR products (within 24 h) using the TOPO TA Cloning for Sequencing kit according to the manufacturer's instruction and transform using TOP10 competent cells (*see* **Note 13**). If ligation within 24 h is not possible, incubate the PCR products with Taq and dATP for 10 min at 72 °C to add the –A overhangs prior to ligation.
3. On each LB-agar plate containing ampicillin, add 40 μL of 20 mg/mL X-gal and spread evenly.
4. Plate recovered TOP10 cells and incubate at 37 °C overnight.
5. Screen for colonies with positive insert using either the insert primers or M13 primers located in the backbone of the vector. Plasmid DNA can be amplified with Templiphi or purified with minipreps according to the manufacturer's instruction.
6. Sequence individual clones. We recommend using M13 reverse primer for sequencing, although primers specific to the insert can also be used (*see* **Note 14**).

3.9 Sequence Analysis

Various software tools are available for bisulfite sequencing analysis, such as BiQ Analyzer [9, 10] and MethylViewer [11]. The theory behind the analysis of M.CviPI-treated sequences is similar to the analysis of regular bisulfite sequencing with the following modifications:

1. GpC dinucleotides that are not occupied by tight binding transcription factors or nucleosomes will be accessible and thus methylated by the enzyme. Following bisulfite sequencing, these GpC sites will remain as "GC," whereas inaccessible and unmethylated sites will be sequenced as "GT." Nucleosome-occupied regions are defined as inaccessible regions of >147-bp DNA, whereas protection due to DNA binding factors generally is smaller (<50 bp).
2. GCG sites are excluded from downstream analysis as we cannot determine whether the methylation status of the cytosines is due to endogenous DNA methylation or due to GpC methyltransferase activity (Fig. 4).
3. Depending on the alignment software used, it may be necessary to reduce the minimal alignment requirement to ~60% to

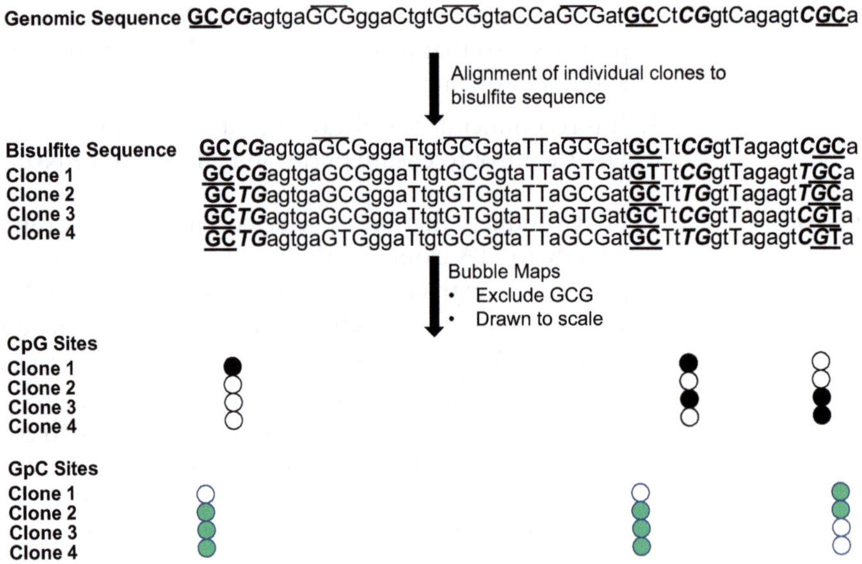

Fig. 4 NOMe-seq sequence analysis. Align sequenced clones to bisulfite converted sequence. When present in GpC (*bold, underlined*) or CpG (*bold, italicized*) context, methylated C will be read as C and an unmethylated C will be read as T. Methylation status is often represented in a bubble map where methylated CpG is shown as *black circles* and methylated GpC is shown as teal circles. Cs present in the GCG context (overlined) are excluded from the analysis as when the C is methylated it cannot be determined whether the methylation status is endogenous, or due to enzyme treatment

compensate for the Cs in the GC context that have not been bisulfite converted.

4. For visualization, we often generate bubble maps to represent the methylation status of CG and GC sites, which are drawn to scale (Fig. 5).

3.10 Whole-Genome Bisulfite Sequencing Library Construction

Before starting, *see* **Note 15**.

1. Sonicate 3–5 μg of enzyme-treated genomic DNA to 150–200 bp (*see* **Note 16**).
2. Ethanol precipitate sonicated DNA and resuspend in 34 μL of nuclease-free water.
3. End-repair sonicated DNA using the End-It DNA End-Repair Kit. In a 1.5 mL microcentrifuge tube, add the following reaction mixture:

DNA (3–5 μg)	34 μL
10× end-repair buffer	5 μL
2.5 mM dNTPs	5 μL
10 mM ATP	5 μL
END-IT enzyme mix	1 μL
Total	50 μL

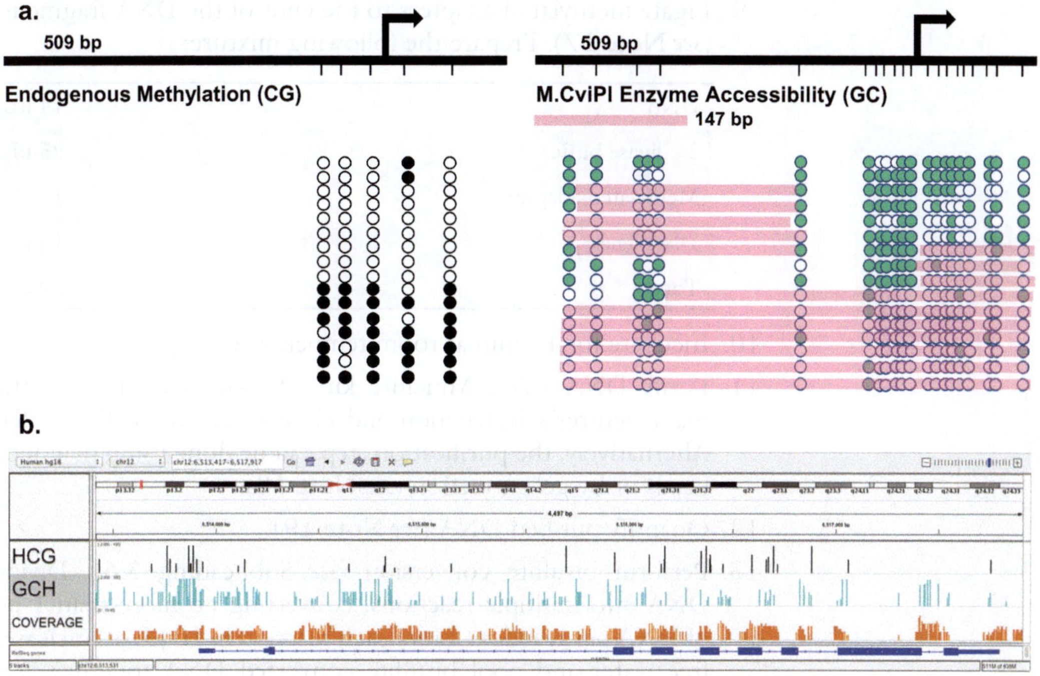

Fig. 5 Visualization of locus-specific and genome-wide NOMe-seq. (**a**) Bubble map chart is shown for the *Xist* locus which is present in multiple chromatin configurations, unmethylated and accessible, and methylated and inaccessible. (**b**) Screenshot of IGV browser showing the average HCG and GCH methylation status in the GAPDH locus following whole-genome bisulfite sequencing

4. Incubate at room temperature for 45 min.
5. Purify DNA using Minelute kit (Qiagen) according to the manufacturer's instruction, and elute in 32 μL of EB buffer. Alternatively, purification step can be done using magnetic beads (AMPure) according to the manufacturer's instruction.
6. Add an A' base to the 3′ end of the DNA in the following reaction mixture:

DNA	32 μL
10× Klenow buffer	5 μL
1 mM dATP	10 μL
Klenow (5 U/μL)	3 μL
Total	50 μL

7. Incubate for 30 min at 37 °C.
8. Purify DNA using Minelute kit (Qiagen) according to the manufacturer's instruction and elute in 14 μL of EB buffer. Alternatively, the purification step can be done using magnetic beads such as AMPure beads.

9. Ligate methylated adapters to the ends of the DNA fragments (*see* **Note 17**). Prepare the following mixture:

DNA	14 μL
2× ligase buffer	25 μL
Methylated adapter	10 μL
DNA ligase	1 μL
Total	50 μL

10. Incubate for 15 min at room temperature.
11. Purify DNA using Minelute kit (Qiagen) according to the manufacturer's instruction and elute in 25 μL of EB buffer. Alternatively, the purification step can be done using magnetic beads or by gel extraction (*see* **Note 18**).
12. Quantify purified DNA (*see* **Note 19**).
13. Perform bisulfite conversion (*see* Subheading 3.6). Divide DNA into multiple reactions, converting no more than 1 μg of DNA per reaction. Elute each reaction in 10 μL of nuclease-free water and pool bisulfite converted DNA together in a microcentrifuge tube.

3.11 Whole Genome Amplification

1. Perform no more than 8 cycles of PCR using high-fidelity Taq polymerase. We routinely used 2× KAPA-Hifi + Uracil master mix to amplify 20 μL of purified bisulfite converted library using a final concentration of 500 nM primers designed to amplify the methylated adapters.
2. Use PCR cycle conditions as recommended by the manufacturer.

   Cycling Condition (*see* **Note 20**):

(a) 98 °C	45 s
(b) 98 °C	15 s
(c) 65 °C	30 s
(d) 72 °C	30 s
Repeat (b–d) for 5–8 cycles	
(e) 72 °C	60 s

3. Add loading buffer to a final concentration of 1× and load all PCR product and DNA ladder to a 2% gel. Run gel at 120 V for 45–60 min or until a clear separation for lower molecular weight products is achieved (*see* **Note 21**).
4. With a clean scalpel, excise bands with fragment size of 200–500 bp (*see* **Note 22**).
5. Purify DNA using a gel extraction kit. Melt the gel slowly at room temperature and proceed with the protocol as indicated

by the manufacturer. Elute in no more than 20 μL of nuclease-free water and store at −20 °C for long-term conservation.

6. Measure the concentration and size distribution of library using Bioanalyzer.
7. Next-generation sequencing is commonly done on HiSeq-2000/2500 or higher using 100 bp PE reads (*see* **Note 23**). Methylation status of CG and GC sites can be extracted using Bis-SNP or other WGBS analysis tools [12].

4 Notes

1. We recommend optimizing the lysis condition for each cell type by adjusting the incubation time and concentration of NP-40 in the lysis buffer as some cells are more easily lysed than the others. For example, fibroblasts such as IMR90 cells require longer incubation time in lysis buffer containing 0.5% NP-40 whereas K562 and GM12878 cells can be lysed in <3 min in 0.25% NP-40. Nuclei can also be isolated by mechanically disrupting the cells using a dounce homogenizer in the presence of NP-40. Monitor lyses by removing a few μL every minute until adequate cell lysis is achieved.
2. For genome-wide application, due to the loss observed during bisulfite conversion, 3–5 μg of footprinted DNA is required to start the library preparation procedure. We recommend performing multiple enzyme reactions for each experimental condition. DNA extracted from multiple reactions can be pooled together as technical replicates.
3. It is critical that the nuclei remain intact throughout the wash and reaction step. Intact nuclei will appear as blue circles under the microscope when visualized with trypan-blue. To resuspend nuclei in the wash buffer, we recommend using a p-1000 pipette tip, which has a wider opening than p-200 tip in order to minimize friction, which may cause the nuclei to burst. Additionally, the wash step can be performed under high-salt conditions (100–400 mM) in order to remove transcription factors.
4. Always include a no-methyltransferase control to help to determine the level of endogenous DNA methylation in ambiguous regions such as the GCG trinucleotides.
5. A highly concentrated version of M.CviPI at 50 U/μL can also be specially ordered from New England Biolabs. If using this version of the enzyme, use the following reaction mixture:

1 M Sucrose	45 μL
10× GpC buffer	5 μL
Nuclei (250,000)	94.5 μL
32 mM SAM	1.5 μL
50 U/μL M.CviPI	4 μL
Total	150 μL

6. If using the highly concentrated enzyme (50 U/μL), boost with the following amount:

32 mM SAM	1.5 μL
50 U/μL M.CviPI	2 μL
Total	150 μL

7. If using the highly concentrated enzyme (50 U/μL) as described in **Note 5**, add 152.75 μL of stop solution.
8. The standard method described in Subheadings 3.1 and 3.2 may also be used for treating nuclei freshly released from single-cell suspension of primary tissue, which is prepared by enzymatic or mechanical disaggregation. However, the nuclei isolation procedure often varies depending on the tissue type and composition and is more laborious compared to isolating nuclei from cell-lines. In order to minimize the technical variability between tissues, we isolated chromatin from cross-linked tissue and treated the chromatin with M.CviPI for a longer incubation time.
9. Use different clean scalpels for each sample when processing multiple tissues at the same time to avoid contamination. If using frozen tissue (−80 °C storage), avoid freeze-thawing the tissue and minimize the time when dissecting.
10. Over-treatment is not a concern when using fixed chromatin, however, it is critical to incubate the reaction for a minimum of 1 h and replenish the SAM periodically when performing longer incubation in order to ensure that all accessible GpC sites are methylated.
11. The primers should not include CpG and GpC sites in the sequence, which will result in a biased PCR amplification. The size of amplicons should be around 350–700 bp so as to provide enough resolution for the region being analyzed. Longer amplicons, on the other hand, are difficult to amplify and sequence due to the fragmentation that occurs during bisulfite conversion.
12. In order to avoid PCR bias, we recommend performing the PCR and cloning in duplicates while combining the

sequencing results. Alternatively, spike-ins of known ratios of methylated and unmethylated amplicons may be included in the PCR reaction.

13. The manufacturer's instructions for the TOPO kit recommend a total reaction volume of 6 μL (4 μL PCR product, 1 μL of TA vector, and 1 μL of 150 mM NaCl). However, we routinely halved the reaction without an adverse effect. We subsequently transformed all 3 μL reactions in 25 μL of competent cells.
14. The number of clones required for clear footprinting pattern will depend on the regions and condition being assessed. We recommend sequencing a minimum of 10 clones to start, which generally suffices for most regions being interrogated.
15. Preparation of WGBS libraries outlined in this section is based on previously described methods [6, 7, 13]. Alternatively, Truseq DNA Methyl Prep kit or other commercially available WGBS library preparation kit can also be used to generate libraries. We also recommend validating a few known control regions at loci-specific level before proceeding to this step in order to confirm that the enzyme treatment was successful.
16. We regularly use the Covaris S2 system to fragment DNA for WGBS in order to generate tightly distributed DNA fragments. DNA is diluted to a total volume of 130 μL and transferred into 6 × 16 mm microtubes, taking care to avoid air bubbles. Sonication is performed at 10% duty cycle, 5 intensity, and 200 cycles per burst for 6 min.
17. The methylated adaptor listed here is the same as the Truseq adaptor used by Illumina. If barcoding libraries, indexed adaptors from Illumina or Bioo Scientific can be used.
18. If performing gel extraction, we recommend using 2% low melting gel or a 50:50 mixture of low and regular melting gel. Excise tightly in order to limit excess gel that may interfere with subsequent purification step and use a clean scalpel each time to avoid cross-contamination between samples. In either case, gel should be melted slowly at room temperature so as to not bias library complexity based on CG density.
19. At this stage, DNA can be safely stored long term to −20 °C if not proceeding to the next step.
20. It is critical to minimize the PCR cycles to less than 10 cycles to avoid excessive PCR duplicates in the libraries, which will bias subsequent analysis.
21. During gel extraction, PCR products may be difficult to see when present in low amount. DNA ladder should be used as a guide when excising size-selected regions.
22. Take care to avoid contamination from self-ligated adapter dimers, which usually appear at ~150 bp.

23. For genome-wide analysis, we generally aim to have a minimum of 300–400 million uniquely mapped sequencing reads. Due to the reduced complexity of bisulfite converted genomes, short single-end reads are likely not sufficient for obtaining a large number of uniquely mappable reads, and thus we recommend sequencing NOMe-seq libraries with 75PE, 100PE or 100SE.

Acknowledgments

This work was funded by 5R37CA082422-13 to PAJ. We thank members of the Jones Lab for helpful discussion.

References

1. Xu M, Kladde MP, Van Etten JL et al (1998) Cloning, characterization and expression of the gene coding for a cytosine-5-DNA methyltransferase recognizing GpC. Nucleic Acids Res 26:3961–3966
2. Kelly TK, Miranda TB, Liang G et al (2010) H2A.Z maintenance during mitosis reveals nucleosome shifting on mitotically silenced genes. Mol Cell 39:901–911
3. Andreu-Vieyra C, Lai J, Berman BP et al (2011) Dynamic nucleosome depleted regions at androgen receptor enhancers in the absence of ligand in prostate cancer cells. Mol Cell Biol 31:4648–4662
4. You JS, Kelly TK, De Carvalho DD et al (2011) OCT4 establishes and maintains nucleosome-depleted regions that provide additional layers of epigenetic regulation of its target genes. Proc Natl Acad Sci U S A 108:14497–14502
5. Taberlay PC, Kelly TK, Liu CC et al (2011) Polycomb-repressed genes have permissive enhancers that initiate reprogramming. Cell 147:1283–1294
6. Kelly TK, Liu Y, Lay FD et al (2012) Genome-wide mapping of nucleosome positioning and DNA methylation within individual DNA molecules. Genome Res 22:2497–2506
7. Hawkins RD, Hon GC, Lee LK et al (2010) Distinct epigenomic landscapes of pluripotent and lineage-committed human cells. Cell Stem Cell 6:479–491
8. Lister R, Pelizzola M, Dowen RH et al (2009) Human DNA methylomes at base resolution show widespread epigenomic differences. Nature 462:315–322
9. Lutsik P, Feuerbach L, Arand J et al (2011) BiQ analyzer HT: locus-specific analysis of DNA methylation by high-throughput bisulfite sequencing. Nucleic Acids Res 39:W551–W556
10. Bock C, Reither S, Mikeska T et al (2005) BiQ analyzer: visualization and quality control for DNA methylation data from bisulfite sequencing. Bioinformatics 21:4067–4068
11. Pardo CE, Carr IM, Hoffman CJ et al (2011) MethylViewer: computational analysis and editing for bisulfite sequencing and methyltransferase accessibility protocol for individual templates (MAPit) projects. Nucleic Acids Res 39:e5
12. Liu Y, Siegmund KD, Laird PW et al (2012) Bis-SNP: combined DNA methylation and SNP calling for bisulfite-seq data. Genome Biol 13:R61
13. Berman BP, Weisenberger DJ, Aman JF et al (2012) Regions of focal DNA hypermethylation and long-range hypomethylation in colorectal cancer coincide with nuclear lamina-associated domains. Nat Genet 44:40–46

Chapter 15

Bisulphite Sequencing of Chromatin Immunoprecipitated DNA (BisChIP-seq)

Clare Stirzaker, Jenny Z. Song, Aaron L. Statham, and Susan J. Clark

Abstract

Epigenetic regulation plays a critical role in gene expression, cellular differentiation, and disease. There is a complex interplay between the different layers of epigenetic information, including DNA methylation, nucleosome positions, histone modifications, histone variants, and other important epigenetic regulators. The different modifications do not act independently of each other and their relationship plays an important role in governing the regulation of the epigenome. Of these, DNA methylation is the best-studied epigenetic modification in mammals. However, the direct relationship between DNA methylation and chromatin modifications has been difficult to unravel with existing technologies, with epigenome-wide integration studies still based on "overlaying" independent chromatin modification and DNA methylation maps. Bisulphite sequencing enables the methylation state of every cytosine residue to be analyzed across a given molecule in a strand-specific context. Here, we describe a direct approach to interrogating the DNA methylation status of specific chromatin-marked DNA, using high-throughput sequencing of bisulphite-treated chromatin immunoprecipitated DNA (BisChIP-seq). This combined approach enables the exquisite relationship between chromatin-modified DNA or transcription factor-associated DNA and the methylation state of each targeted allele to be directly interrogated. BisChIP-Seq can now be widely applied genome-wide to further understand the molecular relationship between DNA methylation and other important epigenetic regulators.

Key words Bisulphite-sequencing, Chromatin immunoprecipitation, BisChIP-seq, DNA methylation

1 Introduction

The genome-wide distribution of DNA methylation and histone modification of chromatin and how this changes during cellular differentiation, in both development and disease, is being revealed by large-scale whole-genome sequencing studies [1, 2]. The many different layers of epigenetic information, including DNA methylation, histone modifications, histone variants, and transcription factor binding sites, do not act independently; rather the complex interplay between the epigenetic layers plays an important role in epigenetic regulation. Of these, DNA methylation is the

Jörg Tost (ed.), *DNA Methylation Protocols*, Methods in Molecular Biology, vol. 1708,
https://doi.org/10.1007/978-1-4939-7481-8_15,

best-studied epigenetic modification in mammals. However, understanding the critical interplay of DNA methylation and histone modifications of chromatin is still a fundamental question in cancer biology. To date, genome-wide studies to address the relationship between histone modification, transcription factor binding and DNA methylation have involved an integrative approach of "overlaying" epigenome maps [3–6]. In contrast, we, and others, have developed an approach to enabling the direct interrogation of DNA methylation across genomic regions associated with a particular histone mark and how to follow how this potentially changes in disease phenotypes [7, 8].

Bisulphite sequencing is the only technique currently described that allows the methylation state of every individual cytosine residue in the target sequence to be defined at a single-base pair resolution [9, 10]. The underlying principle is that DNA is treated with sodium bisulphite to convert cytosine but not 5-methylcytosine to uracil, which is converted to thymine after PCR amplification, whereas 5-methylcytosine residues are read as cytosines [10]. Here, we report the method to perform high-throughput bisulphite sequencing on chromatin immunoprecipitated (ChIP) DNA (BisChIP-seq) to directly address the relationship between histone-marked or transcription factor-associated DNA and DNA methylation (Fig. 1). In this approach, chromatin immunoprecipitation is performed to enrich for a chromatin mark of interest. Specific optimized protocols need to be considered depending on the chromatin mark being interrogated, with histone marks requiring 2–5 $\times$ 10^6 cells, and transcription factors or hormone receptors requiring immunoprecipitation from 10–50 $\times$ 10^6 cells. The most critical step in the chromatin immunoprecipitation being performed, is validation of enrichment of the mark of interest above background (>6–8-fold). Sufficient yield of DNA is also necessary to proceed to successful library preparation and bisulphite sequencing, with ~50 ng DNA required. Following successful chromatin immunoprecipitation, the DNA is subjected to end-repair, adenylation of 3′ ends and ligation of methylated adaptors before bisulphite conversion of DNA is performed. We have tested the optimal bisulphite conditions to use, taking into consideration the small amounts of starting sonicated formaldehyde-fixed ChIP DNA to be converted, to achieve the best DNA yield with the most efficient bisulphite conversion rates (>98%). The final step in the BisChIP-seq protocol is library generation. The adaptor-ligated bisulphite-converted DNA is PCR amplified by 10–14 cycles of PCR to generate a library, which is then subjected to next-generation sequencing. We reference the bioinformatics data analysis site used to identify genomic regions marked by a chromatin modification, while simultaneously interrogating the methylation status after bisulphite treatment.

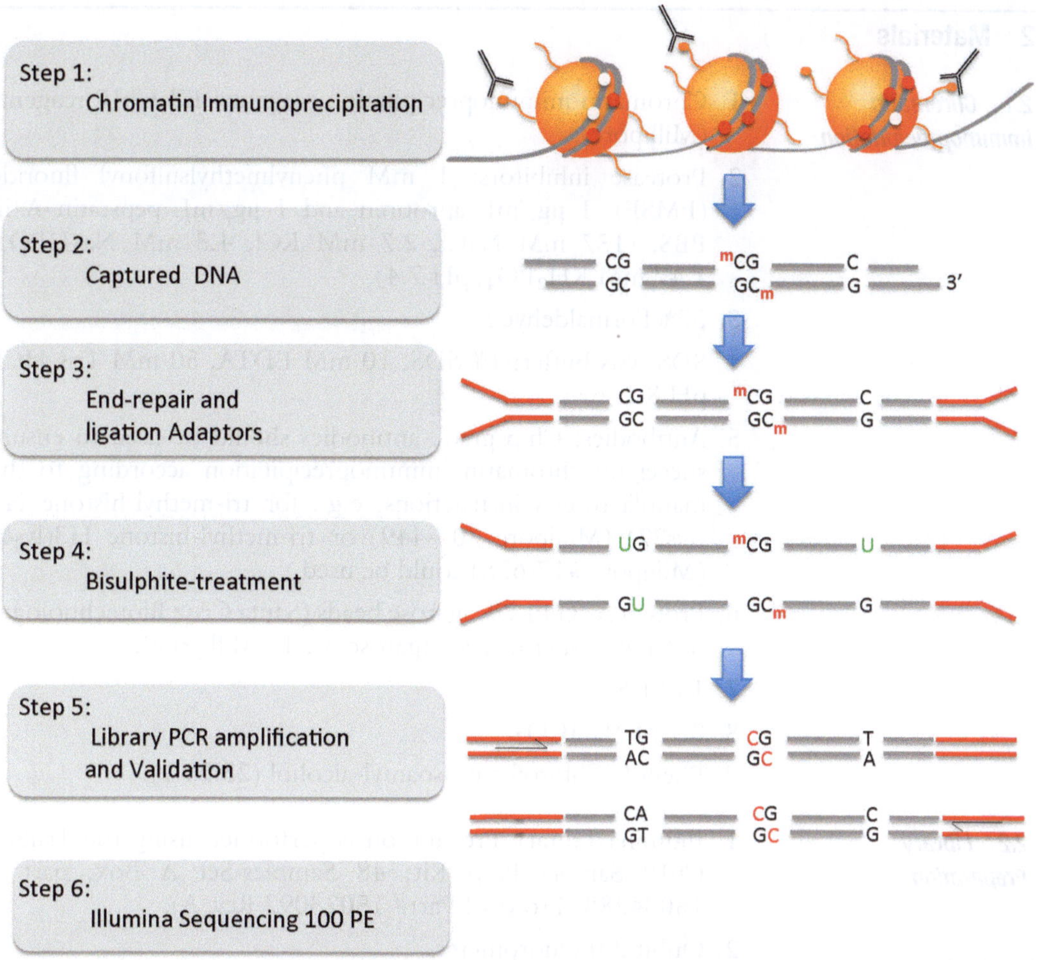

Fig. 1 Flow chart of the BisChIP-seq protocol to perform bisulphite treatment and library preparation on ChIP DNA

BisChIP-Seq is a novel approach that enables the exquisite relationship between DNA methylation and chromatin modified DNA or transcription factor-associated DNA to be directly interrogated. Not only does it allow genome-wide interrogation of two epigenetic marks on the same molecule, but can also reveal allele-specific methylation, as well as single-nucleotide polymorphisms of chromatin-modified DNA. BisChIP-Seq can now be widely applied genome-wide to further understand the complex relationship between DNA methylation and other important epigenetic regulators.

2 Materials

2.1 Chromatin Immunoprecipitation

1. Chromatin immunoprecipitation reagents: EZ-ChIP reagents (Millipore).
2. Protease inhibitors: 1 mM phenylmethylsulfonyl fluoride (PMSF), 1 μg/mL aprotinin and 1 μg/mL pepstatin A in PBS. (137 mM NaCl, 2.7 mM KCl, 4.3 mM Na_2HPO_4, 1.47 mM KH_2PO_4, pH 7.4).
3. 37% Formaldehyde.
4. SDS lysis buffer: 1% SDS, 10 mM EDTA, 50 mM Tris-HCl, pH 8.1.
5. Antibodies: Chip grade antibodies should be used to ensure successful chromatin immunoprecipitation according to the manufacturer's instructions, e.g., for tri-methyl-histone H3 (lys27) (Millipore #07-449) or tri-methyl-histone H3(lys4) (Millipore #17-678) could be used.
6. Protein A/G PLUS agarose beads (Santa Cruz Biotechnology) or Salmon sperm/BSA agarose beads (Millipore).
7. 1% SDS.
8. 0.1 M $NaHCO_3$.
9. Phenol/Chloroform/isoamyl-alcohol (25:24:1).

2.2 Library Preparation

1. Illumina Library Preparation is performed using the TruSeq ChIP Sample Prep Kit, 48 Samples-Set A Box, part # 15034288 (Protocol Part# 15023092 Rev. A).
2. Qubit 2.0 Fluorometer.
3. Qubit dsDNA HS Assay Kit.
4. Qubit assay tubes.
5. Agencourt AMPure XP (Beckman Coulter).
6. DNA LoBind Tube 2.0 mL (for example Eppendorf).
7. 80% Ethanol.
8. Magnetic stand (Thermo Fisher Scientific).

2.3 Bisulphite Treatment

1. EZ DNA Methylation-Gold™ Kit (Zymo Research).

2.4 Amplification and Quantification of Library

1. Pfu Turbo Cx Hotstart DNA Polymerase kit.
2. PCR Primer Cocktail (PPC) from Illumina TruSeq ChIP Sample Prep Kit.
3. 10 mM dNTP kit.
4. Agencourt AMPure XP (Beckman Coulter).
5. 80% Ethanol (EtOH), freshly prepared.

6. Ultra-Pure Water.
7. 2100 Bioanalyzer (Agilent).
8. DNA 1000 Kit (Agilent).
9. High Sensitivity DNA Kit (Agilent).
10. Library Quantification Kit (Kapa Biosystems, Inc.).

2.5 Data Analysis

1. fastqc (http://www.bioinformatics.babraham.ac.uk/projects/fastqc/).
2. trim_galore (http://www.bioinformatics.babraham.ac.uk/projects/trim_galore/).
3. bismark (http://www.bioinformatics.babraham.ac.uk/projects/bismark/) [11].
4. Picard suite (http://picard.sourceforge.net/).
5. UCSC Genome Browser (http://genome.ucsc.edu/) or the Integrated Genomics Viewer (https://www.broadinstitute.org/igv/home).
6. MACS2 (https://github.com/taoliu/MACS/tree/master/MACS2) [12] or PeakRanger (http://www.modencode.org/software/ranger/) [13].

3 Methods

3.1 Chromatin Immunoprecipitation (ChIP)

ChIP assays are carried out according to the protocol provided by the manufacturer (*see* **Note 1**).

1. Fix ~2–5 × 10^6 cells, in a 10 cm dish, by adding formaldehyde to a final concentration of 1% and incubating for 10 min at 37°C.
2. Wash the cells twice with ice-cold PBS containing protease inhibitors, harvest by scraping and treat with SDS lysis buffer for 10 min on ice.
3. Sonicate the resulting lysates to shear the DNA to fragment lengths of 200–500 base pairs (*see* **Note 2**).
4. Immunoprecipitate the chromatin with specific antibodies of interest, for example, antibodies specific for tri-methyl-histone H3(lys27) or tri-methyl-histone H3(lys4). Include a no antibody and IgG controls for each ChIP assay, to facilitate assessment of nonspecific precipitation, during immunoprecipitation by quantitative Real-Time PCR analysis. Process input samples in parallel.
5. Collect the antibody/protein complexes by Protein A/G PLUS agarose beads or Salmon sperm/BSA agarose beads

and wash several times following the manufacturer's instructions.

6. Elute the bound protein/DNA complexes with 250 μL of freshly made 1% SDS and 0.1 M $NaHCO_3$, repeat and combine the eluates. Reverse histone-DNA crosslinks by adding 20 μL 5 M NaCl and heating at 65 °C for 4 h. Treat with 2 μL of 10 mg/mL proteinase K for 1 h and purify the DNA by phenol/chloroform/isoamylalcohol extraction, followed by 70% ethanol precipitation and resuspend in 30 μL H_2O.
7. Perform validation of known candidate genes by qPCR to confirm and assess enrichment of the chromatin immunoprecipitation. This is critical for successful BisChip-Seq (*see* **Note 3**).
8. Quantify the ChIP DNA using Qubit ds (double stranded) DNA High Sensitivity Assay Kit (LifeTechnologies) (*see* **Note 4**).

3.2 Preparation of ChIP DNA for Illumina Library Preparation

Illumina library preparation is performed using the TruSeq ChIP Sample Prep Kit, 48 Samples-Set A Box, according to the ChIP Sample Preparation Protocol.

3.2.1 Perform End Repair

1. Adjust the concentration of ChIP enriched DNA to ~50 ng/ 50 μL in nuclease-free water.
2. Prepare a fresh stock of 80% ethanol.
3. Remove the AMPure XP Beads from 4 °C and let it stand for at least 30 min to bring them to room temperature.
4. Remove Resuspension Buffer and End Repair Mix from −20 °C and thaw them at room temperature, and then place on ice.
5. Use a thermal cycler with a preheat lid option and set to 100 °C, preheat thermal cycler to 30 °C (save as program #CHIP30).
6. Prepare the following reaction mix in a 0.5 mL PCR tube for a single sample:

50 μL/~50 ng	ChIP Enriched DNA
10 μL	Resuspension Buffer
40 μL	End Repair Mix
100 μL	**Total volume**

7. Gently pipette the entire volume up and down ten times to mix thoroughly.
8. Incubate at 30 °C for 30 min in a preheated thermal cycler.
9. Remove the PCR tube from the thermal cycler.

10. Vortex the AMPure XP Beads until they are fully dispersed.
11. Add 160 μL of well-mixed AMPure XP Beads to the reaction mix. Gently pipette the entire volume up and down ten times to mix thoroughly.
12. Incubate the mixture at room temperature for 15 min.
13. Place the PCR tube on a magnetic stand, at room temperature, for 15 min or until the liquid appears clear.
14. Remove and discard supernatant from each tube. DO NOT DISTURB BEADS.

    IMPORTANT: Leave the PCR tube on the magnetic stand while performing the following 80% EtOH wash steps (**steps 15–18**).
15. With the PCR tube on the magnetic stand, add 200 μL of freshly prepared 80% EtOH to each tube without disturbing the beads.
16. Incubate the PCR tube at room temperature for 30 s, then remove and discard all of the supernatant from each tube. Take care not to disturb the beads.
17. Repeat **steps 10** and **11** once for a total of two 80% EtOH washes.
18. Leaving the PCR tube on the magnetic stand, at room temperature for 10 min to allow the pellet to dry.
19. Remove the tube from the magnetic stand and resuspend the dried bead pellet in 16 μL of Resuspension Buffer by pipetting the entire volume up and down 30 times to mix thoroughly.
20. Incubate the PCR tube at room temperature for 2 min.
21. Place the PCR tube on the magnetic stand at room temperature for 5 min or until the liquid appears clear.
22. Transfer 15 μL of the cleared supernatant to a new 0.5 mL PCR tube.

    SAFE STOPPING POINT

    If you do not plan to proceed to Adenylate 3′ Ends immediately, the protocol can be safely stopped here. If you are stopping, store the tube at − 20 °C for up to 7 days.

3.2.2 Adenylate 3′ Ends

1. Remove the A-Tailing Mix from −20 °C and thaw at room temperature.
2. Remove the PCR tube from −20 °C, if it was stored at the conclusion of *Perform End Repair* and let it stand to thaw at room temperature. Centrifuge the thawed PCR tube briefly.
3. Pre-program the thermal cycler with the following program (save as program #CHIP37): Choose the preheat lid option and set to 100 °C, incubate at 37 °C for 30 min followed by 70 °C for 5 min. Hold at 4 °C

4. Prepare the following reaction mix in the following order.

15 μL	End Repaired DNA
2.5 μL	Resuspension buffer
12.5 μL	A-tailing mix
30 μL	**Total volume**

5. Gently pipette the entire volume up and down ten times to mix thoroughly.
6. Place the PCR tube on the pre-programmed thermal cycler. Close the lid and select program #CHIP37.
7. When the thermal cycler has cooled down to 4 °C, remove the PCR tube from the thermal cycler, place on ice and proceed *immediately* to *Ligate Adapters.*

3.2.3 Ligate Adapters

1. Remove appropriate RNA Adapter Indexes (contained within TruSeq ChIP sample Prep Kit) from −20 °C and thaw at room temperature. Place on ice (*see* **Note 5**).
2. Remove Stop Ligation Buffer from −20 °C and thaw at room temperature. Place on ice.

   ***IMPORTANT**: Do not remove the Ligation Mix tube from − 20 °C until instructed to do so in the procedures.*
3. Remove the AMPure XP beads from 4 °C and let it stand for at least 30 min to bring them to room temperature.
4. Preheat the thermal cycler to 30 °C. Choose the thermal cycler preheat lid option and set to 100 °C, saved as program #CHIP30 (*see* **step 5** from Subheading 3.2.1)
5. Briefly centrifuge the thawed RNA Adapter Indexes and Stop Ligation Buffer tubes.
6. Immediately before use, remove the Ligation Mix tube from −20° and place on ice.
7. Prepare the ligation reaction on ice in the following order:

30 μL	Adenylated DNA
2.5 μL	Resuspension buffer
2.5 μL	Ligation mix (***IMPORTANT**: Return the Ligation Mix tube back to −20 °C immediately after use.*)
2.5 μL	RNA adapter index
37.5 μL	**Total volume**

8. Gently pipette the entire volume up and down ten times to mix thoroughly.
9. Briefly centrifuge the PCR tube. Incubate the PCR tube on the preheated thermal cycler at 30 °C for 10 min.

10. Remove the PCR tube from the thermal cycler and add 5 μL of Stop Ligation Buffer to each reaction to inactivate the ligation. Gently pipette the entire volume up and down ten times to mix thoroughly.
11. Vortex the AMPure XP Beads until they are well dispersed.
12. Add 42.5 μL of mixed AMPure XP Beads to each reaction. Gently pipette the entire volume up and down ten times to mix thoroughly.
13. Incubate the binding reaction at room temperature for 15 min.
14. Place the PCR tube on the magnetic stand at room temperature for 5 min or until the liquid appears clear.
15. Remove and discard all of the supernatant from each sample. DO NOT DISTURB BEADS.

    IMPORTANT*: Leave the PCR tube on the magnetic stand while performing the following 80% EtOH wash steps* (**steps 16–18**).
16. With the PCR tube remaining on the magnetic stand, add 200 μL of freshly prepared 80% EtOH to each sample without disturbing the beads.
17. Incubate the PCR tube at room temperature for 30 s, then remove and discard all of the supernatant from each tube without disturbing beads.
18. Repeat **steps 12** and **13** once for a total of two 80% EtOH washes.
19. While keeping the PCR tube on the magnetic stand, let the samples air-dry at room temperature for 10 min and then remove the plate or tube from the magnetic stand.
20. Resuspend the dried bead pellet with 51 μL Resuspension Buffer. Gently pipette the entire volume up and down 30 times to mix thoroughly.
21. Incubate the PCR tube at room temperature for 2 min.
22. Place the PCR tube on the magnetic stand at room temperature for 5 min or until the liquid appears clear.
23. Transfer 50 μL of the clear supernatant to a new 0.5 mL PCR tube.
24. Vortex the AMPure XP Beads until they are well dispersed, then add 50 μL of mixed AMPure XP Beads to each sample for a second cleanup. Gently pipette the entire volume up and down ten times to mix thoroughly.
25. Incubate the PCR tube at room temperature for 15 min.
26. Place the PCR tube on the magnetic stand at room temperature for 5 min or until the liquid appears clear.

27. Remove and discard all of the supernatant from each sample.

    IMPORTANT: Leave the PCR tube on the magnetic stand while performing the following 80% EtOH wash steps (**steps 28–31**).
28. With the PCR tube remaining on the magnetic stand, add 200 μL of freshly prepared 80% EtOH to each tube without disturbing the beads.
29. Incubate the PCR tube at room temperature for 30 s.
30. Remove all of the EtOH from each tube without disturbing the beads.
31. Repeat **steps 24–26** once for a total of two 80% EtOH washes.
32. While keeping the PCR tube on the magnetic stand, let the samples air-dry at room temperature for 10 min and then remove the tube from the magnetic stand.
33. Resuspend the dried pellet with 21 μL Resuspension Buffer. Gently pipette the entire volume up and down 30 times to mix thoroughly.
34. Incubate the PCR tube at room temperature for 2 min.
35. Place the PCR tube on the magnetic stand at room temperature for 5 min or until the liquid appears clear.
36. Transfer 20 μL of the clear supernatant from each PCR tube to a new 1.5 mL microcentrifuge tube.

SAFE STOPPING POINT

If you do not plan to proceed to Bisulphite Treatment of Chip DNA immediately, the protocol can be safely stopped here. If you are stopping, store samples at −20 °C for up to 7 days.

3.3 Bisulphite Treatment of ChIP DNA

To interrogate the methylation status at nucleotide resolution of the chromatin immunoprecipitated DNA, the Chip methyl-adapter ligated gDNA fragments are treated with sodium bisulphite to convert unmethylated cytosine to thymidine, while methylated cytosines are resistant to bisulphite treatment. The procedures are adapted from the EZ DNA Methylation-Gold™ Kit.

1. The **CT Conversion Reagent** supplied within this kit is a solid mixture and must be prepared prior to first use. Prepare as follows: Add 900 μL water, 300 μL of **M-Dilution Buffer**, and 50 μL **M-Dissolving Buffer** to a tube of **CT Conversion Reagent**.
2. Mix at room temperature with frequent vortexing or shaking for 10 min.

   IMPORTANT: It is normal to see trace amounts of undissolved reagent in the ***CT Conversion Reagent****. Each tube of* ***CT***

***Conversion Reagent** is designed for 10 separate DNA treatments* (*see* **Note 6**).

3. *Preparation of M-Wash Buffer:* Add 24 mL of 100% ethanol to the 6 mL **M-Wash Buffer** concentrate before use.
4. Add 130 μL of the **CT Conversion Reagent** to 20 μL of the size selected, methyl-adapter ligated gDNA fragments in a 0.5 mL PCR tube. Mix the sample by flicking the tube or pipetting the sample up and down, then centrifuge the liquid to the bottom of the tube.
5. Place the sample tube in a thermal cycler and perform the following steps:

   Choose the thermal cycler preheat lid option and set to 105 °C, incubate at 98 °C for 10 min, at 64 °C for 2.5 h, and hold at 4 °C storage up to 20 h.
6. Add 600 μL of **M-Binding Buffer** to a **Zymo-Spin™ IC Column** and place the column into a provided **Collection Tube.**
7. Load the sample (from **step 2**) into the **Zymo-Spin™ IC Column** containing the **M-Binding Buffer**. Close the cap and mix by inverting the column several times.
8. Centrifuge at full speed (>10,000 × *g*) for 30 s. Discard the flow-through.
9. Add 100 μL of **M-Wash Buffer** to the column. Centrifuge at full speed for 30 s.
10. Add 200 μL of **M-Desulphonation Buffer** to the column and let it stand at room temperature (20–30 °C) for 15–20 min. After incubation, centrifuge at full speed for 30 s.
11. Add 200 μL of **M-Wash Buffer** to the column. Centrifuge at full speed for 30 s. Add another 200 μL of **M-Wash Buffer** and centrifuge for an additional 30 ss.
12. Place the column into a 1.5 mL microcentrifuge tube. Add 10 μL of **M-Elution Buffer** directly to the column matrix. Centrifuge for 30 s at full speed to elute the DNA.
13. Add another 10 μL of **M-Elution Buffer** directly to the column matrix. Centrifuge for 30 s at full speed to elute the DNA. Total elution volume is 20 μL.

SAFE STOPPING POINT

If you do not plan to proceed to Enrich DNA Fragments immediately, the protocol can be safely stopped here. If you are stopping, store the samples at −20 °C overnight or for up to 4 weeks (Bisulphite treated DNA is not suitable for storage periods longer than 4 weeks.)

3.4 Enrich DNA Fragments

This process uses PCR to selectively enrich those DNA fragments that have methylated adapter molecules on both ends and to amplify the amount of DNA in the library for accurate quantification. The PCR is performed with a PCR primer cocktail that anneal to the ends of the adapters. The number of PCR cycles should be minimized to avoid skewing the representation of the library (*see* **Note 7**).

The volume of adapter ligated fragments added to the PCR reaction below is based on an initial input DNA quantity of 50 ng. With 20 μL of bisulphite-treated elution from the *Bisulphite Treat Ligation Products* procedures, two independent PCR reactions will be set to maximize the library yield and diversity. The final library volume will be 15 μL (*see* **Note 7**).

1. Remove the Pfu Turbo Cx Reaction Buffer, 10 mM dNTP Mix and PCR Primer Cocktail from −20 °C and thaw at room temperature. When thawed, briefly centrifuge and keep the tubes on ice.
2. Remove the AMPure XP beads from 4 °C and let it stand for at least 30 min to bring them to room temperature.
3. Remove Resuspension Buffer from −20 °C and thaw at room temperature, place on ice.
4. Prepare the reaction mix on ice in a 200 μL thin wall PCR tube in the following order:

Reagent	Volume (μL)
DNA	10
Ultra-pure water	27.75
Pfu turbo Cx reaction buffer	5
10 mM dNTP mix	1.25
PCR primer cocktail	5
PfuTurbo Cx Hotstart DNA polymerase	1
Total volume	**50**

5. Vortex the reaction mix briefly to mix thoroughly.
6. Amplify using the following PCR process with a heated lid: Choose the preheat lid option and set to 105 °C, incubate for 5 min at 95 °C, for 30 s at 98 °C. Perform 14 cycles of 10 s at 98 °C, 30 s at 65 °C, 30 s at 72 °C followed by a final extension step for 5 min at 72 °C. Hold at 4 °C (*see* **Note 8** and Fig. 2).
7. Remove the PCR tubes from the thermal cycler and combine the two PCR into one 0.5 mL PCR tube. Each 0.5 mL PCR tube has a total volume of 100 μL of PCR amplified library.
8. Vortex the AMPure XP Beads until they are well dispersed, then add 100 μL of the mixed AMPure XP Beads to each tube containing 100 μL of the PCR amplified library. Return beads to 4 °C.

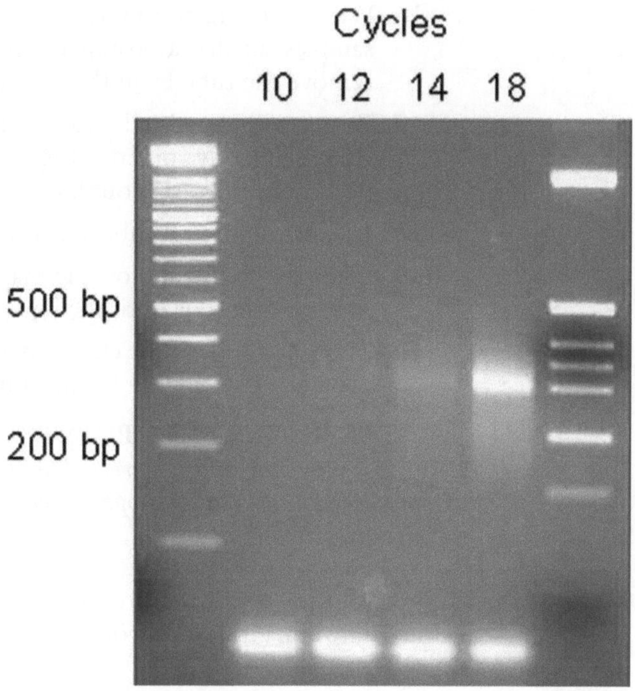

Fig. 2 Optimization of the number of PCR cycles for optimal library preparation testing 10, 12, 14, and 18 cycles, respectively. This shows that the yield was inadequate at 10 cycles, and by 18 cycles there is over-amplification, which may lead to an increase in read duplications. 12–14 cycles achieve a satisfactory yield without over amplification

9. Gently pipette the entire volume up and down ten times to mix thoroughly.
10. Incubate at room temperature for 15 min.
11. Place the PCR tube on the magnetic stand at room temperature for 5 min or until the liquid appears clear.
12. Remove and discard all of the supernatant from each sample. DO NOT DISTURB BEADS.

    IMPORTANT: Leave the PCR tube on the magnetic stand while performing the following 80% EtOH wash steps (**steps 13–16**).
13. With the PCR tube remaining on the magnetic stand, add 200 μL of freshly prepared 80% EtOH to each sample without disturbing the beads.
14. Incubate the PCR tube at room temperature for 30 s, then remove and discard all of the supernatant from each tube.
15. Repeat **steps 10** and **11** once for a total of two 80% EtOH washes.

16. While keeping the PCR tube on the magnetic stand, let the samples air-dry at room temperature for 10 min and then remove the tube from the magnetic stand.
17. Resuspend the dried bead pellet with 16 μL Resuspension Buffer. Gently pipette the entire volume up and down 30 times to mix thoroughly.
18. Incubate the PCR tube at room temperature for 2 min.
19. Place the PCR tube on the magnetic stand at room temperature for 5 min or until the liquid appears clear.
20. Transfer 15 μL of the clear supernatant from each sample to an appropriately labeled 1.5 mL microcentrifuge tube.

SAFE STOPPING POINT

If you do not plan to proceed to Validate Library immediately, the protocol can be safely stopped here. If you are stopping, store the libraries at −20 °C.

3.5 Validate Library

Perform the following procedures for quality control analysis and accurate quantification of the ChIP Bisulphite DNA library templates.

1. **Quantify Libraries**: In order to achieve the highest quality of data on Illumina sequencing platforms, it is important to create optimum cluster densities across every lane of every flow cell. This requires accurate quantification of DNA library templates. Quantify your libraries using qPCR according to the KAPA Library Quantification Kits For Illumina Sequencing Platforms Guide.
2. **Quality Control**: To verify the size of your PCR-enriched fragments, check the template size distribution by running an aliquot of the DNA library on a gel or on an Agilent Technologies 2100 Bioanalyzer using a High Sensitivity DNA chip or DNA 1000 chip. Running samples on a Bioanalyzer should be used for qualitative purposes only. If using the Agilent Bioanalyzer with a High Sensitivity DNA kit, based on yield of libraries, make a 1:5–1:10 dilution of the library using water and load 1 μL of the diluted library on the Agilent High Sensitivity DNA chip. If using the Agilent Bioanalyzer with a DNA 1000 kit, load 1 μL of the library on the Agilent DNA 1000 chip (Fig. 3).
3. When the quality and quantity of the library preparation meet the required criteria, the library is diluted to 10 nM final concentration in preparation for cluster generation and Illumina sequencing. 100 bp PE sequencing to achieve 30 million reads is recommended, leaving 6–10 million reads after mapping and an average coverage of 20–30× of the CpG sites within ChIP peaks.

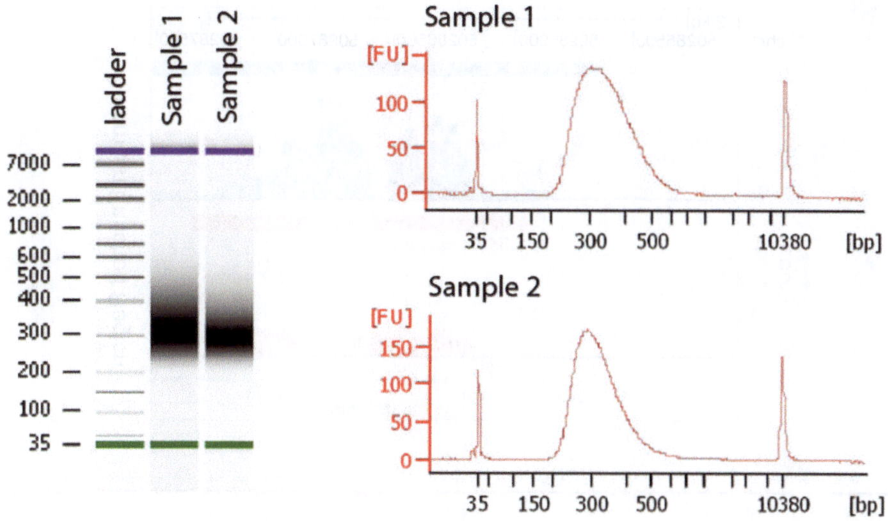

Fig. 3 Agilent Bioanalyzer trace of the bisulphite-treated ChIP DNA library preparation. This shows a library with uniform band and good peak size

3.6 Analysis of BisChIP seq Data

3.6.1 Alignment of Bisulfite-Treated Reads

1. Assess raw reads for quality, bisulfite conversion and adaptor contamination using fastqc.
2. Remove Adaptors and poor quality bases using trim_galore.
3. Align reads to the reference genome using bismark, a bisulfite-aware wrapper for the bowtie2 aligner [11].
4. Remove potential PCR duplicates using MarkDuplicates from the Picard suite.
5. Extract methylation calls at each CpG site using bismark_methylation_extractor.
6. Convert the methylation to bigwig format using R for visualization in either the UCSC Genome Browser or the Integrated Genomics Viewer.

3.6.2 Detection of ChIP-seq-Enriched Regions

The same guidelines that apply to the analysis of native ChIP-seq data apply equally to Bis-ChIP-seq, of which the most important is the choice of peak calling program. Both MACS2 and PeakRanger are widely used, and offer modes for sharp and broad ChIP-seq peak types (Fig. 4).

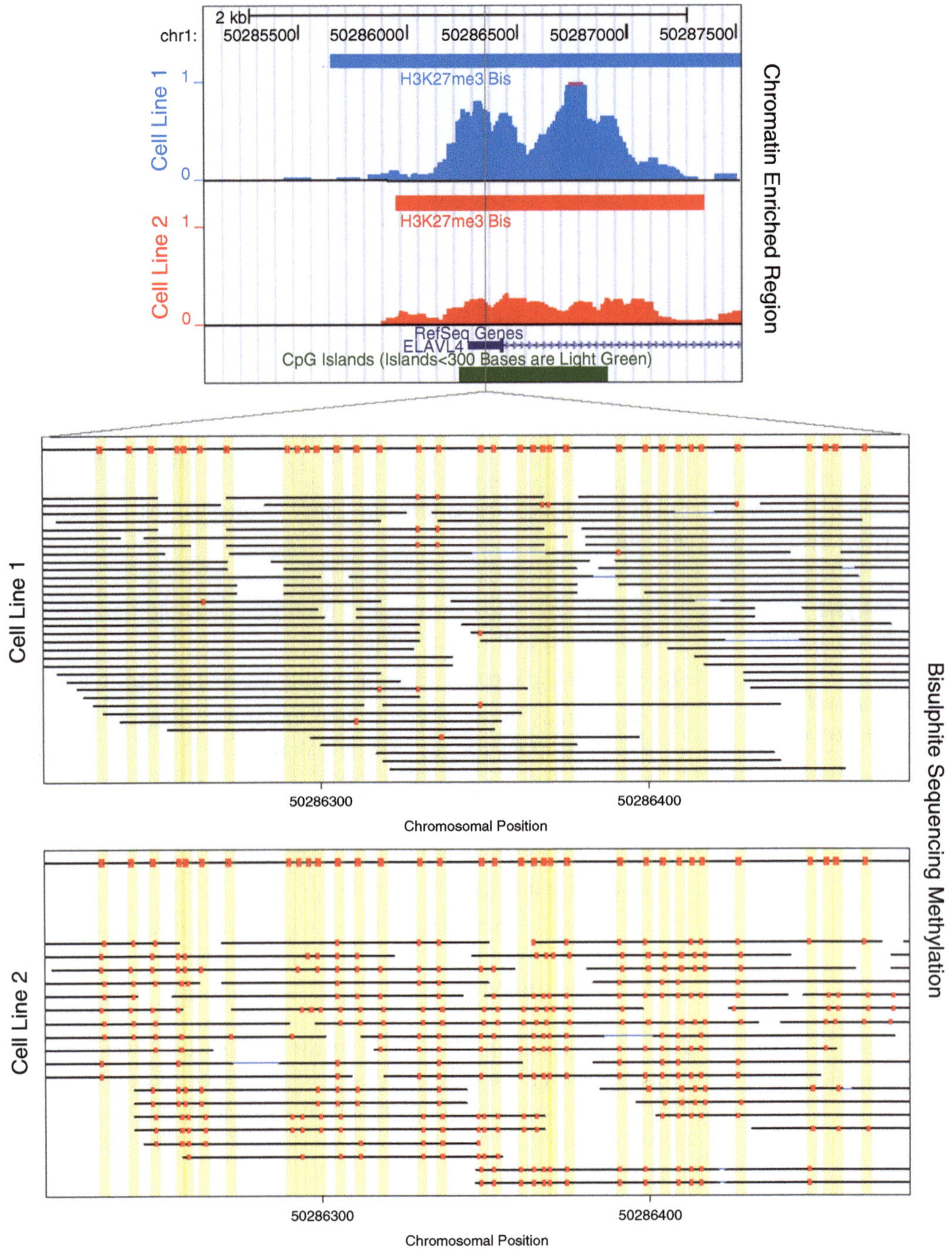

Fig. 4 BisChIP-seq result showing ChIP enrichment and methylation status of the enriched DNA. Individual bisulphite methylation sequencing reads shown with CpG sites (*red circles*) marked in yellow shading for each molecule

4 Notes

1. It is important to consider the most appropriate method to be used for chromatin immunoprecipitation depending on the mark of interest. We describe here the method we have used for the analysis of histone modifications. However, if transcription factors, hormone receptors, or chromatin remodelers are being immunoprecipitated, then the method that allows the best enrichment must be considered and optimized.
2. It is important that the sonicated chromatin DNA falls tightly in the 200–500 base-pair size range to enable uniform library size to be achieved during library preparation.
3. It is critical to validate that the chromatin immunoprecipitation was successful and that there is good enrichment over input. qPCR of known candidate genes should be performed with an enrichment of >6–8-fold of a positive control region over a negative control region.
4. Accurate quantitation of the ChIP DNA is necessary to ensure there is sufficient enriched DNA for bisulphite conversion and subsequent library preparation. For most histone marks, 50 ng is achievable from one chromatin immunoprecipitation and sufficient for one BisChip-Seq experiment. However, for less abundant chromatin marks, or transcription factor-binding sites, it may be necessary to perform the ChIP in triplicate or more and combine ChIP DNAs to obtain the necessary amount of ~50 ng DNA for bisulphite treatment.
5. It is important to consider what indexes to use for multiplexing samples for high-throughput sequencing. Generally for BisChip-Seq 4-6 samples can be multiplexed on one Illumina HiSeq lane, depending on the coverage of the ChIP mark. Consult the Illumina TruSeq Sample Preparation Pooling Guide for appropriate pooling of RNA Adapter indices supplied in the TruSeq ChIP Sample Preparation Kit.
6. The CT Conversion Reagent is light sensitive, so minimize its exposure to light. For best results, the CT Conversion Reagent should be used immediately following preparation. If not used immediately, the CT Conversion Reagent solution will remain usable if stored overnight at room temperature, 1 week at 4 °C, or up to 1 month at −20 °C. Stored CT Conversion Reagent solution must be warmed to 37 °C, then vortexed prior to use.
7. It is important to perform at least 2–3 PCRs in parallel for the library prep amplification step to ensure library complexity.
8. We have optimized the number of PCR cycles to achieve a library that is of sufficient yield for Illumina sequencing. It may be necessary to optimize the number of PCR cycles

depending on the amount of starting material available for library preparation from the chromatin immunoprecipitation and the efficiency and recovery from the bisulphite reaction, to achieve adequate library yield without compromising library complexity due to duplications as a consequence of over-amplification.

Acknowledgments

We thank Dr. Warwick Locke for careful reading of the manuscript. S.J.C. is a National Health and Medical Research Council (NH&MRC) Senior Principal Research Fellow. This work was further supported by an NH&MRC Project Grant (1029584).

References

1. Lister R, Pelizzola M, Dowen RH et al (2009) Human DNA methylomes at base resolution show widespread epigenomic differences. Nature 462:315–322
2. Laurent L, Wong E, Li G et al (2010) Dynamic changes in the human methylome during differentiation. Genome Res 20:320–331
3. Gal-Yam EN, Egger G, Iniguez L et al (2008) Frequent switching of Polycomb repressive marks and DNA hypermethylation in the PC3 prostate cancer cell line. Proc Natl Acad Sci U S A 105:12979–12984
4. Hawkins RD, Hon GC, Lee LK et al (2010) Distinct epigenomic landscapes of pluripotent and lineage-committed human cells. Cell Stem Cell 6:479–491
5. Kondo Y, Shen L, Cheng AS et al (2008) Gene silencing in cancer by histone H3 lysine 27 trimethylation independent of promoter DNA methylation. Nat Genet 40:741–750
6. Coolen MW, Stirzaker C, Song JZ et al (2010) Consolidation of the cancer genome into domains of repressive chromatin by long-range epigenetic silencing (LRES) reduces transcriptional plasticity. Nat Cell Biol 12:235–246
7. Brinkman AB, Gu H, Bartels SJ et al (2012) Sequential ChIP-bisulfite sequencing enables direct genome-scale investigation of chromatin and DNA methylation cross-talk. Genome Res 22:1128–1138
8. Statham AL, Robinson MD, Song JZ et al (2012) Bisulfite sequencing of chromatin immunoprecipitated DNA (BisChIP-seq) directly informs methylation status of histone-modified DNA. Genome Res 22:1120–1127
9. Clark SJ, Harrison J, Paul CL et al (1994) High sensitivity mapping of methylated cytosines. Nucleic Acids Res 22:2990–2997
10. Clark SJ, Statham A, Stirzaker C et al (2006) DNA methylation: bisulphite modification and analysis. Nat Protoc 1:2353–2364
11. Krueger F, Andrews SR (2011) Bismark: a flexible aligner and methylation caller for bisulfite-seq applications. Bioinformatics 27:1571–1572
12. Zhang Y, Liu T, Meyer CA et al (2008) Model-based analysis of ChIP-Seq (MACS). Genome Biol 9:R137
13. Feng X, Grossman R, Stein L (2011) PeakRanger: a cloud-enabled peak caller for ChIP-seq data. BMC Bioinformatics 12:139

Chapter 16

A Guide to Illumina BeadChip Data Analysis

Michael C. Wu and Pei-Fen Kuan

Abstract

The Illumina Infinium BeadChips are a powerful array-based platform for genome-wide DNA methylation profiling at approximately 485,000 (450K) and 850,000 (EPIC) CpG sites across the genome. The platform is used in many large-scale population-based epigenetic studies of complex diseases, environmental exposures, or other experimental conditions. This chapter provides an overview of the key steps in analyzing Illumina BeadChip data. We describe key preprocessing steps including data extraction and quality control as well as normalization strategies. We further present principles and guidelines for conducting association analysis at the individual CpG level as well as more sophisticated pathway-based association tests.

Key words DNA methylation, Epigenome-wide association studies, Hypothesis testing, Normalization, Pathway analysis, Quality control

1 Introduction

DNA methylation is an important epigenetic hallmark associated with gene silencing [1, 2] and involves the addition of a methyl group to the fifth carbon of the cytosines in the human genome. DNA methylation usually occurs at CpG dinucleotides in adult somatic cells, though non-CpG methylation also occurs [3]. The importance of DNA methylation in understanding biological processes, disease etiology, and exposure response has led to the development of a wide range of platforms for profiling DNA methylation, recently including array-based technology and high-throughput sequencing.

The array-based Illumina Infinium HumanMethylation450K BeadChip (Illumina450K) platform [4, 5] is a popular genome-wide CpG methylation profiling technology being used in most large-scale population-based methylation studies of thousands of human individuals due to its comprehensive coverage and high throughput. The Illumina 450K is designed to assess methylation levels at 485,577 individual CpG sites spread across all

Jörg Tost (ed.), *DNA Methylation Protocols*, Methods in Molecular Biology, vol. 1708,
https://doi.org/10.1007/978-1-4939-7481-8_16, © Springer Science+Business Media, LLC 2018

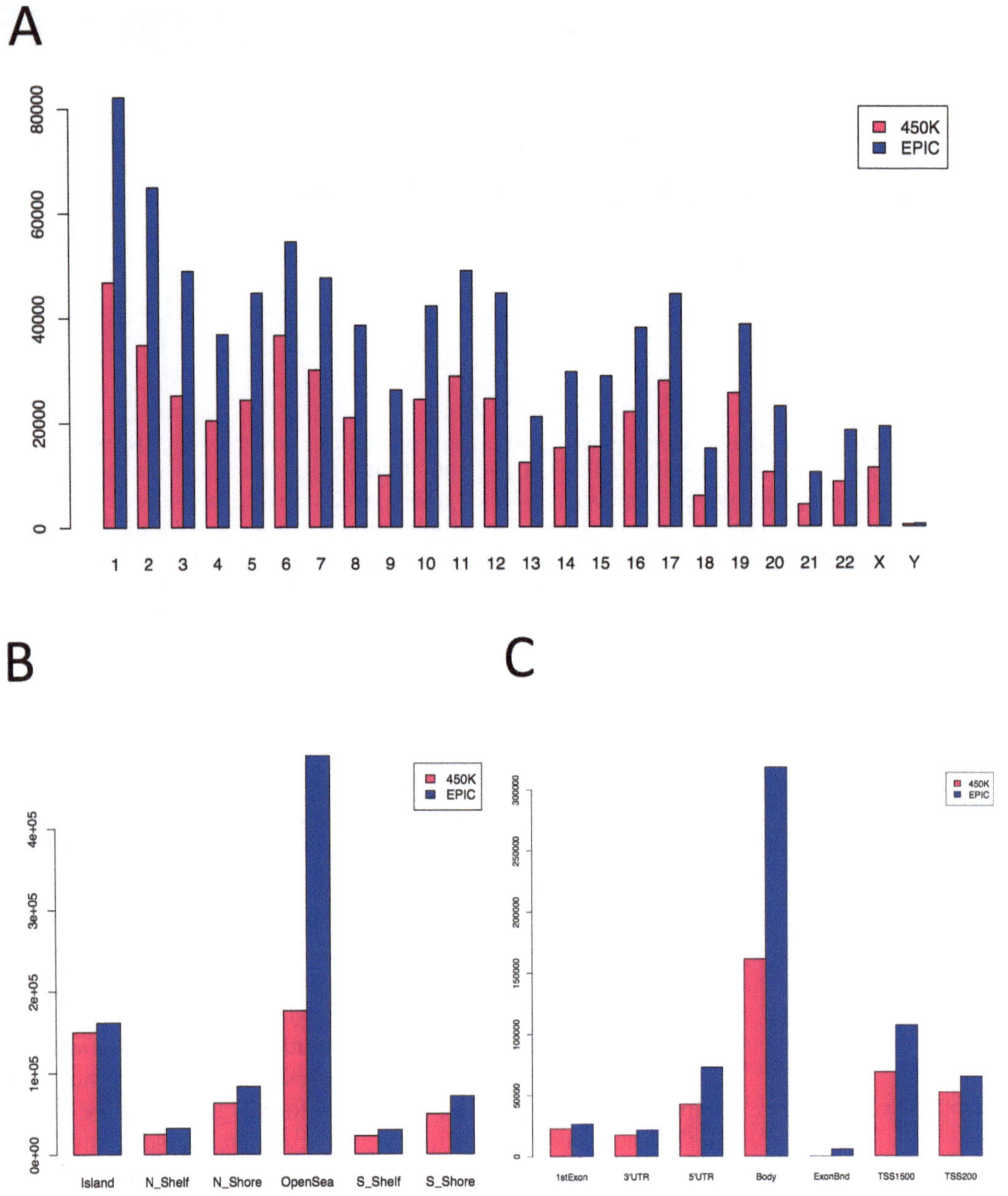

Fig. 1 (**a**) The number of CpGs from each chromosome included on the Illumina 450K and EPIC platform. (**b**) For CpGs associated with islands, the positioning of the CpG relative to the island. (**c**) For CpGs associated with genes, the positioning of the CpG relative to the gene

chromosomes and captures a wide range of important genomic features [4] (Fig. 1). This includes coverage of CpG islands (96%), shores, shelves, and genomic regions including TSS, gene body, first exon, and 3/5′ UTRs of 99% Refseq genes, with an average of 17 CpGs per gene. This platform has been successfully used to identify methylation marks associated with smoking [6–8], cancer [9, 10], aging [11], immune conditions [12], reproductive outcomes [13] and is being employed in studies of a wide range of other exposures, disease outcomes, and conditions. Recently, a new generation of Infinium BeadChips, i.e., MethylationEPIC (EPIC) was released. The EPIC array offers a wider genome-wide coverage at 863,904 CpG and 2932 CNG sites on important regulatory regions, including FANTOM5 enhancers, ENCODE open chromatin and enhancers, DNase hypersensitivity sites, and miRNA promoter regions previously not captured by the 450K array [14]. More than 90% of the 450K array CpG sites are covered by the EPIC array.

Briefly, the Infinium methylation technology is built upon the same technology used for genotyping SNPs. The Infinium assays are based on bisulfite conversion, where unmethylated cytosines are converted to uracils (read of as thymines after PCR) whereas methylated cytosines remain unconverted. Following the Illumina protocol, bisulfite converted DNA are subject to whole genome amplification, enzymatic end-point fragmentation, precipitation and resuspension before hybridizing to the array [5]. The methylation level at each CpG on the array is then measured using one of two probe types (Type I and Type II), which operate using different methods. Type I probes measure methylation at approximately 28% (16%) of the CpGs on 450K (EPIC) and use two probes per CpG corresponding to methylated (M) and unmethylated (U) allele. In contrast, the Type II probes, which cover the remaining 72% (84%) of the CpGs on 450K (EPIC), utilize one probe per CpG and employ different dye colors (red/green) to differentiate M from U alleles. The raw methylation level at each CpG is quantified using the beta-value defined as $\beta = \frac{M}{(M+U+100)}$, where the constant 100 is to stabilize the beta values when the intensities are low. The beta values are bounded between 0 and 1 and correspond approximately to the percentage of cells for which the CpG is methylated with the endpoints representing fully unmethylated and methylated states, respectively.

There are a number of salient features to the data from the Illumina BeadChip (450K and EPIC) platform. First, the distribution of the beta values for a single individual, across all CpGs, is bi-modal with the majority of CpGs close to zero or one (Fig. 2a). Second, although the distribution across CpGs for a single individual is bi-modal, the distribution of beta values for the same CpG across individuals is frequently uni-modal but possibly skewed due to being bounded between zero and one (Fig. 2b). Third, the two different probe types have different properties and exhibit differential behavior (Fig. 3) that may need to be corrected, especially since Type II probes may have lower reproducibility than Type I probes. Finally, accuracy of the particular probes can be influenced by the presence of SNPs (Fig. 4) and cross-hybridization. Combined with

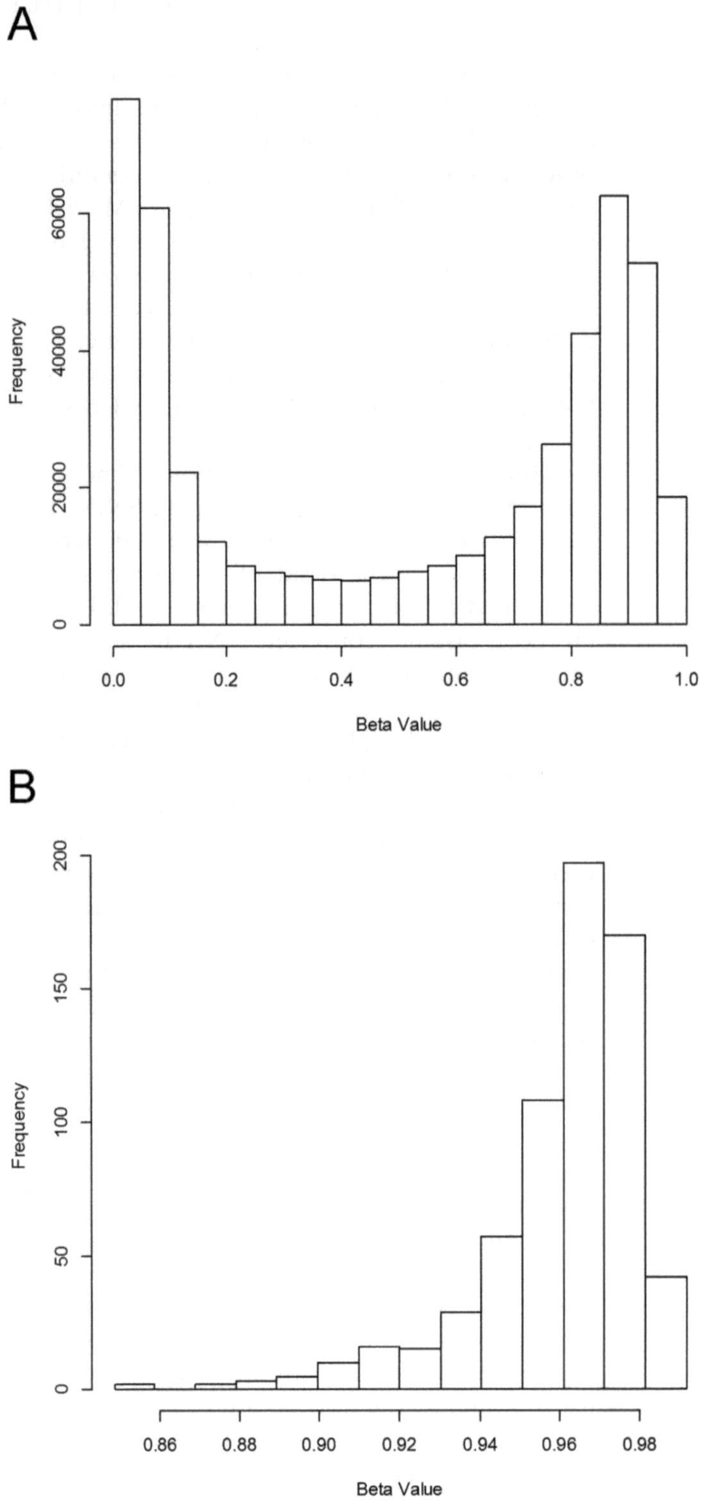

Fig. 2 Using data from a methylation study of aging comprised of 626 individuals [74], we show the distribution of the beta values (**a**) for a single individual across all CpG sites and (**b**) for a single CpG site across all individuals in the study

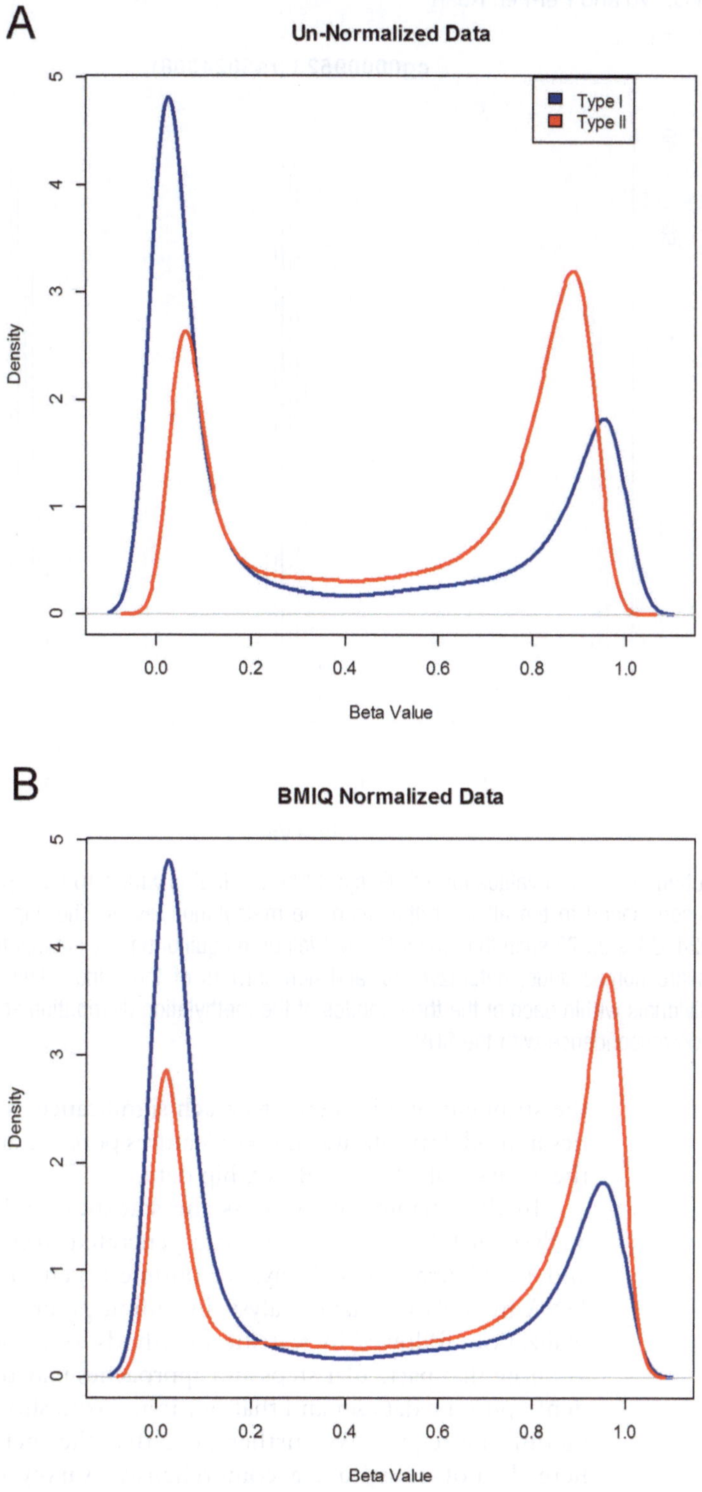

Fig. 3 Comparison of the distributions of the Type I and Type II probes for a single individual without normalization (**a**) and after BMIQ normalization (**b**). The raw data show that the distributions do not align well while BMIQ normalization normalizes Type II probes to the Type I probes and reduces the probe design type differences as evident by the aligned modes

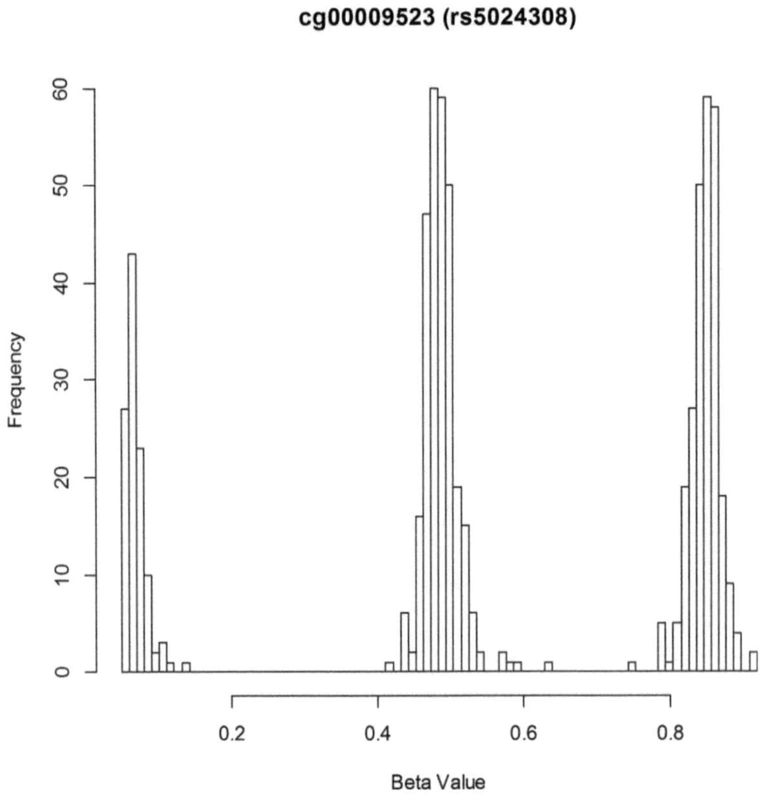

Fig. 4 The distribution of the beta values for a CpG (cg00009523) that is within 10 base pairs of rs5024308. The three modes correspond to the alleles rather than the methylation levels. The population minor allele frequency of rs5024308 is 0.372 such that under Hardy-Weinberg equilibrium we anticipate 14, 47, and 39% of the individuals are homozygous, heterozygous, and non-carriers of the minor allele, respectively. The proportions of individuals within each of the three modes of the methylation distribution are 17, 48, and 35% indicating strong correspondence with the SNP

the stringent requirements to reach significance as well as difficulties in model specification, these features pose a grand challenge for the analysis of Illumina BeadChip data.

In this chapter, we discuss the statistical and computational analysis of DNA methylation data generated from Illumina 450K and EPIC array. Specifically, we outline a general framework for DNA methylation data analysis including preprocessing and normalization, followed by statistical methods for association analysis. We note that particular steps and approaches may be highly dependent upon the data set and that our framework should be viewed as guiding principles. We further note that the methods presented here do not constitute a comprehensive survey of the available methods, which continue to grow rapidly.

2 Materials

2.1 Methylation Measurements

1. *Illumina idat files:* The Illumina Intensity Data (idat) file format is used to summarize the data from the Illumina BeadArray platforms including the 450K and EPIC methylation arrays. For each sample, a "Red" and a "Green" idat file are available representing the intensities of the methylated and unmethylated probes. The files were originally intended for use with GenomeStudio software [15], but other software can also be used to process these files [16]. Although many downstream analysis pipelines do not require direct manipulation of the idat files, these should be retained as analytical best practices are continuing to evolve and emerging methods may require these at a later time.
2. *GenomeStudio Report*: The output from GenomeStudio. At a minimum, the beta value for each probe as well as the corresponding detection *p*-value is necessary for most downstream analyses.

2.2 Software

Most basic analyses can be done within the R statistical computing environment [17] with additional Bioconductor [18] packages. If analyses are starting directly from idat files or the GenomeStudio workspace, then Illumina GenomeStudio software or an appropriate R package for reading the idat files will be necessary.

2.3 Manifest

The Illumina manifest file includes a description of the probes including probe IDs (cg-number), chromosome, location, relation to epigenetically relevant features, nearby SNPs, gene membership, etc. Both the 450K and EPIC array manifests are also available on Bioconductor [19, 20].

2.4 Meta Data

For analyses involving correlation or association with case-control status, exposures or other experimental conditions (referred to as the "variable of interest," for simplicity), information on the variable of interest is needed as meta data. However, since epigenetic variability is influenced by many factors, additional covariates such as age, gender, and other possible confounders should also be collected where possible.

3 Methods

3.1 Data Extraction

1. *Extraction from idat Files*: If starting from the raw Illumina idat files, relevant data entries for each sample must be extracted. This includes both the beta value for each probe and the corresponding detection *p*-value, which represents a confidence measure for the reported beta value. The idat files can be

processed using GenomeStudio software or the minfi package [21] from Bioconductor.

2. *Extraction from Final Report*: Frequently, investigators are provided with a final report in a tab delimited text format from GenomeStudio. These reports generally contain extraneous information; consequently, we still encourage users to separate out the beta values and detection *p*-values from the other information.

3.2 Filtering Probes and Samples

Probes and samples that fail can increase noise as well as lead to spurious results. Failed probes and samples need to be filtered.

1. *Detection p-value*: Within each study, some probes may fail in individual samples. For each sample, set the methylation values of the probes with detection *p*-value greater than 0.05 to be missing values. We recommend that probes failing the detection *p*-value threshold in more than 20% of the samples should be removed from the analysis.
2. *Control Probes*: Control probes should be removed from the analysis, but the measurements can be retained separately in order to conduct quality control and sample checking.
3. *Failed Samples*: Sometimes, entire samples fail due to inadequate DNA concentration or other processing issues. These samples are frequently characterized by large numbers of probes failing detection and serve merely to introduce variability. We recommend that samples with more than 10% of probes failing detection or average detection *p*-value across all probes greater than 0.05 should be removed from the study.

3.3 Flagging Potentially Problematic Probes

Some of the probes on the platform may be capturing effects other than CpG methylation [22], such as underlying single-nucleotide polymorphisms (SNPs) and need to be flagged as potentially problematic. We encourage flagging of the probes rather than outright omission since the degree of the problem is not always clear: for example, SNPs present only in individuals of African ancestry may not be problematic in studies with individuals of Northern European ancestry. If flagged probes reach significance, results should be carefully examined.

1. *Probes with SNPs*: Since the Illumina Infinium BeadChip platform is essentially a genotyping technology, if a SNP is present at or near the CpG site the methylation value may actually be capturing the SNP rather than the CpG methylation (Fig. 4) with SNPs closer to the CpG being more influential. If possible, use the 1000 Genomes Project data [23] or alternative reference to determine whether a SNP lies within 50 base pairs of the CpG site. The reference should be similar in ethnicity and population genetic structure to the study population. Alternatively, the

Illumina Manifest also contains information on whether SNPs lie near probes of interest, but these annotations are incomplete and may not be tailored to the study population. Probes with SNPs near the CpG site should be flagged.

2. *Cross-reactive or Nonspecific Probes*: Recent work has shown that many of the probes are cross-reactive or bind nonspecifically to the target region of interest. Many probes also fall within repetitive elements of the genome. Consequently, these probes may not actually be measuring methylation at the particular target CpG. These CpGs do not affect the statistical validity of any discovered associations and can still serve as useful predictive biomarkers, but any biological interpretations drawn from these CpGs need to be made with caution. We encourage flagging of these probes many of which have been previously annotated [22, 24].
3. *Using Positive Controls*: If positive controls, in which all probes should be fully methylated, are included in the experiment, it is possible to identify individual probes, which do not perform well, e.g. probes with beta value <0.8. These probes may be removed from the analysis for all samples or can be flagged for further examination.

3.4 Sample Checks

Quality control should be conducted, where possible, to check for inconsistencies in the methylation data. Common approaches include looking for gender mismatch as well as correlation between presumed replicates. Large numbers of errors are a cause for concern that sample meta-data may not be correctly linked or that samples may be mislabeled during processing thus requiring further inquiry.

1. *Gender mismatch or outliers*: Multi-dimensional scaling (MDS) plots can be constructed based on the X chromosome data. Males and females should separate into two distinct clusters. Samples separated into the wrong cluster (males in the female cluster or females in the male cluster) should be checked for mislabeling or omitted from the study. Samples not falling into distinct clusters should be omitted.
2. *Genotyping Information:* Many epigenetic studies are conducted within existing cohorts, including cohorts with genotyping information. Since 65 (59) of the control probes on the Illumina 450k (EPIC) BeadChip platform are for direct interrogation of genotype, these can be cross checked against existing genotype data to identify mislabeled samples.
3. *Replicates*: If replicates are available, these should be checked for consistency. In particular, correlation between replicates should be higher than the correlation between non-replicates. Poor correlation is an indication of either data quality issues or

possible mislabeling. If correlation is high, only a single sample needs to be used for all subsequent analyses and other replicates for the same sample can be omitted.

3.5 Normalization

A key objective of normalization is to remove technical and systematic variability from the data in order to make measurements comparable across samples. Within the context of Illumina methylation analysis, a range of normalization procedures have been proposed [25–28] with varying characteristics and performances [29, 30]. In contrast to other types of genomic data, the emphasis has been on within-sample normalization. As noted earlier, the Illumina 450K and EPIC methylation array's probes fall within two different design types (Type I and Type II), which differ in the manner through which they quantify methylation levels [25]. The overall distribution of methylation values measured using Type I and Type II probes is different [4] (Fig. 3a) and Type II probes are sometimes both less reproducible and sensitive [31]. Some of the more popular normalization approaches operate by normalizing the Type II probes to the Type I probes. While reducing differences between different probe types, this can also improve reproducibility due to the higher total reproducibility of Type I probes. In this section, we describe the BMIQ normalization procedure [26] but we also offer some justification for using the raw, un-normalized data.

1. *Raw Data*: The simplest approach is to directly analyze the raw data without application of any normalization procedure. Although normalization is an imperative process for other types of genomic data, the need for normalization for Illumina methylation data varies on a study-by-study basis and may not be necessary. In some studies, the raw data are already highly reproducible with some normalizations potentially decreasing reproducibility [30]. Furthermore, if down-stream analyses are conducted at the individual CpG level, different probe types are not compared such that the analyses are still valid.
2. *BMIQ Normalization*: BMIQ is a popular normalization approach that focuses on transforming the distribution of Type II probes to be similar to the Type I probes (Fig. 3b). Briefly, individual probes (of both types) on a sample are assigned to be methylated, hemi-methylated, and unmethylated. Then the Type II probes classified as methylated or unmethylated are quantile normalized to have identical distribution as the Type I probes of the corresponding class. Hemi-methylated Type II probes are then location and scale adjusted to span the range between the methylated and unmethylated Type II probes. The BMIQ method can be run on a sample-by-sample basis, allowing for improved computational efficiency, and requires the raw methylation values for each probe as well as the probe design

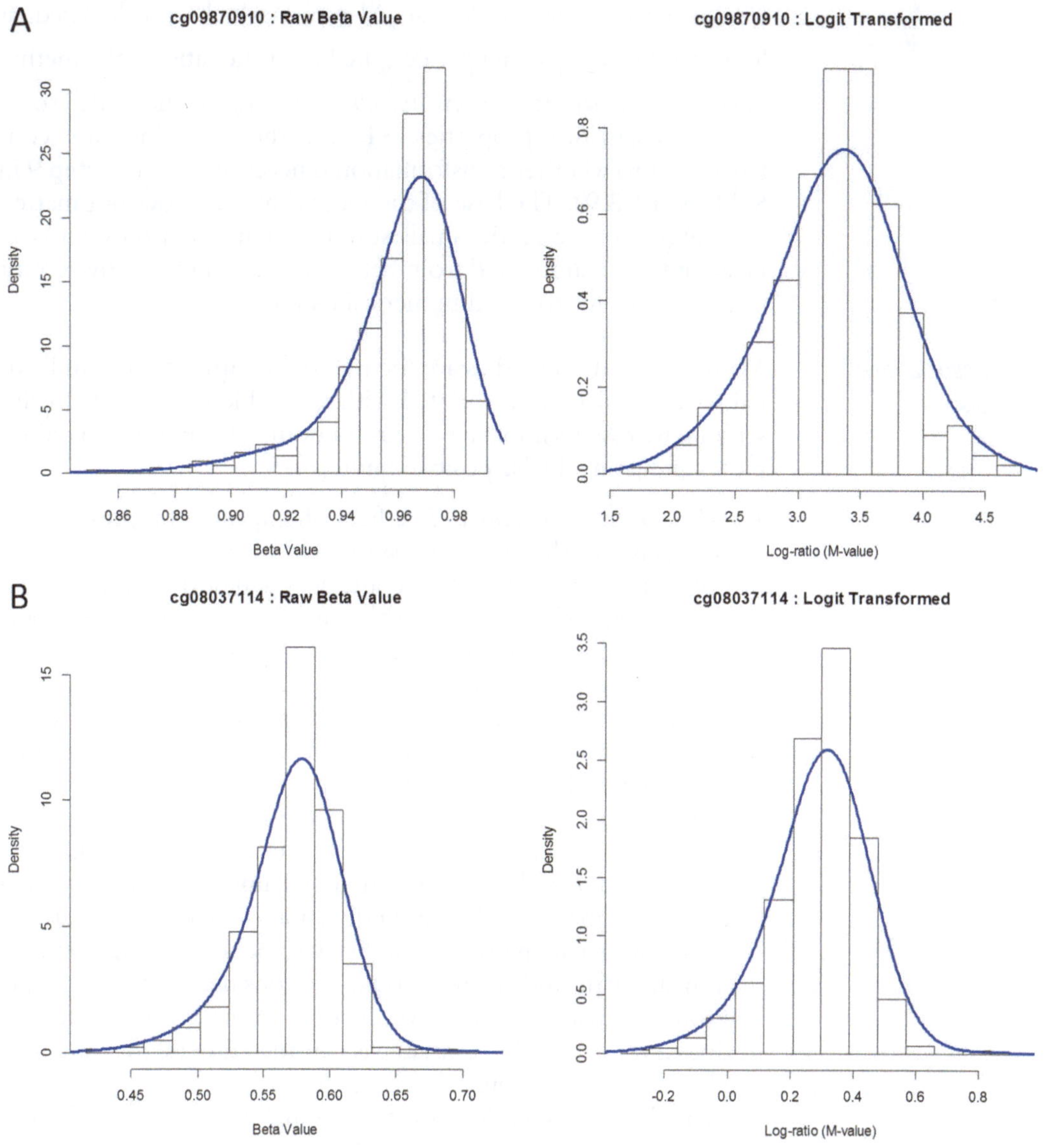

Fig. 5 Comparisons of the distribution of the beta values and log ratios for two different CpGs. For some probes, logit transforming beta values to log ratios can reduce skewness (**a**), but for other CpGs, the transformation does not make much difference (**b**)

types. BMIQ is implemented within the wateRmelon R package [32] and ChAMP pipeline [33].

3.6 Logit Transformation

The beta values are natural measurements of methylation that intuitively correspond, approximately, to the proportion of cells for which the particular CpG is methylated in the sample. However, as a proportion, the values are inherently bounded between 0 and 1 and the distributions can be slightly skewed (Fig. 5). Consequently, the beta values can also be logit transformed to the log-ratio value, also called an M-value, which is calculated as $\text{logit}\,\beta = \ln\left[\frac{\beta}{(1-\beta)}\right] = \ln\left(\frac{M}{U}\right)$, i.e., the log of the ratio of the methylated to unmethylated intensity. Using log-ratios may offer some improved statistical properties [34], but there remains some controversy as to whether transformation is necessary (*see* also **step 9** in Subheading 3.9). The base of the logarithm used does not matter. A small number (e.g., the smallest nonzero methylation value) can be added to or subtracted from beta values equal to exactly zero or one, respectively, to avoid numerical issues.

3.7 Associations Analysis

Many methylation studies are focused on identification of individual CpGs, which are associated with a variable of interest. In this section, we discuss important steps for doing this in order to reduce biases and control false positive findings.

1. *Select Outcome Variable*: The field of epigenetics represents the confluence of the gene expression field with the genetic association study field. Within the context of gene expression studies, the majority of statistical methods model expression as the outcome variable with disease status or experimental condition as the predictor. On the other hand, in genetic association studies, genetic effects are almost always the predictor and a complex trait or phenotype is the outcome. Which variable is the predictor and which variable is the response or outcome depends closely upon the data at hand and should be selected carefully depending on prior knowledge. For example, methylation in cord blood could be expected to be influenced by smoking during pregnancy such that methylation is the outcome variable while asthma at age three is unlikely to be influencing methylation at birth and asthma could be considered the outcome. From a statistical perspective, if there are no covariates then it does not matter which variable is considered to be the outcome. However, if additional covariates are to be adjusted for (as is strongly recommended) then choosing the correct model can yield improved power.
2. *Selecting Covariates*: DNA methylation is known to be influenced by a range of factors such as environmental variables, demographic considerations, and other possible confounders. In traditional epidemiological and statistical analysis, the model would change for each CpG analyzed depending on whether particular covariates meet the criteria to be confounders. However, for genome-wide methylation analysis, the covariates to be included in the model should be selected up-front and just once: applying model selection procedures separately for each CpG is computationally challenging, makes interpretation of final results difficult, and leads to inflated false positive rates due to inference post model selection when testing at lower significance levels.

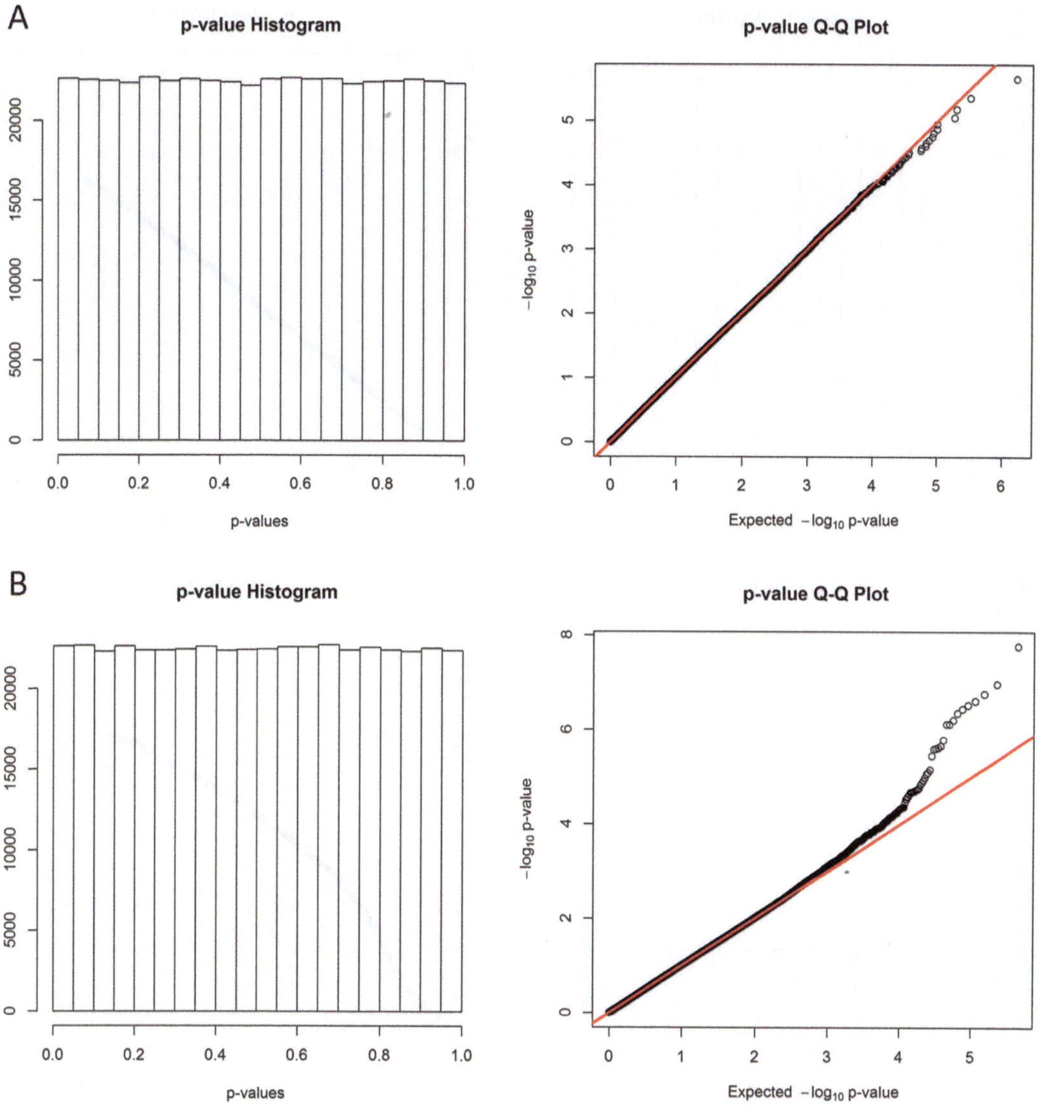

Fig. 6 *p*-value distributions and corresponding Q-Q plots under four different scenarios. (**a**) Under the global null hypothesis that none of the CpGs are related to the variable of interest, the *p*-value distribution is uniform and the Q-Q plot should lie along the red diagonal line. (**b**) When a small proportion of the CpGs are related to the variable of interest, the overall distribution should be near uniform but there should be an inflation of smaller *p*-values as visible in the Q-Q plot. This reflects the GWAS scenario where a few CpGs are related to the outcome. (**c**) The deflated number of small *p*-values and the fact that the Q-Q plot lies below the diagonal is an indication that the model may be incorrectly specified. In this case, an important confounder is omitted. (**d**) In this scenario an important confounder has again been omitted but the Q-Q plot now quickly deviates from the diagonal. This can also happen when large-scale methylation changes are associated with the variable of interest or when there is correlation among the *p*-values (though correlation is frequently induced by missing confounders). Distinguishing modeling issues from large-scale changes is often difficult

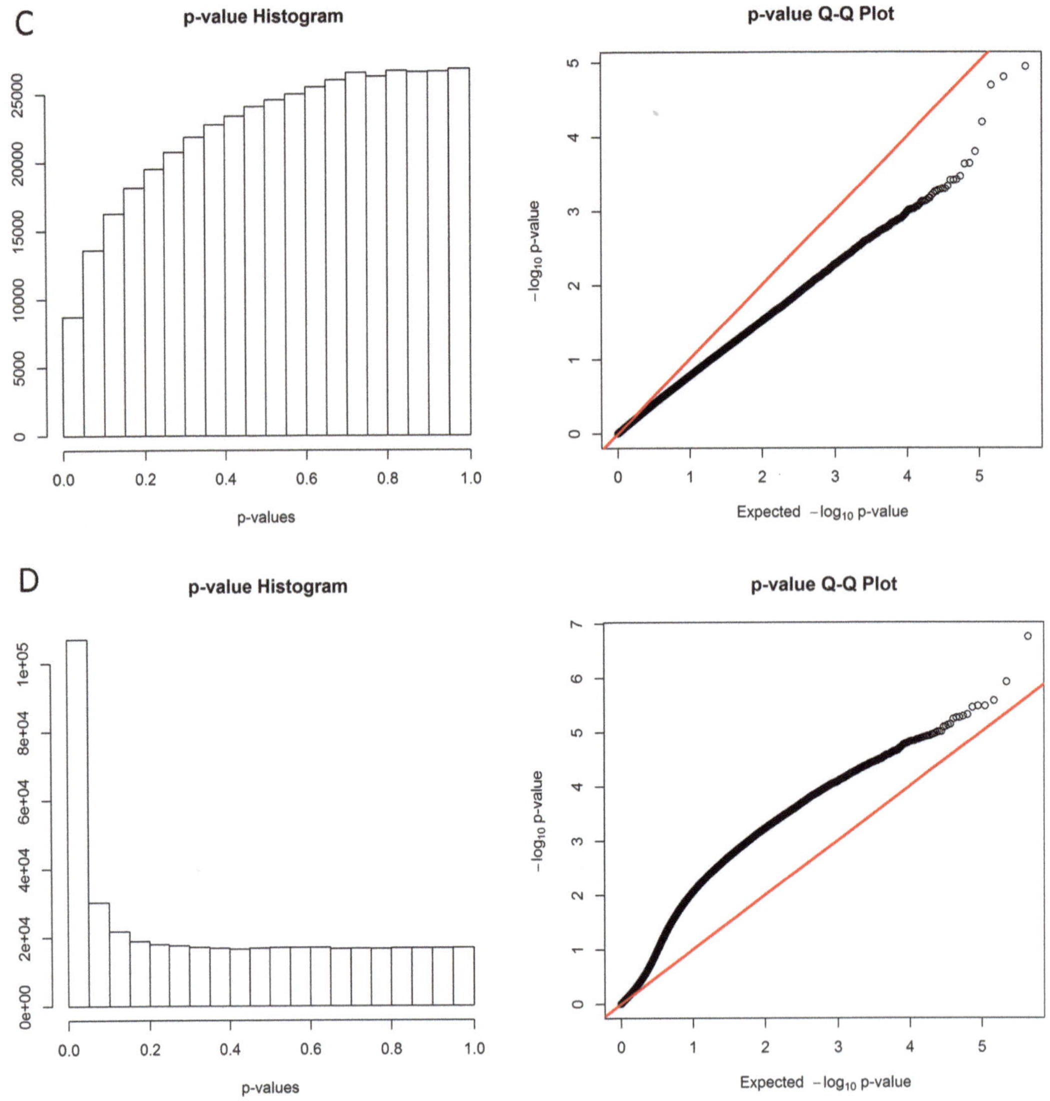

Fig. 6 (continued)

3. *Hypothesis Testing*: Following the determination of the outcome and covariate selection, a univariate testing approach can be used to assess the association between each CpG and the variable of interest while adjusting for additional covariates. If methylation is the outcome variable, then linear regression is typically used to regress methylation on the covariates as well as the variable of interest. Standard Wald tests or likelihood ratio tests can be applied to compute a *p*-value for significance, but we note that score tests sometimes offer improved properties within the context of large-scale hypothesis testing. We further note that adjustments for heteroskedasticity and potential outliers may also be necessary (*see* also Subheading 3.9).

4. *Control for Multiple Testing*: Due to the large number of CpGs being tested, adjustments must be made for multiple testing to control the overall type I error rate. Traditionally, the most common approach was the Bonferroni correction and this remains the de facto standard in genetic association studies. To control the overall type I error rate at the usual $\alpha = 0.05$ level, each of the CpGs is tested for significance at the α/m level where m is the total number of CpGs surviving preprocessing with p-values smaller than α/m called significant. Equivalently, a Bonferroni adjusted p-value can be calculated by multiplying each p-value by m and calling adjusted p-values less than $\alpha = 0.05$ as significant. In applying the Bonferroni correction, the probability of any false positives is then controlled at 5%; however, in practice, many investigators are willing to tolerate some proportion of false positives. Therefore, in such studies, it may be better to choose to control the false discovery rate (FDR) [35], which enables investigators to fix the proportion of significant results, which are actually false. By allowing some false positives, significance threshold can be easier to achieve. Q-values, representing the FDR level at which a CpG would be called significant, are often used analogously with adjusted p-values and can be computed using the p.adjust function within R.
5. *Global Assessment of Results*: Following computation of p-values for each CpG, results should be assessed on a global scale to diagnose pervasive problems. A standard approach is to examine the p-value distribution. Under the global null hypothesis, in which none of the CpGs are associated with the variable of interest, the distribution of p-values should be approximately uniform and the corresponding Q-Q plot should lie along the diagonal. Large deviations away from uniformity often indicate a systemic problem with the data or the analysis. This often means that the model used in the analysis is highly misspecified (sometimes indicating that there are missing confounders or experimental factors), that there are problems with the data such as outliers or heteroskedasticity, or that the statistical methods used for the analysis may not be appropriate (*see* also Subheading 3.9). Deviations from uniformity can be in either direction with too few or too many low p-values. Both are potentially indicative of problems. However, we note that this may be contextually dependent: for some conditions, one could anticipate large, global changes such that very large proportion of the CpGs on the array are associated with disease, e.g., comparing different types of tumors, such that large numbers of small p-values are not unanticipated. We further note that having too few small p-values is frequently more a sign of modeling or data problems than an issue of power (even underpowered studies should still have globally uniform p-value distribution). Figure 6 shows p-value histograms and Q-Q plots under several possible scenarios.

6. *Assessment of Individual Results*: Following global assessment of results, a careful examination of individual results is necessary. In particular, flagged probes, e.g., those containing SNPs that are called significant should be carefully examined. This is contextually specific depending on the results. For example, if the probe contains a SNP, then one could examine the allele frequency within a relevant reference population to determine whether the SNP is of concern. Further, since CpG methylation tends to demonstrate strong regional correlation, one would expect multiple probes to share *p*-values of similar strength as the probe containing a SNP; thus, additional probes in the neighborhood with significant (or even just nearly significant) *p*-values would be suggestive of a true association.

3.8 Pathway and Gene Set Analysis

Pathway and gene set analysis is a commonly used approach to augment individual CpG analysis. For simplicity we use the term "pathway" very loosely to represent any collection of genes (and the CpGs within the genes), which have been grouped together based on prior biological knowledge. By conducting analyses at the pathway level, this allows for reduced multiple testing and potentially improved power. Conversely, in studies in which too many CpGs are called significant, pathway analysis allows for an objective means for drawing conclusions and interpreting results. Furthermore, pathway analyses can identify complex, multi-CpG effects: biological functions require the concerted activity of multiple genes in which individual CpGs have modest effects individually, but large effect when combined.

A wide range of pathway analysis methods have been developed, largely within the context of gene expression profiling or genetic association studies. In this section, we describe some methods, which can be applied to methylation studies. Over-representation analysis and Gene Set Enrichment Analysis test a competitive null hypothesis, in which the significance of the CpGs within the pathway is compared to the CpGs outside of the pathway. On the other hand, averaging methods and regression-based global tests consider self-contained null hypotheses, i.e., whether any of the CpGs within the pathway are significant (*see* **step 20** in Subheading 3.9 for more details).

1. *Mapping CpGs to Genes and Genes to Pathways:* Individual CpGs can be mapped to genes based on their location, which are annotated within the Illumina 450K and EPIC manifests. Genes can then be mapped to pathways or functional groupings using standard databases such as the Kyoto Encyclopedia of Genes and Genomes [36] (KEGG) and the gene ontology [37] (GO) consortium. KEGG includes well-curated pathways with an emphasis on metabolic pathways, which can be limiting, whereas GO contains broad range of pathways (functional

groupings) but with lower quality. Proprietary databases also exist but the specific pathways are not necessarily easily extracted and require use of pathway association analysis methods, which may not be valid.

2. *Simple Over-Representation Analysis:* Many popular pathway analysis methods test whether the pathway is "enriched" for CpGs that are related to the variable of interest, i.e., whether there is an over-representation of significant CpGs (here significance can be based on multiple comparisons corrections or based on looser thresholds) among the CpGs within the pathway. Association between pathway membership and CpG significance is assessed by constructing a contingency table and then applying standard methods such as fisher's exact test, Chi-square test, proportions tests, hypergeometric tests, or other related tests [38]. Although this is the most popular method, it does not generate valid *p*-values [39, 40] and we strongly emphasize that these *p*-values should be taken as descriptive, qualitative measures rather than *p*-values in the usual sense.
3. *Gene Set Enrichment Analysis (GSEA)*: Like the simple over-representation analysis methods, gene set enrichment analysis [41] (GSEA) approach looks for enrichment of significant CpGs among the CpGs in the pathway. However, instead of drawing a threshold for significance, GSEA first ranks each CpG by significance for the ranked CpG list and then computes a 1-sided Kolmogorov-Smirnov (KS) statistic to summarize whether the significant CpGs are closer to the top of the ranked CpG list. Permutation analysis is used to obtain a final *p*-value for significance. *P*-values from GSEA are valid, though permutations can pose problematic due to the presence of covariates (*see* **step 18** in Subheading 3.9).
4. *Averaging Methods:* Averaging methods collapse the CpGs in the pathway into a single number for each individual by taking a weighted average of the CpGs within the pathway. Since the pathway is now summarized by a single number, the collapsed value can be tested for association with the variable of interest. Weights for the collapsing can be estimates without regard to the variable of interest, such as taking the top principal component(s) [42], or estimated with regard to the outcome using supervised learning methods [43]. If the weights are selected with regard to the outcome, then permutation is again necessary to evaluate significance while standard regression-based methods can be used if weights are selected without regard to the outcome.
5. *Regression-Based Global Tests:* As suggested by the name, regression-based global tests regress the variable of interest on all of the CpGs within the pathway to test their cumulative

effect. Testing for the significance of the pathway then corresponds to the regression coefficients of the CpGs, which use recent developments in variance component tests. Several methods including the global test [44] and the kernel machine test [45–48] can be used to do this and obtain accurate p-values for the joint effect of the CpGs. Although regression-based global tests operate under the model in which the variable of interest is the outcome, the p-value is still valid even if methylation is actually dependent upon the variable of interest (same as in the single CpG association testing framework).

3.9 Important Points to Consider

1. *Avoiding Confounding by Batch*: To avoid confounding by batch, samples with similar characteristics in terms of their variables of interest should be spread across plates, arrays, and batches. For example, running all of the cases on the same plate while all of the controls are on a different plate induces perfect confounding such that potential plate effects cannot be distinguished from true differences between cases and controls. Efforts should be made to balance other variables, besides the main variable of interest, across plates as well. Mistakes in running samples in a non-random order have led to serious errors in the past [49–51].
2. *Inclusion of Technical Replicate Samples*: When possible, technical replicates should be included within the study. This includes replicates across plates and also replicates within the same plate. Inclusion of replicates allows for assessment and possible correction of possible plate effects, batch effects, and problematic plates. Replicates can also be used to identify the best preprocessing and normalization choices for the data set by the examination of correlation among replicates after applying different methods.
3. *Inclusion of Positive Controls*: Inclusion of positive controls, i.e., fully methylated samples, can identify some problematic probes. Probes with methylation beta values far from one on the control samples may not be performing well. In principle, negative controls could also be included, but identifying and generating such samples can be challenging.
4. *Filtering Thresholds*: For several steps in the quality control process, thresholds, such as those on the detection p-values, must be specified. The thresholds presented earlier are guidelines with some user flexibility. In particular, more stringent thresholds can also be used, e.g., calling probes with detection p-value greater than 10^{-5} as failed, but we do not recommend using more liberal thresholds than those indicated. More stringent thresholds should be used if sample sizes are larger as such studies can tolerate more data loss and results are interpreted more definitively.

5. *Filtering Based on Non-variable Probes*: Probes with low variation are frequently omitted from the analysis in other data types. For example, rules of thumb for gene expression data include removing the 40% of genes with lowest variability with the understanding that some genes are not expressed. This can also be done for methylation analysis, but the specific thresholds for determining what constitutes a probe with low variability remain unclear. The number of probes is unlikely to change the total number of probes by an order of magnitude (i.e., genome-wide significance levels are not dramatically impacted). Thus, we do not recommend such filtering at this time.
6. *Missing Data*: After filtering probes based on detection *p*-values, many probes may contain some missing values. Although imputation techniques can be used [52], we recommend keeping these values as missing and running a complete case analysis in which the samples with missing CpG value are omitted (when analyzing the particular CpG). The rationale is that if the number of missing values is modest then this will not impact power while larger proportions of missing values may yield unreliable results anyway.
7. *BMIQ Normalization in Practice*: Since aspects of the method are random, different runs can result in slightly different numbers so random seeds should be specified and saved to ensure reproducibility. Occasionally, the algorithm will fail to converge, but this can often be overcome by just specifying a different random seed. Missing data can also introduce problems since early versions of the code did not allow for any of the probe values to be missing, but since the data are analyzed on a sample-by-sample basis, the missing data for each sample can be omitted, the non-missing probes can be normalized, and the missing data can be re-introduced. Newer versions of the code no longer suffer this drawback.
8. *Alternative Normalization Approaches*: A wide range of normalization and automated preprocessing procedures beyond the BMIQ procedure have been devised. Some useful and popular alternatives include the Illumina normalization and preprocessing method implemented in Illumina's GenomeStudio software (Illumina Inc.), the subset-quantile within array normalization (SWAN) method [25], and the complete pipeline (CP) for preprocessing based on subset quantile normalization by Touleimat and Tost [27] as well as others.

   The Illumina normalization method is provided within GenomeStudio and within the minfi [21] Bioconductor package. The approach uses background subtraction to reduce variability between arrays by using negative control probes: probes that are thermodynamically equivalent to regular probes but that do not target specific parts of the transcriptome. Probes are rescaled such that the control probes have a common mean across samples. Methylated and unmethylated probes are normalized separately.

The SWAN method is similar in flavor to the BMIQ procedure and again reduces the bias and variability arising from the different probe types. Specifically, SWAN assumes the distribution of probe intensities should be the same between Type I and Type II probes (conditional on the number of CpGs within the probes) such that subset quantile normalization can be applied to reduce technical differences within each array. No between array normalization is done.

The CP approach is one of the earliest methods that offered the attractive option to comprehensively preprocess the data including background subtraction and color bias correction, and within and across sample quantile normalization. The software also conveniently allows extensive QC including filtering based on the detection p-values, number of beads, and other probe characteristics such as presence of SNPs. The normalization focuses on using a subset quantile normalization to normalize the Type I probes and the Type II probes separately, with probes first divided into categories probe annotations and intelligently selected anchors for computing the reference distributions of the Type I and Type II probes.

9. *Logit Transformation*: There is some debate within the field as to whether beta values or log-ratios (M-values) should be used. Arguments for using the logit transformation are that using log-ratios offers better statistical properties [34]: the log-ratio is no longer constrained within 0 and 1 and can be less skewed than the beta values and with reduced heteroskedasticity. Furthermore, traditional analysis of two-color gene expression array data also used the log-ratio. On the other hand, some favoring the use of beta values argue that log-ratios are artificial and because of the nonlinear transformation, are difficult to interpret. Also, despite the non-normality of the beta values, as long as the sample size is large enough, methods assuming normality such as the t-test are still valid. In practice, it has been observed that choosing to transform or not to transform does not seriously affect analysis results [53], though in the presence of covariates, analysis on the incorrect scale could lead to some bias.

10. *Batch Effects*: As with a wide range of other types of genomic data types, the Illumina BeadChip is sensitive to batch effects, i.e., running arrays or samples at different times, using different batches of reagents, or employing different technicians can result in systematic differences in the data. As noted, batch

effects should be avoided (particularly when confounded with the variable of interest), but in some cases, they may not be preventable. When the batch effects are known, a popular strategy for correcting the batch effects is to apply ComBat [54], which is implemented within the sva package in R [55]. ComBat requires the user to input in the methylation values for all subjects, the batch variable, the variable of interest, as well as additional covariates and returns the batch corrected methylation values. Since ComBat was designed for gene expression data (which is not constrained between 0 and 1), it should be run on the logit transformed data, though the transformation can be reversed prior to analysis if desired. In principle, ComBat should be run each time a different variable of interest is specified and when covariates change, but if the variables of interest and the covariates are balanced across batches, then ComBat can be used to generate a single batch corrected data set. ComBat offers an option for parametric adjustment or nonparametric adjustment. Although nonparametric adjustment may be preferable, the increased computational expense makes parametric adjustment more common. An alternative to ComBat is to include indicator variables within the analysis models for batch; if the sample size is extremely large, then this should be approximately similar to using ComBat. If batch effects are more cryptic in nature, then alternative approaches such as surrogate variable analysis (*see* **step 17**) should be considered.

11. *Sex Chromosomes*: Following common practice in genetic association studies, we recommend omitting sex chromosomes from the analysis when the methylation data is collected on cohorts including subjects with different sexes. If the subjects share the same gender, then the analyses can proceed as earlier.

12. *Complex Variables of Interest*: Broadly speaking, the specific testing approach used will depend strongly on the variable of interest, which may be complex (i.e., not a simple continuous variable or dichotomous variable). For example, if the focus is on understanding the role of methylation in a censored survival problem, then the cox proportional hazards model could be used to regress survival time on the methylation values and additional covariates. In general, this is straightforward; however, we note that many statistical methods rely on obtaining asymptotic null distributions in order to obtain *p*-values. These distributions are often valid when the sample sizes are very large and at relatively large α-levels (e.g., 0.05 level), but may not be accurate at low α-levels used in testing large-scale genetic data—results may be conservative or liberal. At lower α-levels, the sample size required for tests to correctly control type I error may be much larger and careful consideration

should be made in evaluating the appropriateness of the modeling strategies at stringent significance thresholds.

13. *Outliers and Heteroskedasticity*: If methylation is used as the outcome, then some control for heteroskedasticity and outliers may be necessary. Outliers can strongly influence outcomes and could result from data quality issues or even (unknown) rare genetic variants at or near the CpG. Similarly, while logit transformations have been suggested as an adequate means for reducing heteroskedasticity, it is difficult to assess how successful this is on a genome-wide scale. Thus, if sample size is large, we recommend the use of robust regression [56] combined with use of sandwich estimators of covariates in order to obtain significance. The rlm function within the MASS R package [57] can be used to compute the robust linear model and the coeftest function (lmtest R package [58]) with vcov parameter set equal to vcovHC (from the sandwich R package [59, 60]) can be used to obtain a p-value using sandwich estimators of the covariance. This should not be used if the sample size is modest. Applying normal quantile transformations [61] (also called inverse normal transformations) may be an alternative strategy for reducing the effect of outliers and heteroskedasticity.
14. *FDR Levels*: In contrast to the, largely, agreed upon $\alpha = 0.05$ significance threshold for controlling family-wise error rate, there is no consensus regarding the significance level for controlling FDR. Thresholds as high as 0.20 have been suggested as appropriate for identifying suggestive associations [62] while more stringent levels can be selected to identify more confident results. Although the threshold for significance should ideally be selected a priori, in practice, the FDR significance level is sometimes selected post hoc to determine a reasonable number of significant results.
15. *Controlling Cell Type Composition*: Since methylation is tissue and cell type specific, a major concern within the field of methylation analysis pertains to controlling for potential confounding from differential cell type composition across samples. In particular, most samples profiled are composed of a mixture of cell or tissue types. Differences in cell type proportions can result in possibly confounded results. Currently, a popular strategy for correcting differences in cell type involves using a reference data set with sorted cell types in order to estimate the cell type proportions [63], which can then be included as covariates for within regression models. However, this approach requires availability of appropriately sorted reference samples such as the data set of Reinius et al. [64] which profiles sorted blood samples from adult Swedish men which is commonly used despite questionable appropriateness, e.g., in

studies with infant cord blood. Difficulties in identifying appropriate reference sets have led to reference-free approaches [65, 66] and new methods are under development. Surrogate variable analysis (**step 17**) may also be an appropriate strategy for controlling potential confounding from cell type. Currently, it remains unclear which approaches are the best and indeed, there is some question as to whether cell type adjustment is even necessary since cell type may sit along the causal pathway, i.e., adjusting for cell type adjusts out the signal that one is searching for. There is also some concern in the scenario where methylation is the outcome: since cell type composition is estimated based on methylation, then one is regressing methylation on methylation inflated type I error, but given that the cell type estimates are based on a large number of CpGs, this is unlikely to be a major concern. Nonetheless, given the open questions concerning cell type adjustments, a possible strategy is to conduct analyses both without adjustment and then with adjustment as a sensitivity analysis.

16. *Population stratification*: Ancestry and population stratification can be an important confounder that needs to be corrected for in some epigenetic studies. Within the context of genetic association studies, standard approaches for controlling population confounding include computing the top principle components of genetic variability [67, 68]. Recently, recognizing the potential need for correcting for genetic ancestry, methods have been developed for estimating principle components of genetic variability within the context of methylation [69]. These then serve as new variables, which can be adjusted for along with other confounders in the association testing phase.
17. *Un-modeled Confounders and Surrogate Variable Analysis*: Methylation is known to be sensitive to a wide range of factors including technical issues such as batch effects as well as other variables such as age, gender, and environmental exposures. Within the context of association analysis, adjustments for batch (e.g., via ComBat) and for confounders through inclusion of covariates are imperative. In practice, however, it is often difficult to fully identify the confounders or even cryptic batch effects, yet omission of key confounders can lead to inflated type I error rate or decreased power. Surrogate variable analysis (SVA), first developed for gene expression data, is a powerful and popular tool for controlling for un-modeled confounders [55, 70–72]. Essentially, SVA considers methylation at each CpG to be the outcome variable, regresses out the effects of the confounders as well as the variable of interest, and then looks for latent structure within the residual variability by constructing principal components based on the matrix of

residuals. Principal components are selected as surrogate variables representing un-modeled confounders and then can be included in the regression model as covariates. SVA can, in principle, correct for any sources of un-modeled variability and confounding, it may also be useful for controlling batch effects, cell type heterogeneity, and population stratification. Despite being widely used, there are a number of criticisms of SVA. First, interpretation of the surrogate variables is difficult in that it is unclear what the surrogate variables represent. More practically, SVA assumes that methylation is the outcome and it remains unclear how to use SVA when the methylation is treated as a predictor. When one wishes to use the variable of interest as the outcome, it may still be possible to estimate surrogate variables with methylation as the outcome and then include the surrogate variables in the models with methylation as the predictor, though the properties of this analysis have not been well studied. As with the cell type correction methods, there is a possibility that SVA can lead to deflated estimates of standard errors and there is also the possibility that in the case where no confounding is present, SVA can introduce confounding. Finally, since the surrogate variables are not easily understood, then as with cell type adjustments, it may inadvertently adjust out the desired signal in scenarios where the surrogate variables lie in the causal pathway.

18. *Permutation testing methods*: Permutation analysis is a popular strategy for testing due to the lack of assumptions concerning normality and is also commonly used when the null distribution cannot be analytically determined such as in some pathway tests like GSEA. The idea behind permutation is that under the null, the data are exchangeable [73], i.e., interchangeable in a loose sense. Thus, to compute a p-value one first computes a statistic measuring association using the original data set. Then to simulate the null distribution, one permutes the outcome variable, re-computes a statistic measuring association using the permuted data, and repeats this for a large number of times to generate and empirical null distribution. The permutation p-value is given as the proportion of statistics equal to or more extreme than the original statistic. Despite being popular, there are some difficulties related to permutation testing. In particular, the computational expense can be very high. To estimate significance at lower α-levels, a large number of permutations are necessary to obtain accurate p-values. Further, permutation methods need to be used with extreme caution in the presence of covariates. In particular, if methylation is the outcome variable, then independence between the covariates and the variable of interest is necessary in order for permutation to be valid. Similarly, if the variable of interest is the

outcome, then methylation and the covariates need to be independent. Given these challenges, we recommend carefully examining the data and considering alternatives before using permutation testing

19. *Gene Level Analysis*: Since genes contain many CpGs, one can also conduct the analysis at the gene level. In principle, all of the pathway analysis methods can be used to assess the significance of the CpGs within a gene, but in practice, we recommend the use of averaging methods or the regression-based global tests, i.e., the self-contained testing methods. We further note for pathway analysis using over-representation analysis or GSEA, the analysis can be first run at the gene level using a self-contained test and then feeding these into the pathway test to determine if there is enrichment of significant genes within the pathway.

20. *Self-contained* vs. *Competitive Null Hypotheses*: Analyses using self-contained or competitive methods can be statistically valid, but the interpretation of the results differs. In particular, self-contained tests assess whether any of the CpGs within the pathway are related to the outcome whereas competitive tests compare the relative significance of the CpGs within the pathway to the significance of the CpGs outside of the pathway. In theory, self-contained tests should have higher power and are more natural generalizations of single CpG tests (applying a self-contained test to a pathway with only one CpG is usually equivalent to just analyzing the single CpG). Further, when consider multiple pathways, competitive tests cannot be easily used in conjunction with FDR control since competitive tests produce *p*-values that are negatively correlated and FDR is not guaranteed under negative correlation. However, self-contained tests could be called significant if even a single CpG in the pathway is truly related to the outcome; this can be difficult to interpret as a true pathway effect suggesting that competitive tests may be more useful for discerning true pathway level effects.

References

1. Razin A (1998) CpG methylation, chromatin structure and gene silencing-a three-way connection. EMBO J 17:4905–4908
2. Razin A, Riggs AD (1980) DNA methylation and gene function. Science 210:604–610
3. Ramsahoye BH, Biniszkiewicz D, Lyko F et al (2000) Non-CpG methylation is prevalent in embryonic stem cells and may be mediated by DNA methyltransferase 3a. Proc Natl Acad Sci U S A 97:5237–5242
4. Bibikova M, Barnes B, Tsan C et al (2011) High density DNA methylation array with single CpG site resolution. Genomics 98:288–295
5. Sandoval J, Heyn H, Moran S et al (2011) Validation of a DNA methylation microarray for 450,000 CpG sites in the human genome. Epigenetics 6:692–702
6. Joubert BR, Haberg SE, Bell DA et al (2014) Maternal smoking and DNA methylation in newborns: in utero effect or epigenetic

inheritance? Cancer Epidemiol Biomark Prev 23:1007–1017

7. Joubert BR, Haberg SE, Nilsen RM et al (2012) 450K epigenome-wide scan identifies differential DNA methylation in newborns related to maternal smoking during pregnancy. Environ Health Perspect 120:1425–1431
8. Joubert BR, Felix JF, Yousefi P et al (2016) DNA methylation in newborns and maternal smoking in pregnancy: genome-wide consortium meta-analysis. Am J Hum Genet 98:680–696
9. Heyn H, Carmona FJ, Gomez A et al (2013) DNA methylation profiling in breast cancer discordant identical twins identifies DOK7 as novel epigenetic biomarker. Carcinogenesis 34:102–108
10. Shen J, Wang S, Zhang YJ et al (2013) Exploring genome-wide DNA methylation profiles altered in hepatocellular carcinoma using Infinium HumanMethylation 450 BeadChips. Epigenetics 8:34–43
11. Heyn H, Li N, Ferreira HJ et al (2012) Distinct DNA methylomes of newborns and centenarians. Proc Natl Acad Sci U S A 109:10522–10527
12. Liu Y, Aryee MJ, Padyukov L et al (2013) Epigenome-wide association data implicate DNA methylation as an intermediary of genetic risk in rheumatoid arthritis. Nat Biotechnol 31:142–147
13. Engel SM, Joubert BR, Wu MC et al (2014) Neonatal genome-wide methylation patterns in relation to birth weight in the Norwegian mother and child cohort. Am J Epidemiol 179:834–842
14. Pidsley R, Zotenko E, Peters TJ et al (2016) Critical evaluation of the Illumina MethylationEPIC BeadChip microarray for whole-genome DNA methylation profiling. Genome Biol 17:208
15. GenomeStudio® Methylation Module v1.8 User Guide (2011) Illumina Inc., San Diego, CA, USA
16. Smith ML, Baggerly KA, Bengtsson H et al (2013) Illuminaio: an open source IDAT parsing tool for Illumina microarrays. F1000Res 2:264
17. R Development Core Team (2010) R: a language and environment for statistical computing. R Foundation for Statistical Computing, Vienna, Austria
18. Gentleman RC, Carey VJ, Bates DM et al (2004) Bioconductor: open software development for computational biology and bioinformatics. Genome Biol 5:R80
19. Hansen KD (2016) IlluminaHumanMethylation450kanno.ilmn12.hg19: annotation for Illumina's 450k methylation arrays. R package version 0.6.0
20. Hansen KD (2016) IlluminaHumanMethylationEPICanno.ilm10b2.hg19: annotation for Illumina's EPIC methylation arrays. R package version 0.6.0
21. Aryee MJ, Jaffe AE, Corrada-Bravo H et al (2014) Minfi: a flexible and comprehensive bioconductor package for the analysis of Infinium DNA methylation microarrays. Bioinformatics 30:1363–1369
22. Chen YA, Lemire M, Choufani S et al (2013) Discovery of cross-reactive probes and polymorphic CpGs in the Illumina Infinium HumanMethylation450 microarray. Epigenetics 8:203–209
23. Siva N (2008) 1000 Genomes project. Nat Biotechnol 26:256
24. Zhou W, Laird PW, Shen H (2016) Comprehensive characterization, annotation and innovative use of Infinium DNA methylation BeadChip probes. Nucleic Acids Res 45(4):e22
25. Maksimovic J, Gordon L, Oshlack A (2012) SWAN: subset-quantile within array normalization for illumina infinium HumanMethylation450 BeadChips. Genome Biol 13:R44
26. Teschendorff AE, Marabita F, Lechner M et al (2013) A beta-mixture quantile normalization method for correcting probe design bias in Illumina Infinium 450 k DNA methylation data. Bioinformatics 29:189–196
27. Touleimat N, Tost J (2012) Complete pipeline for Infinium((R)) human methylation 450K BeadChip data processing using subset quantile normalization for accurate DNA methylation estimation. Epigenomics 4:325–341
28. Triche TJ Jr, Weisenberger DJ, Van Den Berg D et al (2013) Low-level processing of illumina infinium DNA methylation BeadArrays. Nucleic Acids Res 41:e90
29. Marabita F, Almgren M, Lindholm ME et al (2013) An evaluation of analysis pipelines for DNA methylation profiling using the Illumina HumanMethylation450 BeadChip platform. Epigenetics 8:333–346
30. Wu MC, Joubert BR, Kuan PF et al (2014) A systematic assessment of normalization approaches for the infinium 450K methylation platform. Epigenetics 9:318–329
31. Dedeurwaerder S, Defrance M, Calonne E et al (2011) Evaluation of the Infinium methylation 450K technology. Epigenomics 3:771–784
32. Pidsley R, Y Wong CC, Volta M et al (2013) A data-driven approach to preprocessing illumina

450K methylation array data. BMC Genomics 14:293

33. Morris TJ, Butcher LM, Feber A et al (2014) ChAMP: 450k Chip analysis methylation pipeline. Bioinformatics 30:428–430
34. Du P, Zhang X, Huang CC et al (2010) Comparison of Beta-value and M-value methods for quantifying methylation levels by microarray analysis. BMC Bioinformatics 11:587
35. Benjamini Y, Hochberg Y (1995) Controlling the false discovery rate: a practical and powerful approach to multiple testing. J R Stat Soc Series B Methodol 57:289–300
36. Kanehisa M, Goto S (2000) KEGG: kyoto encyclopedia of genes and genomes. Nucleic Acids Res 28:27–30
37. Ashburner M, Ball CA, Blake JA et al (2000) Gene ontology: tool for the unification of biology. The gene ontology consortium. Nat Genet 25:25–29
38. Khatri P, Draghici S (2005) Ontological analysis of gene expression data: current tools, limitations, and open problems. Bioinformatics 21:3587–3595
39. Goeman JJ, Buhlmann P (2007) Analyzing gene expression data in terms of gene sets: methodological issues. Bioinformatics 23:980–987
40. Wu MC, Lin X (2009) Prior biological knowledge-based approaches for the analysis of genome-wide expression profiles using gene sets and pathways. Stat Methods Med Res 18:577–593
41. Subramanian A, Tamayo P, Mootha VK et al (2005) Gene set enrichment analysis: a knowledge-based approach for interpreting genome-wide expression profiles. Proc Natl Acad Sci U S A 102:15545–15550
42. Tomfohr J, Lu J, Kepler TB (2005) Pathway level analysis of gene expression using singular value decomposition. BMC Bioinformatics 6:225
43. Wu MC, Zhang L, Wang Z et al (2009) Sparse linear discriminant analysis for simultaneous testing for the significance of a gene set/pathway and gene selection. Bioinformatics 25:1145–1151
44. Goeman JJ, van de Geer SA, de Kort F et al (2004) A global test for groups of genes: testing association with a clinical outcome. Bioinformatics 20:93–99
45. Kwee LC, Liu D, Lin X et al (2008) A powerful and flexible multilocus association test for quantitative traits. Am J Hum Genet 82:386–397
46. Liu D, Ghosh D, Lin X (2008) Estimation and testing for the effect of a genetic pathway on a disease outcome using logistic kernel machine regression via logistic mixed models. BMC Bioinformatics 9:292
47. Liu D, Lin X, Ghosh D (2007) Semiparametric regression of multidimensional genetic pathway data: least-squares kernel machines and linear mixed models. Biometrics 63:1079–1088
48. Wu MC, Kraft P, Epstein MP et al (2010) Powerful SNP-set analysis for case-control genome-wide association studies. Am J Hum Genet 86:929–942
49. Baggerly KA, Morris JS, Edmonson SR et al (2005) Signal in noise: evaluating reported reproducibility of serum proteomic tests for ovarian cancer. J Natl Cancer Inst 97:307–309
50. Petricoin EF, Ardekani AM, Hitt BA et al (2002) Use of proteomic patterns in serum to identify ovarian cancer. Lancet 359:572–577
51. Ransohoff DF (2005) Lessons from controversy: ovarian cancer screening and serum proteomics. J Natl Cancer Inst 97:315–319
52. Troyanskaya O, Cantor M, Sherlock G et al (2001) Missing value estimation methods for DNA microarrays. Bioinformatics 17:520–525
53. Bell JT, Pai AA, Pickrell JK et al (2011) DNA methylation patterns associate with genetic and gene expression variation in HapMap cell lines. Genome Biol 12:R10
54. Johnson WE, Li C, Rabinovic A (2007) Adjusting batch effects in microarray expression data using empirical Bayes methods. Biostatistics 8:118–127
55. Leek JT, Johnson WE, Parker HS et al (2012) The sva package for removing batch effects and other unwanted variation in high-throughput experiments. Bioinformatics 28:882–883
56. Huber PJ (1973) Robust regression: asymptotics, conjectures and Monte Carlo. Ann Stat 1:799–821
57. Venables WN, Ripley BD (2013) Modern applied statistics with S-PLUS. Springer, Berlin
58. Zeileis A, Hothorn T (2002) Diagnostic checking in regression relationships. R News 2: 7–10.
59. Zeileis A (2004) Econometric computing with HC and HAC covariance matrix estimators. Journal of Statistical Software 11:1–17.
60. Zeileis A (2006) Object-oriented computation of sandwich estimators. Journal of Statistical Software 16:1–16.
61. Peng B, RK Y, Dehoff KL et al (2007) Normalizing a large number of quantitative traits using empirical normal quantile transformation. BMC Proc 1(Suppl 1):S156

62. Efron B (2007) Size, power and false discovery rates. Ann Stat 35:1351–1377
63. Houseman EA, Accomando WP, Koestler DC et al (2012) DNA methylation arrays as surrogate measures of cell mixture distribution. BMC Bioinformatics 13:86
64. Reinius LE, Acevedo N, Joerink M et al (2012) Differential DNA methylation in purified human blood cells: implications for cell lineage and studies on disease susceptibility. PLoS One 7:e41361
65. Houseman EA, Molitor J, Marsit CJ (2014) Reference-free cell mixture adjustments in analysis of DNA methylation data. Bioinformatics 30:1431–1439
66. Zou J, Lippert C, Heckerman D et al (2014) Epigenome-wide association studies without the need for cell-type composition. Nat Methods 11:309–311
67. Patterson N, Price AL, Reich D (2006) Population structure and eigenanalysis. PLoS Genet 2:e190
68. Price AL, Patterson NJ, Plenge RM et al (2006) Principal components analysis corrects for stratification in genome-wide association studies. Nat Genet 38:904–909
69. Barfield RT, Almli LM, Kilaru V et al (2014) Accounting for population stratification in DNA methylation studies. Genet Epidemiol 38:231–241
70. Leek JT, Scharpf RB, Bravo HC et al (2010) Tackling the widespread and critical impact of batch effects in high-throughput data. Nat Rev Genet 11:733–739
71. Leek JT, Storey JD (2007) Capturing heterogeneity in gene expression studies by surrogate variable analysis. PLoS Genet 3:1724–1735
72. Leek JT, Storey JD (2008) A general framework for multiple testing dependence. Proc Natl Acad Sci U S A 105:18718–18723
73. Good P (2010) Exchangeable random variables. Bioinformatics 26:2214. author reply 2215
74. Hannum G, Guinney J, Zhao L et al (2013) Genome-wide methylation profiles reveal quantitative views of human aging rates. Mol Cell 49:359–367

Part IV

Analysis of Highly Multiplexed Target Regions

Chapter 17

Microdroplet PCR for Highly Multiplexed Targeted Bisulfite Sequencing

H. Kiyomi Komori, Sarah A. LaMere, Traver Hart, Steven R. Head, Ali Torkamani, and Daniel R. Salomon*

Abstract

Many methods exist for examining CpG DNA methylation. However, many of these are qualitative, laborious to apply to a large number of genes simultaneously, or are not easy to target to specific regions of interest. Microdroplet PCR-based bisulfite sequencing allows for quantitative single base resolution analysis of investigator selected regions of interest. Following bisulfite conversion of genomic DNA, targeted microdroplet PCR is conducted with custom primer libraries. Samples are then fragmented, concatenated, and sequenced by high-throughput sequencing. The most recent technology allows for this method to be conducted with as little as 250 ng of bisulfite-converted DNA. The primary advantage of this method is the ability to hand-select the targeted regions covered by up to 10,000 amplicons of 500–600 bp. Moreover, the nature of microdroplet PCR virtually eliminates PCR bias and allows for the amplification of all targets simultaneously in a single tube.

Key words DNA methylation, Epigenetics, CpG methylation, Microdroplet PCR

1 Introduction

Epigenetics has become a fast-growing field of research aimed at enhancing our understanding of the interaction between the genetic code and the environment. Multiple epigenetic mechanisms have been defined in the literature, although in most cases their relationship with gene expression is still generally poorly understood. One of the best-studied epigenetic modifications is DNA methylation, which involves a covalent modification made directly to DNA. This modification is generally deposited onto CG dinucleotides (i.e., CpG residues) and in most cases is symmetrical between DNA strands. Particularly in promoter regions, the presence of DNA methylation is thought to represent a type of

*Deceased

Jörg Tost (ed.), *DNA Methylation Protocols*, Methods in Molecular Biology, vol. 1708,
https://doi.org/10.1007/978-1-4939-7481-8_17,

molecular memory governing repression of gene expression, although more recent evidence has also demonstrated this phenomenon in enhancer regions and gene bodies related to the regulation of alternative splicing in addition to gene expression [1, 2]. Higher density CpG promoters, termed CpG islands, are most commonly unmethylated raising interesting questions about why this feature of the genome has been conserved. On the other hand, differential CpG methylation is typically found in promoters without CpG islands and thus, the importance of CpG methylation in regulating gene expression for these sites is clear [3].

Previous work has demonstrated CpG methylation to be a critical factor in temporal gene expression during embryonic development [4, 5]. Its importance has also been demonstrated in dysregulation of proto-oncogenes and tumor suppressors in cancer studies [6, 7]. Additionally, aberrant DNA methylation has been linked to several chronic diseases, including autoimmune, inflammatory, and metabolic disorders [8, 9].

Numerous methods have been designed to study CpG methylation, although few are able to accomplish large-scale analysis of CpG methylation at single base resolution. Bisulfite conversion has historically been the “gold standard” for studying CpG methylation. This method entails a chemical modification that converts unmethylated cytosines to uracils [10]. Methylated vs. unmethylated cytosines can then be delineated based upon their conversion evaluated by sequencing.

The advent of high-throughput sequencing technologies opened the door for the study of bisulfite conversion of whole genome samples (*see also* Chapters 5–9). However, while costs are rapidly decreasing, whole genome bisulfite sequencing is still cost prohibitive and also presents unique hurdles for bioinformatic analysis. Thus, several methods of large scale DNA methylation analysis have been developed in order to avoid these issues, each with its own distinct set of advantages and drawbacks. One example is MeDIP (*see also* Chapter 12), where an anti-methylcytosine antibody is used to immunoprecipitate methylated CpG residues, followed by high-throughput sequencing. While this method is useful for evaluating CpG methylation on a genome-wide basis, it does not achieve single base methylation resolution. Another such method is MBD-seq (*see also* Chapter 10), which uses methyl-binding protein domains to precipitate methylated DNA. This method achieves similar results to MeDIP, although it does not require DNA denaturation, thus allowing evaluation of symmetry more readily. Additionally, it is reported to be more sensitive than MeDIP, allowing precipitation of regions with lower CpG densities. Finally, methylation-sensitive restriction enzymes (MREs) have frequently been used in combination with next-generation sequencing in order to achieve high-throughput evaluation of DNA methylation at single CpG resolution [11] (*see also* Chapter 12).

While the above methods each have advantages, in many cases the ideal approach to large-scale study of DNA methylation would be more targeted. Strategies for this approach have included the use of PCR, bead arrays (*see also* Chapter 16), or microarrays to target regions of interest. One hurdle with the array-based methods is the inability to examine multiple closely apposed CpG residues within a region. PCR targeting of bisulfite-converted DNA vanquishes this problem, but historically PCR of multiple regions was a time-consuming endeavor [11]. Previous attempts to address these issues involved combinations of array-based capture of bisulfite-converted DNA followed by next-generation sequencing [12]. While this method accomplishes large-scale targeted interrogation of CpG methylation, it also requires large starting amounts of input DNA, making it a less viable option, particularly for anyone wishing to study clinical samples. A modified padlock probe capture approach overcame the problem with starting material, but it also required the synthesis of numerous 150 bp probes, making it a less cost-effective and more time-consuming methodology [13] (*see also* Chapter 19). Finally, previous use of multiplex PCR systems to study low starting amounts of bisulfite-converted DNA had required multiple rounds of PCR amplification, making samples subject to amplification bias and requiring days for processing [14].

We sought to develop a method that would allow high-throughput, targeted bisulfite sequencing. Our method utilizes the microdroplet PCR amplification system from RainDance Technologies [15], which allows for the amplification of 1.5 million parallel microdroplet amplifications in a single PCR reaction within an hour. After combining this method with bisulfite conversion, we demonstrated the efficacy of this method for 50 genes [16] and subsequently expanded this approach to successfully interrogate 57,706 CpGs across 1946 genes [17].

2 Materials

2.1 Primer Library Design

1. Primer design algorithm [16] (available upon request from the corresponding author).
2. Primer3 [18] (http://biotools.umassmed.edu/bioapps/primer3_www.cgi).
3. e-PCR [19] (http://www.ncbi.nlm.nih.gov/tools/epcr/).

2.2 Sample Preparation and Bisulfite Treatment

1. Genomic DNA purification kit.
2. Bisulfite conversion kit.
3. AMPure XP beads (Beckman Coulter Genomics) and magnet.
4. Fresh 70% EtOH.
5. 10 mM Tris–HCl, pH 8.0.
6. Quant-iT ssDNA Assay Kit (Thermo Fisher Scientific).
7. Qubit fluorometer (Thermo Fisher Scientific).

2.3 Preparation of Microdroplet PCR

1. Platinum high-fidelity Taq polymerase and 10× buffer (Thermo Fisher Scientific).
2. dNTPs.
3. Betaine.
4. DMSO.
5. RDT Droplet Stabilizer (RainDance Technologies).
6. Nuclease-free water.
7. RDT1000 (RainDance Technologies).
8. Primer library custom manufactured and supplied by contracting with RainDance Technologies.

2.4 PCR and Library Purification

1. Thermocycler with a heated lid.
2. RDT1000 Droplet Destabilizer (RainDance Technologies).
3. Vortex.
4. Microfuge.
5. Gel loading tips.
6. MinElute kit (Qiagen) or equivalent PCR cleanup kit for small amounts of DNA.
7. Bioanalyzer and DNA assay kit (Agilent).

2.5 Sample Preparation for Sequencing

1. Covaris S2.
2. High sensitivity DNA assay kit (Agilent).
3. Quant-iT dsDNA HS assay kit (Thermo Fisher Scientific).

2.5.1 Fragmentation

2.5.2 Concatenation

1. T4 polynucleotide kinase.
2. Klenow fragment.
3. T4 DNA polymerase.
4. dNTPs.
5. T4 DNA ligase and buffer.
6. DNA clean & concentrator-5 columns (Zymo Research).
7. Rapid DNA ligase and buffer (Enzymatics).

2.5.3 Shearing

1. Covaris S2.
2. High sensitivity DNA Assay chip (Agilent).

2.6 Library Prep

1. NEB Next Ultra DNA Library Prep Kit for Illumina (New England Biolabs).
2. 100% Ethanol.
3. Nuclease-free water.

4. 20× TE: 200 mM Tris–HCl, 20 mM EDTA, pH 8.0.
5. LoBind tubes (Eppendorf).
6. 16-position Magnetic Stand (Thermo Fisher Scientific).
7. NEB Next Multiplex Oligos for Illumina (index primer set 1).
8. Agencourt Ampure XP beads (Beckman Coulter Inc).
9. Freshly prepared 80% ethanol.
10. E-Gel® 2% Agarose Gel (Thermo Fisher Scientific).
11. E-Gel® iBase™ and E-Gel® Safe Imager™ Combo Kit (Thermo Fisher Scientific).
12. Track-It 100 base pair ladder (Thermo Fisher Scientific).
13. Zymoclean™ Gel DNA Recovery Kits.
14. Quant-iT dsDNA Assay Kit (Thermo Fisher Scientific).
15. Qubit fluorometer (Thermo Fisher Scientific).
16. Bioanalyzer and DNA assay kit (Agilent).

2.7 Data Analysis

1. CASAVA (Illumina).
2. Novoalign software (Novocraft).

3 Methods

The workflow for microdroplet PCR-based targeted bisulfite sequencing consists of eight steps (Fig. 1).

3.1 Assay Design

3.1.1 Selection of Genes

Careful selection of targeted genes is a critical step in the development of microdroplet based targeted bisulfite sequencing. In its current capacity, the RainDance platform can accommodate up to 10,000 amplicons in a single library, with the average library size being ~4000 amplicons and the upper limit on amplicon length being 500–600 bp. Thus, in determining how many genes to target, the size of the targeted regions must be taken into consideration. For example, targeting 1 kb regions would minimally require two primer pairs per region and some consideration for variations in primer efficacy. If constructing an average size library, ~4000 amplicons corresponds to 2000 regions. If regions are targeted that are difficult to sequence, have high homology to other sequences, or are A/T or G/C rich, then a larger number of primers are required. This issue is further amplified when designing primers for bisulfite-converted DNA as bisulfite conversion reduces genomic complexity and yields ssDNA, demanding that primer sets be designed to cover both the forward and reverse strands individually. Nonetheless, our experience using the primer design tool we developed for this purpose was that the majority of our targeted regions were successfully sequenced.

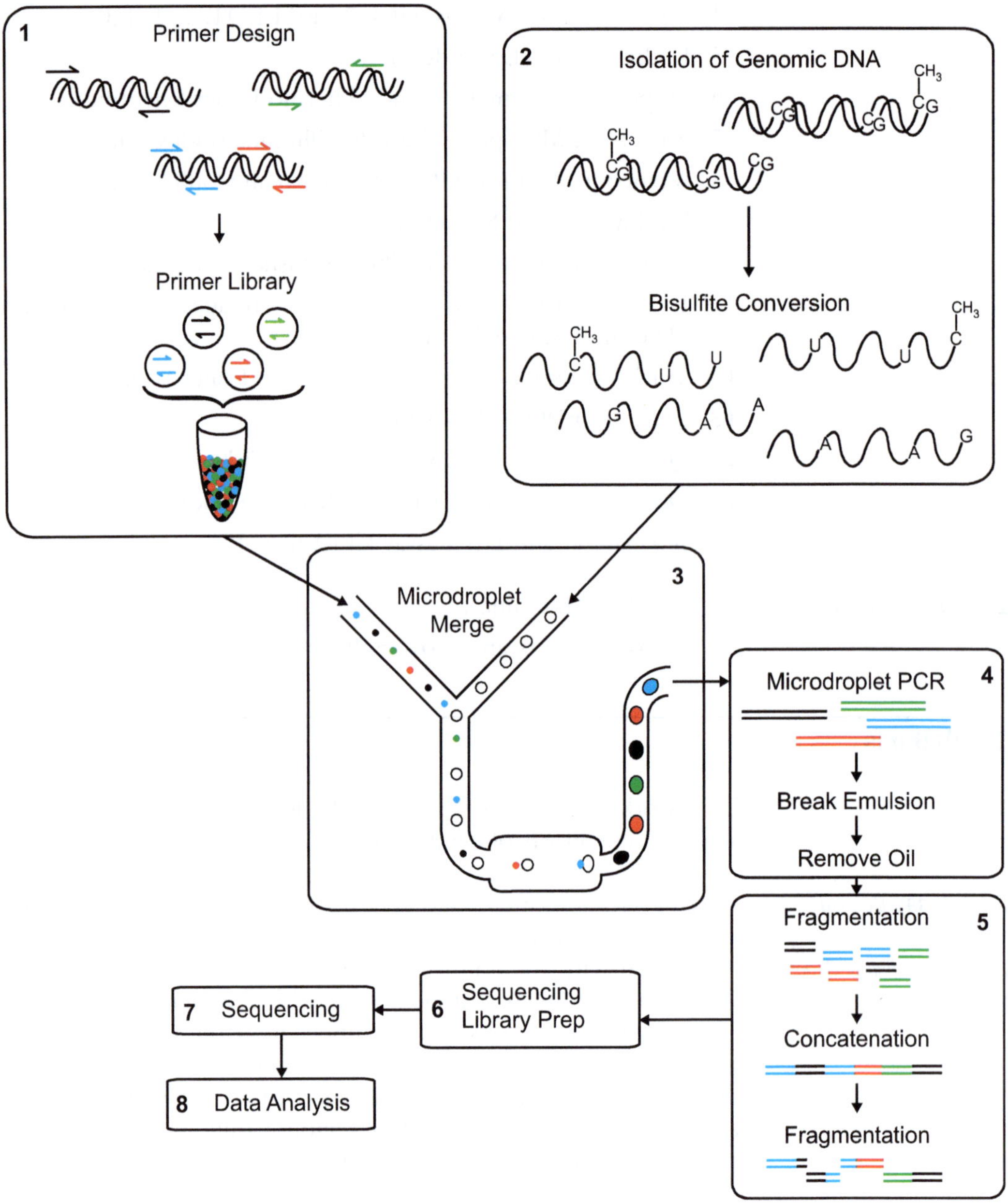

Fig. 1 Microdroplet PCR bisulfite sequencing workflow

3.1.2 Primer Design

A primer design algorithm is utilized in order to minimize the number of CpG containing primers used to cover the region of interest while maintaining ideal design parameters including amplicon length, primer and amplicon Tm, minimal off-target amplifications, and minimal primer-primer interactions. The steps are as follows:

1. *Select the regions of interest*: Extract the targeted region of interest (ROI), plus 450 bp upstream and downstream of the ROI from an in silico bisulfite-converted and SNP-masked genome. In silico bisulfite conversion transforms all non-CpG cytosines to thymine and all CpG cytosines into unknown nucleotides (N). This operation is performed on both the positive and negative strand of the ROI.
2. *Design the initial primer set*: Primer3 [18] is utilized to produce as many primer pairs as possible from the ROI—using the settings in Table 1. Penalty scores from Primer3 are retained for the primer tiling step.

Table 1
Primer3 settings

dffigCondition	Value
Minimum primer length	15 bp
Maximum primer length	36 bp
Optimal primer length	26 bp
Minimum primer tm	56 °C
Maximum primer tm	60 °C
Optimal primer tm	58 °C
Thermodynamic parameters and salt correction	SantaLucia
Monovalent salt concentration	50 mM
Divalent salt concentration	2 mM
DNTP concentration	0.2 mM
Maximum self complementarity	46 °C
Maximum primer pair complementarity	46 °C
Maximum self end complementarity	46 °C
Maximum hairpin stability	46 °C
Maximum primer GC content	65%
Minimum primer GC content	35%
Minimum primer 3′ distance	3 bp
Primer task	Pick detection primers
Primer maximum ns accepted	0
Amplicon size range	200–450 bp
Optimum amplicon size	450 bp

3. *Primer filtration*: Designed primers are filtered for off-target amplification using NCBI reverse e-PCR [19] by scanning them against an in silico bisulfite-converted genome—allowing two mismatches during primer scanning. Any primer pairs mapping within 50–2000 bp of one another at off-target sites are removed from consideration.
4. *Amplicon tiling*: The goal of the amplicon tiling step is to cover the ROI with the least number of primers while selecting for the primers with the best chance of success. Each base in the target region, excluding the flanking nucleotides added to the ROI for primer design, is considered covered if at least one amplicon, from either strand, overlaps the base. Note, a modification to this procedure would require amplicons covering both strands (*see* **Note 1**). The steps are:
   (a) *Initialization*: Amplicon tiling proceeds by first initializing a large selection of potential tiling paths by starting with all possible pairs of amplicons.
   (b) *Filtration*: Potential tiling paths are filtered in order to retain only those paths achieving the highest coverage of the ROI. Note: the primers themselves from low coverage paths are not discarded, just the paths. In the first filtration round, only paths overlapping the flanking nucleotides added for primer design are discarded. In the case that a large number of paths achieve the same coverage, a threshold number of paths are retained by ranking each path based on the average penalty score calculated for each primer.
   (c) *Extension*: The remaining potential tiling paths are then extended by one amplicon by generating all possible combinations of the remaining paths with all primers not already contained within each path.
   (d) *Iterate*: The Filtration and Extension steps are repeated until maximum coverage does not increase or until a path hits 100% coverage of the ROI. By extending all paths in parallel, the shortest possible path attaining 100% coverage is selected. If multiple shortest paths achieving 100% coverage are produced, the path containing primers with the lowest average penalty is selected. If no path achieves 100% coverage, continue to **step 5**.
5. *Iterate primer design and tiling*: If 100% coverage of the ROI is not achieved in the first round of primer design, filtration, and tiling, additional rounds are executed by loosening restrictions on the primer design step. Additional rounds of primer design proceed by iteratively increasing the number of allowable CpGs per primer pair and the maximum amplicon size.

(a) Increased number of allowed CpGs per primer pair by one CpG per primer pair. Filter primers. Extend tiling paths. If 100% coverage is not attained:

(b) Increase number of allowed CpGs per primer pair by one to two CpGs per primer pair. Filter primers. Extend tiling paths. If 100% coverage is not attained:

(c) Increase the maximum amplicon size by 200 bp. Reset allowable CpGs to 0 per primer pair. Filter primers. Extend tiling paths. If 100% coverage is not attained:

(d) Repeat **steps 5a–c** until maximum amplicon size is 1050 bp with two CpGs allowed per primer pair. In general, 100% coverage is achieved before this point.

3.2 Sample Preparation and Bisulfite Treatment

The major consideration in the preparation of genomic DNA for targeted microdroplet PCR and subsequent bisulfite sequencing is the amount of starting material. It is important to recognize that bisulfite conversion results in significant loss of DNA during the conversion and due to DNA destabilization and fragmentation. In testing a number of commercially available bisulfite conversion kits we found that the yield of bisulfite-converted DNA ranged from 40 to 70% of input [16].

1. Purify genomic DNA using commercially available kits or other standard isolation methods.
2. Bisulfite-convert the genomic DNA using a standard sodium bisulfite treatment protocol or a commercially available kit (*see* **Note 2**).
3. Quantify the concentration of the converted DNA using the Quant-it ssDNA assay kit and a Qubit fluorometer.
4. Assess the degree of fragmentation by running a sample of the converted DNA on a 1% agarose gel.
5. If concentration of the bisulfite-converted DNA is required, adjust the salt concentration of the DNA using 10× PCR buffer and dilute to a final concentration of 1× PCR buffer. Agencourt AMPure XP beads are added according to manufacturer's instructions. Incubate samples for 5 min on an orbital shaker or rocker. Place samples on a magnet for 2 min, remove the supernatant and wash the beads with 200 μL fresh 70% EtOH two times. Elute DNA from the beads in 20 μL of 10 mM Tris–HCl, pH 8.0.
6. Quantify DNA using the Quant-iT ssDNA Assay kit and a Qubit fluorometer.

3.3 Preparation of Microdroplet PCR

Microdroplets containing primer pairs specific for a single amplicon are custom manufactured by RainDance and are merged with droplets containing bisulfite-converted genomic DNA and PCR reagents using the RainDance instrument (*see* **Note 3**). At completion of the RainDance run, the resulting emulsion contains over one million singleplex PCR droplets.

1. Preparation of genomic template DNA.
   (a) 2 μg of bisulfite treated genomic DNA is added to 4.7 μL 10× High-Fidelity PCR buffer, 1.8 μL 50 mM $MgSO_4$, 1.7 μL 10 mM dNTPs, 3.6 μL 4 M Betaine, 3.6 μL RDT Droplet Stabilizer, 1.8 μL DMSO, and 0.7 μL of 5 U/μL Platinum High-Fidelity Taq.
   (b) Samples are adjusted to a final volume of 25 μL with nuclease-free water.
2. Microdroplets are generated on a RDT1000 (or the newer instrument platform called ThunderStorm) following the manufacturer's instructions.
   (a) For primers that do not contain a degenerate base (C/T or G/A), each primer droplet contains matched pairs of forward and reverse primers at 5.2 μM per primer yielding a final primer concentration in the merged droplet of 1.6 μM.
   (b) Primers that contain a degenerate base are present at 0.8 μM in each droplet.

3.4 PCR and Library Purification

Amplification of merged droplets is conducted in a standard thermocycler. As each droplet contains a very small amount of DNA and each set of primers has a different priming capacity, a high number of cycles are used to ensure that all PCR reactions proceed to completion, reducing PCR bias.

1. Merged samples are amplified in a thermocycler under the following conditions:
   (a) Initial denaturation at 94 °C for 2 min.
   (b) 55 cycles of 94 °C 15 s denaturation, 56 °C 20 s primer annealing, 68 °C 40 s extension (*see* **Note 4**).
   (c) Final extension at 68°C for 10 min.
   (d) Hold at 4 °C until further processing.
2. Following amplification, the PCR droplet emulsions are broken using an equal volume of RDT 1000 Droplet Destabilizer. Vortex the samples for 15 s and spin in a microcentrifuge at 12,000 × *g* for 5 min.
3. Carefully remove as much oil as possible from below the aqueous phase using a gel loading tip, being careful to not remove any of the PCR product in the aqueous phase.
4. Clean up the aqueous PCR product using a minElute kit (or equivalent) following the manufacturer's instructions.
5. Assess the size distribution and quantity of the resulting product on an Agilent Bioanalyzer by using a DNA chip.

3.5 Fragmentation and Concatenation

Following microdroplet PCR amplification of targeted regions, the resulting DNA ranges in size from 200 to 600 bp (depending on primer design conditions). During preparation of sequencing libraries, ~200 bp products will be selected. As such, prior to preparation of sequencing libraries, the amplified PCR product must be fragmented to 200 bp. Simply fragmenting the microdroplet PCR product before sequencing skewed the sequencing results to the ends of the targeted regions with little coverage through the middle of larger targeted regions. To circumvent this bias, fragmented samples are concatenated and refragmented to 200 bp prior to construction of the sequencing library.

1. Fragment the PCR products in a Covaris S2 using a duty cycle of 10%, an intensity of 5, 200 cycles per burst, twenty 60 s bursts, and total volume of 100 μL.
2. Assess fragmentation with a HS DNA chip on an Agilent Bioanalyzer.
3. Determine the fragmented DNA concentration using a Quant-iT dsDNA HS Assay kit and a Qubit Fluorometer.
4. End-repair 400 ng of fragmented DNA with 50 U T4 polynucleotide kinase, 5 U Klenow fragment, 15 U T4 DNA polymerase, 400 μM dNTPs, and 1× T4 DNA ligase buffer for 20 min at 30°C.
5. Purify end-repaired products on a DNA Clean & Concentrator-5 column or equivalent.
6. Concatenate end-repaired DNA with 600 U rapid T4 DNA ligase in 1× rapid ligation buffer for 15 min at 20 °C.
7. Purify concatenated products on a DNA Clean & Concentrator-5 column or equivalent.
8. Fragment the concatenated products in a Covaris S2 using a duty cycle of 10%, an intensity of 5, 200 cycles per burst, twenty 60 s bursts, and total volume of 100 μL.
9. Assess fragmentation on an Agilent Bioanalyzer using a HS DNA chip.

3.6 Library Prep

While the protocol outlined below is tailored for sequencing on an Illumina HiSeq, sequencing can be conducted on any high-throughput sequencing platform. If longer or shorter reads are used on other platforms the fragmentation steps outlined above can be adjusted.

1. Follow instructions in *NEB Next Ultra Library Preparation kit for Illumina* for the DNA sample *End Prep* reaction.
2. After *End Prep*, follow instructions in the *NEB Next Multiplex Oligos for Illumina (index primer set 1)* kit for adapter ligation.

3. Perform bead-based size selection using Ampure XP beads following the instructions in the *NEB Next Ultra Library Preparation kit for Illumina*. Use the protocol for selecting a total library size of 320 bp (insert size of 200 bp).
4. PCR amplify size-selected product for 12–18 cycles to generate adequate yield for subsequent gel-based size selection.
5. Clean up PCR product using Ampure XP beads using the "1×" protocol. This will not size select the PCR products but will remove PCR primers, nucleotides and salts allowing improved gel migration and subsequent gel-based size selection.
6. Load up to 500 ng per well of cleaned PCR product onto a 2% E-Gel and run for approximately 20 min. Load the Track-It 100 base pair ladder in an adjacent well. Visualize the gel using the E-Gel® Safe Imager™ and excise the PCR products in the size range of 300–350 bp in length.
7. Purify the excised library from the agarose using the Zymo DNA Recover kit following the manufacturer's instructions.
8. Quantitate the library using the Quant-iT dsDNA Assay Kit and Qubit fluorometer.
9. Confirm library size using the Agilent Bioanalyzer and DNA assay kit.
10. Library is now ready for final dilution and loading onto an Illumina Flow Cell following Illumina instructions.

3.7 Sequencing

1. Sequencing is conducted following standard protocols from the sequencer manufacturer.
2. Single end 100 bp reads are collected and base calls are made using the CASAVA software.

3.8 Data Analysis

Mapping sequencing reads of bisulfite-treated DNA to a reference genome presents particular challenges. The goal is to detect the difference between cytosine, which implies protection by a methyl group, and uracil, the product of bisulfite-mediated cytosine deamination, at a given genomic locus. Unless specific allowances are made at the read mapping stage, valid biological signals can be interpreted as reads that are mismatched to the genome. As a single sequencing read could span several genomic CpG loci, particularly in CpG-rich regions, naïve sequence analysis will result in discarding many otherwise valid reads because they exceed some predetermined number of mismatches to the reference genome.

The mapping issue is further complicated by the effective doubling of the genome, as amplified bisulfite-treated DNA no longer forms standard Watson–Crick base pairs (e.g., all single stranded DNA). This is why primers for generating amplicons of bisulfite-treated DNA must be strand-specific; naturally, the resulting

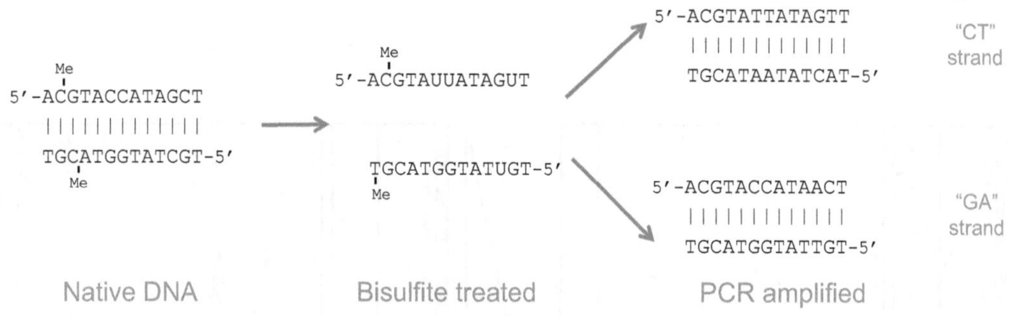

Fig. 2 Mapping bisulfite-converted DNA to the CT or GA strand

sequencing reads must be mapped specifically to the GA strand or the CT strand (Fig. 2).

To address the problem of bisulfite sequence mapping, we turned to a commercial software package from Novocraft called Novoalign (*see* **Note 5**). In bisulfite mode, Novoalign (version 2.7 at time of our analysis) assembles a four-strand index and allows mapping to ambiguous bases in the reference genome without penalty; for example, both CpG and TpG reads will map equally to a CpG locus.

1. Runtime parameters were set in consultation with the vendor; the Novoalign command line was:

```
novoalign -d [genome-index-file] -f [fastq-file] -F ILMFQ -b 4
-c 8 -a -h 120 -t 240 -s 50 -o SAM > [output-file.sam]
```

   where [genome-index-file] was the bisulfite index of the genome generated by novoindex; [fastq-file] is the input file of sequencing reads, and [output-file.sam] is the file to which output is to be written in SAM format. Of particular note is the –s 50 parameter, which we used to recover partial reads that spanned junctions created during the concatenation step described previously. As the software is mapping reads to an ambiguous reference genome that is doubled in size, it is perhaps not surprising that the mapping speed of this platform is 10–100× slower than other contemporary tools such as bowtie.

2. We used the samtools suite and custom Perl scripts to analyze the resulting mapped reads. Reads were divided according to whether they mapped to the GA or CT strand (Fig. 3). Novoalign identifies GA and CT reads with the custom ZB:Z tag in the SAM file.

3. Percent methylation is calculated at each predetermined CpG locus. Using only reads mapped to the CT strand, the number of C and T reads at the C position of the CpG locus is

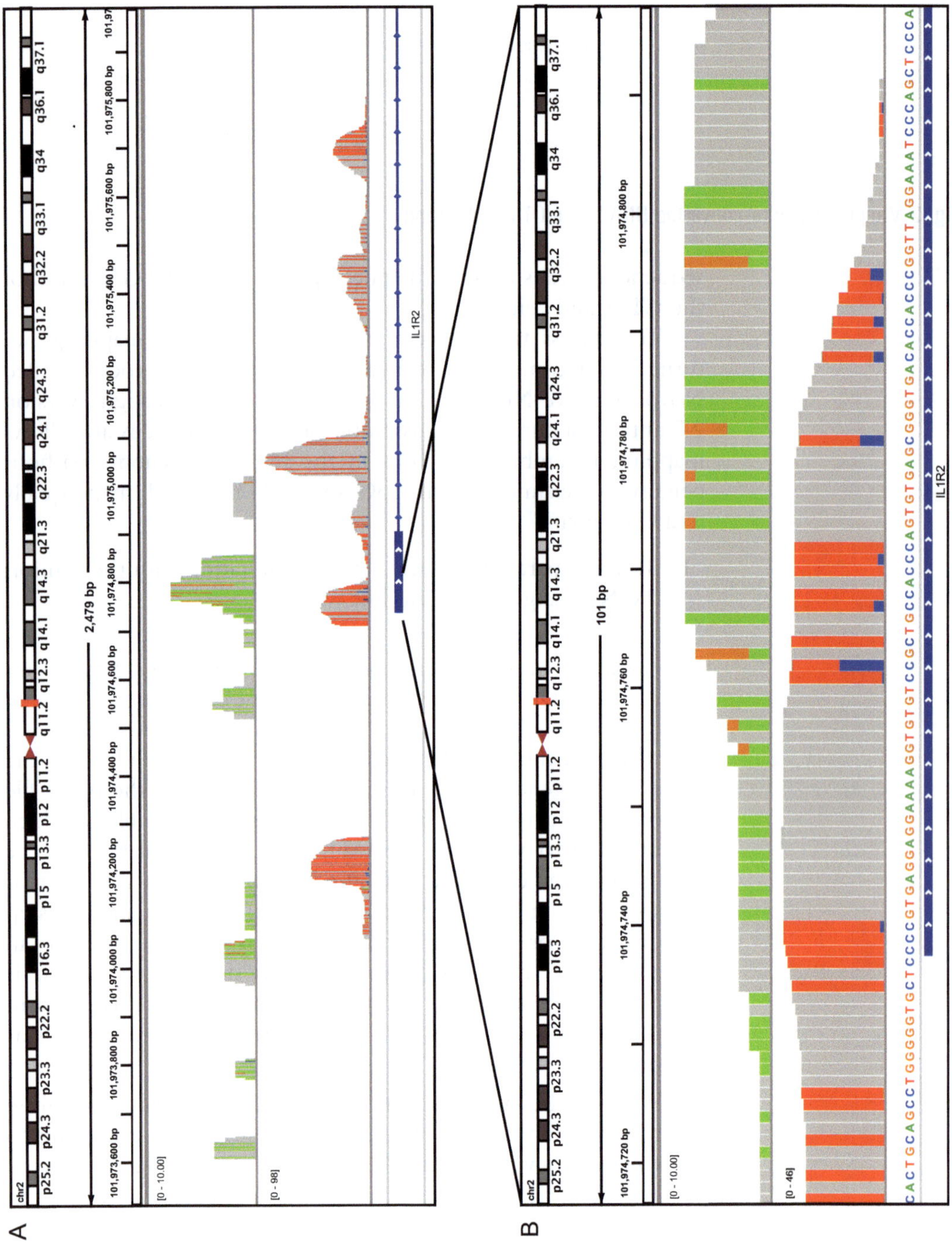

Fig. 3 Integrative genomics viewer pileup demonstrating aligned mapping to the GA and CT strands

determined using samtools, and the percent methylation is then #C reads/(#C + #T reads). If the amplicon is designed against the reverse strand, the percent methylation is calculated using GA-strand reads; the corresponding calculation is #G reads /(#G + #A reads) at the G position of the CpG.

4. As CpG methylation is symmetric, for CpG loci probed by both forward- and reverse-strand amplicons, the % methylation value should be approximately equal; this equality serves as a useful sanity check for resulting values. (*see* **Note 6**).

4 Notes

1. Although bisulfite conversion of double-stranded DNA results in single stranded DNA, the primer algorithm designs primers to cover only a single strand. The rationale behind this design metric is that CpG methylation is symmetric across both strands, so if a CpG is methylated on the 5′ strand then the complementary CpG on the 3′ strand is also methylated. If an investigator is interested in addressing the potential for non-symmetric methylation, then primers would need to be designed to cover both strands.
2. There are many commercially available bisulfite conversion kits. As expected, these kits vary in time required, cost, and the minimum/maximum input required. They also vary in the amount of fragmentation and loss of DNA. We have tested four of these kits and found great variability in the amount of fragmentation seen [16]. Fragmentation is important to consider in relation to the size of the amplicons targeted by your primer library. It is essential that the majority of the bisulfite-converted DNA is larger than the targeted amplicons or successful amplification of the targeted sequences will be impossible to achieve.
3. At the time we developed this protocol the only instrument available from RainDance was the RDT1000. Since that time they introduced the ThunderStorm, which is the same basic technology but capable of doing from 1 to 96 separate samples in a single run.
4. Annealing temperature and extension times are based upon the primer design parameters.
5. Other bisulfite sequence aligners are available for free or commercially.
6. CpG within primer binding regions are subject to large biases in PCR amplification and % methylation at these loci should be viewed with extreme skepticism, if not ignored entirely.

References

1. Deaton AM, Bird A (2011) CpG islands and the regulation of transcription. Genes Dev 25:1010–1022
2. Brown SJ, Stoilov P, Xing Y (2012) Chromatin and epigenetic regulation of pre-mRNA processing. Hum Mol Genet 21:R90–R96
3. Rose NR, Klose RJ (2014) Understanding the relationship between DNA methylation and histone lysine methylation. Biochim Biophys Acta 1839:1362–1372
4. Geiman TM, Muegge K (2010) DNA methylation in early development. Mol Reprod Dev 77:105–113
5. Laurent L, Wong E, Li G et al (2010) Dynamic changes in the human methylome during differentiation. Genome Res 20:320–331
6. Cheung HH, Lee TL, Rennert OM et al (2009) DNA methylation of cancer genome. Birth Defects Res C Embryo Today 87:335–350
7. Ehrlich M (2009) DNA hypomethylation in cancer cells. Epigenomics 1:239–259
8. Grolleau-Julius A, Ray D, Yung RL (2010) The role of epigenetics in aging and autoimmunity. Clin Rev Allergy Immunol 39:42–50
9. Villeneuve LM, Natarajan R (2010) The role of epigenetics in the pathology of diabetic complications. Am J Physiol Renal Physiol 299: F14–F25
10. Shapiro R, Servis R, Welcher M (1970) Reactions of uracil and cytosine derivatives with sodium bisulfite. J Am Chem Soc 92:422–424
11. Fouse SD, Nagarajan RO, Costello JF (2010) Genome-scale DNA methylation analysis. Epigenomics 2:105–117
12. Hodges E, Smith AD, Kendall J et al (2009) High definition profiling of mammalian DNA methylation by array capture and single molecule bisulfite sequencing. Genome Res 19:1593–1605
13. Deng J, Shoemaker R, Xie B et al (2009) Targeted bisulfite sequencing reveals changes in DNA methylation associated with nuclear reprogramming. Nat Biotechnol 27:353–360
14. Varley KE, Mitra RD (2010) Bisulfite patch PCR enables multiplexed sequencing of promoter methylation across cancer samples. Genome Res 20:1279–1287
15. Tewhey R, Warner JB, Nakano M et al (2009) Microdroplet-based PCR enrichment for large-scale targeted sequencing. Nat Biotechnol 27:1025–1031
16. Komori HK, LaMere SA, Torkamani A et al (2011) Application of microdroplet PCR for large-scale targeted bisulfite sequencing. Genome Res 21:1738–1745
17. Komori HK, Hart T, LaMere SA et al (2015) Defining CD4 T cell memory by the epigenetic landscape of CpG DNA methylation. J Immunol 194:1565–1579
18. Rozen S, Skaletsky H (2000) Primer3 on the WWW for general users and for biologist programmers. Methods Mol Biol 132:365–386
19. Schuler GD (1997) Sequence mapping by electronic PCR. Genome Res 7:541–550

Chapter 18

Multiplexed DNA Methylation Analysis of Target Regions Using Microfluidics (Fluidigm)

Martyna Adamowicz, Klio Maratou, and Timothy J. Aitman

Abstract

Whole genome shotgun bisulfite sequencing is a method used to generate genome-wide methylation profiles. There are many available protocols to validate the results of this genome-wide method, but they mostly share the limitation of measuring methylation at a small number of CpG positions in small numbers of samples. We developed a multiplexed DNA methylation analysis protocol, which allows for the simultaneous quantitative measurement of cytosine methylation at single nucleotide resolution in 48 PCR amplicons and 48 samples utilizing the microfluidic system established by Fluidigm. Following bisulfite conversion of 500 ng of the target DNA, a PCR reaction is performed using a 48.48 Access Array, which allows parallel amplification of 48 samples by 48 primer pairs. The products of each reaction are labeled with individual, sample specific tags, pooled in a single library and sequenced using the Illumina MiSeq sequencer. The advantages of this system are: speed, small amount of input material, single nucleotide resolution, high coverage of each locus, low cost of simultaneously assaying multiple CpG loci in multiple DNA samples and high reproducibility.

Key words Bisulfite treatment, Multiplexed targeted sequencing, Microfluidics, 48.48 Access Array, MiSeq

1 Introduction

Traditional, capillary-based Sanger sequencing has been the sequencing method of choice since the 1970s [1] and was the method used in the Human Genome Project, taking over 10 years at a cost of nearly $3 billion. Over the last decade remarkable technological advances have led to dramatic reductions in cost and laboratory time with massive increases in throughput [2]. Whole genome sequencing has become a matter of routine in most technologically advanced institutes. For example, the sequencing technology developed by Illumina tracks the addition of labeled nucleotides in a massively parallel fashion, generating up to 1 terabase of sequence data in a single run. Using Illumina technology we developed and reported a protocol for whole

Jörg Tost (ed.), *DNA Methylation Protocols*, Methods in Molecular Biology, vol. 1708,
https://doi.org/10.1007/978-1-4939-7481-8_18, © Springer Science+Business Media, LLC 2018

genome shotgun bisulfite sequencing (WGSBS) [3] (*see also* protocols reported in Chapter 5–7 and 9). The protocol uses standard bisulfite sequencing in which sodium bisulfite treatment changes the chemical structure of the unmethylated cytosines in DNA into uracil [4, 5] which, following PCR amplification, are read as a thymidine in the sequencing reaction. Methylated cytosines are left unchanged after bisulfite treatment, allowing discrimination of methylated and unmethylated cytosines in the original template DNA. Following ligation of adapters and indices, a DNA library is prepared and sequenced on an Illumina HiSeq sequencer. Paired-end sequence reads, usually around 100 bp in length are mapped to the in silico converted genome as described and percentage methylation is calculated as the number of unmodified cytosines as a proportion of the total number of cytosines and thymidines at each nucleotide location [3]. Carrying out this procedure on genomic DNA allows cytosine methylation levels to be assayed at single nucleotide resolution across the genome [3].

We have adapted this genome-wide bisulfite sequencing protocol to establish a new method of multiplexed DNA methylation analysis of multiple target regions in multiple DNA samples utilizing the Fluidigm 48.48 Access Array microfluidics system and the Illumina MiSeq sequencer. The Access Array System library preparation workflow consists of six steps (Fig. 1): (1) Bisulfite conversion of genomic DNA, (2) sample quality assessment and quantification, (3) target-specific primer design and validation, (4) highly multiplex PCR amplification using the Access Array Integrated Fluidic Circuit (IFC), (5) quality assessment and quantification of the harvested PCR products for sequencing and (6) incorporation of sample-specific, bidirectional barcodes which allows demultiplexing of 100 base pair (bp) paired end reads generated by the sequencer. Amplicon tagging on the Access Array system significantly reduces the time required for enrichment of tagged sequences, by combining amplicon generation with library preparation. The system allows parallel assay of DNA methylation levels at 48 loci in 48 DNA samples. The Access Array protocol, combined with highly parallel Illumina sequencing, provides quantitative, medium-throughput targeted measurement of DNA methylation at single nucleotide resolution, which—in comparison to other existing methods—is easy to use and affordable for the user.

2 Materials

2.1 Assay Design

1. Extracted, high molecular weight DNA.
2. EpiDesigner software (Sequenom, http://www.epidesigner.com/start3.htmL).

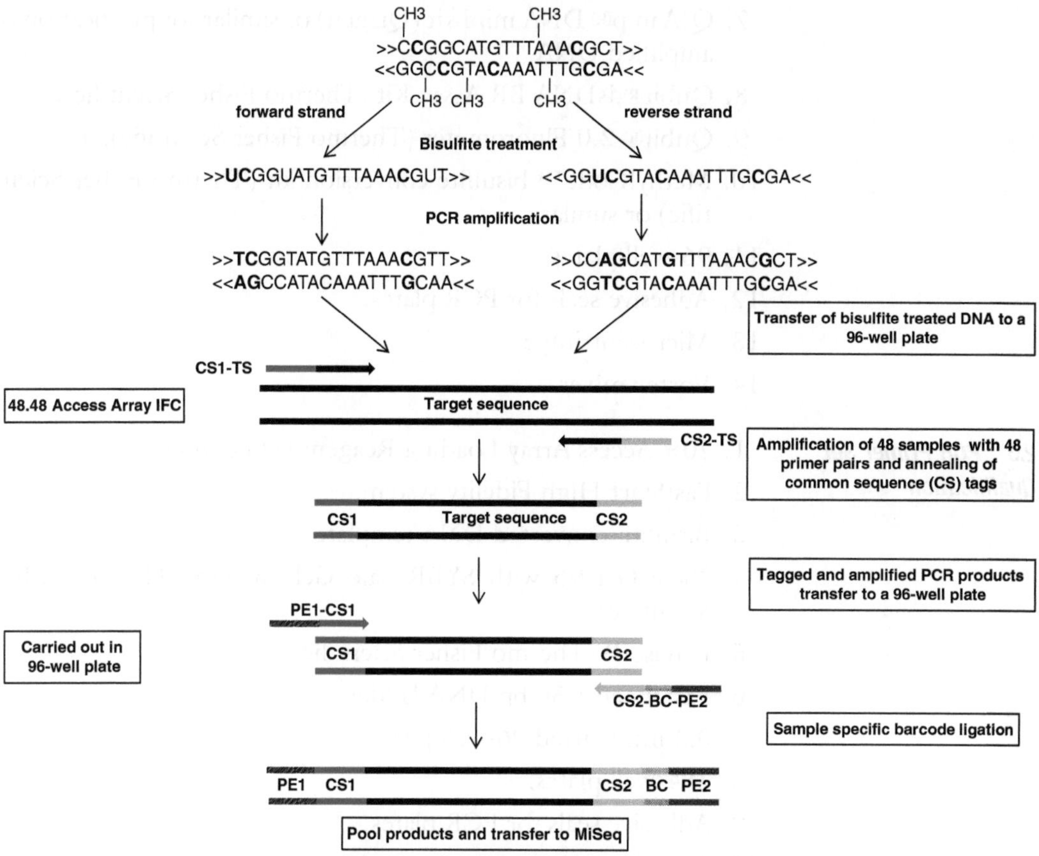

Fig. 1 Overview the protocol of multiplex DNA methylation analysis of target regions using Fluidigm microfluidics. After genomic DNA isolation, DNA is subjected to bisulfite treatment, which converts unmethylated cytosines into uracils. Following PCR reaction they are read as thymines while methylated cytosines remain unchanged and are read as cytosines. DNA is bisulfite-treated followed by site specific tagged primers. Next PCR reaction is performed on the 48.48 Access Array followed by DNA sample-specific bar-code ligation to PCR products. Pooling of all individually bar-coded DNA samples creates a library which is then sequenced on Illumina Mi-seq. *CS1* tag 1, *CS2* tag 2, *TS* target-specific primer sequence, *PE1* paired end sequence 1, *PE2* paired end sequence 2, *BC* barcode sequence, *IFC* integrated fluidic circuit. Adapted from Figures in [9] and the Access Array System for Illumina Platform User Guide by Fluidigm

2.2 Control and Sample Preparation Followed by Bisulfite Treatment

1. Proteinase K for DNA extraction.
2. 0.6% agarose gel for DNA integrity test.
3. Ribonuclease A (RNase).
4. *Sss*I methylase or a commercial fully methylated DNA of a species of interest, e.g., CpG Methylated NIH 3T3 Mouse Genomic DNA (New England Biolabs).
5. MinElute Reaction Cleanup kit.
6. REPLI-g Ultra-Fast kit (Qiagen) or similar for whole genome amplification to create completely unmethylated DNA.

7. QIAamp® DNA mini kit (Qiagen) or similar for purification of amplified DNA.
8. Qubit®dsDNA BR Assay Kit (Thermo Fisher Scientific).
9. Qubit® 2.0 Fluorometer (Thermo Fisher Scientific).
10. MethylCode™ bisulfite conversion kit (Thermo Fisher Scientific) or similar.
11. 96-well plates.
12. Adhesive seals for PCR plates.
13. Microcentrifuge.
14. Vortex mixer.

2.3 PCR Primer Set Qualification

1. 20× Access Array Loading Reagent (Fluidigm).
2. FastStart High Fidelity system.
3. Bisulfite-converted DNA template.
4. 2% E-Gel 96 with SYBR Safe Gels system (Thermo Fisher Scientific).
5. E-Base™ (Thermo Fisher Scientific).
6. Fast-Load® 50 bp DNA ladder.
7. 0.2 mL Skirted 96-well plates.
8. 384-well plates.
9. Adhesive seals for PCR plates.
10. Microcentrifuge.
11. Vortex mixer.
12. Plate centrifuge.

2.4 48.48 Access Array Protocol for PCR Reaction and Barcoding

1. FastStart High Fidelity PCR System, dNTPack.
2. 20× Access Array Loading Reagent (Fluidigm).
3. 1× Access Array Harvest Solution (Fluidigm).
4. Bidirectional 384 Barcode Kit for the Illumina GAII, HiSeq and MiSeq Sequencers (Fluidigm).
5. Target-specific primer pairs tagged with universal tags (CS1 forward tag: ACACTGACGACATGGTTCTACA., CS2 reverse tag: TACGGTAGCAGAGACTTGGTCT): 50 μM CS1-Tagged TS Forward Primer, 50 μM CS2-Tagged TS Reverse Primer.
6. 48.48 syringes of 300 μL of Control Line Fluid (Fluidigm).
7. Template DNA at 50 ng/μL.
8. Agencourt AMPure XP purification system (Beckman Coulter Inc.).
9. High Sensitivity DNA Reagents (Agilent).

10. Qubit® dsDNA HS Assay Kit (Thermo Fisher Scientific).
11. Qubit® 2.0 Fluorometer (Thermo Fisher Scientific).
12. MicroAmp Optical 96-well plate (Thermo Fisher Scientific).
13. Magnetic stand.
14. 0.2 mL Skirted 96-well PCR plate.
15. Adhesive seals for PCR plates.
16. Microcentrifuge.
17. Vortex mixer.
18. Plate centrifuge.
19. 48.48 Access Array Integrated Fluidic Circuits (IFC) (Fluidigm).
20. IFC Controller AX (for Pre-PCR and Post-PCR) (Fluidigm).
21. FC1™ Cycler (Fluidigm).

2.5 MiSeq Reads Alignment and Methylation Calling

1. Hardware: UNIX/Linux workstation with 8GB of RAM and multi-core processor connected to the Internet.
2. Software: UNIX shell commands (mkdir, cd, cat, zcat, echo, wget, grep, unzip, gzip, tar, md5sum), GNU make, GNU Compiler (gcc), Perl, Java, BWA, Samtools, Picard, NxtGenUtils.

3 Methods

3.1 Assay Design

Assay design is the most important step of this targeted highly multiplexed high-throughput sequencing protocol. The software used for primer design, the EpiDesigner from Sequenom, performs in silico bisulfite conversion prior to primer design, which is a very helpful step for the user. It should be noted that primer design is strand specific. If the designed primers are complementary to the DNA reverse strand and the reverse strand sequence serves as an input data, the settings of the software should remain unchanged ("default-forward strand") otherwise the designed primers will be complementary to the forward strand.

3.1.1 Design of Target Specific (TS) Primers

1. Following bisulfite treatment, forward and reverse DNA strands are not complementary and C methylation levels on both strands are assayed separately. This can be useful for example for investigating hemi-methylation (*see also* Chapter 29).
2. Optimal assay TS primers should be 25 base pairs (bp) in length; however a variation of 5 bp is acceptable. It is important to keep the Tm of the primers at 62 °C or at least as close as

possible to it. The product size may vary from 140 to 210 bp with the optimal length 180 bp (*see* **Note 1**).

3. Ideally primers should be placed in the region of at least three cytosines that have been converted during bisulfite treatment.
4. TS primers should not contain any CpG positions and palindromic sequences within primers should be avoided. This provides higher specificity of the primers to the DNA template.
5. Generated amplicons and the sites of primer annealing should be carefully investigated for any polymorphisms to avoid preferential primer attachment to these sequences or assaying only DNA molecules carrying polymorphism.

3.1.2 Design of TS Fluidigm Amplification Primers

1. Generate tagged forward primers by adding the common sequence forward (CS1) tag at the 5′ end of the PCR primer (CS1 sequence: ACACTGACGACATGGTTCTACA).
2. Generate tagged reverse primers by adding the common sequence reverse (CS2) tag at the 5′ end of the PCR primer (CS2 sequence: TACGGTAGCAGAGACTTGGTCT).
3. Primers should be ordered in 96 well plates, diluted in water, at a concentration of 50 μM and volume 100 μL. They should be subjected to desalting purification by the manufacturer following primer synthesis.

3.2 Control and Sample Preparation Followed by Bisulfite Treatment

1. Extract DNA from the samples of interest using either commercially available kits, e.g., Qiagen DNeasy Blood & Tissue Kit or using a standard procedure of phenol/chloroform extraction including a proteinase K digestion step. RNA digestion step by RNase should be performed using for example RNase A.
2. Prepare a fully methylated control DNA. This can be achieved either by ordering a highly methylated DNA of the selected species, e.g., CpG Methylated NIH 3T3 Mouse Genomic DNA or alternatively by incubation of genomic DNA with the CpG methylase SssI. Add 7.5 μL of NEB-buffer 2, 10 nmol of S-adenosylmethionine (SAM), and 6 U of SssI to 4.5 μg of genomic DNA in a final volume of 67.5 μL. Incubate at 37 °C in a thermal cycler. Add additional 10 nmol of SAM and 6 U of SssI twice: first after 3 h of incubation and then after additional 2 h and incubate the solution overnight at 37 °C in a thermal cycler. Inactivate the methylase at 95 °C for 5 min Purify the samples with the MinElute Reaction Cleanup kit following the manufacturer's instructions. Methylated DNA should be stored at −20 °C.
3. Prepare fully unmethylated DNA control using the REPLI-g Ultra-Fast whole-genome amplification kit following the manufacturer's instructions. Purify the amplification product with

QIAamp DNA mini kit following the manufacturer's instructions.

4. Prepare 5 μg of DNA which will be used as a template for the validation of PCR primers. This is a sufficient amount to perform 96 primer reactions.
5. Measure the DNA concentration with the Qubit® dsDNA broad range assay kit and adjust the concentration of all DNA samples and controls to 50 ng/μL.
6. Perform bisulfite conversion of 500 ng of each sample, control DNA and PCR template DNA with MethylCode™ bisulfite conversion kit following the manufacturer's protocol. Make sure there is sufficient DNA to perform the validation of all the PCR primers (*see* **Note 2**).

3.3 PCR Primer Set Validation

This primer validation protocol prepares enough reagents to perform 96 primer and 96 non-template-control (NTC) primer reactions. 20× Access Array Loading Reagent and DMSO should be warmed up to room temperature before use.

Prepare the 5× Target-Specific Primer Solutions for 96 individual primer pairs by mixing:

1. Prepare the 5× Target-Specific Primer Solutions for 96 individual primer pairs by mixing 50 μM CS1-TS Forward Primer and 50 μM CS2-TS Reverse Primer to a concentration of 1 μM. Note that the final TS Forward and Reverse Primer concentrations in this mix are 1 μM and 200 nM in the PCR reaction.
2. Prepare Target-Specific Primer Validation Reaction mix. Reaction conditions are: 1× FastStart High Fidelity Reaction Buffer without $MgCl_2$, supplemented with 4.5 mM $MgCl_2$, 5% DMSO, 200 μM each PCR Grade Nucleotide Mix, 0.05 U/μ L FastStart High Fidelity Enzyme Blend, 1× Access Array Loading Reagent, 8.3 ng/μL bisulfite-converted genomic DNA and water to a total volume of 4 μL per reaction. Values provided here are concentrations in the final solution. All the reagents apart from bisulfite-treated DNA are part of the FastStart High Fidelity PCR System, dNTPack. The mix should be prepared for 120 reactions carried out in a 384-well plate. It is recommended to prepare this amount to provide sufficient reagent to minimize errors due to pipetting. Also, the NTC primer validation reaction components should be prepared at the same time. Use the same reagents diluted to the same final concentrations as for the Target Specific Reactions, but substitute the bisulfite converted genomic DNA with water.
3. Prepare the Primer Validation PCR reactions by mixing 4 μL of Primer Validation Reaction Mix in each well with 1 μL of the 5× Target-Specific Primer Solution for both template and non-template primer control.

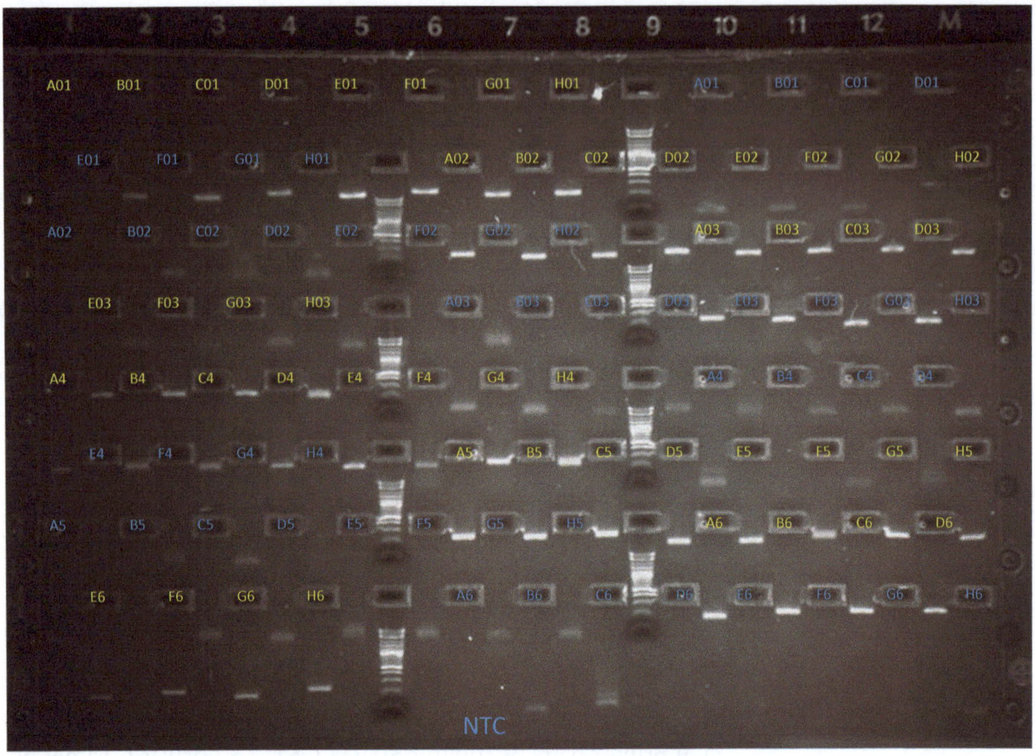

Fig. 2 2% E-Gel 96 with SYBR Safe Gel system. Expected to see product size of 180–210 bp. If primer dimers, multiple bands or bands of a wrong size are visible it is advised to redesign primers

4. The PCR program consists of an incubation step of 70 °C for 20 min, denaturing step of 95 °C for 10 min, followed by 40 cycles of 15 s at 95 °C, 30 s at 57 °C and 1 min at 72 °C, with 2 min at 72 °C of final extension.
5. PCR reaction is evaluated by a visualization step using a 2% E-Gel 96 with SYBR Safe Gel system according to manufacturer's instruction. A band of 180–210 bp in size is expected to be visible on the gel. If the band is not visible, this suggests a wrong annealing temperature, and if more than one band is visible, the designed primers are not specific. In both cased they need to be redesigned to assure the optimization and specificity of the reaction (*see* **Note 3** and Fig. 2).

3.4 Fluidigm 48.48 Access Array Protocol for DNA Methylation Detection

The following reagents stored at −20 °C should be thawed on ice: FastStart High Fidelity PCR System, 20× Access Array Loading Reagent, 1× Access Array Harvest Solution, Bidirectional 384 Barcode Kit for the Illumina GAII, HiSeq and MiSeq sequencers, target-specific primer pairs tagged with universal tags, template DNA at 50 ng/μL, fully methylated and unmethylated control DNA. The following reagents stored at 4 °C should be warmed up to room temperature: Agencourt AMPure XP purification system, Agilent High Sensitivity DNA Reagents and Qubit® dsDNA HS Assay kit.

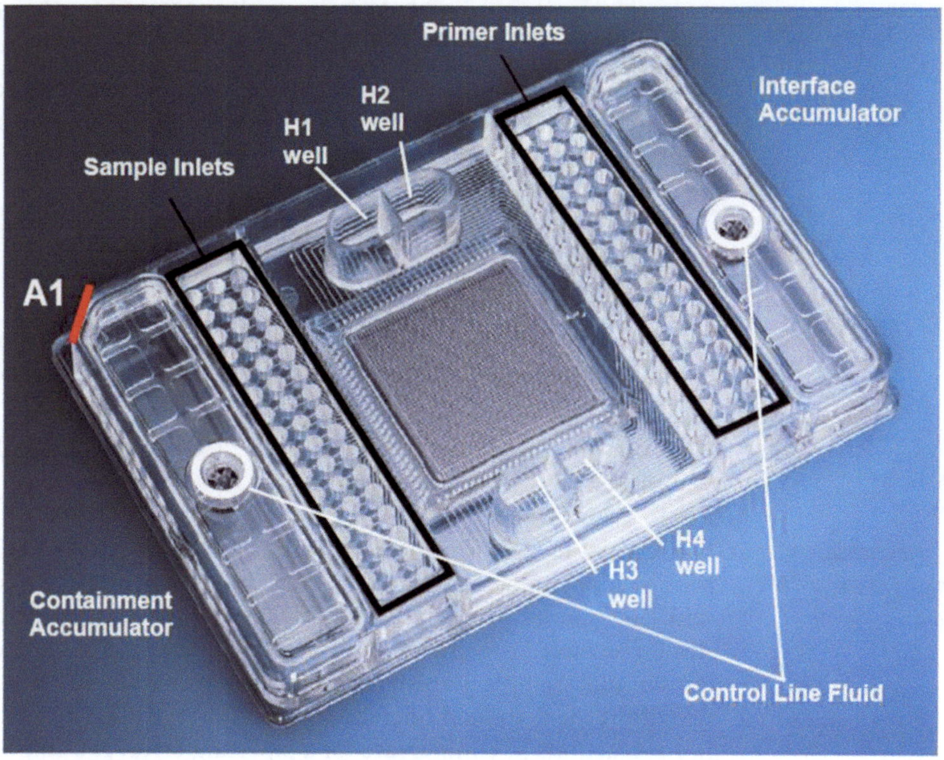

Fig. 3 48.48 Access array integrated fluidic circuits (IFC). The high-throughput 48.48 Access Array IFC enables target enrichment of 48 unique samples at the same time

3.4.1 Preparation of 20× Primer Solutions

1. Prepare the 20× Primer Solutions for 48 individual primer pairs with 50 μM CS1-Tagged TS Forward Primer, 50 μM CS2-Tagged TS Reverse Primer and 20× Access Array Loading Reagent to final concentrations of 1 μM per each primer and 1× for Access Array Loading Reagent.
2. Vortex the 20× Primer Solutions for a minimum of 20 s.
3. Centrifuge for 30 s to spin down all components. These primer solutions will be loaded into the Primer Inlets of a 48.48 Access Array Integrated Fluidic Circuit (IFC).

3.4.2 48.48 Access Array IFC Priming

A picture of a 48.48 Access Array IFC is shown in Fig. 3. The following steps must be performed in the pre-PCR area to avoid contamination with the post-PCR products (*see* **Note 4**). The 48.48 Access Array IFC should be used within 24 h of opening the package. Fluid in the Control Line of the IFC or in the inlets makes the IFC unusable. Only 48.48 syringes with 300 μL of Control Line Fluid should be used.

1. 48.48 Access Array IFC should be primed by injecting Control Line Fluid using Control Line Fluid Syringes into containment and interface accumulators of the IFC. It is crucial to load the

IFC into the Pre-PCR IFC Controller AX within 60 min of priming.

2. 500 μL of 1× Access Array Harvest Solution should be added into the H1–H4 wells (Fig. 3) on the IFC as quickly as possible following the priming.
3. These steps should be performed in the pre-PCR area to prevent contamination. Remove and discard the blue protective film from the bottom of the 48.48 Access Array IFC and load the IFC into the Pre-PCR IFC Controller AX.
4. Place the IFC onto the Pre-PCR IFC Controller AX tray by aligning the notched corner of the IFC to the A1 mark of the controller.
5. To prime the IFC, the "Select Prime (151×)" script should be chosen.

3.4.3 Preparation of Sample Pre-mix and Mix Solutions

It is crucial to perform these steps in a DNA-free area to avoid a possible contamination with DNA molecules, which do not represent the samples of interest, i.e., a DNA-free hood. The reagents should be mixed together in the volume which would allow preparing enough Sample Pre-Mix for 60 reactions. This will make up enough reagent to load one 48.48 Access Array IFC with 16 additional reactions to compensate for a pipetting error. It is important to remember that 48 samples should include water blank, acting as negative control against PCR contamination as well as non-methylated and fully methylated controls. It is essential to vortex all components to ensure complete mixing.

1. Mix together the reagents to the following final concentrations: 1× FastStart High Fidelity Reaction Buffer without $MgCl_2$, 4.5 mM $MgCl_2$, 5% DMSO, 200 μM each nucleotide of PCR Grade Nucleotide mix, 0.05 U/μL FastStart High Fidelity Enzyme Blend, 1× Access Array Loading Reagent and PCR certified water up to a total reaction volume of 4 μL.
2. Prepare the sample mix solutions by adding 4 μL of Sample Pre-Mix to 1 μL of Bisulfite Converted Genomic DNA of your Individual samples and controls.

3.4.4 Samples and Primer Loading onto 48.48 Access Array IFC

An 8-channel pipette is recommended in this step:

1. Load 4 μL of 20× Primer solution into each of the primer inlets.
2. Load 4 μL of Sample Mix into each of the sample inlets.
3. Load the 48.48 Access Array into the Pre-PCR IFC Controller AX. It is important to place the IFC onto the tray by aligning the notched corner of the IFC to the A1 mark in the IFC Controller AX.

4. Run the "Load Mix (151×)" script. These steps should be performed in the pre-PCR area to avoid contamination.
5. Once the reaction is completed, transfer the 48.48 Access Array IFC into the FC1 Cycler to perform amplification reaction according to the PCR program consisting of an incubation step of 70 °C for 20 min, denaturing at 95 °C for 10 min, followed by 40 cycles of 15 s at 95 °C, 30 s at 57 °C and 1 min at 72 °C.

3.4.5 48.48. Access Array IFC Harvesting

All subsequent steps should be performed in the post-PCR area to exclude any contamination of pre-PCR spaces. For these steps only sample inlets are used. To avoid possible mistakes primer inlets may be covered with foil.

1. Remove the 1× Access Array Harvest Reagent from the H1–H4 wells and substitute with 600 μL of fresh reagent followed by addition of 2 μL of 1× Access Array Harvest Reagent into each of the Sample Inlets on the IFC.
2. Place the IFC into the post-PCR IFC Controller AX located in the post-PCR area and align the notched corner of the IFC with the A1 mark of the IFC Controller AX followed by the harvesting program "Harvest (151x)".
3. Finally, label a 96-well plate with the 48.48 Access Array IFC barcode for easier recognition in the future and 10 μL of harvested PCR products from each of the Sample Inlets should be very carefully transferred into columns 1–6 of the plate.
4. Verify the quality of the PCR product from ten random samples described above using an Agilent DNA HS chip following manufacturer's instructions. PCR products can be stored at −20 °C.
5. Depending on the expected size of the PCR products, a smear may be visible on the Agilent DNA HS chip. Comparison of the electropherogram to a histogram of the expected sizes indicates correctness of the product size range. The PCR products should exhibit a band shift of +44 bp when compared to the amplicon insert size as they contain extra 22 bp from the universal forward tag CS1 and 22 bp from the universal reverse tag CS2. An example of Bioanalyzer traces (a) and a histogram (b) are presented in Fig. 4.

3.4.6 PCR Clean-Up Using AMPure Beads

1. Bring the volume of each PCR reaction to 50 μL with PCR grade water and mix very well with 50 μL of resuspended AMPure XP beads. It is important to be very careful with this step to avoid cross-contamination of the PCR products. If needed, an adhesive lid could be applied to pulse spin the plate in order to bring the content down to the bottom of the plate.

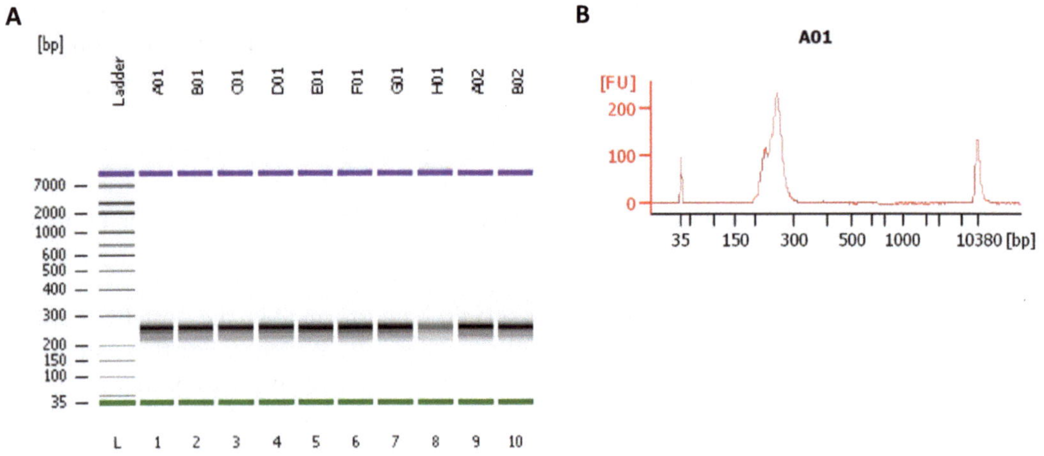

Fig. 4 Example of Bioanalyzer analysis post PCR reaction performed on the 48.48 access array integrated fluidic circuits. (**a**) Bioanalyzer traces of ten samples with the amplicon sizes varying between 200 and 300 bp. They exhibit a band shift of +44 bp when compared to the amplicon insert size as they contain extra 22 bp from the universal forward tag CS1 and 22 bp from the universal reverse tag CS2. (**b**) Electropherogram of 48 PCR products from one sample

2. Incubate the plate at room temperature for 15 min to allow the PCR amplified fragments attach to the beads.
3. Place the plate on the magnetic stand for 10 min, or longer if necessary, until the solution appears clear.
2. Remove 95 μL of the supernatant from each well of the plate taking great care not to disturb the beads. This should be done individually rather than by using an eight channel pipette making sure that the tip is changed between samples.
3. Leaving the plate on the magnetic stand thought entire time, add 200 μL of freshly prepared 80% ethanol to each well without disturbing the beads.
4. Incubate the plate at room temperature for 30 s followed by removal of the supernatant. If the beads are disturbed in the washing step and the solution changes the color to brown, the incubation should be extended until the solution clears.
5. Perform the wash step twice followed by air-drying of the samples for 5 min (*see* **Note 5**).
6. Finally, resuspend the dried pellet with 32.5 μL of water, apply an adhesive seal, mix the plate thoroughly by vortexing, spin down briefly and repeat the vortexing and centrifugation one more.
7. Incubate the plate at room temperature for 10 min and clear the solution by 10 min incubation on the magnetic stand. Transfer 30 μL of the clear supernatant from each well to the corresponding well of a new 96-well plate (*see* **Note 6**).

8. Verify the quality of the purification of the same ten random PCR-amplified samples tested previously together with a water control again using Agilent DNA HS chips following the manufacturer's recommendation.
9. Check the results of the chip to determine if the bead cleanup was successful. The same smear of the PCR products should still be visible, but the primer dimer bands in the samples and the water control should not be no longer detectable.

3.4.7 Bidirectional Amplicon Tagging and Bead Purification of Tagged Samples

1. Add 20 μL of water to the 30 μL of bead-purified samples. Perform this step in the post-PCR area. To remove any possible minute quantity of bead contamination carried over from the previous step, the plate should be placed on the magnetic stand.
2. Meanwhile, prepare in the pre-PCR area the Sample Pre-Mix by combining the components as follows: 1× FastStart High Fidelity Reaction Buffer without $MgCl_2$, 4.5 mM $MgCl_2$, 5% DMSO, 200 μM each nucleotide from PCR Grade Nucleotide Mix, 0.05 U/μL FastStart High Fidelity Enzyme Blend, and water to a total reaction volume of 15 μL. The reaction mix should be sufficient to perform 60 reactions, which is sufficient for 48 PCR product pools harvested from one 48.48 Access Array IFC with 12 additional reactions to compensate for pipetting error.
3. In the pre-PCR area, combine 15 μL of Sample Pre-Mix, 4 μL of unique bidirectional barcode of the barcode kit and 1 μL of bead purified sample in a new 96-well plate resulting in each sample having its own unique barcode. The plate can now be moved to the PCR thermal cycler for the amplification and tagging reaction according to the PCR program consisting of an denaturation step at 95 °C for 10 min, followed by 15 cycles of 15 s at 95 °C, 30 s at 60 °C and 1 min at 72 °C with a 3 min step of final extension at 72 °C.
4. The final step of library preparation involves purification of barcoded samples using AMPure XP beads at room temperature. Combine 4 μL of each sample pool in a new microcentrifuge tube and creating a Harvest Samples' Pool of up to 48 samples.
5. Mix 12 μL of such a pool with 24 μL of water and 36 μL of AMPure XP beads and incubate at room temperature for 15 min followed by a 15 min separation on the magnetic stand.
6. Wash twice with 180 μL of freshly prepared 80% ethanol. Following these washing steps, gently air-dry the beads for 5 min making sure that they are not overdried (*see* **Note 5**).

7. Add 40 μL of water to the microtube and following 5 min incubation at room temperature and 5 min placement on the magnetic separator, the supernatant containing a library ready for sequencing can be removed to a new tube.
8. Final step includes the quality and concentration control using a BioAnalyzer DNA HS chip and Qubit® dsDNA HS Assay following manufacturers' recommendation. The final library size and concentration will vary from 250–400 bp to 8–15 ng/μL respectively. It should be sequenced on an Illumina MiSeq sequencer generating 100 bp pair end reads with 50% PhiX control spike-in (*see* **Note** 7).

3.5 MiSeq Read Alignment and Methylation Calling

The demultiplexing and alignment of sequence reads to the in silico-converted genome has been described elsewhere and follows standard protocols [3, 6]. The main steps in read alignment include: building of the reference genome sequence by downloading the sequence of the chosen species from the UCSC website, in silico bisulfite conversion of the reference genome sequence, preprocessing of the sequencing reads, mapping of the reads to the converted reference genome, postprocessing of the read mappings, filtering, and the pileup of mapped reads. A detailed protocol of read alignment and methylation calling can be found in [3].

4 Notes

1. For successful sequencing using the Illumina MiSeq sequencer combined with 100 bp pair end reads, the insert size should not exceed 140–210 bp.
2. Any method of bisulfite conversion can be used and can be done without a commercially available kit as described previously [7, 8]. Store bisulfite-treated DNA at −20 °C. Bisulfite treatment degrades DNA, do not store bisulfite-treated DNA for more than 3 months.
3. Any given primer pair should only be used if visualization using E-gels indicates there is no primer dimer (Fig. 2).
4. It is essential to use pre- and post-PCR areas in the process of the library preparation to avoid possible contamination and amplification of PCR products.
5. Do not overdry the beads in **step 5** of Subheading 3.4.6 and **step 6** of Subheading 3.4.7. There is a fine balance between over-drying the beads (cracks appear in the pellet) and making sure there is no ethanol carryover in the sample. Cracks in the bead pellet indicate of over-dying, which will result in the lower yield of recovered DNA fragments.

6. If at any stage magnetic beads get carried-over, incubate a plate on a magnetic stand for an extra 10 min before removing sample aliquots.
7. 50% of Phi X is used in Mi-seq sequencing process to balance the base composition of bisulfite-treated DNA.

References

1. Sanger F, Nicklen S, Coulson AR (1992) DNA sequencing with chain-terminating inhibitors. 1977. Biotechnology 24:104–108
2. Institute NHGR (2014) Sequencing costs http://www.genome.gov/sequencingcosts
3. Johnson MD, Mueller M, Game L et al (2012) Single nucleotide analysis of cytosine methylation by whole-genome shotgun bisulfite sequencing. Curr Protoc Mol Biol 99:21–23
4. Hayatsu H, Wataya Y, Kai K et al (1970) Reaction of sodium bisulfite with uracil, cytosine, and their derivatives. Biochemistry 9:2858–2865
5. Shapiro R, Cohen BI, Servis RE (1970) Specific deamination of RNA by sodium bisulphite. Nature 227:1047–1048
6. Illumina Casava package. http://support.illumina.com/sequencing/sequencing_software/casava.ilmn
7. Tost J, El abdalaoui H, Gut IG (2006) Serial pyrosequencing for quantitative DNA methylation analysis. Biotechniques 40:721–722. 724, 726
8. Kwabi-Addo B, Chung W, Shen L et al (2007) Age-related DNA methylation changes in normal human prostate tissues. Clin Cancer Res 13:3796–3802
9. Krueger F, Kreck B, Franke A et al (2012) DNA methylome analysis using short bisulfite sequencing data. Nat Methods 9:145–151

Chapter 19

Large-Scale Targeted DNA Methylation Analysis Using Bisulfite Padlock Probes

Dinh Diep, Nongluk Plongthongkum, and Kun Zhang

Abstract

Bisulfite padlock probes (BSPP) are a method for the targeted quantification of DNA methylation in mammalian genomes. They can simultaneously characterize the level of methylcytosine modification in a large number of targeted regions at single-base resolution. A major advantage of BSPP is that it allows the flexible capture of an arbitrary subset of genomic regions (hundreds to hundreds of thousands of genomic loci) in single-tube reactions. Large number of samples can be processed efficiently and converted into multiplexed sequencing libraries with only three enzymatic steps, without the conventional library preparation procedures. BSPP are applicable to clinical studies, screening cell lines, and for quantifying low abundance regions using deep sequencing.

Key words DNA methylation quantification, Multiplexing, Padlock, Capture, Epigenetics

1 Introduction

A bisulfite padlock probes (BSPP) consist of a common linker sequence connecting two variable capture arms that can anneal to two adjacent genomic regions of bisulfite treated DNA with a gap size up to hundreds of bases [1, 2] (Fig. 1). The capture arms anneal to genomic regions that have been chemically modified by bisulfite such that all unmethylated cytosines are converted to uracils. The first capture arm from the 3′ end anneals in the forward direction, and the linker sequence, which is noncomplementary, provides space to allow the second capture arm to anneal in the reverse direction to a region upstream of the first capture arm. The gap in between the two arms are "captured" by extension with a thermostable polymerase lacking strand displacement and exonuclease activity from the 5′ end of the second capture arm to the 3′ end of the first. After extension, a ligase that is stable and active at high temperatures joins the extension end to the 3′ end of the probe, creating a circular DNA product. The circular DNA product is then amplified by polymerase chain reaction (PCR) and using a

Jörg Tost (ed.), *DNA Methylation Protocols*, Methods in Molecular Biology, vol. 1708,
https://doi.org/10.1007/978-1-4939-7481-8_19,

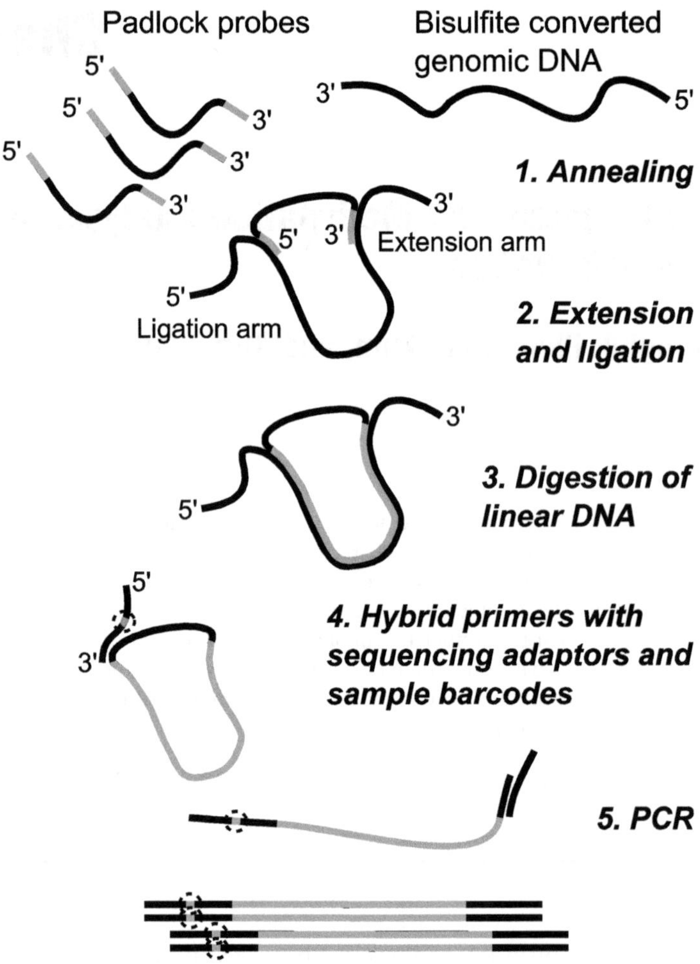

Fig. 1 Bisulfite padlock probe capture. (*1*) The two annealing arms of padlock probes "capture" bisulfite converted DNA during annealing incubation. (*2*) A mixture of dNTPs, polymerase, ligase, and buffer is added to generate a circular product in extension and ligation. (*3*) The genomic DNA and leftover padlock probes are removed by exonucleases. (*4*) Next, hybrid primers are used to label individual samples by appending barcodes (barcode regions are highlighted by *dashed circles*) and to additionally append common adaptors for sequencing. (*5*) PCR amplification is performed on capture products to generate multiple copies of the captured DNA

pair of primers that anneal to the common linker sequence (Fig. 1). The PCR primers contain adaptor sequences and sample barcodes that are compatible with next-gen sequencing instruments. High-throughput sequencing can be performed directly on the capture products, without additional library preparation procedures, to yield information on the methylation of the captured DNA. By design, single molecules such as the Watson and Crick strands are always captured independently. As CpG methylation are expected to be symmetrical, the correlation of CpG sites that are captured on both strands is an estimate of the technical variability within a capture experiment.

Variable capture arm sequences are selected computationally by analyzing the reference genome sequence and fitting user-desired probe constraints. The efficiency of padlock probes, which is the number of molecules captured by each probe sequence, can be estimated using a back-propagation neural network [3]. Efficiency data estimated from previous probe sets was used to train the neural network on the features such as target length, target GC content, binding arm melting temperature, binding arm length, local single-stranded folding energy of the target, and the dinucleotides present at the extension site and ligation sites during probe capture. Padlock probes can be synthesized with standard single columns or on microarrays, with single-columns yielding higher quantities of probes but at a higher up-front cost than array-synthesis. Microarray-based oligo synthesis can produce large numbers (thousands to hundreds of thousands) of oligos but in very limited quantities as a pool. To increase the quantity, bisulfite padlock probes can be synthesized with common adaptors in order to produce functional probes by PCR amplification and enzymatic removal of the adaptors.

Off-target annealing is a primary concern in bisulfite padlock probe capture. During capture, a circular product can only form when a pair of capture arms anneals adjacently to a single molecule with the 5′ and 3′ ends facing each other. Off-targeting annealing primarily leads to lower efficiency when capture arms anneal in multiple locations that do not lead to circular products. If capture arms anneal in off-target locations that leads to a circular product, this would lead to sequences that may not be applicable for analysis, however, this effect can be minimized by ensuring that pairs of capture arm sequences do not overlap with repeats. A minimum length of 25 bp for the capture arm is typically required to obtain melting temperatures above 50 °C, but the combined length of both capture arms should not exceed 60 bp as this would lead to more nonspecific annealing.

Steps for analysis of sequencing reads are similar to the steps taken for whole genome bisulfite sequencing (WGBS, *see also* Chapters 5, 7, and 9) or reduced representation bisulfite sequencing (RRBS, *see* Chapter 8) using readily available tools for next-gen sequence analysis [4, 5]. Similarly, genetic alterations such as single nucleotide polymorphisms (SNPs) can be determined from BSPP data. Furthermore, allele-specific methylation analysis can be performed by capturing regions with heterozygous SNPs [6]. Some considerations for processing BSPP reads include: (1) the necessary trimming from the 5′ end of the reads containing the capture arm sequences and (2) the proportion of clonal reads cannot be identified as all reads map to the same start and end positions. Clonal reads can affect the accuracy of methylation quantification when library complexity is low and too much sequencing was performed on such a library. Appending a unique molecular identifier (UMI) to individual padlock probe oligo is a strategy that allows identification of clonal reads by matching UMIs [7].

2 Materials

2.1 Padlock Probes Design

1. Genome reference sequences in FASTA format are required and can be downloaded from the UCSC Genome Browser. These cannot be in the multi-FASTA format with multiple chromosomes per file.
2. A list of targeted regions in BED format is required to design targets.
3. *ppDesigner* can be downloaded from http://genome-tech.ucsd.edu/public/Gen2_BSPP/ppDesigner/ppDesigner.php.
4. A Mac OS or any other modern Unix-based system is required to run *ppDesigner.*
5. *Perl* is required to run *ppDesigner.* It should already be included in all Unix-based system.
6. *Perl* modules, File::Temp and Sort::Array, are required to run *ppDesigner* and can be downloaded from http://www.cpan.org.
7. *BioPerl* toolkit is required to run *ppDesigner* and can be obtained from http://www.bioperl.org.
8. *Optional*: *UNAfold* software version 3.8. Using *UNAFold* will result in a more accurate prediction of probe efficiency. It is not absolutely required, and it will not change the probe sequence.

2.2 Padlock Probes Preparation (Only for Array-Synthesized Oligonucleotides)

1. Custom synthetic oligonucleotides from service providers such as Agilent, LC Sciences, or CustomArray.
2. 2× KAPA SYBR Fast qPCR master mix (Kapa Biosystems) or similar.
3. Primers for padlock probes amplification (asterisks indicates a phosphorothioate bond).
   (a) pAP1V61U: 5′-G*G*G TCATATCGGTCACTGTU-3′.
   (b) AP2V6: 5′-/5Phos/CACGGGTAGTGTGTATCCTG-3′.
4. Deionized H_2O.
5. 96-well PCR plate.
6. 50 mL disposable reservoir.
7. 200 μL multichannel pipette.
8. Optical adhesive film.
9. 100% Ethanol.
10. 75% Ethanol.
11. 3 M Sodium acetate (NaOAc), pH 5.5.
12. GlycoBlue (15 mg/mL, Thermo Fisher Scientific).
13. TE buffer: 10 mM Tris–HCl, 1 mM EDTA, pH 8.0.

14. QIAquick PCR purification kit (Qiagen).
15. Lambda exonuclease (New England Biolabs, NEB).
16. ssDNA/RNA clean and concentrator column (Zymo Research).
17. USER enzyme mix (NEB).
18. *Dpn*II (50,000 U/mL) (NEB).
19. RE-*Dpn*II-V6 oligo guide: 5′-GTGTATCCTGATC-3′.
20. Real-time thermal cycler.
21. Benchtop centrifuge.
22. 1.5 mL DNA LoBind tube.
23. 0.5 mL microcentrifuge tube.
24. Needle 22 G.
25. 5 mL tube.
26. Nanosep MF 0.2 μm column (PALL).
27. 12-well 6% TBE gel (Thermo Fisher Scientific).
28. 12-well 6% TBE-Urea gel (Thermo Fisher Scientific).
29. 2D-well 6% TBE-Urea gel (Thermo Fisher Scientific).
30. Disposable scalpel #10.
31. 2× TBE-Urea sample buffer (Thermo Fisher Scientific).
32. 10 bp DNA ladder.
33. 10,000× SYBR Gold nucleic acid gel stain (Thermo Fisher Scientific).
34. Benchtop orbital shaker.
35. UV transilluminator.
36. NanoDrop Spectrophotometer (NanoDrop Technologies Inc.) or similar.
37. Qubit fluorometer (Thermo Fisher Scientific).
38. XCell SureLock electrophoresis system (Thermo Fisher Scientific).
39. Electrophoresis power supply.
40. Gel Doc Imager (Bio-Rad) or similar.

2.3 Sample Preparation and Bisulfite Treatment

1. DNeasy blood and tissue kit (Qiagen).
2. TE buffer: 10 mM Tris–HCl, 1 mM EDTA, pH 8.0.
3. Deionized H2O.
4. EZ-96 DNA Methylation-Lightning MagPrep (Zymo Research).
5. 1.5 mL microcentrifuge tube.
6. Thermomixer.

7. Benchtop centrifuge.
8. 0.2 mL PCR tube strips.
9. Qubit dsDNA HS assay kit (Thermo Fisher Scientific).
10. Qubit ssDNA assay kit (Thermo Fisher Scientific).
11. Qubit assay tube (Thermo Fisher Scientific).
12. Qubit fluorometer (Thermo Fisher Scientific).
13. NanoDrop Spectrophotometer (NanoDrop Technologies Inc.) or similar.

2.4 Capture Setup

1. Deionized H_2O.
2. 10× Ampligase Buffer (Epicentre).
3. Functional padlock probes from Subheading 2.2.
4. Bisulfite treated genomic DNA from Subheading 2.3.
5. 96-well PCR plate (Eppendorf).
6. 10 and 20 μL multichannel pipette.
7. Mineral oil.
8. 0.2 mL PCR tube strips.
9. Aluminum seal.
10. Thermocycler.

2.5 Adding KlenTaq, Ligase, and Nucleotides (KLN Mix)

1. Deionized H_2O.
2. 1 mM dNTP mix (NEB).
3. 10× Ampligase Buffer (Epicentre).
4. Hemo KlenTaq (NEB).
5. Ampligase (Epicentre).
6. 0.2 mL PCR tube strips.
7. 10 μL multichannel pipette.

2.6 Exonuclease Digestion

1. Exonuclease I (20 units/μL, Epicentre).
2. Exonuclease III (200 units/μL, Epicentre).
3. 0.2 mL PCR tube strips.
4. 10 μL multichannel pipette.

2.7 Polymerase Chain Reaction

1. Primers for amplification.
   (a) AmpF6.4Sol: 5′-AATGATACGGCGACCACCGAGATC-TACACCACTCTCAGATGTTATCGAGGTCCGAC-3′.
   (b) AmpR6.3 Indexing primers: 5′-CAAGCAGAAGACGG CATACGAGAT<u>XXXXXX</u>GCTAGGAACGATGAGCCT-CCAAC-3′ (<u>XXXXXX</u> is the sample barcode, *see* [2] for details).

2. 2× KAPA SYBR Fast qPCR master mix (Kapa Biosystems).
3. 6× Gel loading dye.
4. 50 mL disposable reservoir.
5. 200 μL multichannel pipette.
6. 10 μL multichannel pipette.
7. 96-well PCR plate.
8. Optical adhesive film.
9. PCR cooler.
10. Real-time thermal cycler.
11. Agencourt AMPure XP beads (Beckman Coulter).
12. SPRI plate 96-ring (Beckman Coulter).
13. 75% Ethanol.

2.8 Size Exclusion by Polyacrylamide Gel

1. 5-well 6% TBE gel (Thermo Fisher Scientific).
2. Low DNA mass ladder (Thermo Fisher Scientific).
3. TE buffer: 10 mM Tris–HCl, 1 mM EDTA, pH 8.0.
4. Nanosep MF 0.2 μm column (PALL).
5. Thermomixer.
6. 100% Ethanol.
7. 75% Ethanol.
8. 3 M Sodium acetate (NaOAc), pH 5.5.
9. GlycoBlue (15 mg/mL, Thermo Fisher Scientific).
10. Deionized H_2O.
11. Benchtop orbital shaker.

2.9 Sequencing

1. SolSeq6.3.3 primer: 5′-TACACCACTCTCAGATGTTATCG AGGTCCGAC-3′.
2. SolSeqV6.3.2r primer: 5′-GCTAGGAACGATGAGCCTCC AAC-3′.
3. AmpR6.3IndSeq primer: 5′-GTTGGAGGCTCATCGTTCC TAGC-3′.
4. Illumina sequencing system. Tested systems are MiSeq, HiSeq2000, HiSeq2500 RapidRun, and GAIIx.
5. Illumina Cluster kit (including flowcells).
6. Illumina sequencing kit.

2.10 Data Analysis

1. *BisReadMapper* is required for analysis and can be downloaded from https://github.com/dinhdiep/BisReadMapper. Otherwise, refer to specific instructions for other software packages designed for the analysis of either Whole Genome Bisulfite

Sequencing (WGBS) or Reduced Representation Bisulfite Sequencing (RRBS) (*see also* Chapters 5–9).

2. Genome reference sequences in FASTA format are required and can be downloaded from UCSC Genome Browser.
3. A modern Unix-based system is required to run *BisReadMapper*.
4. *Perl* is required to run *BisReadMapper*.
5. *Trim Galore!* is required: http://www.bioinformatics.babraham.ac.uk/projects/trim_galore/.

   Note that Cutadapt is required also for Trim Galore!
6. b*amUtils* is required: http://genome.sph.umich.edu/wiki/BamUtil.
7. One of the four supported aligners: Supported aligners are *bowtie2* (version bowtie2–2.1.0), *bwa* (bwa-0.7.5a), *SOAP2* (soap2.21release), *LAST* (last-458), or *GEM* (GEM-binaries-Linux-x86_64-core_i3–20,130,406-045632). Note that BWA is recommended for general usage. Refer to *BisReadMapper* README for further instructions.
8. *Samtools* is required for *BisReadMapper* and can be downloaded from http://sourceforge.net/projects/samtools/files/samtools/. Must be version 1.18 or higher.
9. *OPTIONAL*: *Samtools* v1.08 to run the variant caller. Alternatively, use *BisSNP* from http://epigenome.usc.edu/publicationdata/bissnp2011/.
10. *OPTIONAL*: *Perl* module Statistics::LSNoHistory to calculate the forward and reverse correlation.
11. *OPTIONAL*: Latest SNP database for filtering low confident variants from UCSC Genome Browser, i.e., snp147, is http://hgdownload.soe.ucsc.edu/goldenPath/hg19/database/snp147.txt.gz.

3 Methods

This protocol can be subdivided into three main phases, which are design, experiment, and analysis. In the design phase, the researcher decides on the genomic targets to capture and design padlock probes to capture those targets. Previous works have established the design parameters that work best for bisulfite padlock probes to enable robust and consistent capture results. Padlock probes can then be ordered through a vendor such as IDT DNA, CustomArray, Agilent, LC Sciences and others. Based on the size of the padlock probe set, the researcher is limited to either array-based or single-tube synthesis methods. Single-column oligonucleotide synthesis is more cost-effective for padlock probe sets of sizes

1–1000, while array-based methods are more cost effective for padlock probe libraries of sizes several thousands and above. The cost per padlock probe is higher for single-tube methods; however, the yields and purity are often better and allows bypassing the additional steps required for the preparation of the padlock probes. Padlock probe preparation is required for libraries from array-based synthesis and take up to a week to complete, however, this can be performed in large batches and can be automated. Once padlock probes are ready, it is possible to move to the experiment phase, which involves the processing of samples. Generally, up to 96 or 384 samples may be processed in parallel in 96-well or 384-well plates to minimize batch-to-batch variability. Each sample is prepared for capture in separate tubes. To prepare for the analysis phase, the samples are prepared for sequencing with a two-step PCR protocol that enables multiplexing of individual samples using barcoded primers and pooled size exclusion to enrich for targeted fragment based on the expected fragment sizes. In the analysis phase, the sequencing library is sequenced and then analyzed with next-generation sequencing analysis software.

3.1 Padlock Probes Design

1. Ensure that individual reference sequences (FASTA) files are placed in a common directory.
2. Convert the target list BED file to a target file in the format required. The file should be tab-delimited and have four required columns (1) the unique ID for each target region, (2) the FASTA filename, such as chr22 (for chr22.fa), (3) The starting position, and (4) the final position. The final fifth column can be the required strand to capture. If strand is not indicated, the program will pick the more efficient probes from either strand. An example target file is given in the Example folder.
3. Generate a job file in the format required. An example job file is given in the Example folder. All of the parameters must be given. The *unafold_path* variable can be set to “NA” if *using_-unafold=0*. All paths must be full paths (*see* **Notes 1–5** for important considerations on choosing parameter values).
4. Run the ppDesigner.pl script. See the README for *ppDesigner* for specific usage instructions.
5. Run the primer2padlock.pl script. The *maxH1H2len* is a numerical value and should be the same as the *H1_plus_H2_len* variable from the job file. Indicate “*array*” to generate probes sequences that contain amplification adapters for array synthesis of probes. *See* the README for *ppDesigner* for specific usage instructions.
6. The probe sequences are now ready to be synthesized. It is recommended to randomize the order of the sequences so that

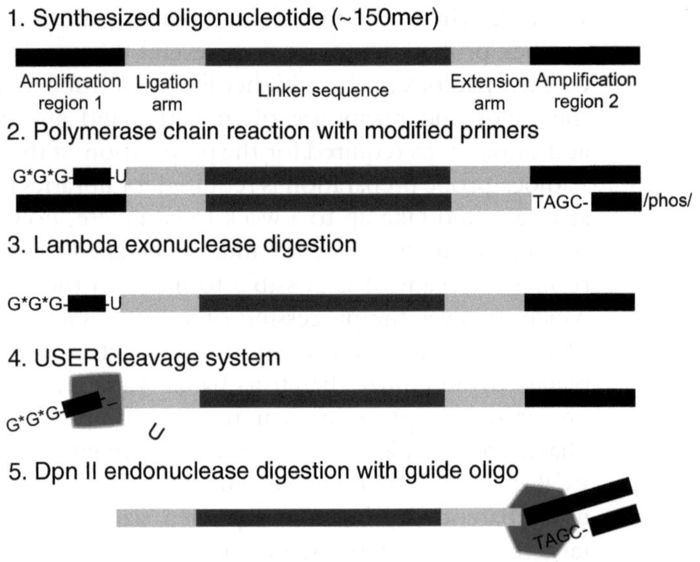

Fig. 2 Bisulfite padlock probe preparation. (*1*) Schematic view of a synthesized oligonucleotide. (*2*) Amplification of oligonucleotides is carried out using a forward primer containing multiple phosphorothioate bonds (marked with a *star*) to prohibit 5′ digestion by lambda exonuclease, and a uracil base at the 3′ end and a reverse primer that is phosphorylated at the 5′ end to promote lambda exonuclease digestion and a *Dpn*II recognition site at the 3′ end. (*3*) Digestion of the reverse complement strand is carried out by lambda exonuclease to generate single stranded probe. (*4*) The 5′ adaptor is cleaved by the USER cleavage system, which contains Uracil DNA glycosylase (UDG, not pictured) that catalyzes the removal of the uracil base and Endonuclease VIII (pictured as a *square*) that breaks the phosphodiester backbone. (5) The 3′ adaptor is cleaved by *Dpn*II (pictured as a *hexagon*) with the help of a guide oligonucleotide to make the cleavage region double-stranded

technical effects such as bad quality spots on the arrays will affect probes in a random manner and do not appear as systematic errors.

3.2 Padlock Probes Preparation

In this section, we describe the preparation of functional padlock probes from libraries of synthetic oligonucleotides (Fig. 2). The libraries of oligonucleotides (typical length of 150 nt and flanked by common amplification adaptors) could be produced by ink-jet printing on programmable microarrays. A first round of PCR, called "expansion PCR," produces the necessary templates for subsequent PCR steps. Amplicons are not functional padlock probes and have to be made functional via a few enzymatic steps that in total can take as little as one full day to accomplish. To purify the functional padlock probe from incompletely digested oligonucleotides, we perform size selection on denaturing 6% TBE-urea PAGE gels.

3.2.1 Expansion PCR

1. Perform the expansion PCR in two reactions (100 μL each) with 1–100 nM template oligonucleotides, 400 nM each of pAP1V61U and AP2V6 primers, and 1× KAPA SYBR Fast

qPCR master mix. The conditions of expansion PCR are as follows: 95 °C for 30 s, 15 cycles of 95 °C for 10 s, 55 °C for 20 s, and 70 °C for 30 s, and final extension at 70 °C for 2 min. The qPCR is monitored and terminated right before fluorescent levels reach plateau (*see* **Note 6**).

2. Purify amplicons with two QIAquick columns following the manufacturer's protocol and elute each column with 50 μL EB buffer (from the QIAquick PCR purification kit).
3. Pool eluted amplicons and measure concentration with NanoDrop.
4. Dilute amplicons to 10 nM and use as template for production PCR.

3.2.2 Production PCR

1. Set up production PCR mix for 100 reactions for 96-well plate (100 μL each) in a 50 mL disposable reservoir. PCR conditions including 0.02 nM of template oligonucleotides, 400 nM each of pAP1V61U and AP2V6, and 1× KAPA SYBR Fast qPCR master mix and amplify at 95 °C for 30 s, 15 cycles of 95 °C for 10 s, 55 °C for 20 s, and 70 °C for 30 s, and final extension at 70 °C for 2 min.
2. Precipitate amplicons in 5 mL tube by pooling 1.2 mL of amplicons to each tube. In each tube, add 120 μL of 3 M NaOAc, pH 5.5 (0.1× volume), 4 μL of GlycoBlue, and 3 mL of 100% ethanol (2.5× volume) and incubate at −80 °C for at least 20 min.
3. Centrifuge at 2700°× *g* for 30 min at 4 °C. Discard supernatant and wash the pellet with 1 mL 75% ice cold ethanol.
4. Dry DNA pellet and resuspend each with 150 μL H_2O.
5. Repurify amplicons with 12 QIAquick columns following the manufacturer's protocol.
6. Elute each column with 50 μL EB buffer and pool the purified amplicons.
7. Measure the concentration with NanoDrop.

3.2.3 Removal of Amplification Adapters

1. Digest the bottom strand of oligonucleotides with lambda exonuclease by incubating 15–20 μg of purified amplicons with 50 units of lambda exonuclease (5 U/μL) in 1× lambda exonuclease buffer in total volume 150 μL at 37 °C for 1 h.
2. Purify the resulting single-strand amplicons with a Zymo ssDNA/RNA clean and concentrator column.
3. Measure the concentration of purified single-strand amplicons with a NanoDrop.
4. To remove the 5′ end amplification adapter, digest 3–5 μg of single-strand amplicons with 5 units of USER (1 U/μL) in 1×

*Dpn*II buffer in total volume 80 μL at 37 °C for 1 h. The USER digested DNA is ~130 nt.

5. To remove the 3′ end amplification adapter, add 5 μL of 100 μM RE-*Dpn*II-V6 guide oligo, 2 μL of 10× *Dpn*II buffer, and 8 μL H_2O to USER digested DNAs, denature at 94 °C for 2 min, decrease temperature to 37 °C for 3 min, add 250 units of *Dpn*II (50 U/μL), incubate at 37 °C for 2 h, and heat inactivate *Dpn*II at 75 °C for 20 min. The USER/*Dpn*II digested DNA size is ~110 nt.
6. Purify USER/*Dpn*II digested DNAs with a Zymo ssDNA/RNA clean and concentrator column.
7. Measure the concentration with a NanoDrop.

3.2.4 Padlock Probe Size Exclusion by PAGE Size Selection

1. Mix 2–3 μg of USER/*Dpn*II digested DNA with 2× TBE-Urea sample buffer in 1:1 ratio in final volume 80–150 μL in a 200 μL PCR tube. Prepare 10 bp DNA ladder in 1× TBE-Urea sample buffer in total volume of 10 μL.
2. Incubate digested DNA and DNA ladder-loading dye mixture at 75 °C for 15 min. At the same time, prerun a 2D-well TBE-Urea gel at 250 V for 15 min.
3. Quick cool digested DNA and ladder on ice, and flush the well with a 1 mL pipette to remove residual urea in the well before loading DNA.
4. Quickly load DNA and DNA ladder into the well.
5. Run at 250 V in 0.5× TBE buffer for 30 min.
6. Remove the gel from the cassette and stain with 1× SYBR Gold in 0.5× TBE buffer for 8 min on an orbital shaker.
7. Place stained gel on a clean Saran wrap on the UV transilluminator with the long wavelength.
8. Locate the position of USER/*Dpn*II digested probe at ~110 bp according to the 10 bp DNA ladder.
9. Cut the gel with scalpel as close as to the band and minimize exposure time to UV light.
10. Move the gel slide to a clean area and chop the gel piece to small pieces with scalpel and transfer to shearing tubes (one gel goes to two tubes).
11. Shearing apparatus preparation for DNA purification from polyacrylamide gel: Place 0.5 mL tube into a 1.5 mL DNA LoBind tube. Punch 0.5 mL tube with a 22 G needle. Treat the tube under UV light for 15 min if possible.
12. Centrifuge the tube at 20,800 × *g* for 2 min at RT and remove 0.5 mL tube.
13. Add 450 μL of 1× TE buffer.

14. Shake vigorously on a shaker for 60 min at 37 °C.
15. Centrifuge the tube at 20,800 × *g* for 2 min at RT.
16. Transfer the clear supernatant to a Nanosep column and centrifuge at 20,800 × *g* for 1 min at RT.
17. Transfer as much as possible the remaining supernatant to a Nanosep column. Centrifuge at 20,800 × *g* for 1 min at RT.
18. Remove the Nanosep column and transfer supernatant to fresh 1.5 mL LoBind tube (volume ~ 400 μL /tube).
19. Precipitate digested DNA with 0.05 μg/μL GlycoBlue, 0.1× volume of 3 M sodium acetate, pH 5.5, and 2.5× volume of 100% ethanol.
20. Mix well and incubate the tube at −80 °C for at least 20 min.
21. Centrifuge at 20,800 × *g* for 30 min at 4 °C.
22. Discard supernatant and wash with 800 μL 75% ice cold ethanol as previously described.
23. Dry the DNA pellet and dissolve with ~10–15 μL H_2O/tube.
24. Pool dissolved padlock probes to the same tube and mix well.
25. Measure DNA concentration with Qubit ssDNA assay.
26. Verify padlock probe preparation on a 12-well 6% TBE-Urea gel using the same procedures as describe above for running a TBE-Urea denaturing gel.

3.3 Sample Preparation and Bisulfite Treatment

1. Extract genomic DNA from cell or tissue samples using a DNeasy blood and tissue kit following the manufacturer's instructions.
2. Measure genomic DNA concentration with Qubit dsDNA HS assay kit and verify genomic DNA quality with NanoDrop.
3. Perform bisulfite conversion on approximately 500 ng to 1 μg of genomic DNA using the EZ-96 DNA Methylation-Lightning Magprep kit in 96-deep-well plate following the manufacturer's protocol and elute bisulfite-treated genomic DNA with 25 μL elution buffer.
4. Measure concentration of bisulfite-converted genomic DNA with Qubit ssDNA assay kit.

3.4 Capture Setup

1. Set up capture reaction in total 20 μL in 96-well plate with the following components: 100–200 ng of bisulfite-treated genomic DNA, normalized amount of BSPP to target based on optimization (*see* **Note 7**), and 1× Ampligase Buffer (*see* **Note 8**).
2. Mix the capture mixture with a 20 μL multichannel pipette and spin down the plate at 1200 × *g* for 1 min.

3. Layer the reaction with 20 μL mineral oil to prevent evaporation during incubation.
4. Seal the plate with aluminum seal and spin down the plate at 1200 × *g* for 1 min.
5. Incubate the capture reactions on thermocycler as the following program: 94 °C for 30 s, gradually lower temperature to 55 °C at 0.02 °C/s in a thermocycler, and let hybridize at 55 °C for 20 h.

3.5 Adding KlenTaq, Ligase, and Nucleotides (KLN mix)

1. Prepare the KLN solution mix as follows: 20% (v/v) Hemo KlenTaq, 0.5 unit/μL Ampligase, 100 μM of dNTP mix, and 1x Ampligase Buffer.
2. Aliquot KLN mix to 8-tube strip and add 2 μL of KLN solution mix to each well with 10 μL multichannel pipette (*see* **Note 9**).
3. Swirl pipette five times to mix the capture reaction without moving the plate from thermal cycler.
4. Continue to incubate at 55 °C for 4–20 h.
5. Heat inactivate enzyme at 94 °C for 2 min.

3.6 Exonuclease Digestion

1. Prepare the exonuclease I/III mix by mixing exonuclease I (20 units/μL) and exonuclease III (200 units/μL) in 1:1 ratio and aliquot equal volumes to a 8-tube strip.
2. Add 2 μL of exonuclease I/III mix to each well with multichannel pipette (*see* **Note 9**).
3. Mix the reaction by swirling pipette around the well five times.
4. Incubate reaction at 37 °C for 2 h.
5. Heat-inactivate exonuclease I/III at 94 °C for 5 min.

3.7 PCR Setup

To enrich for captured DNA, perform PCR to amplify captured DNA in a 100 μL reaction in a 96-well PCR plate.

1. The PCR reaction consists of 10–30 μL of circularized DNA, 200 nM each of AmpF6.4Sol primers and AmpR6.3 indexing primer, and 1× KAPA SYBR Fast qPCR master mix.
2. The PCR program is: 95 °C for 30 s, 8 cycles of 95 °C for 10 s, 58 °C for 30 s, and 72 °C for 20 s, 10–15 cycles of 95 °C for 10 s and 72 °C for 20 s, and a final extension at 72 °C for 3 min. Monitor qPCR during amplification and stop the reaction before the fluorescent curve reaches plateau to prevent overamplification.
3. Verify the amplified amplicons on a 12-well 6% TBE gel by loading 3 μL of PCR products and run at 250 V in 0.5× TBE buffer for 25 min.

4. Stain the gel with 1× SYBR Gold in 0.5× TBE buffer for 4 min and visualize on a Gel Doc.
5. Purify PCR products with 0.7× volume of AMPure magnetic beads using the standard protocol by the manufacturer and elute with 50 μL TE buffer.
6. Measure the concentration of the purified PCR products with a NanoDrop or BioAnalyzer. BioAnalyzer will give a more accurate reading of the amplicon bands.
7. Pool equimolar ratio of each library into single tube for PAGE size selection.

3.8 Size Exclusion by Polyacrylamide Gel

Perform PAGE size selection of expected DNA size on a 5-well 6% TBE gel. The procedure is similar to that described in Subheading 3.2.4 except that DNA is purified on a TBE gel and no DNA denaturation and prerunning of the gel is performed. Approximately 1 μg can be purified on one gel.

1. Mix pooled sequencing libraries in 1× DNA loading dye and load ~250 ng per well of a 5-well 6% TBE gel. Load Low DNA mass ladder in the middle lane for size estimation.
2. Run at 250 V in 0.5× TBE buffer for 25 min.
3. Stain the gel with 1× SYBR Gold in 0.5× TBE buffer for 4 min as describe in Subheading 3.2.4 in a clean container to avoid library contamination.
4. Place stained gel on a clean Saran wrap on the UV transilluminator with the long wavelength.
5. Locate the position of sequencing libraries by comparing to Low DNA mass ladder.
6. Cut the gel with scalpel as close as to the band of sequencing libraries.
7. Follow the procedures for purifying DNA from polyacrylamide gel as described in Subheading 3.2.4.
8. Quantify sequencing libraries with a Qubit dsDNA HS assay.

3.9 Sequencing

1. Sequencing should be performed by a trained technician either within a Core facility or in-house. The adaptors are designed to be compatible with the Illumina sequencing systems. For other sequencing platforms, the PCR primers need to be modified to be compatible.
2. Paired-end sequencing must be performed for target sizes greater than the limit read length of the sequencer to be able to sequence across the entire target inserts.
3. Use the sequencing primer SolSeq6.3.3 for single-end sequencing or the first read of PE sequencing.

4. Use the sequencing primer AmpR6.3IndSeq for the indexing read.
5. Use the sequencing primer SolSeqV6.3.2r for the second read of PE sequencing.

3.10 Data Analysis

1. Use genomePrep.pl to generate the in silico bisulfite converted references, Cytosines (C) converted to Thymines (T) for Watson, and Guanines (G) converted to Adenines (A) for the Crick strand. Note that bisulfite conversion makes the two strands noncomplementary.
2. Use aligner software to create *index* files from the reference sequence for alignment. Both strands (genome.bis.CT and genome.bis.GA) can be concatenated into one file and only one index needs to be created so long as the aligner and computer RAM can support larger index files.
3. Use samtools "faidx" command to generate the fasta index file from the reference sequence.
4. Generate list_fastq_file or a tab separated table of the files to be processed. Each column in this file represents:

   <sample id> <dir> <read1.fq | read1.fq,read2.fq | sampleID.sam> <phred> <clonal method> <adaptor r1 sequence> <adaptor r2 sequence><library type>. Clonal method could be "none" or "samtools". The supported library types are "RRBS," "WGBS," and "BSPP." Trimming must be done specific to each library type for accurate methylation quantitation. For BSPP libraries, trimming from 5′ end of reads is required to remove capture arm sequences (~27 bases).
5. Generate the list_paths file. An example of a list_paths file is in the Example directory.
6. Run MasterBisReadMapper.pl with list_fastq_files and list_-paths as inputs.

4 Notes

1. The maximum target length must take into consideration the desired sequencing platform for the assay. For most applications, a maximum target length of 100 bp should be sufficient. Longer target lengths will require longer sequencing reads or paired-end sequencing to cover the entire targeted region. There is also an inverse correlation between the capture efficiency and the target length.
2. The minimum and maximum target length must be close to enable selection of specific capture products prior to sequencing. Larger differences will lead to more potential nonspecific products being sequenced.

3. The maximum number of CpGs in the capture arm should be limited to 0 or 1. Including more CpGs means that more alternative probes must be synthesized to avoid capture bias towards methylated or unmethylated targets. The capture efficiency of probes can also vary between methylated or unmethylated probes. The presence of CH methylation is negligible in most cell types so we safely assume that they are unmethylated.
4. Minimum H1 and H2 lengths lower than 25 bases should be avoided as this may lead to capture sequences with very low melting temperatures that may not anneal efficiently. H1 plus H2 length should not be longer than 60 bases as this will lead to higher synthesis cost and potentially more nonspecific products.
5. Unique molecular identifiers (UMI) can be attached to padlock probes to allow for single molecules counting and removal of clonal reads [7].
6. There are variations of oligonucleotide quality from different vendors. The concentration of oligonucleotides in expansion PCR should be optimized in small volume (25 μL) by varying concentration between 1, 20, 50, and 100 nM.
7. Probe to target ratio could be varied. Optimization is required to get optimal amount of probes for successful BSPP capture. If the library size is less than ~12,000 probes, the ratio of 1000:1 up to 2000:1 are recommended. If the library size is larger than 12,000 probes, the ratio could be reduced to 500:1 or 200:1.
8. If padlock capture is performed with a small amount of bisulfite converted DNA and a small padlock library size, reducing capture volume to 10 μL can increase capture efficiency.
9. KLN mix and exonuclease I/III mix should be added directly to the capture reaction and not to the layer of mineral oil. This could be done by inserting the pipette tip underneath the mineral oil layer for few seconds before releasing solution mix.

Acknowledgments

We gratefully acknowledge Athurva Gore and Jie Deng for early works on developing the software for padlock probe design and experiments on capture conditions respectively. This work is funded by NIH grants R01GM097253 and R01AG042187. D.D. is supported by a UCSD-CIRM predoctoral fellowship.

References

1. Deng J, Shoemaker R, Xie B et al (2009) Targeted bisulfite sequencing reveals changes in DNA methylation associated with nuclear reprogramming. Nat Biotechnol 27:353–360
2. Diep D, Plongthongkum N, Gore A et al (2012) Library-free methylation sequencing with bisulfite padlock probes. Nat Methods 9:270–272

3. Gore A, Li Z, Fung HL et al (2011) Somatic coding mutations in human induced pluripotent stem cells. Nature 471:63–67
4. Li H, Durbin R (2010) Fast and accurate long-read alignment with burrows-wheeler transform. Bioinformatics 26:589–595
5. Li H, Handsaker B, Wysoker A et al (2009) The sequence alignment/map format and SAMtools. Bioinformatics 25:2078–2079
6. Shoemaker R, Deng J, Wang W et al (2010) Allele-specific methylation is prevalent and is contributed by CpG-SNPs in the human genome. Genome Res 20:883–889
7. Hiatt JB, Pritchard CC, Salipante SJ et al (2013) Single molecule molecular inversion probes for targeted, high-accuracy detection of low-frequency variation. Genome Res 23:843–854

Chapter 20

Targeted Bisulfite Sequencing Using the SeqCap Epi Enrichment System

Jennifer Wendt, Heidi Rosenbaum, Todd A. Richmond, Jeffrey A. Jeddeloh, and Daniel L. Burgess

Abstract

Cytosine methylation has been shown to have a role in a host of biological processes. In mammalian biology these include stem cell differentiation, embryonic development, genomic imprinting, inflammation, and silencing of transposable elements. Given the central importance of these processes, it is not surprising to find aberrant cytosine methylation patterns associated with many disorders in humans, including cancer, cardiovascular disease, and neurological disease. While whole genome shotgun bisulfite sequencing (WGBS) has recently become feasible, generating high sequence coverage data for the entire genome is expensive, both in terms of money and analysis time, when generally only a small subset of the genome is of interest to most researchers. This report details a procedure for the targeted enrichment of bisulfite treated DNA via SeqCap Epi, allowing high resolution focus of next generation sequencing onto a subset of the genome for high resolution cytosine methylation analysis. Regions ranging in size from only a few kb up to over 200 Mb may be targeted, including the use of the SeqCap Epi CpGiant design which is designed to target 5.5 million CpGs in the human genome. Finally, multiple samples may be multiplexed and sequenced together to provide an inexpensive method of generating methylation data for a large number of samples in a high throughput fashion.

Key words Sequencing, NGS, Bisulfite, Targeted enrichment, SeqCap Epi, Hybridization

1 Introduction

The most commonly studied post-replicative and reversible modification of DNA is cytosine methylation [1]. Cytosine methylation has been shown to have a role in a host of biological processes. In mammalian biology these include stem cell differentiation, embryonic development, genomic imprinting, inflammation, and silencing of transposable elements. Given the central importance of these processes, it is not surprising to find aberrant cytosine methylation patterns associated with many disorders in humans, including cancer, cardiovascular disease, and neurological disease. In plant biology, cytosine methylation has been found to be similarly

Jörg Tost (ed.), *DNA Methylation Protocols*, Methods in Molecular Biology, vol. 1708,
https://doi.org/10.1007/978-1-4939-7481-8_20,

important for transposable element and gene silencing, as well as for genome stability. Recent interest among plant biologists in cytosine methylation comes from the commercialization of transgenic crop species since the transgene function can be susceptible to the methylation patterns of DNA adjacent to the integration site. In both plant and animal genomes an understanding of cytosine methylation patterns ("DNA methylation" hereafter) can be essential for understanding gene regulation. DNA methylation patterns are mitotically and meiotically stable, allowing them to be passed both to daughter cells and through the germ line to progeny. The ability to transmit DNA methylation patterns faithfully through multiple cellular generations, and the ability of the daughter cells to consistently interpret the pattern biologically, has led to their exploration as an inherited mechanism of gene regulation. Though heritable, DNA methylation is also reversible and generally considered an "epigenetic" mechanism of gene regulation, that is, regulation imparted "above" that of the nucleotide base sequence itself.

Since the development of robust methods for chemically deaminating cytosine with bisulfite were first published [2] analysis has increasingly been aimed at deciphering the methylation status of individual cytosine bases. Because the fidelity of cytosine methylation pattern maintenance is lower than the fidelity of the intrinsic base sequence [3], careful analysis requires a high depth of coverage to quantify the proportion of methylation occupancy at each cytosine position [4].

Whole genome shotgun bisulfite sequencing (WGBS) has recently become practical to consider as a means to characterize cytosine methylation patterns, as it provides DNA methylation data at single base resolution and allows for the assessment of the percentage of DNA methylation at each position in the genome by analyzing the same position across a population of individual molecules [5]. However, WGBS is time-consuming and costly due to a large amount of sequencing and data analysis required (most vertebrate and plant genomes contain hundreds of millions of cytosines), when generally 65% of the reads do not contain methylate-able sites (predominantly CG or CHG sites), and only a small subset of the genome is of interest to most researchers.

Chemical deamination of cytosine (C) bases to uracil (U) bases with bisulfite treatment brings a series of challenges to variant analysis, since millions of single nucleotide variants (SNVs) are created relative to the reference sequence. Each C base may be converted into a U base dependent upon whether or not it was methylated at the carbon-5 position (5mC). 5mC is substantially resistant to chemical deamination via bisulfite, while unmodified C is not. The U bases introduced by bisulfite are replaced with thymidine (T) bases during conventional DNA amplification and the sequence is analyzed for C or T content at C positions in the reference. The appearance of a T at a particular cytosine location in the reference is interpreted as evidence that the original C was not methylated in the sample. Bisulfite conversion chemistry is effectively chemical mutagenesis ex vivo. Substantial analytical complexity is imposed by converting a large fraction of C bases in a genome, and this happens on two levels. First, bisulfite conversion effective doubles a haploid genome by making the DNA strands noncomplementary to each other. Second, the loss of many Cs from the previously well-balanced four base repertoire of DNA reduces the algorithmic complexity of the new larger genome. Bioinformatic analyses of bisulfite treated DNA sequence including mapping and assembly routines, must take into account that each apparent C or T might have started as a C, a 5mC, or even a T polymorphism relative to the reference, and then analyze the variants across all molecules spanning the same position as a way to characterize the average methylation occupancy of the cellular population.

Whole genome sequencing with current instruments requires that a sequencing library be generated from a DNA sample. A series of laboratory manipulations fragments the DNA, repairs damaged ends, and ligates sequencing adapters onto them. The libraries are often then amplified before they are sequenced. The subsequent data analysis provides a comprehensive genome sequence by ensuring that each base in a genome is represented on multiple molecules in the library and then repeatedly sampling that library. However, only a fraction of the genome is typically of interest to most researchers. Sequence Capture [6] technology provides a means to focus the depth of coverage available with current sequencing systems over genomic regions of interest [7]. This is accomplished by hybridizing a sequencing library to immobilized biotinylated oligonucleotide probes, which are designed to be complementary to the regions of interest in a genome. Library fragments that are not complementary to probes can be washed away, leaving the regions of interest over-represented in the resulting captured library.

To integrate Sequence Capture technology with a bisulfite sequencing approach, at least two workflows are possible. A universal constraint on both workflows is that, since the methylation patterns on DNA molecules recovered from a biological sample are not copied with currently available DNA amplification systems, any necessary amplification steps must only occur after bisulfite treatment of the DNA sample is completed. In the first workflow, a sequencing library is constructed, then library fragments from the regions of interest are captured, bisulfite treated, amplified and sequenced. In the second workflow, a sequencing library is constructed, then bisulfite treated *before* the fragments of interest are captured, amplified and sequenced (Fig. 1).

The first workflow (capture then convert) has the advantage of allowing the capture probes to be designed in native (four base)

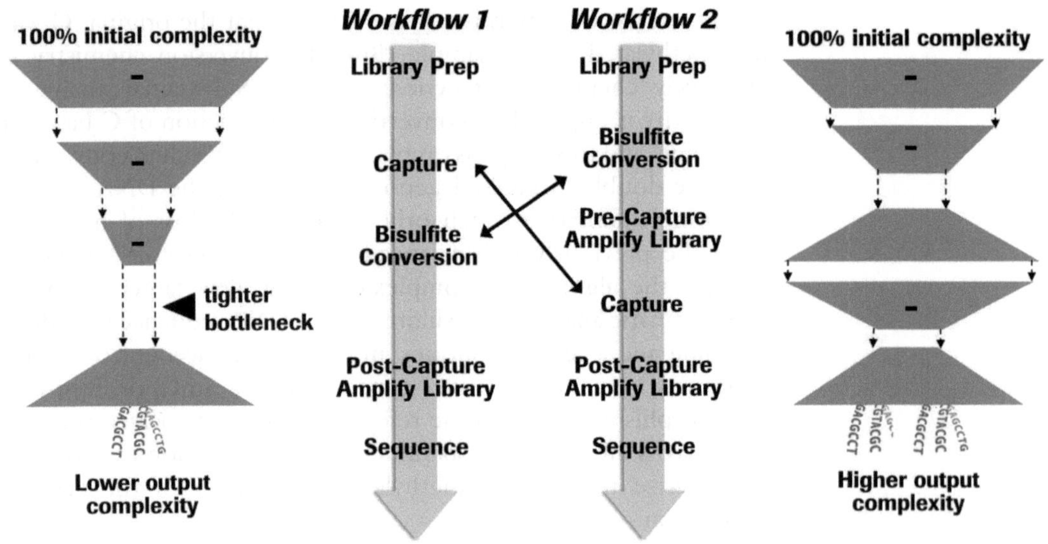

Fig. 1 Alternate workflows for integrating Sequence Capture and bisulfite sequencing. Library preparation, bisulfite conversion, and sequence capture are all steps where significant molecular complexity and information can be lost through inefficiency or bias. If those three steps are arranged in an uninterrupted series, as shown for Workflow 1, the effects are compounded and the information complexity bottleneck is tighter. More sample is required to offset the bottleneck effect and a higher rate of duplicate sequencing reads is expected. In Workflow 2 (SeqCap Epi), the three steps are arranged so that an intervening pre-capture amplification increases the total mass of DNA prior to the capture step to alleviate the bottleneck effect

sequence space, which requires fewer probes and should allow for higher capture specificity manifest as a higher on-target read rate. This method suffers from a hidden disadvantage, however, namely that the small molecular population recovered from capture of the sample library must be bisulfite converted before it can be amplified and sequenced. Bisulfite conversion is a harsh process that can damage as much as 90% of the library fragments to the extent that they cannot be amplified. Chaining each of these low yield steps together increases the potential that output sequencing libraries will suffer from low molecular diversity. Sequencing of a low diversity library yields highly redundant and more readily biased data, which may not be reflective of the actual methylation pattern in the sample.

The second capture workflow (convert then capture) does not suffer the same limitation, since an amplification step prior to capture is possible, which can mitigate diversity losses without affecting the integrity of the methylation pattern on each molecule. However, workflow 2 does have a distinct disadvantage relative to workflow 1 in terms of probe design; since the methylation occupancy of the relevant Cs are not known prior to capture, the probe design must take into account the myriad of possible methylation states for each C in each probe target. The practical effect of this disadvantage is that probe design and synthesis must address a

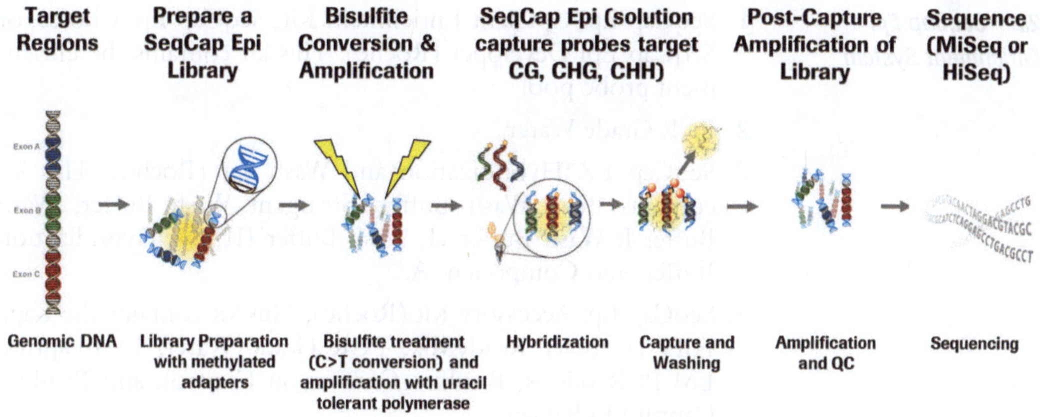

Fig. 2 SeqCap Epi workflow. The SeqCap Epi workflow incorporates bisulfite treatment and amplification steps before the hybrid capture step to reduce information loss through compounded molecular bottleneck effects

much larger possible sequence space to ensure high efficiency of molecule recovery in a manner irrespective of the actual native methylation occupancy of each molecule.

The SeqCap Epi Enrichment System, employing workflow 2, has overcome these challenges by utilizing a design algorithm that creates probes complementary to the possible methylation configurations on both strands from a bisulfite converted genomic template, and a manufacturing process which can create DNA oligonucleotide probe pools with tens of millions of oligos. This manufacturing process and the innovations in probe design have made the second workflow (convert then capture; Fig. 2) not only technically feasible, but effective. The SeqCap Epi Enrichment system enables the reproducible targeting of selected genomic regions, up to 210 Mb in cumulative size, from bisulfite treated genomic DNA, in a single workflow. The system has been validated in both plant (maize) and mammalian genomes (mouse and human) [8–10].

2 Materials

2.1 DNA Sample Library Preparation and Bisulfite Conversion

1. KAPA Library Prep Kit Illumina (Roche) or similar.
2. TE Buffer (1×): 10 mM Tris–HCl, pH 8.0, 0.1 mM EDTA.
3. SeqCap Adapter Kit A (Roche).
4. SeqCap Adapter Kit B (Roche).
5. Agencourt AMPure XP (Beckman Coulter).
6. Ethanol, 200 proof (absolute), for molecular biology.
7. EZ DNA Methylation-Lightning Kit (Zymo Research or similar).

2.2 SeqCap Epi Enrichment System

1. SeqCap Epi CpGiant Enrichment Kit, SeqCap Epi Choice, or SeqCap Epi Developer (Roche). This kit contains the enrichment probe pool.
2. PCR Grade Water.
3. SeqCap EZ Hybridization and Wash Kit (Roche). This kit contains Bead Wash Buffer, Stringent Wash Buffer, Wash Buffer I, Wash Buffer II, Wash Buffer III, 2× Hybridization Buffer, and Component A.
4. SeqCap Epi Accessory Kit (Roche); This kit contains the Kapa HiFi HotStart ReadyMix, PCR Grade Water, Post-capture LM-PCR oligos, Bisulfite Conversion Control, and Bisulfite Capture Enhancer.
5. SeqCap HE-Oligo Kit A (Roche).
6. SeqCap HE-Oligo Kit B (Roche).
7. SeqCap EZ Pure Capture Bead Kit (Roche). This kit contains Purification Beads and Capture Beads.
8. High Sensitivity DNA Kit (Agilent).
9. DNA 1000 Kit (Agilent).

3 Methods

3.1 Storing the SeqCap Epi Probe Pool, Accessory Kit, Adapter Kits A and B, HE-Oligo Kits A and B, and Pure Capture Bead Kits

1. Upon receipt, undertake the following steps to ensure the highest performance of the SeqCap Epi probe pool to avoid multiple freeze/thaw cycles or potential accidental contamination:
   (a) If frozen, thaw the tube of SeqCap Epi probe pool on ice.
   (b) Vortex the SeqCap Epi probe pool for 3 s.
   (c) Centrifuge the tube of SeqCap Epi probe pool at 10,000 × *g* for 30 s to ensure that the liquid is at the bottom of the tube before opening.
   (d) Aliquot the SeqCap Epi probe pool into single-use aliquots (4.5 μL/aliquot) in 0.2 mL PCR tubes (or 96-well plates if following the higher throughput protocol) and store at −15 to −25 °C until use. The presence of some residual volume after dispensing all single-use aliquots is normal.
   (e) When ready to perform the experiment, thaw the required number of single-use SeqCap Epi probe pool aliquots on ice.
2. Upon receipt, store the SeqCap Epi Accessory Kit, SeqCap Hybridization and Wash Kit, SeqCap Adapter Kits A and B, and SeqCap HE-Oligo Kits at −15 to −25 °C.
3. Upon receipt, store the SeqCap Pure Capture Bead Kit at +2 to +8 °C.

3.2 Preparing the Sample Library and Performing Bisulfite Conversion

1. Resuspend the SeqCap adapters.
   (a) Spin the lyophilized index adapters, contained in the SeqCap Adapter Kit A and B, briefly to allow the contents to pellet at the bottom of the tube.
   (b) Add 50 μL cold, PCR-grade water to each of the 12 tubes labeled "SeqCap Index Adapter" in the SeqCap Adapter Kit A or B. Keep adapters on ice.
   (c) Briefly vortex the index adapters plus PCR-grade water and spin down the resuspended index adapter tubes.
   (d) The resuspended index adapter tubes should be stored at −15 to −25 °C.
2. Preparation of the bisulfite conversion control.
   (a) Briefly spin down the tube containing the Bisulfite conversion control to ensure that the entire tube contents are in the bottom of the tube.
   (b) Add 1 mL of PCR-grade water directly to the tube containing the Bisulfite conversion control.
   (c) Vortex for 10 s to mix.
   (d) Briefly spin down.
   (e) Aliquot 990 μL of the PCR-grade water into a new 1.5 mL tube.
   (f) Add 10 μL of the diluted Bisulfite conversion control to the tube containing 990 μL of PCR-grade water.
   (g) Vortex for 10 s to mix.
   (h) Briefly spin down. The Bisulfite conversion control is now ready for use. The diluted Bisulfite conversion control should be stored at −15 to −25 °C and repeated freeze/thaws should be avoided by creating aliquots of this diluted reagent.
3. Preparing the sample library.
   (a) Add 5.8 μL of the diluted Bisulfite conversion control to 1 μg of the gDNA sample of interest.
   (b) Vortex for 10 s to mix.
   (c) Adjust the volume of the combined gDNA sample and Bisulfite conversion control to a total volume of 52.5 μL using 1× TE and transfer to a Covaris microTUBE for fragmentation.
   (d) Fragment the DNA pool so that the average DNA fragment size is 180–220 bp.
   (e) Following fragmentation, transfer 50 μL of the fragmented DNA to a 0.2 mL PCR tube and construct the sample library following the procedure described in the KAPA

Library Preparation Kit Technical Data Sheet, v1.14 (or later).

- Follow the protocol beginning with "End Repair Reaction Setup" and "End Repair Cleanup", **steps 2** and **3** of the KAPA Library Preparation Kit Technical Data Sheet.
- Proceed to "A-Tailing Reaction Setup" and "A-Tailing Cleanup", **steps 4** and **5** of the KAPA Library Preparation Kit Technical Data Sheet.
- Proceed to "Adapter Ligation Reaction Setup", **step 6** of the KAPA Library Preparation Kit Technical Data Sheet. Use 5 μL of a 10 μM solution of the SeqCap Library Adapter of choice in the "Adapter Ligation Reaction Setup".
- Proceed to "Adapter Ligation Cleanup", **step 7** of the KAPA Library Preparation Kit Technical Data Sheet, and perform the "First Post-Ligation Cleanup". Skip the "Second Post-Ligation Cleanup".
- Proceed to the "Dual-SPRI Size Selection", **step 8** of the KAPA Library Preparation Kit Technical Data Sheet.

(f) Once the sample library is constructed, proceed with the bisulfite conversion.

4. Bisulfite conversion of the sample library.

(a) Follow the steps in the Zymo Research "EZ DNA Methylation-Lightning Kit" for the bisulfite conversion of the DNA sample libraries.

- Note: Some thermal cycles have a maximum volume capacity of 100 μL. When using a thermal cycler that is not designed to work with volumes greater than 100 μL, split the bisulfite conversion reaction into two separate 0.2 mL PCR tubes (75 μL per tube). Once the thermal cycler program has completed, the two-like samples can be purified using a single Zymo EZ DNA Methylation-Lightning Kit column.
- Elute the purified bisulfite-converted sample library using 20 μL of PCR-grade water.

(b) Once the bisulfite conversion of the DNA sample library is complete, proceed with amplification of the sample.

3.3 Amplifying the Bisulfite-Converted Sample Library Using LM-PCR

1. Resuspend the SeqCap Pre-LM-PCR oligos.

(a) Spin the lyophilized primers briefly to allow the contents to pellet at the bottom of the tube. Please note that both primers are contained within a single tube.

(b) Add 550 μL PCR-grade water to the tube labeled "Pre-LM-PCR Oligos 1 & 2" from the SeqCap Adapter Kit A or B.

(c) Briefly vortex the primers plus PCR-grade water and spin down the resuspended oligo tube.

(d) The resuspended oligo tube should be stored at −15 to −25 °C.

2. Preparing the LM-PCR reaction.

(a) Prepare the LM-PCR Master Mix in a 1.5 mL microcentrifuge tube (or 15 mL conical) tube on ice. The amount of each reagent needed is listed below (if desired, increase all Master Mix volumes by 10% to account for pipetting variance). It is recommended to include both a negative (water) and positive (previously amplified library) controls in this step.

Pre-capture LM-PCR master mix	Per individual sample library or negative control (μL)	24 Sample libraries (μL)	96 Sample libraries (μL)
KAPA HiFi HotStart Uracil + ReadyMix (2×)[a]	25	600	2400
PCR grade water	2	48	192
Pre LM-PCR Oligo 1 and 2, 5 μM	3	72	288
Total	30	720	2880

[a]Note: The Kapa HiFi HotStart Uracil + Ready Mix (2×) is contained within the SeqCap Epi accessory kit

(b) Pipette 30 μL of LM-PCR Master Mix into each PCR tube or well.

(c) Add the 20 μL of the bisulfite-converted sample library (or PCR grade water for negative control) to the PCR tube, or each well of the 96-well plate, containing the LM-PCR Master Mix. Mix well by pipetting up and down five times. Do not vortex.

3. Performing the PCR amplification.

(a) Place the PCR tube (or 96-well PCR plate) in a thermal cycler.

(b) Amplify the bisulfite-converted sample library using the following Pre-Capture LM-PCR program: 2 min at 95 °C, followed by 12 cycles of 30 s at 98 °C, 30 s at 60 °C and 4 min at 72 °C, followed by 10 min at 72 °C, hold at 4 °C (*see* **Note 1**).

(c) Store the reaction at +2 to +8 °C (for up to 72 h) until ready for cleanup.

4. Purifying the amplified bisulfite-converted sample library using DNA purification beads (*see* **Note 2**).

    (a) Allow the DNA Purification Beads to warm to room temperature for at least 30 min prior to use.

    (b) Transfer each amplified, bisulfite-converted sample library into a separate 1.5 mL microcentrifuge tube (approximately 50 μL). Process the negative control in exactly the same way as the amplified, bisulfite-converted sample library.

    (c) Vortex the beads for 10 s before use to ensure a homogenous mixture of beads.

    (d) Add 90 μL (1.8× volume) DNA Purification Beads to the 50 μL of amplified, bisulfite-converted sample library.

    (e) Vortex briefly and incubate at room temperature for 15 min to allow the DNA to bind to the beads.

    (f) Place the tube containing the bead bound DNA in a magnetic particle collector.

    (g) Allow the solution to clear.

    (h) Once clear, remove and discard the supernatant being careful not to disturb the beads.

    (i) Add 200 μL of freshly prepared 80% ethanol to the tube containing the beads plus DNA. The tube should be left in the magnetic particle collector during this step.

    (j) Incubate at room temperature for 30 s.

    (k) Remove and discard the 80% ethanol and repeat **steps i–j.** for a total of two washes with 80% ethanol.

    (l) Following the second wash, remove and discard all of the 80% ethanol.

    (m) Allow the beads to dry at room temperature with the tube lid open for 15 min (or until dry) (*see* **Note 3**).

    (n) Remove the tube from the magnetic particle collector and resuspend the DNA using 52 μL of PCR-grade water (*see* **Note 4**).

    (o) Vortex for 10 s to mix to ensure that all of the beads are resuspended.

    (p) Incubate at room temperature for 2 min.

    (q) Place the tube back in the magnetic particle collector and allow the solution to clear.

    (r) Remove 50 μL of the supernatant that now contains the amplified, bisulfite-converted sample library and aliquot into a new 1.5 mL tube.

5. Checking the quality of the amplified, bisulfite-converted sample library.

   (a) Measure the A_{260}/A_{280} ratio of the amplified, bisulfite-converted sample library to quantify the DNA concentration using a NanoDrop spectrophotometer and determine the DNA quality.

   - The A_{260}/A_{280} ratio should be 1.7–2.0.
   - The sample library yield should be ≥1.0 μg.
   - The negative control yield should be negligible. If this is not the case, the measurement may be high due to the presence of unincorporated primers carried over from the LM-PCR reaction and not an indication of possible contamination between amplified sample libraries; however, this should be verified using an 2100 Bioanalyzer instrument as follows.

   (b) Run 1 μL of each amplified, bisulfite-converted sample library (and negative/positive controls) on a High Sensitivity DNA chip (Fig. 3). Run the chip according to the manufacturer's instructions. If the amplified, bisulfite-converted sample library has a concentration above 20 ng/μL consider diluting the sample 1:10 using PCR-grade water prior to running it on the High Sensitivity DNA chip. The Bioanalyzer trace should indicate that the average DNA fragment size falls between 150 and 500 bp. The negative control should not show any significant signal within this range, which could indicate

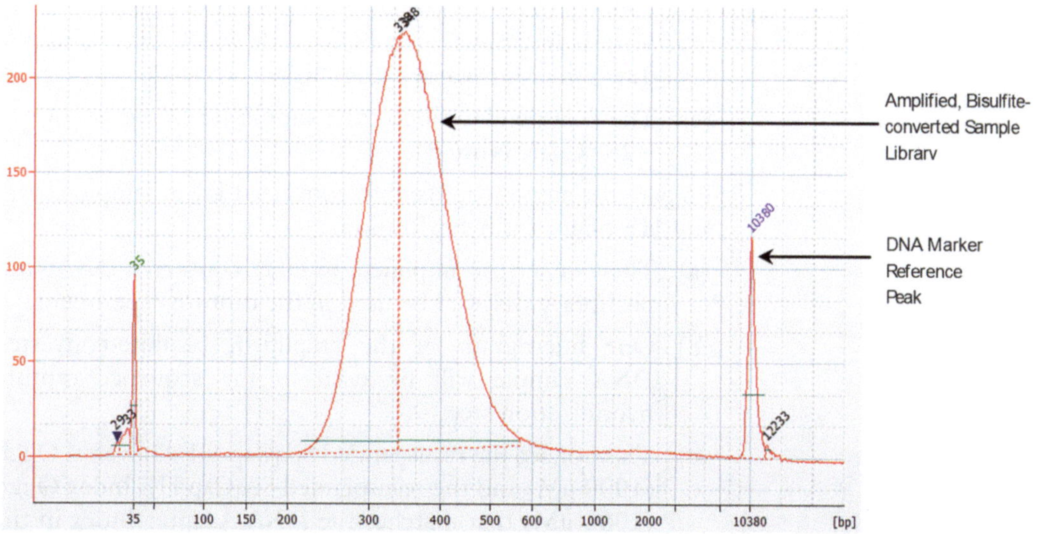

Fig. 3 Pre-Capture LM-PCR. Example of an amplified, bisulfite-converted sample library analyzed using an Agilent high sensitivity DNA chip

contamination between amplified sample libraries, but it may exhibit sharp peaks visible below 150 bp. If the negative control reaction shows a positive signal by NanoDrop spectrophotometer, but the Bioanalyzer trace indicates only the presence of a sharp peak below 150 bp in size, then the negative control should not be considered contaminated.

(c) If the amplified, bisulfite-converted sample library meets these requirements, proceed to hybridization. If the amplified, bisulfite-converted sample library does not meet these requirements, reconstruct the sample library.

3.4 Hybridizing the Amplified, Bisulfite-Converted Sample Library and SeqCap Epi Probe Pool

1. Preparing for hybridization.
   (a) Turn on a heat block to +95 °C and let it equilibrate to the set temperature.
   (b) Remove the appropriate number of 4.5 μL SeqCap Epi probe pool aliquots (one per bisulfite-converted sample library) from the −15 to −25 °C freezer and allow them to thaw on ice.
2. Resuspend the SeqCap HE universal and SeqCap HE Index oligos.
   (a) Spin the lyophilized oligo tubes, contained in the SeqCap HE-Oligo Kits A or B, briefly to allow the contents to pellet to the bottom of the tube.
   (b) Add 120 μL PCR-grade water to the SeqCap HE Universal Oligo tube (1000 μM final concentration).
   (c) Add 10 μL PCR-grade water to each SeqCap HE Index Oligo tube (1000 μM final concentration).
   (d) Vortex the primers plus PCR-grade water for 5 s and spin down the resuspended oligo tube.
   (e) The resuspended oligo tube should be stored at −15 to −25 °C (*see* **Note 5**).
3. Prepare the amplified, bisulfite-converted DNA sample library and HE oligos for hybridization.
   (a) Thaw on ice the amplified, bisulfite-converted DNA sample library that will be used in the capture experiment.
   (b) One microgram of the amplified, bisulfite-converted DNA sample will be used in the sequence capture hybridization step.
   (c) Thaw on ice the resuspended SeqCap HE Universal Oligo (1000 μM) and the resuspended SeqCap HE Index Oligo (1000 μM) that matches the DNA Adapter Index in the amplified, bisulfite-converted DNA sample library.

4. Prepare the hybridization sample.
   (a) Add 10 μL of Bisulfite Capture Enhancer (of the SeqCap Epi Accessory Kit) to a new 1.5 mL tube.
   (b) Add 1 μg of the amplified, bisulfite-converted DNA sample library to the 1.5 mL tube containing the 10 μL aliquot of Bisulfite Capture Enhancer.
   (c) Add 1 μL of the SeqCap HE Universal Oligo and 1 μL of the appropriate SeqCap HE Index Oligo to the amplified, bisulfite-converted DNA sample plus Bisulfite Capture Enhancer.
   (d) Close the tube's lid and make a hole in the top of the tube's cap with an 18–20 G (or smaller) needle. The closed lid with a hole in the top of the tube's cap is a precaution to suppress contamination in the DNA vacuum concentrator.
   (e) Dry the amplified, bisulfite-converted DNA sample/Bisulfite Capture Enhancer/Hybridization Enhancing Oligos in a DNA vacuum concentrator on high heat (+60 °C) (*see* **Note 6**).
      - To each dried-down amplified, bisulfite-converted DNA sample/Bisulfite Capture Enhancer/Hybridization Enhancing Oligos, add 7.5 μL of 2× Hybridization Buffer and 3 μL of Hybridization Component A (*see* **Note 7**).
      - The tube with the amplified, bisulfite-converted DNA sample/Bisulfite Capture Enhancer/Hybridization Enhancing Oligos should now contain the following components:

Component	Solution capture
Bisulfite capture enhancer	100 μg
Amplified, bisulfite-converted DNA sample	1 μg
Hybridization enhancing Oligos	2000 pmol[a]
2X hybridization buffer	7.5 μL
Hybridization component A	3 μL
Total	*10.5 μL*

[a]Composed of 50% (1000 pmol) SeqCap HE Universal Oligo and 50% (1000 pmol) of the appropriate SeqCap HE Index oligo

- Cover the hole in the tube's cap with a sticker or small piece of laboratory tape.
- Vortex the amplified, bisulfite-converted DNA sample/Bisulfite Capture Enhancer/Hybridization Enhancing Oligos plus

Hybridization Cocktail (2× Hybridization Buffer + Hybridization Component A) for 10 s and centrifuge at maximum speed for 10 s.

- Place the amplified, bisulfite-converted DNA sample/Bisulfite Capture Enhancer/Hybridization Enhancing Oligos/Hybridization Cocktail in a + 95 °C heat block for 10 min to denature the DNA.
- Centrifuge the amplified, bisulfite-converted DNA sample/Bisulfite Capture Enhancer/Hybridization Enhancing Oligos/Hybridization Cocktail at maximum speed for 10 s at room temperature.
- Transfer the amplified, bisulfite-converted DNA sample/Bisulfite Capture Enhancer/Hybridization Enhancing Oligos/Hybridization Cocktail to the 4.5 μL aliquot of SeqCap Epi probe pool in a 0.2 mL PCR tube (the entire volume can also be transferred to one well of a 96-well PCR plate).
- Vortex for 3 s and centrifuge at maximum speed for 10 s.
- The hybridization sample should now contain the following components:

Component	Solution capture
Bisulfite capture enhancer	100 μg
Amplified, bisulfite-converted DNA sample	1 μg
Hybridization enhancing Oligos	2000 pmol[a]
2× hybridization buffer	7.5 μL
Hybridization component A	3 μL
SeqCap Epi probe pool	4.5 μL
Total	*15 μL*

[a]Composed of 50% (1000 pmol) SeqCap HE Universal Oligo and 50% (1000 pmol) of the appropriate SeqCap HE Index Oligo

- Incubate in a thermal cycler at +47 °C for 64–72 h. The thermal cycler's heated lid should be turned on and set to maintain +57 °C (10 °C above the hybridization temperature).

3.5 Washing and Recovering the Captured Bisulfite-Converted DNA Sample

1. Preparing sequence capture and bead wash buffers.
   (a) Dilute the 10× Wash Buffer (I, II, III, and Stringent) and the 2.5× Bead Wash Buffer to create 1× working solutions (all of the buffers can be found in the SeqCap Hybridization and Wash Kit).

Amount of concentrated buffer	Amount of PCR grade water (μL)	Total volume of 1× buffer[a] (μL)
40 μL—10× stringent wash buffer	360	400
30 μL—10× wash buffer I	270	300
20 μL—10× wash buffer II	180	200
20 μL—10× wash buffer III	180	200
200 μL—2.5× bead wash buffer	300	500

[a]Store working solutions at room temperature (+15 to +25 °C) for up to 2 weeks. The volumes in this table are calculated for a single experiment; scale up accordingly if multiple samples will be processed

(b) Preheat the following wash buffers to +47 °C in a water bath:
- 400 μL of 1× Stringent Wash Buffer.
- 100 μL of 1× Wash Buffer I.

2. Preparing the capture beads.
   (a) All the Capture Beads to warm to room temperature for 30 min prior to use.
   (b) Mix the beads thoroughly by vortexing for 15 s.
   (c) Aliquot 100 μL of bead for each capture into a single 1.5 mL tube (i.e., for one capture use 100 μL of beads and for four captures using 400 μL of beads, etc.).
   (d) Place the tube in a magnetic particle collector. When the liquid becomes clear, remove and discard the liquid being careful to leave all of the beads in the tube. Any remaining traces of liquid will be removed wit subsequent wash steps.
   (e) While the tube is in the magnetic particle collector, add twice the initial volume of beads of 1× Bead Wash Buffer (i.e., for one capture use 200 μL of 1× Bead Wash Buffer and for four captures use 800 μL of 1× Bead Wash Buffer, etc.).
   (f) Remove the tube from the magnetic particle collector and vortex for 10 s.
   (g) Place the tube back in the magnetic particle collector to bind the beads. Once clear, remove and discard the liquid.
   (h) Repeat **steps e–g** for a total of two washes.
   (i) After removing the buffer following the second wash, resuspend by vortexing the beads in 1× the original volume using 1× Bead Wash Buffer (i.e., for one capture using 100 μL 1× Bead Wash Buffer and for four captures use 400 μL 1× Bead Wash Buffer, etc.).

(j) Aliquot 100 μL of resuspended beads into new 0.2 mL tubes (one tube per captured sample).

(k) Place the tube in the magnetic particle collector to bind the beads. Once clear, remove and discard the liquid.

(l) The Capture Beads are now ready to bind the captured DNA. Proceed immediately to "**step 3**" (below).

3. Binding DNA to the capture beads.

(a) Transfer the hybridization samples to the tubes containing the Capture Beads prepared above.

(b) Mix thoroughly by pipetting up and down ten times.

(c) Bind the captured sample to the beads by placing the tubes containing the beads and DNA in a thermal cycler at +47 °C for 45 min (heated lid set to +57 °C). Mix the samples by vortexing for 3 s at 15 min intervals to ensure that the beads remaining in suspension. It is helpful to have a vortex mixer located close to the thermal cycler for this step.

4. Washing the capture beads plus bound DNA.

(a) After the 45-min incubation, add 100 μL of 1× Wash Buffer I heated to +47 °C to the 15 μL of Capture Beads plus bound DNA.

(b) Mix by vortexing for 10 s.

(c) Transfer the entire content of each 0.2 mL tube to 1.5 mL tubes.

(d) Place the tubes in a magnetic particle collector to bind the beads. Remove and discard the liquid once clear.

(e) Remove the tubes from the magnetic particle collector and add 200 μL of 1× Stringent Wash Buffer heated to +47 °C. Pipette up and down ten times to mix. Work quickly so that the temperature does not drop much below +47 °C.

(f) Incubate at +47 °C for 5 min.

(g) Repeat **steps d–f** for a total of two washes using 1× Stringent Wash Buffer heated to +47 °C.

(h) Place the tubes in the magnetic particle collector to bind the beads. Remove and discard the liquid once clear.

(i) Add 200 μL of room temperature 1× Wash Buffer I and mix by vortexing (continually) for 2 min. If liquid has collected in the tube's cap, tap the tube gently to collect the liquid into the tube's bottom before continuing to the next step.

(j) Place the tubes in the magnetic particle collector to bind the beads. Remove and discard the liquid once clear.

(k) Add 200 μL of room temperature 1× Wash Buffer II and mix by vortexing (continually) for 1 min.

(l) Place the tubes in the magnetic particle collector to bind the beads. Remove and discard the liquid once clear.

(m) Add 200 μL of room temperature 1× Wash Buffer III and mix by vortexing (continually) for 30 s.

(n) Place the tubes in the magnetic particle collector to bind the beads. Remove and discard the liquid once clear.

(o) Remove the tubes from the magnetic particle collector and add 50 μL of PCR-grade water to each tube of bead-bound captured DNA sample.

(p) Store the beads plus captured samples at −15 to −25 °C or proceed with the next step (*see* **Note 8**).

3.6 Amplifying the Captured, Bisulfite-Converted DNA Sample Using LM-PCR and Sequencing the Captured Samples

1. Resuspend the post-LM-PCR oligos.
   (a) Spin the lyophilized oligos, contained in the SeqCap Epi Accessory Kit, briefly to allow the contents to pellet at the bottom of the tube. Please note that both oligos are contained within a single tube.
   (b) Add 480 μL PCR-grade water to the tube labeled "Post-LM-PCR Oligos 1 & 2" from the SeqCap Epi Accessory Kit.
   (c) Briefly vortex the primers plus PCR-grade water and spin down the resuspended oligo tube.
   (d) The resuspended oligo tube should be stored at −15 to −25 °C.
2. Preparing the LM-PCR reaction.
   (a) Prepare the LM-PCR Master Mix in a 1.5 mL microcentrifuge tube (or 15 mL conical) tube on ice. The amount of each reagent needed for two reactions (one captured DNA sample) is listed below (if desired, increase all Master Mix volumes by 10% to account for pipetting variance). It is recommended to include both a negative (water) and positive (previously amplified library) controls in this step.

Pre-capture LM-PCR master mix	Two reactions (for one captured DNA sample)
KAPA HiFi HotStart ReadyMix (2×)[a]	50 μL
Post LM-PCR Oligo 1 & 2, 5 μM	10 μL
Total	*30 μL*

[a]Note: The Kapa HiFi HotStart Ready Mix (2×) is contained within the SeqCap Epi accessory kit

(b) Pipette 30 μL of LM-PCR Master Mix into the PCR tubes or wells.

(c) Vortex the bead-bound captured DNA to ensure a homogenous mixture of beads.

(d) Aliquot 20 μL of the bead-bound captured DNA as template into each of the two PCR tubes or wells. Add 20 μL of PCR-grade water to the negative control.

(e) Mix well by gently pipetting up and down five times.

(f) Store the remaining bead-bound captured DNA at −15 to −25 °C.

3. Performing the PCR amplification.

   (a) Place the PCR tubes (or 96-well PCR plate) in a thermal cycler.

   (b) Amplify the captured DNA using the following Post-Capture LM-PCR program: 45 s at 98 °C, followed by 16 cycles of 45 s at 98 °C, 30 s at 60 °C and 30 s at 72 °C, followed by 1 min at 72 °C, hold at 4 °C.

   (c) Store the reaction at +2 to +8 °C (for up to 72 h) until ready for clean-up.

4. Purifying the amplified, captured bisulfite-converted sample using DNA purification beads (*see* **Note 9**).

   (a) Allow the DNA Purification Beads to warm to room temperature for at least 30 min prior to use.

   (b) Pool the like-amplified, captured bisulfite-converted DNA samples into a 1.5 mL tube (approximately 100 μL). Process the negative control in exactly the same way as the amplified, captured sample.

   (c) Vortex the beads for 10 s before use to ensure a homogenous mixture of beads.

   (d) Add 180 μL (1.8 × volume) DNA Purification Beads to the 100 μL pooled amplified, captured bisulfite-converted DNA samples.

   (e) Vortex briefly and incubate at room temperature for 15 min to allow the DNA to bind to the beads.

   (f) Place the tube containing the bead bound DNA in a magnetic particle collector.

   (g) Allow the solution to clear.

   (h) Once clear, remove and discard the supernatant being careful not to disturb the beads.

(i) Add 200 μL of freshly prepared 80% ethanol to the tube containing the beads plus DNA. The tube should be left in the magnetic particle collector during this step.

(j) Incubate at room temperature for 30 s.

(k) Remove and discard the 80% ethanol and repeat **steps i–j** for a total of two washes with 80% ethanol.

(l) Following the second wash, remove and discard all of the 80% ethanol.

(m) Allow the beads to dry at room temperature with the tube lid open for 15 min (or until dry) (*see* **Note 3**).

(n) Remove the tube from the magnetic particle collector and resuspend the DNA using 52 μL of PCR-grade water.

(o) Pipette up and down ten times to mix to ensure that all of the beads are resuspended.

(p) Incubate at room temperature for 2 min.

(q) Place the tube back in the magnetic particle collector and allow the solution to clear.

(r) Remove 50 μL of the supernatant that now contains the amplified, bisulfite-converted sample library and aliquot into a new 1.5 mL tube.

(s) Store the purified DNA sample at −15 to −25 °C.

5. Checking the quality of the amplified, captured bisulfite-converted DNA sample.

(a) Analyze 1 μL of the amplified, captured bisulfite-converted DNA sample and negative control using a DNA 1000 chip (Fig. 4) and measure the A_{260}/A_{280} ratio using a NanoDrop spectrophotometer to quantify the concentration of DNA and to determine the DNA quality. The negative control should not show significant amplification, which could be indicative of contamination. Amplified, captured bisulfite-converted DNA should exhibit the following characteristics:

- The A_{260}/A_{280} ratio should be 1.7–2.0.
- The LM-PCR yield should be ≥500 ng.
- The average DNA fragment length should be between 150 and 500 bp.

(b) The amplified, captured bisulfite-converted DNA meets the requirements, proceed to sequencing. If the amplified, captured bisulfite-converted DNA does not meet the A_{260}/A_{280} ratio requirement, purify again using the DNA Purification Beads (or alternatively, a second Qiagen QIAquick PCR Purification column).

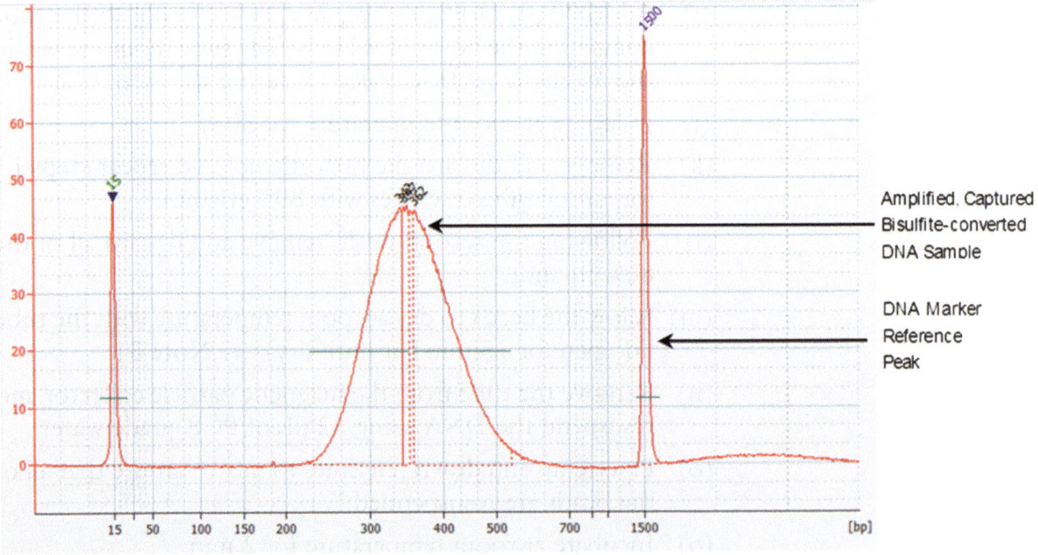

Fig. 4 Post-Capture LM-PCR. Example of an amplified, captured bisulfite-converted DNA sample analyzed using an Agilent DNA 1000 chip

6. Sequence the amplified, captured bisulfite-converted DNA sample (*see* **Note 10**).

3.7 Bisulfite Conversion and Capture Control (See Note 11)

1. Sequence reads should be aligned against the reference genome of the sample as well as against the lambda genome (NC_001416.1) (*see also* Chapter 5).
2. Once reads have been mapped, process the mapped reads using a methylation analysis package. The BSMAP alignment application (https://code.google.com/p/bsmap/) contains the methratio.py script, which determines methylation calls. A sample usage is:
   python2.6 methratio.py -d hg19_plus_NC001416.fa -s /usr/local/samtools -m 1 -z -i skip -o Sample.methylation_results.txt Sample.bam.
   This will produce a file with methylation calls for all C positions covered by mapped reads. Filter the result file to only include the methylation calls for NC_001416 between positions 4500 and 6500 (the region targets for capture).
   The first eight columns of the resulting file should look like the following:

Chr	Pos	Strand	Context	Ratio	Eff_CT_count	C_count	CT_count
NC_001416	4500	–	GCGGG	0.005	388	2	388
NC_001416	4501	–	CGGGT	0	392	0	392
NC_001416	4502	–	GGGTT	0	399	0	399
NC_001416	4512	–	TTGTG	0.007	437	3	437
NC_001416	4514	–	GTGCG	0.002	449	1	449
NC_001416	4515	+	TGCGC	0.007	538	4	538
NC_001416	4517	+	CGCTT	0	555	0	555
NC_001416	4521	+	TGCAG	0.003	575	2	575
NC_001416	4523	–	CAGGC	0.002	490	1	490
NC_001416	4524	–	AGGCC	0	497	0	497

Use the C_count and CT_count columns to calculate the percent conversion:

conversion rate = 1−(sum(C_count)/sum(CT_count).

For the above example, the conversion rate = 1−(13/4720) = 0.9972 = 99.72%.

3. Bisulfite conversion rates should generally be above 99.5% to be considered successful.

4 Notes

1. For the hybridization set up, 1 μg of amplified, bisulfite-converted sample library is recommended. Therefore, depending on the quality and quantity of the gDNA used during sample library construction and the results of the bisulfite conversion step, it may be necessary to increase the total number of PCR cycles to more than 12 cycles. Increasing the PCR cycles could result in an increase in the PCR duplicate rate (observed following sequencing). Additionally, it is possible to decrease the total number of PCR cycles if more than 1 μg of amplified, bisulfite-converted sample library is routinely obtained following the Pre-Capture LM-PCR.
2. Alternatively, samples can be purified using the Qiagen QIAquick PCR Purification Kit. If this purification method is chosen, follow the protocol detailed in the Qiagen QIAquick PCR Purification Kit guide with the following exception: Elute the amplified, bisulfite-converted sample library using PCR-grade water instead of Qiagen buffer EB.

3. Over drying of the beads can result in yield loss.
4. It is critical that the amplified, bisulfite-converted sample library is eluted with PCR-grade water and not Qiagen buffer EB or TE.
5. To prevent damage to the Hybridization Enhancing (HE) Oligos due to multiple freeze/thaw cycles, once resuspended the oligos can be aliquoted into smaller volumes to minimize the number of freeze/thaw cycles.
6. Denaturation of the DNA with high heat is not problematic after linker ligation because the hybridization utilizes single-stranded DNA.
7. These two reagents can be found in the SeqCap Hybridization and Wash Kit.
8. There is no need to elute the DNA off the beads. The beads plus captured DNA will be used as template in LM-PCR, as detailed below.
9. Alternatively, samples can be purified using the Qiagen QIAquick PCR Purification Kit. If this purification method is chosen, follow the protocol detailed in the Qiagen QIAquick PCR Purification Kit guide.
10. When working with bisulfite-converted sample libraries, some considerations need to be taken into account for sequencing when using an Illumina sequencing instrument. Due to the decreased nucleotide complexity in bisulfite-treated DNA, it may be necessary to perform one of the following options:
    (a) Designate a "control" lane when using a HiSeq2000 sequencing instrument. This lane should contain a DNA sample with a complex mixture of all four dinucleotides. Do not put your bisulfite-converted DNA sample in the control lane.
    (b) Mix in a sample with a more diverse representation of all four dinucleotides (with a different barcode index) into the bisulfite-converted DNA being sequenced. The diverse sample should be greater than or equal to 10% of the entire sample mixture.
    (c) In addition, due to the lower diversity, the input amount should be lowered slightly to generate a lower cluster count than standard. This will increase sequencing quality and result in more useful sequencing data.

    For best results, contact Illumina Technical Support prior to sequencing bisulfite-converted DNA.
11. This *Note* describes a procedure for using the Bisulfite conversion and Capture Control (BCC) that is provided in the SeqCap Epi Accessory Kit. This control consists of genomic DNA

from the Enterobacteria phage lambda, which is not naturally methylated at CpG dinucleotides in the *E. coli* host organism (*see also* Chapter 5). Bisulfite conversion of the phage lambda genomic DNA should thus result in all cytosines (C), including those in a CpG context, being converted to thymidines (T) after PCR amplification and sequencing. The completeness of C to T conversion in lambda DNA is used as a proxy for the completeness of bisulfite conversion of the experimental sample DNA in the mixture. If conversion of C to T in the lambda DNA is nearly complete, then observation of a CpG in the sample DNA is more likely to reflect true methylation status (i.e., true positives) rather than bisulfite conversion inefficiency (i.e., false positives). Although completeness of bisulfite conversion of non-CpG cytosines in the sample DNA itself (e.g., mitochondrial DNA) can also be used to estimate the completeness of bisulfite conversion using a reference sequence as a baseline, the use of a spike-in control will avoid variability related to sample quality and sequence polymorphism. Probes targeting a two-kilobase region between coordinates 4500 and 6500 bp of the lambda genome (NC_001416.1) are included by default in every SeqCap Epi design to specifically capture this region from the BCC. This streamlines the process of calculating bisulfite conversion by focusing analysis on a small and well-defined target.

References

1. Baubec T, Schübeler D (2014) Genomic patterns and context specific interpretation of DNA methylation. Curr Opin Genet Dev 25:85–92
2. Frommer M, McDonald LE, Millar DS et al (1992) A genomic sequencing protocol that yields a positive display of 5-methylcytosine residues in individual DNA strands. Proc Natl Acad Sci U S A 89:1827–1831
3. Ushijima T, Watanabe N, Okochi E et al (2003) Fidelity of the methylation pattern and its variation in the genome. Genome Res 13:868–874
4. Genereux DP, Miner BE, Bergstrom CT et al (2005) A population-epigenetic model to infer site-specific methylation rates from double-stranded DNA methylation patterns. Proc Natl Acad Sci U S A 102:5802–5807
5. Lister R, Pelizzola M, Dowen RH et al (2009) Human DNA methylomes at base resolution show widespread epigenomic differences. Nature 462:315–322
6. Hodges E, Xuan Z, Balija V et al (2007) Genome-wide in situ exon capture for selective resequencing. Nat Genet 39:1522–1527
7. Parikh H, Deng Z, Yeager M et al (2010) A comprehensive resequence analysis of the KLK15-KLK3-KLK2 locus on chromosome 19q13.33. Hum Genet 127:91–99
8. Li Q, Suzuki M, Wendt J et al (2015) Post-conversion targeted capture of modified cytosines in mammalian and plant genomes. Nucleic Acids Res 43(12):e81
9. Li Q, Eichten SR, Hermanson PJ et al (2014) Genetic perturbation of the maize methylome. Plant Cell 26:4602–4616
10. Allum F, Shao X, Guenard F et al (2015) Characterization of functional methylomes by next-generation capture sequencing identifies novel disease-associated variants. Nat Commun 6:7211

Chapter 21

Multiplexed and Sensitive DNA Methylation Testing Using Methylation-Sensitive Restriction Enzymes "MSRE-qPCR"

Gabriel Beikircher, Walter Pulverer, Manuela Hofner, Christa Noehammer, and Andreas Weinhaeusel

Abstract

DNA methylation is a chemically stable key-player in epigenetics. In the vertebrate genome the 5-methyl cytosine (5mC) has been found almost exclusively in the CpG dinucleotide context. CpG dinucleotides are enriched in CpG islands very frequently located within or close to gene promoters. Analyses of DNA methylation changes in human diagnostics have been conducted classically using methylation-sensitive restriction enzymes (MSRE). Since the discovery of bisulfite conversion-based sequencing and PCR assays, MSRE-based PCR assays have been less frequently used, although especially in the field of cancer epigenetics MSRE-based genome-wide discovery and targeted screening applications have been and are still performed successfully. Even though epigenome-wide discovery of altered DNA methylation patterns has found its way into various fields of human disease and molecular genetics research, the validation of findings upon discovery is still a bottleneck. Usually several multiples of 10 up to 100 candidate biomarkers from discovery have to be confirmed or are of interest for further work. In particular, bisulfite PCR assays are often limited in the number of candidates which can be analyzed, due to their low multiplexing capability, especially, if only small amounts of DNA are available from for example clinical specimens. In clinical research and diagnostics a similar situation arises for the analyses of cell-free DNA (cfDNA) in body fluids or circulating tumor cells (CTCs). Although tissue- or disease- (e.g., cancer) specific DNA methylation patterns can be deduced very efficiently in a genome-wide manner if around 100 ng of DNA are available, confirming these candidates and selecting target-sequences for studying methylation changes in liquid biopsies using cfDNA or CTCs remains a big challenge. Along these lines we have developed MSRE-qPCR and introduce here method details, which have been found very suitable for the efficient confirmation and testing of DNA methylation in a quantitative multiplexed manner (e.g., 48–96 plex) from ng amounts of DNA. The method is applicable in a standard qPCR setting as well for nanoliter scaled high-throughput qPCR, enabling detection of <10 copies of targets, thus suitable to pick up 0.1–1% of specific methylated DNA in an unmethylated background.

Key words Methylation-sensitive restriction enzyme, Multiplex, qPCR, MSRE-qPCR, DNA methylation

Jörg Tost (ed.), *DNA Methylation Protocols*, Methods in Molecular Biology, vol. 1708, https://doi.org/10.1007/978-1-4939-7481-8_21, © Springer Science+Business Media, LLC 2018

1 Introduction

Epigenetics, in particular DNA methylation, has become one of the major research subjects since its fundamental role in a huge variety of different cellular and developmental processes has been discovered, as well as its contribution in the formation of various diseases such as different cancer types or age related impairments and diseases [1]. Nevertheless, DNA methylation serves also as potential diagnostic tool for such diseases, whereby the focus is laid especially on the analysis of 5-methylcytosines (5mC) within the context of CpG dinucleotides [2].

The development of DNA methylation-based biomarkers, which may be used in a diagnostic test typically starts with a genome-wide screening. Various techniques are available for such a biomarker screening experiment including different array and next generation sequencing- based methods targeting either single CpG dinucleotides or certain gene regions [3] (*see also* Chapters 5–16). These screenings typically result in the definition of a large number of potential candidate biomarkers. Therefore further techniques are needed to confirm the findings, to reduce the number of significant targets and to validate markers in a large study cohort. To achieve these three major approaches can be applied including bisulfite deamination-, affinity- and restriction enzyme-based methylation testing [4] (*see also* Chapters 17–20). DNA methylation patterns have been investigated already very early with restriction enzymes and are still used nowadays, typically using methylation-sensitive restriction enzymes (MSREs), which contain CpG motifs in their recognition site. Since the activity of these enzymes is blocked by 5mC, only unmethylated sites are digested while methylated regions remain unaffected. Thus, upon successful digestion and after PCR amplification only methylated DNA should result in a detectable PCR product [5]. These amplification products can be simply detected by quantitative polymerase chain reaction (qPCR) and thus enable a statement about the methylation level of the targets of interest. Moreover, a calibration curve can be used for either relative or absolute quantification of the amount of methylated DNA at the investigated genomic regions [6]. The combination of MSRE digestion and qPCR-based detection offers the opportunity to use high-throughput PCR platforms and thus enables methylation testing of large study cohorts. In addition, since the available amount of DNA is often a limiting factor, the use of high-throughput instruments such as Fluidigm's Biomark™ System can constitute an optimal alternative not only to safe costs and time, but also to significantly reduce the amount of DNA needed per assay, which is crucial to be able to investigate a large number of targets [7].

Successful MSRE-qPCR methylation testing is characterized by one major prerequisite, which is the design of assays with respect to restriction sites as well as the completeness of the enzymatically digestion, to avoid false-positive results [8]. Design of MSRE-qPCR assays is usually very straightforward and less complicated than the design of bisulfite-based assays, as "native" DNA is targeted; however, primers must cover target regions presenting at least one MSRE-cut-site. To avoid technical difficulties and guarantee complete digestion of the targeted regions, we recommend using PCR assays providing at least three MSRE–cut-sites within their amplicon sequence. CpG-rich regions, which are mostly also candidate regions for methylation differences, are perfect targets for MSRE-qPCR assay design, as they typically contain a large number of MSRE cut-sites. Furthermore, the use of more than one MSRE is strongly recommended, in particular restriction enzymes *AciI*, *Hin6I*, *HpyCH4IV*, and *HpaII* usually provide a very good coverage of CpG-rich sequences [9]. We routinely use a combination of all four MSREs, which increases the probability of available cut sites, completeness of digestion as well as the multiplexing capabilities for targeting various regions in parallel. Although various primer design programs are available, we have set up a "high-throughput" MSRE-qPCR primer design tool freely available at http://sourceforge.net/projects/msrehtprimer/ [10, 11].

After successful design, primers should be tested and validated according to the specifications detailed in the MIQE guidelines by Bustin et al. [12] (Minimum Information for Publication of Quantitative Real-Time Experiments [12]). MIQE guidelines define the criteria to successfully evaluate qPCR assays and specify parameters, which are associated with highly reliable and bias-free qPCR data. By means of a calibration curve different parameters, as stated in the MIQE guidelines, including PCR efficiency, limit of detection (LOD), limit of quantification (LOQ), and the correlation coefficient can be calculated and allow an assessment of the performance and quality of the qPCR assay [12]. For MSRE-qPCR these technical parameters can be determined using undigested genomic DNA.

As mentioned above, the availability of sample is often a limiting factor, but can be overcome by using preamplification upon MSRE digestion, where only a little fraction of sample material is required to analyze a large number of targets. However, in order to ensure that sufficient DNA template is available for qPCR detection a certain minimum amount of starting material is required for preamplification. Even if assays are well optimized and can detect targets at single copy sensitivity, detection of for example 0.1% circulating tumor DNA in cell-free DNA from plasma or bodily fluids is only feasible when the theoretical input of 1000 genome equivalents (corresponds to 6.5 ng diploid human DNA) is available for amplification. MSRE-qPCR, targeting native DNA, does enable successful implementation of preamplification for DNA methylation analyses. To our knowledge, this is less easily achieved using bisulfite-based PCR. In our experience, preamplification can be integrated in the experimental design, which greatly facilitates the investigation of up to 100 target regions in parallel [13]. Thus, samples are preamplified in a multiplex reaction prior to the qPCR evaluation, where one reaction contains all assays being under investigation. This approach allows increasing the number of targets and technical replicates, which can be evaluated from one single sample [14]. Although the sensitivity and the precision can be significantly improved by adding such an additional amplification step, the number of cycles should be limited to avoid unintended bias and over-amplification of the most abundant targets [15]. Another disadvantage of such high multiplex reaction is that some of the assays might be underrepresented in the reaction mix and might lead to poor or loss of the amplification signal in the subsequent qPCR reaction, thus assays should be tested on standard qPCR experiments before clinical samples are analyzed. The combination of preamplification followed by "single-gene"- specific qPCR readout via double strand intercalating fluorescence dyes such as SYBR- or Eva-Green is well suited for high-throughput qPCR analyses using Fluidigm's nano-liter-scaled qPCR arrays as well as using for example 384-well qPCR platforms available from several vendors. Both strategies are described with all necessary technical details in the methods section.

The experimental design of the MSRE-qPCR methylation study has a direct impact on the different options for the analysis of the results. Basically, assays can be run in a "paired manner" to directly compare undigested and MSRE digested aliquots of each sample, or relative methylation differences can be deduced using internal standards. Absolute quantification is best performed using a calibration series of undigested DNA. Similar to other qPCR analyses, a minimum number of controls and control PCR reactions need to be integrated for MSRE-qPCR. For methylation studies these would be amplicons targeting known differentially methylated regions, unmethylated regions and amplicons where no restriction sites are present. Standards enable normalization of qPCR Ct values as well as delta-Ct or delta-delta-Ct-based analyses. In this respect the use of SYBR- and Eva-Green enables also to control the specificity of amplicons by melting curve analyses. Similar to established methods for, for example, mRNA expression qPCR data, absolute quantification via calibration series and internal standards can be conducted and a large number of software tools are available [16]. The main strategies based on our own experiences are detailed in the methods section.

Although MSRE-qPCR enables very efficient analyses from small amounts of DNA in a highly multiplexed manner and is well suited for clinical samples from various biological sources—which are often available in very limited quantities only—the approach is not well suited for the analysis of FFPE fixed tissue. For MSRE-qPCR examination of FFPE-derived DNA the variance between different sample handling routines and laboratories has to be considered and ideally examined in a proof-of-principle setting before putting much effort into such a study. Alternative fixatives—also useful for pathological examinations of tissue—like RCL2, urea or guanidinium- and methanol/acetic acid fixatives are, however, well suited. MSRE-qPCR can also be applied to more special applications making use of the specificity of restriction enzymes, e.g., *BstNI* and *PspGI* used for studying non-CpG methylation, 5-hydroxymethylation (upon glycosylation [17]) as well as to digest exclusively methylated DNA and amplify then the unmethylated DNA using enzymes like *McrBC*. Although the methods and analysis principles described in this chapter are focused on CpG methylation, these could be also of use for studying adenosine methylation in bacterial DNA using differential digestion of the DNA template with appropriate restriction cut sites (e.g., *MboI*, *BsaBI*).

In summary, MSRE-qPCR can be applied to a wide range of clinical and molecular biological epigenetic analyses, when multiple targets/regions of interest have to be studied in parallel and in an analytically sensitive manner.

2 Materials

2.1 Assay Design

1. PCR primer design using the design engine Primer3 from the University of Massachusetts (*http://biotools.umassmed.edu/bioapps/primer3_www.cgi*).
2. UCSC browser (http://genome.ucsc.edu/).
3. MSRE-HT primer design tool (software download and instructions available via: http://sourceforge.net/projects/msrehtprimer/).

2.2 MSRE-qPCR Testing

2.2.1 Sample and Calibration Standard Preparation

1. Genomic DNA extraction kit (e.g., High Pure PCR Template Preparation Kit, Roche Applied Science).
2. Quant-iT™ PicoGreen® dsDNA Assay Kit (Thermo Fisher Scientific).
3. Real-time thermocycler (e.g., LightCycler® 480 II).
4. Fluorospectrometer (e.g., NanoDrop 3300).

2.2.2 MSRE Digestion

1. *AciI* (New England Biolabs).
2. *HpyCH4IV* (New England Biolabs).
3. *Hin6I* (Thermo Fisher Scientific).

4. *HpaII* (Thermo Fisher Scientific).
5. Tango Buffer 10× (Thermo Fisher Scientific).
6. Thermocycler.

2.2.3 Digestion Control PCR Gel Electrophoresis

1. Digestion control PCR primers.
2. HotStarTaq DNA polymerase (Qiagen).
3. dNTPs.
4. DMSO (≥99.9%).
5. Thermocycler.
6. Agarose.
7. Loading dye.
8. DNA Ladder (100 bp).
9. Large-scale horizontal gel electrophoresis system (e.g., PerfectBlue Gel System Maxi S Plus (Peqlab)).
10. 10× Tris–Borate–EDTA (TBE) buffer: 890 mM Tris–borate and 20 mM EDTA, pH 8.3.
11. Gel stain (e.g., SYBR® Safe DNA Gel Stain).
12. Imaging System (e.g., BioSpectrum® 310, UVP or similar).
13. Microcon®-10 Centrifugal Filters Ultracel® YM-10 (Merck Millipore).
14. PCR certified water.

2.2.4 Preamplification

1. MSRE-qPCR primer pool.
2. HotStarTaq DNA polymerase (Qiagen).
3. dNTPs.
4. DMSO (≥99.9%).
5. Thermocycler.

2.2.5 MSRE-qPCR (Micro-Fluidic or 384-Well Format)

1. LightCycler® 480 II (Roche Applied Science).
2. 384 well PCR plate (e.g., FrameStar, 4titude Ltd).
3. 96 × 96 Dynamic Array IFC (Fluidigm).
4. Control Line Fluid (Fluidigm).
5. dNTPs.
6. 25 mM $MgCl_2$.
7. DMSO.
8. 20× EvaGreen Binding Dye.
9. HotStarTaq DNA polymerase (Qiagen).
10. ROX Reference Dye.
11. 20× DNA binding Dye Loading Reagent (Fluidigm).

12. 2× Assay Loading Reagent (Fluidigm).
13. HX IFC Controller HX (Fluidigm).
14. Biomark™ System (Fluidigm).

3 Methods

3.1 Assay Design

The development of MSRE-qPCR assays for methylation testing is based mostly on defined candidate regions, which are of interest for analyses or have been identified for example in screening experiments such as genome-wide approaches using microarray or NGS. Since the MSRE-qPCR-based approach is strongly dependent on the successful digestion of the unmethylated DNA fraction, the assay design must consider the restriction map of the target region. Primer pairs should be placed to encompass the genomic location of the selected target as well as several MSRE cut sites.

Well-designed and optimized qPCR assays allow detection of DNA in the single digit copy or even in the single copy range. Due to this analytical sensitivity complete cleavage of unmethylated DNA by the MSREs is crucial. To ensure complete cleavage of unmethylated DNA we recommend combining different MSREs to increase the number of possible targets for the MSREs, and to ensure that at least three cut sites from different MSREs are present in the final assay. In addition, an over digest by applying an excess amount of MSREs and an extended isothermal incubation time of up to 16 h is useful.

Before starting the assay design, which makes use of the design tool Primer 3 of the University of Massachusetts (http://biotools.umassmed.edu/bioapps/primer3_www.cgi), various design parameters must be specified including intended melting temperature, GC content, primer and product size. Here we describe the assay design, as well as the entire procedure, which is optimized for medium (384-well) and the high-throughput Biomark™ qPCR system (Fluidigm).

1. Following the identification of the target regions or CpGs of interest, define the genomic region around the targets or CpGs, respectively. This can be done by for example using the UCSC genome browser (http://genome.ucsc.edu/) to extract the sequence upstream and downstream the predefined target region or CpG in FASTA file format. An identical number of base pairs should be extracted on both sides, upstream as well as downstream to ensure that the region of interest is exactly in the middle of the deduced sequence to facilitate the later assay design. The final sequence should be adjusted to a length of approximately 300 bp.

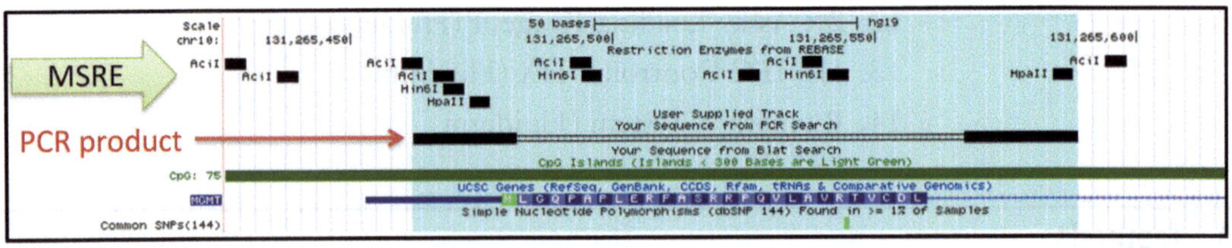

Fig. 1 UCSC-Browser image illustrating a MSRE-PCR design for the *MGMT* region exemplifying the location of MSRE sites (*HpaII*, *Hin6I*, *HypyCH4IV*, *AciI; green arrow*) within the CpG island (*dark green bar*) covering also the 1st exon of *MGMT* (*blue tracks*). The PCR product lane is indicated by a *red arrow*, primers are depicted as *black bars*; the PCR covered region highlighted in *blue*. (Region *chr10:131,265,400-131,265,651*; UCSC Genome Browser on Human Feb. 2009-GRCh37/hg19—Assembly)

2. Import the generated FASTA file into the Primer 3 online tool and design the primers with the following design parameters: product size region 65–125 base pairs (bp), primer size 18–25 bp, primer melting temperature between 68 and 72 °C and GC content between 20 and 100%. In addition, the melting temperature for the forward and reverse primer should not differ by more than 2 °C and the primer or product sequence should not contain any common repeats or single nucleotide polymorphisms (SNPs). The design parameters can be adapted to different experimental requirements.
3. Check the location and the length of the amplicons by in silico PCR using the UCSC genome browser. Make sure that there at least three restriction enzyme cut sites in the amplification product (Fig. 1).

For assisting the design process of multiple assays in parallel, we have developed the MSRE-HTPrimer program, which is an open-source, portable, web-based and easy-to-use pipeline, which facilitates the paralleled design of MSRE assays from a table of candidate regions. Software download and detailed instructions are available online (http://sourceforge.net/projects/msrehtprimer/). This high-throughput primer design tool is using reference genomes and genomic information including SNPs, RefSeq genes, repeats and CpG islands. MSRE-HTPrimer has also integrated a customized automatic hierarchical filtering option covering all parameters given in Primer3, considers annotated genomic features to avoid SNPs, and repeats, as well as warrants MSRE sites in the designed assays [10, 11].

3.2 MSRE-qPCR Assay

The workflow of the MSRE-qPCR assay is depicted in Fig. 2. Isolated DNA is subjected to MSRE digestion using a combination of enzymes; before the digestion is stopped we recommend conducting a PCR to control for completion of digestion. If DNA concentrations are too low, volume-reduction of the digestion should be performed by ultrafiltration (spin columns to enable a

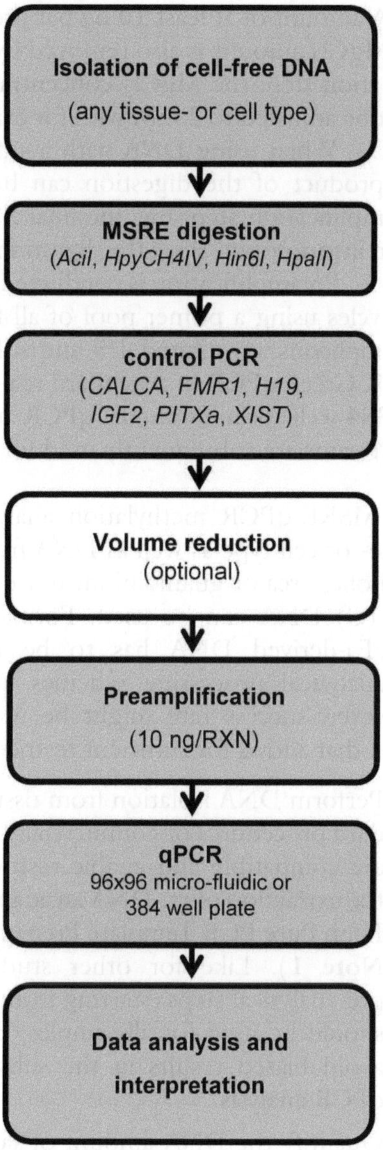

Fig. 2 Workflow for MSRE-qPCR analyses: Isolated DNA is subjected to MSRE digestion using a combination of enzymes; before the digestion is stopped, we recommend conducting a PCR to control for completion of digestion. Optionally when DNA concentrations are too low volume-reduction of the digestion should be done by ultrafiltration (Microcon®) spin columns to enable a DNA amount of at least 10 ng per preamplification reaction. Since the $MgCl_2$ amount is also removed by the purification and volume reduction step, the $MgCl_2$ concentration for the preamplification must be adjusted to 2 mM with for example PCR buffer containing $MgCl_2$. When using DNA with sufficient high concentration, it can be directly transferred to the preamplification, but the final $MgCl_2$ concentration must be taken into account, since the digestion mix already contains 10 mM $MgCl_2$. Preamplification is conducted on a standard PCR cycler for 22 cycles using a primer pool of all forward and reverse primers. Preamplicons are diluted 1:5 and subjected then to single Eva/SYBR-Green qPCR in standard well (preferably 384 well) or microfluidic qPCR arrays. Ct and Tm values from qPCR runs are then used for data analysis

DNA amount of at least 10 ng per preamplification reaction. Since the $MgCl_2$ amount is also removed by the purification and volume reduction step, the $MgCl_2$ concentration for the preamplification must be adjusted to 2 mM with for example PCR buffer containing $MgCl_2$. When using DNA with a sufficiently high concentration, the product of the digestion can be directly transferred to the preamplification step, but the final $MgCl_2$ concentration must be taken into account, since the digestion mix already contains 10 mM $MgCl_2$. Preamplification is conducted on a standard PCR cycler for 22 cycles using a primer pool of all forward and reverse primers. Preamplicons are diluted 1:5 and subjected then to single Eva-/SYBR-Green qPCR in a standard real-time Thermocycler (preferably 384 well) or microfluidic qPCR arrays. Ct and Tm values from qPCR runs are subsequently used for data analysis.

3.2.1 Sample Preparation and Calibration Standards

For MSRE-qPCR methylation analysis DNA isolated from any tissue- or cell type as well as DNA from fixed tissue (using RCL2, alcoholic, urea or guanidinium-based fixatives), or circulating cell-free (cf) DNA can be used. Formalin fixed paraffin embedded (FFPE)-derived DNA has to be tested with respect to the pre-analytical processing schemes in the different laboratories. However, success rate might be moderate for obtaining FFPE DNA that allows for sufficient restriction digestion.

1. Perform DNA isolation from tissue or cells using various standard procedures or commercially available isolation kits, which are compatible and enable restriction enzyme digestion. For the extraction of cf.-DNA an adapted protocol using the Roche High Pure PCR Template Preparation Kit is recommended (*see* **Note 1**). Like for other studies, the same protocol for pre-analytical steps (starting from biological sample collection) should be used for all samples/different biological groups to avoid biased results in the subsequent comparative MSRE-qPCR analysis.
2. Quantify the DNA amount of samples. Especially for samples with low DNA concentration quantification using the Quant-iT™ Picogreen® dsDNA Assay Kit is recommended. Perform the measurement according to the manufacturers' protocol. The DNA quantification can be performed in a real-time PCR system or by using a fluorospectrometer.
3. Prepare calibration standards by serial dilutions of a purified or commercially available DNA of high quality and known concentration. The concentration of the calibration curve should cover the range of expected samples' concentrations to ensure a reliable quantification.

3.2.2 MSRE Digestion

The digestion step is performed with methylation-sensitive restriction enzymes (MSREs), which contain CpG motifs in their respective recognition sites (*see* **Note 2**). Thus only unmethylated DNA is digested by the enzymes whereas methylated DNA remains unaffected. In our assay a cocktail of four different methylation-sensitive restriction enzymes is used to ensure complete digestion.

1. Digest 10 ng of input DNA with 9 U of each MSRE (*AciI*, *HpyCH4IV*, *Hin6I*, *HpaII*) in 1× Buffer Tango (containing 10 mM $MgCl_2$) in a total reaction volume of 65 μL. In order to ensure reliable results the calibration standards should be treated like the investigated DNA samples, prepare a so-called "mock-digestion" by replacing the volumes of the enzymes with DNase-free water.
2. Incubate the digestion mix at 37 °C for 16 h in a standard thermocycler. After the incubation step, inactivate the enzymes at 65 °C for 20 min.

3.2.3 Digestion Control PCR

The completeness of the digestion step is the major pre-requisite for the success of the MSRE-qPCR-based methylation analysis as even traces of undigested unmethylated DNA can lead to false-positive results [8]. Therefore, the completeness of the restriction digestion needs to be controlled by an endpoint PCR containing targets with known methylation status, as previously described [18]. The multiplex PCR contains a total of 6 targets including *IGF2* [insulin-like growth factor 2 (somatomedin 2)], *H19* [imprinted maternally expressed transcript], *XIST* [X inactive specific transcript], *PITX2* [paired-like homeodomain 2], *FMR1* [fragile X mental retardation 1] and *CALCA* [calcitonin-related polypeptide alpha]. *H19* and *IGF2* are known imprinted genes, which are differentially methylated on the paternal and maternal chromosome. Digestion of DNA with MSREs results in fragmentation of unmethylated DNA, while the DNA on the methylated chromosome remains uncut and can be amplified in subsequent PCR reaction. Transcription factor *PITX2* as well as *CALCA* are genes, which are unmethylated in normal (non-cancerous) DNA on both parental chromosomes. Completeness of digestion reaction is assured when no PCR product of these genes can be detected after PCR amplification (*see* **Note 3**).

The *FMR1* locus is located on the X chromosome and results in a sex-specific signal upon MSRE digestion, since only females carry a methylated allele on the inactive X chromosome, while males carry a single unmethylated allele. Therefore PCR amplification fails in DNA samples derived from male individuals after complete digestion, while DNA remains uncut on the methylated allele in female-derived samples and can be amplified. *XIST located on the chromosome X (chrX)* is unmethylated on the inactive chrX (present in females only) and methylated on the active chrX (present in both

Table 1
Primer sequences and PCR product length for the "Control multiplex-PCR"

PCR Product	Length (bp)	Sense Primer	Antisense Primer
CALCA	381	ccgccgctgccaccgcctctg	acgccgtccctgatcgcaccgtctg
FMR1	230	cccggcccgctgcgggtgtaa	cgacctgtcaccgcccttcagccttcc
H19	528	ggagtctggcaggagtgatgacgggtgga	tgccctcttgctctttctgcctggaacgg
IGF2	462	tgtcaggtcaggcgggcggctcag	gggaggcggggcacggggact
PITX2	347	tgctcctgctcctcggttggctcctaagt	ctttctcgctcggtctcgcactctctcct
XIST	301	gcagcaatccagcactgtccatcccacct	tgggcgaggctttgttaagtggttagcatggt

male and female DNA). Thus upon digestion there should be always an *XIST* amplicon derived from the active chrX independent of the gender. Thus *XIST* will give a similar MSRE-PCR-product pattern as *H19* and *IGF2.*

Digestion control PCR products are then evaluated by conventional gel electrophoresis.

1. One μL of digested or *mock*-digested DNA sample is amplified in a multiplex PCR reaction including all six primer pairs (Table 1). Perform PCRs in a 20 μL reaction volume containing 1× PCR buffer (containing 15 mM $MgCl_2$), 0.17 mM dNTPs, 5% DMSO, 0.65 U HotStarTaq DNA Polymerase, 10 pmol of each forward and reverse primer for *XIST, PITX2*, and *CALCA* and 40 pmol of each forward and reverse primer for *IGF2*, *H19*, and *FMR1.* The thermal cycling program starts with a heat activation step at 95 °C for 15 min, followed by 40 cycles of a two-step amplification protocol with a denaturation step at 94 °C for 30 s and an annealing step at 68 °C for 80 s, followed by a final elongation step at 72 °C for 7 min.
2. Prepare a 2% agarose gel with 1× TBE buffer and add a DNA gel stain such as SYBR® Safe DNA Gel Stain to the molten gel. Combine 2 μL aliquot of the digested or mock digested amplified DNA with 8 μL water and 2 μL loading dye. Pipette the DNA samples and the DNA ladder on the gel and run it for 75 min at 160 V. Visualize the amplification product with the UV trans-illuminator and take pictures using a gel documentation system.

3.2.4 Purification and Volume Reduction

After the MSRE digestion, samples have to be purified from the excess of $MgCl_2$ as well as reduced with respect to the sample volume in order to transfer the entire 10 ng of input DNA to the following preamplification step (*see* **Note 4**). If the DNA concentration is high enough, digested DNA can also be transferred directly to the preamplification step without this purification step. However, the $MgCl_2$ concentration of the digested DNA must be taken into consideration, since a final concentration of 2 mM should be in the preamplification reaction.

1. Load the entire digestion mix volume (except 1 μL for the digestion control PCR) onto a Microcon®-10 Centrifugal filter and centrifuge to dryness for 30 min at 14,000 × *g*.
2. Pipette 10 μL PCR certified water onto the membrane and incubate for 5 min at room temperature.
3. Elute the DNA by turning the filter upside down at 1000 × *g* for 3 min according to the manufacturer's protocol.

3.2.5 Preamplification

Since the amount of DNA is often limited for analysis of clinical samples, the detection of rare mutations or the methylation level at a given loci can be very challenging [19]. Therefore, an additional amplification step the so-called preamplification ensures that enough starting material for subsequent single-target qPCR analysis on microfluidic platforms or conventional real-time instruments is available. The samples are preamplified in a high multiplex reaction, containing a pool of all primer pairs in one single primer pool; afterwards targets are read out individually in subsequent single-target qPCR reactions. For an alternative amplification protocol with a different reagent setup follow the manufacturer's protocol (*see* **Note 5**).

1. Prepare the preamplification reaction mix by combining 9 μL purified DNA sample (10 ng/reaction) from the digestion or *mock* digestion reaction with 2 μL 10× PCR buffer containing 15 mM $MgCl_2$, 0.16 mM dNTPs, 0.6 U HotStarTaq DNA Polymerase, 5% DMSO, 50 nM pooled primer pairs and 0.4 μL 25 mM $MgCl_2$ in a total volume of 20 μL.
2. Amplify samples in the thermocycler with the following program: 95 °C for 15 min followed by 22 cycles of 95 °C for 40 s, 65 °C for 40 s and 72 °C for 80 s, finalized by 7 min at 72 °C. After amplification, dilute the samples 1:5 with DNase-free water and proceed to the next step.

3.2.6 MSRE-qPCR

In function of the number of samples and targets of interest, the preamplified amplicons can be either evaluated on conventional real-time PCR instruments or on high-throughput microfluidic devices. We describe first (Subheading 3.2.6.1) the setup using 384-well format qPCR, followed by the high-throughput microfluidic qPCR setup (Subheading 3.2.6.2). The big advantage of the high-throughput machines such as Fluidigm's Biomark™ System is the extremely reduced time and cost per sample, as up to 9216 qPCRs can be conducted in a single run. The individual qPCR reactions take place in nano-liter volumes on so-called Dynamic

Array IFCs; an input of only 10 ng total DNA per sample is enough to evaluate the methylation level up to 96 different targets in a single run. Different array formats are available and enable high flexibility in experimental design, allowing for example parallel analysis of 48 different samples with 48 different assays (48 × 48 array), 96 different samples with 96 different assays (96 × 96 array) or 192 samples with 24 different assays (192 × 24 array, *see* **Note 6**).

qPCR Analysis in 384-Well Plates

1. Combine 1× PCR buffer containing 15 mM $MgCl_2$ with 0.16 mM dNTPs, 5% DMSO, 0.5× EvaGreen Binding Dye, 0.3 U HotStarTaq DNA Polymerase 1.6 pmol of forward and reverse primer, and 2 μL of the 1:5 diluted preamplified DNA in a total volume of 10 μL.
2. Run the following cycling program: heat activation for 15 min at 95 °C, 45 cycles with 40 s at 95 °C, 40 s at 65 °C, and 80 s at 72 °C, followed by 7 min at 72 °C and the melting curve analysis with 2 °C/s increasing temperature from 65 to 95 °C.

qPCR Analysis on 96 × 96 Dynamic Arrays

For an alternative amplification protocol with a different reagent setup follow the manufacturer's protocol (*see* **Note 7**).

1. Prepare 6 μL sample mix per sample containing 1× PCR buffer with 15 mM $MgCl_2$, 2× DNA binding Dye Loading Reagent, 0.16 mM dNTPs, 1.1× EvaGreen Binding Dye, 5% DMSO, 0.18 U HotStarTaq DNA Polymerase, 0.004 μL ROX Reference Dye, and 1.5 μL 1:5 diluted preamplified sample.
2. Prepare the assay master mix by mixing 3 μL 2× Assay Loading Reagent with 0.3 μL water and 2.7 μL 20 μM pooled forward and reverse primer of the targets of interest.
3. Prime the 96.96 chip by pipetting 150 μL control line fluid into both accumulators of the dynamic array and run the priming program on the HX IFC Controller, where the control fluid is pumped into the chip.
4. After priming, load the sample master mix and the assay master mix on the respective position on the micro-fluidic array (*see* also the manufacturer's protocol) and run the loading program on the HX IFC Controller to load the reaction mixes in the respective reaction chambers.
5. Run the loaded Fluidigm array on the Biomark™ System with the following program: thermal mixing step with 2 min at 50 °C, 30 min at 70 °C and 10 min at 25 °C, followed by a heat activation step for 15 min at 95 °C, the cycling program with 45 cycles for 40 s at 95°c, 40 s at 65 °C and 80 s at 72 °C. The amplification program is finished after a final elongation step for 7 min at 72 °C.

6. Confirm the specificity of the amplification by a melting curve analysis starting from 65 to 95 °C with a temperature increase of 2 °C/s.

3.3 Analysis

Once the qPCR run—independently of the used platform (either 384-well plate or microfluidics)—has been successfully finished, the generated data have to be analyzed.

3.3.1 Prefiltering

1. In order to prepare the data for statistical analysis, export the raw Ct values including the corresponding Tm values to text files.
2. Set samples with missing values (i.e., samples are completely digested) to a Ct value of 40 or 45 (depending on the total number of conducted cycles). This corresponds to a completely unmethylated state of the investigated region.
3. Use Tm analysis to monitor the specificity of the reactions and filtering of artifacts and primer dimers. Eliminate each data point with a deviation of the Tm value of ±1.5 °C compared to a reference Tm value for the same assay.

3.3.2 Normalization

1. Normalize the resulting data matrix with samples in columns and assays in rows to prevent bias and to correct for diverging input amounts of DNA and different DNA qualities.
2. Calculate the delta-Ct (dCt) value, using either a set of control genes or—if the experimental setup allows for it—the undigested sample.
3. For the first approach, calculate the geometric mean of a set of control assays (assays target regions either without cut sites, or make use of imprinted or known methylated gene regions). We usually use two assays without cut sites and two assays targeting imprinted regions, having a shift of one in Ct values when the digested sample is compared to the undigested. The geometric mean of the calculated control assays is subtracted from sample Ct values ($dCt = Ct_{Sample} - Ct_{Controls}$).
4. For the second approach the sample is applied two times to the PCR, digested und undigested, and the Ct value for the undigested sample is subtracted from the digested sample ($dCt = Ct_{Sample1\_digested} - Ct_{Sample2\_undigested}$). Both methods compensate for differences in applied DNA starting material and DNA quality.

3.3.3 Statistical Analysis

For the statistical analyses of the generated data a variety of tools are available. Advanced users may benefit most using the command line-based open source tool R (https://www.r-project.org/), which is an integrated software suite for data manipulation. A large repository of R-packages to analyze and prepare the data is available on *www.bioconductor.org*. For "non-bioinformaticians" we recommend BRB Array tools (https://linus.nci.nih.gov/BRB-ArrayTools.html), combining the flexibility of R with a graphical user interface, which can easily be accessed from within Excel as AddIn. A third tool, also aiming on users not familiar with command line-based software is the *Qlucore Omics Explorer*, which is under constant development and offers a growing usability as well as a growing repertoire of statistical and graphical tools. Basic analysis of the data should include comparative analyses using a random-variance t-test to identify statistically significant differentially methylated regions. In addition, different prediction algorithms to allocate samples to predefined classes are available.

4 Notes

1. For the cfDNA isolation with the Roche High Pure PCR Template Preparation Kit an adapted protocol according to Wielscher et al. [20] can be used, where multiple sample loading steps increase the DNA extraction yield significantly.
2. To obtain a high frequency of cutting sites in the genome we used the four enzymes *AciI*, *Hin6I*, *HpaII*, and *HpyCh4IV*. These four enzymes cover 39% of the CpGs in the genome. The cut site frequencies of the enzymes were taken from Schumacher et al. [9]. In this work the best-suited enzymes for methylation analyses have been investigated when studying high-throughput DNA methylation profiling with methylation-sensitive restriction enzymes, and interrogated on microarrays.
3. Especially in cancer-settings *PITX2* and/or *CALCA* can be methylated and might therefore yield amplification products in some samples.
4. Alternatively, the volume reduction step can also be performed right after the DNA isolation (before the quantification) in order to reduce the volume of the digestion mix. However, please keep the $MgCl_2$ concentration for the preamplification step in mind, since the digestion mix already contains 10 mM $MgCl_2$.
5. The preamplification step can also be performed with the TaqMan® Pre-amp Master Mix (Life Technologies).
6. In recent publications [18, 21] we have shown, that the MSRE method showed a high correlation with other quantitative—even single CpG-based—approaches like Pyrosequencing (*see also* Chapter 22) or Agena's (formerly known as Sequenom) EpiTYPER/MassARRAY-based approach (*see also* Chapter 26). Moreover, we have shown in these publications the applicability of the assay format to clinical diagnostics and

that reliable quantification down to the one-digit copy number range is feasible and accurate.

7. The MSRE-qPCR experiment on Fluidigm's Biomark™ System can also be performed with TaqMan® Gene Expression Assays from Applied Biosystems.

Acknowledgments

We thank former colleagues Rudolf Pichler, Markus Sonntagbauer, Elisabeth Reithuber, and Matthias Wielscher involved in developing and improving the described MSRE-qPCR approach. Part of this work was supported by: European Community's Seventh Framework program EurHEALTHAgeing HEALTH-F2-2011-277849 and RESOLVE FP7-HEALTH-F4-2008-202047.

References

1. Daniel M, Tollefsbol TO (2015) Epigenetic linkage of aging, cancer and nutrition. J Exp Biol 218:59–70
2. Shridhar K, Walia GK, Aggarwal A et al (2016) DNA methylation markers for oral pre-cancer progression: a critical review. Oral Oncol 53:1–9
3. Lendvai Á, Johannes F, Grimm C et al (2012) Genome-wide methylation profiling identifies hypermethylated biomarkers in high-grade cervical intraepithelial neoplasia. Epigenetics 7:1268–1278
4. Noehammer C, Pulverer W, Hassler MR et al (2014) Strategies for validation and testing of DNA methylation biomarkers. Epigenomics 6:603–622
5. Olkhov-Mitsel E, Bapat B (2012) Strategies for discovery and validation of methylated and hydroxymethylated DNA biomarkers. Cancer Med 1:237–260
6. Ståhlberg A, Zoric N, Åman P et al (2005) Quantitative real-time PCR for cancer detection: the lymphoma case. Expert Rev Mol Diagn 5:221–230
7. Szita N, Polizzi K, Jaccard N et al (2010) Microfluidic approaches for systems and synthetic biology. Curr Opin Biotechnol 21:517–523
8. Egger G, Wielscher M, Pulverer W et al (2012) DNA methylation testing and marker validation using PCR: diagnostic applications. Expert Rev Mol Diagn 12:75–92
9. Schumacher A, Kapranov P, Kaminsky Z et al (2006) Microarray-based DNA methylation profiling: technology and applications. Nucleic Acids Res 34:528–542
10. Pandey RV, Pulverer W, Kallmeyer R, Beikircher G, Pabinger S, Kriegner A, Weinhäusel A (2016) MSRE-HTPrimer: a high-throughput and genome-wide primer design pipeline optimized for epigenetic research. Clin Epigenet 8(1):84
11. Pandey RV, Pulverer W, Kallmeyer R, Beikircher G, Pabinger S, Kriegner A, Weinhäusel A (2016) MSP-HTPrimer: a high-throughput primer design tool to improve assay design for DNA methylation analysis in epigenetics. Clin Epigenet 8(1):101
12. Bustin SA, Benes V, Garson JA et al (2009) The MIQE guidelines: minimum information for publication of quantitative real-time PCR experiments. Clin Chem 55:611–622
13. Devonshire AS, Elaswarapu R, Foy CA (2011) Applicability of RNA standards for evaluating RT-qPCR assays and platforms. BMC Genomics 12:118
14. Jackson JB, Choi DS, Luketich JD et al (2016) Multiplex preamplification of serum DNA to facilitate reliable detection of extremely rare cancer mutations in circulating DNA by digital PCR. J Mol Diagn 18:235–243
15. Ståhlberg A, Kubista M (2014) The workflow of single-cell expression profiling using quantitative real-time PCR. Expert Rev Mol Diagn 14:323–331
16. Pabinger S, Rödiger S, Kriegner A et al (2014) A survey of tools for the analysis of quantitative PCR (qPCR) data. Biomol Detect Quantif 1:23–33

17. Wielscher M, Liou W, Pulverer W et al (2013) Cytosine 5-Hydroxymethylation of the LZTS1 gene is reduced in breast cancer. Trans Oncol 6:715–721
18. Weinhaeusel A, Thiele S, Hofner M et al (2008) PCR-based analysis of differentially methylated regions of GNAS enables convenient diagnostic testing of pseudohypoparathyroidism type Ib. Clin Chem 54:1537–1545
19. Andersson D, Akrap N, Svec D et al (2015) Properties of targeted preamplification in DNA and cDNA quantification. Expert Rev Mol Diagn 15:1085–1100
20. Wielscher M, Pulverer W, Peham J et al (2011) Methyl-binding domain protein-based DNA isolation from human blood serum combines DNA analyses and serum-autoantibody testing. BMC Clin Pathol 11:11
21. Pulverer W, Hofner M, Preusser M et al (2014) A simple quantitative diagnostic alternative for MGMT DNA-methylation testing on RCL2 fixed paraffin embedded tumors using restriction coupled qPCR. Clin Neuropathol 33:50–60

Part V

Locus-Specific DNA Methylation Analysis

Chapter 22

Quantitative DNA Methylation Analysis at Single-Nucleotide Resolution by Pyrosequencing®

Florence Busato, Emelyne Dejeux, Hafida El abdalaoui, Ivo Glynne Gut, and Jörg Tost

Abstract

Many protocols for gene-specific DNA methylation analysis are either labor intensive, not quantitative and/or limited to the measurement of the methylation status of only one or very few CpG positions. Pyrosequencing is a real-time sequencing technology that overcomes these limitations. After bisulfite modification of genomic DNA, a region of interest is amplified by PCR with one of the two primers being biotinylated. The PCR generated template is rendered single-stranded and a pyrosequencing primer is annealed to analyze quantitatively cytosine methylation. In comparative studies, pyrosequencing has been shown to be among the most accurate and reproducible technologies for locus-specific DNA methylation analyses and has become a widely used tool for the validation of DNA methylation changes identified in genome-wide studies as well as for locus-specific analyses with clinical impact such as methylation analysis of the *MGMT* promoter. Advantages of the Pyrosequencing technology are the ease of its implementation, the high quality and the quantitative nature of the results, and its ability to identify differentially methylated positions in close proximity.

Key words Pyrosequencing, Real-time synthesis, Bisulfite, Epigenotyping, Quantification, Heterogeneous DNA methylation, Biomarker

1 Introduction

Pyrosequencing is a sequencing-by-synthesis method, in which the incorporation of nucleotides complementary to a template strand is monitored bioluminometrically (Fig. 1) [1, 2]. A history of the development of the technology has recently been published [3]. Pyrophosphate (PP_i) is released upon incorporation of the nucleotide(s) by the Klenow fragment of the *E. coli* DNA polymerase I and is used by an ATP sulfurylase to produce ATP from adenosine phosphosulfate [4]. This ATP provides the necessary energy for the luciferase to oxidize D-luciferin. The product oxyluciferin is generated in an excited state, which decays to the ground state with the emission of a photon that can be detected by a

Jörg Tost (ed.), *DNA Methylation Protocols*, Methods in Molecular Biology, vol. 1708,
https://doi.org/10.1007/978-1-4939-7481-8_22, © Springer Science+Business Media, LLC 2018

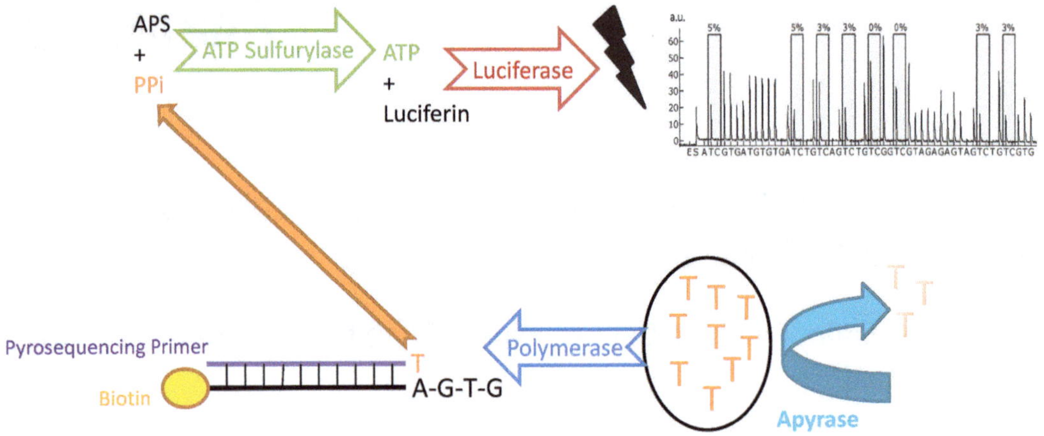

Fig. 1 Principle of the pyrosequencing technology. A complementary nucleotide (T in this figure) to the template strand is added and incorporated by the polymerase while non-incorporated nucleotides are degraded by the Apyrase. The pyrophosphate released by nucleotide incorporation is converted into ATP by an ATP sulfurylase using adenosine phosphosulfate (APS). This ATP provides the necessary energy for the luciferase to oxidize D-luciferin. The product oxyluciferin is generated in an excited state, which decays to the ground state with the emission of a photon that can be detected by a charge-coupled device (CCD) camera. All enzymes are included in the pyrosequencing Enzyme mixture, and APS, luciferin, and some other substrates are included in the pyrosequencing Substrate mixture

charge-coupled device (CCD) camera. In contrast to conventional Sanger sequencing that uses a mixture of the four fluorescently labeled terminating ddNTPs and strand elongating dNTPs [5], only one nucleotide is dispensed at a time by an inkjet-type cartridge in pyrosequencing reactions. Unincorporated nucleotides as well as excess ATP are degraded prior to addition of the next nucleotide by an apyrase. The carefully optimized interplay of the kinetics of the four enzymes ensures that at most one nucleotide is present at any time in the reaction mixture permitting clear assignment of the light signal to the incorporation of a specific nucleotide and thereby reconstruction of the sequence synthesized by the iterative addition of nucleotides. The dispensation order of nucleotides can either be predefined for the analysis of known sequences, which is the most frequently used method for the analysis of DNA methylation patterns, or consists of cyclic ACGT dispensations for de novo sequencing. The procedure of the pyrosequencing assay is simple and robust and results are highly reproducible. The instrument provides a great flexibility performing either different assays (up to 96) in parallel or analyzing up to 96 samples with one specific assay. These properties have made pyrosequencing a widely used analysis platform for various biological and/or diagnostic applications such as routine (multiplex) genotyping of single nucleotide polymorphisms (SNPs), bacterial typing or mutation detection [6]. The intensity of the bioluminometric response is directly proportional to the amount of incorporated nucleotides, i.e., a peak corresponding to the incorporation of two consecutive (and identical) nucleotides will have the double height compared to the signal of a single nucleotide incorporation. The peak heights in the resulting output format, termed Pyrogram, thus inform on the extent of homopolymeric sequences and proportions of alleles can be deduced directly from the relative height of the peaks corresponding to variable nucleotide positions. The quantitative nature of the results is the most important characteristic of the pyrosequencing technology—especially with regard to DNA methylation analysis. The bioluminometric response is linear for the sequential addition of up to five identical nucleotides (dCTP, dGTP, TTP) or three α-S-dATPs. The latter is used instead of dATP which serves as a direct—though less efficient—substrate for the luciferase and would therefore uncouple signal detection from nucleotide incorporation. Pyrosequencing has therefore been used in variety of applications where quantitative assessment of the relative abundance of two individual nucleotides or short sequences is required such as determination of SNP allele frequencies in pooled samples [7], analysis of copy number variation [8], karyotyping [9], abundance of treatment relevant somatic mutations [10] and DNA methylation analysis [11–15].

For DNA methylation analysis, genomic DNA extracted from a tissue sample is treated with sodium bisulfite to "freeze" the methylation status of the cytosines and to translate the epigenetic modification into sequence information [16]. A small region of interest is amplified in a methylation-independent PCR reaction with one of the two amplification primers being biotinylated. This label is subsequently used to generate a single-stranded template to which the pyrosequencing primer is annealed. DNA methylation analysis by pyrosequencing thus permits simultaneous analysis and quantification of the methylation status of several CpG positions in close proximity. This point is of particular interest as successive CpGs might display significantly different levels of methylation as demonstrated in the differentially methylated region of imprinting genes as well as at promoter CpG islands. Such heterogeneous DNA methylation patterns might lead to the failure of assays relying on methylation dependent-amplification such as methylation-specific PCR (*see* Chapters 23–25), while pyrosequencing when performed with amplification primers located in CpG-free sequence stretches will yield accurate quantification of DNA methylation levels despite variability in DNA methylation levels between CpGs in close proximity [17, 18]. Pyrosequencing combines the advantages of sequence-based approaches, such as in-built quality control and resolution of individual CpG sites, with the possibilities of medium to high-throughput (for DNA methylation analysis) and the advantages of PCR-based technologies. It features a limit of detection of ~3% for the minor component of a quantitative signal and a quantitative resolution of 5% or better [13, 19]. Of course, performance depends on assay design and the specific region targeted and should be carefully evaluated [20]. In several multilaboratory comparisons pyrosequencing was among the most accurate and reproducible technologies for locus-specific DNA methylation analysis and is one of the most suitable technologies for DNA methylation-based diagnostic tests [18, 21].

In addition to the frequent use as method for the analysis of candidate genes and validation of DNA methylation changes identified in genome-wide DNA methylation studies performed using Illumina BeadArrays (*see* Chapter 16) or other microarray and sequencing-based methods [22–24], pyrosequencing can also be used for the determination of the global DNA methylation content of a sample [25–27]. Combining pyrosequencing with allele-specific PCR amplification or allele-specific pyrosequencing primers permits the analysis of allele-specific DNA methylation patterns [28–30]. Similar to the genome-wide approach described in Chapter 15, pyrosequencing can also be used to analyze the co-occurrence of different levels of epigenetic modifications such as histone modifications and DNA methylation by analyzing the DNA methylation patterns on immunoprecipitated DNA [31] as well as DNA methylation patterns on single molecules or in a few cells [32, 33]. Furthermore, using specifically developed algorithms for the design of the dispensation order and the analysis of the Programs enable the analysis of cytosine methylation in other sequence contexts than CpGs, which is particular useful for the analysis of DNA methylation in plants or embryonic stem cells [34]. As the template strand used for pyrosequencing is not altered during the pyrosequencing reaction, it can be recovered after the sequencing run for the successive use of several sequencing primers on the same DNA template (serial pyrosequencing) improving throughput while reducing cost, labor and analysis time and saving precious DNA samples [35]. The broad range of applications combined with the above-described advantages has made pyrosequencing a widely used analysis method. However, there are some inconveniences associated with the technology, mainly concerning the size of the amplicon and the sequencing read-length. Due to the thermal instability of the enzymes (especially the luciferase) pyrosequencing has to be carried out at 28 °C, which limits the size of the amplified amplicon to 300 bp or less as the formation of secondary structures can complicate annealing of the sequencing primer or increase background signals. The limitation in the read-length is mainly due to dilution effects and increasing background due to frame shifts of subpopulations of sequenced molecules. Furthermore, in contrast to next generation sequencing-based methods analyzing DNA methylation patterns of target regions at the level of individual molecules [36], routine pyrosequencing will yield only the average methylation level of a large number of molecules at a given CpG position, which will be, however, sufficient for most applications. In the protocol described in this chapter, which is an updated version of the publication in the previous edition of this book [37], we describe in detail all the different steps necessary for the analysis of DNA methylation patterns and provide a number of tips and tricks, which will help in the successful development and application of pyrosequencing assays.

2 Materials

2.1 Assay Design

1. Design of PCR primers using MethPrimer [38] (http://www.urogene.org/methprimer/index.html).
2. Nucleic Acid Sequence Massager (http://www.cybertory.org/resources/sequenceTools/seqMassager.html).
3. Design of pyrosequencing primers manually or using the commercial PSQ assay design software (Qiagen).

2.2 Sample Preparation, Calibration Standards and Bisulfite Treatment

1. Qubit® dsDNA HS Assay Kit (Thermo Fisher Scientific).
2. Qubit fluorometer (Thermo Fisher Scientific).
3. EpiTect Fast DNA Bisulfite Kit (Qiagen) or the respective high-throughput kits converting 96 samples at a time EpiTect Fast 96 DNA Bisulfite Kit (*see* **Note 1**).
4. Methylated and unmethylated standards, e.g., EpiTect Control DNA (Qiagen) or Human Methylated & Non-Methylated DNA Set (Zymo Research).

2.3 PCR Amplification

1. Primers for PCR amplification.
2. Biotinylated primers for PCR amplification (*see* **Note 2**).
3. HotStar Taq DNA polymerase (Qiagen) or Phusion U Hot Start PCR Master mix (Thermo Fisher Scientific).
4. dNTP Mix (8 mM).
5. PCR plates, e.g., AB800 (ABgene) or Thermo-Fast PCR Plate, 24-well (Thermo Fisher Scientific).
6. 96 Gradient Mastercycler.

All reagents should be stored at −20 °C.

2.4 Sample Preparation for Pyrosequencing Analysis

1. Streptavidin Sepharose HP.
2. PyroMark Q96 Vacuum Workstation or PyroMark Q24 Vacuum Workstation [39] (Qiagen) using PyroMark Vacuum Prep Filter Probe (Qiagen).
3. PyroMark Q96 Vacuum Prep Trough or PyroMark Q24 Vacuum Prep Troughs (Qiagen).

4. Binding buffer: 10 mM Tris–HCl, 2 M NaCl, 1 mM EDTA, 0.1% Tween 20; pH 7.6 (*see* **Note 3**).
5. Denaturing solution: 0.2 M NaOH.
6. Wash buffer: 10 mM Tris–acetate, pH 7.6.
7. Annealing buffer: 20 mM Tris–acetate, 2 mM Mg–Acetate, pH 7.6.
8. Thermowell sealing tape.
9. Thermomixer or similar (room temperature).
10. Heating device, for example heating plate or thermoblock.
11. PyroMark Q96 HS Sample Prep Thermoplate (Qiagen).
12. Primers for pyrosequencing.
13. Plate for pyrosequencing analysis, PyroMark Q96 HS Plate or PyroMark Q24 Plate (Qiagen).

All reagents used for this step should be stored at room temperature except for the streptavidin coated sepharose beads (+4 °C) and the pyrosequencing primer (−20 °C).

2.5 Analysis by Pyrosequencing Reaction

1. Pyrosequencer PyroMark Q96 MD or Q24. For long sequence runs on the Q24, we recommend the PyroMark Q24 Advanced pyrosequencer.
2. Cartridge for reagent dispensation: PyroMark Q96 HS Dispensing Tip Holder (to use with NDT tips) or PyroMark Q96 HS Capillary Tip Holder (to use with CDT tips) or PyroMark Q24 Cartridge.
3. PyroMark Q96 HS Reagent Tips.
4. Nucleotide Dispensing Tips (NTDs): PyroMark Q96 HS Nucleotide Tips.
5. Capillary Dispensing Tips (CDTs): PyroMark Q96 HS Capillary Tips.
6. PyroMark Gold Q96 Reagents or PyroMark GOLD Q96 SQA Reagents or PyroMark Gold Q24 Reagents (*see* **Notes 4** and **5**).
7. PyroMark Q-CpG Software or PyroMark Q24 Advanced Software.

All Pyrosequencing Accessories and Reagents for both Q96MD and Q24 are sold by Qiagen.

2.6 Serial Pyrosequencing

All required reagents are listed under Subheadings 2.4 and 2.5.

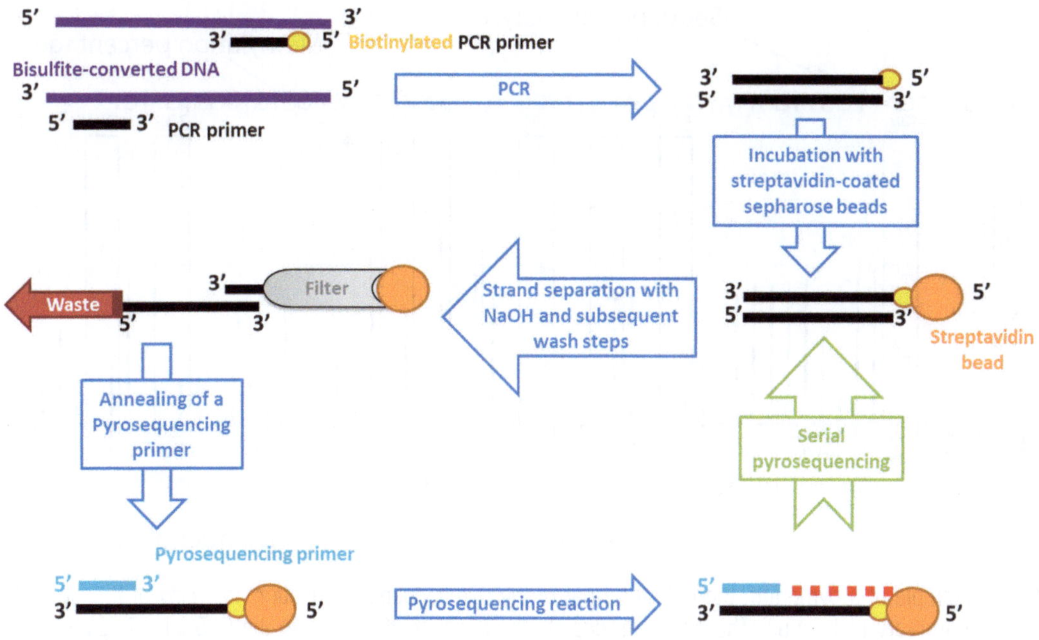

Fig. 2 Outline of the different steps of the pyrosequencing procedure. The target region is amplified by PCR with two primers, one of them being biotinylated. The resulting biotinylated PCR product is incubated with streptavidin-coated sepharose beads: once the strand is captured by streptavidin, the bead prevents the strand to pass through the filter while the non-biotinylated strand is washed off under denaturing conditions. Then, a sequencing primer is annealed to the template strand for the pyrosequencing reaction. After the pyrosequencing run, the biotinylated strand can be purified again by discarding the de novo synthesized complementary strand according to the same principle. The same template strand is then available for the sequencing of another part of the target region with another pyrosequencing primer leading to additional CpG positions analysis (serial pyrosequencing)

3 Methods

The protocol for DNA methylation analysis using the pyrosequencing technology can be subdivided into seven steps (Fig. 2):

1. *The assay design* for a quantitative gene- or promoter-specific analysis of CpG methylation.
2. *The bisulfite treatment* of samples and standards, if the latter have not been purchased already bisulfite converted.
3. *The PCR amplification* of the target sequence performed with one of the PCR primers biotinylated.
4. *The sample preparation for pyrosequencing analysis,* which requires rendering the PCR product single-stranded and subsequent annealing of the pyrosequencing primer to the single-stranded DNA template.

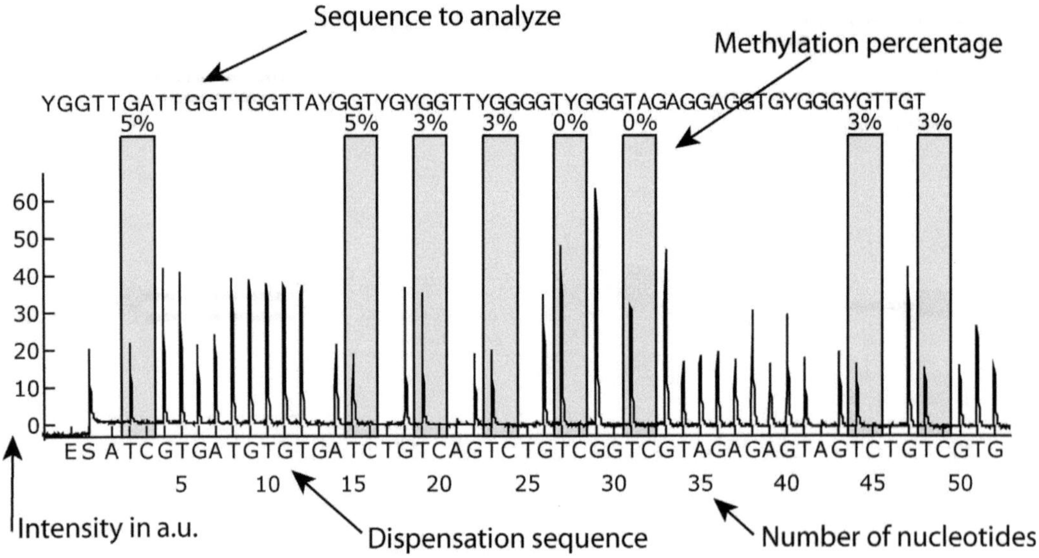

Fig. 3 Example of a pyrogram, the output format of the pyrosequencing reaction. Intensity of the signals is given in arbitrary units (a.u.). The pyrogram depicts the analysis of 8 CpGs in the CpG island spanning the transcription start of *CDKN2A*

5. *The pyrosequencing reaction* synthesizes the complementary strand to the single-stranded DNA template and analyzes quantitatively the CpG (and other polymorphic) positions in the sequence (Fig. 3).
6. *Serial pyrosequencing* enables the analysis of additional CpGs that have not been analyzed in the first pyrosequencing run on the previously used template strand.

3.1 Assay Design

The assay design is probably the most critical step for pyrosequencing-based DNA methylation analysis. Great attention should therefore be paid to the design as this will critically influence the successful outcome of the assay.

Many standard software tools developed for conventional PCR cannot handle primer design on bisulfite converted DNA because of the lower complexity of bisulfite treated DNA. However, some commercial and freely available software have been especially designed for this purpose such as MethPrimer [38], Bisearch [40], or MethylPrimer Express (Thermo Fisher Scientific). In our laboratory, we use MethPrimer and the procedure is described in Subheading 3.1.1. PCR primers are positioned to amplify the target region regardless of its methylation status. Then, one or several pyrosequencing primers are necessary to cover the target region: a manual design method is described in Subheading 3.1.2, which is in our hands as successful as the commercial design software (*see* **Note 6**).

3.1.1 Design of Amplification Primers

1. As the two strands of genomic DNA are no longer complementary after bisulfite treatment, both strands can be analyzed for possible amplification products. MethPrimer takes only the forward strand into account and the reverse strand has to be created with a variety of software tools such as Nucleic Acid Sequence Massager.
2. Primers length should be between 15 and 35 bp and the amplification product should be between 150 and 380 bp, although we successfully analyzed DNA methylation in PCR products up to 400 bp. With increasing length of the amplification product, there is an increasing risk of hairpin loop formation during pyrosequencing, but since the loop strongly depends of the amplified sequence, only an experimental verification can determine the success of the pyrosequencing reaction (*see* **Note 7**).
3. Place preferably primers in a region containing four or more cytosines that have been converted during bisulfite treatment to ensure complementarity to completely converted DNA.
4. Primers should preferentially contain no CpG positions. If it cannot be avoided the maximum number of CpGs included in the primer should be restricted to one, and the CpG should not be included in the last five bases from the 3′ terminus to avoid preferential amplification of specific DNA methylation patterns.
5. To ensure specificity, palindromes within primers and complementary sequences between primers as well as degenerated bases and inosine should be avoided.
6. The generated amplicon should be verified for the presence of potential polymorphic positions such as SNPs in the population under investigation. We strongly recommend the redesign of amplification primers annealing to a potentially polymorphic sites, which has in the past lead to erroneous results [41].

3.1.2 Design of Pyrosequencing Primers

1. Sequences are identified where a sequencing primer could be positioned and at least the last five bases from the 3′ terminus do not overlap with any other potentially variable position including CpGs and SNPs that are retained after bisulfite treatment.
2. The last four or five bases from the 3′ terminus should be verified to be unique in the amplicon by using, for example, the MS Word find tool. As few as four consecutive nucleotides complementary to a sequence in the amplification product might add to background signal impeding precise quantification.
3. Successfully designed primers should also be checked for primer dimers and possible hairpin structures.

4. As read-lengths of up to 120 bases can be achieved with the PyroGold SQA kit, primers can also be positioned in non-polymorphic regions next to the variable region using part of the nucleotide dispensations to approach the region of interest.
5. At least one cytosine not followed by a guanine should be included in the dispensation order to control for complete bisulfite conversion.
6. The direction of the pyrosequencing primer defines which of the amplification primers needs to be biotinylated. This primer should be checked carefully not to form any hairpin structure.
7. If several pyrosequencing primers are required to cover the region of interest, analysis of a few CpG positions by more than one pyrosequencing primer improves confidence into the acquired results and helps to detect potential technical artifacts.
8. Indicate potential polymorphic positions such as SNPs included in the sequence to analyze in the dispensation order with degenerated bases.

3.2 Sample Preparation, Calibration Standards, and Bisulfite Treatment

A high number of CpG positions in the amplified fragment results in a large sequence difference between completely methylated and unmethylated molecules after bisulfite treatment. This difference strongly influences the melting and annealing properties during PCR amplification and predisposes for preferential amplification of one allele (PCR bias) [42]. As a PCR bias is both sequence and strand specific, it is almost impossible to predict. PCR bias can be detected by using samples with a known degree of methylation or mixtures of completely methylated and unmethylated DNA prepared as described below.

1. Verify the concentrations of the samples of interest by using the Qubit fluorometer (*see* **Note 8**).
2. Treat samples and genomic DNA standards (which are not sold already bisulfite treated) standards with sodium bisulfite using the EpiTect kit following the manufacturer's instructions (*see* **Note 1**). Bisulfite-treated standards can be amplified directly.
3. We recommend to bisulfite-convert at least 200 ng of genomic DNA to have sufficient amount of DNA after the conversion available for PCR amplification. Carrier RNA might be added to improve DNA recovery, but should be verified to not yield an amplification signal in the target regions.

3.3 PCR Amplification

A strong and specific PCR amplified product with one of the two primers being biotinylated is required for a successful pyrosequencing reaction.

1. An increased random amplification during PCR is observed with decreasing amounts of template DNA used in the PCR amplification [14]. A minimum of 10 ng DNA is therefore necessary to ensure high reproducibility (*see* **Note 9**).
2. Typical reaction conditions using the HotStar Taq are 1× HotStar Taq buffer supplemented with 1.6 mM $MgCl_2$, 0.4 mM dNTP Mix, and 2.0 U of HotStar Taq polymerase, and 5 pmol of forward and reverse primer in a 25 μL volume, one of them being biotinylated (*see* **Note 10**).
3. Reaction conditions with the Phusion U polymerase are 2× Master Mix and 4 pmol of forward and reverse primer in a 20 μL volume, one of them being biotinylated.
4. The HotStar PCR program consists of a denaturing step of 15 min at 95 °C followed by 50 cycles of 30 s at 95 °C, 30 s at the respective annealing temperature determined beforehand by a temperature gradient, and 20 s at 72 °C, with a final extension of 5 min at 72 °C. Perform 50 cycles of amplification to ensure complete exhaustion of the free biotinylated primer and to yield a strong PCR product (*see* **Note 11**).
5. The PCR program using the Phusion polymerase consists of a denaturing step of 15 min at 98 °C followed by 50 cycles of 30 s at 98 °C, 30 s at the respective annealing temperature determined beforehand by a temperature gradient, and 20 s at 72 °C, with a final extension of 5 min at 72 °C.
6. Deposit 5 μL of the PCR product as well as the positive and negative controls (*see* **Note 12**) on an agarose gel. The entire PCR amplification should be repeated if the slightest signal in the negative controls is detected.
7. PCR products can be stored at +4 °C for several days or at −20 °C for several months.

3.4 Sample Preparation for Pyrosequencing Analysis

In this step, the previously amplified double-strand DNA is rendered single-stranded to enable annealing of the sequencing primer since the pyrosequencing reaction takes place at 28 °C due to the thermal instability of the enzyme mix. Template preparation can be applied to a complete 96 well plate, a complete 24 well plate, as well as to a single well. Use the PyroMark Q96 Vacuum Workstation for 96 well plates intended for PyroMark 96 and the PyroMark Q24 Vacuum Workstation for 24 well plates intended for the PyroMark Q24.

1. Transfer 5 μL of the PCR product into a new PCR plate, add 40 μL of binding buffer and 1 μL sepharose beads and complete to 80 μL with water. Cover the plate with a sealing tape and incubate the reaction mixture for 10 min at room

temperature under constant mixing (1400 rpm). It is crucial that the beads do not sediment (*see* **Note 12**).

2. During this incubation step prepare the pyrosequencing plate by diluting 4 pmol of the pyrosequencing primer into 12 μL of annealing buffer into the respective wells of the 96 plate or 4 pmol in 20 μL of annealing buffer in the 24 well plate. One or several different pyrosequencing primers can be used on the same plate (*see* **Note 13**).
3. Fill the four troughs of the vacuum preparation tool with approximately 180 mL (40 mL for the Q24 Vacuum Workstation) of 70% ethanol and denaturing solution (0.2 M NaOH) and 200 mL (50 mL for Q24 Vacuum Workstation) of wash buffer and water, respectively. This different level assures that the NaOH is completely washed off in the following washing step as it might otherwise inhibit the ensuing pyrosequencing reactions.
4. Turn on the workstation, create vacuum in the aspiration device and clean the tips of the filters by immersion in water for several seconds (*see* **Note 14**). Remove the PCR plate from the mixer and aspirate the binding mix.
5. Immerse the tips of the filters for 5 s in the successive baths of ethanol 70%, and denaturing solution followed by 10 s in the wash buffer. Turn over the tool and release the vacuum.
6. Immerse the tip of the filters in the annealing mix of the sequencing plate and shake gently to release the beads into the wells.
7. Check the plate used for incubation with the binding buffer for the presence of remaining beads. The presence of beads indicates incomplete retrieval of the PCR product, which might lead to failure of the subsequent pyrosequencing analysis due to insufficient quantity of template material (*see* **Note 15**).
8. *Optional*: Add to one well of the freshly prepared pyrosequencing plate a single-stranded synthetic template and the corresponding pyrosequencing primer to detect potential problems during the pyrosequencing run (e.g., failed dispensations due to blocked tips).
9. Prepared plates can be stored at this point for 2 days at 4 °C prior to pyrosequencing analysis.
10. Incubate the sequencing plate for 2 min at 80 °C on the thermoplate placed on a heating device. Sealing of the plate is not necessary. Allow the plate to cool down to room temperature.

3.5 Pyrosequencing Reaction

1. During cooling of the sequencing plate (Subheading 3.4, **step 9**), program the sequencing run on the Pyrosequencer.
2. The software calculates the quantity of reagents necessary to perform the run. Dispense the reagents and enzyme mix in the appropriate tips placed in the cartridge (*see* **Note 16**). Wear a lab coat and powder-free gloves when handling the dispensing cartridge to prevent dust particles from clogging the tip heads of the NDTs and be careful not to create air bubbles when filling the tips. For sequencing runs of up to 30 nucleotides on the PyroMark Q96MD the PyroGold Q96 Reagents kit can be used. The PyroGold SQA Reagents kit enables longer read-length of up to 120 nucleotides. A PyroMark Q24 Advanced (and not only a PyroMark Q24) is required to use the PyroMark Q24 Advanced CpG Reagents, which are suitable for longer runs.
3. Deposit the sequencing plate and the reagents' cartridge in the Pyrosequencer.
4. Perform the dispensation test (no dispensation test is necessary on the Q24) with a sealed pyrosequencing plate to verify that the dispensing tips are properly working. Droplets should be clearly visible and homogeneous. Change the tips if necessary.
5. Start the pyrosequencing run. The length of a run is proportional to the number of dispensations (1 per min). Remaining time is indicated on the Q24 screen.
6. After the end of the sequencing run, analyze the results with the PyroMark Q96 MD Software or the PyroMark Q24 Advanced Software (Fig. 4). In some cases the signals corresponding to part of the sequence might be missing. This is in most cases due to a problem of nucleotides dispensation (tips blocked). Check the point of the tips for large droplets that might have accumulated at a dust particle or due to an adverse electrostatic environment (humidity too low).
7. If the plate will be used for serial pyrosequencing (*see* below), seal the plate and store the plate at 4 °C for not more than 3 days.
8. The PyroMark Q96 MD software is able to export the results, so that they can be treated with statistical or graphical software such as Excel®.
9. Remove the cartridge and clean the tips well with pure water.

3.6 Serial Pyrosequencing

If the volume of the PCR product is not sufficient to perform all pyrosequencing reactions on a target region amplified in a single PCR reaction, it is possible to recycle the template strand for additional analyses as the biotinylated strand is not altered during pyrosequencing. In case of dispensation problems during the run,

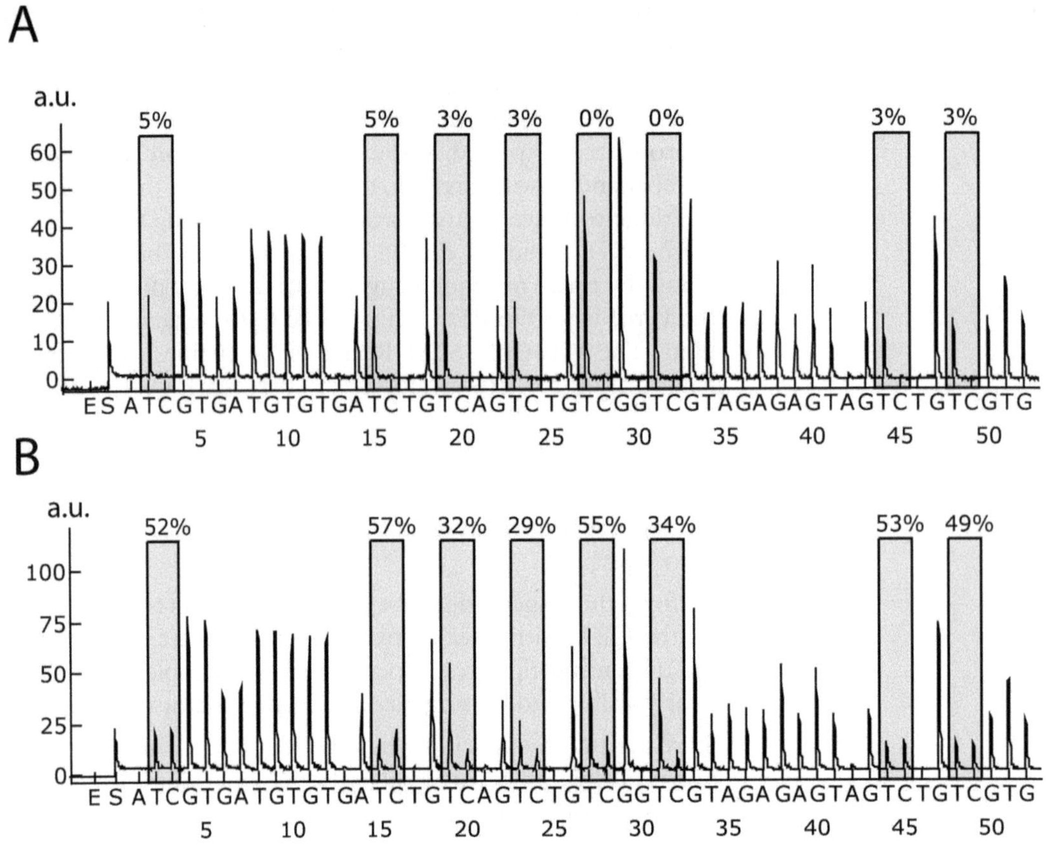

Fig. 4 Analysis of DNA methylation patterns in the *CDKN2A* promoter in human hepatocellular carcinoma. Panel A depicts the peritumoral liver tissue corresponding to the tumor shown in panel B. Hardly any methylation is detected in the peritumoral tissue (average 2.7%) while on average 45% methylation is found in the tumor

this method also allows to repeat the pyrosequencing run without redoing a PCR amplification and using more of the sample DNA. Although there is a slight loss in intensity due to incomplete recovery of the biotinylated strand, quantitative results are unaltered for several cycles of pyrosequencing on the same template [35].

1. Prepare a pyrosequencing plate with a new pyrosequencing primer as described in Subheading 3.4, **step 2** and prepare the workstation as described in Subheading 3.4, **step 3**.
2. Add 20 μL binding buffer to the completed sequencing reaction (from Subheading 3.5, **step** 7) and resuspend the sepharose beads by pipetting up and down several times.

Purify the mixture without further incubation as described in Subheading 3.4, **steps 4–10** and perform a new pyrosequencing reaction as described in Subheading 3.5.

4 Notes

1. A detailed protocol for the bisulfite conversion of genomic DNA without the use of commercial kits for bisulfite conversion has been previously published by us and others [15, 43] and is also described in other chapters of this volume (*see* for example Chapter 23). The mentioned bisulfite kit from Qiagen is routinely used in our laboratory, but other kits are commercially available from, Zymo Research, Human Genetic Signatures, or Thermo Fisher Scientific probably yielding similar results.
2. Biotinylated primers are more sensitive to storage conditions than unmodified primers and therefore dilutions as well as stock tubes should be kept at −20 °C. Dilutions should be aliquoted and not subjected to more than five cycles of freezing and thawing. It should also be noted that the quality might differ substantially from provider to provider. Biotinylated primers should always be ordered HPLC-purified to remove free biotin, which competes with the PCR product for the binding sites of the streptavidin-coated beads used for template preparation.
3. The different buffers used for template preparation are also commercially available from Qiagen.
4. The PyroMark Gold Q96 SQA Reagents (972812) contain more nucleotides than PyroMark Gold Kit (972804) to support longer sequencing. The ratio between E/S-Mix to nucleotides in the PyroMark Gold Q96 CDT reagents (972824) is intermediate compared to the two other kits. For six runs with a full plate of the same assay it will support an assay length with max. ten dispensations of one nucleotide species. The alternative for long sequencing with CDTs is to use the PyroMark Gold Q96 SQA Reagents (972812) and dilute the nucleotides 1:1 in TE buffer.
5. Protocols have been published to avoid the commercial reagent kits and prepare the enzyme mix in-house [44]. We have, however, not tested these protocols and can therefore not comment on the respective performance.
6. The commercial pyrosequencing design software has been developed for the design of SNP genotyping assays and its quality criteria for successful assay design might be too stringent. Despite a low quality score and the display of a number of error messages, primers designed by the software do often work in practice, but should be carefully verified using the appropriate controls and calibration standards. Manual design of the pyrosequencing primers is in our hands as successful.

7. The recommended length is applicable to the analysis of DNA methylation patterns in DNA extracted from cell lines, blood or frozen tissue. For the analysis of DNA extracted from FFPE tissues or cell-free circulating DNA, the size of the amplification product should be restricted to 80–120 bp.
8. Alternative quality controls based on qPCR to estimate the total amount of amplifiable DNA, the total amount of amplifiable bisulfite-converted DNA as well as the conversion status have been developed and are described in detail in Chapter 25. They provide useful alternatives for estimation of the quality and amount of bisulfite converted DNA, especially for laboratories with little experience in bisulfite-based methods.
9. We normally use 1 μL containing at least 10 ng of bisulfite treated DNA for a total reaction volume of 25 μL. Higher volumes of bisulfite-treated DNA have shown to inhibit PCR amplification in some cases probably to some reagents carried over from the bisulfite conversion.
10. Avoid a higher amount of primers: excess of biotinylated primer may decrease the capture of the amplification product on the streptavidin-coated beads and lead to high background signals during the pyrosequencing reaction.
11. The high number of amplification cycles could result in amplification of very small amounts of contaminating DNA. Therefore it is important to always include appropriate controls, especially several negative controls. We also strongly recommend a spatial separation of pre- and post-PCR manipulations to reduce the risk of contaminations. Conventional decontamination methods such as dUTP incorporation can only be applied to bisulfite treated DNA under special conditions [45] as unmethylated cytosines are converted to dUTPs during bisulfite treatment.
12. In most cases, 5 μL of PCR product are sufficient to obtain a strong pyrosequencing signal. But depending on the intensity of the PCR product on the agarose gel, it may be necessary to use more PCR product. Generally, we never use less than 5 μL as it leads to insufficient peak height and inaccurate quantification in the resulting pyrogram. However, using too much PCR product might lead to premature loss of signal due to depletion of the reagents. In this case, it is possible to launch the same plate again after modifying the sequences to analyze starting from the first missing nucleotide for each well. Another alternative is to purify the plate following the instructions in the serial pyrosequencing paragraph (Subheading 3.6).
13. Each primer should be checked for background signals in the absence of a DNA template by preparing a well of the plate containing annealing buffer and the respective pyrosequencing primer.

14. It is useful to check the proper working of the filter tips by aspirating liquid from a PCR plate filled with water. Any remaining liquid indicates a sub-optimal performance of the filter tip and the filter tip should be changed.
15. If no liquid has been aspirated in a well, the respective filter tip must be changed. If the liquid has been correctly aspirated, resuspend the beads by adding 20 μL of binding buffer and follow the purification procedure described in Subheading 3.4, **steps 3–6** above.
16. Take the pyrosequencing kit out of the fridge 30 min prior to use. Otherwise small air bubbles might form in the cartridge. If problems with blocked cartridges are frequent, it might help to centrifuge the nucleotides before use. If multiple consecutive runs are planned in the same day, the dispensing tips of the cartridge can be filled in advance with a sufficient volume for all runs.

Acknowledgments

Work in the laboratory of JT is supported by grants from the ANR (ANR-13-EPIG-0003-05 and ANR-13-CESA-0011-05), Aviesan/INSERM (EPIG2014-01 and EPIG2014-18) and INCa (PRT-K14-049), a Sirius research award (UCB Pharma S.A.), a Passerelle research award (Pfizer), iCARE (MSD Avenir), and the institutional budget of the CNRGH.

References

1. Ronaghi M, Uhlen M, Nyren P (1998) A sequencing method based on real-time pyrophosphate. Science 281(363):365
2. Ronaghi M (2001) Pyrosequencing sheds light on DNA sequencing. Genome Res 11:3–11
3. Nyren P (2015) The history of pyrosequencing ((R)). Methods Mol Biol 1315:3–15
4. Ahmadian A, Ehn M, Hober S (2006) Pyrosequencing: history, biochemistry and future. Clin Chim Acta 363:83–94
5. Sanger F, Nicklen S, Coulson AR (1977) DNA sequencing with chain-terminating inhibitors. Proc Natl Acad Sci U S A 74:5463–5467
6. Lehmann U, Tost J (2015) Pyrosequencing-methods and protocols, Methods Mol Biol, vol 1315. Humana Press, Springer, New York
7. Wasson J (2007) Allele quantification and DNA pooling methods. Methods Mol Biol 373:63–74
8. Pielberg G, Andersson L (2007) Gene copy number detection in animal studies. Methods Mol Biol 373:147–156
9. Deutsch S, Choudhury U, Merla G et al (2004) Detection of aneuploidies by paralogous sequence quantification. J Med Genet 41:908–915
10. Sefrioui D, Mauger F, Leclere L et al (2017) Comparison of the quantification of KRAS mutations by digital PCR and E-ice-COLD-PCR in circulating-cell-free DNA from metastatic colorectal cancer patients. Clin Chim Acta 465:1–4
11. Uhlmann K, Brinckmann A, Toliat MR et al (2002) Evaluation of a potential epigenetic biomarker by quantitative methyl-single nucleotide polymorphism analysis. Electrophoresis 23:4072–4079
12. Colella S, Shen L, Baggerly KA et al (2003) Sensitive and quantitative universal pyrosequencing methylation analysis of CpG sites. Biotechniques 35:146–150
13. Tost J, Dunker J, Gut IG (2003) Analysis and quantification of multiple methylation variable positions in CpG islands by pyrosequencing. Biotechniques 35:152–156

14. Dupont JM, Tost J, Jammes H et al (2004) De novo quantitative bisulfite sequencing using the pyrosequencing technology. Anal Biochem 333:119–127
15. Tost J, Gut IG (2007) Analysis of gene-specific DNA methylation patterns by pyrosequencing technology. Methods Mol Biol 373:89–102
16. Frommer M, McDonald LE, Millar DS et al (1992) A genomic sequencing protocol that yields a positive display of 5-methylcytosine residues in individual DNA strands. Proc Natl Acad Sci U S A 89:1827–1831
17. Alnaes GI, Ronneberg JA, Kristensen VN et al (2015) Heterogeneous DNA methylation patterns in the GSTP1 promoter lead to discordant results between assay technologies and impede its implementation as epigenetic biomarkers in breast cancer. Genes (Basel) 6:878–900
18. consortium B (2016) Quantitative comparison of DNA methylation assays for biomarker development and clinical applications. Nat Biotechnol 34:726–737
19. Dejeux E, Audard V, Cavard C et al (2007) Rapid identification of promoter hypermethylation in hepatocellular carcinoma by pyrosequencing of etiologically homogeneous sample pools. J Mol Diagn 9:510–520
20. Lehmann U (2015) Quantitative validation and quality control of pyrosequencing (R) assays. Methods Mol Biol 1315:39–46
21. Quillien V, Lavenu A, Ducray F et al (2016) Validation of the high-performance of pyrosequencing for clinical MGMT testing on a cohort of glioblastoma patients from a prospective dedicated multicentric trial. Oncotarget 7:61916–61929
22. Irizarry RA, Ladd-Acosta C, Wen B et al (2009) The human colon cancer methylome shows similar hypo- and hypermethylation at conserved tissue-specific CpG island shores. Nat Genet 41:178–186
23. Miceli-Richard C, Wang-Renault SF, Boudaoud S et al (2016) Overlap between differentially methylated DNA regions in blood B lymphocytes and genetic at-risk loci in primary Sjogren's syndrome. Ann Rheum Dis 75:933–940
24. Morales E, Vilahur N, Salas LA et al (2016) Genome-wide DNA methylation study in human placenta identifies novel loci associated with maternal smoking during pregnancy. Int J Epidemiol 45:1644–1655
25. Yang AS, Estecio MR, Doshi K et al (2004) A simple method for estimating global DNA methylation using bisulfite PCR of repetitive DNA elements. Nucleic Acids Res 32:e38
26. Luttropp K, Sjoholm LK, Ekstrom TJ (2015) Global analysis of DNA 5-methylcytosine using the luminometric methylation assay, LUMA. Methods Mol Biol 1315:209–219
27. Tabish AM, Baccarelli AA, Godderis L et al (2015) Assessment of changes in global DNA methylation levels by pyrosequencing(R) of repetitive elements. Methods Mol Biol 1315:201–207
28. Wong HL, Byun HM, Kwan JM et al (2006) Rapid and quantitative method of allele-specific DNA methylation analysis. BioTechniques 41:734–739
29. Busato F, Tost J (2015) SNP-based quantification of allele-specific DNA methylation patterns by pyrosequencing(R). Methods Mol Biol 1315:291–313
30. Kristensen LS, Johansen JV, Gronbaek K (2015) Allele-specific DNA methylation detection by pyrosequencing(R). Methods Mol Biol 1315:271–289
31. Moison C, Assemat F, Daunay A et al (2015) DNA methylation analysis of ChIP products at single nucleotide resolution by pyrosequencing (R). Methods Mol Biol 1315:315–333
32. Hajj NE, Kuhtz J, Haaf T (2015) Limiting dilution bisulfite pyrosequencing(R): a method for methylation analysis of individual DNA molecules in a single or a few cells. Methods Mol Biol 1315:221–239
33. Huntriss J, Woodfine K, Huddleston JE et al (2015) Analysis of DNA methylation patterns in single blastocysts by pyrosequencing(R). Methods Mol Biol 1315:259–270
34. How-Kit A, Daunay A, Mazaleyrat N et al (2015) Accurate CpG and non-CpG cytosine methylation analysis by high-throughput locus-specific pyrosequencing in plants. Plant Mol Biol 88:471–485
35. Tost J, El abdalaoui H, Gut IG (2006) Serial pyrosequencing for quantitative DNA methylation analysis. BioTechniques 40:721–722, 724, 726
36. Masser DR, Berg AS, Freeman WM (2013) Focused, high accuracy 5-methylcytosine quantitation with base resolution by benchtop next-generation sequencing. Epigenetics Chromatin 6:33
37. Dejeux E, El abdalaoui H, Gut IG et al (2009) Identification and quantification of differentially methylated loci by the pyrosequencing technology. Methods Mol Biol 507:189–205
38. Li LC, Dahiya R (2002) MethPrimer: designing primers for methylation PCRs. Bioinformatics 18:1427–1431
39. Dunker J, Larsson U, Petersson D et al (2003) Parallel DNA template preparation using a

vacuum filtration sample transfer device. BioTechniques 34:862–868
40. Aranyi T, Varadi A, Simon I et al (2006) The BiSearch web server. BMC Bioinformatics 7:431
41. Tost J, Jammes H, Dupont JM et al (2007) Non-random, individual-specific methylation profiles are present at the sixth CTCF binding site in the human H19/IGF2 imprinting control region. Nucleic Acids Res 35:701
42. Warnecke PM, Stirzaker C, Melki JR et al (1997) Detection and measurement of PCR bias in quantitative methylation analysis of bisulphite-treated DNA. Nucleic Acids Res 25:4422–4426
43. Kwabi-Addo B, Chung W, Shen L et al (2007) Age-related DNA methylation changes in normal human prostate tissues. Clin Cancer Res 13:3796–3802
44. Ye H, Wu H, Liu Y et al (2015) Prenatal diagnosis of chromosomal aneuploidies by quantitative pyrosequencing(R). Methods Mol Biol 1315:123–132
45. Tetzner R, Dietrich D, Distler J (2007) Control of carry-over contamination for PCR-based DNA methylation quantification using bisulfite treated DNA. Nucleic Acids Res 35:e4

Chapter 23

Methylation-Specific PCR

João Ramalho-Carvalho, Rui Henrique, and Carmen Jerónimo

Abstract

Cytosine methylation is a DNA modification generally associated with transcriptional silencing. Factors that regulate methylation have been linked to human disease, yet how they contribute to malignancies remains largely unknown. Methylation of DNA can change the functional state of regulatory regions, but does not change the Watson–Crick base pairing of cytosine. Moreover, sequence symmetry of CpGs enables propagation of the methyl mark through cell division. This potential for inheritance coupled with the fact that DNA methylation patterns change during development and disease partially explains the interest in DNA methylation as a memory module. DNA methylation analysis also provides an opportunity to discover new and more powerful biomarkers that can help in clinical practice.

Methylation-Specific PCR (MSP) is likely the most widely used technique to study DNA methylation of a locus of interest. MSP can rapidly detect the methylation status of any group of CpG sites within a CpG island, not requiring methylation-sensitive restriction enzymes. It also requires minute amounts of DNA, is very sensitive as it can detect <0.1% of methylated alleles in a specific locus, and can be used in different samples, including bodily fluids, and paraffin-embedded samples.

Key words Methylation-specific PCR, DNA Methylation, Epigenetics, Prostate Cancer, Biomarkers, Gene regulation, MSP, qMSP, Cancer

1 Introduction

DNA methylation is probably the best studied and the best understood epigenetic modification, being conserved among plants, fungi, and animals, and it is also the only epigenetic mark for which a detailed mechanism of mitotic inheritance has been clearly depicted [1, 2]. In mammals, the most common form of DNA methylation results from the addition of methyl group to the five position of cytosine resulting in 5-methylcytosine (5mC), which is mostly restricted to the symmetrical CpG context [3, 4]. High levels of 5mC in CpG-rich promoters are strongly associated with transcriptional repression [5]. However, in CpG-poor genomic regions this dynamics is more complex and depends on the relationship between DNA methylation and transcriptional activity [5].

Jörg Tost (ed.), *DNA Methylation Protocols*, Methods in Molecular Biology, vol. 1708, https://doi.org/10.1007/978-1-4939-7481-8_23,

Deposition and maintenance of 5mC is essential for normal development and these processes are catalyzed by the DNA methyltransferases (DNMTs). DNMT1 is the most abundant in mammalian cells and methylates hemimethylated CpGs, preserving methylation patterns during DNA replication [6]. DNMT3a and DNMT3b are de novo methyltransferases that establish DNA methylation patterns during development and they can methylate unmethylated and hemimethylated DNA [7].

The mammalian genome contains approximately 28 million CpGs and 70–80% of these are methylated due to the high proportion of repetitive elements [8]. Approximately 10% of all CpGs are located within CpG-rich regions—CpG islands—and the remaining genome is depleted for CpGs [4]. CpG-rich regions are often associated with the transcription start sites of genes as well as in gene bodies and intergenic regions [4]. More than half of the housekeeping and developmental regulatory genes have CpG islands encompassing their transcription start sites [8].

Physiologic DNA methylation has been comprehensively studied for its role in various biological processes such as genomic imprinting, transposable elements silencing, stem cell differentiation, embryonic development, and inflammation [8]. Moreover, aberrant DNA methylation (hypomethylation and hypermethylation) has been reported for many pathologies, including cancer [9].

Over the years, several methods have been developed to map 5-methylcytosine, providing unparalleled, genome-wide coverage of DNA methylation changes [9] (*see also* Chapters 5–13). The experimental confirmation and biological reproducibility of DNA methylation for a given candidate gene, is usually done using locus-specific DNA methylation assays, such as methylation-specific PCR (MSP). Indeed, MSP was first established for the evaluation of the methylation status of gene promoters in cell lines and fresh frozen tissues [10]. MSP allows for highly sensitive detection (detection level of 0.1% of the alleles, with full specificity) of locus-specific DNA methylation, using PCR amplification of bisulfite-converted DNA [10]. It is a simple, fast and cost-effective method to analyze DNA methylation status of virtually any group of CpG dinucleotides within a CpG island. The power of MSP to discriminate methylated from unmethylated cytosine, however, depends on sodium bisulfite DNA conversion. During this chemical modification, unmethylated cytosines in DNA samples are converted to uracils, whereas methylated cytosines are resistant to this modification and remain unmodified [11]. Then, bisulfite-modified DNA is PCR amplified using specific primers sets, one that binds specifically to the methylated sequence and other that only binds to the unmethylated sequence.

MSP results are obtained using gel electrophoresis, not requiring further restriction or sequencing analysis. Due to its relative simplicity, safety (radioactivity labeling is not required) and sensitivity, the MSP assay is one of the most widely used methods to assess promoter methylation. Additionally, nested-based approaches, for the analysis of individual (nested MSP) or multiple (multiplex-nested MSP) promoters were developed to assess the methylation status of samples with low DNA quantity and/or quality, being suitable for analysis of DNA extracted from archival and microdissected specimens.

Nevertheless, MSP has been modified to allow for quantification of methylated alleles of a single region among unmethylated DNA (real-time quantitative MSP, QMSP) (*see also* Chapters 24 and 25). QMSP allows for a rapid analysis of multiple CpG sites in a large number of samples, not requiring subsequent manipulation such as electrophoretic gels.

Because the analysis of multiple molecular markers is more likely to provide clinically relevant information than single marker analysis and these assays are amenable to large-scale screening of several target genes in multiple samples, QMSP might be particularly important for validation and standardization of promoter methylation as a powerful biomarker in clinical practice.

2 Materials and Reagents

2.1 DNA Samples

1. Virtually, all samples can be tested by MSP-based methods. The most commonly used are cell pellets, fresh-frozen and formalin-fixed paraffin-embedded tissues.
2. For control purposes, commercially available unmethylated and methylated DNA is the best option. However, normal lymphocytes are an alternative as unmethylated control whereas in vitro methylated DNA (IVD) serves as methylated control. An alternative control approach is the use of DNA from cell lines in which the methylation status of particular genes is known, although it must be confirmed for each gene tested by the user.
3. To prepare the IVD control it is appropriate to use a commercially available IVD kit which includes all the buffers, 2, -S-adenosylmethionine (SAM) ,and the M.SssI methylase.

2.2 Cell Culture

1. Cell culture flasks.
2. Trypsin.
3. Cell culture media.
4. PBS.
5. CO_2 incubator and related equipment.
6. Pipettes, falcons, and other materials routinely used in cell culture.

2.3 Deparaffinization of Slides of Formalin-Fixed Paraffin-Embedded Tissues

1. Xylene.
2. Oven (set approximately at 60 °C).
3. Ethanol (EtOH): 100%, 95%, 70%, and 50% EtOH, at room temperature.

2.4 Reagents and Equipment to be Used in DNA Extraction

1. SE buffer: 75 mM Sodium Chloride, 25 mM EDTA Buffer (4.38 g NaCl, 50 mL of 0.5 M EDTA, add Distilled water to 1 L, autoclave).
2. Sodium dodecyl sulfate (SDS), to be used at 10%: Dissolve 10 g of SDS in 80 mL of H_2O, and then add H_2O to 100 mL. This stock solution is stable for 6 months at room temperature.
3. Proteinase K 10 mg/mL.
4. Digestion buffer FFPE: Combine 5 mL of 1 M Tris–HCl, 1 mL of 0.1 M EDTA, 500 μL of Tween 20 and distilled H_2O up to 100 mL.
5. Water bath or a thermomixer.
6. Phenol and chloroform, pH 8: Mix phenol (e.g., Sigma) with the supplied equilibration buffer and wait approximately 10 min until the solution reaches pH 8. In a flask, combine 400 mL of chloroform and the previously equilibrated phenol solution with the same volume. This solution is pH 8. Mix and store overnight at 4 °C. The next day, aliquot it in 50 mL falcons. Store at −20 °C until use.
7. *Phase lock* gel 1.5 mL, 2.0 mL and 15 mL tubes, according to the needs.
8. Ethanol.
9. Centrifuge.
10. 7.5 M Ammonium acetate.
11. Glycogen.
12. DEPC-treated (nuclease free) H_2O (or another elution buffer e.g., 1 M Tris–HCl, pH 8.0).
13. AKE buffer, 10×: 82.9 g NH_4Cl; 10 g $KHCO_3$; 0.37 g EDTA. Add 800 mL of distilled H_2O and homogenize. Adjust to pH 7.4 (using or NaOH or HCl). Then, add the remaining volume of H_2O to 1 L. Autoclave before use.
14. Phosphate buffered saline (PBS), 10×: Combine 1.48 g of Na_2HPO_4, 0.495 g $NaH_2PO_4 \times H_2O$, 7.2 g NaCl. Add distilled H_2O up to 1 L.
15. Scalpels/razor blades.
16. NanoDrop or another type of spectrophotometer.

2.5 Bisulfite Treatment

The reagents included in this particular section are those required to perform the in-house protocol [12]. The DNA bisulfite treatment may also be carried out using commercially available kits according to the manufacturers' protocol (*see* **Note 1**). Up to 2 μg of DNA can be used as starting material.

1. 2 M and 3 M NaOH.
2. Bisulfite Mix: For 20 samples, mix 4.75 g sodium bisulfite dissolved in 6.25 mL H_2O with 1.25 mL 1 M hydroquinone (prepared by adding 0.275 g to 2.5 mL H_2O and heating to 55 °C for 10 min) and 1.75 mL 2 M NaOH.
3. DNA clean-up kit (e.g., Promega Wizard DNA clean-up system or similar).
4. 7.5 M Ammonium acetate.
5. Glycogen.
6. EtOH (100 and 75%).
7. Vacuum manifold.
8. 80% isopropanol.
9. DEPC-treated (nuclease free) H_2O.

2.6 Reagents and Materials for PCR Amplification (See Note 2)

1. HotStart Taq polymerase and associated buffers for MSP:
   (a) Thermo Scientific Maxima Hot Start Taq DNA Polymerase (Thermo Fisher Scientific).
   (b) AmpliTaq Gold DNA Polymerase (Thermo Fisher Scientific).
   (c) Platinum *Taq* DNA Polymerase (Thermo Fisher Scientific).
   (d) KAPA SYBR FAST Universal 2× qPCR Master Mix (KAPA Biosystems).
2. dNTPs.
3. DS buffer: Combine 1660 μL 1 M $(NH_4)_2SO_4$, 3350 μL 2 M Trizma Preset pH 8.8, 670 μL 1 M $MgCl_2$, 70 μL 2-mercaptoethanol, 100 μL DMSO and 4150 μL DEPC H_2O.
4. DEPC-treated (nuclease free) H_2O.
5. ROX™ Passive Reference Dye (if necessary, *see* **Note 3**).
6. UCSC Genome browser (http://www.genome.ucsc.edu) or similar.
7. CpG island searcher (http://www.uscnorris.com/cpgislands2/cpg.aspx).
8. Primer design programs, such as Methyl Primer Express (https://www.thermofisher.com/order/catalog/product/4376041), the MSPPrimer (http://www.mspprimer.org),

which are also suitable to design MSP, nested, and bisulfite-sequencing primers.

9. Primers (standard desalting, 100 μM scale).
10. Universal Methylated DNA (methylation-positive control for gene methylation studies, e.g., CpGenome Universal Methylated DNA, Merck Millipore or similar).
11. Universal Unmethylated DNA (unmethylated DNA control for gene methylation studies (e.g., CpGenome Universal Unmethylated DNA, Merck Millipore or similar).
12. PCR tubes.
13. Thermocycler or a real time PCR system.
14. PCR hood for post-PCR manipulation and nested setup.

2.7 Agarose Gel Detection

1. Agarose.
2. TBE buffer, 10×: 1 M Tris–Boric Acid, 20 mM EDTA, pH ~ 8.3.
3. Ethidium bromide or equivalent (e.g., GreenSafe Premium or Sybr Safe) to a final concentration of 0.2–0.5 μg/mL (usually about 2–3 μL of lab stock solution per 100 mL gel).
4. Micropipettes.
5. DNA ladder.
6. Gel imaging system.

2.8 Nondenaturing Polyacrylamide Gel Electrophoresis (PAGE) Gel Detection

1. Acrylamide–bisacrylamide (29:1) (30% w/v).
2. Ammonium persulfate (APS) (10% w/v).
3. TBE buffer, 1×: 0.1 M Tris-Boric Acid, 2 mM EDTA.
4. *N*,*N*,*N*′,*N*′-Tetramethylethylenediamine (TEMED) 10% (w/v).
5. ddH_2O.
6. Vertical gel electrophoresis tank for polyacrylamide gels.
7. PAGE plates, spacers, and cast.
8. Loading Buffer.
9. DNA ladder.
10. Ethidium bromide (to be used at 1 μg/mL) or SYBR Gold (to be used as 1×, from the stock).
11. Gel imaging system.

3 Methods

3.1 DNA Extraction

3.1.1 DNA Extraction from Cell Lines, Urine Samples and Fine-Needle Aspirations Biopsies

1. Wash the cells-containing flask twice with cold PBS and trypsinize. The cell number differs depending on the cell type, but 5 million cells render a good amount of starting material.
2. Pellet cells by centrifugation for 5 min at 2000 × *g*. Remove supernatant.
3. If the samples are not from cell lines, pellet down cells by centrifugation (10 min at 12,000–16,000 × *g*).
4. Add 500 μL of SE buffer to pellet and vortex.
5. Add 30 μL of 10% SDS and 20 μL proteinase K (10 mg/mL) to each tube and homogenize.
6. Incubate samples overnight on a dry block or in a water bath set at 55 °C.
7. The next day, inactivate proteinase K through incubation at 95 °C for 10 min.
8. Centrifuge the phase lock gel tube at 12,000–16,000 × *g* for 20–30 s.
9. Transfer the content of each sample (sample volume 100–750 μL to a 2 mL phase lock gel tube and add an equal volume of phenol–chloroform pH 8; mix by inverting the tubes several times.
10. Centrifuge tubes at 12,000–16,000 × *g* for 15 min. The upper aqueous phase is separated from the bottom organic phase by the phase lock gel.
11. Prepare a new phase lock gel tube and add phenol–chloroform, previously prepared. Right after **step 10**, add the aqueous phase (containing the DNA) to the new tube. Centrifuge tubes at 12,000–16,000 × *g* for 15 min.
12. Transfer the upper aqueous phase containing DNA into a new 2 mL tube.
13. Precipitate DNA by adding 7.5 M ammonium acetate (1/10 of the total volume), 2 μL of carrier (e.g., glycogen, which is optional for cell lines), and two volumes of ice cold 100% EtOH. Mix by inverting the tubes.
14. Let stand overnight at −20 °C. Alternatively, an incubation of 2–3 h at −80 °C is sufficient to precipitate DNA.
15. Pellet the DNA by centrifugation at 12,000–16,000 × *g* for 20 min.
16. Discard the supernatant and wash the pellet with 1 mL of ice cold 70% EtOH.
17. Pellet DNA by centrifugation for 20 min at 12,000–16,000 × *g*.

18. Repeat **steps 16** and **17** and air-dry the pellet.
19. Resuspend the pellet in 20–100 μL elution buffer or nuclease-free water. To improve DNA elution, samples may be incubated at 50 °C for 10–20 min.
20. Quantify DNA using a NanoDrop or other spectrophotometer.

3.1.2 DNA Extraction from Blood and Fresh-Frozen Tissues

1. Centrifuge the blood-containing tubes at 1500 × *g* for 10 min.
2. Transfer the plasma to a 1.5 mL tube.
3. Transfer the remaining sample to a 50 mL tube and fill up to 50 mL with AKE.
4. Incubate on ice for 30 min.
5. Centrifuge at 1500 × *g* for 10 min.
6. Discard the supernatant and add 50 mL of AKE.
7. Centrifuge at 1500 × *g* for 10 min.
8. Discard the supernatant and add 50 mL of PBS.
9. Centrifuge at 1500 × *g* for 10 min.
10. Discard the supernatant.
11. Proceed to DNA extraction or store both plasma and leucocytes at −80 °C.
12. Add 2700 μL of SE buffer to the cells/tissue sections. Mix by vortexing.
13. Add 300 μL of SDS and 25 μL of proteinase K (10 mg/mL). Invert the tubes to mix.
14. Incubate samples overnight in a dry block or in a water bath, set at 55 °C. If the samples are not digested, add more 20 μL of proteinase K until digestion is complete.
15. Prepare the 15 mL phase lock gel by centrifuging them at 1500 × *g* for 1–2 min.
16. Add a volume of 3 mL of phenol–chloroform, previously prepared.
17. Transfer the content of each sample tube to a 15 mL phase lock gel tube and mix by inverting the tubes several times.
18. Centrifuge tubes at 1500 × *g* for 20 min. The upper aqueous phase is separated from the bottom organic phase by the phase lock gel.
19. Prepare a new phase lock gel tube and add 3 m L phenol–-chloroform previously prepared. Right after **step 18** centrifugation, add the aqueous phase (contains the DNA) to the new tube. Centrifuge tubes at 1500 × *g* for 20 min.
20. Transfer the upper aqueous phase containing DNA into a new 15 mL tube.

21. Precipitate DNA by adding 7.5 M ammonium acetate (1/3 of the total volume), and two volumes of ice cold 100% EtOH. Mix by inverting the tubes.
22. Incubate overnight at −20 °C. Alternatively, an incubation of 2–3 h at −80 °C is sufficient to precipitate DNA.
23. Pellet DNA by centrifugation for 30 min at 3000 × *g*.
24. Discard the supernatant and wash the pellet with 6 mL of ice cold 70% EtOH.
25. Pellet DNA by centrifugation for 20 min at 3000 × *g*.
26. Repeat the **steps 24** and **25**. Air-dry the pellet.
27. Rehydrate the pellet in 50–150 μL of elution buffer or nuclease-free water. To improve DNA elution, samples may be incubated at 50 °C for 10–20 min.
28. Quantify DNA using a Nanodrop or other spectrophotometer.

3.1.3 DNA Extraction from Paraffin-Embedded Samples

Deparaffinization is critical and mandatory to extract DNA from paraffin-embedded tissue slides.

1. Cut 3–5 sections 5–10 μm thick and place them in a single 1.5 mL tube.
2. Add 1 mL of xylene and invert several times (do not use vortex). Incubate at RT for a minimum of 15 min.
3. Spin down at 12,000 × *g* in a microcentrifuge for 5 min.
4. Discard the xylene and add fresh xylene (1 mL), repeating **steps 2** and **3** two times (this is very important to ensure complete removal of paraffin).
5. Discard the final volume of xylene, add 1 mL of 100% EtOH, mix (do not vortex) and incubate at RT for 15 min.
6. Spin at 12,000 × *g* for 5 min, discard EtOH, add fresh 100% EtOH, and incubate for 5 min.
7. Discard the 100% EtOH, add 1 mL of 70% EtOH, mix (do not vortex) and incubate at RT for 5 min.
8. Spin at 12,000 × *g* for 5 min, discard EtOH, add 1 mL of 50% EtOH, and incubate for 5 min.
9. Spin at 12,000 × *g* for 5 min, discard EtOH, and dry in a heat block for 15 min at 90 °C.
10. Add 1000 μL of digestion buffer FFPE to each specimen and 35 μL of proteinase K. Adjust the volume of digestion buffer and proteinase K according to sample volume.
11. Incubate at 55 °C overnight. A clear liquid indicates complete digestion. If tissue is still visible, add more proteinase K and repeat digestion at 55 °C until all lysate becomes clear.
12. Heat to inactivate proteinase K (10 min at 95 °C) (optional).

13. Spin down briefly.
14. Samples can be stored at this stage at −80 °C, or proceed to next step.
15. Continue with **step 8** of Subheading 3.1.1.
16. After DNA quantification, store it at 4 °C (if used on a regular basis) or −20 °C (longer storage).

3.1.4 DNA Extraction from Macrodissected Paraffin-Embedded Samples

1. Place the paraffin-embedded slides vertically on a Kimwipes tissue and incubate them in an oven, at 60 °C, until the paraffin has melted.
2. Carefully wipe the melted paraffin off the slide using a Kimwipes.
3. Wash the slides 2–3 times for 10 min, with fresh xylene.
4. Rehydrate slides by incubating for 5 min in each of the following solutions: 100%, 95%, 70%, and 50% EtOH (use Coplin jars).
5. Finally, rinse slides in ddH_2O twice for 2 min and allow slides to dry.
6. Label areas of interest on the back of the deparaffinized slide (as previously determined by the pathologist on the corresponding H&E slide).
7. Put a drop of digestion buffer FFPE into each spot (about 10 μL, but adjust according to the scrapping area).
8. Scrape the area of interest using a scalpel, aspirate using a pipette and transfer into a clean tube containing 500 μL of digestion buffer FFPE and 15 μL of proteinase K. Clear liquid indicates complete digestion. If tissue is still visible, add more proteinase K and repeat digestion at 55 °C until all lysate becomes clear.
9. Continue with **step 8** from Subheading 3.1.1. Alternatively, use the whole volume of the digested sample for bisulfite treatment.
10. Quantify DNA using a Nanodrop.
11. Store DNA as previously described.

3.2 Bisulfite Conversion [12]

1. Dilute up to 2 μg of DNA in 20 μL of nuclease-free water.
2. Add 2 μL of 3 M NaOH and incubate 20 min at 50 °C (This step creates single-stranded DNA, which is sensitive to sodium bisulfite conversion).
3. In the meantime, prepare the bisulfite reaction mix,
4. Add 450 μL of reaction mix to each sample and mix gently.
5. Incubate at 70 °C for 3 h.

6. Connect columns to a vacuum manifold.
7. Add 1 mL of Promega DNA wizard clean-up resin to each sample. Mix and transfer to the columns. This step allows for the separation of the bisulfite-treated DNA from the sodium bisulfite solution.
8. Apply vacuum.
9. Fill column with 80% isopropanol. Apply vacuum until column is empty.
10. Wash again with 80% isopropanol. Apply vacuum until the column is empty.
11. To allow the column to dry, apply vacuum continuously for 10 min.
12. Transfer empty column into labeled 1.5 mL tube. Turn off vacuum.
13. Apply 45 μL of heated ddH_2O (60–70 °C) (warm water facilitates DNA elution from the resin).
14. Spin tube/column for 5 min at 12,000 × *g* and discard columns.
15. Add 5 μL of 3 M NaOH to the DNA and incubate at RT for 10 min (this completes the chemical modification by desulfonating the bisulfite-treated DNA).
16. For each tube add 1 μL of glycogen, 75 μL of 7.5 M ammonium acetate and 350 μL of ice cold 100% EtOH.
17. Mix and incubate tubes overnight at −20 °C.
18. Spin tubes for 20 min at 12,000 × *g* at 4 °C, discard supernatant, and add 500 μL of ice cold 70% EtOH.
19. Repeat **steps 17** and **18**. Centrifuge the tubes.
20. Dry pellet and resuspend in 60 μL of ddH_2O (1 μg of DNA/60 μL of H_2O).

3.3 Primer Design

Primer design is a common challenge for all PCR-based methods used to analyze bisulfite-converted DNA and it is a critical step to accurately determine the methylation status of a gene promoter [10]. After bisulfite modification, the two strands of DNA are no longer complementary. Thus, strand-specific primers must be used for PCR amplification. Because any strand can be selected for methylation studies, it is possible to design suitable primers for one strand when the other does not allow for appropriate primer design.

The chemical conversion of all unmethylated Cs to Ts, leads to a biased base composition of the bisulfite-converted DNA, which is then comprised mostly of only three of the possible four bases. The original (bisulfite-converted) strands of DNA are T, G, A-rich, and the reverse complement is A, C, T-rich. Thus, primers designed for amplification of bisulfite-converted DNA frequently have multiple annealing sites, leading to alternative unspecific amplicons, and are more prone to primer-dimer formation (*see* **Note 4**).

3.3.1 Manual Design of MSP Primers

1. The target DNA promoter sequence to be analyzed might be obtained from the UCSC Genome browser or a similar public database.
2. Select approximately 1000–2000 base pairs (bp) of genomic sequence upstream of the transcription start site. Nonetheless, include approximately 200 bp of sequence following the transcription start site as CpG islands often continue after the transcription start site.
3. Confirm the presence of a CpG island by pasting the sequence into CpG island searcher using default settings.
4. Transfer the sequence into a text document. Find and replace in the whole document CG by X, then C by T and finally X by CG. This is the methylated-modified sequence from which the methylated set of MSP primers will be designed.
5. Paste the same original sequence in a different page of the text document. Find C and replace by T in the whole document. This is the unmethylated-modified sequence from which unmethylated MSP primers sets will be designed. Coloring CGs in a distinct color allows for easier visualization of CpG islands and helps with primer design.
6. MSP primers are designed using the same strand (either sense or antisense strand).
7. Methylated MSP primers should have the following characteristics:
   (a) Use the methylated-modified sequence.
   (b) Methylated amplicon products should be less than 200 bp.
   (c) Methylated primers should amplify a region in the CpG island rather than on the edge of the island.
   (d) Methylated primers must contain a minimum of 2–3 CpG dinucleotides and at least 1–2 CpG located at the 3′ of the primer sequences.
   (e) In the methylated primers, the forward must terminate with a C and the reverse has to end with a G.
   (f) The remaining unconverted cytosine (C), which will appear as T in the converted sequence, should be located 3′ to ensure specific amplification of bisulfite-converted DNA.
   (g) Melting temperature should be approximately 60–65 °C.

8. Unmethylated MSP primers should have the following characteristics:
   (a) Use the unmethylated-modified sequence.
   (b) Unmethylated amplicon products should be less than 200 bp.
   (c) The target sequence must be a similar region as that of the methylated set of primers.
   (d) Terminate with a T for the forward unmethylated primer and with an A for the reverse unmethylated primer.
   (e) Melting temperature should be approximately 60–65 °C.
   (f) If some adjustments in the melting temperatures of unmethylated primer sequences are required, extend the 5′ end. The melting temperatures for unmethylated and methylated MSP primers should be within a 5 °C range.

3.3.2 Manual Design of Nested Primers

Nested primers should amplify the region where the methylated MSP primers have been designed. Nested and MSP primers may overlap a few bases (approximately 10 bp).

Nested primers should not have CG residues so that the amplified products generated using these primers are neither specific for the methylated nor the unmethylated sequence. If this cannot be avoided, minimize the number of CG and replace in the primer sequence the C of the CG by a Y (an equimolar mixture of pyrimidines C and T) in the forward nested primer, and the G of the CG by an R (an equimolar mixture of purines A and G) in the reverse nested primer sequence.

3.3.3 Primer Design Using Methyl Primer Express® Software

The Methyl Primer Express® Software, developed by M.P. Fraga, C. Ferrero, and M. Esteller, allows for the design of bisulfite sequencing (BSP) and methylation-specific PCR (MSP) primers. The software examines DNA sequences for CpG islands, simulates bisulfite modification on the CpG-containing sequences, and it finally generates primer pairs for MSP or BSP. Primer design for either method can be adjusted to accommodate variations in experimental design.

Methyl Primer Express software includes specific roles for primer design to guarantee the ability of the primers to form a stable duplex strand with the specific site on the target DNA:

1. In case multiple islands are found, any of the predicted islands is a target region for amplification.
2. If a CpG island is smaller than the minimal amplicon size, the primer pair should span the whole island.
3. If a CpG island is greater than the maximal amplicon size, the primer pair should be within the island.

4. If a CpG island is between the minimal and maximal amplicon size, the primer pair should cover at least two-thirds of the island.
5. Two primer sets are required for MSP: one assuming a fully methylated sequence and the other a fully unmethylated sequence. The recommendations for picking MSP primers must observe the following conditions:
6. Each primer should contain at least three CpG sites in its sequence, and one of the CpG sites should be at the 3′ end of the sequence to maximally discriminate between methylated DNA and unmethylated DNA.
7. Each primer should have a maximum number of non-CpG Cytosines in its sequence to amplify only the bisulfite-modified DNA.
8. The primer pair for the methylated DNA (M pair) and the primer pair for the unmethylated DNA (U pair) should contain the same CpG sites within their sequences. For example, if the forward primer for an M pair has the sequence ATTAGTTTCGTTTAAGGTTCGA, then the forward primer for the U pair must also contain the two CpG sites, although they may differ in length and start position.

3.3.4 Probe Design for QMSP

Real-time detection is based on the recognition of a fluorescent signal produced proportionally during the amplification of a PCR product. A fluorogenic probe is designed to anneal to a target sequence between forward and reverse primers. The 5′ end is labeled with a reporter fluorochrome (normally FAM) and the 3′ end is labeled with a quencher (TAMRA or MGB). The following rules must be observed when designing a QMSP probe:

1. Keep the GC content close to 50%.
2. Avoid long stretches of a single nucleotide.
3. The probe should not hybridize with the primers.
4. The probe must have a minimum of 2–3 CG dinucleotides and at least 1–2 CGs located at the 3′ of the primer sequence.
5. It is suggested that the probe sequence ends with a C.
6. The melting temperature of the probe should be 10 °C higher than that of the primers.
7. The probe may be designed in either forward or reverse orientation.

3.4 PCR Amplification

Before analyses in the samples of interest, the primers and PCR conditions (for any form of PCR described in this chapter) must be previously optimized (*see* **Note 4**).

3.4.1 Methylation-Specific PCR

1. Thaw the 10× PCR buffer, dNTP mix, and primers (*see* **Notes 2** and **5**).

2. Scale the master mix according to the number of samples to be analyzed taking the unmethylated and the methylated amplifications as well as a negative template control into account according to the following table:

10× PCR buffer	1.88 μL
dNTP mix (2 mM)	1.88 μL
Primer F and R (10 μM)	1.88 μL
MaximaHotStart TaqDNA polymerase (5 U/μL)	0.15 μL
Nuclease-free H_2O	12.21 μL

3. Pipette 18 μL of each PCR master mix into separate PCR tubes. Make sure the tubes are correctly labeled.
4. Add 2 μL of bisulfite-modified DNA template.
5. Initiate the PCR with a 10 min denaturation step at 95 °C, followed by 35 cycles: 95 °C—30 s (denaturation); AT°C—30 s (annealing: primer specific); 72 °C—1 min (elongation) and a final elongation step at 72 °C for 7 min (*see* **Note 6**).
6. Store at 4 °C until analysis.

3.4.2 Nested-MSP and Multiplex Nested-MSP

1. Thaw the 10× PCR buffer, dNTP mix, and primers.
2. Scale the master mix according to the number of samples to be analyzed taking the unmethylated and the methylated amplifications as well as a negative template control into account according to the following table:

10× PCR buffer	1.88 μL
dNTP mix (2 mM)	1.88 μL
Primer F and R (10 μM)	1.88 μL
MaximaHotStart TaqDNA polymerase (5 U/μL)	0.15 μL
Nuclease-free H_2O	12.21 μL

3. Pipette 18 μL of each PCR master mix into separate PCR tubes. Make sure the tubes are correctly labeled.
4. Add 2 μL of bisulfite-modified DNA template.
5. Continue as described in **step 4** of Subheading 3.4.3.

3.4.3 Multiplex Nested-MSP Stage 1 Amplification Setup

It is convenient to prepare a mix for multiplex-nested primers (e.g., if six promoters will be analyzed, combine 50 μL of each of the forward- and reverse-nested primers (100 pmol/μL) for a total volume of 600 μL) (*see* **Note 7**).

1. Prepare the following mix (for one sample: quantities should be multiplied by the number of samples and controls analyzed):

10× PCR buffer	1.88 μL
dNTP mix (2 mM)	1.88 μL
Nested primer Mastermix (10 μM)	10 μL
MaximaHotStart TaqDNA polymerase (5 U/μL)	0.2 μL
Nuclease-free H_2O	4.19 μL

2. Pipette 18 μL of each PCR master mix into separate PCR tubes. Make sure the tubes are correctly labeled.
3. Add 2 μL of bisulfite-modified DNA template.
4. Initiate the PCR with a 10 min denaturation step at 95 °C, followed by 35 cycles: 95 °C—30 s (denaturation); AT °C—30 s (annealing: primer specific); 72 °C—1 min (elongation) and a final elongation step at 72 °C for 7 min (*see* **Note 6**).
5. Store at 4 °C until analysis.
6. Dilute nested amplification products, at least the least 1/500, with nuclease-free ddH_2O and use 2 μL as template for conventional MSP.

   Each of the promoters amplified in the multiplex-nested reaction will be assayed individually in an MSP reaction.

3.5 Quantitative MSP

The QMSP assay allows for the quantification of methylated alleles of a given region among unmethylated DNA. The sensitivity of QMSP is very similar to conventional MSP (1:1000–1:10,000). There are two commons chemistries used to perform QMSP: the nucleic acid SYBR® Green dye and the oligonucleotide fluorescent probes. SYBR® Green binds to double-stranded DNA and, during PCR, DNA polymerase amplifies the target sequence, creating PCR products to which SYBR® Green dye then binds. As PCR progresses, more PCR product is created and more SYBR® dye binds to all double-stranded DNA, resulting in increased fluorescence intensity proportional to the amount of PCR product synthesized. The main advantages of SYBR green are: any double-stranded DNA target sequence can be analyzed; reduced assay setup and running costs, assuming that your PCR primers are well designed. The main disadvantage is that it may generate false positive signals.

The oligonucleotide probes act as a third primer in the reaction and detect only the DNA sequence complementary to the probe. Thus, use of the reporter probe significantly increases specificity due to more stringent amplification. The advantages of probes are: specific hybridization between probe and target is required to generate fluorescent signal; probes can be labeled with different reporter dyes allowing amplification and detection of distinct sequences in one reaction tube and post-PCR processing is excluded, which reduces assay labor and material costs. The main disadvantage is that different sequences require the synthesis of specific probes.

3.5.1 QMSP Using SYBR Green Reagents

Determine the number of samples to be used, as it might involve different PCR plates to be used (*see* **Note 5**).

1. Thaw the 10× PCR buffer, dNTP mix and primers.
2. Scale the master mix according to the number of samples to be analyzed taking the unmethylated and the methylated amplifications as well as a negative template control into account.
3. Prepare a standard curve dilution of the positive control (Universal Methylated DNA).
4. Prepare the following mix (for one sample, quantities should be multiplied by the number of samples and controls analyzed): combine 10 μL of SYBR green master mix, 1 μL of primers (0.5 μL of each primer) and complete with 7 μL DEPC H_2O. Rox dye usage depends on the equipment to be used (*see* **Note 3**). Please keep in mind that these volumes might be scaled down.
5. Pipette 18 μL of mix to each well.
6. Add 2 μL of sample.
7. Centrifuge the plate at 1200 × *g* for 1 min.
8. The PCR cycling conditions are those provided by the manufacturer. The annealing and elongation step might need optimization (*see* **Notes 4–6**).
9. Include a melting curve analysis according to the thermocycler specifications to assess the specificity of PCR amplicons produced [13].

3.5.2 QMSP Using Oligonucleotide Probes (with AmpliTaq Gold)

1. Thaw the 10× PCR buffer, dNTP mix, $MgCl_2$, and primers.
2. Scale the master mix according to the number of samples to be analyzed taking the unmethylated and the methylated amplifications as well as a negative template control into account.
3. Prepare a standard curve dilution of positive control (Universal Methylated DNA).
4. Prepare the following mix (for one sample, quantities should be multiplied by the number of samples and controls analyzed [14]. Please keep in mind that these volumes might be scaled down.

DEPC H_2O	10.6 μL
$MgCl_2$	4.4 μL
10× buffer	2 μL
dNTPs	0.16 μL
Primer F (100 μM)	0.12 μL
Primer R (100 μM)	0.12 μL
Rox dye (*see* **Note 3**)	0.4 μL
Probe (100 μM)	0.04 μL
AmpliTaq gold	0.2 μL

5. Pipette 18 μL of mix to each well.
6. Add 2 μL of sample.
7. Centrifuge the plate at 1200 × *g* for 1 min.
8. Initiate the PCR with a 10 min denaturation step at 95 °C followed by 45 cycles: 15 s at 95 °C (denaturation) and 1 min at 60 °C (annealing and elongation). The annealing and elongation step might need optimization (*see* **Notes 4–6**).

3.5.3 QMSP Using Oligonucleotide Probes (with PlatinumTaq)

1. Thaw the 10× PCR buffer, dNTP mix, $MgCl_2$, and primers.
2. Scale the master mix according to the number of samples to be analyzed taking the unmethylated and the methylated amplifications as well as a negative template control into account.
3. Prepare a standard curve dilution of positive control (Universal Methylated DNA).
4. Prepare the following mix (for one sample, quantities should be multiplied by the number of samples and controls analyzed [15]. Please keep in mind that these volumes might be scaled down.

DEPC H_2O	14.96 μL
10× DS buffer	2 μL
dNTPs	0.16 μL
Primer F (100 μM)	0.12 μL
Primer R (100 μM)	0.12 μL
Rox dye (*see* **Note 3**)	0.4 μL
Probe (100 μM)	0.04 μL
Platinum Taq	0.2 μL

5. Pipette 18 μL of mix to each well.
6. Add 2 μL of sample.

7. Centrifuge the plate at 1200 × *g* for 1 min.
8. Initiate the PCR with a 10 min denaturation step at 95 °C followed by 45 cycles: 15 s at 95 °C (denaturation) and 1 min at 60 °C (annealing and elongation). The annealing and elongation step might need optimization (*see* **Notes 4–6**).

A representative result from qMSP assay is depicted in Fig. 1a (reference gene, e.g., β-actin) and 1B (target gene, e.g., *GSTP1*).

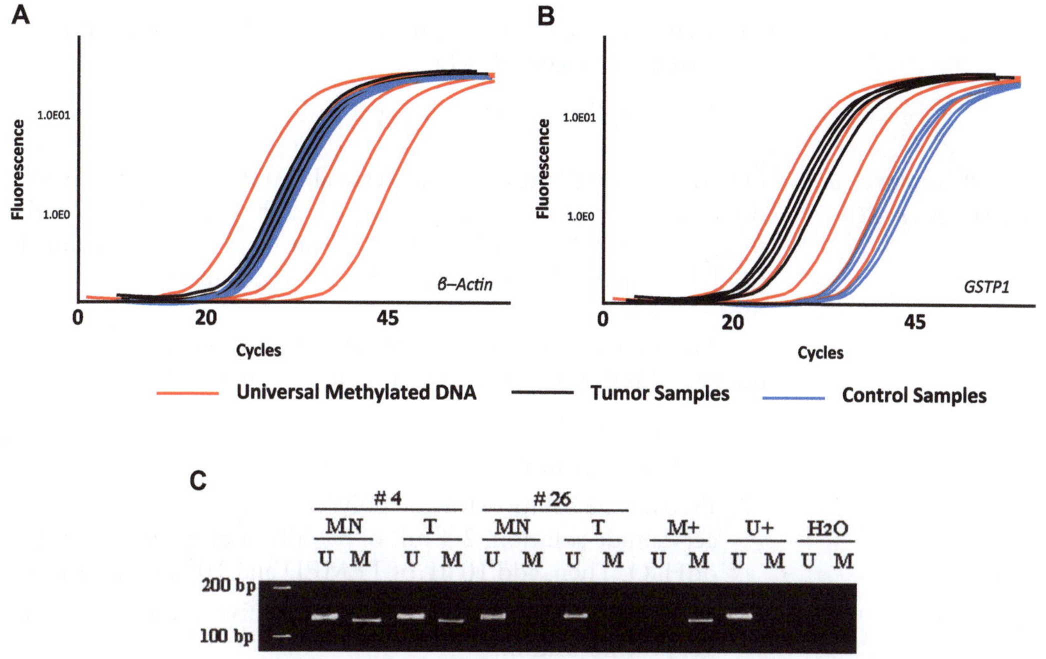

Fig. 1 Schematic representation of MSP and QMSP results for a given locus. (**a**) Ideal amplification curves for the reference gene (e.g., *β–Actin).* The *red lines* represent the calibration curves constructed using serial dilutions of bisulfite-modified Universal Methylated DNA. In dark are depicted the tumor and in blue the control samples. In this case, the tumor samples amplification is similar to the control samples, indicating that the number of alleles is comparable between the tumor and the control samples. The reference gene (e.g., *β–Actin*) for QMSP must ideally not vary dramatically between samples type, as it is for example the case for the target gene depicted in Fig. (**b**). Please note that any units on both X-and Y-axis are arbitrary. (**b**) Amplification signal for multiple samples, including tumor and control samples, for a potential hypermethylated locus in Prostate Cancer (e.g., *GSTP1*). The *red lines* represent the calibration curves constructed using serial dilutions of bisulfite-modified Universal Methylated DNA. In dark are depicted the tumor and in blue the control samples. In this case, the tumor samples amplification is earlier than control samples, indicating that the number of methylated alleles is higher in tumor than in control samples. Please note that any units on both X-and Y-axis are arbitrary. (**c**) Representation of an agarose gel electrophoresis for MSP. The picture contains the result for 2 samples (#4 and #26), a positive control for methylated (M+) and for unmethylated DNA (M−). In the column H_2O is depicted the result for controlling the reaction mix for any PCR unintended amplification (unspecific PCR product, or other contaminations)

3.6 Gel Detection

3.6.1 Agarose Gel Resolution

1. Prepare a 2% agarose (*see* **Note 8**) gel combining agarose and TBE buffer. Add the appropriate dye to the gel (*see* **Note 9**).
2. Load 10 μL of the PCR product from the amplification for unmethylated molecules so that every other lane contains unmethylated samples.
3. Load 10 μL of the methylated product into the remaining lanes (confirm that MSP reactions have been done simultaneously).
4. Load the DNA ladder in the first lane and apply constant voltage until required resolution is completed.
5. Visualize amplification products using UV in a gel imaging system (*see* **Notes 9–11**).

An example is depicted in Fig. 1c.

3.6.2 Polyacrylamide Gel Electrophoresis (PAGE)

One of the advantages of using vertical PAGE is the better resolution obtained compared to agarose gels. Polyacrylamide gels is able to separate DNA that differs by 0.2% in length, well beyond the resolving capabilities of agarose (2% difference in DNA length (*see* **Notes 8** and **9**).

The preparation and loading are more difficult. As a consequence, PAGE is not amenable for running many samples.

1. Mount PAGE plates, spacers, and cast together. Make sure that there is no leakage.
2. Prepare the solution for the polyacrylamide: combine 3 mL of acrylamide solution, 2.4 mL of 5× TBE buffer, and 6.5 mL of ddH_2O. Then, add 10 μL of TEMED and 100 μL of 10% APS.
3. Dispense the mixture into cast and leave for 15 min to allow for gel polymerization.
4. Transfer the cast gel into the PAGE electrophoresis system, add the running buffer, and load the samples carefully to prevent cross-contamination.
5. Run the gel until the desired resolution is obtained.
6. Disassemble cast and transfer the gel into a reservoir containing ethidium bromide.
7. Prepare a solution 0.5 μg/mL ethidium bromide in water or buffer to completely submerge the gel (*see* **Notes 8** and **9**).
8. Keep the gel in the staining solution for 15–30 min (depending upon gel thickness).
9. Wash with ddH_2O. Visualize using UV in a gel imaging system (*see* **Notes 9–11**).

3.7 Data Analysis

The identification of a specific band representing a PCR product with adequate molecular weight reveals the presence of either unmethylated or methylated alleles (Fig. 1). During the analysis of clinical samples, and if studying a tumor-suppressor gene, the expected scenario is that normal tissue are fully unmethylated, whereas primary tumor might be also unmethylated, fully methylated or display an unmethylated and a methylated band (due to intratumoral heterogeneity and/or contamination with normal cells from the stroma). The latter result might also derive from the analysis of cells containing both unmethylated and methylated alleles of a given target gene. To minimize contamination with normal cells, macrodissection or laser-capture microdissection is highly recommended. For troubleshooting *see* **Notes 10** and **11**.

3.7.1 QMSP Data Analysis

Because QMSP depends on the input of template DNA, an internal reference should be used to normalize the assay. A housekeeping gene (e.g., *ACTB*) without CpG islands is the most frequent choice. Several strategies might be employed to express this relative quantity and the most common is a ratio calculated using the values obtained for the target gene and for the reference gene in the same sample. This represents the relative level of methylation of a given gene sequence in that particular sample. Moreover, the quantitative assay using a probe was shown to reliably detect promoter methylation in the presence of 10.000-fold excess of unmethylated alleles, i.e., a tenfold increase in sensitivity over conventional MSP.

If the QMSP is based on SYBR green detection, it allows for the identification of specific amplification of DNA fragments through the analysis of their melting temperature as this is specific for each amplified PCR product. The results are acquired by comparing the melting curves for each sample. If different peaks from the expected amplification product are observed, it is suggested that the alternative peak might result from primer-dimer formation (*see* **Notes 10** and **11**).

To reliably quantify the gene methylation of each sample it is recommended to ensure that some real time PCR parameters are met:

1. R = (0.97–1).
2. Slope = (−3.30; −3.92); ensuring that at least 80% efficiency is attained.
3. The inter-plate difference in efficiency must be less than 2%.
4. The standard deviation of replicates of each sample should be below 0.35.
5. To determine the ratio of methylation: (relative quantity of target gene/relative quantity of reference gene), the following formula can be used: Methylation level = *(Target gene/ Housekeeping gene)* × *1000* [16].

4 Notes

1. The suggested bisulfite-conversion protocol has been used for a long time in our laboratory. However, commercially available kits (such as EZ DNA Methylation-Gold™ Kit, Zymo Research) are now being increasingly used. Those kits are faster, the conversion efficiency is more accurate, and the performance in multiple downstream applications is more reliable. Indeed, the batch effects are limited using the available kits. However, both the in-house protocol and commercial solutions are effective in the conversion of most of the unmethylated cytosines to uracils, maintaining good DNA integrity. Additionally, some adjustments can be made to the protocol (e.g., increased duration of incubation steps, and temperature >50 °C and a pH < 5.0 will produce more complete conversion). Nonetheless, increased degradation of template DNA is more likely to occur. Independently of the used protocol, DNA degradation due to bisulfite treatment is rather common. If DNA is degraded, the amplification of products larger than 1 kb will be very difficult and amplification of products in the 500–1000 bp range will be less efficient. This might be particularly problematic for bisulfite sequencing, although it has less impact on MSP.
2. Good laboratory practices are strongly recommended. PCR product contaminations generated in previous experiments are of great concern. It is advocated to have separate micropipette sets for sample and reagent preparation and for reaction setup. Moreover, these steps should be separated from reaction analyses. Filtered micropipette tips should be used for all pre-PCR manipulations to minimize potential contaminations.
3. ROX™ Passive Reference Dye is formulated for use on Applied Biosystems (ABI)/Thermo Fisher Scientific and Stratagene real-time PCR instruments. This inert dye, whose fluorescence does not change during the reaction, may be added to quantitative, real-time PCR reactions to normalize the well-to-well differences that may occur due to artifacts such as pipetting errors or instrument limitations. ROX Passive Reference Dye is composed of a 25 μM glycine conjugate of ROX in 10 mM Tris–HCl, pH 8.6, 0.1 mM EDTA, and 0.01% Tween® 20. *See* specific instrument instructions for further details on passive dye usage. As indication, ROX™ concentrations for different instruments are given below:
   (a) ABI 7000, 7300, 7700, 7900HT and 7900HT Fast:
      - Amount per 50 μL reaction: 1.0 μL (0.6–1.0 μL).
        - Final ROX Concentration: 500 nM (300–500 nM).
        - Dilution Factor: 50×.

(b) ABI 7500 and ABI 7500 Fast; Stratagene Mx3000™, Mx3005P™, and Mx4000™:
   - Amount per 50 μL reaction: 0.1 μL (0.06–0.1 μL).
     - Final ROX Concentration: 50 nM (30–50 nM).
     - Dilution Factor: 500×.

4. PCR amplification must be very stringent to assess the methylation status of each sample. PCR optimization is a critical step: our experience dictates that annealing temperatures must be optimized for each gene using a temperature gradient. Sometimes, the annealing time needs to be optimized as well. Due to the base bias created by the bisulfite-modification, the unmethylated primers are rich in A/T and, consequently, such primers are longer than the corresponding methylated set. If possible, annealing temperatures must be optimized so that both primers sets have a similar annealing temperature. Usually, increasing the annealing temperature helps to obtain a specific PCR product. In our experience, there is no need to use PCR additives (DMSO) to the reaction, because the bisulfite treatment reduces the CG content and, therefore, the amplification is easier. Include adequate controls in every batch for both the unmethylated and methylated reactions.
5. If a large number of samples will be analyzed, it is highly recommend using PCR plates (either 96 or 384 wells). For qMSP both plate systems are equally valuable. The volumes need to be scaled accordingly.
6. Avoid excessive number of cycles that frequently lead to unspecific amplification.
7. For multiplex-nested PCR, the primer combinations used need to have similar annealing temperatures to allow for amplification in the same reaction.
8. For preparing the gel, consider the following table:

Agarose Gels		**Polyacrylamide Gels**	
% agarose	**Fragment Size (bp)**	**% acrylamide**	**Fragment Size (bp)**
0.5	1000–30,000	3.5	1000–2000
0.7	800–12,000	5	80–500
1.0	500–10,000	8	60–400
1.2	400–700	12	25–150
1.5	200–500	15	25–150
2	100–400	20	6–100

Please note that to ensure adequate buffering power during vertical electrophoresis, TBE Buffer is used for polyacrylamide gel electrophoresis at a working strength of 1×. Lower dilutions of the buffer or the use of TAE Buffer may cause gels to overheat and result in band "smiling" throughout the gel.

9. Ethidium bromide is, probably, the most frequently used stain for detecting DNA/RNA. It yields low background and has a detection limit of 1–5 ng. The major drawback to ethidium bromide is that it is a potent mutagen. It is a DNA intercalating dye, inserting itself into the free gaps between the base pairs. Ethidium bromide displays UV absorbance between 300 and 360 nm and reemits this energy at 590 nm. Besides the sensitivity, the simplicity and nondestructive nature of ethidium bromide staining have made it the standard stain for double-stranded DNA. Please note that ethidium staining is strongly enhanced by the double-stranded structure of native DNA. Staining of denatured ssDNA or RNA is rather insensitive, requiring some tenfold more nucleic acid for equivalent detection. Another limitation is that the fluorescence of ethidium is quenched by polyacrylamide, reducing sensitivity by 10–20 fold in PAGE gels. During its manipulation extreme caution is advised: handle only with gloves and proper protective equipment; Solutions must be handled with extreme caution, and decontaminated prior to disposal.

10. *No amplification*: If no amplification is observed with unmethylated or methylated primers, the first recommendation it to optimize the primers. Nonetheless, DNA from paraffin-embedded samples, due to degradation, often fails to amplify by MSP. A different MSP-based approach, such as nested-MSP, might overcome this limitation. To determine if bisulfite-modified DNA has a sufficient quality for further analysis, it is strongly recommended to use an established control MSP primer set which amplifies a sequence devoid of cytosines (e.g., *ACTB*, primer sequences: *ACTB*_F: TGGTGATGGAGGAGGTTTAGTAAGT; *ACTB*_R: AACCAATAAAACCTACTCCTCCCTTAA.) under conditions previously shown to provide adequate results. If control primers fail to amplify a specific PCR band, the DNA template is the most likely source of the problem. If high quality genomic DNA was initially used, it is plausible that problems occurred during bisulfite treatment or storage after the chemical modification. However, if a PCR using control primers produces acceptable results, a problem with DNA may be excluded, and a problem in the PCR reaction itself is the most likely explanation. Primer design and annealing temperature are some of the most critical aspects of the MSP-based techniques. An easy and fast experiment is to test the primers with decreased (by a few degrees) annealing temperature in a proper control sample. If no amplification is detected, it is necessary to check the primer design: be sure that the amplicon generated is

below 200 bp, that the primers specifically recognize bisulfite-modified DNA and that they do recognize the same DNA strand (sense or antisense) after bisulfite modification, make certain that the primers have equal annealing temperatures and that they are designed in the 5′-3′ direction. Ultimately, even if applying all these parameters a specific band is not detected, design new PCR primers.

11. *Methylation-specific amplification in all samples*: The most common cause for detection of methylation products in all samples, including negative control, is the contamination with previously amplified PCR products. It is mandatory to work with clean micropipettes, bench and racks. The use of DNA degradation solutions helps preventing those contaminations. Avoid working with PCR products in the PCR-setup area. If the negative control has no amplification, but all samples are positive for methylation, this might be due to either the presence of methylated alleles in all samples or to primers not specific enough to discriminate between unmethylated and methylated alleles. Although some optimization can be made (e.g., changing the annealing temperature), it is preferable to design new primers and include up to two extra CpG dinucleotides in the newly designed primer to increase discriminatory power.

Acknowledgments

This work was supported by Federal funds through Programa Operacional Temático Factores de Competitividade (COMPETE) with co-participation from the European Community Fund (FEDER) and by national funds through Fundação para a Ciência e Tecnologia (FCT) under the project EXPL/BIM-ONC/0556/2012 (CJ). João Ramalho-Carvalho is supported by a fellowship from FCT (SFRH/BD/71293/2010).

References

1. Feng S, Jacobsen SE, Reik W (2010) Epigenetic reprogramming in plant and animal development. Science 330:622–627
2. Bird A (2002) DNA methylation patterns and epigenetic memory. Genes Dev 16:6–21
3. Ziller MJ, Muller F, Liao J et al (2011) Genomic distribution and inter-sample variation of non-CpG methylation across human cell types. PLoS Genet 7:e1002389
4. Jones PA (2012) Functions of DNA methylation: islands, start sites, gene bodies and beyond. Nat Rev Genet 13:484–492
5. Bock C (2012) Analysing and interpreting DNA methylation data. Nat Rev Genet 13:705–719
6. Vertino PM, Sekowski JA, Coll JM et al (2002) DNMT1 is a component of a multiprotein DNA replication complex. Cell Cycle 1:416–423
7. Okano M, Bell DW, Haber DA et al (1999) DNA methyltransferases Dnmt3a and Dnmt3b are essential for de novo methylation and mammalian development. Cell 99:247–257

8. Smith ZD, Meissner A (2013) DNA methylation: roles in mammalian development. Nat Rev Genet 14:204–220
9. Heyn H, Esteller M (2012) DNA methylation profiling in the clinic: applications and challenges. Nat Rev Genet 13:679–692
10. Herman JG, Graff JR, Myohanen S et al (1996) Methylation-specific PCR: a novel PCR assay for methylation status of CpG islands. Proc Natl Acad Sci U S A 93:9821–9826
11. Clark SJ, Statham A, Stirzaker C et al (2006) DNA methylation: bisulphite modification and analysis. Nat Protoc 1:2353–2364
12. Costa VL, Henrique R, Ribeiro FR et al (2007) Quantitative promoter methylation analysis of multiple cancer-related genes in renal cell tumors. BMC Cancer 7:133
13. Savva-Bordalo J, Ramalho-Carvalho J, Pinheiro M et al (2010) Promoter methylation and large intragenic rearrangements of DPYD are not implicated in severe toxicity to 5-fluorouracil-based chemotherapy in gastrointestinal cancer patients. BMC Cancer 10:470
14. Jeronimo C, Usadel H, Henrique R et al (2001) Quantitation of GSTP1 methylation in non-neoplastic prostatic tissue and organ-confined prostate adenocarcinoma. J Natl Cancer Inst 93:1747–1752
15. Ramalho-Carvalho J, Pires M, Lisboa S et al (2013) Altered expression of MGMT in high-grade gliomas results from the combined effect of epigenetic and genetic aberrations. PLoS One 8:e58206
16. Henrique R, Ribeiro FR, Fonseca D et al (2007) High promoter methylation levels of APC predict poor prognosis in sextant biopsies from prostate cancer patients. Clin Cancer Res 13:6122–6129

Chapter 24

Quantitation of DNA Methylation by Quantitative Multiplex Methylation-Specific PCR (QM-MSP) Assay

Mary Jo Fackler and Saraswati Sukumar

Abstract

The defining feature of the Quantitative Multiplex Methylation-Specific PCR (QM-MSP) method to sensitively quantify DNA methylation is the two-step PCR approach for a multiplexed analysis of a panel of up to 12 genes in clinical samples with minimal quantities of DNA. In the first step, for up to 12 genes tested, one pair of gene-specific primers (forward and reverse) amplifies the methylated and unmethylated copies of the same gene simultaneously and in multiplex, in one PCR reaction. This methylation-independent amplification step produces amplicons of up to 10^9 copies per μL after 36 cycles of PCR. In the second step, the amplicons of the first reaction (STEP 1) are quantified with a standard curve using real-time PCR and two independent fluorophores to detect methylated/unmethylated DNA of each gene in the same well (e.g., 6FAM and VIC). One methylated copy is detectable in 100,000 reference gene copies. Methylation is reported on a continuous scale. For the gene panel, the highest level of normal DNA methylation above which a sample would be called positive is derived by using Receiver Operating Characteristic (ROC), maximizing assay specificity and sensitivity to distinguish between normal/benign versus tumor DNA. QM-MSP can be applied to clinical samples of fresh or fixed ductal cells, ductal fluid, nipple fluid, fine needle aspirates, core biopsies, and tumor tissue sections.

Key words QM-MSP, Quantitation, Tissue, Cells, DNA methylation, Bisulfite, Methylation-specific PCR

1 Introduction

Quantitative multiplex methylation-specific PCR (QM-MSP) [1, 2] is a highly sensitive, specific, and quantitative methylation assay used by our laboratory and others [1–10] (Fig. 1a). It combines the principles of conventional gel-based MSP [11] (*see also* Chapter 23), quantitative real-time MSP (qMSP) [12, 13] (*see also* Chapters 23 and 25) and multiplexed gel-based MSP [14] into one format developed to enable quantification of methylated gene panels in clinical samples with limited DNA quantities available (pg-ng; 50–1000 cells). The assay is easily performed on fresh or fixed cytological samples including ductal lavage/ductoscopy fluids and cells, nipple fluids, and fine needle aspirates as well as tissues

Jörg Tost (ed.), *DNA Methylation Protocols*, Methods in Molecular Biology, vol. 1708,
https://doi.org/10.1007/978-1-4939-7481-8_24,

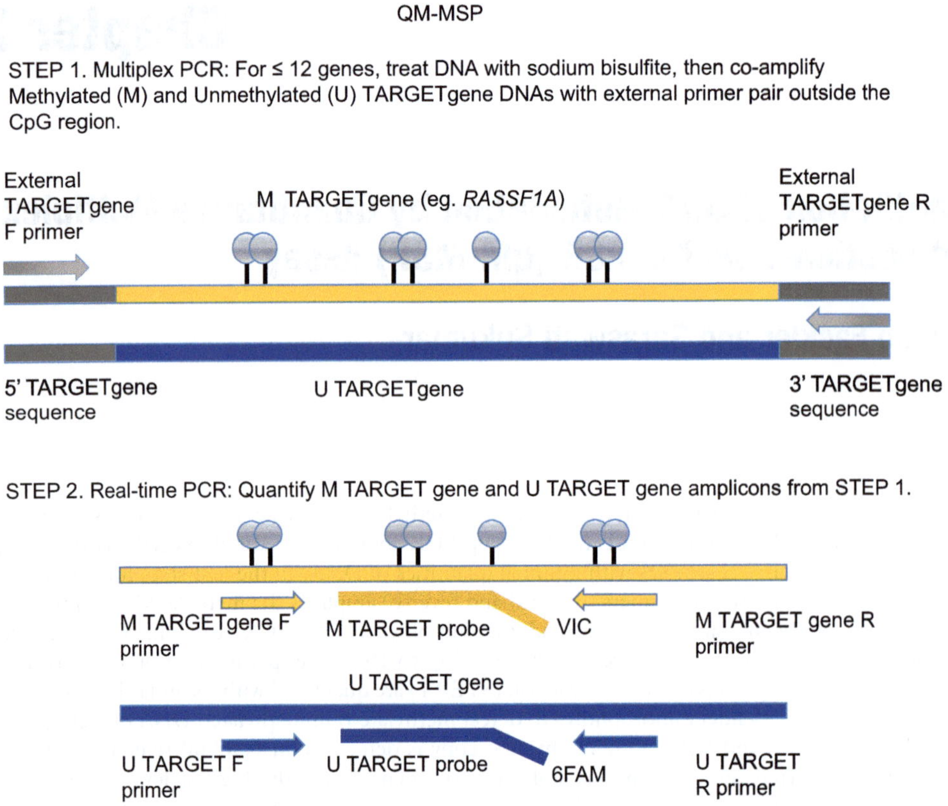

Fig. 1 Schema of QM-MSP. A two-step PCR is performed where the first PCR reaction (STEP 1) contains one pair of external primers per gene (forward and reverse) that co-amplifies DNA from the methylated gene of interest (TARGETgene; M) and the gene-specific unmethylated allele (U). In the second PCR reaction (STEP 2), amplicons of **step 1** are assayed by absolute quantitative real-time PCR with specific sets of primers (*forward and reverse*) and hydrolysis probes (in two colors) recognizing methylated TARGETgene. Figure adapted by permission from the American Association for Cancer Research: Fackler et al., Quantitative multiplex methylation-specific PCR assay for the detection of promoter hypermethylation in multiple genes in breast cancer. Cancer Res. 2004; 64:4442–52 [2]

and core biopsies. The QM-MSP method has an analytical sensitivity of 1 methylated copy in 100,000 unmethylated copies [2, 3, 10], nearly tenfold higher than qMSP, and 100-fold higher than gel-based MSP techniques. With more than double the clinical sensitivity for detection of cancer compared to cytology alone [2, 10], a panel of genes can be co-amplified and quantified on a continuous scale, discriminating normal or benign versus cancer samples using an ROC-derived threshold.

In the protocols described in this chapter, we present methods for QM-MSP as well as instructions for preparation and handling of genomic DNA extracted from fresh cells or formalin-fixed, paraffin-embedded tissue, samples obtained from fine needle aspirate, ductoscopy, ductal lavage, core needle biopsy, excisional biopsy tissue and tumor. We have included all the different steps necessary for the processing of patient samples, sodium bisulfite treatment of DNA, construction and calibration of standard curves, primer design, setup and execution of multiplex PCR, and setup and execution of real-time PCR.

2 Materials

All chemicals and solutions must be handled with PCR safe techniques. Unless otherwise specified (*see* **Note 1**), stock chemicals are stable indefinitely at room temperature.

2.1 Buffers

1. 1 M NH_4SO_4, prepare fresh each time.
2. 1 M Tris–HCl, pH 8.0.
3. 2 M Tris–HCl, pH 8.8.
4. 1 M $MgCl_2{\cdot}6H_2O$.
5. TE: 10 mM Tris–HCl, pH 8.0, 1 mM EDTA, pH 8.5.
6. 5 M NaCl.
7. 500 mM EDTA, pH 8.0.
8. DTT: 1 M dithiothreitol.
9. 10% SDS.
10. β-mercaptoethanol (Sigma; 14.3 M), keep stock at −20 °C, replace every couple of months.
11. DMSO, PCR grade.
12. Isopropanol.
13. Ethanol (100%, 70%, 50%).
14. Xylene.
15. ddH_2O.
16. Proteinase K (20 mg/mL).
17. RNase A-, DNase-, and protease-free (10 mg/mL).
18. TNES: 10 mM Tris, pH 8.0, 150 mM NaCl, 2 mM EDTA, 0.5% SDS.
19. TNES/SS: 1 mL TNES, 5 μL Salmon sperm DNA (10 mg/mL).
20. TNES/SS/PK: 900 μL TNES/SS, 100 μL Proteinase K (20 mg/mL).
21. 10× MSP Buffer: 166 mM NH_4SO_4, 670 mM Tris–HCl, pH 8.8, 67 mM $MgCl_2$, 100 mM β-mercaptoethanol, 1% DMSO (*see* **Note 2**).
22. 1× Dilution Buffer: 100 μL 10× MSP buffer, 5 μL salmon sperm DNA 10 mg/mL, 5 μL tRNA, 10 mg/mL, 890 μL ddH_2O).

23. DNA Hydrating Solution (Low TE): 1 mM Tris–HCl, pH 8.0, 0.1 mM EDTA.
24. PBS (1×): 10 mmol Na_2HPO_4, 1.76 mmol KH_2PO_4, 137 mmol NaCl, 2.7 mmol KCl without calcium or magnesium, pH 7.4 (e.g., Corning, 21-040-CV).

2.2 Controls

1. M.SssI methylase (New England Biolabs).
2. PUREGENE DNA Purification Kit (Qiagen) or similar.
3. NanoDrop (Thermo Fisher Scientific).

2.3 Sample Preparation

1. PAP jar.
2. Temperature-controlled oven.

2.4 PCR Reagents

1. Forward and reverse gene-specific external primers.
2. Gene-specific forward and reverse internal primers and probes (FAM- or VIC-labeled fluorophore and TAMRA as a quencher).
3. Platinum Taq polymerase (Thermo Fisher Scientific).
4. dNTPs.
5. ROX 50× (Thermo Fisher Scientific).
6. tRNA (10 mg/mL).
7. RAMP Taq 5 U/μL (Denville Scientific).
8. 7500 Real-time PCR System (Thermo Fisher Scientific) or similar.

3 Methods

3.1 Primer Design

Software is available from several commercial sources that can be used to design primers and probes for methylation-specific PCR. However, we recommend the manual design of primers/probes. By convention, forward primers/probes use sense sequences and reverse primers use the reverse complement of the sense sequence. Primer sequences are written in the 5′-3′ orientation. For each gene, primers for a methylated TARGETgene (M), and a REFERENCEgene [unmethylated TARGETgene (U) is used as reference for QM-MSP] are designed. For example, M $TARGET_{RASSF1}$, $REFERENCE_{RASSF1}$ (U $TARGET_{RASSF1}$ for QM-MSP).

1. Find an area that is rich in CGs overall. When studying the role of gene promoter hypermethylation in silencing gene expression, usually choose an area to situate primers/probes near the translational start site of the gene. When studying genes discovered by molecular methods such as methylome array (*see*

also Chapter 16), choose a location for QM-MSP primers/probes within 100 bp of the array probe, if possible (*see* **Note 3** for an example analyzing *AKR1B1*).

2. Copy/paste the genomic RefSeq nucleotide sense strand sequence for the CpG region of interest plus several hundred upstream and downstream bases, into a word processing document. Make all bases lower case. Find/replace all CG dinucleotides with bold upper case letters. Find/replace all lower case "c" with "t." This is now the predicted sequence of bisulfite-converted methylated DNA. Save.
3. Copy/paste the predicted sequence of bisulfite-converted methylated DNA to a new section of the document and within the pasted text find/replace all bold C residues with bold T residues. This is now the predicted sequence of bisulfite-converted unmethylated DNA. Save.
4. Design of the Multiplex methylation-independent external primers for each gene for STEP 1: Primers specifically co-amplify bisulfite converted DNA template for both the TARGETgene as well as the REFERENCEgene. Up to 12 TARGET/REFERENCE genes can be amplified in one tube. Select sequences for external primers based on the following:
   (a) Position each external primer sequence in a CpG poor stretch of bases flanking or within the CG rich area. Ideally, external primers should be positioned to amplify 150–250 bp DNA and be 20–25 bp each in length, with a predicted melting point (Tm) of 56–58 °C (based on the C or G = 4 °C, A, or T = 2 °C rule).
   (b) Several internal stand-alone C residues should be present.
   (c) Avoid repetitive nucleotides of >4 bases. Try to have the last base or two at the 3′ end of the primers finish with a C or G. Between the binding sites for forward and reverse primers there should be a sufficient number of bases, including CpG dinucleotides, to accommodate the internal region of the next PCR step.
   (d) Primers must not be complementary to each other or contain palindromes.
   (e) External primers should amplify the target sequences independent of the methylation status and therefore should not contain CpG dinucleotides. If this cannot be avoided they should be synthesized to have equimolar degenerate nucleotides for cytosine at the CG position, designated as "Y" (C + T) in the forward primer and as "R" (G + A) in the reverse primer rendering the amplification methylation-independent (*see* **Note 4**).

5. Design of Real-time STEP 2 Methylation-specific Internal Primers/Probes that co-amplify diluted amplicons from STEP 1. Select sequences for the methylation-specific inner primers/probes based on the following criteria:
   (a) Methylated target gene set: To design the methylated TARGETgene primer/probe set, begin with the sequence predicted for methylated bisulfite converted DNA and select a forward and reverse primer pair positioned to amplify 80–100 bp DNA within the region amplified by the external primers in **step 4** above. Internal primers should be 20–25 bp each, have a predicted Tm = 64-66 °C, several independent stand-alone C residues, 2–3 CpG dinucleotides per primer, and end with a 3′ C or G if possible.
   (b) Avoid repetitive stretches of the same nucleotide, as well as CGCG regions and palindromes. Primers must not be complementary to each other (*see* **Note 4**) or the hydrolysis probe.
   (c) The hydrolysis probe is usually designed in the reverse orientation, should be 25–30 bp, have a predicted Tm = 74–76 °C, several independent stand-alone C residues, 3–4 CpG dinucleotides, and must not have a 5′ G residue in the first or second position. Ideally the probe will terminate with a 3′ C or G. The 5′ end should be linked to an intense fluorophore such as 6-FAM, although others such as VIC are adequate, with the 3′ end linked to a quencher such as TAMRA (*see* **Note 5**).
   (d) Unmethylated target gene set: To design the unmethylated REFERENCEgene primer/probe set use the sequence predicted for unmethylated bisulfite converted DNA and perform the same steps as above. Try to match the Tm of the unmethylated primers to that of the methylated primers. Situate the forward primer, reverse primer and probe overlapping the sites for methylated primers/probe. Unmethylated primers/probe should be 2–3 nucleotides longer than the methylated set to compensate for the conversion of CG to TG and thus the change in predicted Tm.

3.2 Universally Methylated DNA Stock (M.SssI CpG Methylase Reaction)

Fully methylated control DNA is used as "M" control in QM-MSP STEP 1, and subsequently in STEP 2, the M amplicons are used as an M control. Another use is to establish the gene-specific M standard curve for real-time PCR. There are several commercial sources of universally methylated DNA. However to prepare it yourself, purify genomic methylated cancer cell line DNA (e.g., MDA-MB-453), and then methylate this DNA by treating the

DNA with the CpG methylase enzyme M.SssI. SssI enzyme methylates cytosine residues in the sequence CpG. The protocol described uses the reagents supplied by New England Biolabs, but the M.SssI enzyme is commercially available from several sources.

Definition: 1 U of SssI methylates 1 μg of DNA (lambda) in 1 h at 37 °C in the presence of 160 μM S-adenosyl methionine (SAM). To drive the forward reaction, high amounts of fresh SAM must be used and SAM/SssI must be replenished during the incubation with the methylase.

1. Assemble the reaction mix by adding 2 μg of DNA, 5 μL NEB 2 buffer (500 mM NaCl, 100 mM Tris-HCl, 100 mM $MgCl_2$, 10 mM DTT, pH 7.9 , or similar buffer), 1 μL SAM (supplied stock: 32 mM stored at −80 °C, freshly thawed,) and 5 μL SssI enzyme (4 U/μL).
2. After 3 h at 37 °C add another 50 μL of freshly prepared reaction mix (no DNA), continue to incubate an additional 3 h 37 °C.
3. Heat inactivate at 95 °C for 5 min.
4. Perform the sodium bisulfite DNA conversion reaction directly after this step or store sample frozen at −80 °C. There is no need to clean up the DNA prior to the sodium bisulfite reaction step.

3.3 Unmethylated DNA Stock

Human sperm DNA, leukocyte DNA or other cell source lacking methylation at the locus of interest may be used as a source of control unmethylated DNA. Unmethylated control DNA is used as “U” control in the STEP 1 multiplex PCR, and subsequently in STEP 2 the U amplicons are used as an U control. Another use is make the gene-specific U standard curve for real-time PCR. For leukocytes, prepare genomic DNA using routine phenol/chloroform extraction or commercially available kits.

Human sperm DNA (HSD) can be purified using the PUREGENE DNA Purification Kit:

1. Dilute 5 mL fresh seminal fluid to 25 mL with TE and place on ice.
2. Centrifuge the specimen at 2100 × *g* for 15 min at 4 °C.
3. Discard the supernatant and vortex the cellular pellet rigorously for 1 min, then resuspend in 6 mL of PureGene Cell Lysis Solution, 240 μL 1 M dithiothreitol, and 30 μL Proteinase K (20 mg/mL).
4. Mix the sample by inverting 25 times and incubated at 55 °C overnight.
5. Add 30 μL RNase A solution (10 mg/mL, provided in the kit), to the cell lysate and mix the sample by inverting another 25 times, and then incubate at 37 °C for 60 min.

6. Cool the sample to room temperature; add 2 mL of Protein Precipitation Solution and vortex the lysate vigorously for 20 s.
7. Place the sample in an ice bath for 5 min, and then centrifuge at 2000 × *g* for 10 min.
8. Transfer the supernatant (~8 mL) to a 50 mL conical tube.
9. Precipitate the DNA in the supernatant by adding 2 mL 10 M ammonium acetate, and 25 mL absolute ethanol, then invert 50 times, and centrifuge at 3000 × *g* for 15 min. Wash the DNA pellet in 80% ethanol. Air dry, and then rehydrate the pellet in 400 μL DNA Hydrating Solution. Incubate at 65 °C 1 h until completely dissolved. Store at 4 °C.

3.4 Standard Curve Stocks for STEP 2, Real-Time PCR

For each gene, master PCR stocks of (1) methylated TARGETgene ("M") and (2) REFERENCEgene (U) for the real-time standard curves, are made ahead and frozen (−80 °C) indefinitely. To make the working standard curve stock, small aliquots of target and reference DNAs are mixed in equimolar ratios and these "Standard Curve" stocks are kept frozen as a gene-specific master stock. Note that each gene has its own standard curve stock where M and U are in equimolar amounts. Because the stock is too concentrated to use directly, a dilution of 1:10 is prepared from the master standard curve stock. From the 1:10 dilution a standard curve with dynamic range of 10^{-2}–10^{-8} is prepared each time the real-time run is performed. The 1:10 dilution is refrozen and stored at −80 °C.

1. Prepare single-gene reaction mixes using the recipe shown in Subheading 3.7 (STEP 1 multiplex PCR recipe) using 20–40 ng DNA of *either* the universally methylated (M) *or* the unmethylated (U) DNA and *2 μL each* of *external* forward/reverse primers (stock is 100 ng/μL).
2. Perform a PCR reaction in a 500 μL PCR microfuge tube using a final volume of 50 μL volume. Thermocycler settings are 95 °C for 5 min, followed by 36 cycles of 95 °C for 30 s, 56 °C for 45 s, and 72 °C for 45 s, with a final extension of 72 °C for 7 min, and a 4 °C hold.
3. Visualize 3 μL of the PCR product using 3% agarose gel electrophoresis. Products should be single bands of the expected size. Store the stocks undiluted at −80 °C.
4. Final "Standard Curve Stocks" have equimolar M TARGET-gene and REFERENCE gene (U) mixed together in one tube for a single gene. To prepare this stock for a single gene, begin by combining equal volumes of the M only and U only stock PCR reactions above (e.g., 3 μL each in one 500 μL tube, only 1 gene per tube).
5. Make serial 1:10 dilutions (e.g., 3 μL DNA + 27 μL 1× dilution buffer) to prepare seven dilutions ranging from 10^{-2} to 10^{-8} of the M/U mix.

6. Assay 4 μL of each dilution per well by real-time PCR (M target and reference DNAs assayed in the same well) and determine the cycle threshold (Ct) at each dilution for M and U.
7. Calculate the average ΔCt between M and U at each point over the entire curve (10^{-2}–10^{-8}). Based on this ΔCt, calculate the fold difference in copies: for example if average ΔCt = 1.5 cycles, then $2^{1.5}$ = 2.8-fold difference on average over the entire curve.
8. From aliquots of the master PCR stocks, remix the M only and U only controls to make an equimolar mixture of M and U (the real-time PCR amplification plots will overlap), taking into account the fold difference in copies and adding more of the lowest concentration control, i.e., 3 μL of M + 8.4 μL (3 μL × 2.8-fold) of U, if U was at the lowest concentration.
9. Retest the new mix to ensure that M and U curves are now overlapping (≤0.5 cycles difference). Store the Master PCR stocks (M or U) at −80 °C (stable indefinitely).
10. Dilute the M/U stock 1:10 in 1× dilution buffer and this will be the working M/U curve stock. Freeze these stocks at −80 °C indefinitely (*see* **Note 6**).
11. In **step 6** (Subheading 3.8) serial 1:10 dilution curves will be made from the working M/U curve stock for a dynamic range of 10^{-2}–10^{-8} for every real-time PCR. If the working M/U curve stock begins to drift after several freeze/thaw cycles, remake the 1:10 dilution from the M/U master standard curve stock.

3.5 Samples

As few as 50–100 cells can be assayed by QM-MSP, but 1000–20,000 cells are preferred.

3.5.1 Fresh Cells

1. Centrifuge freshly isolated cells in phosphate buffered saline (PBS), pH 7.4.
2. Discard the supernatant, transfer cells to a 500 μL microcentrifuge tube, briefly centrifuge to pellet the cells, discard the supernatant and resuspend the cells in 30 μL TNES/SS/PK (for cell numbers greater than 20,000 use up to 100 μL TNES/SS/PK).
3. Incubate the sample at 56 °C overnight, add 1/10th volume of additional proteinase K (20 mg/mL), continue to incubate for 4 h, heat inactivate at 95 °C for 10 min and store the sample at 4 °C.

3.5.2 Frozen Tissue Embedded in OCT Compound

1. Thaw frozen tissue mounted on a glass slide and dissolve the embedding compound in 70% ethanol for 1 min, then 50% ethanol for 5 min.

2. Wash tissue by dipping the slide up and down gently until all the OCT rinses off.
3. Then dip the slide five times in 70% ethanol, drain and air dry the sample.
4. Harvest, at a minimum, a 4–5 mm^2 area of tissue.
5. With a fresh single-edge razor or pipet tip, scrape the tissue into a microfuge tube containing 27 μL TNES/SS (*see* **Notes 7** and **8**).
6. Add 3 μL proteinase K (20 mg/mL) and continue with the proteinase K digestion as described in Subheading 3.5.1.
7. For slices of tissue in OCT not on slides place the cryosection/s into a 1.5 mL microfuge tube and incubate in 1 mL 50% ethanol for 5 min, pulse vortex 1 min, centrifuge briefly to pellet the tissue, remove the supernatant.
8. Repeat with 70% ethanol. Carefully remove the supernatant and pellet the sample.
9. Add 30 μL TNES/SS/PK (for a large tissue section use up to 100 μL of TNES/SS/PK) and digest the tissue as described in Subheading 3.5.1 (*see* **Note 7**).

3.5.3 Formalin-Fixed, Paraffin-Embedded (FFPE) Tissue Sections

QM-MSP requires at least 4–5 mm^2 tissue depending on the quality of the DNA, which is routinely fragmented by formalin and further damaged by prolonged fixation.

1. Paraffin-embedded tissue on glass slides should be incubated in 40 mL fresh xylene for 10 min at room temperature in a disposable PAP jar, and the slide dipped up and down ten times.
2. Repeat once with new xylene.
3. Air-dry the slide in a chemical hood.
4. With a fresh sterile flat single-edge razor blade, scrape the tissue into a 500 μL microcentrifuge tube containing 30 μL TNES/SS/PK (*see* **Notes 7** and **8**). For a large tissue section, use up to 100 μL of TNES/SS/PK.
5. Process the sample overnight as described for proteinase K digestion in Subheading 3.5.1.

3.5.4 FFPE Tissue Plugs/Cores

1. Deparaffinize FFPE cores or slices from tissue blocks with xylene in a 1.5 mL microfuge tube.
2. Incubate the tissue in 1 mL fresh xylene for 10 min, vortex for 1 min, centrifuge it for 5 min at full speed and remove the supernatant.
3. Repeat twice more with 1 mL fresh xylene, drain.
4. Add 1 mL of absolute ethanol to wash off the xylene from the tissue, centrifuge and drain well.

5. Add 100 μL TNES/SS/PK (*see* **Note 7**).
6. Digest the sample as described in Subheading 3.5.1.
7. Use 5–10 μL of the lysate for the QM-MSP assay and store the rest of the lysate at 4 °C or colder.

3.5.5 Stained Cytology Samples

PAP stained cellular samples such as those prepared using a cytocentrifuge or ThinPrep device, mounted with a coverslip can be assayed (*see* **Note 9**):

1. Dissolve the mounting media holding the coverslip in place by incubating in xylene for 2–3 days or longer until the coverslip falls off on its own. Do not force it off or some cells may adhere.
2. Rinse the slide by dipping in fresh xylene and air-dry.
3. Wet the section with 20–40 μL of TNES. With a clean pipette tip or single edge razor blade, scrape cells into a 500 μL microcentrifuge tube taking care to rinse the cells repeatedly with TNES/SS (*see* **Note 7**).
4. Briefly centrifuge the sample to pellet the cells.
5. Remove all TNES/SS and resuspend the cell pellet in 30 μL TNES/SS/PK. Proteinase K-treat the sample as described in Subheading 3.5.1.

3.6 Protocol for Sodium Bisulfite Conversion of the DNA Template

Sodium bisulfite chemically converts unmethylated cytosine residues in DNA to uracil, which is read as thymidine during PCR. When cytosine is protected by DNA methylation (i.e., contains a 5′ methyl group, ^{m}CG) it resistant to sodium bisulfite and there will be no nucleotide sequence change of the ^{m}CG after incubation with sodium bisulfite. This is the basis of the methylation specific PCR (MSP) method described by Herman et al. [11]. Primers that are predicted to differentially hybridize to either methylated or unmethylated DNA after the sodium bisulfite reaction are referred to as methylated (M) or unmethylated (U) methylation-specific PCR (MSP) primers, respectively. Many commercial sodium bisulfite DNA conversion kits are available, including the EZ DNA Methylation Kit (D5001; Zymo Research) described in this subheading. This protocol is a modified version of the manufacturer's instructions.

1. Denature the double stranded DNA: Mix DNA (up to 3 μg purified DNA or 40 μL of lysate in TNES/SS/PK) + ddH$_2$O to make a final volume of 42.5 μL in a 500 μL microcentrifuge tube.
2. On ice, add 7.5 μL M-dilution buffer (supplied with the kit). Heat samples at 42 °C 15 min to denature the DNA (*see* **Note 10**). Chill samples on ice.
3. Perform the bisulfite conversion: During the incubation in **step 1**, prepare the CT reagent (supplied) and use it immediately

after the 42 °C incubation is finished: On a per vial basis (small vial sufficient for 10.5 samples), add 750 μL ddH_2O + 185 μL M-dilution buffer to the vial containing 567 mg of CT reagent. On a per sample basis, combine 71.4 μL water + 17.6 μL M-dilution buffer and 54 mg of conversion reagent. Dissolve the CT conversion reagent by rotating it in the dark at room temp for 10 min. When the denaturing step is completed (**step 2**), add to each sample 97.5 μL CT conversion reagent. Pulse vortex. Centrifuge 10 s.

4. Incubate the sample overnight in a thermocycler: 16 cycles of 95 °C 30 s, 50 °C 1 h (*see* **Notes 11** and **12**). The final volume is 150 μL. Hold the sample at 4 °C until the DNA clean-up step.
5. DNA clean-up: Using a 1 mL pipet, add four volumes of M-Binding Buffer (supplied) into a Zymo-Spin IC column seated in a 2 mL collection tube (supplied) and then add the converted DNA (i.e., 150 μL bisulfite reaction and 600 μL M-Binding Buffer).
6. Use the pipet tip to gently mix the sample with the binding buffer pipetting up and down 8–10 times (*see* **Note 13**).
7. Centrifuge at full speed 30 s at room temperature and transfer the column to a clean collection tube (*see* **Note 11**).
8. Add 100 μL M-Wash Buffer (supplied) to the column and centrifuge full speed for 30 s (*see* **Note 11**).
9. Add 200 μL M-Desulfonation Buffer (supplied) to the column, incubate 15 min at room temperature, and centrifuge full speed for 30 s (*see* **Note 11**).
10. *Immediately* neutralize the column by adding 200 μL M-Wash Buffer and centrifuging it at full speed for 30 s [it is important to wash away the Desulfonation buffer promptly, (*see* **Note 11**)].
11. Change to a new collection tube. Add 200 μL M-Wash Buffer to the column and centrifuge full speed for 1 min.
12. Transfer the column to a new collection tube.
13. Add 12 μL of warm water (70 °C).
14. Incubate the column with the water for 5–10 min at room temperature and centrifuge the column 1 min at full speed to recover the DNA. Transfer tubes to ice.
15. Use immediately for STEP 1 Multiplex PCR (20 ng/reaction is ideal) or store at −80 °C. For small amounts of DNA use the entire sample (10 μL) for the STEP 1 reaction.

3.7 Protocol for STEP 1: Multiplex PCR Reaction for ≤12 Genes

1. Perform the STEP 1 multiplex reaction in a 500 μL PCR microfuge tube using a final 50 μL reaction volume (Table 1) containing 16.6 mM NH_4SO_4, 67 mM Tris pH 8.8, 6.7 mM $MgCl_2$, 1.25 mM of each dNTP 10 mM β-mercaptoethanol,

Table 1
Preparation of Multiplex STEP 1 Reaction Mix for One Sample, 12 genes

Multiplex STEP 1 reaction mix	1× Master mix (μL)	Final concentration
10× MSP	5.0	1×
25 mM dNTP	2.5	1.25 mM
Platinum Taq (5 U/μL)	2.0	0.2 U/μL
Water	6.5	
Subtotal μL	16.0	
For each of 12 genes:		
External primer 100 ng/μl forward	1.0 × 12 genes	2 ng/μL per gene
External primer 100 ng/μl reverse	1.0 × 12 genes	2 ng/μL per gene
Subtotal μL	24.0	
DNA (bisulfite-treated) ≤ 40 ng total	10.0	
Total μL	50	

0.1% DMSO, 2 ng/μL per external primer and 10 U/50 μL Platinum Taq polymerase (*see* **Note 14**). Thermocycler settings are 95 °C for 5 min, followed by 36 cycles of 95 °C for 30 s, 56 °C for 45 s, and 72 °C for 45 s, with a final extension of 72 °C for 7 min.

2. After completion of PCR, the reaction mix is diluted 1:5 by adding 200 μL 1× Dilution Buffer before performing the quantitative real-time PCR assay (STEP 2).
3. Include controls such as 100% methylated TARGETgene DNA (e. g., universally methylated DNA), unmethylated DNA (e.g., leukocyte or human sperm DNA), and water (no template control).
4. Optimal DNA input concentration is 20 ng, however up to 1000-fold less input DNA can be amplified if the sample has good integrity of the DNA (not formalin fixed).
5. The reaction mix can be stored frozen at −20 °C indefinitely. Prior to real-time PCR (STEP 2) each sample is further diluted 1:10^2 and/or 1:10^4 (**step 1** in Subheading 3.8).

3.8 Protocol for STEP 2: Quantitative Real-Time MSP (Two-Color qMSP)

The amplicons produced in STEP 1 are assayed in the same well in duplex using methylated (M) target and reference (U) sets of primers situated internally with respect to the multiplex primers. For each gene, the methylated target and reference U amplicons are tested with a unique set of primers (forward and reverse) and hydrolysis probes specific for M gene target or U reference gene.

1. Preparation of Sample & Control Dilution for STEP 2:
   (a) Stocks of STEP 1 multiplexed patient DNA samples and U or M controls are stored frozen at −20 °C at a dilution of 1:5 in 1× Dilution Buffer.
   (b) Thaw each sample at room temperature, and prepare a working stock by further diluting the sample/control $1{:}10^{-2}$ and/or $1{:}10^{-4}$. These working stocks can be preassembled in a 96-well "template" plate, where the samples are organized in the same layout as in the STEP 2 real-time PCR plate.
   (c) Apply an adhesive foil cover to the template plate, freeze at −20 °C.
   (d) Before using, thaw and centrifuge the template plate (115 × *g* for 5 min) then remove the foil adhesive cover carefully in order to prevent cross-contamination of the wells. Repeated freeze/thaws will not affect the samples/controls.
2. Preparation of a master stock of primers/probes for both the target and reference sets: Keep the probes wrapped in foil or otherwise in the dark.
   (a) Thaw individual primers and probes, vortex, combine and place on ice.
   (b) Make enough for an entire run (e.g., 99× plus extra = 110× for one full plate) and label this tube "primers/probes". For example, for the 110× stock combine 352 μL of each of the four primers with 4.4 μL of each of the two probes. This stock is **Tube A**. Refreeze master stocks.
3. Preparation of Master Reaction Mix: On ice, prepare sufficient amount of reaction master mix for one entire plate plus extra (e.g., 110× per plate). Following the recipe shown in Table 2 below, combine 10× MSP, dNTP, ROX, tRNA, water and Taq polymerase sufficient for 110× (*see* **Note 14**). For example, standards require 3× reaction mix for each of the seven dilutions = 21×, a full plate of 76 samples requires 76×, plus 6× for six controls = 106×, including extra = 110×. Controls include M only, U only, multiplex water, real-time PCR water, and known samples. Set aside.
4. For sample/control wells: Label a 1.5 mL centrifuge tube "All" and transfer into it the amount of primers/probes prepared in **step 2** needed for wells B3–12, and C1 through H12 (Fig. 2).
5. Add the appropriate amount of reaction mix prepared in **step 3** of this subheading. Example for 85×: Combine 1095 μL of primer/probe + 265 μL of reaction mix. Pulse vortex, briefly centrifuge. Keep on ice. This stock is **Tube B**.

Table 2
Preparation of Real-time MSP Reaction Mix Sufficient for 72 sample wells and U, M and Water Controls

Real-time MSP (STEP 2) Reaction Mix		1× (μL)	110× (μL)	Final concentration
Reaction mix	10× MSP buffer	2.00	220.0	1×
	25 mM dNTP (Denville Scientific)	0.16	17.6	200 μM
	ROX 50× (Life Technologies)	0.24	26.4	300 nM
	10 mg/mL tRNA (Roche)	0.10	11.0	50 μg/mL
	5 U/μL RAMP Taq (Denville Scientific)	0.20	22.0	1.0 U
	Water	0.42	46.2	
	Reaction mix subtotal μL	3.12	343.2	
Primes/probes	Internal M primer, 5 μM forward	3.20	352	700 nM
	Internal M primer, 5 μM reverse	3.20	352	700 nM
	Internal U or primer, 5 μM forward	3.20	352	700 nM
	Internal U or primer, 5 μM reverse	3.20	352	700 nM
	M probe, 100 μM (FAM/TAMRA)	0.04	4.4	200 nM
	U or probe, 100 μM (VIC/TAMRA)	0.04	4.4	200 nM
	Primer/probe mix subtotal μL	12.88	1416.8	
DNA	DNA (diluted $1{:}10^{-2}$ or $1{:}10^{-4}$)	4.00		
	Total μL	20.0		

1 Full plate	Calculated amount	Reaction mix (Rxn)	Primers/Probes
Standard curve	(7 dilutions)(3× rxn) = 21×	(21×)(3.12) = 66 μL	(14×)(12.88) = 180 μL
Samples	(82×) + (3× extra) = 85×	(85×)(3.12) = 265 μL	(85×)(12.88) = 1095 μL
Total	Amount needed	106× + extra = 110×	99× + extra = 110×

QM - MSP Plate Set-Up

	1	2	3	4	5	6	7	8	9	10	11	12
A	10^{-2}	10^{-2}	10^{-3}	10^{-3}	10^{-4}	10^{-4}	10^{-5}	10^{-5}	10^{-6}	10^{-6}	10^{-7}	10^{-7}
B	10^{-8}	10^{-8}	U only Control Mult	M only Control Mult.	PCR #1 Water Mult.	PCR #2 Water	#1 $1{:}10^2$	#2 $1{:}10^2$	#3 $1{:}10^2$	#4 $1{:}10^2$	#5 $1{:}10^2$	#6 $1{:}10^2$
C	#7 $1{:}10^2$	#8 $1{:}10^2$	#9 $1{:}10^2$	#10 $1{:}10^2$	#11 $1{:}10^2$	#12 $1{:}10^2$	#13 $1{:}10^2$	#14 $1{:}10^2$	#15 $1{:}10^2$	#16 $1{:}10^2$	#17 $1{:}10^2$	#18 $1{:}10^2$
D	#19 $1{:}10^2$	#20 $1{:}10^2$	#21 $1{:}10^2$	#22 $1{:}10^2$	#23 $1{:}10^2$	#24 $1{:}10^2$	#25 $1{:}10^2$	#26 $1{:}10^2$	#27 $1{:}10^2$	#28 $1{:}10^2$	#29 $1{:}10^2$	#30 $1{:}10^2$
E	#31 $1{:}10^2$	#32 $1{:}10^2$	#33 $1{:}10^2$	#34 $1{:}10^2$	#35 $1{:}10^2$	#36 $1{:}10^2$	#1 $1{:}10^4$	#2 $1{:}10^4$	#3 $1{:}10^4$	#4 $1{:}10^4$	#5 $1{:}10^4$	#6 $1{:}10^4$
F	#7 $1{:}10^4$	#8 $1{:}10^4$	#9 $1{:}10^4$	#10 $1{:}10^4$	#11 $1{:}10^4$	#12 $1{:}10^4$	#13 $1{:}10^4$	#14 $1{:}10^4$	#15 $1{:}10^4$	#16 $1{:}10^4$	#17 $1{:}10^4$	#18 $1{:}10^4$
G	#19 $1{:}10^4$	#20 $1{:}10^4$	#21 $1{:}10^4$	#22 $1{:}10^4$	#23 $1{:}10^4$	#24 $1{:}10^4$	#25 $1{:}10^4$	#26 $1{:}10^4$	#27 $1{:}10^4$	#28 $1{:}10^4$	#29 $1{:}10^4$	#30 $1{:}10^4$
H	#31 $1{:}10^4$	#32 $1{:}10^4$	#33 $1{:}10^4$	#34 $1{:}10^4$	#35 $1{:}10^4$	#36 $1{:}10^4$	#37 $1{:}10^2$	#37 $1{:}10^4$	#38 $1{:}10^2$	#38 $1{:}10^4$	Control - Replicate	Control - Replicate

Fig. 2 The QM-MSP plate setup schema

6. The stocks for Standard Curves for Absolute Quantitation were prepared in Subheadings 3.2–3.4 and are stored at −80 °C. They were prepared separately for each gene and they consist of equimolar amounts of fully methylated target DNA and fully unmethylated DNA. A 1:10 M/U or M gene curve stock is kept as the working curve stock at −80° C and is thawed to prepare dilutions of the standard curve for STEP 2 real-time PCR.
   (a) Set up a series of seven 500 μL tubes and place 27 μL of water containing salmon sperm DNA (50 μg/μL) into each tube. Perform a series of serial 1:10 dilutions of the Working Standard Curve stock by transferring 3 μL DNA to the first tube, mixing well, and transferring 3 μL of that DNA to the next tube using a new pipette tip, continuing in this way for all seven tubes. This will make 10^{-2}, 10^{-3}, 10^{-4}, 10^{-5}, 10^{-6}, 10^{-7} and 10^{-8} dilutions. Vortex, keep these tubes on ice.
   (b) Set up another set of seven 500 μL tubes and place 9.36 μL reaction mix in each tube. Beginning with the first DNA dilution, transfer 12 μL of each dilution to respective tubes of reaction mix, cap, and vortex and keep these tubes on ice. These tubes will be used directly for STEP 2. Label these **Tubes C 10^{-2}–C 10^{-8}**. Refreeze the working M/U or M 1:10 curve stock.
7. Verify the following tubes and samples are ready, keep on ice:
   (a) **Sample DNAs and controls** diluted 10^{-2} and/or 10^{-4} (**step 1**).
   (b) **Tube A**: "Primers/probes" for target and reference sets combined (**step 2**).
   (c) Master Reaction Mix (**step 3**).
   (d) **Tube B**: "All", primers/probes (from Tube A) + reaction mix in sufficient amount for all samples and controls (**step 5**).
   (e) **Tube C** tubes 10^{-2}, 10^{-3}, 10^{-4}, 10^{-5}, 10^{-6}, 10^{-7}, 10^{-8} "10^{-2}–10^{-8}": Working mix of 3× volume of curve DNA + 3× volume of reaction mix per tube (**step 6**).
8. Assemble the 96-well plate on ice using plates recommended by the manufacturer of the real-time PCR instrument.
9. Pipet 12.88 μL of **Tube A** (Primers/probe) into each well of Row A, and the first two wells of Row B, and then pipet 7.12 μL of **Tube C 10^{-2}, 10^{-3}, 10^{-4}, 10^{-5}, 10^{-6}, 10^{-7}, 10^{-8}** (curve DNA + reaction mix) into the wells as indicated (Fig. 2).
10. Pipet 16 μL of **Tube B** primers/probes/reaction mixture into all empty wells (Row B3–B12, Rows C-H, Fig. 2), and then pipet 4 μL diluted **sample or control DNA** into these wells, as indicated.

3.9 Performing the Real-Time MSP run

1. The real time PCR reaction is run a final 20 μL volume containing 16.6 mM NH_4SO_4, 67 mM Tris pH 8.8, 6.7 mM $MgCl_2$, 10 mM β-mercaptoethanol, 0.1% DMSO, 300 nM ROX, 200 μM dNTP 5 μg/mL tRNA, 1.0 U RAMP Taq polymerase, 700 mM each forward and reverse primer, and 200 nM for each hydrolysis probe.
2. On ice, secure an optical adhesive cover on the 96-well PCR plate, taking special care to seal all edges well to prevent evaporation.
3. Centrifuge the plate in the cold at 250 × *g* for 3 min.
4. Do not contaminate the underside of the plate with any laboratory dyes such as ethidium bromide or the PCR block will become contaminated.
5. Using the software provided by the manufacturer, specify the PCR as the absolute quantitation method, label wells for standards, unknown samples, and controls, identify which probe in conjugated to which of the two fluorophores, and indicate the presence of the ROX passive dye.
6. Specify parameters as 95 °C for 10 min, then 40 cycles of 95 °C for 30 s and 65 °C for 1 min (*see* **Note 15**).
7. Specify the volume in the wells as 20 μL.
8. Transfer the plate to the real-time PCR machine and start the run. Approximate time to completion is 90 min. Before the run the plates may be stored refrigerated in the dark up to 24 h without any loss of signal.

3.10 Analysis: Calculating Gene Methylation

Using the Applied Biosystem's 7500 Real-time PCR System, sample values are extrapolated from the standard curve for target and reference DNAs. This is called absolute quantitation. It is performed according to the instructions from the manufacturer, with several modifications. Use the software supplied to analyze the data.

1. With the baseline set at automatic, manually set a single threshold for all wells at the point of maximal overlap of the curves for methylated TARGETgene (M) and REFERENCE (U gene) (usually ΔRn between 0.02 and 0.04).
2. For the standard curve, assign the Ct for the 10^{-2} dilution = 200,000,000 copies. The Ct for the 10^{-3} dilution is assigned 20,000,000 copies, the 10^{-4} Ct is assigned 2,000,000 copies, etc. with the final 10^{-8} Ct assigned 200 copies.
3. If M and U do not perfectly overlap over the entire range of the standard curve, the copy number is adjusted according to the average ΔCt of the two curves.

4. Calculate the ΔCt between Ct of methylated TARGETgene and Ct of unmethylated REFERENCE gene DNAs at each dilution.
5. Calculate the overall average ΔCt between the two curves, and then transform the ΔCt value to fold-difference in copies (*see* example below).
6. Adjust the copy number of the highest dilution curve based on this factor. Example, if the methylated TARGETgene curve was at the lowest dilution (had the lower Ct values) and the average ΔCt between both curves was 1.5 cycles (considering all dilutions), the difference would be $2^{1.5} = 2.828$ fold.
7. Assign 200,000,000 copies to the 10^{-2} dilution of the methylated TARGETgene and 70,700,000 copies to the 10^{-2} dilution of the REFERENCE gene (200,000,000 copies ÷ 2.828 fold = 70,700,000 copies). After the standard curve Ct values are assigned copy numbers, use the software to extrapolate from the curve to the samples the copy number of M and U gene DNA.
8. Use the formulas below to calculate the extent of methylation for each gene separately, and cumulatively for the gene panel (Examples in Figs. 3 and 4): **QM-MSP**: Percent Methylation (%M) = [copies Methylated TARGETgene ÷ (copies Methylated TARGETgene + copies unmethylated TARGETgene) copies] [100]; CMI = the sum of all %M values within the panel.
9. If the run fails to meet minimum criteria, rerun STEP 2 after making appropriate corrections.
   (a) Standard curve: $R^2 = \geq 0.99$, efficiency = 90 ± 10% and slope = −3.33 ± 10%. TARGETgene and REFERENCE gene C_t overlap ≤1.5 cycles, averaged over the whole curve.
   (b) For any sample, copy number of methylated TARGET gene or REFERENCE gene must not exceed the upper range of the standard curve (200,000,000 copies, 10^{-2} dilution of the curve) and the copy number of the REFERENCE gene must be at least 20,000 copies.
   (c) Control water (after both multiplex and real-time PCR), $C_t \geq 38$.
   (d) Control fully methylated TARGETgene DNA (universally methylated DNA template, after multiplex and real-time PCR), %M or MI = 100.
   (e) Control REFERENCE DNA (fully unmethylated allele of TARGETgene after multiplex and real-time PCR), %M or MI = 0.

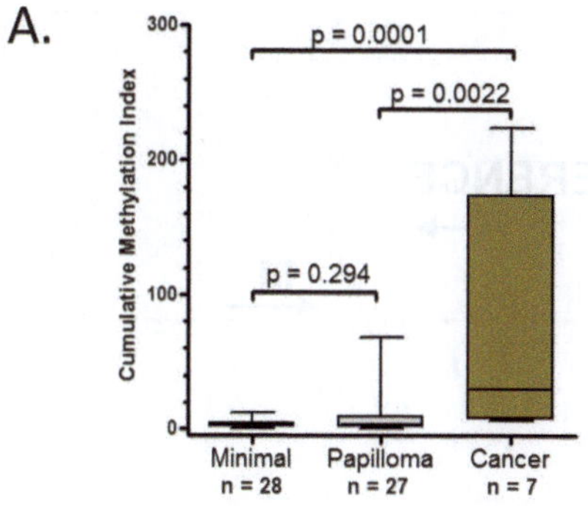

6-Gene Panel	Cumulative Methylation		
	Minimal	Papilloma	Cancer
Number of values	28	27	7
Minimum	0.13	0.03	5.54
25% Percentile	1.02	0.95	7.02
Median	2.43	2.68	29.75
75% Percentile	3.88	8.72	173.60
Maximum	12.24	68.23	223.90
Mean	3.05	8.60	83.13
Lower 95% CI of mean	1.97	2.74	0.76
Upper 95% CI of mean	4.13	14.46	165.50

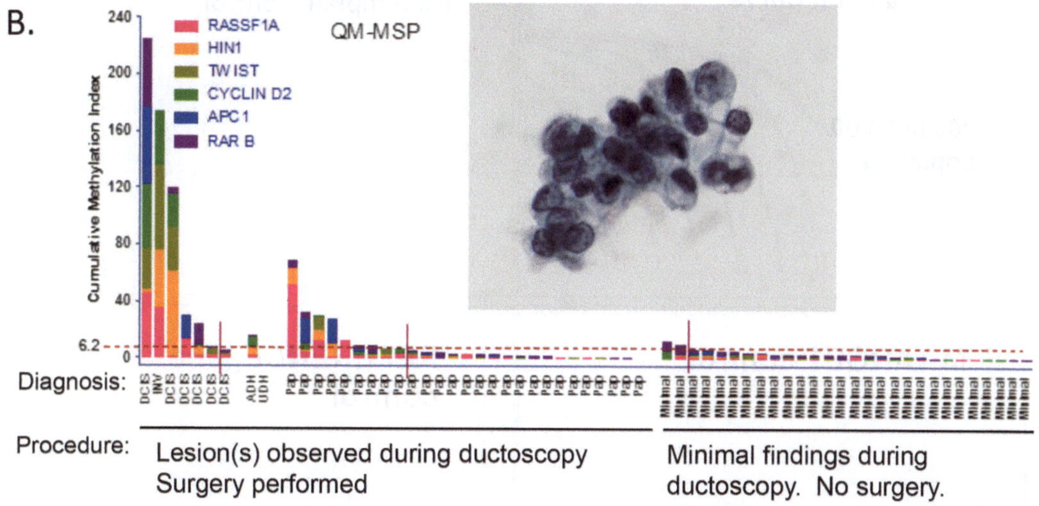

Fig. 3 QM-MSP Assay of Breast Cells collected during ductoscopy. (**a**) In women presenting with spontaneous nipple discharge, QM-MSP was performed for a panel of six genes on DNA from ductal cells retrieved during ductoscopy. (**b**) Lesions of concern observed during ductoscopy were surgically excised and classified (*X*-axis) according to the histological diagnosis: DCIS (ductal carcinoma in situ), INV (invasive ductal carcinoma, ADH (atypical ductal hyperplasia), UDH (usual ductal hyperplasia), or Pap (papilloma). Procedure: Ducts with minimal findings (Minimal) during ductoscopy were not subjected to surgery; in cells derived from these ducts methylation was low (90th percentile = 6.2 units of cumulative methylation). Methylation was significantly higher in cells derived from ducts containing DCIS and INV cancer compared to those cells obtained from ducts with from papilloma (benign) or ducts with minimal findings (Mann–Whitney, $p = 0.002$ and $p = 0.0001$, respectively). Each colored segment represents the % methylation of a different gene, and cumulative methylation is represented as the sum of methylation of all genes (*Y*-axis). (Inset) Approximately 30 cells were adequate to quantitate the level of methylation for the six-gene panel in a DCIS in (**b**). Figure adapted by permission from the American Association for Cancer Research: Fackler et al., Hypermethylated genes as biomarkers of cancer in women with pathologic nipple discharge. Clin Cancer Res. 2009; 15:3802–11 [10]

(f) Average duplicate samples for patient samples, duplicate aliquots are assayed and averaged.

(g) Implausible extreme values (e.g., due to contamination) are discarded/repeated.

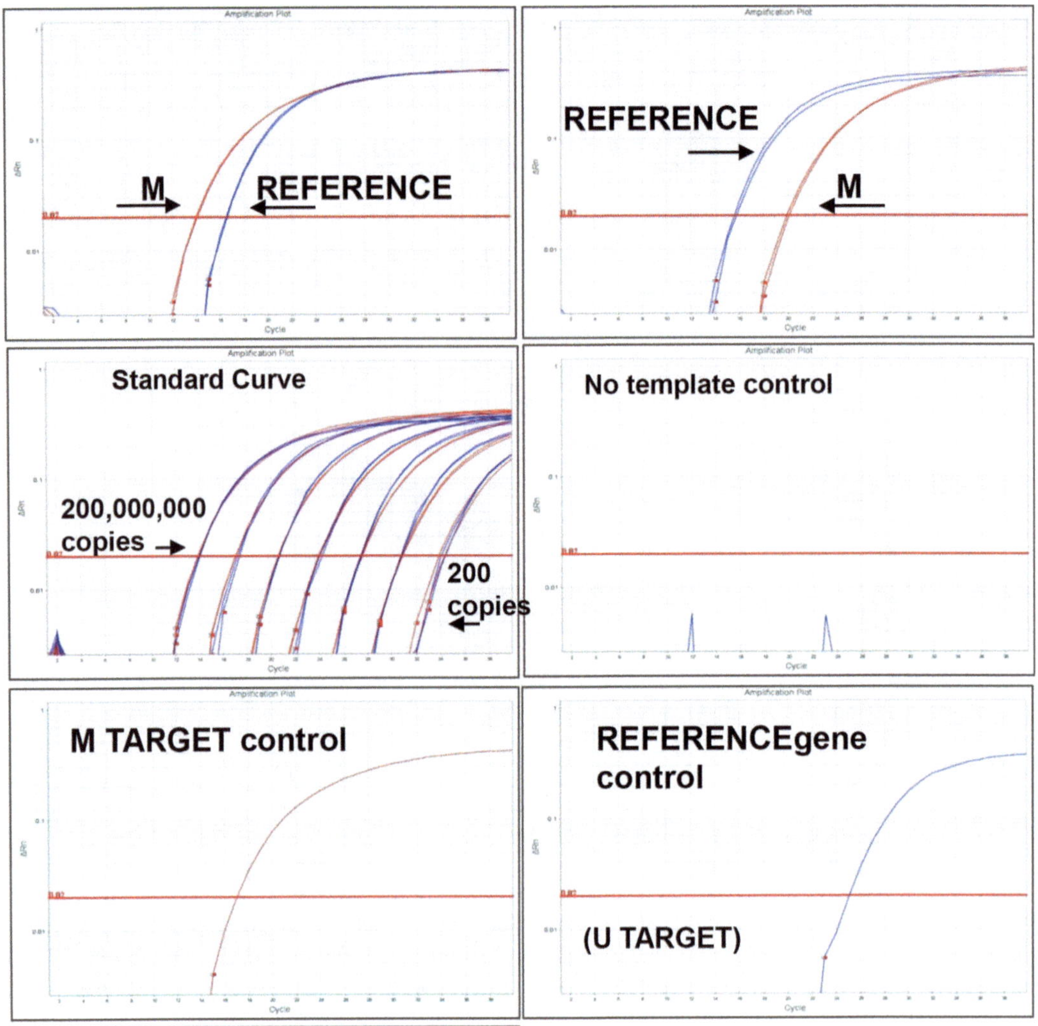

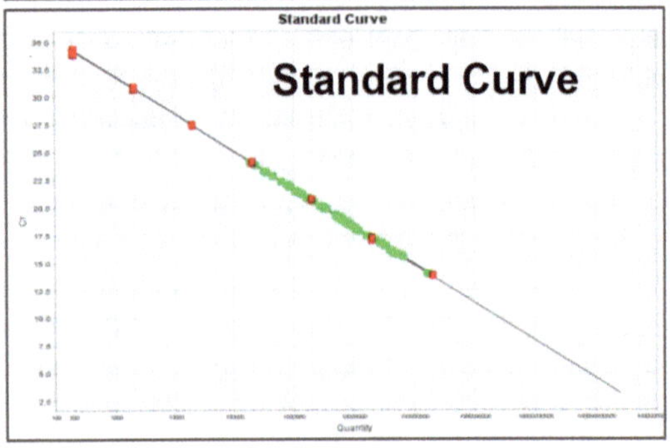

	Slope	R^2	Eff%
M TARGET	-3.381	1	97.612
REFgene	-3.408	1	96.521

Fig. 4 STEP 2 amplification curves and analytical parameters. For the *RASSF1* gene, amplification plots are shown for methylated TARGET (M) and REFERENCE DNAs, as indicated by *arrow*. Standard curve amplification plot demonstrates linear overlapping M and REFERENCE U curves ranging from 200,000,000 to 200 copies, with amplification efficiency of 97.6% and 96.5%, respectively. Figure adapted by permission from the American Association for Cancer Research: Fackler et al., Quantitative multiplex methylation-specific PCR assay for the detection of promoter hypermethylation in multiple genes in breast cancer. Cancer Res. 2004;64:4442–52 [2]

4 Notes

1. Stocks of 1 M Tris–HCl (pH 8.0), 5 M NaCl, 500 mM EDTA (pH 8.0), 10% SDS, TNES, 1 M $MgCl_2$ are stable at room temperature indefinitely. Salmon sperm and tRNA stocks are stable at −20 °C indefinitely. Proteinase K should be kept at −20 °C until the expiration date specified on the tube. 10× MSP buffer is only stable for 1–2 months at −80 °C. 1 M NH_4SO_4 should be prepared immediately before using to make 10× MSP buffer. Purchase fresh beta mercaptoethanol every 6–12 months and keep it in the −20 °C freezer.
2. Keep the 10× MSP buffer stocks at −80 °C. This buffer goes "off" after a few months.
3. Example: Design of *AKR1B1* QM-MSP primers/probes.

The *AKR1B1* gene was first reported to be hypermethylated in breast cancers based on our previous methylome array studies [15, 16]. The methylation status was verified by QM-MSP assays using primers/probes located within 100 bp of the array probe [15, 16].

Wild type genomic DNA sequence of AKR1B1

gcagctgaggaactcctttctgccaCGCGgggCGCGggCGagCGttgggg
gCGgaaagaatcCGctgccactaggaccaggCGgaagaagcatcccCGcC
GacccttggggaaggcCGcCGCGgcacccccagCGcaaccaatcagaagg
ctcctttCGCGcagCGgCGCGccaacCGcaggCGcccttttctgcCG
acctc
aCGggctatttaaaggtaCGCGcCGCGgccaaggcCGcacCGtactgggC
GggggtctggggagCGcagcagccatggcaagcCGtctcctgctcaacaa
CGgCGccaagatgcccatcctggggttgggtacctggaaggtaggtgctC
GtggggggCGCGggccCGgggctCGcctcacactctcCGCGCGgc
ctgtat
tggCGagggacccCGagtgaccctgagcagctCGcccCG
CGgaCGccCGg

Sequence of bisulfite-treated methylated AKR1B1 DNA

gtagttgaggaattttttttgttaCGCGgggCGCGggCGagCGttgggg
gCGgaaagaattCGttgttattaggattaggCGgaagaagtattttCGtC
GatttttggggaaggtCGtCGCGgtattttttagCGtaattaattagaagg
ttttttCGCGtagCGgCGCGttaatCGtaggCGtttttttgtCGatttt
aCGggttatttaaaggtaCGCGtCGCGgttaaggtCGtatCGtattgggC
GggggtttggggagCGtagtagttatggtaagtCGtttttgtttaataa
CGgCGttaagatgtttattttggggttgggtatttggaaggtaggtgttC
GtggggggCGCGggttCGgggttCGttttatatttttCGCGCGgtttgtat
tggCGagggatttCGagtgattttgagtagttCGtttCGCGgaCGttCGg

Sequence of bisulfite-treated unmethylated AKR1B1 DNA

gtagttgaggaattttttttgttaTGTGgggTGTGggTGagTGttgggg gTGgaaagaattTGttgttattaggattaggTGgaagaagtattttTGtT GatttttggggaaggtTGtTGTGgtattttagTGtaattaattagaagg ttttttTGTGtagTGgTGTGttaatTGtaggTGtttttttgtTGatttt aTGggttatttaaaggtaTGTGtTGTGgttaaggtTGtatTGtattgggT GggggtttggggagTGtagtagttatggtaagtTGtttttgtttaataa TGgTGttaagatgtttattttggggttgggtatttggaaggtaggtgttT GtggggGTGTGggttTGgggttTGttttatattttTGTGTGgtttgtat tggTGagggatttTGagtgattttgagtagttTGtttTGTGgaTGttTGg

AKR1B1_F_Ext	gYGtaattaattagaaggtttttt	Tm 58	216 bp
AKR1B1_R_Ext	aacacctaccttccaaatac	Tm 56	
(Gtatttggaaggtaggtgtt sense)			
AKR1B1_FM	gCGCGttaatCGtaggCGttt	Tm 64	84 bp
AKR1B1_RM	cccaataCGataCGaccttaac	Tm 64	
(gttaaggtCGtatCGtattggg sense)			
AKR1B1_FUM	TGgTGTGttaatTGtaggTGtttt	Tm 64	86 bp
AKR1B1_RUM	cccaataCAataCAaccttaacC	Tm 64	
(GgttaaggtTGtatTGtattggg sense)			
AKR1B1_M_Probe	CGtacctttaaataaccCGtaaaatCGa	Tm 76	
(tCGattttaCGggttatttaaaggtaCG sense)			
AKR1B1_U_Probe	ACAtacctttaaataaccCAtaaaatCAac	Tm 76	
(gtTGattttaTGggttatttaaaggtaTGT sense)			

4. Often, not all design criteria can be met, so factors must be balanced against the sequence in the location that is targeted. CG content and Tm are the most critical primer criteria.
5. TARGETgene and REFERENCEgene for one analyzed region will be amplified in the same well as a duplex reaction requiring thus a FAM/TAMRA labeled hydrolysis probe for the TARGETgene and a VIC/TAMRA labeled hydrolysis probe for the REFERENCEgene (or vice versa, *see* also Fig. 1).
6. For prolonged storage, these DNAs are routinely diluted in 1X Dilution Buffer containing a large excess of salmon sperm and tRNA in Tris buffer, pH 8.8. This stabilizes the DNA, and prevents losses due to degradation and adsorption to the plastic microcentrifuge tube.
7. One of the most important issues when working with small amounts of DNA sample is to add carrier such as salmon sperm DNA and/or tRNA to the lysis buffer to prevent nonspecific losses of template DNA. However, if the sample will also be used for methylation array or DNA sequencing do not use carrier.

8. Before scraping, prewet the tissue with a couple microliters of TNES so it stays stuck to the slide.
9. Samples stained with Wright's stain may not amplify well by MSP.
10. For the 42 °C incubation with M-dilution buffer in the sodium bisulfite conversion protocol, prepare the samples on ice and at the end of the incubation transfer the samples back to ice. This reduces sample degradation. Do not incubate samples longer than specified by the protocol. Similar precautions need to be taken during desulfonation of samples. Take into consideration the amount of time needed to process all samples when timing incubations. Neutralize the reaction with M-wash buffer immediately after removal of Desulfonation Buffer from the column.
11. Do not centrifuge the columns longer than specified times or the column may become overly dry. This may reduce DNA recovery.
12. Small bisulfite crystals may form during the overnight sodium bisulfite conversion step, but they will not interfere with DNA binding to the column after the addition of M-binding buffer.
13. For clean-up of DNA after the sodium bisulfite conversion step the DNA sample must be thoroughly mixed with the M-binding buffer before centrifugation of the column, otherwise the recovery of DNA may be low due to poor loading of the column.
14. Vortex all reagents, primers, and probes after thawing. However, do not vortex stocks of Taq polymerase. In STEP 1, gently mix by stirring with a pipette tip between the addition of all components, followed at the end with a pulse-vortex and brief centrifugation of the master mix.
15. Annealing/elongation temperatures may need to be optimized (ranging from 60 to 65 °C) depending on the primers.

Acknowledgments

We thank Sidra Hafeez for critical review of the manuscript. This work was supported by grants to SS from AVON Research Foundation, the Rubenstein family, John A. Sellon Charitable Trust, the Department of Defense Center of Excellence on "Targeting Metastatic Breast Cancer" grant W81XWH-04-1-0595, and SKCCC Core grant P30 CA006973.

References

1. Swift-Scanlan T, Blackford A, Argani P et al (2006) Two-color quantitative multiplex methylation-specific PCR. Biotechniques 40:210–219
2. Fackler MJ, McVeigh M, Mehrotra J et al (2004) Quantitative multiplex methylation-specific PCR assay for the detection of promoter hypermethylation in multiple genes in breast cancer. Cancer Res 64:4442–4452
3. Fackler MJ, Malone K, Zhang Z et al (2006) Quantitative multiplex methylation-specific PCR analysis doubles detection of tumor cells in breast ductal fluid. Clin Cancer Res 12:3306–3310
4. Euhus DM, Bu D, Ashfaq R et al (2007) Atypia and DNA methylation in nipple duct lavage in relation to predicted breast cancer risk. Cancer Epidemiol Biomarkers Prev 16:1812–1821
5. Lee JS, Lo PK, Fackler MJ et al (2007) A comparative study of Korean with Caucasian breast cancer reveals frequency of methylation in multiple genes correlates with breast cancer in young, ER, PR-negative breast cancer in Korean women. Cancer Biol Ther 6:1114–1120
6. Locke I, Kote-Jarai Z, Fackler MJ et al (2007) Gene promoter hypermethylation in ductal lavage fluid from healthy BRCA gene mutation carriers and mutation-negative controls. Breast Cancer Res 9:R20
7. Euhus DM, Bu D, Milchgrub S et al (2008) DNA methylation in benign breast epithelium in relation to age and breast cancer risk. Cancer Epidemiol Biomarkers Prev 17:1051–1059
8. Suijkerbuijk KP, Fackler MJ, Sukumar S et al (2008) Methylation is less abundant in BRCA1-associated compared with sporadic breast cancer. Ann Oncol 19:1870–1874
9. Wu JM, Fackler MJ, Halushka MK et al (2008) Heterogeneity of breast cancer metastases: comparison of therapeutic target expression and promoter methylation between primary tumors and their multifocal metastases. Clin Cancer Res 14:1938–1946
10. Fackler MJ, Rivers A, Teo WW et al (2009) Hypermethylated genes as biomarkers of cancer in women with pathologic nipple discharge. Clin Cancer Res 15:3802–3811
11. Herman JG, Graff JR, Myohanen S et al (1996) Methylation-specific PCR: a novel PCR assay for methylation status of CpG islands. Proc Natl Acad Sci U S A 93:9821–9826
12. Eads CA, Danenberg KD, Kawakami K et al (2000) MethyLight: a high-throughput assay to measure DNA methylation. Nucleic Acids Res 28:E32
13. Lehmann U, Langer F, Feist H et al (2002) Quantitative assessment of promoter hypermethylation during breast cancer development. Am J Pathol 160:605–612
14. Palmisano WA, Divine KK, Saccomanno G et al (2000) Predicting lung cancer by detecting aberrant promoter methylation in sputum. Cancer Res 60:5954–5958
15. Fackler MJ, Umbricht CB, Williams D et al (2011) Genome-wide methylation analysis identifies genes specific to breast cancer hormone receptor status and risk of recurrence. Cancer Res 71:6195–6207
16. Fackler MJ, Lopez Bujanda Z, Umbricht C et al (2014) Novel methylated biomarkers and a robust assay to detect circulating tumor DNA in metastatic breast cancer. Cancer Res 74:2160–2170

Chapter 25

MethyLight and Digital MethyLight

Mihaela Campan, Daniel J. Weisenberger, Binh Trinh, and Peter W. Laird

Abstract

MethyLight is a quantitative, fluorescence-based, real-time PCR method to sensitively detect and quantify DNA methylation of candidate regions of the genome. MethyLight is uniquely suited for detecting low-frequency methylated DNA regions against a high background of unmethylated DNA, as it combines methylation-specific priming with methylation-specific fluorescent probing. The quantitative accuracy of real-time PCR and the ability to design bisulfite-dependent, DNA methylation-independent control reactions together allow for a quantitative assessment of these low frequency methylation events. Here we describe the experimental steps of MethyLight analysis in detail. Furthermore, we present principles and design examples for three types of quality control reactions. QC-1 reactions are methylation-independent reactions to monitor sample quantity and integrity. QC-2 reactions are bisulfite-independent reactions to monitor recovery efficiencies of the bisulfite conversion methodology used. QC-3 reactions are bisulfite-independently primed reactions with variable bisulfite-dependent probing to monitor completeness of the sodium bisulfite treatment. We show that these control reactions perform as expected in a time course experiment interrupting sodium bisulfite conversion at various timepoints. Finally, we describe Digital MethyLight, in which MethyLight is combined with Digital PCR, for the highly sensitive detection of individual methylated molecules, with use in disease detection and screening.

Key words DNA Methylation, Real-time PCR, TaqMan, Bisulfite, Epigenetics, Cancer, Quantitative, Methylation-specific PCR, Digital PCR

1 Introduction

MethyLight is a sodium-bisulfite-dependent, quantitative, fluorescence-based, real-time PCR method to sensitively detect and quantify DNA methylation in genomic DNA [1–5]. MethyLight relies on methylation-specific priming [6], combined with methylation-specific fluorescent probing [1–5]. This combination of methylation-specific detection principles results in a highly methylation-specific detection technology, with an accompanying ability to sensitively detect very low frequencies of hypermethylated alleles. The high sensitivity and specificity of MethyLight make it uniquely suited for the detection of low-frequency DNA methylation

Jörg Tost (ed.), *DNA Methylation Protocols*, Methods in Molecular Biology, vol. 1708,
https://doi.org/10.1007/978-1-4939-7481-8_25,

biomarkers as evidence of disease [7]. At the same time, the quantitative accuracy of real-time PCR and the flexibility to design bisulfite-dependent, methylation-independent control reactions [5] allows for a quantitative assessment of these low frequency methylation events.

In addition to discussing in detail how to perform the experimental steps of MethyLight analysis, we present here how template, primer and probe design flexibility can be used to develop quality control reactions. The major principles are presented in Fig. 1. Methylation-independent, bisulfite-dependent reactions can be used as quality controls of sample quantity and integrity, as illustrated in Fig. 1: QC-1 [5]. We have developed a series of reactions to monitor recovery and completeness of the sodium bisulfite conversion step. One of the challenges in monitoring recovery during the conversion step is that the sequence changes as a result of the conversion. We therefore selected a region of the genome that does not contain any cytosine residues on one DNA strand over a short stretch (Fig. 1: QC-2), and thus remains unaffected by treatment with sodium bisulfite. Therefore, we can use a reaction targeting this locus (C-LESS) to monitor DNA recovery at any step during the sodium bisulfite conversion. Figure 2, right panel, shows a time course of a sodium bisulfite conversion reaction. It is evident that the C-LESS reaction is relatively impervious to the effects of sodium bisulfite. We also designed reactions to monitor the efficacy and completeness of sodium bisulfite conversion for a given sample (Fig. 1: QC-3). For this purpose, we selected a locus in the human genome for which the primer locations do not cover any cytosines in one strand of the DNA. Thus, the amplification of this strand would be bisulfite-conversion-independent. The region covered by the probe contains four cytosine residues. We designed and tested all 16 permutations of the probe, assuming either conversion to uracil, or lack of conversion at each cytosine (Fig. 2). Experimental results with the probes indicated by arrows on the left are shown for the bisulfite conversion time course on the right. We recommend using these probes to monitor completeness of the reaction. A threshold for bisulfite conversion quality control can be implemented simply as a delta cycle threshold (Δ-C(t)) for each of these reactions, compared to the methylation-independent ALU-C4 QC-1 quantity control.

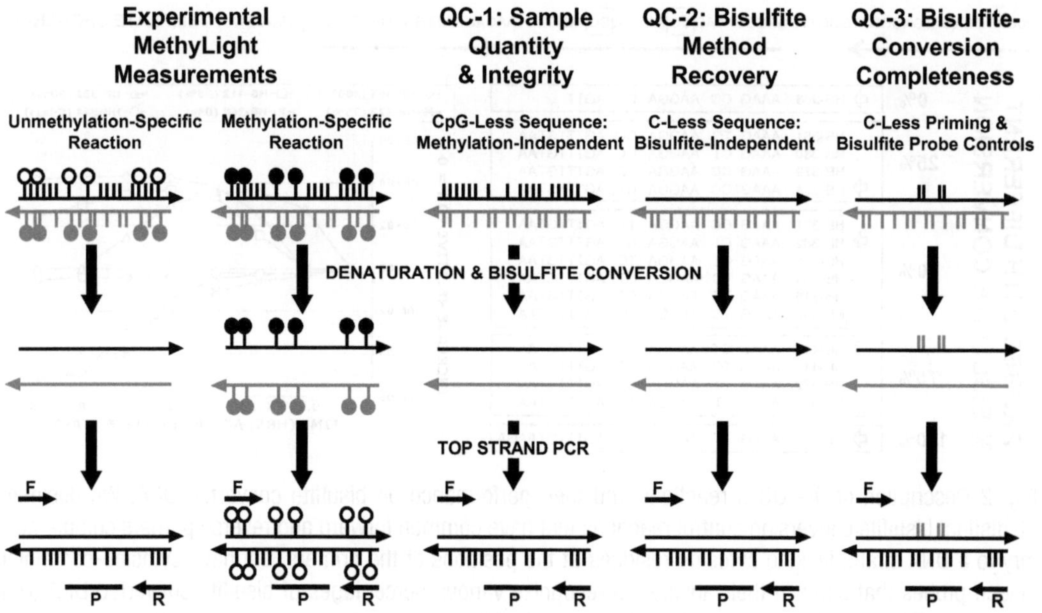

Fig. 1 TaqMan-based MethyLight experimental and quality control PCR reactions used for quantitative methylation analysis of bisulfite converted DNA. Several types of TaqMan-based PCR reactions are used for the quantitative DNA methylation analysis of bisulfite converted DNA. Following bisulfite conversion, methylated cytosines remain unchanged, while unmethylated cytosines are deaminated to uracils. During subsequent PCR amplification of the bisulfite-converted DNA, thymine is incorporated in the place of uracil. The two strands of the bisulfite-converted DNA are no longer complementary, such that separate PCR reactions can be designed to amplify either the *top* or the *bottom* DNA strands. Each *horizontal line* represents one DNA strand. The *tick marks* represent cytosines not in the context of CpG dinucleotides, while the *lollipops* represent cytosines in the context of CpG dinucleotides. The methylated cytosines are depicted as *solid black* while the unmethylated ones are *open white*. The bottom strand does not participate in the further analysis after bisulfite conversion, and is depicted in *gray*. The experimental MethyLight reactions are specific for unmethylated or methylated DNA sequences. These reactions (the first 2 panels) are designed to include cytosines within CpG sequences (methylation-specific) as well as cytosines located outside the CpG context (bisulfite conversion-specific). Reactions towards unmethylated DNA sequences are designed to amplify TG-containing sequences, while the reactions toward methylated DNA sequences are designed to amplify CG-containing sequences. Three methylation-independent reactions are used as quality controls (QC) to monitor the sample quantity and integrity (QC-1), as well as bisulfite conversion recovery (QC-2) and bisulfite conversion completeness (QC-3). The QC-1 reaction is designed towards a CpG-less sequence that still contains cytosines outside the CpG context (bisulfite conversion-specific). QC-2 is a bisulfite-independent reaction in which both primers and the probes are designed towards a DNA region that does not contain any cytosines on one of the DNA strands (C-LESS). QC-3 reactions comprise a panel of 16 different reactions designed towards a single DNA sequence that have the same primer sequences but distinct probes. The DNA sequence covered by the primers lacks cytosine residues on one of the strands, while the DNA sequence covered by the probes contains four cytosines outside the CpG context (Fig. 2)

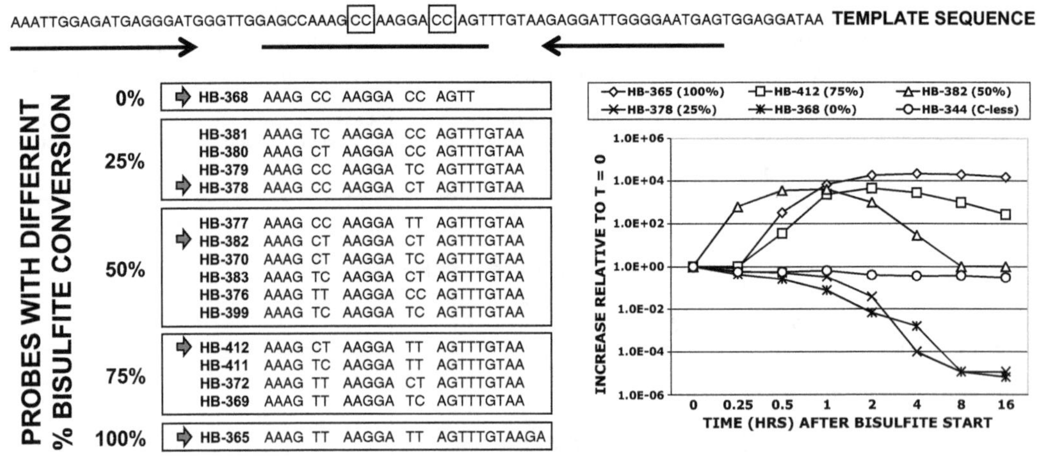

Fig. 2 Description of the QC-3 reactions and their performance on bisulfite converted DNA. We designed 16 distinct bisulfite conversion control reactions that have common forward and reverse primers complementary to a DNA strand lacking cytosine residues at the positions of the primers, but have cytosine-containing unique probes that differ in their abilities to recognize various percentages of bisulfite converted DNA (0%, 25%, 50%, 75%, and 100% conversion). The genomic DNA sequence covered by these probes contains four cytosines that are normally modified to uracils after bisulfite conversion and then to thymines after a subsequent PCR amplification. The degree of conversion of these residues can be monitored by these probes, since they contain 16 different permutations of these residues to thymine reflecting possible changes that could occur in case of complete or incomplete bisulfite conversion. We tested the performance of these reactions in a time course experiment where DNA was either left unconverted or was bisulfite converted for different periods of time (0.25 h, 0.5 h, 1 h, 2 h, 4 h, 8 h and 16 h). The ability of the 0% and 100% conversion reaction as well as the best 25%, 50%, and 75% conversion reactions to detect various degrees of bisulfite converted DNA is presented in the right panel of this figure along with the performance of the C-LESS reaction that is not affected by the bisulfite conversion process since it is designed towards a DNA sequence that contains no cytosines on one strand

2 Materials

2.1 M.SssI Modification

1. M.*Sss*I enzyme supplied with 10× buffer and 32 m*M* *S*-adenosyl methionine (SAM) (e.g., New England Biolabs).
2. Peripheral blood leukocyte (PBL) DNA.
3. DNA extraction kit for circulating nucleic acids (e.g., Qiagen QIAamp Circulating Nucleic Acid Kit or similar).

2.2 Bisulfite Conversion and Recovery

1. EZ DNA Methylation kit (Zymo Research) or similar.

2.3 MethyLight PCR

1. AmpliTaq Gold DNA Polymerase with Buffer II and $MgCl_2$ (Thermo Fisher Scientific). The kit contains the AmpliTaq Gold Enzyme, 10× reaction buffer II (100 mM Tris–HCl, pH 8.3, 500 mM KCl), and the 25 m*M* $MgCl_2$ stock.

2. Deoxynucleotide triphosphates (dNTPs) are combined and diluted to a stock concentration of 10 m*M* for each nucleotide.
3. Primers and Black-Hole Quencher containing probes are obtained from Biosearch Technologies Inc. (Novato, CA). The primers and probes are prepared as 300 μM and 100 μM solutions, respectively, in H_2O. The probes containing the Minor Groove Binder Non Fluorescent Quencer (MGBNFQ) are obtained from Thermo Fisher Scientific, and are prepared as 100 μM solution in H_2O.

3 Methods

3.1 DNA Isolation

Highly purified DNA is not a requirement for MethyLight analysis. Crude DNA extraction protocols involving lysis of the cells or tissues followed by DNA precipitation, or just crude lysates alone, can be used in conjunction with MethyLight analysis. These approaches are desirable when limited quantities of DNA are available, such as in tissues embedded in paraffin slides, tissues from biopsies, or in body fluids such as blood (plasma/serum) that contain small amounts of free circulating DNA. After biopsy tissues or microdissected cells from paraffin slides are lysed, an aliquot of this lysis solution can be directly used in bisulfite conversion. For the sensitive detection of DNA methylation in plasma or serum, the plasma/serum DNA needs to be concentrated from a larger initial volume. This can be achieved by using commercially available kits for DNA extraction from blood or various biological fluids or by precipitation of the DNA.

3.2 M.SssI Modification

M.*Sss*I is a CpG methylase and therefore each CpG dinucleotide is a target of the enzyme, which uses *S*-adenosyl methionine (SAM) as a methyl donor. M.*Sss*I treated DNA is used as a universally methylated reference sample in most MethyLight implementations. PBL DNA is used as a substrate in this protocol. A dilution of bisulfite-converted M.*Sss*I-DNA will be used for normalization and is the basis for the ALU-C4 standard curves.

1. Carry out the M.*Sss*I treatment overnight at 37 °C in a solution containing H_2O, 0.05 μg/μL of PBL DNA, 0.16 m*M* of SAM, 1× reaction buffer, and 0.05 units/μL of M.*Sss*I enzyme.
2. On the next day, add an extra boost of M.*Sss*I enzyme and SAM (1/3 of the original volume) for both components together with H_2O in a total volume representing 1/50 of the initial treatment volume.
3. In order to achieve complete methylation at all the genomic CpG sites, multiple rounds of M.*SssI* treatment can be performed.
4. Store M.*SssI* treated DNA at +4 °C, and use 20 μL (~1 μg) for each bisulfite conversion.

3.3 Bisulfite Conversion and Recovery

1. Before starting, prepare the CT Conversion Reagent and M-Wash Buffer included in the Zymo EZ DNA Methylation kit. Add 750 μL of water and 210 μL of M-Dilution Buffer to one tube of CT Conversion Reagent and mix by vortexing every 1–2 min for a total of 10 min. Each tube of CT Conversion Reagent is designed to treat 10 DNA samples. For best results the prepared CT Conversion Reagent should be used immediately. Add 24 mL of 100% ethanol to the M-Wash Buffer Concentrate.
2. Start the bisulfite conversion protocol by adding 5 μL of the M-Dilution Buffer to the DNA sample and adjust the total volume to 50 μL with sterile H_2O. Mix the sample by flicking or pipetting up and down.
3. Incubate the sample at 37 °C for 15 min.
4. Add 100 μL of the prepared CT Conversion Reagent to each sample and vortex gently.
5. Incubate the sample in the dark at 50 °C for 12–16 h.
6. Incubate the sample on ice for 10 min.
7. Add 400 μL of M-Binding buffer to the sample and mix by pipetting up and down.
8. Load the sample onto a Zymo-Spin I column and place the column into a 2 mL collection tube.
9. Centrifuge at full speed (>10,000 × *g*) for 30 s. Discard the flow-through.
10. Add 200 μL of M-Wash Buffer to the column. Spin at full speed for 30 s.
11. Add 200 μL of M-Desulfonation Buffer to the column and let column stand at room temperature for 15 min. After the incubation, spin at full speed for 30 s.
12. Add 200 μL of M-Wash Buffer to the column. Spin at full speed for 30 s.
13. Add another 200 μL of M-Wash Buffer and spin at top speed for 2 min.
14. Add 10 μL of M-Elution Buffer directly to the column matrix. Place into a 1.5 mL tube. Spin at top speed for 1 min to elute the DNA.
15. Bisulfite converted DNA is stored at −20 °C.

3.4 MethyLight Assay Design Guidelines

1. The size of the MethyLight PCR amplicon should not be larger than 130 base pairs (bp) for optimal performance (shorter amplicons of 80–100 bp are preferred).
2. Each primer and probe oligomer should contain at least two CpGs.

Exceptions: MethyLight reactions containing only one CpG in each primer and probe oligomer can be used when there is no alternative, such as when the sequence is located in a CpG poor region. In this situation, the CpG should be positioned at the 3′ end in each primer and towards the middle of the probe oligomer in order to maximize discrimination between methylated and unmethylated loci. Additional specificity can be obtained by using shorter oligomers with minor groove binding (MGB) annealing enhancing modified nucleotides. This approach increases the relative contribution of each individual CpG. Further methylation specificity can be obtained by the use of blocking oligonucleotides, an approach referred to as HeavyMethyl [8].

3. In general, CpGs should be located at or near the 3′ end of the primer sequences, however, the final five bases at the 3′ end of the primer oligomer should not contain more than three guanines and/or cytosines.
4. In general, CpGs should be in the middle part of the probe oligomer sequence in order to maximize the melting point differences between methylated and unmethylated template molecules.
5. Avoid mononucleotide repeats (e.g., AAAAAAA) in the primer and probe sequences, as these may decrease priming specificity. With respect to cytosine and guanines mononucleotide repeats longer than three bases should be avoided in assay design.
6. The melting temperature (Tm) of the probe oligomer should be 10° higher than that of the primers. The Tm for each primer should be at 59 °C ± 2 °C, and should not be more than one degree different between each primer for an assay.
7. Avoid using probe sequences in which the 5′ end begins with a guanine. Moreover, probe sequences should contain fewer guanines than cytosines. This can be achieved by using the complementary strand for probe design.
8. Probe oligomers should not be greater than 30 bp in length.

3.5 TaqMan PCR Reaction Setup for MethyLight Analysis

The MethyLight assay makes use of the TaqMan PCR principle, which requires forward and reverse primers as well as a nonextendable oligomeric probe which emits fluorescence only after it is degraded by the 5′ → 3′ exonuclease activity of the *Taq* polymerase.

1. Each PCR reaction uses the same basic reaction setup. The choice of primer/probe sets is the only variable in these reactions.
2. Each individual PCR reaction contains 10 μL DNA, 15.4 μL PreMix Solution, 4.5 μL OligoMix Solution (1.5 μL of each

primer and probe) and 0.1 μL *Taq* polymerase in a 30 μL total PCR volume.

3. The PreMix Solution contains all the TaqMan components except Taq polymerase. These components and their final concentration in a PCR reaction are: $MgCl_2$ (3.5 m*M*), 1× TaqMan Buffer II (10 mM Tris–HCl, pH 8.3, 50 mM KCl), and 0.2 m*M* of each dNTP. Each AmpliTaq Gold DNA Polymerase Kit is sufficient for 350 MethyLight reactions. Therefore, to prepare a PreMix Solution for 350 reactions, mix 1.47 mL of the 25 mM $MgCl_2$ stock with 1.05 mL of 10× TaqMan Buffer, 210 μL of 10 m*M* combined dNTPs and 2.66 mL of H_2O. Small aliquots are stored at +4 °C. For each PCR reaction use 15.4 μL of the PreMix Solution in a 30 μL total PCR volume.
4. The OligoMix Solution is specific for each MethyLight and quality control reactions and represents a mixture of both primers and the probe. From the working stock of primers (300 μM) and probe (100 μM), prepare an OligoMix Solution by combining both primers and the probe in one tube. The concentrations of each the forward and reverse primers in the OligoMix Solution are 2 μM, and the probe concentration is 0.67 μM. For each PCR reaction use 4.5 μL of the OligoMix Solution in a 30 μL total PCR volume.
5. The combined PreMix Solution, OligoMix Solution and *Taq* Gold polymerase for each reaction is referred to as the MasterMix Solution (*see* **Note 1**). Load 10 μL of bisulfite converted DNA and 20 μL of the MasterMix Solution in each PCR well.
6. For example, to determine the DNA methylation status of a specific gene of interest, such as *MLH1*, first prepare an *MLH1* OligoMix Solution by combining 2 μL of the *MLH1* forward primer (300 μM), 2 μL of the *MLH1* reverse primer (300 μM) and 2 μL of the *MLH1* probe (100 μM) with 294 μL water. The MethyLight primers and probe sequences for *MLH1* have previously been published [9]. In each individual *MLH1* PCR reaction, combine 4.5 μL of this *MLH1* OligoMix Solution with 15.4 μL of PreMix Solution, 0.1 μL *Taq* polymerase and 10 μL of the bisulfite converted DNA sample to be analyzed.
7. Individual OligoMix Solutions are prepared for any other gene investigated by MethyLight or any other Quality Control reactions used in the analysis, and 4.5 μL aliquots are then combined with the PreMix Solution, *Taq* polymerase and bisulfite converted DNA as described above.
8. Each MethyLight-based data point is the result of the combined analysis of a methylation-dependent PCR reaction (Experimental MethyLight reaction, *see* Fig. 1) and methylation-independent PCR reaction (CpG-less sequence,

see Fig. 1) on reference (M.*Sss*I-treated DNA) and experimental DNA samples. The MethyLight assay setup is described in Subheading 3.7.

9. All PCR reactions are carried out as follows: one cycle at 95 °C for 10 min followed by 50 cycles at 95 °C for 15 s and 60 °C for 1 min.

3.6 Initial Quality Control

1. *QC-1*: Sample Quantity and Integrity. Samples vary in the initial template quantity and integrity. The most reliable measure of amplifiable DNA quantities after bisulfite conversion is a bisulfite-dependent, methylation-independent reaction for a multi-copy number sequence well distributed throughout the genome. For this purpose, use the ALU-C4 bisulfite control reaction (5) (*see* Table 1 for primer and probe sequences) to perform a preliminary TaqMan PCR test to check the C (t) value of each sample. Following the purification of bisulfite treated DNA using the Zymo kit, the DNA is contained in a 10 μL volume. Dilute the sample 1:10 (final volume 100 μL) and test 2 μL by PCR using the ALU-C4 bisulfite control reaction and the PCR conditions described in Subheading 3.4. The The C(t) value generated from this 1:5 dilution will give an indication of the amount of bisulfite converted DNA available for further analysis (*see* **Note 2**).
2. *QC-2*: Sample Recovery. If problems are regularly encountered with recovery of samples after bisulfite conversion, then C-less bisulfite-independent reactions can be used to monitor recovery of each step. Primer and probe sequences of the C-less reaction are given in Table 1, and the cycling conditions are described in Subheading 3.4.
3. *QC-3*: Bisulfite Conversion Completeness. The efficacy and completeness of the bisulfite conversion of the sample can be assessed using a panel of bisulfite conversion reactions (Fig. 2) (*see* Table 1 for the primer and probe sequences of these reactions and Subheading 3.4 for cycling PCR conditions). These reactions are specific for unconverted DNA (0% conversion), fully converted DNA (100% conversion) or DNA with various degrees of conversion (25%, 50%, and 75% conversion) (*see* **Note 3**).

3.7 MethyLight Reactions

Two types of reactions are used in the MethyLight protocol that both use the bisulfite converted DNA as a substrate: methylation-dependent reactions (CpG-based) specific for methylated DNA and methylation-independent control reactions (CpG-less), as described for the QC-1 quality control reaction above.

1. The methylation-dependent reactions are both bisulfite- and methylation-specific reactions, i.e., they cover CpGs as well as

Table 1
Primer and probes sequences for the quality control PCR reactions used for the analysis of bisulfite treated DNA

HB-Number	Reaction ID	Description	Forward Primer Sequence (5′ to 3′)	Reverse Primer Sequence (5′ to 3′)	Probe Sequence (5′ to 3′)
HB-313	ALU-C4	QC-1	GGT TAG GTA TAG TGG TTT ATA TTT GTA ATT TTA GTA	ATT AAC TAA ACT AAT CTT AAA CTC CTA ACC TCA	CCT ACC TTA ACC TCC C
HB-344	C-LESS-C1	QC-2	TTG TAT GTA TGT GAG TGT GGG AGA GA	TTT CTT CCA CCC CTT CTC TTC C	CTC CCC CTC TAA CTC TAT
HB-365	CONV-C1	QC-3: 100%	AAA TTG GAG ATG AGG GAT GGG T	TTA TCC TCC ACT CAT TCC CCA A	TCT TAC AAA CTA ATC CTT AAC TTT
HB-368	CONV-C2	QC-3: 0%	AAA TTG GAG ATG AGG GAT GGG T	TTA TCC TCC ACT CAT TCC CCA A	AAC TGG TCC TTG GCT TT
HB-369	CONV-C3	QC-3: 75%	AAA TTG GAG ATG AGG GAT GGG T	TTA TCC TCC ACT CAT TCC CCA A	TTA CAA ACT GAT CCT TAA CTT T
HB-370	CONV-C4	QC-3: 50%	AAA TTG GAG ATG AGG GAT GGG T	TTA TCC TCC ACT CAT TCC CCA A	TTA CAA ACT GAT CCT TAG CTT T
HB-372	CONV-C5	QC-3: 75%	AAA TTG GAG ATG AGG GAT GGG T	TTA TCC TCC ACT CAT TCC CCA A	TTA CAA ACT AGT CCT TAA CTT T
HB-376	CONV-C9	QC-3: 50%	AAA TTG GAG ATG AGG GAT GGG T	TTA TCC TCC ACT CAT TCC CCA A	TTA CAA ACT GGT CCT TAA CTT T
HB-377	CONV-C10	QC-3: 50%	AAA TTG GAG ATG AGG GAT GGG T	TTA TCC TCC ACT CAT TCC CCA A	TTA CAA ACT AAT CCT TGG CTT T

HB-378	CONV-C11	QC-3: 25%	AAA TTG GAG ATG AGG GAT GGG T	TTA TCC TCC ACT CAT TCC CCA A	TTA CAA ACT AGT CCT TGG CTT T
HB-379	CONV-C12	QC-3: 25%	AAA TTG GAG ATG AGG GAT GGG T	TTA TCC TCC ACT CAT TCC CCA A	TTA CAA ACT GAT CCT TGG CTT T
HB-380	CONV-C13	QC-3: 25%	AAA TTG GAG ATG AGG GAT GGG T	TTA TCC TCC ACT CAT TCC CCA A	TTA CAA ACT GGT CCT TAG CTT T
HB-381	CONV-C14	QC-3: 25%	AAA TTG GAG ATG AGG GAT GGG T	TTA TCC TCC ACT CAT TCC CCA A	TTA CAA ACT GGT CCT TGA CTT T
HB-382	CONV-C16	QC-3: 50%	AAA TTG GAG ATG AGG GAT GGG T	TTA TCC TCC ACT CAT TCC CCA A	TTA CAA ACT AGT CCT TAG CTT T
HB-383	CONV-C17	QC-3: 50%	AAA TTG GAG ATG AGG GAT GGG T	TTA TCC TCC ACT CAT TCC CCA A	TTA CAA ACT AGT CCT TGA CTT T
HB-399	CONV-C15	QC-3: 50%	AAA TTG GAG ATG AGG GAT GGG T	TTA TCC TCC ACT CAT TCC CCA A	TTA CAA ACT GAT CCT TGA CTT T
HB-411	CONV-C20	QC-3: 75%	AAA TTG GAG ATG AGG GAT GGG T	TTA TCC TCC ACT CAT TCC CCA A	TTA CAA ACT AAT CCT TGA CTT T
HB-412	CONV-C21	QC-3: 75%	AAA TTG GAG ATG AGG GAT GGG T	TTA TCC TCC ACT CAT TCC CCA A	TTA CAA ACT AAT CCT TAG CTT T

All probes contain a 6FAM fluorophore at the 5′ end and a minor groove binder (MGB) nonfluorescent quencher at the 3′ terminus. The genomic coordinates for the QC-2 reaction are; chromosome 20, 19199387-19199455, and for the QC-3 reactions; chromosome 11, 17649485-17649620 obtained from the NCBI Build 36.2. The QC-1 reaction is based on an ALU consensus sequence and therefore does not have precise genomic coordinates

Cs not in a CpG context in their sequence (methylated CpGs will remain CpGs, other Cs and unmethylated CpGs will become Ts or TpGs, respectively, after bisulfite conversion and PCR).

2. The methylation independent control reaction (CpG-less) (ALU-C4) is used to normalize for differing quantities and quality of DNA samples (*see* **Note 4**). This reaction is not methylation-specific since there are no CpGs in the primers/probe sequences, but is specific for bisulfite converted DNA since it covers Cs not in a CpG context.

3.8 MethyLight Assay Setup

In order to determine the methylation status of a specific gene using the MethyLight assay, four PCR reactions are required. Two types of samples are needed: the bisulfite converted DNA of the sample of interest and the M.*Sss*I-converted DNA as a reference sample. For each of these DNA samples, we perform a PCR reaction for the gene of interest (Experimental MethyLight measurement, *see* Fig. 1) and one control PCR reaction to measure the amount of amplifiable DNA sample (QC-1, ALU-C4 reaction, *see* Fig. 1). The use of M.*Sss*I-converted DNA as a reference helps to control for variations in reagent batches, including primers and probes, reaction efficiency, machine performance, and various other sources of batch effects (*see* **Note 5**).

1. Dilute the bisulfite converted M.*Sss*I-DNA (1:100) and use 10 μL of this sample for each PCR reaction. Use 10 μL of the experimental sample (diluted based on the ALU-C4 C(t) value from the 1:5 dilution test, *see* Subheading 3.5, **step 1**). Perform each MethyLight reaction as well as each control reaction in duplicate.
2. Two independent four-point standard curves using the ALU-C4 control reaction and bisulfite converted M.*Sss*I-modified DNA (diluted 1:100) are required for quantification. From this initial stock of bisulfite converted M.*Sss*I-modified DNA, perform 1:25 serial dilutions. A volume of 10 μL of each dilution should be used for each serially diluted sample.

3.9 MethyLight Percentage of Methylated Reference (PMR) Calculations

1. The formula to calculate PMR values represents the quotient of two ratios (multiplied by 100). Thus, the formula is: $100 \times [(\textit{GENE-X}\ \text{mean value})_{sample}/(\textit{ALU-C4}\ \text{mean value})_{sample}]/[(\textit{GENE-X}\ \text{mean value})_{\text{M.}\textit{Sss}\text{I}}/(\textit{ALU-C4}\ \text{mean value})_{\text{M.}\textit{Sss}\text{I}}]$.
2. Once the real-time PCR program is finished, the C(t) values are converted to mean values/copy numbers using the standard curve for each plate (*see* **Note 6**).
3. One PMR value per sample will be calculated based on the mean values derived from each of the two standard curves.

The two PMRs obtained will be averaged at the end of the procedure.

4. Using the data generated with the first standard curve, divide the mean/copy value for the methylation reaction of the sample of interest by the mean/copy value of the ALU-C4 reaction for the very same sample.
5. Divide the mean/copy value for M.*Sss*I sample for the same methylation reaction as in **step 4** by the mean/copy value for the ALU-C4 reaction of the M.*Sss*I sample. Average this quotient for duplicate reactions.
6. Divide the value from **step 4** by the value from **step 5** and multiply that value by 100. This is the first PMR value.
7. Calculate the second PMR value by the same procedure using the data generated based on the second standard curve. This can be achieved by simply reassigning the second ALU-C4 standard curve wells as standards. Then redetermine the values from **steps 4** and **5**. The PMR values from each quantitation can then be averaged to generate the final PMR value for each sample.

3.10 Digital MethyLight

We have also applied MethyLight technology to digital PCR applications in developing Digital MethyLight, an ultra-sensitive method for detecting and quantitating individual methylated DNA regions in biological fluids, such as plasma or serum [10]. Digital MethyLight involves distributing a MethyLight reaction across a 96- or 384-well plate or higher in a microfluidic device, such that the mean initial template DNA concentration is less than one molecule per reaction compartment. Amplification of methylated DNA molecules occurs in a small minority of PCR wells, and therefore represents a digital readout of the original number of template molecules in each sample (Fig. 3).

1. Isolate genomic DNA from a specific volume of plasma, serum, etc., using the QIAamp Circulating Nucleic Acid as described by the manufacturer.
2. Convert genomic DNA from each sample with sodium bisulfite using the EZ DNA methylation kit as described by the manufacturer.
3. Mix bisulfite-converted DNAs, representing a specific volume of plasma or serum, with 200 μM dNTPs, 0.3 μM forward and reverse MethyLight primers, 0.1 μM MethyLight probe, 3.5 mM $MgCl_2$ and 50 units of AmpliTaq GOLD polymerase in a 2.88 mL volume and distribute in 30 μL aliquots in each well of a 96-well plate. Since the amount of DNA in these samples is very small, and in order to minimize costs, we recommend scaling down the components of the assay by

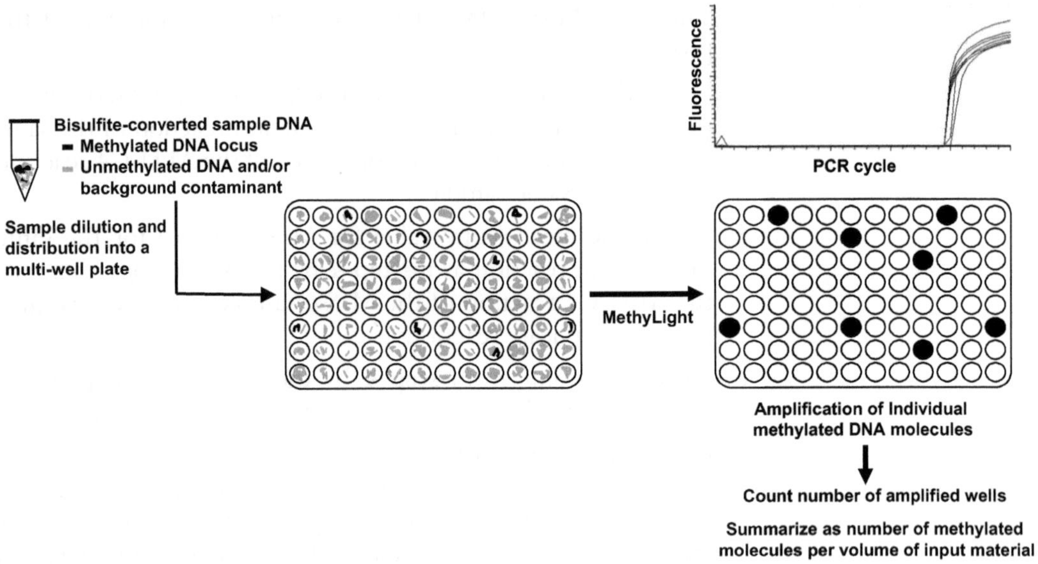

Fig. 3 Digital MethyLight Technology Overview. Genomic DNA is isolated from a specific volume (usually 100–1000 μL) of plasma or serum. Genomic DNA is then bisulfite converted, mixed with MethyLight assay reagents and distributed across a multi-well PCR plate such that each well contains less than one methylated DNA molecule. Sample distribution also minimizes the amount and influence of unmethylated DNA and/or background contaminants that may affect the performance of classic MethyLight assays. MethyLight is performed on the entire multi-well plate, and the PCR-amplified reaction wells are counted. Each sample/assay is scored as the number of amplifications (number of methylated DNA molecules) per volume of input material

1/3 and distribute the total volume of 0.996 mL (2.88 mL/3) in 10 μL aliquots in each well of a 96-well plate.

4. Perform digital MethyLight PCRs using the exact same PCR program as with Classic Digital MethyLight. Each well that shows a PCR amplification is scored as positive, and the total number of amplifications per 96-well plate is later normalized for the volume of input material.

3.11 Multiplexing in Digital MethyLight Assays

Digital MethyLight assays can be performed with multiplexed markers. Multiplexing provides the most cost-effective way of analyzing multiple markers, while maximizing the simultaneous use of precious samples such as plasma or serum from patients. Presently, we have succeeded in combining up to three sets (triplex) of markers. Each probe is labeled with one of three different fluorophores (Fam, Hex and Quasar), so that three candidate gene regions can be concurrently and simultaneously interrogated. In order to develop multiplexed reactions with minimal impact on the performance of each individual reaction, combinations of the three primer/probe sets need to be evaluated to determine the impact of spurious primer annealing events by noncognate primers for each labeled probe.

1. Set up each MethyLight reaction containing OligoMix Solutions with combinations of the three sets of primers and probes. Each PCR reaction uses the same reaction setup (discussed in Subheading 3.4).
2. Prepare OligoMix Solutions for Reactions 1, 2, and 3 independently.
3. To check for any interference of Probe 1 by any of the Primer Sets, prepare one OligoMix Solution with Primers for reactions 1 and 2 with the Probe for reaction 1, and another OligoMix solution with Primers for reactions 1 and 3 with the Probe for reaction 1.
4. To check for any interference of Probe 2 by any of the Primer Sets, prepare one OligoMix Solution with Primers 1 and 2 with the Probe for reaction 2, and another OligoMix solution with Primers 2 and 3 with the Probe for reaction 2.
5. To check for any interference of Probe 3 by any of the Primer Sets, prepare one OligoMix Solution with Primers for reactions 1 and 3 with Probe 3, and a second OligoMix solution with Primers 2 and 3 with Probe for reaction 3.
6. Finally, prepare: (1) OligoMix Solution with all three primers sets and the Probe for reaction 1; (2) an OligoMix solution with all three primers sets and Probe for reaction 2, and (3) an OligoMix solution with all three primers sets and the Probe for reaction 3.
7. Prepare all of the above OligoMix Solution from a diluted (1:3) working stock of primers (100 μM) instead of 300 μM. The probe working stock concentration of the probe will remain 100 μM. The concentrations of each the forward and reverse primers as well as the probe in the OligoMix Solution is 0.67 μM. For each PCR reaction use 4.5 μL of the OligoMix Solution in a total PCR volume of 30 μL.
8. Set up PCR reactions (in triplicate) with these 12 different OligoMix Solutions and the C(t) values of these primer/probes combinations will be compared. Any differences in C(t) values between the individual reactions and the Probes in combination with other primers will be indicative that the multiplex it is not possible for that particular Probe.
9. The interference of any of the Probes by the other Probes can only be determined after each Probe is labeled with one of two different fluorophores (Hex or Quasar). Similar OligoMix Solutions can be made this time using combinations of two primers with two probes and three primers and all three probes. The C(t) values of these reactions can be compared with those of the individual reactions to determine if the Probes are compatible with each other.

4 Notes

1. Uracil DNA glycosylase (AMPerase) should not be included in MethyLight reactions. This is a component used in some TaqMan reaction kits, and this poses a complication in MethyLight reactions as uracil is a product following bisulfite conversion.
2. It should be noted that low C(t) values are always preferred to achieve the best possible data. An ALU-C4 C(t)$\leq$17 is usually desirable, but data can also be generated from samples with C(t) $\leq$ 22 when the samples are precious.
3. The lack of complete decline of the mean values obtained for the 75% and 50% reactions is likely due to some degree of cross-hybridization with fully converted sequence. Nevertheless, these two reactions are likely to be the most sensitive probes for detecting incomplete bisulfite conversion of a sample. It should be noted that these reactions assess completeness of bisulfite conversion only at this one locus. To the extent that sequences differ in their resistance to denaturing or bisulfite conversion, this locus may not be representative for other parts of the genome.
4. Cancer genomic DNA samples can contain copy number alterations, which can affect the quantitation of both the locus of interest as well as the methylation-independent reaction. Ideally, one would design a CpG-less methylation-independent reaction as close to the MethyLight reaction as possible to adjust for such events. This significantly increases the cost of reaction design and experimental implementation. As a next-best solution, we avoid the influence of copy-number alterations of single loci for the methylation-independent QC-1 control reaction by using a repetitive element. Even in samples with aneuploidy, this control reaction will yield a reasonable approximation of total DNA quantities due to the distributed nature of the target sequence.
5. We describe here the standard procedure in our laboratory to calculate the PMR value as a universal measure for the percentage of fully methylated alleles of a DNA sample, regardless of origin or locus being assessed. We find this a useful measure for most instances. Although our implementation is based on a comparison to a reference sample, it utilizes the absolute method of quantitation for real-time PCR, which is based on mean values derived from a standard curve of defined initial template quantities. By comparing to a control reaction and to a reference sample, we turn this absolute method into a relative method. We also sometimes implement relative methods of analysis, including the calculation of Δ-C(t) values. We use this most frequently, when we are interested in a separate

within-sample comparison, such as the comparison of different bisulfite conversion control reactions. We prefer the PMR method as a general measure of DNA methylation, since it controls for many other sample-independent sources of experimental variation and error.

6. Under usual real-time PCR conditions, the standard curve is based on dilutions of known absolute quantities of template. Although this could be implemented for each reaction, using synthetic or cloned template, we prefer to avoid this, in part to limit sources of high-concentration PCR contamination. Since the PMR calculation is a relative measure, it is sufficient to use unknown quantities of standard DNA, but with precisely defined dilutions. This will yield mean values that do not have any absolute meaning, but which can be used to derive the ratios in the PMR calculation. In other words, the unknown quantity of template DNA in the standards is divided away.

References

1. Eads CA, Danenberg KD, Kawakami K et al (1999) CpG island hypermethylation in human colorectal tumors is not associated with DNA methyltransferase overexpression. Cancer Res 59:2302–2306
2. Eads CA, Danenberg KD, Kawakami K et al (2000) MethyLight: a high-throughput assay to measure DNA methylation. Nucleic Acids Res 28:e32
3. Trinh BN, Long TI, Laird PW (2001) DNA methylation analysis by MethyLight technology. Methods 25:456–462
4. Ogino S, Kawasaki T, Brahmandam M et al (2006) Precision and performance characteristics of bisulfite conversion and real-time PCR (MethyLight) for quantitative DNA methylation analysis. J Mol Diagn 8:209–217
5. Weisenberger DJ, Campan M, Long TI et al (2005) Analysis of repetitive element DNA methylation by MethyLight. Nucleic Acids Res 33:6823–6836
6. Herman JG, Graff JR, Myohanen S et al (1996) Methylation-specific PCR: a novel PCR assay for methylation status of CpG islands. Proc Natl Acad Sci U S A 93:9821–9826
7. Laird PW (2003) The power and the promise of DNA methylation markers. Nat Rev Cancer 3:253–266
8. Cottrell SE, Distler J, Goodman NS et al (2004) A real-time PCR assay for DNA-methylation using methylation-specific blockers. Nucleic Acids Res 32:e10
9. Fiegl H, Gattringer C, Widschwendter A et al (2004) Methylated DNA collected by tampons-a new tool to detect endometrial cancer. Cancer Epidemiol Biomarkers Prev 13:882–888
10. Weisenberger DJ, Trinh BN, Campan M et al (2008) DNA methylation analysis by digital bisulfite genomic sequencing and digital MethyLight. Nucleic Acids Res 36:4689–4698

Chapter 26

Quantitative Region-Specific DNA Methylation Analysis by the EpiTYPER™ Technology

Sonja Kunze

Abstract

DNA methylation plays a profound role in development and health as well as development and progression of disease. High-throughput quantitative DNA methylation analysis is therefore crucial for the study of the normal physiology of the epigenome and its dysregulation in disease. Many target areas are identified by a range of emerging genome-wide cytosine methylation techniques, but these whole genome scans usually only provide methylation data for a few individual CpG sites (CpGs) within a region. The EpiTYPER™ assay is a region-specific method for the detection and quantitative analysis of DNA methylation that allows performing a high-resolution scan of selected regions. It thus enables a more detailed analysis of single CpGs and the surrounding area and can, in addition to candidate gene methylation analysis, be used to validate CpGs detected by genome wide techniques. The EpiTYPER™ assay allows a fast and reproducible targeted quantification of individual CpGs in a high throughput manner and is based on base-specific cleavage of bisulfite-converted genomic DNA and matrix-assisted laser desorption/ionization time-of-flight mass spectrometry (MALDI-TOF MS). Up to 85% of the CpGs within a target region can be analyzed and the detection precision allows quantifying methylation differences as low as 5–7%.

Key words EpiTYPER, Base-specific cleavage, MALDI-TOF MS, DNA methylation pattern, Quantitative, Bisulfite

1 Introduction

The identification, quantification, and mapping of DNA methylation within the genome can be conducted by a wide range of available methods and approaches [1]. The method of choice depends on the intended application and required level of information. Since the methods for DNA methylation analysis differ in their coverage and sensitivity, they enable global profiling of DNA methylation as well as the measurement of the degree of methylation at specific target regions or the creation of genome-wide methylation maps.

For region-specific analysis numerous methods have been reported. While the earlier ones relied exclusively on restriction enzymes, utilization of bisulfite conversion has revolutionized the

Jörg Tost (ed.), *DNA Methylation Protocols*, Methods in Molecular Biology, vol. 1708,
https://doi.org/10.1007/978-1-4939-7481-8_26, © Springer Science+Business Media, LLC 2018

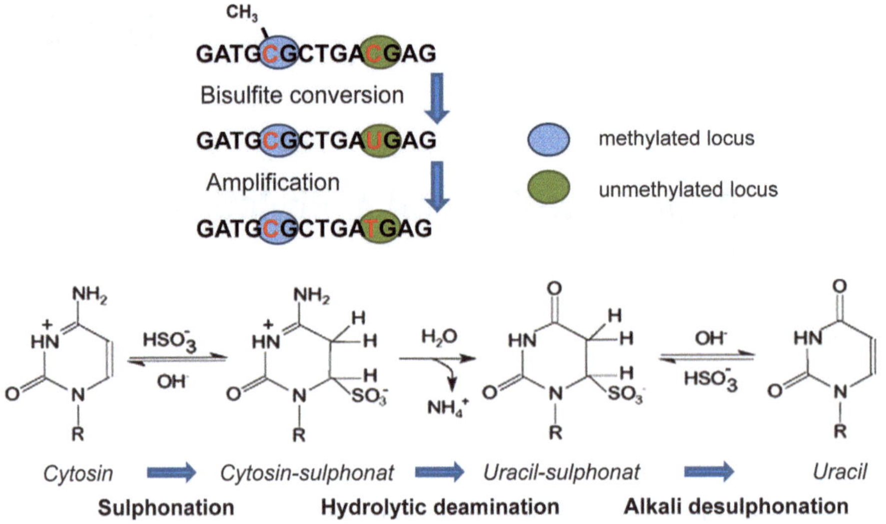

Fig. 1 Principle of the bisulfite conversion and chemical background. The chemistry of cytosine deamination by sodium bisulfite involves three steps: sulfonation (the addition of bisulfite to the 5–6 double bond of cytosine), hydrolic deamination of the resulting cytosine-bisulfite derivative to give a uracil-bisulfite derivative and alkli desulfonation (the removal of the sulfonate group by an alkali treatment) to give uracil

field. Treatment of denatured single-stranded genomic DNA with sodium bisulfite converts unmethylated cytosines to uracils, whereas methylated cytosine bases are protected from conversion and remain unchanged. The uracils are amplified in subsequent polymerase chain reactions (PCR) as thymines, producing methylation-dependent sequence variations of C to T for the unmethylated locus (Fig. 1). One method based on the bisulfite treatment of genomic DNA is, besides for example pyrosequencing (*see* Chapter 22), the EpiTYPER assay from Agena Biosciences formerly Sequenom. It is used to carry out region-specific high-throughput DNA methylation analysis for the targeted quantitation of individual CpGs. In short, the assay detects and quantitatively analyzes DNA methylation using base-specific cleavage of bisulfite-converted genomic DNA and matrix-assisted laser desorption/ionization time-of-flight mass spectrometry (MALDI-TOF MS) [2]. An overview of the concept of the EpiTYPER method is given in Fig. 2. The bisulfite treatment of genomic DNA is followed by PCR amplification in which a T7- promoter tag is introduced by a T7-promoter tagged reverse primer. Also, a 10mer tag sequence is added to the forward primer to balance the PCR primer length. The PCR primers are independent of the methylation state of the genomic DNA, i.e., they bind to both methylated and nonmethylated template, and should be designed to yield a product within a 200–600 base pair (bp) range (henceforth referred to as amplicon). After an incubation with Shrimp Alkaline Phosphatase (SAP) reaction, that removes unincorporated dNTPs, in vitro RNA

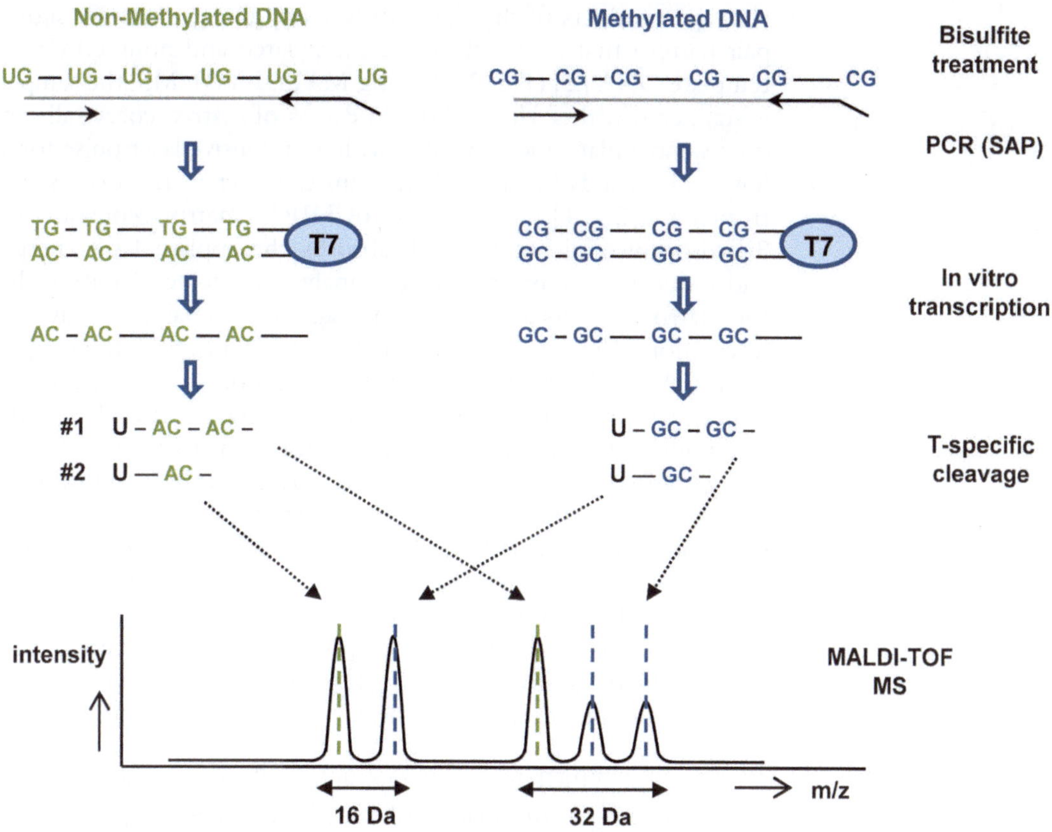

Fig. 2 Scheme of the EpiTYPER technique. Bisulfite-treated DNA is amplified by PCR using primers located around the CpG site of interest, one of them being tagged with a T7-promoter sequence. The PCR product is transcribed into a RNA transcript and cleaved in a base-specific manner after each T (thymin). The cleavage products are then analyzed by matrix-assisted laser desorption/ionization time-of-flight (MALDI-TOF) mass spectrometry (MS) and a characteristic mass signal pattern can be obtained. The *green* signal represents nonmethylated DNA and the *blue* one represents methylated DNA. This figure was modified after Ehrich et al. [2]

transcription is performed on the reverse strand, followed by base-specific T-cleavage using RNase A. Within this step, cleavage products are generated for the reverse transcription reaction for U (thymidine (T)). Since T is found in DNA it is used from this point forward, however, cleavage actually happens at U in an RNA molecule. Both methylated and nonmethylated regions are cleaved at every T to produce fragments of different length and mass. For a specific CpG site the cleavage products resulting from methylated and nonmethylated DNA respectively have the same length and differ only in their nucleotide composition. The methylation dependent C/T sequence changes introduced by bisulfite treatment are presented as G/A changes on the reverse strand and therefore result in a mass difference of 16 Da (Dalton) for each CpG site enclosed in the cleavage products generated from the RNA transcript. MALDI-TOF MS allows measurement of the cleavage products of the EpiTYPER assay, giving a distinct signal pair pattern that results from the methylated and nonmethylated template DNA [2] (Fig. 2). During MALDI-TOF MS, the sample is staggered with a 100–1000 fold excess of matrix, cocrystallized on a sample plate and irradiated with an intensive laser pulse for a few nanoseconds in the high vacuum chamber of the mass spectrometer [3, 4]. The SpectroCHIP® matrix consists of 3-hydroxypicolinic acid, which absorbs the applied laser energy and induces the ionization of the analyte molecule. Additionally, the matrix prevents a photolytic damage of the analyte and avoids interaction of analyte molecules with each other or with the sample carrier. After electric field-induced acceleration in the mass spectrometer source region, the gas phase ions travel through a field-free region at a velocity inversely proportional to their mass-to-charge ratios (m/z), until they hit the detector. Ions with low m/z values are faster than ions with higher m/z values and reach the detector earlier. The TOF-analyzer measures exactly the time until the ions hit the detector [5, 6]. The resulting time-resolved mass spectrum is finally translated into a mass spectrum upon comparison with known calibration standards and further processed and analyzed by the software EpiTYPER (Agena Biosciences) that performs baseline correction, peak identification and quality assessments.

The EpiTYPER assay as a region-specific method is especially recommended for the analysis of a small number of amplicons in a large sample cohort. It is commonly used in candidate gene methylation analysis and is a great tool to validate single CpGs and the surrounding area detected by, for example, genome wide methods (e.g., Illumina 450K or EPIC, *see* also Chapter 16) [7]. The advantages of the EpiTYPER assay are the fast, targeted quantification of individual CpGs in a high-throughput manner. The time needed for bisulfite-treated DNA to data is approx. 8 h and the detection precision allows to quantify methylation differences as low as 5–7% within a methylation range of 10–90% [2]. Disadvantages of the EpiTYPER assay are for example difficulties to measure DNA methylation across CpG islands and that only the average methylation data per cleaved fragment is examined. Ideally, a fragment contains a single isolated CpG site and has a unique molecular weight. However, a fragment may also contain multiple CpGs or may have a molecular weight that overlaps with another CpG-containing fragment and then the constituent CpGs have to be measured in aggregates. Or the fragment molecular weight is too small or too large and lies outside the testable mass window. For these reasons, depended on the sequence content and the distribution of CpGs, up to 85% of the CpGs within a target region are analyzed. Another point is that polymorphisms in the DNA sequence analyzed can be a problem for the MassCLEAVE assay and obscure the analysis of the degree of DNA methylation

[8, 9]. Overall the EpiTYPER assay is an accurate, reproducible, and quantitative tool to analyze genomic DNA methylation. It has been, for example, applied to determine cardiovascular disease-related and cancer-associated DNA methylation patterns [10–12] and to validate single CpGs and the surrounding area detected by genome-wide methods [7, 13].

Another possible approach for DNA methylation analysis using MALDI-TOF MS is the primer extension method. This method as usually employed for the analysis for single nucleotide polymorphisms (SNPs) and mutations [14, 15] can be adapted and used for the analysis of genomic DNA methylation within single CpGs that have been found to be relevant within the T-cleavage or other methods. In brief, the C/T sequence change as a result of the bisulfite treatment can be treated as a SNP. After PCR of a target region harboring the CpG site of interest a primer is annealed immediately adjacent to the CpG site and, depending on the methylation status of the desired CpG site, extended by one of four possible terminators (ddNTPs) with the use of a DNA polymerase (C/T on the forward strand ($\Delta m = 15$ Da) or G/A on the reverse strand ($\Delta m = 16$ Da)). The peak area ratio of mass signals derived from the single primer extension products representing methylated and unmethylated DNA can be used to estimate the degree of methylation of the respective CpG site [16]. The primer extension method allows for multiplexing and, therefore, minimizes the amount of DNA needed as well as the costs to run selected single CpGs in a large sample size.

Although this chapter will only address the protocol for the EpiTYPER assay, the primer extension method is a great tool on top of the EpiTYPER assay. The EpiTYPER assay allows the simultaneous analysis of multiple CpGs within a specific region, whereas the primer extension method allows the analysis of selected CpGs with proven functional relevance of different regions, both within a single reaction.

2 Materials

2.1 Software

1. Primer Design software: www.epidesigner.com (Agena Biosciences) or MethPrimer (http://www.urogene.org/cgi-bin/methprimer/methprimer.cgi).

2.2 Sample Preparation and Bisulfite Treatment

1. NanoDrop spectrophotometer.
2. EZ-96 DNA Methylation™ Kit (Zymo Research).
3. Nanopure Water (ddH_2O).
4. 100% ethanol.
5. Centrifuge for 96-well microplates.

6. Standard thermocycler, for cycling of 96-well microplates.
7. Plate-sealing film.
8. Methylation standards (e.g., EpigeneDX).

2.3 PCR, SAP and MassCLEAVE™ Protocol

1. EpiTYPER T Reagent Set—Complete PCR Reagent Set (includes: HotStart Buffer (10×; containing 15 mM $MgCl_2$), HotStart Taq DNA Polymerase (5 U/μL) and dNTP Mix (25 mM each); Agena Biosciences formerly Sequenom).
2. Nanopure Water (ddH_2O).
3. PCR primers (designed during Assay Design step, *see* Subheading 3.2 of this chapter).
4. EpiTYPER T Reagent Set—MassCLEAVE T7 Kit (includes: RNase-free ddH_2O, SAP (Shrimp Alkaline Phosphatase; 1.7 U/μL), T7 polymerase buffer (5×), T-cleavage Mix, DTT (dithiothreitol, 100 mM), T7 RNA & DNA Polymerase (50 U/μL) and RNase A (10 mg/mL); Agena Biosciences).
5. 384-well microplates.
6. Centrifuge for 384-well microplates.
7. Standard thermocycler, for cycling of 384-well microplates (with temperature gradient functionality if possible).
8. Plate-sealing film.

2.4 Conditioning, Nanodispensing and MALDI-TOF MS Measurements

1. EpiTYPER T Reagent Set—SpectroCHIP II Arrays & Clean Resin Kit (includes: CLEAN Resin and SpectroCHIP® II G384; Agena Biosciences).
2. Rotator.
3. 4-Pt Calibrant (included in the MassCLEAVE T7 Kit; Agena Biosciences).
4. MassARRAY™ Nanodispenser (Agena Biosciences).
5. MassARRAY™ Compact MALDI-TOF MS (Agena Biosciences).
6. MassARRAY™ EpiTyper software v1.2 (Agena Biosciences).

3 Methods

3.1 Bisulfite Treatment

Protocols are performed according to the manufacturer's instructions (Zymo Research EZ DNA methylation kit) with slight modifications as recommended by Agena Biosciences.

The bisulfite conversion can be carried out in a 96-well plate format or in single tubes, depending on the sample size (*see* **Note 1**). The protocol differs only slightly; in the following protocol the steps for the conversion in 96-well plate format are listed.

1. Prepare a 45 μL DNA mix in a Conversion Plate (96-well format) for each of the DNA samples by adding a certain amount of genomic DNA to the respective volume of nanopure water, depending on the DNA concentration. For one reaction a minimum amount of 10 ng of bisulfite-treated DNA is needed. Therefore the amount of genomic DNA added depends on the number of planned reactions. Usually a range between 500–1000 ng is recommended, however if DNA is limited, the input can be reduced to 200 ng (*see* **Note 2**).
2. Add 5 μL M-Dilution Buffer to each DNA mix, mix by pipetting up and down, seal the plate and incubate the samples at 37 °C for 15 min.
3. During the incubation time of the samples, prepare the CT Conversion Reagent by adding 7.5 mL of nanopure water and 2.1 mL of M-Dilution Buffer to one bottle of CT Conversion Reagent. Mix the solution with frequent vortexing for at least 10 min at room temperature (fix the tube to the vortexer by adhesive tape; prolong the vortex step if the solution is not clear of crystals) (*see* **Note 3**).
4. After incubation, add 100 μL of the CT Conversion Reagent to each sample, mix by pipetting up and down.
5. Incubate the Conversion Plate on a thermal cycler using the following program: 20 cycles of 95 °C for 30 s and 50 °C for 15 min, hold at 4 °C.
6. After incubation, keep the samples on ice for 10 min, prepare the M-Wash Buffer by adding 144 mL of 100% ethanol to the 36 mL M-Wash Buffer concentrate and set the centrifuge to a temperature of 25 °C.
7. Stack a Silicon-ATM Binding Plate on top of a Collection Plate and add 400 μL of M-Binding Buffer to each well.
8. Centrifuge the samples which have been on ice for 10 min, load them into the wells of the Silicon-ATM Binding Plate containing the M-Binding Buffer, and mix by pipetting up and down as long as smear is observable.
9. Centrifuge the plate for 3 + 3 min at $\geq 3000 \times g$ (RCF 4500) (*see* **Note 4**) and discard the flow-through to prevent contamination of the column contents.
10. Add 500 μL of M-Wash-Buffer to each well and centrifuge for 3 + 3 min at $\geq 3000 \times g$. Discard flow-through.
11. Add 200 μL of M-Desulfonation Buffer to each well and incubate the samples at room temperature (20–30 °C) for 15–20 min.
12. Centrifuge for 3 + 3 min at $\geq 3000 \times g$. Discard flow-through.

13. Add 500 μL of M-Wash Buffer to each well and centrifuge for 3 + 3 min at ≥3000 × *g*. Discard flow-through.
14. Add another 500 μL of M-Wash Buffer and centrifuge for 6 + 6 min at ≥3000 × *g*. Discard flow-through.
15. Place the Silicon-ATM Binding Plate onto an Elution Plate and pipette an appropriate volume of nanopure water (*see* **Note 5**) directly to the binding matrix in each well.
16. Incubate the plate for about 2–5 min and centrifuge for 3 + 3 min at ≥3000 × *g*.
17. The bisulfite-converted DNA can be used immediately stored at 4 °C for 1 week or at or below −20 °C for later use (*see* **Note 6**).
18. Since the DNA is single stranded due to the conversion process with limited nonspecific base-pairing at room temperature, an absorption coefficient of 260 nm, which resembles that of RNA, has to be used to determine the concentration of the recovered bisulfite-treated DNA (value of 40 μg/mL for Abs260 = 1.0), if desired.

A complete bisulfite conversion is essential for the DNA methylation analysis, as incomplete conversion of cytosines to uracils can result in false-positive methylation signals. Incomplete bisulfite conversion can arise from incomplete denaturation before bisulfite treatment or reannealing during the bisulfite conversion, as the conversion only takes place within single stranded DNA. Thus repeated denaturation cycles, as stated in **step 5**, are part of the bisulfite conversion process. On the other hand overtreatment with bisulfite degrades DNA and may lead to an increased incidence of methylated cytosines converting to thymine residues, which results in under-reporting of DNA methylation (*see* **Note** 7).

3.2 Assay Design

The design of the assays is the most crucial step for a successful outcome of the methylation analysis. In our laboratory primers are designed using Agena Biosciences's Primer Design software as we usually have good experiences with this tool. But also other commercial and freely available software's can handle primer design on bisulfite-converted DNA such as MethPrimer.

The genomic sequence is pasted into the EpiDesigner, the PCR parameters adjusted or left as default (*see* recommended settings in Table 1). The optimal size range for PCR amplicons is 200–600 bp, depending on the quality of bisulfite-treated DNA. A good DNA quality allows successful amplification of sequences exceeding 600 bp, whereas highly degraded DNA will only yield stable results for shorter sequences. The strand can be selected (or both) as well as the reaction type (T-cleavage). The primers of the amplicons are designed to contain at least four non-CpG cytosines within the 1st Primer that will be replaced by Ts, or guanines within the 2nd Primer that will be replaced by adenines, as a consequence, these

Table 1
Settings for Primer design using the Agena Biosciences Primer Design software

Setting	Minimum	Optimum	Maximum
Primer T_m (°C)	56	62	64
Primer size (bp)	20	25	30
Product size (bp)	100	300	600
Primer non-CpG 'C's	4	–	–

primers will only bind to fully converted templates and should yield a defined, single band on an agarose gel. Also an assay name can be added and the run can be saved for a later time in "My Primers" (*see* **Note 8**).

The advantage of the EpiDesigner is that one can review primer locations and sequences in the results window. By moving the mouse pointer over the potential amplicons displayed in the results chart one can also verify which CpGs within the sequence will be covered by each corresponding amplicon and which ones will be not (e.g., due to low or high mass). If the detection of a certain CpG site is required (e.g., to validate the methylation difference of a CpG detected with the Illumina 450K) the amplicon pairs can be selected accordingly to certainly cover the desired CpG (*see* **Note 9**).

Also, the T7 promoter tag including an 8 bp insert ("cagtaatacgactcactatagggagaaggct") and the 10mer tag ("aggaagagag") are added automatically by the software upon export of the designed PCR primers (*see* **Note 10**). The T7-promoter tagged reverse primer is incorporated into the amplification product for in vitro transcription during PCR amplification and the 10mer-tag sequence to the forward primer is added to balance the PCR primer length.

3.3 Polymerase Chain Reaction (PCR)

3.3.1 Optimizing Polymerase Chain Reaction (PCR)

To obtain a defined, single band on the agarose gel (without any additional nonspecific bands) and thus the ideal annealing temperature of the primers, PCR incubation conditions can be optimized by test PCR.

1. Prepare 1 μM PCR primer mixes in 1.5 mL tubes by adding 4 μL of both forward and reverse primers to 392 μL of nanopure water for each assay.
2. Prepare PCR cocktails for 20 reactions in 1.5 mL tubes according to Table 2 including the Primermix (the 20 reactions include 8 bisulfite-treated DNA samples for the different temperatures, 8 negative controls without adding bisulfite-treated DNA for the different temperatures and overhang).

Table 2
PCR cocktail for EpiTYPER methylation analysis

Reagent	Volume for single reaction in μL	Volume for test PCR in μL	Volume for 384-well plate[a] in μL
ddH_2O	1.42	28.4	654.34
10× PCR-buffer (containing 15 mM $MgCl_2$)	0.50	10	230.40
dNTP (25 mM each)	0.04	0.8	18.43
PCR enzyme (Agena biosciences, 5 U/μL)	0.04	0.8	18.43
Mastermix volume:	2.00	40	921.60
Primermix (1 μM) (T7Reverse & 10mer forward tags)	2.00	40	–
Bisulfite-treated DNA (10 ng/μL)	1.00	–	–
Total volume	5.00	–	–

[a]Volumes for a 384-well plate include 20% overhang to account for possible pipetting loss

3. Dispense 4 μL of the PCR—Primermix cocktail into the wells of a 384-well plate in a way that allows running one sample and one negative control each with the different annealing temperatures, seal and centrifuge the plate.
4. Dispense 1 μL of the bisulfite-converted DNA (10 ng/μL) to the PCR—Primermix cocktail (*see* **Note 11**), but leave out the negative controls and add 1 μL of nanopure water instead.
5. Run a Gradient-PCR (56–64 °C) to determine the optimum annealing temperature using the following program: 94 °C for 4 min, 45 cycles of 94 °C for 20 s, 56–64 °C for 30 s and 72 °C for 1 min. Terminate the program at 72 °C for 3 min and hold at 4 °C (*see* **Note 12**).
6. After the incubation step, run all samples on a 3% agarose gel.
7. Choose the annealing temperature that shows the best result with a defined, single band on the gel. If primers do not show good PCR products, it is necessary to design new assays and test again (*see* **Note 13**).

3.3.2 Preparing Polymerase Chain Reaction (PCR)

1. Design a Plate Layout to simplify the pipetting of the PCR plate. In addition to the bisulfite-converted DNA samples, 0%, 25%, 50%, 75%, and 100% methylation standards should be included within each run as an indicator of the intra-run variation in the degree of bisulfite-related bias (if there are not enough free wells available on the plate to include all methylation standards at least the 25% and 75% ones should be

a Amplicons

	1	2	3	4	5	6	7	8	9	10	11	12	13	14	15
A	1	1	1	2	2	2	3	3	3	4	4	4	5	5	5
B	1	1	1	2	2	2	3	3	3	4	4	4	5	5	5
C	1	1	1	2	2	2	3	3	3	4	4	4	5	5	5
D	1	1	1	2	2	2	3	3	3	4	4	4	5	5	5
E	1	1	1	2	2	2	3	3	3	4	4	4	5	5	5
F	1	1	1	2	2	2	3	3	3	4	4	4	5	5	5
G	1	1	1	2	2	2	3	3	3	4	4	4	5	5	5
H	1	1	1	2	2	2	3	3	3	4	4	4	5	5	5
I	1	1	1	2	2	2	3	3	3	4	4	4	5	5	5
J	1	1	1	2	2	2	3	3	3	4	4	4	5	5	5
K	1	1	1	2	2	2	3	3	3	4	4	4	5	5	5
L	1	1	1	2	2	2	3	3	3	4	4	4	5	5	5
M	1	1	1	2	2	2	3	3	3	4	4	4	5	5	5
N	1	1	1	2	2	2	3	3	3	4	4	4	5	5	5
O	1	1	1	2	2	2	3	3	3	4	4	4	5	5	5
P	1	1	1	2	2	2	3	3	3	4	4	4	5	5	5

Amplicon 1: cg05575921 Amplicon 2: cg21566642 Amplicon 3: cg03636183 Amplicon 4: cg06126421 Amplicon 5: cg19572487

b Samples

	1	2	3	4	5	6	7	8	9	10	11	12	13	14	15
A	1	17	33	1	17	33	1	17	33	1	17	33	1	17	33
B	2	18	34	2	18	34	2	18	34	2	18	34	2	18	34
C	3	19	35	3	19	35	3	19	35	3	19	35	3	19	35
D	4	20	36	4	20	36	4	20	36	4	20	36	4	20	36
E	5	21	37	5	21	37	5	21	37	5	21	37	5	21	37
F	6	22	38	6	22	38	6	22	38	6	22	38	6	22	38
G	7	23	39	7	23	39	7	23	39	7	23	39	7	23	39
H	8	24	40	8	24	40	8	24	40	8	24	40	8	24	40
I	9	25	41	9	25	41	9	25	41	9	25	41	9	25	41
J	10	26	42	10	26	42	10	26	42	10	26	42	10	26	42
K	11	27	0%	11	27	0%	11	27	0%	11	27	0%	11	27	0%
L	12	28	25%	12	28	25%	12	28	25%	12	28	25%	12	28	25%
M	13	29	50%	13	29	50%	13	29	50%	13	29	50%	13	29	50%
N	14	30	75%	14	30	75%	14	30	75%	14	30	75%	14	30	75%
O	15	31	100%	15	31	100%	15	31	100%	15	31	100%	15	31	100%
P	16	32	NTC	16	32	NTC	16	32	NTC	16	32	NTC	16	32	NTC

0%: unmethylated DNA; 25 - 100%: methylated DNA; NTC: Negative control

Fig. 3 Example of a (**a**) amplicon layout and (**b**) sample layout for the preparation of the PCR plate. Only the first 15 columns are shown. The amplicon layout shows where the amplicons analyzed within the plate (1–5) should be located and the sample layout shows the location of the corresponding samples including the bisulfite-converted DNA samples (1–42), methylation standards (0–100%) and negative controls (NTC)

included). The DNA methylation standards are only included to show if the experiment worked in principle and if the trend given for each CpG site is correct. They should not be used as a standard curve for adjustment of the methylation data. It is important to also include negative controls (NTC's, by adding nanopure water instead of bisulfite-treated DNA) to verify the occurrence of PCR contamination and formation of primer dimers (Fig. 3) (*see* **Note 14**).

2. Prepare PCR cocktails as displayed in Table 2 without adding the bisulfite-treated DNA.
3. Dispense 2 μL of the PCR cocktail into the wells of a 384-well plate; seal and centrifuge the plate (*see* **Note 15**).
4. Dispense 2 μL of the PCR primer mixes prepared in Subheading 3.3.1 of this chapter to each well as per the layout in Fig. 3a, seal and centrifuge the plate.
5. Dispense 1 μL of the bisulfite-converted DNA (10 ng/μL) to each well as per layout in Fig. 3b.

6. Properly seal the plate with plate sealing film, vortex and centrifuge at 540 × *g* for 1 min.
7. Incubate the 384-well plate on a thermal cycler using the following program: 94 °C for 4 min, 45 cycles of 94 °C for 20 s, 57 °C* for 30 s and 72 °C for 1 min. Terminate the program at 72 °C for 3 min and hold at 4 °C (*Annealing temperature adjusted according to test PCR).
8. After the incubation step, run all 384 PCR products on a 3% agarose gel. As primers will only bind to fully converted templates a defined, single band on the agarose gel for each PCR product suggests that the corresponding sample is fully converted (*see* **Note 7**).

3.4 SAP Reaction

To reduce the presence of by-products, the amplification products have to be treated with shrimp alkaline phosphatase (SAP), which cleaves phosphates of unincorporated dNTPs, rendering them unavailable for future polymerase reactions.

1. Prepare the SAP enzyme solution in a 1.5 mL tube according to Table 3.
2. Thoroughly mix the SAP enzyme solution by vortexing and centrifuge.
3. Centrifuge the 384-well PCR plate after completion of the PCR.
4. Add 2 μL SAP enzyme solution to each well of the 384-well PCR plate using a 12-channel pipettor or liquid handling robot. The SAP enzyme solution is moderately viscous, so this has to be done with care to minimize loss of solution due to adhesion to the pipette tips.
5. Properly seal the 384-well PCR plate with plate sealing film, vortex and centrifuge at 3000 × *g* for 1 min.
6. Incubate the plate on a thermal cycler at 37 °C for 20 min and inactivate the enzyme at 85 °C for 5 min, hold at 4 °C.

Table 3
SAP enzyme solution

Reagent	Volume for single reaction in μL	Volume for 384-well plate[a] in μL
RNase-free ddH_2O	1.70	783.36
Shrimp alkaline phosphatase (SAP)	0.30	138.24
Total volume	2.00	921.60

[a]Volumes for a 384-well microtiter plate include 20% overhang to account for possible pipetting loss (*see* **Note 16**)

Table 4
T-cleavage transcription/RNase cocktail

Reagent	Volume for single reaction in μL	Volume for 384-well plate[a] in μL
RNase-free ddH_2O	3.21	1479.16
5× T7-polymerase buffer	0.89	410.11
T-cleavage mix	0.22	101.38
DTT, 100 mM	0.22	101.38
T7-RNA & DNA polymerase	0.40	184.32
RNase A (10 mg/mL)	0.06	27.65
Total volume	5.00	2304.00

[a]Volumes for a 384-well microtiter plate include 20% overhang to account for possible pipetting loss (*see* **Note 16**)

3.5 MassCLEAVE Reaction

In this step, in vitro RNA transcription is performed on the reverse strand followed by base-specific T-cleavage using RNase A (*see* **Note 17**). This yields to fragmented RNA molecules that can be analyzed by MALDI-TOF MS.

1. Prepare the T-cleavage transcription/RNase A cocktail according to Table 4.
2. After SAP treatment, transfer 2 μL of each PCR product of the 384-well PCR/SAP plate "one to one" into a new 384-well T-cleavage plate using a 12-channel pipettor or liquid handling robot. Leftovers of the PCR/SAP products can be stored at −20 °C or discarded.
3. Add 5 μL of the T-cleavage transcription/RNase cocktail per well.
4. Properly seal the plate with plate sealing film, vortex and centrifuge at 540 × *g* for 1 min.
5. Incubate the plate on a thermal cycler at 37 °C for 3 h.
6. Upon completion, condition the sample plate immediately (next step) or store it at −20 °C. Do not leave the plate on 4 °C or room temperature.

3.6 Conditioning

After the MassCLEAVE reaction a cation exchange resin (Clean Resin) has to be added to the products to remove residual salt that could interfere with mass spectrometry analysis.

1. Spread out 6 mg/well Clean Resin on a 384-well Clean Resin plate (also called dimple plate) with the help of the spoon and scraper.

2. Scrape excess Clean Resin off the plate and return it to its container.
3. Allow the 384-well Clean Resin plate to stand for about 15–25 min (*see* **Note 18**).
4. Meanwhile, centrifuge the 384-well T-cleavage plate at 540 × *g* for 1 min after completion of the MassCLEAVE reaction.
5. Dilute the samples with 20 μL of nanopure water, best prepared in a reservoir, using a 12-channel pipettor.
6. Seal the plate and centrifuge at 540 × *g* for 1 min.
7. To add the Clean Resin to the products gently place the 384-well T-cleavage plate upside-down onto the Clean Resin plate directly after the centrifugation step.
8. Hold the two plates together and flip them over so that the Clean Resin plate is on top and the Clean Resin falls into the T-cleavage plate.
9. Make sure all the Clean Resin fell into the T-cleavage plate and tap the Clean Resin plate to help the Clean Resin fall into the T-cleavage plate if necessary.
10. Properly seal the T-cleavage plate and rotate it 360° perpendicular to its long axis for about 30–60 min at room temperature using a rotator at lowest speed.
11. Centrifuge the plate at 3200 × *g* for 5 min.

The reaction products are now ready for transfer to a SpectroCHIP® (*see* **Note 19**).

3.7 Nanodispensing

The reaction products need to be transferred from the 384-well plate format to a 384-element chip array (SpectroCHIP®) using a Nanodispenser.

1. Perform a volume check on the Nanodispenser to determine, which dispense speed is required to spot an optimal product volume of 15–25 nL. This step is necessary, since the volume deposited on the chip is affected by the low viscosity of the reactions.
2. Dispense the 4-point calibrant, a mix of four oligonucleotides with known masses, to the SpectroCHIP®.
3. Dispense the reaction products from the T-cleavage plate to the SpectroCHIP®.

3.8 MALDI-TOF MS Measurements

MALDI-TOF MS analyzes the mixture of cleavage products differing in length and mass. A distinct signal pair pattern results from the mass signals representing methylated and nonmethylated template DNA, which is representative for the CpGs within the analyzed sequence substring.

1. Set up the plate in the "Plate Editor" of the EpiTYPER software by defining the three constituents (T-cleavage reaction, amplicons and samples) for each well according to the Plate Layout designed for pipetting of the PCR plate (*see* Subheading 3.3.2 of this chapter, Fig. 3) and export it to the RT Workstation (check the "Duplicate Wells Report" under "View" before the export as plates with incomplete or duplicate wells cannot be exported. Correct any wells if necessary and resave the experiment).
2. Load the SpectroCHIP® into a MassARRAY mass spectrometer.
3. Set the MassARRAY Workstation Compact to MassCLEAVE (MassCleave.par) for both FlexControl and SpectroAcquire.
4. Load the plate into the MassARRAY Workstation.
5. Before starting the run, control the 4-point calibrant manually on stage position "F0" (alternatively G0 or H0). The solution has four standardized spectra mass signals for calibrating the analysis system (in Da) at 1479.0, 3004.0, 5044.4, and 8486.6, which should not be shifted for more than 2–3 Da.
6. Run the SpectroCHIP®.

3.9 Data Analysis

The EpiTYPER software automatically calculates the amount of methylated DNA for each CpG position by the ratio of the intensities of the distinct signal pairs resulting from the mass signals representing methylated and nonmethylated template DNA. Quantitative results can be obtained for each sequence-defined CpG unit, which contains either one individual CpG site or an aggregate of subsequent CpG sites (CpG Unit). The resulting mass spectra, the nucleotide sequences that produced the spectrum and the amplicon-specific methylation data can be viewed in the EpiTYPER Analyzer program. The so-called EpiGram tab allows a graphical representation for better visualization then given by the result table alone (Fig. 4). The EpiTYPER methylation data can then be exported as desired and analyzed with various statistical packages. One example of how the EpiTYPER as a region-specific method can be used as a validation technique for single CpGs and the surrounding area detected by genome-wide methods (e.g., Illumina 450K) is given in Table 5.

3.9.1 Test for Complete Bisulfite Conversion and SNP Prediction Analysis

It is important to test for complete bisulfite conversion, as incomplete conversion of cytosine to uracil can result in false-positive methylation signals. The R package "MassArray" offers a bisulfite conversion efficiency calculation by measuring levels of unconverted non-CpG cytosines in a given sample [8]. As cytosine methylation in mammals usually occurs in the context of CpG, unmethylated cytosines outside this context can be used to measure

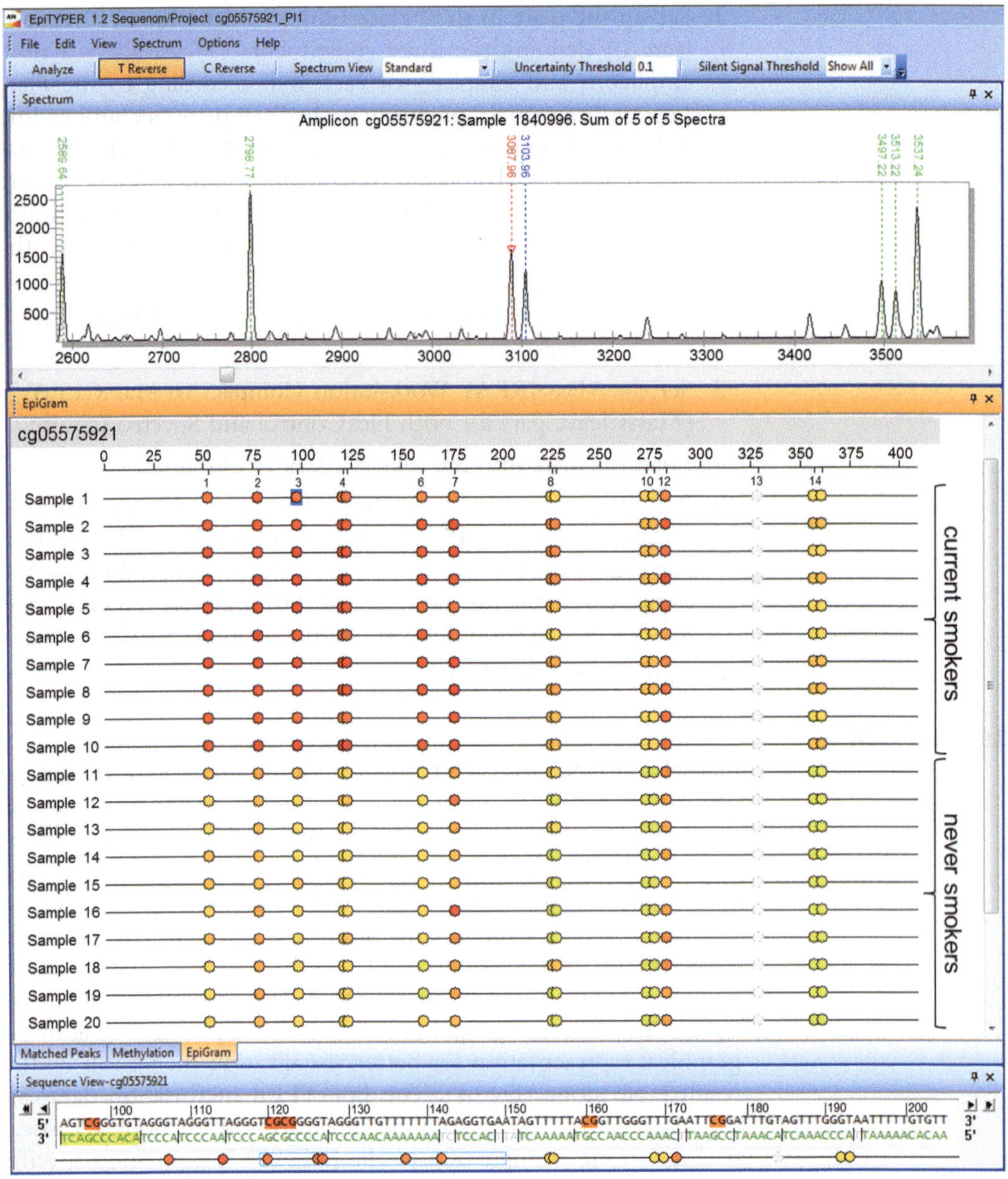

Fig. 4 EpiTYPER methylation data of genomic DNA samples from human whole blood for the smoking-associated candidate gene *AHRR*. The figure shows the varying DNA methylation levels of 10 current compared to 10 never smokers in a 375 bp long amplicon located within the *AHRR* gene as it can be viewed in the EpiGram tab of the EpiTYPER™ Analyzer program. The EpiGram tab is a graphical representation of methylation ratio, each circle displays a CpG site and the color code within the circles denotes the level of methylation found at this particular site in the selected sample. The default color spectrum ranges from *red* (0% methylated) to *yellow* (100% methylated). Above the set of samples the nucleotide position and the CpG site number are given. The amplicon includes an Illumina 450K site (cg05575921; CpG site number 3 in the EpiGram) repeatedly identified to be differentially methylated between never and current smokers [7]. The Spectrum panel displays the mass spectrum for all analyzed amplicons and the Sequence View panel shows the forward and reverse nucleotide sequence for the selected amplicon. This screenshot was made from the EpiTYPER™ Analyzer program of the EpiTYPER Version 1.2 software (Agena Biosciences)

Table 5
Technical validation of Illumina 450K BeadChip methylation intensities by the EpiTYPER assay

Gene CpG	450K: Median ß-value methylation difference	Median ß-value methylation difference in %	*p*-value	Pearson r^2	ß-value as median (first quartile–third quartile) Never smokers	Current smokers
AHRR						
CpG1*	CpG3	−42.50	8.52E−07		0.820 (0.798–0.823)	0.395 (0.295–0.565)
CpG2*	cg05575921:	−24.50	3.53E−04		0.640 (0.610–0.710)	0.395 (0.308–0.488)
CpG3*	−24.40%	−38.00	3.46E−07	0.976456	0.820 (0.800–0.833)	0.440 (0.305–0.560)
CpG4.5*		−40.00	6.11E−06		0.835 (0.783–0.850)	0.435 (0.305–0.555)
CpG6*		−40.00	2.41E−08		0.885 (0.878–0.893)	0.485 (0.388–0.613)
CpG7		−26.50	4.11E−02		0.650 (0.420–0.713)	0.385 (0.260–0.485)
CpG8.9		−12.50	1.63E−02		0.935 (0.855–0.960)	0.810 (0.675–0.900)
CpG10.11*		−8.00	2.05E−04		0.930 (0.920–0.940)	0.850 (0.808–0.870)
CpG12		−5.50	1.24E−01		0.615 (0.565–0.670)	0.560 (0.443–0.608)
CpG14.15		−7.00	6.71E−03		0.930 (0.920–0.940)	0.860 (0.815–0.890)

Displayed is the median ß-value methylation difference between 41 current and 41 never smokers measured by the EpiTYPER, and the corresponding result of the 450K analysis based on 749 never and 262 current smokers. The calculation was done with a linear model adjusted for age, sex, BMI, alcohol, and white blood cell count. For the EpiTYPER validation, the correlation coefficient (Pearson r^2) between the Illumina array derived *b*-value and the EpiTYPER methylation is reported for the associated CpG site cg05575921. *Significant after Bonferroni $p \leq 0.05/28 = 0.0018$ [7]

bisulfite treatment efficiency, acting as "conversion controls". It is essential to measure the extent of bisulfite conversion, because even though the primers are designed to enrich for completely converted sequences (they contain at least 4 non-CpG "C"s), some proportion of amplicons may contain remnant unconverted cytosines, which may cause a molecular weight shift in the subsequent fragmentation profile. The disadvantage of this approach is that it can only be implemented after the EpiTYPER methylation analysis has been carried out, as an output file of the methylation data for each amplicon has to be generated in order to run the pipeline.

Also SNPs within the target sequence of an amplicon could have an effect on the DNA methylation analysis, whether in the form of a fragment mass shift or a different fragmentation pattern, which can lead to misinterpretation of the results [9]. Each novel peak among the MassArray spectrum can be explained by any number of potential SNPs. By using an exhaustive string substitution approach as implemented in the R package "MassArray" [8], putative SNPs can be identified by comparing expected and observed data (*see* **Note 20**).

4 Notes

1. Since the 96-well plate format compared to the tube format is much easier to handle in the lab and therefore less prone to handling errors as well as more cost effective, it is

recommended to favor the plate format when exceeding a certain sample size.

2. It is strongly recommended to convert enough DNA to be able to repeat the PCR step or the entire procedure for all amplicons if necessary (e.g., calculate for three times as many reactions as originally planned). Leftover bisulfite-treated DNA can be stored at −20 °C/−80 °C for a couple of months, depending on the manufacturer of the kit, and used as a test DNA to optimize amplicon annealing temperatures for other EpiTYPER projects (*see* **Note 11**).
3. As the CT Conversion reagent is light sensitive, the exposure to light has to be minimized by darkening the lab room. For best results the prepared CT Conversion Reagent should be used immediately after preparation for all experiments. Alternatively, the CT Conversion Reagent solution can be stored overnight at room temperature, 1 week at 4 °C or up to 1 month at −20 °C. Stored CT Conversion Reagent solution must be warmed to 37 °C, then vortexed well prior to use.
4. The following is relevant for all centrifugation steps within the cleanup of the bisulfite-treated DNA: After each step all wells have to be completely dry, but in our lab usually a few are not even after an extended centrifugation time, which can be seen by swaying the plate under the bench light. This problem can be solved by rotating the plate 180° in between the centrifugation step (thus 3 + 3 min), probably due to a slight change of the centrifugal force. This might not be necessary for all users, if all wells are dry after the usual 5 min centrifugation step.
5. The volume of nanopure water used for the elution depends on the amount of DNA added. If 500 ng DNA is added, the elution step can be carried out with 40 μL of nanopure water as the following PCR step should be carried out with at least 10 ng of bisulfite-treated DNA and a loss of about 20% of the input DNA has to be expected during the bisulfite conversion process. The elution volume should not be too small to assure sufficient recovery of the bisulfite-treated DNA. If the amount of input DNA is small, it might be therefore necessary to add more than 1 μL of the bisulfite-treated DNA to the PCR-Mix to gain the 10 ng needed for the PCR reaction.
6. It is best to use the bisulfite-treated DNA fresh and to avoid freezing/thawing. If it is not possible to do all PCRs within a week (the period within the bisulfite-treated DNA can be kept at 4 °C) it is recommended to make aliquots and only thaw them once.
7. Some working groups carry out a test PCR with one primer pair specific for bisulfite-converted DNA that covers a short sequence that contains many non-CpG cytosines, which are

converted to thymines (Ts) during the bisulfite treatment. But unfortunately, even though resulting DNA bands are as expected and seem to confirm a complete bisulfite conversion, we observed that incomplete conversion can still be detected within the Bisulfite Treatment Control Probes of the Illumina 450K BeadChip in some cases. For that reason we think that a single test PCR covering only a small region of the genome is not really trustworthy, as even though the test PCR shows good results, other parts of the genome than the one covered by the test primers, might still lack complete conversion. We therefore check completion of the bisulfite treatment for each sample within the actual amplicons designed for the corresponding project (Subheading 3.2 of this chapter) after PCR by running an agarose gel with the PCR products (*see* Subheading 3.3.2 of this chapter). Also, the bisulfite conversion efficiency can be tested with the generated data of the MassArray platform by using the corresponding analytical tool of the MassArray package as addressed in Subheading 3.9.1 of this chapter.

8. For primers saved in "My Primers": Because of a bug in the EpiDesigner software the "detail file" that has to be exported for the EpiTYPER process is incorrect concerning the target sequences with an "R"-letter in the "Direction" column. These sequences should be displayed as reverse complement but are incorrectly displayed as forward sequences. Therefore it is important to change these target sequences from forward to reverse complement in order to correct the file. For this purpose the freely available web-based program "Reverse Complement" (http://www.bioinformatics.org/sms/rev_comp.html) can be used for example.

9. Nevertheless, with the EpiDesigner it is not possible to see if the covered CpGs within the desired amplicon will be analyzed as single or in units and if for example silent peaks will overlap the methylated peak. It is possible to reutilize a SpectroChip in order to carry out a "Ghost Run" in the laboratory and get more information about the expected mass spectrum. In principle, an already used SpectroCHIP® has to be loaded into the MassARRAY mass spectrometer. A plate has to be set up by defining the three constituents (T-cleavage reaction, the desired amplicons and any samples) in the "Plate Editor", exported to the RT Workstation and run as described in Subheading 3.8. Also, it can be tricky to figure out which CpG site within the amplicon is the actual desired one. Keep in mind that you look at a forward or revers sequence depending on the direction of the sequence loaded into the EpiDesigner and the direction of the selected amplicon.

10. If alternative software's such as MethPrimer are used for the design of the primers, it is important to keep in mind that the T7 promoter tag ("cagtaatacgactcactataggg") including an 8 bp insert ("agaaggct") for reducing abortive cycling during transcription (5′–cagtaatacgactcactatagggagaaggct + gene-specific sequence–3′) and the 10mer tag (5′–aggaagagag + gene-specific sequence–3′) have to be added manually.
11. Since it is best to use fresh bisulfite-treated DNA for the actual later experiments, stored bisulfite-treated DNA of former experiments can be used for this test process (*see* **Note 2**).
12. If no thermal cycler with temperature gradient functionality is available, the primers can also be tested with selected annealing temperatures. In our hands most primers have an optimum annealing temperature of 57–60 °C, so one test run with 57 °C and another one with 60 °C might already give enough information on which annealing temperature might be best to yield good PCR products.
13. For time-critical projects we recommend to order two different sets of primers for each amplicon and run both with the test PCR. Usually at least one of them works properly, which may avoid the need of redesigning, ordering and testing new primers and therefore save time.
14. If only a small number is analyzed it is recommended to run assays in duplicate or even triplicate for all samples and to take the mean of methylation value for each sample/CpG site. It is also possible to pool genomic DNA [17] in order to use less amount of DNA and minimize costs and then to only run differentially methylated amplicons with single samples in a second run.
15. Depending on the number of amplicons and samples within one plate, the pipetting process can be quite time-consuming. The preparation of the 384-well plate for PCR should then be done on ice.
16. If a liquid handling robot is used it might be necessary to increase the excess volume to over 30%.
17. The C-specific cleavage reaction theoretically provides specific results for selected CpGs but is to our knowledge practically not used by most working groups. Its usability is limited by low CpG coverage and at least within our projects we could not find a use-case for it that would have brought additional value in comparison to the T-cleavage reaction. Moreover, Agena Biosciences has discontinued this kit recently.
18. Do not let the Clean Resin overdry, only wait as long as necessary to be able to transfer it into the T-cleavage plate (*see* Subheading 3.6, **steps 8** and **9**). The color should not

appear too light (yellowish) and the Clean Resin should remain as soggy as possible. The time therefore depends on the relative room temperature and humidity.

19. If the reaction products cannot be immediately transferred to a SpectroCHIP®, the conditioned T-cleavage plate can be tightly sealed by an adhesive sealing foil and stored at −20 °C for up to 2 weeks. If so, thaw and centrifuge the plate at 540 × *g* for 3 min before the transfer.
20. It is of note that these analyses cannot directly confirm or deny the presence or absence of a SNP. However, if samples show a putative high-confidence SNP that maps to a fragment containing one or more CpGs, methylation data from that site should be interpreted with caution.

References

1. Umer M, Herceg Z (2013) Deciphering the epigenetic code: an overview of DNA methylation analysis methods. Antioxid Redox Signal 18:1972–1986
2. Ehrich M, Nelson MR, Stanssens P et al (2005) Quantitative high-throughput analysis of DNA methylation patterns by base-specific cleavage and mass spectrometry. Proc Natl Acad Sci U S A 102:15785–15790
3. Karas M, Hillenkamp F (1988) Laser desorption ionization of proteins with molecular masses exceeding 10,000 daltons. Anal Chem 60:2299–2301
4. Kirpekar F, Nordhoff E, Larsen LK et al (1998) DNA sequence analysis by MALDI mass spectrometry. Nucleic Acids Res 26:2554–2559
5. Hillenkamp F, Karas M, Beavis RC et al (1991) Matrix-assisted laser desorption/ionization mass spectrometry of biopolymers. Anal Chem 63:1193A–1203A
6. Griffin TJ, Smith LM (2000) Single-nucleotide polymorphism analysis by MALDI-TOF mass spectrometry. Trends Biotechnol 18:77–84
7. Zeilinger S, Kuhnel B, Klopp N et al (2013) Tobacco smoking leads to extensive genome-wide changes in DNA methylation. PLoS One 8:e63812
8. Thompson RF, Suzuki M, Lau KW et al (2009) A pipeline for the quantitative analysis of CG dinucleotide methylation using mass spectrometry. Bioinformatics 25:2164–2170
9. Coolen MW, Statham AL, Gardiner-Garden M et al (2007) Genomic profiling of CpG methylation and allelic specificity using quantitative high-throughput mass spectrometry: critical evaluation and improvements. Nucleic Acids Res 35:e119
10. Lim SP, Wong NC, Suetani RJ et al (2012) Specific-site methylation of tumour suppressor ANKRD11 in breast cancer. Eur J Cancer 48:3300–3309
11. Vanaja DK, Ehrich M, Van den Boom D et al (2009) Hypermethylation of genes for diagnosis and risk stratification of prostate cancer. Cancer Invest 27:549–560
12. Breitling LP, Salzmann K, Rothenbacher D et al (2012) Smoking, F2RL3 methylation, and prognosis in stable coronary heart disease. Eur Heart J 33:2841–2848
13. Christensen BC, Kelsey KT, Zheng S et al (2010) Breast cancer DNA methylation profiles are associated with tumor size and alcohol and folate intake. PLoS Genet 6:e1001043
14. Tost J, Gut IG (2005) Genotyping single nucleotide polymorphisms by MALDI mass spectrometry in clinical applications. Clin Biochem 38:335–350
15. Werner M, Sych M, Herbon N et al (2002) Large-scale determination of SNP allele frequencies in DNA pools using MALDI-TOF mass spectrometry. Hum Mutat 20:57–64
16. van den Boom D, Ehrich M (2009) Mass spectrometric analysis of cytosine methylation by base-specific cleavage and primer extension methods. Methods Mol Biol 507:207–227
17. Docherty SJ, Davis OS, Haworth CM et al (2010) DNA methylation profiling using bisulfite-based epityping of pooled genomic DNA. Methods 52:255–258

Chapter 27

Methylation-Specific Multiplex Ligation-Dependent Probe Amplification (MS-MLPA)

Cathy B. Moelans, Lilit Atanesyan, Suvi P. Savola, and Paul J. van Diest

Abstract

This chapter describes a method for the rapid assessment of promoter hypermethylation levels or methylation of imprinted regions in human genomic DNA extracted from various sources using methylation-specific multiplex ligation-dependent probe amplification (MS-MLPA).

Multiplex ligation-dependent probe amplification (MLPA) is a powerful and easy-to-perform PCR-based technique that can identify gains, amplifications, losses, deletions, methylation and mutations of up to 55 targets in a single reaction, while requiring only minute quantities of DNA (about 50 ng) extracted from blood, fresh frozen or formalin-fixed paraffin-embedded materials. Methylation-specific MLPA (MS-MLPA) is a variant of MLPA, which does not require sodium bisulfite conversion of unmethylated cytosine residues, but instead makes use of the methylation-sensitive endonuclease HhaI. MS-MLPA probes are designed to contain a HhaI recognition site (GCGC) and thus target one CpG dinucleotide within a CpG island. If the HhaI recognition site is not methylated, HhaI will cut the probe–sample DNA hybrid and no PCR product will be formed. If the target DNA is methylated, HhaI is not able to cut, and the fragment will be amplified during subsequent PCR. For data analysis, MS-MLPA peak patterns of the HhaI-treated and -untreated reactions are compared, leading to calculation of a methylation percentage. The methylation profile of a test sample is assessed by comparing the probe methylation percentages obtained on the test sample to the percentages of the reference samples. MS-MLPA can be combined with copy number and point mutation detection in the same reaction.

Key words MS-MLPA, MLPA, Methylation, Genomic DNA, Methylation-sensitive restriction enzymes paraffin, FFPE, Coffalyser.Net

1 Introduction

The MLPA technique was first described in 2002 by Schouten et al. [1] and an overview is shown in Fig. 1. Up to 55 probe targets can be analyzed in one reaction. MLPA reactions are easy to perform and require short hands-on time. Up to 96 samples can be handled simultaneously and results can be obtained within 24 h. MLPA has been used to assess changes in gene copy number [2–4], gene expression [5, 6], point mutation status [7–9] and DNA methylation [10–13]. Due to the short lengths of the target sequences of

Jörg Tost (ed.), *DNA Methylation Protocols*, Methods in Molecular Biology, vol. 1708,
https://doi.org/10.1007/978-1-4939-7481-8_27,

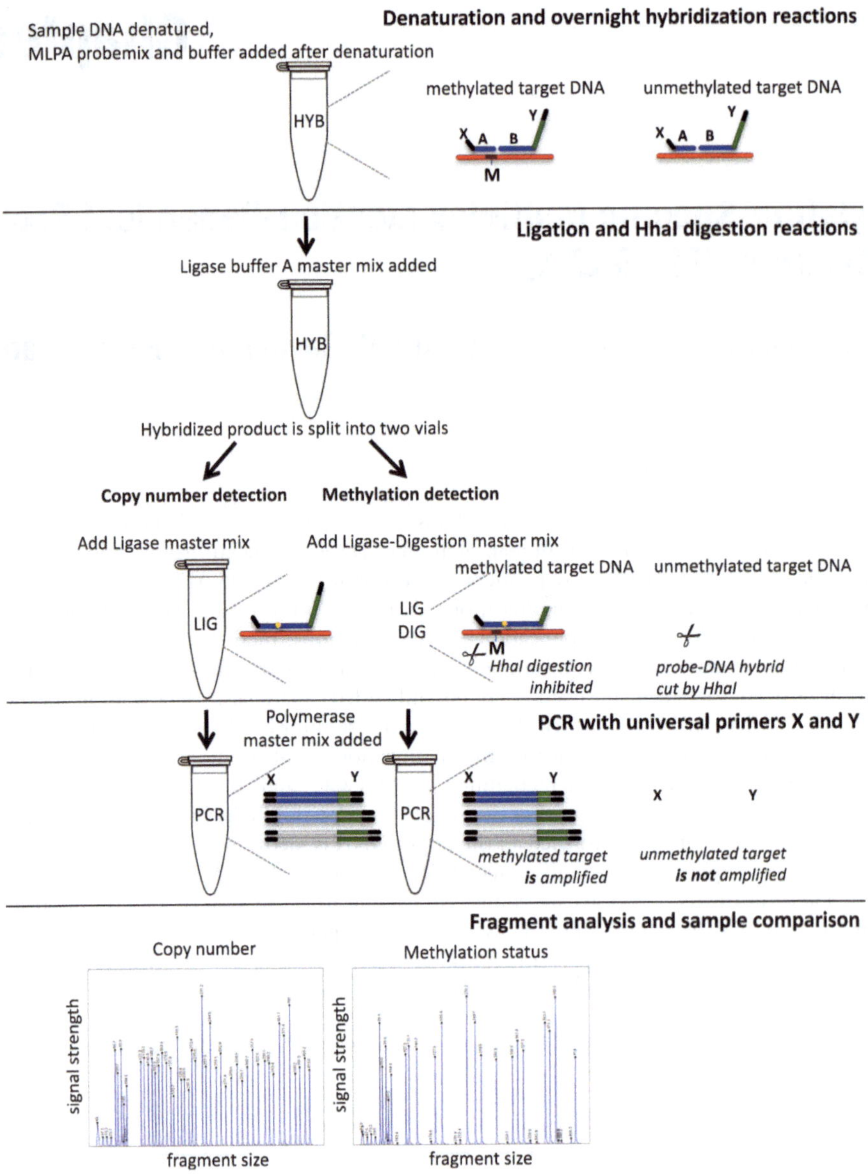

Fig. 1 Principle of methylation-specific multiplex ligation-dependent probe amplification (MS-MLPA). MLPA uses a mixture of hemiprobe sets that consist of two oligonucleotides, both having PCR primer sequences (X/Y) on the outer ends, while both on the inner ends having a sequence complementary to a part of the target sequence (A or B). One of the hemiprobes has a stuffer (in *green*) of variable length in between the PCR primer sequence and the complementary target sequence. When the complementary target sequences of both hemiprobes hybridize adjacent to each other on the target sequence, they can be ligated to each other and subsequently amplified using the PCR primer binding sequences. However, methylation-specific MLPA probes are designed to target DNA sequences which contain a restriction site for the methylation-sensitive endonuclease HhaI (indicated by "M"). After probe hybridization, the MS-MLPA reaction is split into two parts. One part of the MS-MLPA reaction is processed as a normal MLPA reaction (LIG), providing information on copy number status of the target DNA. The other part is also treated with the HhaI endonuclease (LIG, DIG), which provides information on methylation status of the target DNA. When hybridizing to an unmethylated DNA target, the methylation-specific probes will be ligated and simultaneously digested by HhaI. A digested

hemiprobes, MLPA can not only be applied to DNA isolated from fresh-frozen materials, but is also suitable for the more fragmented DNA from formalin-fixed, paraffin-embedded (FFPE) materials. Depending on the quality of the DNA, 50–200 ng of DNA suffice. The ability to carry out a multiplex copy number and methylation assessment on small amounts of FFPE materials makes MLPA a very attractive method in pathology.

A variant of classic MLPA called "methylation-specific MLPA (MS-MLPA)" allows semiquantitative detection of the methylation status of genes and their copy number simultaneously. The key component in MS-MLPA is the methylation-sensitive endonuclease HhaI. Technically, MS-MLPA is similar to MLPA, with the exception that the samples are divided into two reactions after the ligation step (*see* Fig. 1). One of the reactions is carried out in the presence of HhaI and a ligase, the other reaction is carried out with the ligase only. If the HhaI recognition site is not methylated, HhaI will cut the probe–sample DNA hybrid, and no PCR product will be formed. If the sample DNA is methylated, HhaI digestion is prevented and the fragment will be amplified in the subsequent PCR. For data analysis, MLPA peak patterns of the HhaI-treated and -untreated reactions are compared, providing an estimate of the methylation percentage within a given sample. The disadvantage of using the HhaI endonuclease is that the design of MS-MLPA probes is confined to its recognition sites (GCGC). The major advantage of MS-MLPA compared with other methods for methylation detection such as methylation-specific PCR (MSP, *see* also Chapters 23–25), is that it does not require sodium bisulfite conversion of unmethylated cytosine residues, a step that is often difficult to standardize and leads to degradation of the DNA. In addition, MS-MLPA can analyze up to 55 probes simultaneously and it can be combined with copy number and point mutation detection in the same reaction [7, 14]. Previous studies have shown a good correlation between MS-MLPA, (quantitative multiplex) MSP and bisulfite (pyro)sequencing [11, 15–18].

Obviously, MS-MLPA is a nonmorphological method that requires proper evaluation by histopathology on the input material. MS-MLPA can be used in samples with lower tumor percentages, as long as 30% of methylated DNA / tumor DNA is present in the sample, the methylation status will be recognized correctly [14]. In cases with a lower percentage of relevant materials, macrodissection or laser microdissection may be necessary [19, 20].

Fig. 1 (continued) MS-MLPA probe will therefore not generate PCR product and subsequently no peak signal will be detected. In contrast, when the target sequence of the MS-MLPA probe is methylated, the methyl group will prevent HhaI-digestion. An undigested, ligated probe can thus be amplified during PCR, resulting in a peak signal. Comparing the electrophoresis patterns of the undigested MS-MLPA reactions allows for the detection of copy number changes. Comparing the peak patterns of digested reactions may reveal methylation of DNA target sequences

2 Materials

2.1 Sample Preparation

1. Proteinase K (10 mg/mL): dissolve 1 g Proteinase K in 100 mL distilled water and store 1 mL aliquots at −20 °C.
2. Lysis buffer: 50 mM Tris–HCl buffer, pH 8.0, 0.5% Tween 20.

   Dissolve 0.61 g Tris in 80 mL molecular biology grade water, add 20 mL molecular biology grade water and adjust pH to 8.0 with HCl (37%). Add Tween 20 until final concentration of 0.5%.
3. Xylene.
4. Ethanol: 100, 85 and 70% EtOH.

2.2 MS-MLPA

1. MS-MLPA probemix and reagents (*see* MRC-Holland website www.mlpa.com) (*see* **Note 1**).
2. HhaI endonuclease (e.g., Promega 10 units/μL).
3. TE buffer: 10 mM Tris–HCl, 1 mM EDTA, pH 8.2.

   Dissolve 1.21 g Tris and 0.37 g Sodium-EDTA in 900 mL molecular biology grade water, adjust the pH to 8.2 with HCl and add up to 1 L with molecular biology grade water.
4. Thermocycler with heated lid and capillary electrophoresis equipment.
5. Labeled size standard (ROX-500/LIZ-500) (*see* **Note 2**).
6. Deionized formamide (*see* **Note 3**).
7. Performance optimized polymer (POP, Thermo Fisher Scientific) (*see* **Note 2**).
8. Fragment analysis software (Genescan, Thermo Fisher Scientific).
9. Coffalyser.Net MLPA analysis software (freeware at http://coffalyser.wordpress.com/)

3 Methods

The preparation of DNA samples, and the optimum reaction conditions for MS-MLPA are described. A set of guidelines specific for software analysis of MS-MLPA assays is included.

3.1 DNA Extraction from Formalin-Fixed Paraffin-Embedded Material

1. Cut 3–4 μm sections to perform a routine H&E staining and mark the appropriate tumor area to be macrodissected.
2. For DNA extraction cut 8–10 μm thick sections, mount on sequential slides, deparaffinize in xylene for 10 min and rehydrate through graded ethanol (100, 85 and 70% for 1 min each) (*see* **Notes 4–6**).

3. Scrape off the appropriate tumor area using a scalpel.
4. Add 50 μL lysis buffer to the tube containing the macrodissected material (*see* **Note 7**).
5. Add 10 μL Proteinase K (10 mg/mL) and incubate at 56 °C for at least 1 h, then heat inactivate at 80 °C for 10–15 min. Cool the lysate immediately on ice (*see* **Note 8**).
6. Centrifuge 2 min at room temperature at full speed and carefully pipet the DNA-containing supernatant into a clean tube (*see* **Note 9**).

3.2 MS-MLPA Assay

The thermocycling programs used for this subheading are summarized in Table 1.

1. Dilute the DNA sample (optimal range 50–100 ng) with TE and add 5 μL to 0.2 mL PCR tubes. Process appropriate positive/negative controls along with your samples (*see* **Notes 10–15**).
2. Heat 5 min at 98 °C and cool to 25 °C before opening the thermocycler (denaturation of sample DNA).
3. Add a mixture of 1.5 μL SALSA MLPA Probemix (black cap) + 1.5 μL MLPA buffer (yellow cap, vortex before use) to each tube. Mix with care (*see* **Notes 16** and **17**).
4. Incubate 1 min at 95 °C (denaturation), followed by 16 h at 60 °C (hybridization overnight) (*see* **Note 18**).
5. Prepare a ligase master mix (less than 1 h before use), containing for each reaction 8.25 μL molecular biology grade water,

Table 1
Methylation-specific multiplex ligation-dependent probe amplification (MS-MLPA) thermocycler conditions

DNA denaturation reaction	98 °C 25 °C	5 min Pause
Hybridization reaction	95 °C 60 °C 20 °C	1 min Pause Pause
Ligation (±digestion) reaction	48 °C 48 °C 98 °C 20 °C	Pause 30 min 5 min Pause
PCR reaction	95 °C[a] 60 °C[a] 72 °C[a] 72 °C 15 °C	30 s 30 s 1 min 20 min Pause

[a]35 cycles

1.5 μL Ligase buffer B (white cap) and 0.25 μL Ligase-65 enzyme (green cap). Mix by pipetting and store at room temperature (*see* **Notes 19** and **20**).

6. Prepare a ligase-digestion master mix, containing for each reaction 7.75 μL molecular biology grade water, 1.5 μL Ligase buffer B (white cap), 0.25 μL Ligase-65 enzyme (green cap), and 0.5 μL HhaI endonuclease. Mix by pipetting and store on ice (*see* **Note 20**).
7. Continue the thermocycler program by pausing at 20 °C.
8. Prepare a Ligase buffer A master mix, containing 3 μL Ligase buffer A (transparent cap) and 10 μL molecular biology grade water for each reaction. Add 13 μL of this master mix to each tube, mix and transfer 10 μL to a new (second) tube.
9. Continue the thermocycler program by pausing at 48 °C. While at this temperature, add 10 μL ligase master mix to the first tube (prepared in **step 5**) and 10 μL ligase-digestion mix (prepared in **step 6**) to the second tube and mix well.
10. Incubate 30 min at 48 °C (for ligation and HhaI digestion), then heat for 5 min at 98 °C and pause at 20 °C. At this point, the tubes can be removed from the thermocycler (*see* **Note 21**).
11. Prepare a Polymerase master mix by adding for each reaction 3.75 μL molecular biology grade water, 1 μL SALSA PCR primer mix (brown cap) and 0.25 μL SALSA Polymerase (orange cap). Mix well by pipetting up and down. Do not vortex and store on ice until use (*see* **Note 22**).
12. At room temperature, add 5 μL Polymerase master mix to each tube containing 20 μL ligation(-digestion) reaction, mix by pipetting and start the PCR reaction with 35 cycles of 30 s at 95 °C, 30 s at 60 °C followed by 60 s at 72 °C. The PCR ends with 20 min incubation at 72 °C and pauses at 15 °C (*see* **Note 23**).
13. Separate the amplification products by a capillary sequencing system with fragment analysis software: in our case (ABI-3730), we mix 20 μL deionized formamide, 0.5 μL ROX-500 and 1 μL PCR product, denature at 98 °C for 3 min, and keep the plate at 4 °C until the start of the run (*see* **Note 24**).

3.3 Software Analysis and Interpretation of the Data

1. To analyze MS-MLPA data, we recommend using Coffalyser.Net software (freeware) developed by MRC-Holland. A slideshow and pdf tutorial is available at http://coffalyser.wordpress.com/tutorials/ (*see* **Note 25**).
2. Open Coffalyser.Net, create a new project or add an experiment to an existing project. Choose the experiment type

(DNA/MS-MLPA). Set the CE device (in this case ABI-3730; this has to be set only once).

3. Choose the correct 6-FAM probemix. A new probemix can be added by clicking the right button in the "select a sheet" field. Then right click on the "Select Coffalyser Work Sheets" window, select "Add" and choose from the drop-down list. Select the size marker (in this case ROX GS500-250) (*see* **Note 26**).
4. Click "next", right click in the next window, select "Add (from file)" and browse to find the folder containing your MLPA raw data (.fsa files). Select "Import".
5. In the column "sample type", choose which samples are to be used as sample, reference sample, no DNA control or digested sample (just click in the box). Then click "start fragment analysis". Perform a quality control by checking whether all probes are present in all (undigested) samples except the no DNA control. Check DNA concentration (Q-fragments), denaturation (D-fragments), digestion and gender (X and Y fragments) (*see* **Note 27**).
6. Click "Next" without changing the automatic settings. Select the samples that need to be further analyzed (based on the quality control from the previous step). You can do this by right clicking somewhere in the sheet, select "Select Samples For Analysis", and then "All Samples Suitable For Analysis". Un-tick all boxes that need no analyzing. For MS-MLPA, right click in the sheet and select "Digested Samples (For MS-MLPA)" and then "Match Samples Automatically".
7. Click "Start Comparative Analysis". When the analysis is finished, the last three columns indicate the signal sloping after correction (FSLP), the reference sample quality (RSQ) and the reference probe quality (RPQ) respectively (*see* **Note 28**).
8. To further analyze the results, right click somewhere in the sheet, and select "Open Experiment Results". Go to the tab "Statistics" and check whether the standard deviations in the references samples are ok. If not, delete the reference sample that is deviating, and repeat the analysis omitting this sample as reference. If OK, then go to the tab "ratio overview", right click somewhere in the page and click "Export Grid Data". For MS-MLPA it is possible to export copy number data and methylation data separately. To do this, right click in the sheet and select "Show Data Type", then choose "DNA" or "MS" (*see* **Notes 29–33**).

4 Notes

1. In case MRC-Holland cannot offer an (MS-)MLPA probemix for your application, you can design your own synthetic MLPA probes according to a protocol provided by the manufacturer. Synthetic probes differ from MRC-Holland probes in that the latter consist of a synthetic and a clone-derived part. The use of cloned oligonucleotides allows preparation of longer probes and inclusion of up to 55 different probes in one MLPA reaction. As an extra control, MLPA probemixes often contain more than one probe per region of interest. In addition, it is in some cases possible to develop methylation-specific probemixes for a certain tumor or disease in collaboration with MRC-Holland.
2. Size standard, polymer and other specifications on capillary electrophoresis depend on the capillary electrophoresis instrument type.
3. Nowadays, most capillary sequencers no longer use Deionized Formamide. In these cases, formamide is replaced by distilled water. Formamide can become acidic when stored for longer periods. This can result in depurination and fragmentation of the DNA upon heating. We recommend storing formamide aliquots at −20 °C.
4. DNA extracted from blood or fresh frozen tissue can be also used for (MS-)MLPA.
5. The most recent version of the protocol is provided on the MRC-Holland website. It is recommended to check it for updates.
6. Chemical removal of paraffin by means of xylene or other commercially available deparaffinization solvents (e.g., Deparaffinization Solution, QIAGEN) can be replaced by a physical method, such as melting the paraffin at 90 °C for 15 min after scraping off from glass slide and adding lysis buffer.
7. We generally add 50 μL of lysis buffer, but it is also possible to use less or more dependent on the amount of tissue that was macrodissected.
8. If necessary, the incubation time can be increased to an overnight step, in most cases this will lead to a higher DNA concentration. Always use a lysis buffer/proteinase K proportion of 5/1.
9. The extracted DNA is stable for 2–3 weeks at 4–10 °C. This easy, fast and cheap DNA isolation method leads to lower quality and fragmented DNA, but is sufficient for (MS-)MLPA analysis due to the very short hybridizing probe sequences of MLPA probes. MLPA does not require a special

method or kit for DNA extraction. However, it is important that the extraction method does not leave a high concentration of contaminants such as EDTA and salt (use with caution: Qiagen EZ1, M6, M48, and M96 systems, unless an alternative protocol (can be obtained from MRC-Holland) is applied). We have tried different column precipitation isolation methods that also work well and generally lead to a higher DNA quality, but at the same time to a lower DNA quantity.

10. The EDTA concentration in the DNA sample should not exceed 2.5 mM. Concentrating samples by evaporation, e.g., using SpeedVac may result in high EDTA concentrations that can influence MLPA results.

11. The sample volume should be 5 μL. The volume of the reaction is important for the hybridization speed and the Tm (melting temperature), which are both probe- and salt concentration-dependent. If the DNA concentration of a sample is very low, add 5 μL of DNA without TE buffer. Please consider that a minimum of 5 mM (preferably 10 mM) Tris–HCl with a pH between 8.0 and 9.0 should be present in the 5 μL DNA sample before heating to 98 °C for DNA denaturation in order to prevent DNA depurination.

12. Both fresh frozen and highly fragmented DNA samples yield reliable results with MLPA. The MLPA results are not strongly influenced by variations in the amount of DNA between samples because of the intrinsic comparison with the internal reference probes and limited use of PCR primers (*see* **Note 14**). This is of special interest for crude lysates from FFPE specimens as a reliable measurement of the sample DNA concentration can be difficult because of the presence of contaminants. In our experience, optical density (260 nm) measurements often overestimate the DNA concentration, especially in the case of direct lysis DNA isolation methods (five to tenfold overestimation). Therefore, we recommend to use 1–5 μL of direct lysis sample per MLPA reaction. Whether the DNA quantity was sufficient can be estimated on the basis of the Q-fragments, as explained in **Note 26**.

13. Negative (and optional positive) control samples should be run simultaneously with the test samples. It is highly recommended to compare reference and tumor samples extracted by the same method and derived from the same source (e.g., blood versus blood) to minimize the variation caused by possible contaminants. Use at least 3 reference samples in each MLPA experiment. When using more than 21 samples, add 1 additional reference sample for each 7 samples. Reference samples should be spread randomly over the sample plate to avoid technical bias and thus to minimize variation. For MS-MLPA, we also

process a nonmethylated sample (human sperm DNA) and fully methylated cell line- or blood-derived DNA treated with the M.SssI CpG methyltransferase enzyme in parallel. It is also recommended to include a no-DNA (5 μL water or TE) reaction in each experiment, as it will reveal contamination of water, MLPA reagents, electrophoresis reagents or capillaries.

14. As the primers are almost 100% consumed in the PCR reaction, the absolute amounts of amplicons formed is not dependent on the amount of sample DNA used. As a result it is not needed to use the exact same amount of sample DNA in each MLPA reaction.

15. DNA samples diluted with water may give aberrant results due to depurination in the 5 min 98 °C DNA denaturation step. In case the ionic composition of the sample is not known, it is recommended to add Tris-HCl pH 8.5 to the sample, resulting in a 5–10 mM final concentration.

16. From January 2nd, 2013 onwards, a newly developed SALSA MLPA buffer became available. The new buffer, containing a short blocking oligonucleotide that reduces potential hairpin formation of probes, makes MLPA more robust. Its main advantages are reduced standard deviation within one experiment, reduced sensitivity to variation in pipetting volumes, reduced sensitivity to evaporation during PCR, reduced sensitivity to differences in thermocycler ramping speeds and considerably improved results especially for MS-MLPA experiments.

17. The SALSA MLPA buffer is usually frozen at −20 °C, so thawing is necessary before use. Furthermore, the MLPA buffer is viscous and does not mix easily. Always vortex thawed buffers and probemixes before use, as salts and other constituents are concentrated at the bottom of the tube.

18. The hybridization time can be anywhere between 12 and 24 h (16–18 h is recommended, but hybridization should be nearly complete at 12 h). Following the overnight incubation at 60 °C, the volume at the bottom of the tubes should be at least 5.5 μL. Be sure that there is no excessive evaporation, if so try to optimize the pressure from the heated lid, a different brand of tubes or try using mineral oil on top.

19. Never vortex solutions that contain enzymes. Master mixes containing enzymes should be mixed gently by pipetting up and down.

20. If the Ligase Mastermix is prepared more than 1 h before use and placed on ice, it should be warmed to room temperature before adding it to the MLPA reactions, because adding it cold causes increased nonspecific peaks in the no DNA reaction.

21. Ligation reaction products can be stored at room temperature for several hours or at 4 °C for up to 1 week.
22. Vortex the SALSA PCR primer mix before use. Warm the polymerase for 10 s in your hand in order to reduce viscosity.
23. The recommended number of PCR cycles is 35, however the number of cycles can be reduced to 30 and, in the case of small DNA amounts, the number of cycles can be increased up to 37. The sequence (5′-3′) of the labeled forward PCR primer is GGGTTCCCTAAGGGTTGGA and of the unlabeled reverse primer GTGCCAGCAAGATCCAATCTAGA. The PCR product can be stored in the dark at 4 °C for up to 1 week. For longer periods, storage at −20 °C is recommended.
24. The amount of the PCR product required for analysis by capillary electrophoresis depends on the instrument and fluorescent label used. The recommended settings per instrument can be found at the MRC-Holland website in the fragment separation section under MLPA procedure.
25. Coffalyser.Net is free software designed by MRC-Holland and was made specifically for the analysis of (MS-)MLPA data. Coffalyser.Net uses a SQL client–server database model to store all project/experiment- related data. The client-server model has one main application (server) that deals with one or several slave applications (clients). Clients may communicate to a server over the network, allowing data sharing within and even beyond their institutions. The software is compatible with binary data files produced directly by all major capillary electrophoresis systems including: ABIF files (*.FSA, *.AB1, *. ABI) produced by Applied Biosystems devices, RSD files produced by Megabace™ systems (Amersham) and ESD files produced by CEQ systems (Beckman).
26. A library containing all probemix datasheets can be downloaded directly from MRC-Holland's server. The probemix datasheet library is updated regularly and the datasheets may include important notifications about probes.
27. Coffalyser.Net displays several quality check points: the number of found probes, the number of found reference probes, whether the 92 bp benchmark probe was found, whether the sample was male or female (X and Y probe at 100 and 105 bp), whether there was enough DNA (by estimation of the relative signal of Q-fragments (64, 70, 76, 82 bp) present in every MLPA mix) and whether the DNA was denatured completely (relative signals of denaturation fragments (88, 96 bp) present in every probemix). Relatively low 88 and 96 bp fragment signals might indicate presence of excess salt in DNA sample. Therefore sample DNA salt content should be low in order to allow DNA denaturation of those genomic regions that have a

very high CG content; e.g., CpG islands with a GC content above 80% will not be completely denatured in samples that contain 40 mM NaCl or KCl, or more than 80 mM Tris–HCl. Additional purification of sample DNA by ethanol precipitation or column-based methods is beneficial for removing salts or other contaminants. The MRC-Holland website is providing up-to-date methods for sample DNA purification.

28. Optimally, all values should be green or orange but in most cases, when using FFPE tissue, the RSQ and RPQ values of some of the samples will be red (inevitably).
29. To determine the methylation status of each sample, the peak pattern of each digested DNA sample is compared to the corresponding undigested sample. The approximate methylation percentage is calculated by dividing the normalization constant (NC) for the MS-MLPA probe in digested sample A by the NC for the MS-MLPA probe in the undigested sample A, and multiplying this ratio by 100%. The NC is a result of intra-sample normalization, which is achieved by dividing the signal of each MS-MLPA probe by the signal of every reference probe in that sample (reference probes detect sequences that are expected to have a normal copy number in the condition/disease of interest and they do not contain a HhaI restriction site), and calculating the median of all these ratios per probe.
30. Most MS-MLPA probes detect methylation of the first cytosine nucleotide in a single HhaI restriction site within the sequence detected by the probe (GC^{m}GC). One CpG site may not be representative of the entire CpG island.
31. A point mutation or a SNP in the HhaI restriction site of a MS-MLPA probe may result in a false positive methylation signal.
32. MS-MLPA probes detecting sequences in CpG islands of promoter regions often reproducibly yield a low residual signal in the digested reference samples. This might be due to methylation of the sequence in a small percentage of the cells tested. This background signal is often higher in probes detecting sequences near the edge of the CpG island. Note also that methylation patterns vary between different tissue types and even between age groups.
33. The optimal cutoff values for detecting a significant change in methylation of a sequence is probe-, sample type-, and application-dependent. A 15% cutoff to define presence of methylation is often used [13, 21].

References

1. Schouten JP, McElgunn CJ, Waaijer R et al (2002) Relative quantification of 40 nucleic acid sequences by multiplex ligation-dependent probe amplification. Nucleic Acids Res 30:e57
2. Moelans CB, de Weger RA, van Blokland MT et al (2010) Simultaneous detection of TOP2A and HER2 gene amplification by multiplex ligation-dependent probe amplification in breast cancer. Mod Pathol 23:62–70
3. Moelans CB, de Weger RA, van Diest PJ (2010) Absence of chromosome 17 polysomy in breast cancer: analysis by CEP17 chromogenic in situ hybridization and multiplex ligation-dependent probe amplification. Breast Cancer Res Treat 120:1–7
4. Vorstman JA, Jalali GR, Rappaport EF et al (2006) MLPA: a rapid, reliable, and sensitive method for detection and analysis of abnormalities of 22q. Hum Mutat 27:814–821
5. Eldering E, Spek CA, Aberson HL et al (2003) Expression profiling via novel multiplex assay allows rapid assessment of gene regulation in defined signalling pathways. Nucleic Acids Res 31:e153
6. Hess CJ, Denkers F, Ossenkoppele GJ et al (2004) Gene expression profiling of minimal residual disease in acute myeloid leukaemia by novel multiplex-PCR-based method. Leukemia 18:1981–1988
7. Bunyan DJ, Skinner AC, Ashton EJ et al (2007) Simultaneous MLPA-based multiplex point mutation and deletion analysis of the dystrophin gene. Mol Biotechnol 35:135–140
8. Bergval IL, Vijzelaar RN, Dalla Costa ER et al (2008) Development of multiplex assay for rapid characterization of Mycobacterium tuberculosis. J Clin Microbiol 46:689–699
9. Gutierrez-Enriquez S, Balmana J, Baiget M et al (2008) Detection of the CHEK2 1100delC mutation by MLPA BRCA1/2 analysis: a worthwhile strategy for its clinical applicability in 1100delC low-frequency populations? Breast Cancer Res Treat 107:455–457
10. Dikow N, Nygren AO, Schouten JP et al (2007) Quantification of the methylation status of the PWS/AS imprinted region: comparison of two approaches based on bisulfite sequencing and methylation-sensitive MLPA. Mol Cell Probes 21:208–215
11. Nygren AO, Ameziane N, Duarte HM et al (2005) Methylation-specific MLPA (MS-MLPA): simultaneous detection of CpG methylation and copy number changes of up to 40 sequences. Nucleic Acids Res 33:e128
12. Procter M, Chou LS, Tang W et al (2006) Molecular diagnosis of Prader-Willi and Angelman syndromes by methylation-specific melting analysis and methylation-specific multiplex ligation-dependent probe amplification. Clin Chem 52:1276–1283
13. Moelans CB, Verschuur-Maes AH, van Diest PJ (2011) Frequent promoter hypermethylation of BRCA2, CDH13, MSH6, PAX5, PAX6 and WT1 in ductal carcinoma in situ and invasive breast cancer. J Pathol 225:222–231
14. Homig-Holzel C, Savola S (2012) Multiplex ligation-dependent probe amplification (MLPA) in tumor diagnostics and prognostics. Diagn Mol Pathol 21:189–206
15. Suijkerbuijk KP, Pan X, van der Wall E et al (2010) Comparison of different promoter methylation assays in breast cancer. Anal Cell Pathol 33:133–141
16. Jeuken JW, Cornelissen SJ, Vriezen M et al (2007) MS-MLPA: an attractive alternative laboratory assay for robust, reliable, and semiquantitative detection of MGMT promoter hypermethylation in gliomas. Lab Investig 87:1055–1065
17. Hess CJ, Ameziane N, Schuurhuis GJ et al (2008) Hypermethylation of the FANCC and FANCL promoter regions in sporadic acute leukaemia. Cell Oncol 30:299–306
18. Furlan D, Sahnane N, Mazzoni M et al (2013) Diagnostic utility of MS-MLPA in DNA methylation profiling of adenocarcinomas and neuroendocrine carcinomas of the colon-rectum. Virchows Arch 462:47–56
19. Moelans CB, de Weger RA, Ezendam C et al (2009) HER-2/neu amplification testing in breast cancer by multiplex ligation-dependent probe amplification: influence of manual- and laser microdissection. BMC Cancer 9:4
20. Moelans CB, de Weger RA, van Diest PJ (2011) Amplification testing in breast cancer by multiplex ligation-dependent probe amplification of microdissected tissue. Methods Mol Biol 755:107–118
21. Gylling A, Abdel-Rahman WM, Juhola M et al (2007) Is gastric cancer part of the tumour spectrum of hereditary non-polyposis colorectal cancer? A molecular genetic study. Gut 56:926–933

Chapter 28

Methylation-Sensitive High Resolution Melting (MS-HRM)

Dianna Hussmann and Lise Lotte Hansen

Abstract

Methylation-Sensitive High Resolution Melting (MS-HRM) is an in-tube, PCR-based method to detect methylation levels at specific loci of interest. A unique primer design facilitates a high sensitivity of the assays enabling detection of down to 0.1–1% methylated alleles in an unmethylated background.

Primers for MS-HRM assays are designed to be complementary to the methylated allele, and a specific annealing temperature enables these primers to anneal both to the methylated and the unmethylated alleles thereby increasing the sensitivity of the assays. Bisulfite treatment of the DNA prior to performing MS-HRM ensures a different base composition between methylated and unmethylated DNA, which is used to separate the resulting amplicons by high resolution melting.

The high sensitivity of MS-HRM has proven useful for detecting cancer biomarkers in a noninvasive manner in urine from bladder cancer patients, in stool from colorectal cancer patients, and in buccal mucosa from breast cancer patients. MS-HRM is a fast method to diagnose imprinted diseases and to clinically validate results from whole-epigenome studies. The ability to detect few copies of methylated DNA makes MS-HRM a key player in the quest for establishing links between environmental exposure, epigenetic changes, and disease.

Key words MS-HRM, PCR bias, Primer design, DNA methylation, Heterogeneous methylation

1 Introduction

The principle behind detecting DNA methylation at specific loci of interest using the Methylation-Sensitive High Resolution Melting (MS-HRM) method is based upon a specific primer design and measurement of the difference in melting properties between the methylated and unmethylated alleles after bisulfite conversion.

Bisulfite treatment of DNA is used to preserve the methylation marks at cytosines when preceding guanine, which are normally erased during the PCR, as there are no enzymes or methyl donors in the reaction mix to replace the methyl groups during the amplification. Bisulfite treatment will leave methylated cytosines unchanged, whereas unmethylated cytosines are converted to uracils, which eventually will be replaced by thymines after amplification by PCR [1]. The resulting two DNA strands from each double

Jörg Tost (ed.), *DNA Methylation Protocols*, Methods in Molecular Biology, vol. 1708,
https://doi.org/10.1007/978-1-4939-7481-8_28, © Springer Science+Business Media, LLC 2018

helix are no longer complementary, and the allelic regions in the genome may have a different DNA sequence, depending on the methylation status before bisulfite conversion.

After bisulfite conversion followed by PCR, a fully methylated double stranded DNA region comprises a higher number of the stronger cytosine/guanine bonds, than the same unmethylated cytosine-poor DNA region. These properties are used to separate the methylated from the unmethylated alleles by the difference in the amount of energy necessary to separate the DNA strands of the double helix.

Worm et al. initially reported the principle of separating methylated from unmethylated DNA by melting analysis [2]. The development of MS-HRM is based upon a new primer design, further improvement of the fluorescent dye chemistry, the development of a new generation of instruments, and software capable of performing high resolution melting (HRM) after real-time PCR [3].

HRM was initially used to assess the genotype of single nucleotide polymorphisms (SNP) and to detect other minor sequence variations, based upon the difference in DNA sequences and thereby in melting properties of the resulting amplicons. This principle was transferred to assess the methylation status of a specific genomic region of interest.

After bisulfite conversion the DNA is subjected to a PCR, in which a saturating fluorescent dye is present. This dye will omit fluorescence when bound to double-stranded DNA, making it possible to follow the amplification process during PCR. After a certain number of cycles, preferably when the amplification process has reached the plateau phase, the temperature is increased to dissociate the amplicons. The following decrease in temperature will allow heteroduplex formation, and when the temperature is raised gradually (0.1 °C/s) the double helix will dissociate and release the intercalating dye and thereby promote a decrease in fluorescence. The software generates melting curves showing the relation between temperature and measured fluorescence from the end of amplification until all amplicons are fully dissociated and are present as single DNA strands. Amplicons from templates with a high level of methylation, containing a high proportion of cytosine and guanine after bisulfite conversion, will melt at a higher temperature than templates from unmethylated DNA having a high adenine and thymine content.

MS-HRM relies on controlled amplification of all templates regardless of methylation status. The PCR bias introduced by the difference in base composition between methylated and unmethylated templates after bisulfite conversion, facilitating amplification of AT-rich over CG-rich templates, is overcome by a specific primer design [4–7].

The principle is to design the primers to be complementary to the methylated templates and when choosing the temperature at which both primers anneal to the methylated and unmethylated templates with similar efficiency, the primers are in principle methylation independent (MIP).

The methylation status of a single CpG can be detected by MS-HRM, and preferably a limited number of CpG dinucleotides should be included in the amplicon to avoid additional melting domains, which may interfere with the melting profile and eventually interpretation of the results. Designing amplicons not exceeding 130 base pairs (bp) has proven successful when analyzing degraded DNA material, though larger amplicons have been reported [8].

Assessment of the methylation level is semiquantitative and performed by comparing the melting profile of the test sample to a standard dilution range of fully methylated DNA in unmethylated DNA [9]. Quantification of global methylation of LINE-1 elements assessed by MS-HRM was performed using a standard curve comprising serial dilutions of fully methylated DNA in unmethylated DNA, thereby determining the linear relationship between the standard curves and percentage of methylation [10].

The specific primer design favoring amplification of the methylated allele have resulted in a high sensitivity of MS-HRM assays, and the ability to detect methylation levels between 0.1% and 1% [6, 11].

The high sensitivity is preserved even when using FFPE-tissue derived DNA, which has been stored for up to 30 years [12]. This may be explained by a combination of the ability to detect even a few copies of methylated DNA in a sample with the assay design spanning less than 90 bp.

MS-HRM is performed in a closed tube format limiting the risk of contamination. Analyses can be performed at high-throughput using a robot for pipetting and a HRM platform capable of analyzing 384 well plates. The closed tube format combined with the high sensitivity, the ability to robustly detect methylation status of few copies of DNA derived from highly degraded material, holds a potential for MS-HRM to be used for non- or limited invasive methylation-specific biomarker detection and clinical validation of potential biomarkers identified by epigenomic screening [11, 13, 14].

Noninvasive detection of potential biomarkers using MS-HRM based assays was used to assess the methylation status of four genes, *ZNF154, POUF42, HOXA9*, and *EOMES*, which were significantly differentially methylated in urine derived DNA from bladder cancer patients when compared to urine from healthy controls [14]. Likewise, MS-HRM reproducibly detected methylation levels down to 1% of *SFRP2* and *VIM* in stool from colorectal cancer patients (CRC), in patients with advanced adenomas (AA) [8], and

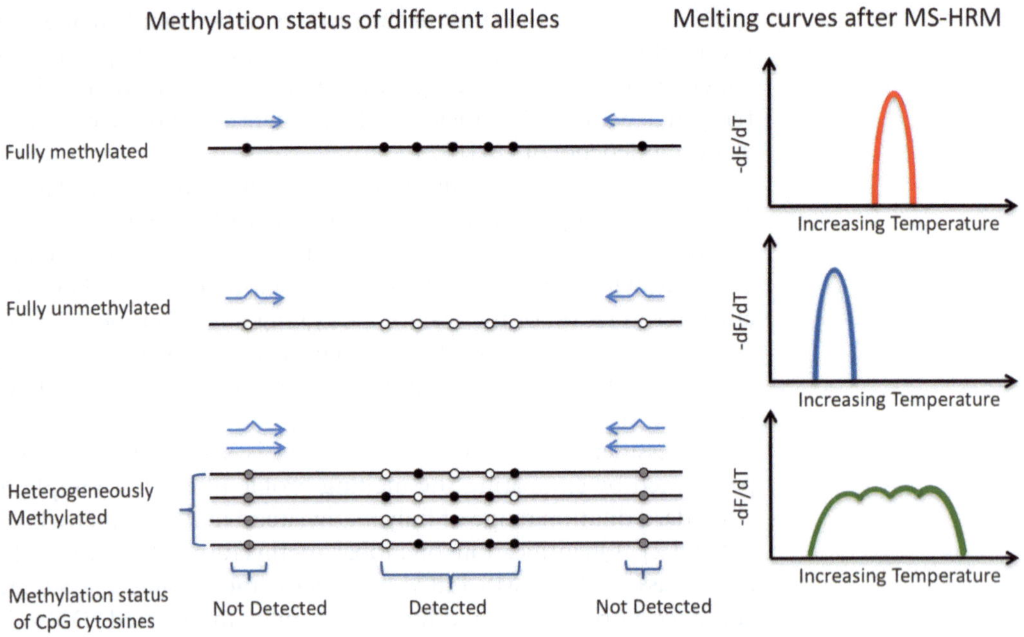

Fig. 1 Schematic illustration of the principle behind MS-HRM. Primers designed for MS-HRM are complementary to the methylated template, but a specific annealing temperature ensures binding to both the methylated and the unmethylated templates. Only the CpG sites between the primers are assessed for their methylation status. A schematic illustration of the resulting melting curve is shown for a fully methylated template (*red curve*), an unmethylated template (*blue*), and a heterogeneously methylated template (*green*). Primers (*blue arrows*); *filled black circles* (methylated cytosines), *filled gray circles* (methylated or unmethylated cytosines), *open circles* (unmethylated cytosines)

BRCA1 promoter methylation was assessed in buccal mucosa from a breast cancer patient [15].

Limited invasive detection of methylation levels by MS-HRM was used to assess *AKAP12* promoter methylation in serum from CRC patients [16], and cancer related genes (*BRCA1, BRCA2, MGMT, MLH1, APC, RASSF1A*) in peripheral blood leucocyte DNA from breast and colorectal cancer patients [15, 17–19].

Specific MS-HRM assays targeting imprinted regions can be used as a diagnostic screen for Prader–Willi and Angelman syndromes when assaying the methylation level of *SNRPN* [20] or Beckwith–Wiedemann and Russell Silver syndromes when investigating the *H19/IGF2* imprinting center [9].

The MS-HRM assay measures the cytosine methylation level between the primers, as there is no template selection by the primers (Fig. 1). Therefore, MS-HRM assays are able to reliably detect heterogeneous methylation, which presents with a highly distinct melting curve generated by the complex mixture of alleles containing different patterns of methylated and unmethylated cytosines [11, 21] (Fig. 1). Being a fast and relatively inexpensive method, MS-HRM can be used to screen for heterogeneous methylation, followed by further detailed analysis by for example pyrosequencing [22]. The phenomenon of heterogeneous methylation and especially its implication in disease need further investigation, as most methylation detection methods will assess heterogeneous methylation as fully methylated DNA.

2 Materials

The following materials are adjusted to a protocol using the Lightcycler®480 (Roche) as instrumentation.

2.1 Instrumentation

1. Ideally, an instrument capable of performing real-time PCR coupled with subsequent high resolution melting such as: Lightcycler®480 (Roche), the Qiagen Rotor-Gene Q (the former Corbett Rotor-Gene 6000), and the Applied Biosystems® 7577 Fast or 7900HT Fast Real-Time PCR systems. If this kind of instrumentation is not available, real-time PCR and high resolution melting can be performed separately using a conventional qPCR instrument and for example the LightScanner® System (BioFire Defense).
2. Thermocycler.
3. Multiwell plate centrifuge.
4. NanoDrop 1000 Spectrophotometer or DNA quantification: e.g., Quant-iT™ PicoGreen® dsDNA Assay Kit or similar using a Fluorometer.
5. For post MS-HRM analysis: 3130xl Genetic Analyzer (ThermoFisher Scientific) or equivalent capillary electrophoresis system.

2.2 Materials for Sample Preparation

1. Methylated control DNA: CpG Methyltransferase (*M.Sss*I) (e.g., CpG Methyltransferase Kit, Thermo Scientific), store at −20 °C or commercially produced methylated DNA (e.g., Methylated Human Genomic DNA, Zymo Diagnostics).
2. Unmethylated control DNA: Whole Genome Amplification (WGA) of genomic DNA (e.g., Illustra GenomiPhi V2 DNA Amplification Kit, GE Healthcare), store at −80 °C, or commercially produced DNA (e.g., Epitech unmethylated human control DNA, Qiagen).
3. QIAquick PCR Purification Kit (Qiagen) or similar.
4. DNA quality control: Genomic DNA LabChip® GX (Perkin Elmer, protocol: http://www.perkinelmer.com/CMSResources/Images/44-151452CLS136856_Rev_02_User_Guide_Genomic_DNA.pdf).

5. EZ DNA Methylation-Gold™ kit (Zymo Research) for bisulfite conversion using one column per sample or EZ-96 DNA Methylation Gold Kit (Zymo Research) for bisulfite conversion of up to 96 samples in 96-well plates or similar products for bisulfite conversion (*see* Chapter 23 for a bisulfite conversion protocol not using commercial kits).

2.3 Software for Assay Design

1. Microsoft Word, or other text programs/web-accessible software, for in silico bisulfite treatment.
2. USCS Genome Browser: provides DNA sequence and location of CpG islands (http://genome.ucsc.edu).
3. OligoAnalyzer: determines secondary structures of the primers. (http://eu.idtdna.com/analyzer/Applications/OligoAnalyzer/).
4. Oligo Calc: calculates melting temperature, lengths, etc. of the primers. (http://www.basic.northwestern.edu/biotools/oligocalc.html).
5. In silico PCR: determines secondary structures of the primers and unspecific annealing (http://genome.ucsc.edu/cgi-bin/hgPcr).

2.4 Materials for PCR Amplification and High-Resolution Melting

1. High Resolution Melting (HRM) Master Mix (e.g., MeltDoctor™ HRM master mix, ThermoFisher Scientific; or LightCycler® 480 High Resolution Melting Master, Roche Diagnostics).
2. MS-HRM primers.
3. Multiwell plate-96 or Multiwell plate-384, white.
4. Optical adhesive covers.

2.5 Data Analysis

1. LightCycler® 480 Software (SW release 1.5.1), Roche. Required modules provided with the instrument: Gene scanning for normalized melting curves and MeltCurve Genotyping for melting peaks.

2.6 Post MS-HRM Analyses

1. Dideoxysequencing: alkaline phosphatase (FastAP) and Exonuclease 1 (Exo1) for purification (ThermoFisher Scientific).
2. BigDye® Terminator v1.1 Cycle Sequencing kit (store at −20 °C) (ThermoFisher Scientific).
3. Cloning: Zero Blunt® TOPO® PCR Cloning Kit (ThermoFisher Scientific) or similar.

3 Methods

3.1 General Principle

The workflow of optimizing and analyzing samples by Methylation-Sensitive High-Resolution Melting (MS-HRM) is as follows:

1. In silico bisulfite treatment of the region of interest for assay design.
2. Primer design following the specific rules for MS-HRM.
3. Preparation of methylated and unmethylated DNA for standard dilutions
4. Bisulfite modification of standards and preparation of standard dilution series.
5. Optimization of the MS-HRM assay using the standard dilution series (Adjusting the annealing temperature (T_A) to obtain a high assay specific sensitivity).
6. Bisulfite treatment of test DNA samples.
7. Performing MS-HRM analysis in duplicates or triplicates.
8. Data analysis.

3.2 Primer Design

1. Select the specific region of interest. The optimal MS-HRM assay design includes from 3–6 CpG sites between the primers (*see* **Note 1**). The target sequence can be obtained from the USCS Genome Browser database, which also allows to assess regional CpG islands. No web-based software takes into consideration the specific primer design principles of MS-HRM.
2. Copy the selected genomic region into Microsoft Word (or a similar text program/web-based tool) for in silico bisulfite treatment. Choose either the sense or antisense DNA strand (*see* **Note 2).** Ensure that "match cases" is turned on, and that the sequence is in lower case. Use the "find and replace"--function to substitute "cg" with "CG" followed by replacement of "c" with "T". Capital "T" illustrates all non-CpG cytosines. This sequence now reflects the 100% methylated DNA strand.
3. Design the primers to target the region of interest following the MS-HRM primer design rules:
   (a) Include one or two CpG dinucleotide close to the 5′ end of each primer. An additional CpG dinucleotide can be included if a higher sensitivity of the assay is required (*see* **Note 3**).
   (b) A thymine, originating from a non-CpG cytosine, may be included at the 3′ end of one or both primers, to prevent amplification of unconverted templates after bisulfite treatment.

(c) Inclusion of a cytosine or guanine at the 3′ end of the primer will facilitate specific primer annealing.

(d) Match the primer melting temperature within 1 °C of difference and preferably at the range of 56–60 °C.

(e) Select an amplicon length of approximately 100 bp (ranging from 60 to 130 bp). Short amplicons limits the risk of including multiple melting domains, which may interfere with data interpretation, and when analyzing highly degraded DNA, e.g., FFPE- tissue, the limited size facilitates amplification of the templates (*see* **Note 4).**

(f) Apply standard guidelines for primer design avoiding: secondary structure, self-dimers and heterodimers, and ensuring a GC content in the range of 35–65%. Standard tools as OligoAnalyzer or the in silico PCR option from UCSC are recommended. During the initial search for the optimal primer pair the simple "2+4 rule" (2 °C are added for each A-T and 4 °C for each C-G bond) works as well in our hands (*see* **Note 5**).

4. The low complexity of the bisulfite-converted template may prove a challenge for proper primer design (*see* **Note 6**).

3.3 Control DNA for Standard Dilution Series

The standard dilution series are obtained from 100% methylated DNA diluted into a background of unmethylated DNA, and serves as methylation level standards. The methylated and unmethylated DNA can be provided commercially or manually produced. Hernández et al. provide a comparison of different available kits [23].

1. Produce unmethylated standard DNA by a two-step whole genome amplification (WGA treatment) (*see* **Note** 7). The general protocol of the illustra GenomiPhi V2 DNA Amplification Kit from GE Healthcare is followed with few exceptions:

(a) Mix on ice 1 μL genomic DNA (10 ng/μL) and 9 μL Sample buffer.

(b) Incubate at 95 °C for 3 min. Cool and keep on ice.

(c) Add 9 μL RxN buffer and 1 μL enzyme mix.

(d) Incubate at 30 °C for 120 min followed by 65 °C for 10 min using a thermocycler.

(e) Store the product at 4 °C (WGA 1).

(f) For the second round of WGA, dilute the WGA product (WGA 1) 1:10 in H_2O and repeat the first round of WGA.

(g) Purify the WGA product (WGA 2) using the QIAquick PCR purification kit or similar.

(h) Elute in 2 × 50 μL Elution Buffer.

(i) Measure the DNA concentration (preferred concentration approximately 50 ng/μL). The samples are ready for bisulfite treatment.

2. Produce methylated DNA using the CpG Methyltransferase (*M.SssI*). One batch of M. SssI treated DNA provides approximately 500 ng of fully methylated DNA.
   (a) Mix at room temperature, 500 ng DNA, 2 μL 10× M.SssI buffer, 0.4 μL 50× SAM, and 1 μL M.SssI in a total volume of 20 μL.
   (b) Mix gently and spin down for a few seconds.
   (c) Incubate at 37 °C for 15 min.
   (d) Inactivate the enzyme by incubating at 65 °C for 20 min.
   (e) Purify the DNA by phenol and chloroform extraction followed by ethanol precipitation or by using a spin column kit.
   (f) Store at −20 °C.
3. Subject one batch of fully methylated DNA (approx. 500 ng) and one batch (approx. 2000 ng) of fully unmethylated DNA to bisulfite modification with 500 ng input DNA per bisulfite conversion reaction (Subheading 3.5).
4. Mix the resulting 100 μL of 100% methylated standard and 400 μL of unmethylated standard into the dilutions comprising e.g.: 100, 10, 1, 0.1, and 0% of methylated DNA. The standards have a theoretical concentration of 5 ng/μL and assay optimization performed in duplicates requires 4 μL of each standard (*see* **Note 8**). Thus one batch of standards can approximately be used for 20 analyses. It is important to note that bisulfite treated DNA has a limited lifespan of approximately 1 month when stored at 4 °C and longer if stored at −20 °C or −80 °C. Thus, when producing the standards, the amount should be adapted in regard to the experiment.

3.4 DNA Input

1. Purify DNA from tissue samples, cells or cell lines and from frozen or formalin fixed paraffin embedded (FFPE) tissues or other fixation methods, using a method of choice.
2. Measure the quantity of the input DNA prior to bisulfite treatment, to ensure a correct amount of DNA. Quant-iT™ PicoGreen® dsDNA Assay Kit has shown higher accuracy in our hands than the NanoDrop 1000 Spectrophotometer, and is therefore recommended.
3. The quality of the DNA can be verified using Genomic DNA LabChip® GX. If the DNA quality is high and the designed amplicons are of shorter length (approx. 100 base pairs) this step may be omitted.

3.5 Bisulfite Modification

1. Perform bisulfite conversion of the DNA. To optimize the MS-HRM assays or if few samples are to be analyzed the EZ-DNA Methylation-Gold Kit works well in our hands. When a large number of samples are to be analyzed use a 96-well plate based conversion kit such as the EZ-96 DNA Methylation-Gold Kit. It is important to use either the columns or 96-well plates for bisulfite treatment of standards and test DNA for each project (*see* **Note 9**).
2. Follow the protocol of the manufacturer with the exception that 500 ng DNA is used as input, and the modified DNA is eluted in water.
3. 500 ng of fully methylated DNA and 4 × 500 ng of fully unmethylated DNA are bisulfite-modified prior to preparing the standard dilutions (Subheading 3.3).
4. Elute the bisulfite-converted DNA depending on the quality of the DNA. DNA obtained from poor quality DNA as from FFPE tissue demands a higher theoretical final DNA concentration, and is therefore eluted in 52 μL sterile H_2O, providing a theoretical concentration of 10 ng/μL. For high quality DNA, as DNA isolated from blood or fresh frozen tissue, the DNA can be eluted in 104 μL sterile H_2O, resulting in a theoretical concentration of 5 ng/μL. A theoretical concentration of 5 ng/μL is sufficient for the DNA used in the optimization procedure.
5. Verify the quality of a new batch of standards of fully methylated and unmethylated DNA after bisulfite conversion, or when obtaining unexpected results after MS-HRM analysis. An already optimized MS-HRM assay can be used for validation processing in parallel previously used standards with known performance status (*see* **Note 10**).

3.6 Optimizing a MS-HRM Assay Using Standard Dilutions

1. Prepare the set of methylation standards by serially diluting 100% methylated DNA into 100% unmethylated DNA as follows: 100, 10, 1, 0.1 and 0% (Subheading 3.3).
2. A standard PCR protocol is used as default: Prepare the PCR in duplicates (2 × 6 reactions) including each of the bisulfite modified standard dilutions (100, 10, 1, 0.1, and 0%) and a no template control (NTC). Genomic DNA may be included ensuring no amplification of unconverted DNA.
3. Mix for each reaction, omitting the DNA in NTC: 10 ng DNA, 5.0 μL of MeltDoctor™ HRM master mix, 300 nM of each primer, H_2O to a total volume of 10 μL (*see* **Note 10** if using the HRM Master mix).
4. Seal the plates with optical adhesive covers.
5. Spin the tubes or plates briefly before placing them in the instrument.

6. Initiate the cycling process with one cycle at 95 °C for 10 min, followed by 40 cycles at 95 °C for 15 s, and the primer specific T_A (e.g., 60 °C) for 1 min.
7. Initiate the melting process with a denaturing step at 95 °C for 1 min followed by 55 °C for 1 min, and last by a stepwise increase in temperature from 55 °C to 95 °C, with a rate of 0.1 °C/s.
8. Initiate the optimization by selecting the deduced primer annealing temperature (T_A) for the PCR (*see* **Note 5**).
9. Analyze the MS-HRM results using the concomitant software following the instrument (Subheading 3.7).
10. All parameters of the PCR setup can be adjusted during assay optimization; however the annealing temperature is the single, most easily adjustable parameter to modify as it determines the PCR bias, followed by numbers of cycles, primer concentration and the $MgCl_2$ concentration.
11. It is recommended to change the T_A by increasing or decreasing it by 1 °C in consecutive runs dependent on the performance of the assay (Fig. 2). The assay sensitivity can be improved by adjusting the T_A accordingly (*see* **Note 11**).
12. Perform the optimization assays in duplicates or triplicates to ensure a reliable outcome (*see* **Note 12**).
13. It is highly important to be aware that external conditions (e.g., temperature fluctuations) may affect the performance of the laboratory equipment (*see* **Note 13** and Fig. 3).
14. Some assays may be difficult to interpret, but if the melting curves are appropriately separated, the assay can be accepted (Fig. 4). The stability of the amplicons can be sought balanced by applying a clamp to the primers (*see* **Note 4**).
15. The aim of the optimization is to ensure proper amplification and a high sensitivity. Ideally, the amplicons should only have a single melting domain, resulting in clear and distinguishable melting curves.

3.7 MS-HRM Analyses of Test Samples

1. Analysis of test samples is performed equivalent to the optimization procedure (Subheading 3.6) along with the standard dilution series and controls preferrably in triplicates.
2. Twenty-eight test samples can be analyzed in parallel with the standard dilution series and the NTC in triplicates in a 96-well plate, and 124 samples in triplicates in a 384 well plate. It is recommended to use a pipetting robot to set up the 384 well plates.
3. Seal the plate and centrifuge briefly before placing it into the instrument.

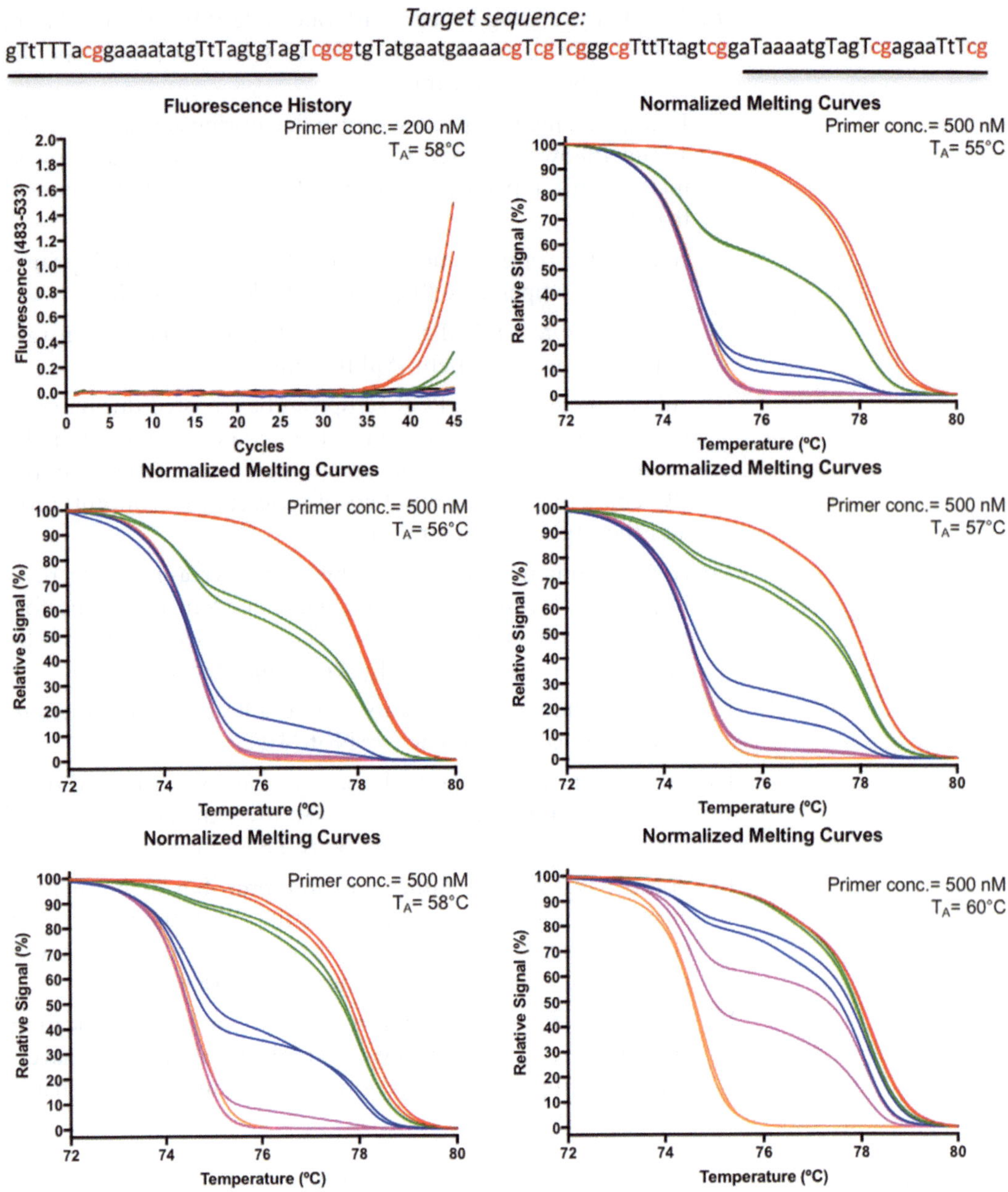

Fig. 2 Optimization procedure. The optimization procedure of the assay targeting the CpG Island of the *CDH13* promoter region is illustrated by the normalized melting curves of the dilution standards. An increase in primer concentration and a stepwise increase of the annealing temperature (T_A) from 55 °C to 60 °C results in a higher sensitivity of the assay. The target sequence is presented in which CpG sites are marked in *red*, non-CpG cytosines as T and the primer locations are underlined. Standards: 100% (*red*), 10% (*green*), 1% (*blue*), 0.1% (*turquoise*), and 0% (*orange*)

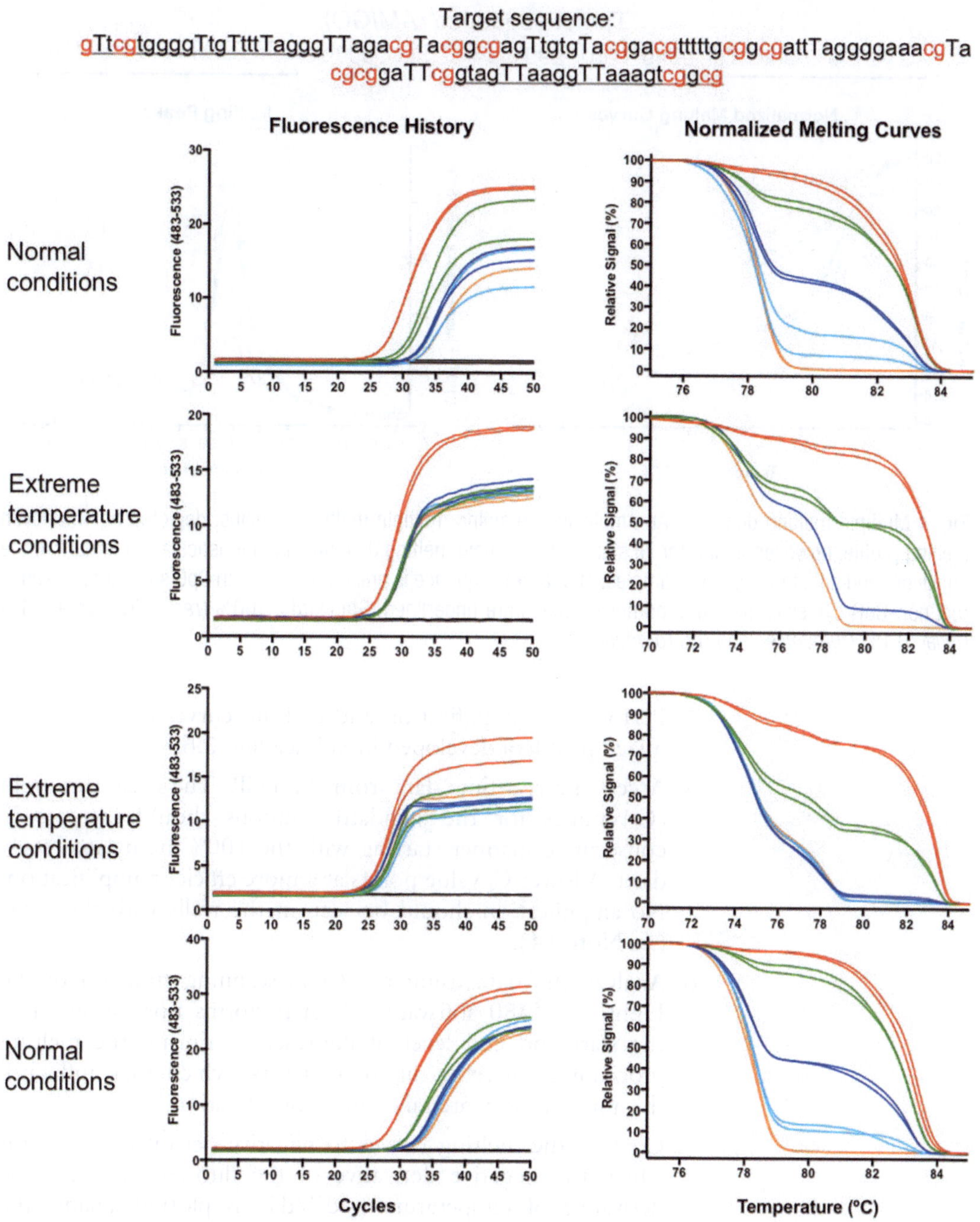

Fig. 3 Sensitivity to external condition. Extraordinary fluctuation of the room temperature affected the instrument performance resulting in abnormal amplification and melting curves. Direct sequencing of the MS-HRM amplicons revealed that the correct target region was assessed and that the bisulfite conversion was successful. The target sequence is presented in which CpG sites are marked in red, non-CpG cytosines as T and the primer locations are underlined. Standards: 100% (*red*), 10% (*green*), 1% (*blue*), 0.1% (*turquoise*), and 0% (*orange*)

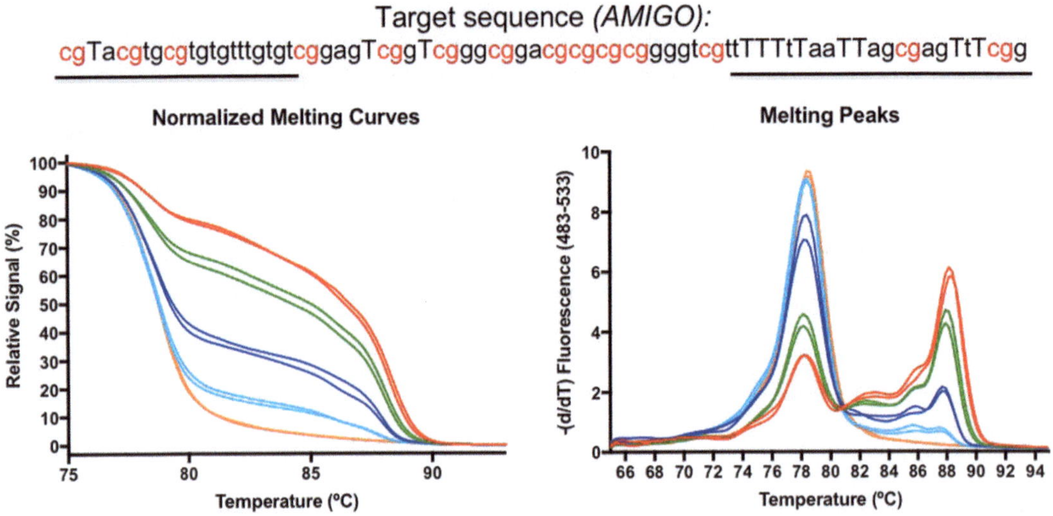

Fig. 4 Multiple melting domains. An amplicon containing multiple melting domains, depicted as an uneven melting profile, however similar for all standards. Multiple melting domain characteristics are a feature of the amplicon, and not of the specific samples. The target sequence is presented, in which CpG sites are marked in red, non-CpG cytosines as T and primer locations are underlined. Standards: 100% (*red*), 10% (*green*), 1% (*blue*), 0.1% (*turquoise*), and 0% (*orange*)

4. Perform the amplification and melting curve using the optimized protocol developed in Subheading 3.6.
5. Access the real-time data from the PCR. The cycle threshold (C_t) values for the standard dilutions should increase in consecutive manner starting with the 100% methylated standard. A lower C_t value points at a more efficient amplification. No amplification should be seen in the wells with the NTC (*see* **Note 14**).
6. Analyze the data using the Gene scanning function of the Lightcycler®480 software, which performs normalization of the start and end level of fluorescence during the melting procedure. Thereby, different factors, which may influence the amplification, are taken into consideration.
7. Convert the melting curves to negative derivative peaks, for which the negative derivative of the fluorescence over the derivative of temperature (−dF/dT) is plotted against the increasing temperature. The top point of the resulting melting curve represents the highest drop in fluorescence, indicating separation of the two DNA strands of the amplicons. This point is designated the melting temperature for the amplicon.
8. Ensure a uniform melting profile at separate temperatures of the 100% methylated and unmethylated standards, preferably using both the normalized melting curves and the negative derivative peaks. The remaining standards should contain two melting domains divided between the two melting temperatures,

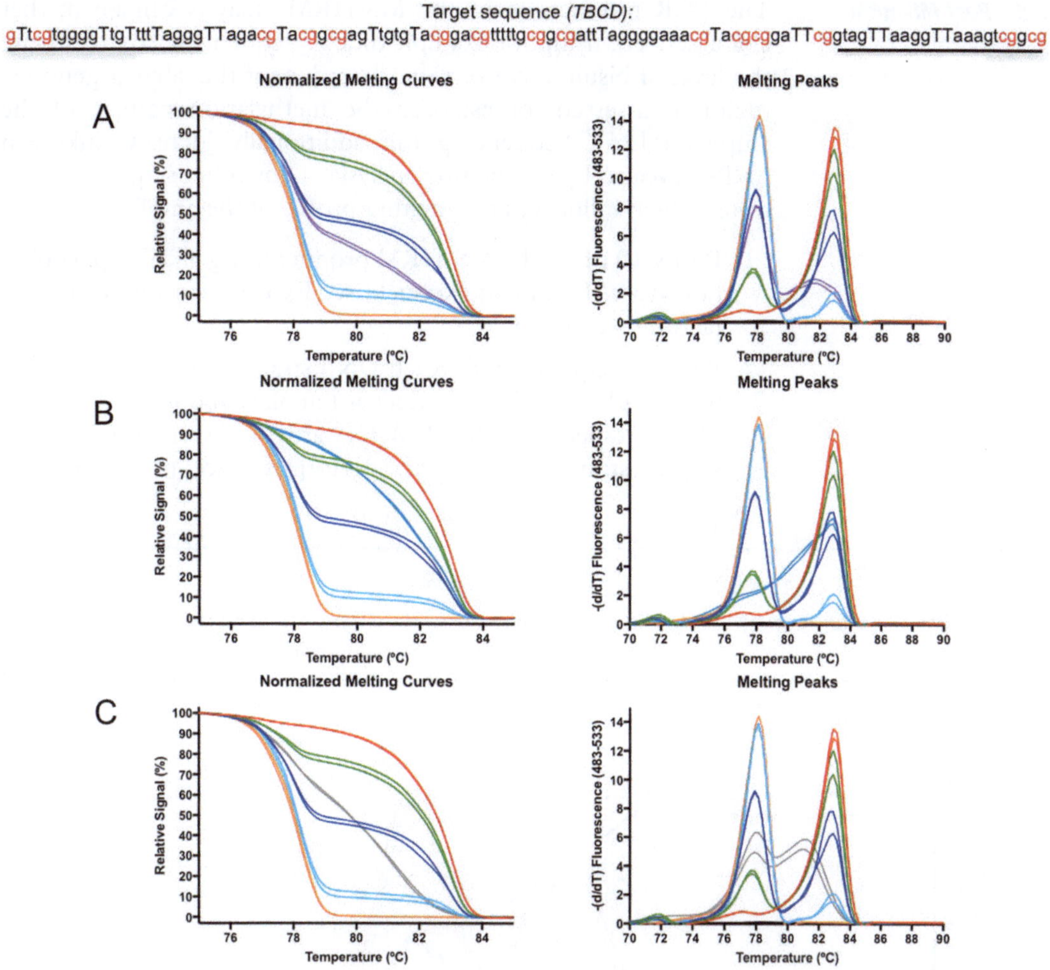

Fig. 5 Examples of heterogeneous melting profiles. Heterogeneously methylated samples: (**a**) (*Purple*), (**b**) (*Light blue*), and (**c**) (*Dark green*). The target sequence is presented in which CpG sites are marked in red, non-CpG cytosines as T and the primer locations are underlined. Standards: 100% (*red*), 10% (*green*), 1% (*blue*), 0.1% (*turquoise*), and 0% (*orange*)

according to the different amounts of methylated and unmethylated template. A pronounced difference in melting temperature between the unmethylated and the methylated amplicons ensures wider separation of the melting curves.

9. Heterogeneously methylated amplicons will not melt uniformly, and are easily detectable (Fig. 5 and **Note 15**). Heterogeneous methylation can only be estimated in a qualitative manner, but can be further studied through additional analyses (Subheading 3.8).
10. When test samples of unknown methylation levels are analyzed in parallel with the control standards, the level of methylation can be semiquantitatively estimated from the direct comparison of melting profile between the test sample and the standards (*see* **Note 16**).

3.8 Post MS-HRM Analyses

The PCR products from the MS-HRM analysis can be further characterized, using for example dideoxysequencing to investigate the level of bisulfite conversion, determine if the correct genomic region is assessed, or establish the methylation pattern of the amplicon [24]. Sequencing can additionally identify unknown SNPs, revealed by the melting analysis or any other sequence variations, which influences the melting profile of the amplicon.

1. Purify 10 μL of the MS-HRM product using alkaline phosphatase (FastAP) and an exonuclease (ExoI) according to manufacturer's protocol.
2. Perform sequencing for each DNA strand (forward or reverse) using 1 μL of purified product as template and 0.5 μL (10 μM) of the respective MS-HRM forward or reverse primer. The single stranded PCR is performed according to the

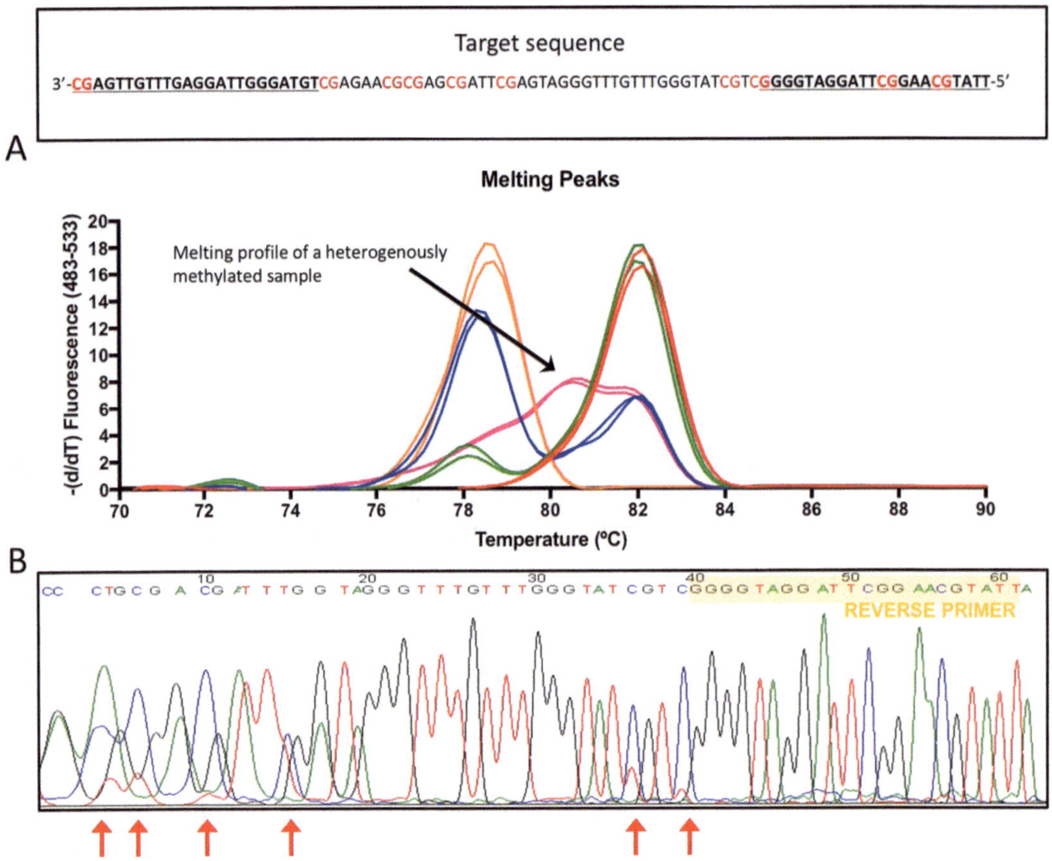

Fig. 6 Direct sequencing of a MS-HRM product. The target sequence is presented in which CpG sites are marked in red, non-CpG cytosines as T and primer locations are underlined. (**a**) Illustrates the uniform melting profiles of the standards (100% (*red*), 10% (*green*), 1% (*blue*), 0.1% (*turquoise*), and 0% (*orange*)) and a heterogeneously methylated sample (*pink*). (**b**) Subsequent direct sequencing of the MS-HRM product from the sample (*pink*), the reverse primer is marked in yellow, and red arrows indicate CpG sites

manufactures protocol (BigDye Terminator v 1.1) and submitted to capillary electrophoresis.

3. An example of heterogeneously methylation verified by direct sequencing of the MS-HRM products is illustrated in Fig. 6.
4. To obtain a detailed picture of the methylation pattern, as for heterogeneous methylated templates, cloning of the MS-HRM products can be performed and single clones sequenced to provide the exact position of each methylated CpG in each allele.
5. Another feature to be combined with MS-HRM is digital PCR (dMS-HRM) performed prior to high resolution melting [15], which may substitute cloning of the MS-HRM products followed by dideoxysequencing. The principle of digital MS-HRM is to dilute the sample into a series containing 0, 1 or occasionally 2 amplifiable templates in each well. The subsequent HRM can then identify the methylation pattern of each allele, and pooling all the dilutions will provide a quantitative measure for the methylation of the initial sample.

4 Notes

1. An assay designed to target the methylation status of only one CpG site results in melting curves, which are hardly separated, due to the limited differences in melting properties between the methylated and unmethylated amplicons. The difficulties in separating the different standard dilution curves may result in assays only able to discriminate between the unmethylated and the fully methylated amplicons (Fig. 7).
2. After bisulfite modification the two DNA strands are no longer complementary. A primer set is designed to target and amplify only one of the DNA strands. Generally, MS-HRM assays are designed to target the sense strand, but if primer design proves to be difficult or unspecific amplification occurs, the antisense strand can be used. Therefore, the first round of amplification will only target the DNA strand, which is complementary to one of the primers. After the first cycle of amplification both primers are able to anneal and prime synthesis of both DNA strands. Note that the strand has to be chosen *before* the in silico bisulfite treatment is made.
3. Including a CpG dinucleotide in the primer sequence will make the primer more specific for the methylated allele. The closer the CpG are to the 3′ end of the primer and if an additional CpG is included, the higher specificity of the primer for the methylated allele.

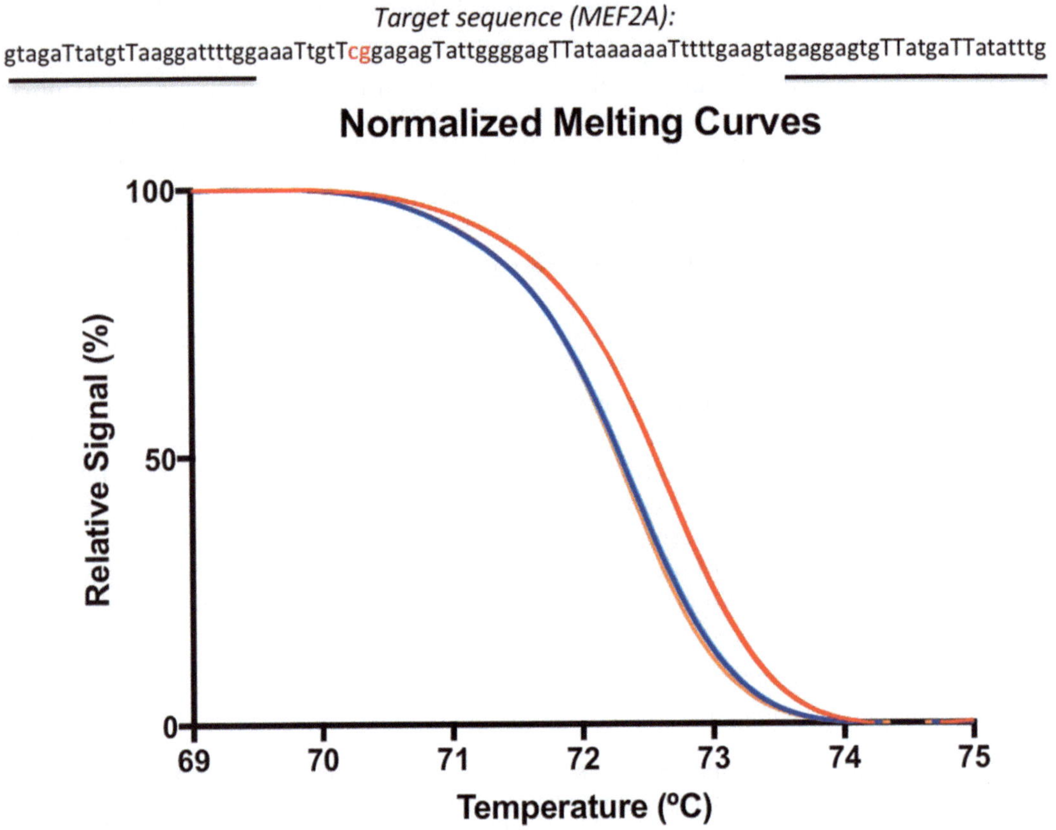

Fig. 7 Low sensitivity assay. MS-HRM assay targeting one CpG site resulting in a low sensitivity of the assay, in which only the 100% methylated standard can be distinguished from the remaining standards, which melts simultaneously. The target sequence is presented in which CpG sites marked in red, non-CpG cytosines as T and the primer locations are underlined. Standards: 100% (*red*), 10% (*green*), 1% (*blue*), 0.1% (*turquoise*), and 0% (*orange*)

4. A long amplicon may contain multiple melting domains, which can interfere with interpretation of the HRM results (Fig. 3). A clamp can be attached to the primers to stabilize the amplicon and make it melt more uniformly. The clamp consists of 3–4 cytosines and/or guanines attached to the 5′ end of the primer, creating an amplicon that is 6–8 bases longer than the region of interest. The primer melting temperature should be adapted accordingly.
5. The primer melting temperature is used to identify the annealing temperature (T_A) of the PCR amplification. Usually, T_A will be a few degrees lower than the primer melting temperature, yet this is dependent on the algorithm used for the calculation. The "2+4 rule" generally results in a higher T_A than most available software and one should aim for primer melting temperature of around 65 °C. In addition to each specific base in the primer, OligoCalc also takes the specific *order* of bases into

consideration. This generally results in a lower melting temperature compared to the "2+4 rule". If melting temperatures are calculated by OligoCalc, a temperature of 60 °C should be aimed at.

6. When using only one strand of bisulfite treated DNA as template for designing MS-HRM primers, it is highly important to take the sequence differences after bisulfite treatment into consideration.
7. The first round of WGA treated DNA results in a final volume of 200 μL and can be stored at −20 °C until further use. One batch of second-round WGA treated DNA yields approx. 2000–3000 ng DNA in 100 μL. Multiple batches can be treated simultaneously, and kept as stock at 4 °C.
8. One batch of standards can approximately be used for 20 analyses. It is important to note that bisulfite-treated DNA has a limited lifespan of approximately 1 month when stored at 4 °C and longer if stored at −20 °C or −80 °C. Thus, when producing the standards, the amount should be adapted in regard to the experiment.
9. The resulting amplification curves and HRM peaks can be slightly different depending on whether the bisulfite treatment has been performed using columns or 96-well plates. Therefore, it is important to use the same system for bisulfite treatment for all standards and test DNA to be analyzed simultaneously. The optimization procedure can be performed using DNA purified using columns, but when applying test samples, the same bisulfite conversion system has to be used for both test samples and corresponding standard.
10. The optimization procedure for the assay targeting the CpG island located in the promoter region of *CDH13* is illustrated in Fig. 2. The final, optimized assay produces an amplicon of 93 base pairs, using the primers: Forward: 5′ −gTtTTTacg-gaaaatatgTtTagtgTag – 3′ and Reverse: 5′ – cgAaAttctcgActA-cattttAtcc – 3′. Set up the reaction mixture with: 20 ng DNA, 500 nM of each primer, 3 mmol/L $MgCl_2$, 4.8 μL of High Resolution Melting Master Mix and sterilized H_2O to a final volume of 10 μL. The cycling protocol comprises: initial denaturation at 95 °C for 10 min, followed by a cycling process with 50 cycles of 95 °C for 5 s, $T_A = 57$ °C for 10 s and 72 °C for 10 s. Melting process: initial denaturing step at 95 °C for 1 min, followed by 65 °C for 1 min and finally a stepwise increase in temperature from 65 °C to 95 °C, with a rate of 0.1 °C/s.
11. Increasing or decreasing the annealing temperature affects the binding properties of the primers. Increasing the temperature results in more specific binding to the methylated template, which are complementary to the primers, and may in addition

eliminate potential unspecific binding and primer-dimer formation, which may be observed in the gDNA control or in the NTC.

12. The number of PCR cycles can be increased up to 60; this may though enhance the risk of unspecific primer annealing. If multiple assays are optimized to work under uniform condition this can minimize machine and hands-on time, necessary prerequisites for a high-throughput setting.
13. External conditions as a highly elevated room temperature have been found to influence assay performance. An example of the influence of increased room temperature on the HRM instrument performance is shown in Fig. 3.
14. The real-time data provides a quality control step for PCR amplification, which should be uniform and reach plateau for all standards. Generation of an amplification product in NTC may reflect contamination, but if it melts prior to the unmethylated amplicons it may reflect self-priming of the primers. Raising the T_A can eliminate unspecific binding.
15. Heterogeneously methylated templates will melt uneven in-between the methylated and unmethylated templates often resulting in broader "bumpy" peaks (Fig. 5). This is due to heteroduplex formation between the heterogeneously methylated templates (Fig. 1). Visualized using the negative derivative peaks heterogeneous methylation can easily be detected (Fig. 5).
16. The melting profiles of biological samples may be easier to interpret when compared to the standards using both the melting peaks and normalized melting curves.

Acknowledgments

The authors thank T.K. Wojdacz and I.L. Daugaard for their critical review of the manuscript.

Competing interests

LLH is an inventor on a patent and co-founder of the company MethylDetect on the aspects of the MS-HRM technology.

References

1. Frommer M, McDonald LE, Millar DS et al (1992) A genomic sequencing protocol that yields a positive display of 5-methylcytosine residues in individual DNA strands. Proc Natl Acad Sci U S A 89:1827–1831
2. Worm J, Aggerholm A, Guldberg P (2001) In-tube DNA methylation profiling by fluorescence melting curve analysis. Clin Chem 47:1183–1189
3. Wojdacz TK, Dobrovic A (2007) Methylation-sensitive high resolution melting (MS-HRM): a new approach for sensitive and high-throughput assessment of methylation. Nucleic Acids Res 35:e41

4. Warnecke PM, Stirzaker C, Melki JR et al (1997) Detection and measurement of PCR bias in quantitative methylation analysis of bisulphite-treated DNA. Nucleic Acids Res 25:4422–4426
5. Wojdacz TK, Hansen LL (2006) Techniques used in studies of age-related DNA methylation changes. Ann N Y Acad Sci 1067:479–487
6. Wojdacz TK, Borgbo T, Hansen LL (2009) Primer design versus PCR bias in methylation independent PCR amplifications. Epigenetics 4:231–234
7. Wojdacz TK, Hansen LL, Dobrovic A (2008) A new approach to primer design for the control of PCR bias in methylation studies. BMC Res Notes 1:54
8. Xiao Z, Li B, Wang G et al (2014) Validation of methylation-sensitive high-resolution melting (MS-HRM) for the detection of stool DNA methylation in colorectal neoplasms. Clin Chim Acta 431:154–163
9. Wojdacz TK, Dobrovic A, Hansen LL (2008) Methylation-sensitive high-resolution melting. Nat Protoc 3:1903–1908
10. Tse MY, Ashbury JE, Zwingerman N et al (2011) A refined, rapid and reproducible high resolution melt (HRM)-based method suitable for quantification of global LINE-1 repetitive element methylation. BMC Res Notes 4:565
11. Wojdacz TK, Windelov JA, Thestrup BB et al (2014) Identification and characterization of locus specific methylation patterns within novel loci undergoing hypermethylation during breast cancer pathogenesis. Breast Cancer Res 16:R17
12. Kristensen LS, Wojdacz TK, Thestrup BB et al (2009) Quality assessment of DNA derived from up to 30 years old formalin fixed paraffin embedded (FFPE) tissue for PCR-based methylation analysis using SMART-MSP and MS-HRM. BMC Cancer 9:453
13. Oster B, Thorsen K, Lamy P et al (2011) Identification and validation of highly frequent CpG island hypermethylation in colorectal adenomas and carcinomas. Int J Cancer 129:2855–2866
14. Reinert T, Modin C, Castano FM et al (2011) Comprehensive genome methylation analysis in bladder cancer: identification and validation of novel methylated genes and application of these as urinary tumor markers. Clin Cancer Res 17:5582–5592
15. Snell C, Krypuy M, Wong EM et al (2008) BRCA1 promoter methylation in peripheral blood DNA of mutation negative familial breast cancer patients with a BRCA1 tumour phenotype. Breast Cancer Res 10:R12
16. Liu W, Guan M, Su B et al (2010) Rapid determination of AKAP12 promoter methylation levels in peripheral blood using methylation-sensitive high resolution melting (MS-HRM) analysis: application in colorectal cancer. Clin Chim Acta 411:940–946
17. Wojdacz TK, Thestrup BB, Cold S et al (2011) No difference in the frequency of locus-specific methylation in the peripheral blood DNA of women diagnosed with breast cancer and age-matched controls. Future Oncol 7:1451–1455
18. Wojdacz TK, Thestrup BB, Overgaard J et al (2011) Methylation of cancer related genes in tumor and peripheral blood DNA from the same breast cancer patient as two independent events. Diagn Pathol 6:116
19. Li X, Wang Y, Zhang Z et al (2013) Correlation of and methylation levels between peripheral blood leukocytes and colorectal tissue DNA samples in colorectal cancer patients. Oncol Lett 6:1370–1376
20. White HE, Hall VJ, Cross NC (2007) Methylation-sensitive high-resolution melting-curve analysis of the SNRPN gene as a diagnostic screen for Prader-Willi and Angelman syndromes. Clin Chem 53:1960–1962
21. Candiloro IL, Mikeska T, Hokland P et al (2008) Rapid analysis of heterogeneously methylated DNA using digital methylation-sensitive high resolution melting: application to the CDKN2B (p15) gene. Epigenetics Chromatin 1:7
22. Candiloro IL, Mikeska T, Dobrovic A (2011) Assessing combined methylation-sensitive high resolution melting and pyrosequencing for the analysis of heterogeneous DNA methylation. Epigenetics 6:500–507
23. Hernandez HG, Tse MY, Pang SC et al (2013) Optimizing methodologies for PCR-based DNA methylation analysis. BioTechniques 55:181–197
24. Wojdacz TK, Moller TH, Thestrup BB et al (2010) Limitations and advantages of MS-HRM and bisulfite sequencing for single locus methylation studies. Expert Rev Mol Diagn 10:575–580

Chapter 29

Hairpin Bisulfite Sequencing: Synchronous Methylation Analysis on Complementary DNA Strands of Individual Chromosomes

Pascal Giehr and Jörn Walter

Abstract

The accurate and quantitative detection of 5-methylcytosine is of great importance in the field of epigenetics. The method of choice is usually bisulfite sequencing because of the high resolution and the possibility to combine it with next generation sequencing. Nevertheless, also this method has its limitations. Following the bisulfite treatment DNA strands are no longer complementary such that in a subsequent PCR amplification the DNA methylation patterns information of only one of the two DNA strand is preserved. Several years ago Hairpin Bisulfite sequencing was developed as a method to obtain the pattern information on complementary DNA strands. The method requires fragmentation (usually by enzymatic cleavage) of genomic DNA followed by a covalent linking of both DNA strands through ligation of a short DNA hairpin oligonucleotide to both strands. The ligated covalently linked dsDNA products are then subjected to a conventional bisulfite treatment during which all unmodified cytosines are converted to uracils. During the treatment the DNA is denatured forming noncomplementary ssDNA circles. These circles serve as a template for a locus specific PCR to amplify chromosomal patterns of the region of interest. As a result one ends up with a linearized product, which contains the methylation information of both complementary DNA strands.

Key words Hairpin bisulfite sequencing, DNA strands, Chromosome, Restriction, Covalent linking, ssDNA circles

1 Introduction

Hairpin Bisulfite Sequencing (HBS) is a method to detect DNA methylation on both complementary DNA strands of individual DNA molecules [1]. HBS allows to discriminate if both strands are methylated or if a hemimethylation is present on only one of the two complementary DNA strands (upper or lower strand) or if both strands are symmetrically unmethylated. It also allows to discriminate a true non-CpG methylation from a genetic polymorphic (mutated) site. Hairpin bisulfite sequencing is more powerful and appropriate compared to conventional bisulfite sequencing

Jörg Tost (ed.), *DNA Methylation Protocols*, Methods in Molecular Biology, vol. 1708,
https://doi.org/10.1007/978-1-4939-7481-8_29, © Springer Science+Business Media, LLC 2018

when one needs to detect the symmetry of DNA methylation patterns on both DNA strands, i.e., when analyzing active demethylation, de novo methylation or maintenance methylation events during cell replication or stages of reprogramming [2–4].

For the use of the HBS method the following general steps have to be considered. A standard HBS approach starts with the digestion of DNA by a defined restriction enzymes (usually four base cutter) that is not sensitive to DNA methylation, followed by covalent linking of the DNA fragments (upper and lower DNA strand) to a short hairpin DNA oligonucleotide using conventional ligation (Fig. 1). Restriction enzymes generating "sticky ends" should be preferred, since this will increase the efficiency of linker ligation. However, in our experience also the use of enzymes creating non-overhanging "blunt ends" is possible. The ligation is carried out using T4 DNA ligase. To ensure a high yield of DNA

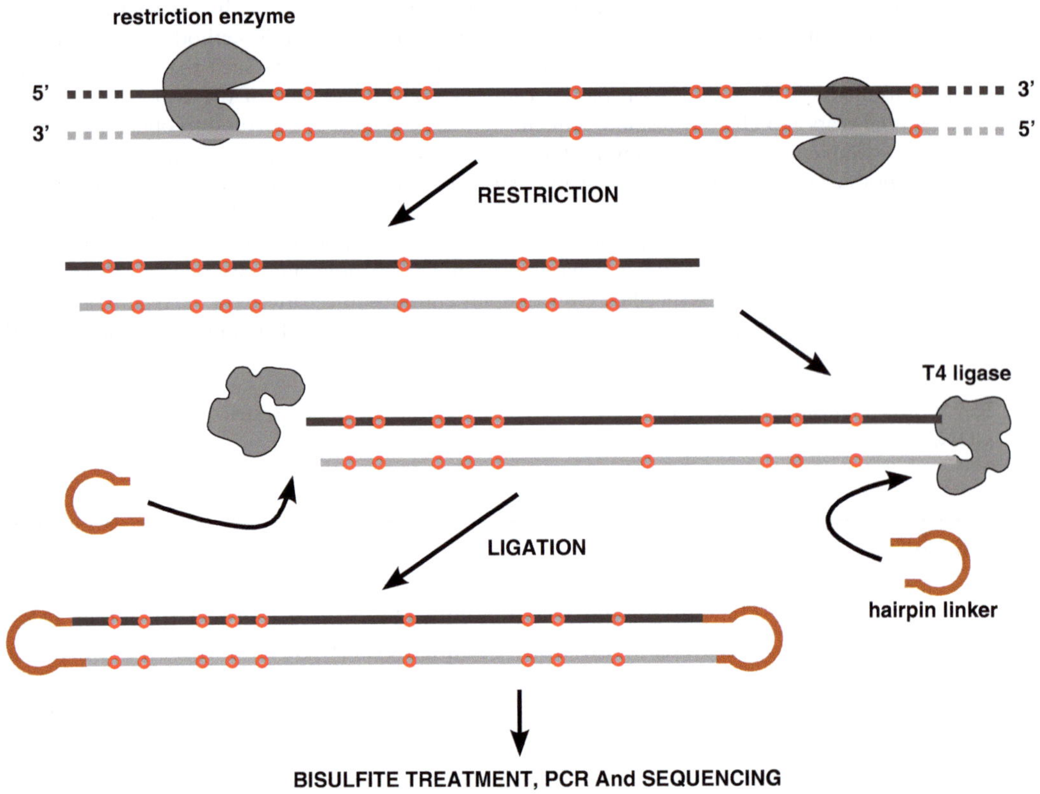

Fig. 1 Workflow of the Hairpin Bisulfite Sequencing protocol; genomic DNA is cut using a restriction enzyme, which is unaffected by DNA methylation. A complementary hairpin oligonucleotide is ligated to link upper and lower strand covalently together. The constructs are in the next step subject to bisulfite treatment resulting in single stranded circular DNA. After treatment the converted DNA serves as a template for a locus-specific PCR. PCR Products are then purified and sequenced. *Straight* and *dashed lines* indicate DNA strands, *red circles* illustrate CpG positions

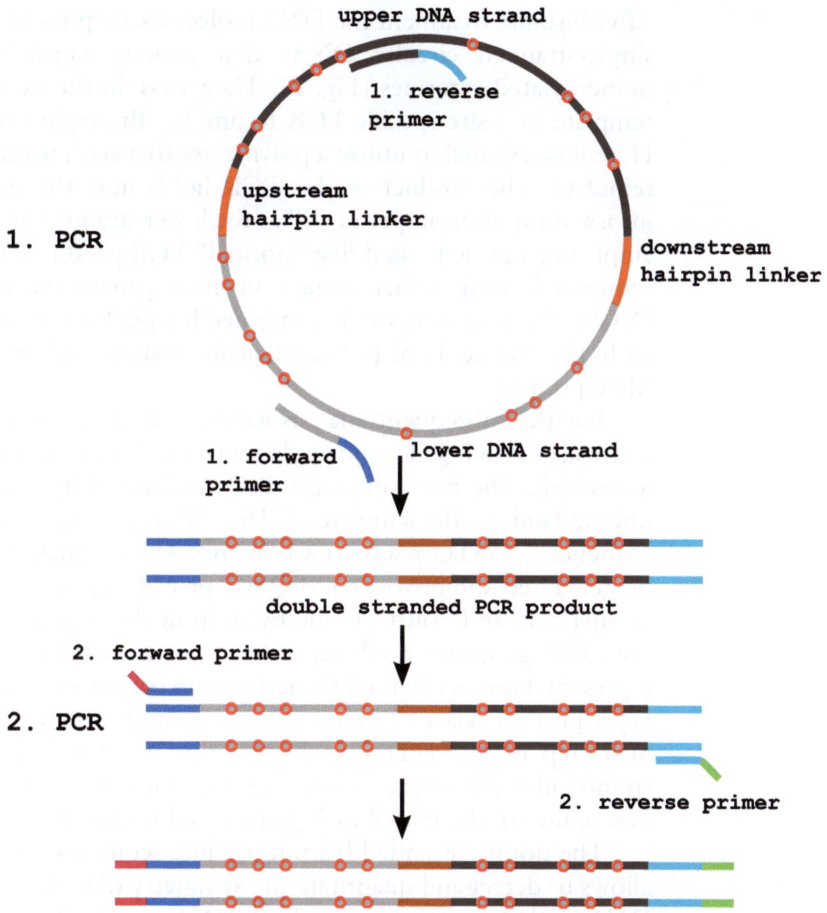

Fig. 2 Workflow after bisulfite treatment; bisulfite treated circular hairpin constructs are amplified in two consecutive PCRs; the fusion primers used in the first amplification step carry on the 5′ end parts of the sequencing adapters which will become part of the PCR product. In the second PCR the rest of the adapter sequence is introduced to the amplicon. *Lines* indicate DNA strands, *red circles* illustrate CpG positions

hairpin constructs the hairpin oligonucleotide is provided in excess to minimize the likelihood of religation of DNA fragments.

The overhang of the linker is designed with complementary overhangs. For example, the restriction enzyme MspI will leave a 5′ CG overhang, accordantly the hairpin linker also has to have a 5′ "CG" overhang (Fig. 2). Because of the enzymatic steps we recommend to use high quality (nondegraded) double stranded DNA (dsDNA). The circular constructs obtained after ligation are then subject to bisulfite treatment, in which all cytosines are converted to uracils. The steric closeness (intertwined ssDNA rings) of the complementary DNA strands favors a quicker renaturation to dsDNA. Hence the bisulfite conversion of hairpin constructs is more challenging than that of normal DNA. To avoid a reannealing we recommend using cycling bisulfite protocols with additional denaturation steps or alternatively higher incubation temperatures.

After bisulfite treatment the DNA molecules are present in form of single-stranded circular DNAs that contain uracil instead of unmethylated cytosines (Fig. 2). They serve in the next step as a template in a site specific PCR to amplify the region of interest. Here it is essential to utilize a polymerase that accepts uracils in the template. The product of this PCR holds now the methylation information of the upper as well as the lower strand. The generated amplicons can be treated like "normal" PCR products and can be sequenced using either Sanger or next generation sequencing (NGS). We have successfully combined hairpin bisulfite sequencing with the Roche FLX pyrosequencing system and the Illumina MiSeq system.

For the subsequent analysis we use the trimmed and quality-controlled FASTQ files and apply two bioinformatics tools develop in our lab. The first tool is the BiQ Analyzer (http://epigenetik.uni-saarland.de/de/software/) [5]. The program aligns the sequenced FASTQ reads to a reference HBS sequence (needs to be generated and provided). BiQ will provide an overview of CpG methylation and non-CpG methylation in the sequences. As outputs BiQ generates a tab separated table and different graphical representations, such as CpG methylation pattern map and quantitative pearl-necklace diagrams. The table output is then used in the next step by the Hairpinanalyzer script to back-fold the single strand information into a double stranded format. A more detailed description of the workflow is given in subsection 3.

The double stranded hairpin bisulfite sequencing output now allows to detect and quantitate the symmetry of CpG and CpNpG DNA methylation patterns on both DNA strands of one individual chromosome and to unambiguously identify nonsymmetrical cytosine methylation. We have applied the method to identify the massive occurrence of hemimethylated sites and general loss of methylation in Dnmt1KO ES-cells [2]. The hairpin bisulfite sequencing method provides matched stranded information, which directly shows the hemimethylated pattern formation.

The use of a hairpin linker also offers additional technical advantages. Since the added linker contains unmodified cytosines, it is possible to directly calculate the true conversion rate obtained during bisulfite treatment and at the same time can estimate the true amount of methylated non-CpG positions [3]. Further, in the loop sequences of the linker variable nucleotides can be introduced creating a barcoding for each DNA molecule. This allows to identify duplicated sequences generated during PCR and to exclude them from further analysis.

Despite of the many advantages of the Hairpin Bisulfite Sequencing method HBS also has some experimental limitations. In a locus-specific HBS analysis the obligate use of a suitable restriction enzyme (absence of recognitions site, distance to the analyzed region) can become a limitation. Moreover, the restriction enzymes must not be affected by DNA methylation, which would otherwise lead to a massive underrepresentation of methylated sites. Further, the size of the region that can be analyzed is limited. Based on the fact, that the PCR product contains the information of both upper and lower strands the product will be double the size of the genomic region. However, this is only a small disadvantage, since modern sequencing techniques allow to analyze sequences with a size over 500 base pairs (bp) in length (particularly on a FLX or MiSeq sequencer).

2 Materials

2.1 Experimental Design

1. Identification of suitable restriction sites close to the region of interest using NEBcutter (http://tools.neb.com/NEBcutter2/) or WatCut restrictions analysis (http://watcut.uwaterloo.ca/template.php).
2. Design of the reference sequence (upper strand-linker-lower strand) for primer search and for the BiQ subsequent methylation analysis. The precise outline will be described in subsection 3.
3. Primers for PCR are either designed "by eye" or with the help of online tools like Primer3Plus (http://www.bioinformatics.nl/cgi-bin/primer3plus/primer3plus.cgi/).
4. Hairpin linkers are designed according to match the restriction enzyme ends and to containing indices/bar codes.

2.2 Sample Preparation

1. Qubit® BR assay kit for measurement of the DNA concentration (Thermo Fisher Scientific).
2. Restriction enzymes.
3. T4 DNA Ligase (New England Biolabs).
4. EZ DNA Methylation™ Kit; EZ DNA Methylation-Gold™ Kit (Zymo Research) or manual protocol.
5. HOT FIREPol® DNA Polymerase (Solis BioDyne) or HotStarTaq DNA polymerase (Qiagen) or similar.
6. Agencourt® AMPure® XP beads (Beckman Coulter).
7. Gel purification kit (e.g., AveGene or similar).
8. Primers and hairpin linkers.
9. dNTPs.
10. 25 mM $MgCl_2$.
11. 1× TE buffer: 10 mM Tris–HCl, pH 7.4; 1 mM EDTA, pH 8.0.

2.3 Sequencing and Data Analysis

1. Genome Sequencer FLX 454 System (Roche) or MiSeq Desktop Sequencer (Illumina)
2. For Methylation analysis BiQ Analyzer HT (http://biq-analyzer-ht.bioinf.mpi-inf.mpg.de/) was used followed by the use of Hairpinanalyzer. The python script of the hairpin analyzer can be received from http://epigenetik.uni-saarland.de/de/software/).

3 Methods

3.1 Experimental Design

A proper planning and design of the experiment for Hairpin Bisulfite Sequencing includes three main steps.

3.1.1 Selecting Restriction Enzymes

The first step when designing the Hairpin Bisulfite experiment is the search for suitable restriction enzymes within the region of interest. When selecting enzymes a few things should be considered.

1. First, the restriction enzyme should not be affected by DNA methylation to ensure that unmethylated, hemimethylated, and fully methylated regions are equally represented in the later results.
2. The use of enzymes, which create sticky ends, should be preferred since the ligation will work more efficient. Nevertheless, the use of blunt end creating enzymes is possible.
3. There are several online tools, which are suitable for the search of restriction enzymes for example "NEBcutter" or "WatCut." Both tools also provide the information on methylation sensitivity and sometimes even other DNA modifications like 5hmC, 5fC, and 5caC.

3.1.2 Hairpin Linker and Reference Sequence Design

The hairpin linker itself can be divided into three sections.

1. The first part is variable and depends on the used restriction enzyme (Fig. 3). For example MspI will create a 5′-CG overhang; therefore the linker must contain a 5′-CG overhang including a free phosphate group to allow ligation to the DNA.
2. The second part always consists of the same sequence and facilitates the formation of the hairpin structure (Fig. 3). The use of unmodified cytosine within this linker part allows later an exact and unbiased calculation of the conversion rate during the bisulfite treatment and permits a more accurate detection of non-CpG methylation.
3. The last part is forming the loop structure of the linker. It contains a unique sequence that cannot form any double strand structures (Fig. 3).

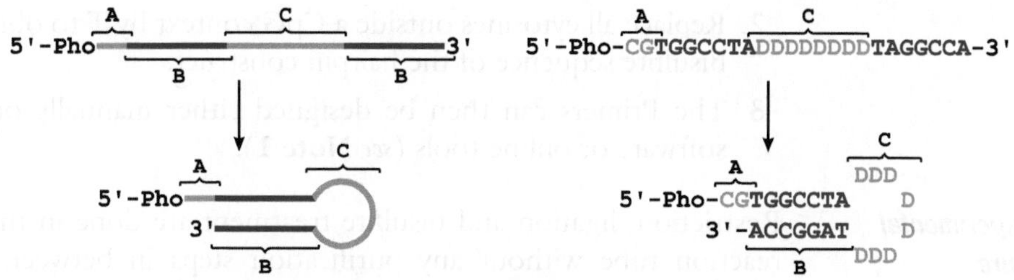

Fig. 3 Schematic illustration of the hairpin linker; *A*, variable, restriction enzyme dependent sequence; *B*, constant, hairpin formation facilitating sequence; *C*, variable loop sequence; as an example the structure and sequence of a hairpin oligo for MspI restriction is shown in a denatured, single strand and annealed, folded state

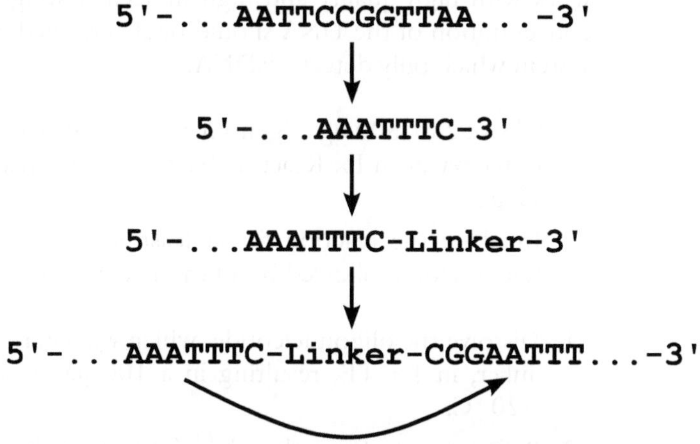

Fig. 4 Example of the design of a hairpin construct; The left part next to the restriction site is removed. The linker sequence is pasted followed by the reverse complement sequence of the right part of the DNA sequence

4. As shown in Fig. 3 the loop can obtain 6–8 variable nucleotide positions. This allows to create an individual barcode for each DNA molecule and to exclude duplicates of the PCR in further computational analysis. Using eight of these positions it is theoretically possible to distinguish between 6561 (3^8) sequences.

3.1.3 Reference Sequence and Primer Design

The next step in the experimental design is the construction of the reference sequence. This sequence is needed in order to design primers and also for later sequencing and methylation analysis.

The reference sequence can easily be designed with any software that can handle text files.

1. Download the genomic sequence of the region, mark the restriction site used, delete the sequence in front or after the restriction site, add the linker sequence and finally paste the reverse complement of the remaining sequence on the other site of the linker (Fig. 4).

2. Replace all cytosines outside a CpG context by T to obtain the bisulfite sequence of the hairpin construct.
3. The Primers can then be designed either manually or using software or online tools (*see* **Note 1**).

3.2 Experimental Procedure

Restriction, ligation and bisulfite treatment are done in the same reaction tube without any purification steps in between which minimizes the loss of DNA. The described protocol below is the standard protocol used in our lab, however depending on the amount of DNA or the restriction enzyme used in the reaction an optimization or adjustments might be necessary. As starting material, genomic DNA from phenol–chloroform or various extraction kits can be used. Due to the nature of the method it is important to work with high quality and high molecular weight dsDNA. The concentration of the DNA should be determined using the Qubit system which only detects dsDNA.

3.2.1 Restriction

1. Cleave 200–500 ng of genomic DNA with 10 units of a restriction enzyme in 1× Reaction buffer in a total reaction volume of 17 μL.
2. Incubate the reaction for at least 3 h at the recommended temperature, followed by a heat inactivation (*see* **Note 2**).

3.2.2 Preparation of Hairpin Linker and Ligation

1. Dissolve the oligonucleotide, which will later form the hairpin linker, in 1× TE, resulting in a 100 μM stock and store at −20 °C.
2. Before usage, form the oligo into the right structure. Heat 50 μL of the 100 μM solution to 98 °C for 15 min followed by cool down using the slowest cooling rate of a thermocycler until 20 °C is reached. In this form the hairpin linker is rather stable but should be stored at −20 °C.
3. Add 1 μL of a 100 μM hairpin linker solution to the reaction together with 200 U T4 DNA ligase and 2 μL of 10 mM ATP.
4. Incubate the reaction for at least 3 h at 16 °C. The ligation can also be performed overnight.

3.2.3 Bisulfite Treatment

As mentioned before, there are several bisulfite kits, which can be used for the conversion of hairpin constructs. Since hairpin DNA molecules tend to fold back rather fast and the conversion of cytosine can only occur on single stranded DNA, we recommend a protocol with higher incubation temperature or additional denaturation steps. Kits successfully used in our lab are listed above in subsection 2. When using a homemade protocol, the conversion rate can be verified by looking into the conversion of cytosines included in the hairpin linker. A manual protocol used in our lab has been described previously [2]. In addition, the hairpin protocol is also extendable to oxidative bisulfite sequencing (*see also* Chapter 34) or other "chemical" forms of sequencing [6].

Table 1
Examples of PCR protocols to amplify Hairpin Bisulfite molecules

HotFirePol PCR	HotStarTaq PCR
3 μL 10× reaction buffer	3 μL 10× reaction buffer
3 μL 25 mM $MgCl_2$	1.2 μL 25 mM $MgCl_2$
2.4 μL 10 mM dNTPs	2.4 μL 10 mM dNTPs
0.5 μL 10 μM forward primer	0.5 μL 10 μM forward primer
0.5 μL 10 μM reverse primer	0.5 μL 10 μM reverse primer
0.5 μL HotFirePol DNA polymerase	0.3 μL HotStarTaq DNA polymerase
Add to 30 μL ddH_2O	Add to 30 μL ddH_2O

Table 2
Temperature profile of HotFire/HotStarTaq

95 °C	15 min	
95 °C 50–62 °C 72 °C	1 min 1 min 1 min	35–45 cycles
72 °C	5 min	
4 °C	Hold	

3.2.4 PCR

Because of the bisulfite treatment it is essential to use a polymerase that recognizes uracil in the template strand for the PCR. The best results in our lab were achieved using the HotFirePol from Solis-BioDyne or HotStarTaq from Qiagen. We recommend a total reaction volume 30 μL and to perform multiple PCRs in parallel to ensure a low number of duplicated reads.

1. Pipette the reaction mixture according to Table 1 using either enzyme.
2. Both enzymes share similar temperature characteristics during PCR. A typical cycler protocol for both enzymes is given in Table 2.
3. Purification of the PCR products is performed using 27 μL AMPureXP Beads (0.9×).

3.2.5 Sequencing

Hairpin bisulfite is compatible with both Sanger and next generation sequencing (NGS). To prepare the samples for NGS one has to introduce adapter on each side of the amplicon, which are compatible with the sequencing platform to be used. These adapter bind to the sequencing platform and are the start point of the sequencing process. In addition each adapter carries a unique sequence ID, which allows sequencing of multiple samples at the same time. The adapter sequence is introduced by the use of fusion primers in two consecutive PCRs (Fig. 2).

In the first PCR the primers consist at their 3′ end of the target specific sequence complementary to the bisulfite treated DNA and carry at their 5′ end the first part of the adapter sequence resulting in an amplification of the target sequence and the introduction of the first part of the adaptor (Table 1 for PCR conditions).

In the second PCR the primer will bind the adaptor part introduced during the first PCR step. These primers carry the sequence, which later bind to the sequencing platform and in addition carry a sample specific sequence ID. Table 3 provides all primer and adapter sequences needed for sequencing on the Illumina MiSeq platform.

The second amplification can be performed as a multiplex PCR where several amplicons of distinct genomic regions can be prepared for sequencing at the same time. For this the concentration of each amplicons must be adjusted to 5 nM and pooled into one reaction.

1. Pipette components and perform the second PCR amplification PCR as outlined in Table 4 (*see* **Note 3**).

2. After incubation, clean-up the reaction again using 55 μL AMPureXP beads (1.1×).

Table 3
Illumina adapter and primer sequences; i5/i7 = index; grey = flow cell binding sequence; Oligonucleotide sequences © 2016 Illumina, Inc. All rights reserved

Primer	Primer Sequence
1.PCR Forward	TCTTTCCCTACACGACGCTCTTCCGATCT_Amplicon Specific
1.PCR Reverse	GTGACTGGAGTTCAGACGTGTGCTCTTCCGATCT_Amplicon Specific
2.PCR Forward	AATGATACGGCGACCACCGAGATCTACAC[i5]ACACTCTTTCCCTACACGACGCTCTTCCGATCT
(i5, 6bp index e.g.: CGTGAT)	
2.PCR Forward	GATCGGAAGAGCACACGTCTGAACTCCAGTCAC[i7]ATCTCGTATGCCGTCTTCTGCTTG
(i7, 6bp index e.g.: AAGCTA)	

Table 4
List of chemicals and cycler condition for the second PCR

Enrichment PCR	Cycler conditions	
25.0 μL 5 nM amplicon pool		
5.0 μL 10× buffer HotStarTaq	95 °C—15 min	
2.0 μL 25 mM $MgCl_2$ 4.0 μL 10 mM dNTPs 2.5 μL 10 μM index primers 2.5 μL 10 μM universal primers	95 °C—30 s 60 °C—30 s 72 °C—30 s 72 °C—5 min	5 cycles
0.6 μL HotStarTaq	4 °C—Hold	
8.4 μL ddH_2O		

3. Adjust to a final concentration of 10 nM.
4. Prepare the final amplicon library by pooling all enrichment PCRs into one reaction.
5. Following the MiSeq preparation protocol from Illumina, dilute the library stepwise to a final concentration of 18 pM.

3.3 Data Analysis

To obtain the information of the symmetric DNA methylation two separate steps are necessary. First, the methylation information has to be obtained from the sequenced PCR product. Second, the methylation information has to be translated back to the DNA double strand. For this we developed in our lab two different bioinformatics tools.

3.3.1 Methylation Analysis

The methylation analysis is performed using the BiQ HT Analyzer [2]. The BiQ HT is a Java based program designed for locus-specific DNA methylation analysis of high-throughput bisulfite sequencing data (*see* **Note 4**). The program aligns the sequenced reads to a reference sequence. Thereby comparing all positions where a cytosine is expected it detects the methylation status of the sequenced loci. Both, reference sequence and sequencing data has to be provided in a FASTA file format. The program calculates different quality scores including alignment score, sequence identity, bisulfite conversion rate and number of missing sites. The different quality scores can also be used to filter the data, for example against low quality reads or low sequence identity. BiQ HT will automatically use the default settings for filtering, but each value can be adjusted manually by the user. Besides the analysis of CpGs, BiQ is also able to detect methylation in a non-CpG context (CpHpG; CpHpH). After analysis the data is presented and can be stored in different ways. A typical output consist of a tab stop separated table which includes all quality and methylation values and different types of methylation diagrams such as pattern maps and pearl-necklace diagram.

For the Hairpin analysis, three independent analyzing steps with BiQ HT are necessary. The first step is the detection of CpG methylation. The reference sequence used in this step consists of the sequence of both DNA strands and the linker sequence in between.

1. Replace in the linker sequence all cytosines by thymines because the linker will be analyzed later in an independent step. Depending on the type of loci, adjust the filter sequence identity. When analyzing repetitive elements for example, a lower sequence identity (80%) should be chosen because of the variability of the sequence of those elements.
2. Calculate the non-CpG methylation. Here it can be advantageous to replace all CpGs in the sequencing data by NpN to avoid confounding by possible mutations or sequencing failures, which create new CpG positions and lead to wrong estimations of non-CpG methylation. Again this is especially important when analyzing repetitive elements, due to their sequence variability. Note that also all CpGs in the reference should be replaced by NpN to allow an accurate alignment of the sequenced reads.
3. Estimate the conversion rate using the unconverted linker sequence as the reference sequence. Like in the non-CpG analysis the method to analyze Cs has to be chosen. When using wobble-position, within the linker loop, lower the sequence identity to 70 or even 60%, otherwise most of the sequences will be filtered out.

For each of the three analysis steps a separate folder has to be created in which the results are stored. The information in the different folders is then used to reconstruct the double strand information. Only the CpG folder is needed to reconstruct the double strand. However, conversion rate and non-CpG positions will then not be analyzed (*see* **Note 5**).

3.3.2 Reconstructing the Double Strand

The four folders containing the information about CpG methylation, non-CpG methylation, conversion rate and optionally SNPs are then used to reconstruct the double strand and calculate the amount of both strands methylated, only upper strand methylated, only lower strand methylated or both strands unmethylated.

For this purpose we have developed in our group a script we called Hairpinanalyzer. The program is based on python and is able to translate the single strand information, created by the BiQ HT,

Fig. 5 Example of a methylation pattern map created by the Hairpinanalyzer; each column represents one CpG dyad (1–5) and one row a sequenced read; *red* = fully methylated CpG dyads; *light* and *dark green* = hemimethylated CpG dyads; *blue* unmethylated CpG dyads; *white bars* indicate mutated or not analyzable CpGs; the *bar* on the *left* site shows a summary of the methylation over all CpG positions analyzed (*M*)

into a double strand output. The Hairpinanalyzer has no graphical interface and has to be run via command line and configuration of the source code. Like BiQ the Hairpinanalyzer creates an output in form of a tab separated table and a methylation pattern map. An example of a pattern map created by the Hairpinanalyzer is given in Fig. 5. Each column of the map represents a CpG position whereas each row indicates one sequenced read. The different methylation statistics include, both strands methylated, left (of the linker) strand methylated, right strand methylated, both strands unmethylated and mutated or not detectable. The colors can be chosen manually. The tab separated table contains information about number of reads, number of analyzed CpGs, Methylation status, Mutations or Sequencing errors as well as the information about SNPs and non-CpG methylation (*see* **Note 6**).

4 Notes

1. Even though both strands are no longer perfectly complementary after bisulfite treatment due to the conversion of cytosine; there are still regions that are complementary to some extent. This makes the primer design sometimes difficult because of dimer formations. To avoid this, it is sometimes necessary to use relatively short primers. This on the other hand will result in relatively low annealing temperatures. In our experience primers with a size of 23 bases and an annealing temperature over 50 °C are working fine. However, if possible a higher temperature should always be preferred, since it will increase the specificity of the PCR.
2. The restrictions conditions described in this protocol were suitable for all restriction enzymes used in our lab so far (BsaWI, DdeI, Eco47I, MspI, and TaqI). However, depending on the amount of DNA and the used restriction enzyme it might be necessary to change the parameters to obtain optimal reaction conditions.
3. When using next generation sequencing, the fusion primers of the PCR have to be adjusted to the sequencing system. In our lab we have successfully used FLX as well as Illumina systems for the sequencing of Hairpin Bisulfite amplicons.
4. The BQ Analyzer HT can be downloaded from http://biq-analyzer-ht.bioinf.mpi-inf.mpg.de. There is also a Java Web Start version available as well as detailed documentation.
5. It is also possible to include a fourth folder that contains the information about SNPs and the barcode of the linker. For that a fourth analysis with BiQ has to be performed. By repeating the analysis for CpG including the option output SNPs.
6. The Hairpinanalyzer was programmed by Mathias Bader in our lab. Unfortunately this script is not available online, but can be obtained upon request from the authors.

References

1. Laird CD, Pleasant ND, Clark AD et al (2004) Hairpin-bisulfite PCR: assessing epigenetic methylation patterns on complementary strands of individual DNA molecules. Proc Natl Acad Sci U S A 101:204–209
2. Arand J, Spieler D, Karius T et al (2012) In vivo control of CpG and non-CpG DNA methylation by DNA methyltransferases. PLoS Genet 8: e1002750
3. Ficz G, Hore TA, Santos F et al (2013) FGF signaling inhibition in ESCs drives rapid genome-wide demethylation to the epigenetic ground state of pluripotency. Cell Stem Cell 13:351–359
4. Seisenberger S, Andrews S, Krueger F et al (2012) The dynamics of genome-wide DNA methylation reprogramming in mouse primordial germ cells. Mol Cell 48:849–862
5. Lutsik P, Feuerbach L, Arand J et al (2011) BiQ Analyzer HT: locus-specific analysis of DNA methylation by high-throughput bisulfite sequencing. Nucleic Acids Res 39:W551–W556
6. Giehr P, Kyriakopoulos C et al (2016) The influence of hydroxylation on maintaining CpG methylation patterns: a hidden Markov model approach. PLoS Comput Biol 12(5):e1004905

Chapter 30

Helper-Dependent Chain Reaction (HDCR) for Selective Amplification of Methylated DNA Sequences

Susan M. Mitchell, Keith N. Rand, Zheng-Zhou Xu, Thu Ho, Glenn S. Brown, Jason P. Ross, and Peter L. Molloy

Abstract

Over the last few years a number of restriction enzymes that cut DNA only if cytosines within their recognition sequences are methylated have been characterized and become commercially available. Cleavage with these enzymes to release DNA fragments in a methylation-dependent manner can be combined with a novel method of amplification, *H*elper *D*ependent *C*hain *R*eaction (HDCR), to selectively amplify these fragments. HDCR uses "Helper" oligonucleotides as templates for extension of the free 3′ end of target fragments to incorporate tag sequences at the ends of fragments. These tag sequences are then used for priming of amplification of target fragments. Modifications to the amplification primers (Drivers) and the Helpers ensure that there is selection for the sequences within target fragments with each cycle of amplification. The combination of methylation-dependent enzymes and HDCR allows the sensitive and selective amplification of methylated DNA sequences without the need for bisulfite treatment.

Key words DNA methylation, Methylation-dependent restriction enzymes, GlaI, Multiplex, DNA amplification, Cancer detection, Diagnostic, Epigenetics

1 Introduction

The frequent occurrence of changes in DNA methylation in many cancers has stimulated the development of assays that can selectively amplify methylated DNA sequences, both for the quantification of the proportion of methylated DNA in a sample or for the sensitive detection of methylated DNA sequences, for example in a blood [1] or fecal [2] sample. Most methods for the selective amplification of methylated DNA rely on initial treatment of the DNA with sodium bisulfite. Selective amplification methods include Methylation-Specific PCR (MSP) [3], HeavyMethyl PCR [4], and Headloop PCR [5]. MSP assays (including MethyLight [6] and ConLight [7] assays that use fluorescent probes, *see also* Chapters 23–25) can be used in combination with a reference, non-methylation-sensitive PCR, to quantify the proportion of

Jörg Tost (ed.), *DNA Methylation Protocols*, Methods in Molecular Biology, vol. 1708,
https://doi.org/10.1007/978-1-4939-7481-8_30, © Springer Science+Business Media, LLC 2018

methylated DNA (PMR) within a sample [8] (*see also* Chapter 25). Assays that use methylation-sensitive restriction enzymes such as HpaII and HhaI have also been developed. These assays use primers flanking one or more methylation-sensitive restriction sites. After digestion with the methylation-sensitive enzyme, uncut, methylated DNA can be amplified, but not DNA that has been cut at any unmethylated CpG site between the primers. Failure to digest or incomplete digestion of unmethylated DNA can lead to false positive signals, and therefore such assays have not become widely used, especially in a clinical context.

We devised a new method that combines the use of methylation-dependent, (as opposed to methylation-sensitive), restriction enzymes and a selective amplification technique, Helper Dependent Chain Reaction (HDCR) to provide sensitive, selective amplification of methylated DNA sequences [9]. Over the last few years a number of restriction enzymes [10, 11] have become available that cut DNA only when cytosines within their recognition sites are methylated (Table 1). Thus, after enzyme digestion, free DNA ends are only produced when a particular site is methylated, providing the opportunity to positively select for amplification based on these cut ends. When combined with HDCR the use of methylation-dependent enzymes provide a means for selective amplification of methylated DNA without the requirement for

Table 1
Methylation-dependent restriction enzymes (*see* Note 1)

Enzyme	Supplier	Target sites	Mammalian (MG) target sites
GlaI	SibEnzyme	5′-RMGY 3′-YGMR	5′-GMGM(G)-3′ 3′-MGMG(C)-5′
PcsI	SibEnzyme	WMG$(N)_7$ MGW	WMG$(N)_7$ MGW
KroI	SibEnzyme	5′-GCMGGC 3′-CGGMCG	5′-GCMGGC 3′-CGGMCG
BlsI	SibEnzyme	5′-RYNRY 3′-YRNYR	
LpnP1	New England Biolabs (NEB)	5′-CMDG—10/14	14/10—CMGG—10/14
MspJI	NEB	5′-MNNR—9/13	MGNR—(9/13) forward (13/9) YNMG reverse (MGNR on opposite strand)
FspEI	NEB	CM—11/15	CMG (11/15) Forward (15/11) MGG Reverse (CMG on reverse strand)

D: A or G or T; M: methylated cytosine; N: any base; R: A or G; W: A or T; Y: C or T; e.g., 9/13: restriction enzyme cutting 9 bases from the restriction enzyme recognition site on the forward strand, 13 bases away on the reverse strand

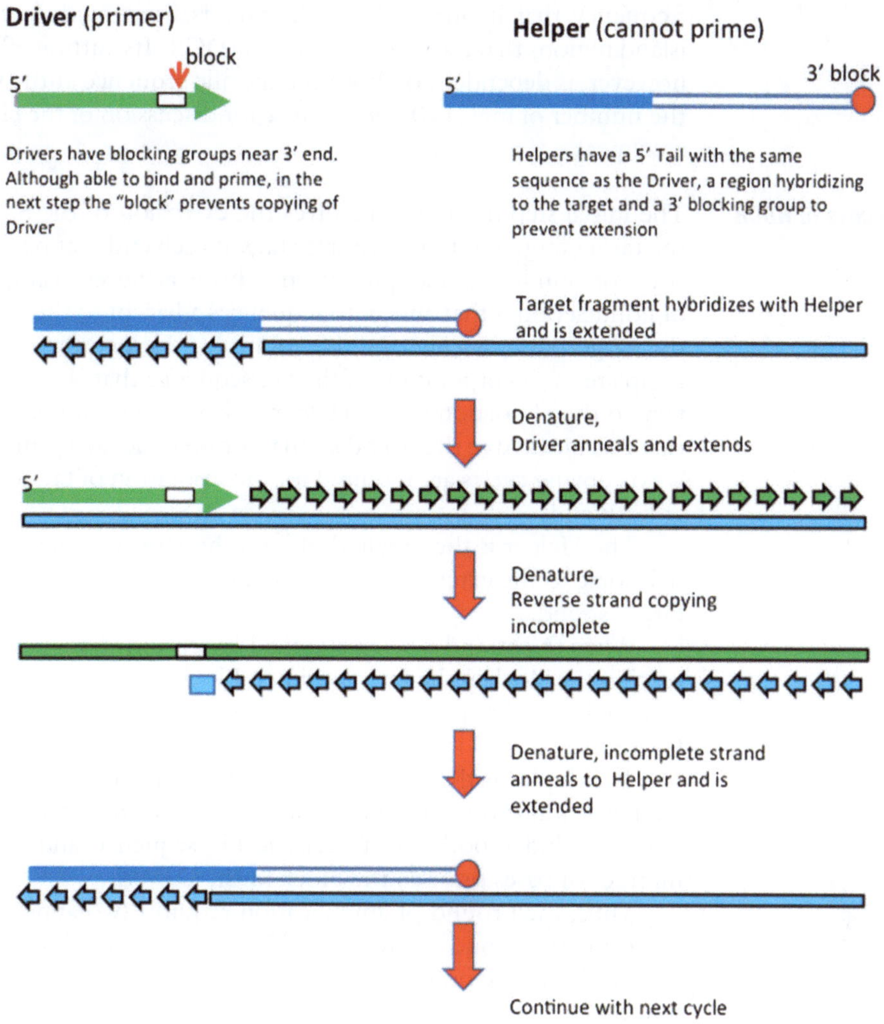

Fig. 1 Steps of HDCR

bisulfite treatment and utilizing only nanogram amounts of DNA. HDCR is based on the presence of a free 3′ end adjacent to a target site that is used for amplification and it also has the desirable property that, unlike in standard PCR amplification, this sequence selectivity is reinforced in every amplification cycle. The principle of HDCR is outlined in Fig. 1 and individual steps discussed below.

1.1 Digestion of DNA with Methylation-Dependent Enzymes

Firstly, DNA is digested with a restriction enzyme that is dependent on the presence of 5-methylcytosine (abbreviated as meC or M) to release fragments of suitable size for HDCR, generally 40–120 bp. A number of such enzymes have become commercially available in recent years and those suitable for analysis of CpG methylation are summarized in Table 1. Of the available enzymes, we have found GlaI to be the preferred option in many cases. It cuts with sufficient frequency that it often releases fragments from methylated CpG island regions that are of ideal size for HDCR. Its cutting efficiency, however, is dependent on both the specific sequence at the site and the number of meC [10] (*see* **Note 1** for discussion of the choice of enzymes).

1.2 Process of HDCR

The initial step of HDCR involves the extension of the 3′ ends of the target fragment to incorporate tags at each end that become the sites for priming and amplification. This is achieved using Helper oligonucleotides that anneal to sequences adjacent to the 3′ ends of the target strands (Fig. 1). The annealed Helper then acts as a template for incorporation of the tag sequence that is complementary to the 5′ portion of the Helper. A key feature of the Helper is that it is blocked at the 3′ end so that it cannot act as a primer itself. Two copying cycles are required for incorporation of tag sequences at both ends.

The Helper is then melted off and the Driver primer is able to hybridize to the extended strand and prime copying of the original target strand.

When this strand is reverse-copied, extension is stopped when it reaches the block in the Driver primer. Blocking of extension can be achieved by the presence of a terminal dideoxynucleotide, a C3-spacer or 2′-*O*-methyl nucleotides.

The duplex is then melted and the incomplete strand extended after hybridization with the helper. At this stage the hybridizing region includes both the target-specific sequence and the small number of bases prior to the block in the Driver.

Thus, each round of amplification requires two annealing and extension steps, one involving the Helper and a strand of the target amplicon and the other involving the Driver and the extended tag sequence.

1.3 Features and Variations of HDCR

The key advantages of methylation-dependent HDCR are:

1. HDCR does not require treatment of DNA with bisulfite and can be applied to nanogram quantities of DNA such as might be available from tissue biopsies.
2. Amplification is dependent on positive cutting by the methylation-dependent restriction enzyme. Thus it is not prone to false positives that can arise through lack of cutting by methylation-sensitive enzyme. Incomplete cutting by a methylation-dependent enzyme will result in a lessening of signal rather than a false one.
3. The requirement for hybridization of the template with Helpers in each amplification cycle means that there is continued selection for target sequences. Thus, any mispriming events are copied at only a linear rate in subsequent reactions, instead of the exponential rate observed with standard PCR.

4. The presence of blocking groups in the Drivers (and Helpers) minimizes the potential for formation of primer dimers or other amplification artifacts.
5. The ability to multiplex different target molecules using common 5′ regions for Helpers and common Drivers.

1.3.1 Competition Between Helpers and Drivers

Both the Driver and Helper are complementary to the extended 3′-end of tagged target molecules, with complementarity of the Helper also including the region of the target molecule adjacent to the 3′ end. The efficiency of HDCR is very dependent on the ability of the Driver to compete with the Helper for annealing to the extended 3′ target. Having the Drivers at a higher concentration than the Helpers and incorporation of modified bases to adjust their Tm routinely allows reactions with efficiencies of >0.5 per cycle; the efficiency of the *NEUROG1* HDCR (Fig. 2) is ~0.8.

1.3.2 Single or Double-Ended HDCRs

Selectivity for methylated sequences is determined by the recognition sequence at target sites and the number of methylated bases that are required for digestion by the particular enzyme (*see* Table 1). Since cutting at two sites to release a suitable fragment provides greater methylation selectivity, we normally prefer to use

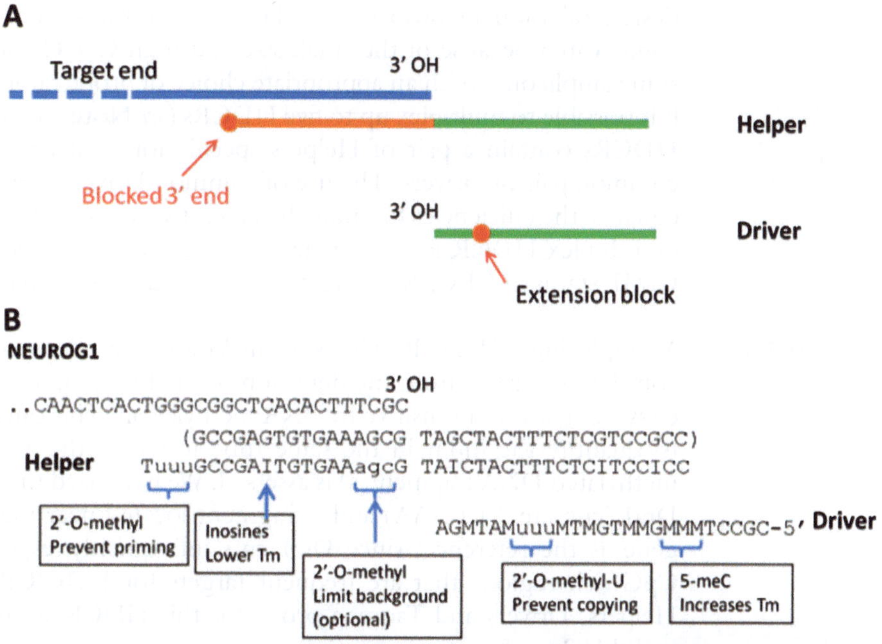

Fig. 2 Helper and Driver design. Panel **a**. Schematic representation of a Helper and Driver in relation to the target sequence. Panel **b**. Design for *NEUROG1* reverse Helper and Driver. The *top* sequence is that of the 3′ end of the *NEUROG1* target. Shown in *brackets* between the target and the Helper is the normal nucleotide sequence complementary to the extended target. In the Helper and the Driver inosines are shown as *I*, 5-methylcytosines as *M* and 2′-*O*-methyl nucleotides in *lower case*

double-ended HDCRs. However, it is not always possible to identify suitable fragments in a target region and a Helper and Driver at one end can be used in combination with a normal PCR primer at the other. Because cutting by GlaI is dependent on the number of methylated cytosines within its cutting site and sequences including two CpG sites such as GMGMG are among the most efficiently cut, it is also advantageous for single-ended HDCR. Other enzymes with longer recognition sequences, such as PcsI (Table 1) that rarely have sites close together can also be candidates for single-ended HDCR. Single-ended HDCR is suitable in regions of low CpG density, or where methylation is localized to a restricted region or one or two CpG sites. An example of a single-ended HDCR for the *SLC6A4* gene is shown in Fig. 3; this gene is among a set identified as good targets for determining the CpG island methylator phenotype (CIMP) in colorectal cancer [12].

1.3.3 Multiplexed HDCR

Common sequences can be used for the 5′ ends of Helper oligonucleotides in combination with the specific sequences in the 3′ portion that are complementary to different target molecules. This enables use of a single pair of Driver oligonucleotides for priming the amplification of different target sequences as described in Rand et al. [9]. Individual Taqman probes are used for the detection of the different target amplicons. Commonly applied rules for the design of Taqman probes are followed, but the design can be complicated because of the small size and high G + C content of some amplicons. With an appropriate choice of probe fluorophores it is possible to multiplex up to five HDCRs (*see* **Note 2**). Multiplex HDCRs contain a pair of Helpers specific for each target and a common pair of Drivers. The use of common Drivers also helps to equalize the efficiencies for individual target sequences. An example of a duplex HDCR for two genes also suggested for determining CIMP status in CRC, *FSTL1* and *KCNC1* [12], is shown in Fig. 3.

1.3.4 Reference HDCR

Multiplexing of HDCRs allows for inclusion of a reference amplicon that is derived from the digestion of the DNA sample with an enzyme that is not sensitive to DNA methylation. This can be used to quantify the input in the same tube in which the amount of methylated DNA fragment(s) is assessed. We have used the enzyme DraI (cut site TTT′AAA) and a fragment located near the *SEPT9* gene as the reference, since DraI cuts infrequently, especially in CpG-rich regions that are frequent targets for HDCR [9]. The Helpers, Drivers and Taqman probe for this HDCR are included in Table 2.

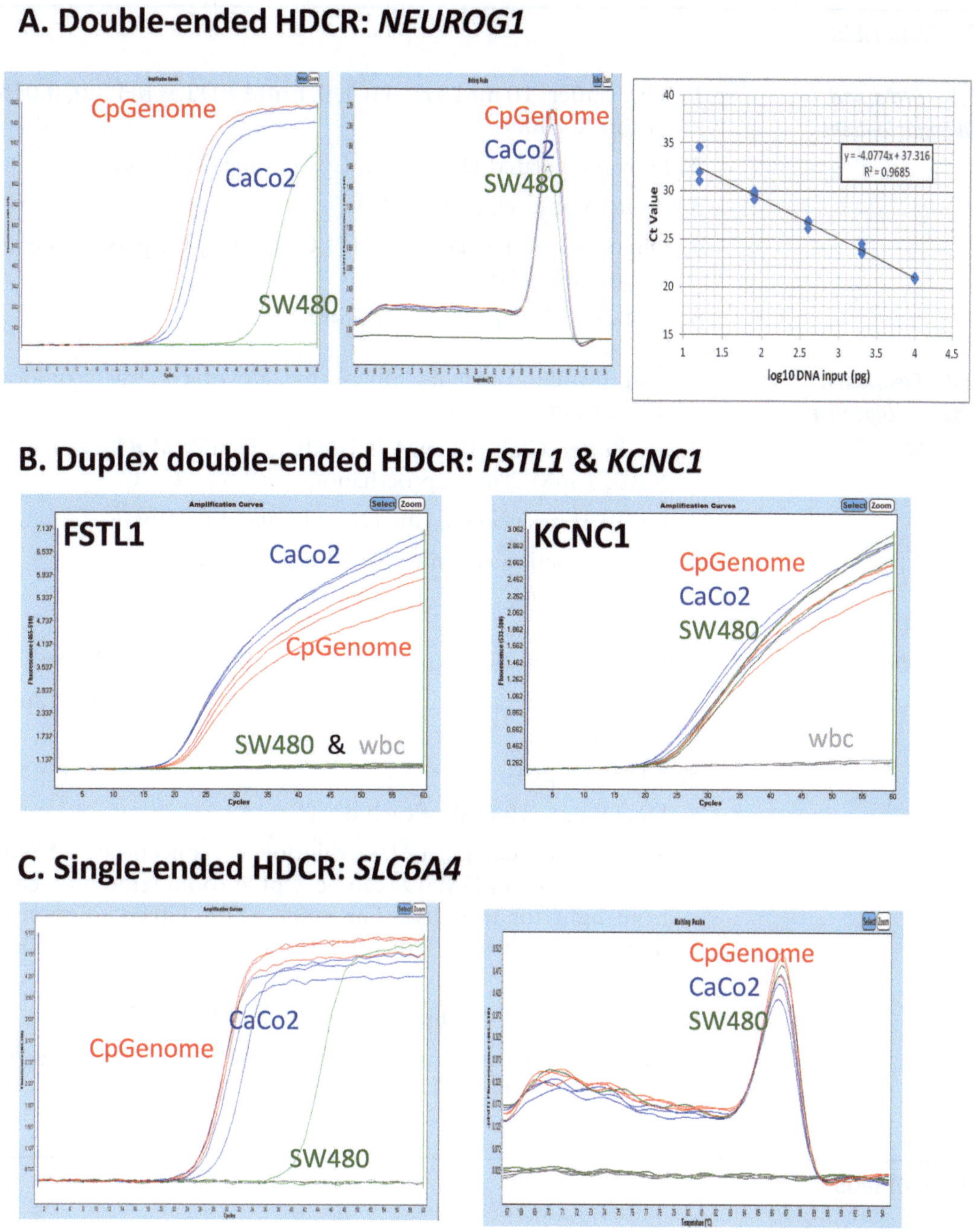

Fig. 3 Examples of HDCRs. Input DNAs are fully methylated DNA (CpGenome, *red*), control wbc DNA (*grey*), CaCo2 cell line DNA (*blue*) and SW480 cell line DNA (*green*). Panel **a**. *NEUROG1* HDCR using conditions as in Subheading 3. *Left* and *centre*—amplification and amplicon melting curves with SYBR Green detection and 100 pg DNA input. *Right*—Standard curve with inputs of 16, 80, 400, 2000, and 10,000 pg fully methylated DNA. Detection was performed using the *NEUROG1* Taqman probe. Panel **b**. Duplex HDCRs with input of 100 pg DNAs and detection with fluorophores FAM (*FSTL1*) and YAK (*KCNC1*). Helpers and Probes (Table 2) were included at 100 nM and Drivers at 400 nM. Panel **c**. Single-ended HDCR for *SLC6A4*, with 100 pg input DNA and detection using SYBR Green. *Left*, amplification curves; *right*, melt curves. Reactions included reverse Helper SLC6A4 G64 R1 at 100 nM, the reverse Driver UDr3H at 400 nM and forward primer SLC6A4 64F1 at 25 nM (*see* **Note 10**)

2 Materials

2.1 DNAs and Oligonucleotides

1. TEX buffer: 10 mM Tris–HCl, 0.1 mM EDTA, pH 8.0, 0.01% Triton X-100.
2. Fully methylated DNA (e.g., CpGenome, Millipore).
3. White blood cell (wbc) DNA.
4. Oligonucleotides: Drivers, HPLC purified; Helpers, sequencing grade (*see* **Note 2**).
5. Probes: HPLC purified (*see* **Note 2**).

2.2 Restriction Enzyme Digestions (See Note 1)

1. GlaI restriction enzyme (Sibenzyme, Novobirsk, Russia or local distributors).
2. SE Buffer GlaI: 10 mM Tris–HCl, 5 mM $MgCl_2$, 10 mM NaCl, 1 mM 2-mercaptoethanol, pH 8.5 at 25 °C.
3. LpnP1 restriction enzyme (New England Biolabs).
4. DraI restriction enzyme (New England Biolabs).
5. 50 mM EDTA.

2.3 HDCR

1. Polymerase buffer, e.g., Go Taq Flexi Buffer (5×) (Promega).
2. DNA polymerase, e.g., Platinum Taq enzyme (5 U/μL) (Promega).
3. 50 mM $MgCl_2$.
4. DNTP mix: 10 mM of each dNTP.
5. Stock SYBR Green: 1/100 dilution of concentrated SYBR Green diluted in DMSO (can be kept at room temp protected from light for months). The stock SYBR Green solution is diluted 1:10 in water just prior to use, to provide the 1:1000 working stock.
6. Driver and Helper Oligonucleotides and Probes.
7. PCR machine (e.g., Ligthcycler480 (Roche) or Rotor-Gene Q (Qiagen), (*see* **Note 3**)).

3 Methods

3.1 Design of Helper and Driver Oligonucleotides

Key features of Helper and Driver oligonucleotides are shown in Fig. 2. Design for the right end of the *NEUROG1* GlaI fragment HDCR is used as a specific example. *NEUROG1* has previously been identified as part of a panel of five genes commonly used for determining the CIMP status in CRC [8]. Sequences of Helpers, Drivers, and probes used for reactions as well as those for the DraI control reaction are shown in Table 2.

Table 2
Oligonucleotide sequences

Drivers		
Universal	UDr2H forward driver	CCCGTMGMMGTMMMTMuuuMTAMAG
Universal	UDr3H reverse driver	CGCCTMMMGMMTGMTMuuuMATMGA
Universal (control)	UDrCT1L control forward driver	CGTCTMMTMMMTMMTMuuuMTTMMA
Universal (control)	UDrCA1L control reverse driver	CCTTCMAMAMMAMMMAuuuMAAMMA
Helpers		
NEUROG1	NEUROG1 G91 F1	CICCITCCCTCTTTCTACAITGcgaCCACTAAICAGCauuT
NEUROG1	NEUROG1 G91 R1	CCICCTICTCTTTCATCIATGcgaAAGTGTIAGCCGuuuT
KCNC1	KCNC1G 57 F1	CICCITCCCTCTTTCTACAITGuguICGIGCCGCTCTuuuA
KCNC1	KCNC1 G57 R1	CCICCTICTCTTTCATCIAT GucgGTCTCGTCGATuuuT
FSTL1	FSTL1 G60-62 F1	CICCITCCCTCTTTCTACAITGCCGAGITAGGCGAGuuuT
FSTL1	FSTL1 G60-62 R1	CCICCTICTCTTTCATCIATGCGcagACCCAAGAGuuuT
SEPT 9 (control)	P2CHF387	CCTCCCTCCTCTTTCTTCCATAaaaGTAAATCTCTTACTTAAGAaaaT
SEPT 9 (control)	P2CHR388	CACACCACCCATTTCAACCATAaaaATCCATCCATTGTTTcgaA
SLC6A4	SLC6A4 G64 R1	CCICCTICTCTTTCATCIATGcggCCGTATTTGTACuuuT
Primer		
SLC6A4	SLC6A4 64 F1	TCTTTGGCGGCGGCTATCT
Probes		
NEUROG1	NEUROG1 G91 LC610	LC610-ACCTGGAGGGTACTTGGAGT-BBQ
FSTL1	FSTL1G60-62 FAM	FAM-CGGGGACCGAGTTCGGACTCCT-BBQ
KCNC1	KCNC1 G57 YAK	YAK-CAGAAGGCCAGCTCCTCCTCG-BBQ
Control	Control Cy5	Cy5-CTTATGGTTCCTCATCAAAAATAAAGG-BBQ

Lower case a, u, c, g: 2′-*O*-methyl nucleotides; BBQ: BlackBerry® quencher; LC610, FAM, YAK: fluorescent dyes

1. *Helpers*

The 3′ portion of a Helper oligonucleotide comprises sequences that are complementary to the 3′ end of the target sequences, while the 5′ portion contains sequences that correspond to the sequence of the Driver primer. The key feature of the Helper is that its 3′ end is blocked so that when it anneals to the target sequence it cannot be extended. In the *NEUROG1* example shown, the nonextendable 3′ end contains three mismatching 2′-*O*-methyl uracils and a mismatched thymine. We normally use this type of blocked end for ease of synthesis. Other 3′ options include use of a 3′-C18 spacer or a dideoxynucleotide.

The region of the Helper complementary to the target sequence is normally 14–17 nucleotides. Since the Helper and Driver compete with each for binding to the tagged target, modifications are introduced into the Helper to lower its Tm and permit better competition by the shorter Driver. We aim for this annealing portion to have a Tm of 50 °C, determined by counting 4 °C for a G/C pair and 2 °C for an A/T pair. Inosines can be incorporated instead of guanines, with a C/I pair counting as 2 °C. The sequence of the 5′ portion (tag) of the *NEUROG1* corresponds to the sequence of the Driver, except that a number of guanines have been replaced with inosines to lower its Tm relative to the Driver. Since common Drivers can be used for different target sequences, this 5′ portion of the Helpers is common to multiple targets (Table 2).

Also included in the *NEUROG1* Helper are 2′-*O*-methyl modifications to the nucleotides two to four bases from the site of priming by the target DNA on the Helper (Fig. 2). This is not an essential feature, but can limit any potential mispriming and copying of the Helper, further limiting potential background. The position of the 2′-*O*-methyl nucleotides does not interfere with priming and tagging of the target.

2. *Drivers*

The Drivers are similar to normal PCR primers except that they contain modifications that prevent them being completely copied. We use a pair of drivers as universal drivers for many different target sequences (UDr2H-F and UDr3H-R, Table 2). They both contain three 2′-*O*-methyl uracils six bases from their 3′ ends. This positioning is sufficient to allow for efficient priming while acting as an effective block to chain elongation of the reverse strand. To initiate priming, the Driver must compete with the longer Helper oligonucleotides for hybridization with the tagged target sequence. The use of meC in place of C in the Drivers increases their relative Tm and allows more efficient competition. We use a separate pair of

Drivers and matching Helpers (Table 2) for amplification of the control methylation-independent fragment when this is included in the same HDCR reaction to measure the level of input DNA.

3.2 HDCR

1. Preparation of oligonucleotide stocks: Prepare stock solutions of Helper and Driver oligonucleotides at 200 μM in TEX buffer and dilute to working stocks of 2 μM and 5 μM respectively. Prepare stocks of probes at 100 μM and dilute to 5 μM in TEX buffer for working stocks.
2. Preparation of standard curves: Prepare standard curves for either fully methylated DNA alone or fully methylated DNA mixed with wbcDNA (unmethylated for most targets) made to a final amount of 10 ng per reaction. A suitable dilution series contains 10 ng, 2 ng, 400 pg, 80 pg, and 16 pg of fully methylated DNA (*see* **Note 4**).
3. DNA Digestion: Digest 1 μg DNA in 50 μL SE Buffer GlaI, with 4 units of GlaI for 2 h at 30 °C. Stop the reaction by heating to 70 °C for 15 min. Add 5 μL of 50 mM EDTA and dilute to 5 ng/mL by addition of 145 μL of TEX buffer, aliquot and store at −20 °C (*see* **Notes 5** and **6** for more information on control reactions and digests of clinical samples).
4. HDCR set-up
   (a) Set-up reactions in 384-well plates. For a standard reaction components are added as shown in Table 3. The second column shows components for a reaction using predigested DNA, while the third column combines DNA digestion in 5 μL followed by addition of reagents for HDCR in 15 μL. Individual reactions may be optimized by titrating the $MgCl_2$ concentration or the concentrations of Helper oligonucleotides (*see* **Note 7**). In the example shown in Fig. 3 Helper oligonucleotides and the TaqMan Probe were used for the *NEUROG1* gene in combination with driver oligonucleotides UdrF1 and UdrR1 (Table 2).
   (b) HDCR cycling conditions
   Typical cycling conditions (as used for *NEUROG1* HDCR shown in Fig. 2) are shown below in Table 4, and representative data are shown in Fig. 3. For many HDCRs 10 mM $MgCl_2$, and simplified cycling conditions (95 °C, 60 sec, 50 cycles - 95 °C, 5 sec, 60 °C, 20 sec, 76 °C, 15 sec, 67 °C (Read) , 20 sec) give improved results.

Table 3
HDCR set-up

	Standard HDCR	One tube digest then HDCR	Final
Restriction digest in SE GlaI buffer, (*see* **Note 6**)		5	
GlaI digested DNA (up to 20 ng)	1		
5× Promega Go Taq Flexi Buffer	2	3	1×
50 mM $MgCl_2$	1.2/2.0	1.3/2.2	6 mM/10mM
10 mM dNTPs	0.2	0.3	200 nM
SYBR Green (1/1000) or TaqMan Probes (5 μM)	0.2 each	0.3 each	1/200,000 or 100 nM
5 μM F driver universal	0.8	1.2	400 nM
5 μM R driver universal	0.8	1.2	400 nM
2 μM F Helpers (gene specific)	0.5 each	0.75 each	100 nM (*see* **Note 7**)
2 μM R Helpers (gene specific)	0.5 each	0.75 each	100 nM
Platinum Taq 5 U/μL	0.2	0.3	0.1 U/μL
Water	To 10 μL	To 15 μL	

Table 4
HDCR cycling conditions

Denature	95 °C/60s				×1
Tagging/selection (*see* **Note 8**)	95 °C/15 s	50 °C/20 s	76 °C/15 s	67 °C/20s	×15 cycles
Cycle/read (*see* **Note 9**)	95 °C/5 s	60 °C/20 s	76 °C/15 s	67 °C/20s Read	×50 cycles
Melt	67–95 °C cont.				×1
Cool	50 °C/30 s				×1

4 Notes

1. Choice of restriction enzymes and sites: For GlaI, the efficiency of cleavage is variable (6–100%) based on the quantity and location of 5meCs, as well as the composition of the enzyme recognition sequence. (GlaI)Data from SibEnzyme states:

• Optimal substrate (100% activity):	5′-GMGM-3′/3′-MGMG-5′.
• Good substrates (>25% activity):	5′-RMGM-3′/3′-YGMG-5′ 5′-AMGT-3′/3′-TGMA-5′.
• Medium substrates (>6% activity):	5′-GMRM-3′/3′-MGYG-5′ 5′-GMGT-3′/3′-CGMA-5′.
• Poor substrates (6% activity):	5′-GMGC-3′/3′-CGMG-5′.
(*M* methylated cytosine, *R* A or G).	

For all enzymes, especially GlaI, it is valuable to experimentally determine the efficiency of cutting at particular sites. Test substrates of either PCR products or clones covering the target region can be prepared by in vitro methylation using M.SssI methyltransferase.

The recognition site for LpnP1 is asymmetric with cutting occurring 10/14 bases 3′ to its recognition sequence. However, in the context of CpG methylation the recognition site is equivalent to a methylated HpaII/MspI site with cutting on either side. While LpnP1 can release suitable size fragments and we have performed successful HDCRs for some [9], its utility is limited by intrinsic secondary cutting of DNA. We have found this secondary cutting is dependent on the density of LpnPI sites (more sites, more secondary cutting) as well as the time of digestion. For enzymes such as LpnPI that cut to leave a 5′ overhang it is preferable to end-fill after cutting to produce blunt-ended fragments as this simplifies the HDCR design.

2. Various suppliers are able to provide oligonucleotides with modifications and alternate bases used in Helpers and Drivers. These include 2′-*O*-methyl nucleotides, inosine, 5-methyl cytosine, D-spacers, dideoxy nucleotides, and 3′ C-spacers. We have purchased probes from TIB Molbiol (www.tib-molbiol.com) because they provide a range of fluorophores that enable multiplexing up to six reactions. These include Cyan 500, LC640, LC610, Cy5, YAK, and FAM. For multiplexed reactions use the FAM fluorophore for the least abundant HDCR target or least sensitive single-plex HDCR. CYAN 500 is the least stable fluorophore.
3. The examples HDCRs shown were run in a Lightcycler (Roche Molecular Diagnostics), but we have also used the Rotor-Gene Q with a 72-place rotor [5]. We have found overlaying samples with mineral oil can improve consistency, but do not routinely do this when using the Lightcycler.
4. For some genes or target sites there will be a background of methylation so digested wbcDNA cannot be used for preparing the standard curve. An alternative is to use wbcDNA either undigested or digested with another enzyme.

5. When the DraI control reaction is included in a multiplex PCR, DraI is included (0.25 units or 0.05 units/μL) in the digestion of DNA with GlaI as it cuts efficiently in SE Buffer GlaI.
6. For establishing and optimizing HDCR conditions, or when samples are being used for multiple reactions, we use aliquots of predigested DNAs (second column, Table 2). For clinical samples we digest up to 10 ng of DNA in 5 μL reactions in SE buffer GlaI, with 0.4 units of GlaI for 2 h at 30 °C. A mix of HDCR components is then added (third column, Table 3), bringing the reactions to 15 μL, prior to HDCR cycling. We note that GlaI is also active in the GoTaq PCR buffer, so it is possible to combine all reagents in a single tube and proceed directly from restriction digestion to HDCR cycling. However, we have seen that some oligonucleotides can inhibit GlaI digestion, so each assay should be verified to see if this is possible.
7. HDCR efficiency is significantly affected by the relative concentrations of Helpers and Drivers. For a double-ended HDCR, a matrix of concentrations of each Helper in the range 40–200 nM should be evaluated. For multiplex(ed) reactions Helpers can initially be tested at the same concentrations as in singleplex reactions. Adjustments can subsequently be made to equalize the efficiencies of HDCR for each of the targets.
8. In the tagging step conditions are optimized for hybridization of the Helper with the target sequences. An annealing temperature equivalent to the estimated Tm of the complementary region of the Helper (50 °C) is used for initial evaluation. If amplification products are seen, higher annealing temperatures to increase the specificity of tagging can be evaluated. Conversely, if there is no amplification product, or amplification is delayed, lowering of the annealing temperature should be evaluated. The tagging step can potentially be inefficient because of the relatively short annealing regions and potential secondary structure at 3′ ends of CpG-rich target strands, therefore we normally allow 10–15 cycles at this lower annealing temperature.
9. After the preliminary tagging cycles the annealing temperature is raised to 60 °C since the products of incomplete extension contain additional bases complementary to the 3′ end of the Driver, and thus the annealing region with the Helper is longer than at the initial tagging step.
10. For single-ended HDCRs the concentration of the standard primer should be lower, 25–50 nM, than the Helper and Driver.

References

1. Lofton-Day C, Model F, Devos T et al (2008) DNA methylation biomarkers for blood-based colorectal cancer screening. Clin Chem 54:414–423
2. Imperiale TF, Ransohoff DF, Itzkowitz SH et al (2014) Multitarget stool DNA testing for colorectal-cancer screening. N Engl J Med 370:1287–1297
3. Herman JG, Graff JR, Myöhänen S et al (1996) Methylation-specific PCR: a novel PCR assay for methylation status of CpG islands. Proc Natl Acad Sci U S A 93:9821–9826
4. Cottrell SE, Distler J, Goodman NS et al (2004) A real-time PCR assay for DNA-methylation using methylation-specific blockers. Nucleic Acids Res 32:e10
5. Rand KN, Ho T, Qu W et al (2005) Headloop suppression PCR and its application to selective amplification of methylated DNA sequences. Nucleic Acids Res 33:e127
6. Eads CA, Danenberg KD, Kawakami K et al (2000) MethyLight: a high-throughput assay to measure DNA methylation. Nucleic Acids Res 28:e32
7. Rand K, Qu W, Ho T et al (2002) Conversion-specific detection of DNA methylation using real-time polymerase chain reaction (ConLight-MSP) to avoid false positives. Methods 27:114–120
8. Weisenberger DJ, Siegmund KD, Campan M et al (2006) CpG island methylator phenotype underlies sporadic microsatellite instability and is tightly associated with BRAF mutation in colorectal cancer. Nat Genet 38:787–793
9. Rand KN, Young GP, Ho T et al (2013) Sensitive and selective amplification of methylated DNA sequences using helper-dependent chain reaction in combination with a methylation-dependent restriction enzymes. Nucleic Acids Res 41:e15
10. Tarasova GV, Nayakshina TN, Degtyarev SKH (2008) Substrate specificity of new methyl-directed DNA endonuclease GlaI. BMC Mol Biol 9:7–18
11. Cohen-Karni D, Xu D, Apone L et al (2011) The MspJI family of modification-dependent restriction endonucleases for epigenetic studies. Proc Natl Acad Sci U S A 108:11040–11045
12. Hinoue T, Weisenberger DJ, Lange CPE et al (2012) Genome-scale analysis of aberrant DNA methylation in colorectal cancer. Genome Res 22:271–282

Part VI

DNA Methylation Analysis of Specific Biological Samples

Chapter 31

DNA Methylation Analysis from Blood Spots: Increasing Yield and Quality for Genome-Wide and Locus-Specific Methylation Analysis

Akram Ghantous, Hector Hernandez-Vargas, and Zdenko Herceg

Abstract

Blood represents an easily accessible human tissue for numerous research and clinical applications, including surrogate roles for biomarkers of other tissues. Dried blood spots (DBS) are space- and cost-efficient storage forms of blood while stably retaining many of its chemical constituents. Consequently, neonatal DBS are routinely collected in many countries, and their biobanks represent gold mines for research. However, the utility of DBS is restricted by the limited amount and quality of extractable biomolecules (including DNA), especially for genome-wide profiling. In particular, DNA methylome analysis in DBS has proven to be technically challenging, mainly due to the requirement for stringent preprocessing, such as bisulfite conversion. Moreover, DNA amplification, required to increase its yield, often leads to a bias in the analysis, particularly in methylome profiles. Thus, it is important to develop methodologies that maximize both the yield and quality of DNA from DBS for downstream analyses. Using a combination of in-house-derived and modified commercial extraction methods, we developed two robust protocols that produced increased DNA yield and quality from DBS. Though both protocols are more efficient relative to other published methods, one protocol yields less DNA compared to the other, but shows improved 260/280 spectrophotometric ratios, which are useful for sample quality assessment. Both protocols consist of two sequential phases, each involving several critical steps. Phase I comprises blood extraction off the filter papers, cell lysis, and protein digestion. Phase II involves DNA precipitation, purification, and elution. Results from subsequent locus-specific and genome-wide DNA methylation analyses demonstrate the high quality, reproducibility, and consistency of the data. This work may prove useful to meet the increased demand for research on DBS, particularly with a focus on the epigenetic origins of human diseases and newborn screening programs.

Key words DNA methylation, Blood spot, DNA extraction, Epigenetics, Illumina HM450 arrays, Pyrosequencing, Whole bisulfitome amplification

1 Introduction

Blood represents an easily accessible human tissue for numerous research and clinical applications. Moreover, it is often used as a surrogate tissue for the identification of metabolic, genetic and epigenetic biomarkers of other tissues [1]. When collected and

Jörg Tost (ed.), *DNA Methylation Protocols*, Methods in Molecular Biology, vol. 1708,
https://doi.org/10.1007/978-1-4939-7481-8_31,

dried on filter paper, in the form of dried blood spots (DBS), blood retains many of its chemical constituents for many years and can be stored with space- and cost-efficient advantages. The use of filter paper for blood collection and analysis was implemented as early as the 1960s by Guthrie et al. using dried-blood samples for newborn phenylketonuria detection [2]. "Guthrie cards" (Whatman 903) are widely used in many types of tests, including chemical, serological, and genetic applications [3]. More recently, Flinders Technology Associates chemically treated filter papers (FTA cards) were specifically developed for DNA/RNA analyses [4]. These chemically treated cards allow long-term storage of DNA at room temperature and are impregnated with denaturants that guard against oxidation, nuclease and ultraviolet damage, and both bacterial and fungal degradation. Other types of DBS filter papers are available (IsoCode, FTA-DMPK-A, FTA-DMPK-B, FTA-DMPK-C, etc.) [5, 6].

Neonatal DBS, routinely collected in many countries, represent gold mines for epidemiology, research, and screening, particularly for exposure and inherent markers of metabolism, infection, genetics, and epigenetics. With the advent of high-throughput technologies and with the general lack of biospecimen collection in observational human studies, many large human cohorts rely on the use of DBS obtained soon after birth as the main source of biological information [7]. For example, the International Childhood Cancer Cohort Consortium (I4C), which is the largest prospective investigation into childhood cancer, currently enrolls a growing number of more than 400,000 births with continuous follow-up [7]. However, the utility of DBS is restricted by the limited amount and quality of extractable biomolecules (including DNA), especially for genome-wide profiling. In particular, DNA methylome analysis in DBS has proven to be technically challenging, especially because such techniques often require stringent bisulfite preprocessing that can further degrade DNA. Other limitations include the variable degradation of DNA and the identification of technical artifacts due to storage and extraction. Moreover, whole-genome amplification to increase DNA yield often leads to a bias in the analysis, particularly for methylome profiles. For example, whole bisulfitome amplification causes considerable deviations in the middle range of DNA methylation levels in locus-specific and genome-wide methylation analyses [8–10]. Therefore, it is important to develop methodologies that maximize both the yield and quality of DNA from DBS for downstream analyses. This also represents one of the important needs of I4C, in the context of which the presented work has been designed.

We have previously tested several genomic DNA extraction methods on DBS, including resin-based, lysis-based, and magnetic bead-based methods [10, 11]. Lysis- and bead-based (ChargeSwitch Forensic DNA Purification kit, Invitrogen) methods were the best, but the latter were not suitable for Illumina BeadChip methylation profiling, leading to a loss of up to 16% of detectable probes on the Infinium HumanMethylation27 arrays (Illumina Inc.) [12]. Therefore, we have recently optimized and compared several lysis-based DNA extraction methods from DBS using combinations of in-house-derived and modified commercial extraction kits [10]. We used a homogenous set of DBS samples, spotted on the same day (approximately 2 years before extraction) and stored in a similar manner, to provide a common platform for cross-protocol comparisons. These DBS were obtained from the National Children Study (NCS)—USA and were FTA-type, which preserves well DNA relative to other types of neonatal cards, hence, allowing comparisons across a wide range of DNA extraction protocols. The best protocols were then tested on Guthrie cards from the Tasmanian Infant Health Survey (TIHS), Australia, which were more than 20 years old, and protocols maintained a robust performance [10].

In this work, we describe two robust and efficient protocols that increase DNA yield from DBS while minimizing DNA degradation. Both protocols consist of two sequential phases, each comprising several steps that were systematically optimized (Fig. 1). Critical steps in Phase I include blood extraction off the filter papers, cell lysis and protein digestion. Phase II involves DNA precipitation, purification and elution (Fig. 1). One protocol, termed GenSolve-Qiagen, represents a combination of (a) GenSolve material in Phase I, (b) Qiagen material in Phase II and (c) in-house modifications. This protocol extracted on average 808 ± 376 ng per 9 mm (diameter) punch from NCS samples (equivalent to 12.7 ng/mm^2, mean 260/280 ratio of 1.83 ± 0.14) and 66 ± 15 ng per two 1 mm punches from TIHS samples (equivalent to 42.0 ng/mm^2, mean 260/280 ratio of 1.66 ± 0.02). The other protocol, termed Qiagen-Ethanol, involves (a) Qiagen consumables in phases I and II, (b) ethanol as a precipitation buffer and (c) in-house modifications. This protocol extracts approximately twice the amount of DNA relative to GenSolve-Qiagen, as quantified by NanoDrop, a spectrophotometry-based method, and Qubit, a fluorescent-based tool. However, the Qiagen-Ethanol protocol does not yield reliable 260/280 ratios, which often represent a criterion for the selection of good quality samples, particularly for costly or laborious downstream applications (260/280 and 260/230 ratios should, however, be treated with caution; for example, different contaminants can compensate for each other's deviations, resulting in misleadingly optimal 260/280 ratios). Notably, though GenSolve-Qiagen protocol yields less DNA than the Qiagen-Ethanol approach, the former still represents a highly efficient protocol with yields exceeding other methods tested in our studies [10] or published protocols (outlined in the next paragraph) [11, 13–15]. Accordingly, either

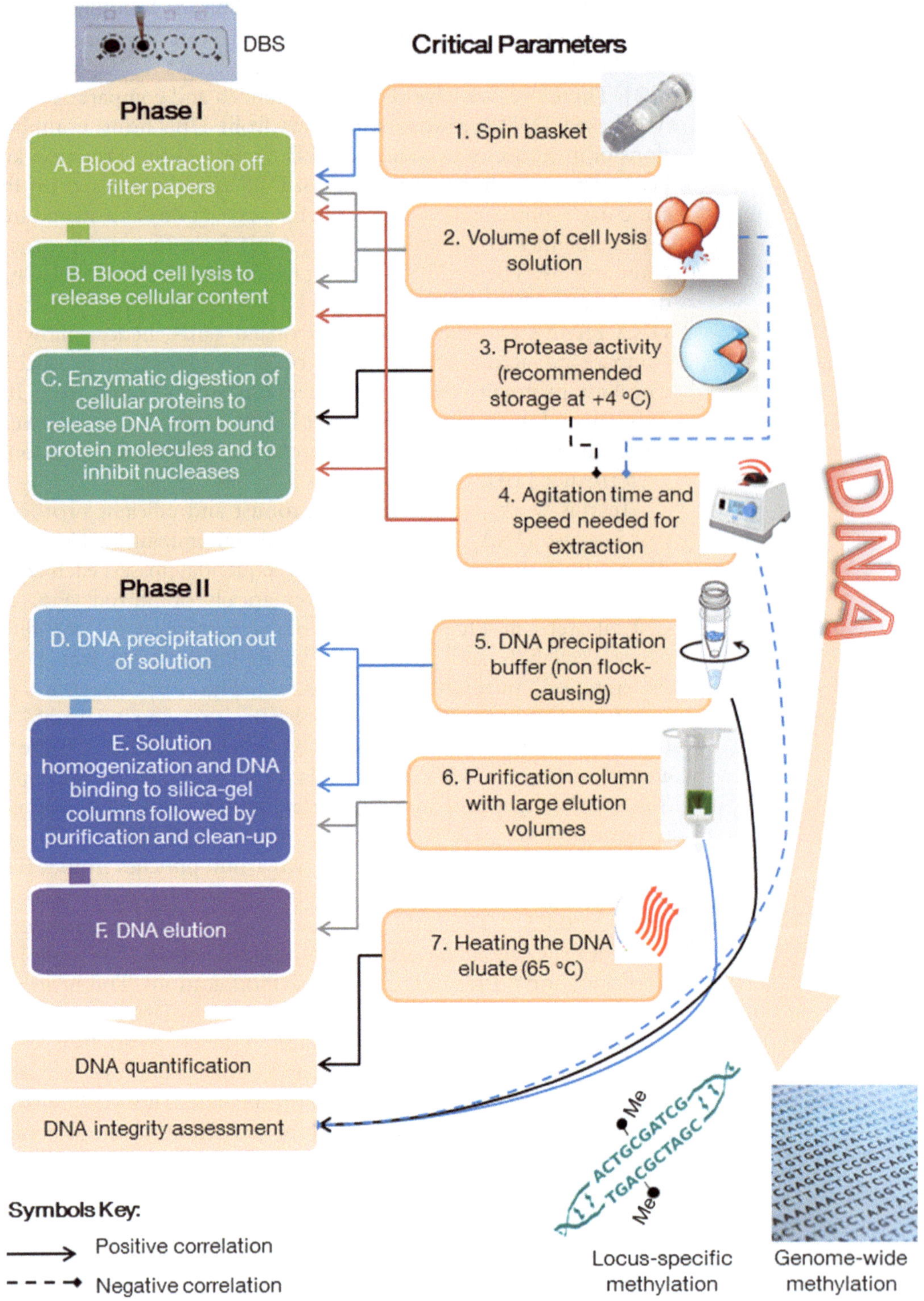

Fig. 1 Phases and critical parameters of DNA extraction protocols optimized for DNA methylation studies based on dried blood spots (DBS). Two sequential phases, each encompassing three steps (*A–C* and *D–F*), are outlined for an optimized protocol leading to increased DNA yield and quality suitable for locus-specific and

of these two protocols could be used according to specific needs. Because TIHS samples were insufficient in size compared to NCS spots, the latter were selected for further downstream applications, using the GenSolve-Qiagen protocol, which enables priori elimination of bad quality samples based on 260/280 ratios. Gel electrophoresis and BioAnalyzer size distribution analyses showed that the average peak size of DNA was greater than 1 kbp in these protocols and often ranging between 4.9 and 9.7 kbp across tested NCS samples. Moreover, the obtained DNA led to reproducible, consistent and high-quality data from locus-specific (by Pyrosequencing®, *see also* Chapter 22) and genome-wide (by Illumina Human Methylation 450 BeadChips, *see also* Chapter 16) DNA methylation analyses [10].

Other studies have compared DNA extraction protocols from DBS, but irrespective of epigenetic applications [13–15]. It is sometimes difficult to compare DNA extraction protocols across different studies due to many reasons, such as differences in the structure of filter papers, on which blood was soaked, the storage conditions and year-durations, and DNA quantifications methods used. However, these studies used reference protocols identical to some that were investigated in ours, hence, providing a common platform for cross-comparisons. Accordingly, the protocols outlined in this work are expected to perform better than other reported methods, including (a) QIAamp DNA Mini Kit (Qiagen), EZNA (Omega Bio-Tek), Chelex 100 (AmershamBiosciences) and alkaline lysis (GenomiPhi DNA Amplification Kit, AmershamBiosciences), as reported by

Fig. 1 (continued) genome-wide methylation studies. In addition, seven critical parameters positively or negatively influencing each phase are distinctly highlighted (*solid arrows*: positive correlation; *dashed line*: negative correlation); some parameters also influence each other, as indicated. All these parameters, however, positively correlate with increased DNA yield and purity, as represented by a *growing arrow*, though some of them compromise DNA integrity, as indicated. Briefly, spin baskets increase the efficiency of blood extraction from filter papers. A spin basket is shown next to parameter 1 and consists of a tube with an embedded perforated basket used to separate blood solutions from the filter papers from which they were extracted. High volumes of cell lysis solution (parameter 2) and efficient proteases (parameter 3) positively influence several protocol steps and decrease the need for high-speed and long-duration agitations (parameter 4), which often compromise DNA integrity, as indicated. Ethanol, as a precipitation buffer (parameter 5), causes fewer flocks of precipitates than other buffers, allowing easier solution homogenization without the need of vortexing that often damages DNA integrity. The absence of flocks also minimizes, in subsequent steps, the clogging of the DNA purification columns, hence, possibly leading to increased DNA yield in some protocols. A silica-gel column with a funnel-shape design is shown next to parameter 6 and often used to elute small volumes (5–30 μL). However, we recommend using large volumes of eluent (greater than 30 μL) on QIAamp spin columns followed by heating to partially evaporate the eluate and concentrate the DNA (*see* **Note 7**). The type of spin column is crucial to maintaining DNA integrity, a criterion satisfied by the QIAamp spin columns. Heating the eluate at 65 °C (parameter 7) does not denature DNA but denatures proteins, including nucleases, and often enhances 260/280 and 260/230 spectrophotometric absorbance ratios and eliminates some solvents that compromise downstream applications, including PCR

Sjoholm et al. [15], (b) Instagene Matrix (Biorad), Instagene Matrix Dx (Biorad), Extract-N-Amp (Sigma), and ReadyAmp Genomic DNA Purification System (Promega), as reported by Wong et al. [11], and (c) in-house methods by Hue et al. [14] and Hollegaard et al. [13].

Our findings support the good performance of our optimized methods relative to many in-house and commercial DNA extraction protocols from DBS. The suggested protocols were cited in other independent studies, including a new application in DBS using methylated DNA immunoprecipitation coupled with next-generation sequencing (MeDIP-seq), which generated high-quality data using a single 3.2 mm DBS punch (60 ng DNA) from filter cards archived for up to 16 years [16]. Moreover, the scale of our suggested protocols can be increased by implementing the QIAcube technology (Qiagen). With scalable designs, laser cutting of DBS punches would be appropriate to eliminate cross-contamination, as recently reported [17]. Interestingly, the reproducibility at a genome-wide level and the repeatability at single CpG sites of HM450-based methylome data have been recently addressed in DBS, in comparison with peripheral blood mononuclear cells, demonstrating ongoing interest in technical applications in this field [18].

2 Materials

The described protocols require simple equipment and few commercial kits and material. However, not all reagents in the suggested kits will be used, and replacements will be suggested. Therefore, individual components of the kits may be purchased separately. The most crucial kit components are the protease and DNA purification columns. The DNA precipitation buffers supplied in the kits should be replaced by ethanol, as suggested.

2.1 Equipment

1. Card puncher.
2. Scissors.
3. Forceps.
4. Micropipettes and tips (to avoid cross-contamination, we recommend pipet tips with aerosol barriers).
5. Vortex.
6. Centrifuge (up to 16,000 × *g*) with rotor for 2 mL microcentrifuge tubes.
7. NanoDrop, Qubit™ dsDNA High Sensitivity Assay, Quant-iT™ PicoGreen® dsDNA Assay Kit or other spectrophotometry- or fluorescent-based DNA quantification equipment.
8. Thermomixer (up to 70 °C and 1400 rpm).

2.2 Kits

1. QIAamp DNA Micro kit (Qiagen).
2. GenSolve (GenVault, GVR).

2.3 Other Material

1. Nuclease free water (pH at least 7.0 to minimize degradation of DNA by acid hydrolysis).
2. Absolute ethanol (96–100%).
3. Spin basket (GVSPIN250).
4. Microcentrifuge tubes (1.7 and 2.0 mL).
5. Blood spots (we used dried neonatal blood spots from NCS and TIHS cohorts; these samples were heel-pricked without anticoagulants added. Permissions from the ethical committees of the International Agency for Research on Cancer (IARC), NCS and TIHS were obtained.)
6. The lysis buffers can be prepared in-house, if needed. One possible recipe for lysis buffer is: 10 mM Tris–HCl, 5 mM $MgCl_2$, 1% v/v Triton X100, 1% w/v SDS, and 10 mM EDTA, pH 8.0. However, the lysis buffer supplied with the kits may be optimal for the enzymatic activity of the supplied protease.
7. Personal protection equipment (lab coat, gloves, and goggles).

3 Methods

The following DNA extraction protocols are described for a DBS punch having a 9 mm diameter. The provided solution volumes may be tailored proportionally to the DBS area in use, although the volume provided for the starting solution, the lysis-protease mix, is given in excess to ensure complete soaking of DBS pieces and was set to 620 μL for both protocols. Use sterile equipment and avoid cross-contamination between samples by cleaning equipment with deionized water followed by 70% ethanol after each use [19]. Cut the DBS punches into small pieces using sterile scissors, over a clean filter paper, and transfer the pieces into a 2 mL microcentrifuge tube (replace the filter paper between samples). The protocols described below proceed with the microcentrifuge tube containing the DBS pieces. The measured reproducibility exhibited <15% CV between duplicate samples and between assays.

3.1 Protocol GenSolve-Qiagen

The following steps were adapted and modified from the GenSolve (GenVault) and QIAamp DNA Micro (Qiagen) kits.

Phase I (use GenSolve reagents):

1. Reconstitute Solution A powder with 16 mL of 1% LiDs to prepare the lysis solution. Vortex to completely resuspend the lyophilized reagent (*see* **Note 1**). Add 620 μL of the

lysis-protease mix, which is composed of 600 μL of reconstituted Solution A + 20 μL protease, per microcentrifuge tube containing the DBS pieces (*see* **Note 2**).

2. Incubate at 65 °C with shaking at 1400 rpm for 1.5 h. Inspect during the first 5 min that all pieces are completely submerged (*see* **Note 3**).
3. Briefly spin to collect liquid off cap.
4. Add 20 μL Recovery Solution B to an empty spin basket and then transfer the solution from **step 3** into the basket, making sure to transfer along the DBS pieces by scooping them with a pipette tip.
5. Spin at 16,300 × *g* for 2 min. Discard basket and paper pieces. Pulse vortex each tube (*see* **Note 4**).

Phase II (use QIAamp reagents):

6. Add 600 μL of 100% ethanol to each microcentrifuge tube. Pulse-vortex each sample (avoid excessive vortexing, which may cause DNA fragmentation), then briefly spin afterward to collect liquid from the caps (*see* **Note 5**).
7. Load 600 μL of the sample onto a QIAamp spin column placed in a collection tube; avoid touching the column with the pipet tip. Spin at 6000 × *g* for 1 min. Discard the eluate and tap the top of the collection tube on clean tissue. Place the column back into the collection tube.
8. Repeat step 7 until the entire sample has been applied onto the column.
9. Add 25 mL ethanol (96–100%) to the bottle containing 19 mL Buffer AW1; mix by shaking. Reconstituted AW1 can be stored at room temperature for up to 1 year. Add 500 μL of AW1 onto each column/collection tube and spin at 6000 × *g* for 1 min (*see* **Note 6**). Discard the eluate and tap the collection tube on tissue. Place the column back into the collection tube.
10. Add 30 mL ethanol (96–100%) to the bottle containing 13 mL Buffer AW2; mix by shaking. Reconstituted AW2 can be stored at room temperature for up to 1 year. Add 500 μL AW2 onto each column/collection tube and spin at 16,300 × *g* for 4 min (*see* **Note 6**). Discard collection tube and replace with a new one.
11. Spin again at 16,300 × *g* for 1.5 min.
12. Place the column in a new 1.7 mL microcentrifuge tube.
13. Add 50 μL of nuclease-free water (preheated at 70 °C) to the spin column to elute DNA (*see* **Notes 7** and **8**). Incubate at room temperature for 5 min. Spin at 6000 × *g* for 1 min.

14. Repeat elution using an additional 30–50 μL nuclease-free water on the same column/microcentrifuge tube set-up, resulting in total eluate volume up to 80–100 μL (two serial elutions will produce higher DNA yields though lower DNA concentrations. However, the DNA solution can be concentrated as needed using the heating step below).
15. Keep the DNA-containing microcentrifuge tube and discard the column.
16. Optional: treat the DNA solution with RNAse to remove possible contaminating RNA (supplied in the Qiagen kit).
17. Use 1 μL of the eluate in a standard PCR reaction to verify the presence of DNA amplification products.
18. Quantify DNA concentrations using NanoDrop, Qubit™ dsDNA High Sensitivity Assay, Quant-iT™ PicoGreen® dsDNA Assay Kit or other spectrophotometry- or fluorescent-based DNA quantification methods.
19. Open the covers of the microcentrifuge tubes containing the DNA eluates, place the tubes in a heat block, cover them with an aluminum foil to minimize aerosol contamination and heat them at 65 °C for at least 10 min (*see* **Note 9**).
20. Requantify DNA and store at −20 °C.
21. Optional: the obtained DNA can be used for downstream methylation studies, including locus-specific (e.g., by Pyrosequencing®, *see also* Chapter 22) and genome-wide (by Illumina Human Methylation 450 BeadChips, *see also* Chapter 16) methylation analyses (*see* **Note 10**).

3.2 Protocol Qiagen-Ethanol

The following steps were adapted and modified from the QIAamp DNA Micro kit.

Phase I (use QIAamp reagents):

1. Add 620 μL of lysis-protease mix, which is composed of 560 μL Buffer ATL + 60 μL proteinase K, per microcentrifuge tube containing the DBS pieces (*see* **Note 2**).
2. Incubate at 56 °C with shaking at 1400 rpm for 1.5 h. Inspect during the first 5 min that all pieces are completely submerged (*see* **Note 3**).
3. Briefly spin to collect liquid off cap.
4. Transfer the solution from **step 3** into an empty spin basket, making sure to transfer along the DBS pieces by scooping them with a pipette tip.
5. Spin at 16,300 × *g* for 2 min. Discard basket and paper pieces. Pulse vortex each tube (*see* **Note 4**).

Phase II (use QIAamp reagents):

6. Add 600 μL of 100% ethanol to each microcentrifuge tube (replacing precipitation buffer AL of the QIAamp DNA Micro kit; AL, unlike ethanol, causes flocks of solid precipitates that can hinder the later steps of DNA purification, leading to decreased DNA yield).
7. Pulse-vortex each sample (avoid excessive vortexing which may cause DNA fragmentation), then briefly spin afterward to collect liquid from the caps (*see* **Note 5**). Heating to homogenize the solution, as in the conventional protocol of the QIAamp DNA Micro kit, is no longer necessary here.
8. Load 600 μL of the sample onto a QIAamp spin column placed in a collection tube; avoid touching the column with the pipet tip. Spin at 6000 × *g* for 1 min. Discard the eluate and tap the top of the collection tube on clean tissue. Place the column back into the collection tube.
9. Repeat **step** 7 until the entire sample has been applied onto the column.
10. Add 25 mL ethanol (96–100%) to the bottle containing 19 mL Buffer AW1; mix by shaking. Reconstituted AW1 can be stored at room temperature for up to 1 year.
11. Add 500 μL of AW1 onto each column/collection tube and spin at 6000 × *g* for 1 min (*see* **Note 6**). Discard the eluate and tap the collection tube on tissue. Place the column back into the collection tube.
12. Add 30 mL ethanol (96–100%) to the bottle containing 13 mL Buffer AW2; mix by shaking. Reconstituted AW2 can be stored at room temperature for up to 1 year. Add 500 μL AW2 onto each column/collection tube and spin at 16,300 × *g* for 4 min (*see* **Note 6**). Discard collection tube and replace with a new one.
13. Spin again at 16,300 × *g* for 1.5 min.
14. Place the column in a new 1.7 mL tube.
15. Add 50 μL of nuclease-free water (preheated at 70 °C) to the spin column to elute DNA (*see* **Notes** 7 and **8**). Incubate at room temperature for 5 min. Spin at 6000 × *g* for 1 min.
16. Repeat elution using an additional 30–50 μL nuclease-free water on the same column/microcentrifuge tube set-up, resulting in total eluate volume up to 80–100 μL (two serial elutions will produce higher DNA yields though lower DNA concentrations. However, the DNA solution can be concentrated as needed using the heating step below).
17. Keep the DNA-containing microcentrifuge tube and discard the column.

18. Optional: treat the DNA solution with RNAse to remove possible contaminating RNA (supplied in the Qiagen kit).
19. Use 1 μL of the eluate in a standard PCR to verify the presence of DNA amplification products.
20. Quantitate DNA concentrations using NanoDrop, Qubit™ dsDNA High Sensitivity Assay, Quant-iT™ PicoGreen® dsDNA Assay Kit, or other spectrophotometry- or fluorescent-based DNA quantification methods.
21. Open the covers of the microcentrifuge tubes containing the DNA eluates, place the tubes in a heat block, cover them with an aluminum foil to minimize aerosol contamination, and heat them at 65 °C for at least 10 min (*see* **Note 9**). Requantify DNA and store at −20 °C.
22. Optional: the obtained DNA can be used for downstream methylation studies, including locus-specific (e.g., by Pyrosequencing®, *see also* Chapter 22) and genome-wide (by Illumina Human Methylation 450 BeadChips, *see also* Chapter 16) methylation analyses (*see* **Note 10**).

4 Notes

1. The performance of the reconstituted solution (before combining it with protease) was tested after storage for up to 1 month at +4 °C and showed no loss of lysis potential relative to a freshly prepared solution [10].
2. The protease and lysis solutions for several samples may be proportionally mixed in one tube and then distributed to each microcentrifuge tube sample. However, it is recommended to add the mixture to the samples immediately after its preparation because the protease may otherwise self-digest in the absence of substrate. Moreover, we recommend protease stocks that are stable at +4 °C instead of (a) room temperature, which is more favorable for the growth of possible contaminating microorganisms, and (b) −20 °C, which necessitates repeated freeze-thaw cycles that diminish protease activity. Accordingly, we prefer using the protease by GenSolve rather than the one supplied by Qiagen in both of the two presented protocols.
3. Incubation in lysis solution for extended durations (1–2 h) with increased shaking speeds (1400 rpm) is critical to dissolve blood material off the filter paper. The filter papers should not retain any observable color after drying using spin baskets (**step 4** of the protocol). Otherwise, it is recommended to increase lysis duration to 3 h, beyond which proteases often lose activity. Some protocols recommend overnight incubation at 700 rpm,

which we have observed to increase DNA yield only if the lysis volume was not in excess, which is not the case in the presented protocols.

4. To fully extract blood material from filter papers, soaking DBS in large lysis buffer volumes, incubation in lysis solution for extended durations with increased shaking speeds and using spin baskets at high speed are recommended. TissueLyser (Qiagen) no longer increases the DNA yield above that obtained using these modifications.
5. DNA precipitation buffers that create flocks of precipitants should be avoided. Ethanol, unlike many buffers used to precipitate DNA out of solutions, causes less precipitates in lysed DBS solutions. Therefore, prolonged vortexing after ethanol addition is not necessary to ensure homogenization and should be replaced by pulse-vortexing or mixing by inversion to prevent DNA fragmentation. By comparing several protocols and their combinations, we observed that ethanol is critical in some protocols to increase the DNA yield, but not necessarily the DNA fragment sizes. However, the nature of purification column was crucial to ensure less DNA fragmentation, and in this regard the recommended Qiagen columns performed well.
6. After binding DNA to the spin column and adding each wash buffer, turn the column inside the centrifuge by 180° compared to the loading position used for binding allowing optimal washing efficiency.
7. With spin columns, two sequential elutions each using the same volume, X μL, of eluent in the same column generally yields more DNA than using $2X$ μL of eluent in one shot. However, this would not be the case if X is less than the minimum elution capacity of the column because then a single elution with X may not be sufficient to completely soak the column resin. In the presented protocols, X should not be less than 30 μL. Specific column designs (termed extra-small, XS) with low elution capacities (as low as 5 μL) were recently developed by Macherey-Nagel (NucleoSpin gDNA clean-up XS); however, we observed that these columns did not consistently improve DNA quantity or quality, whether combined or not with other kits, the DNA precipitation buffer changed to ethanol, or the washing volume and frequency increased [10].
8. Preheating the nuclease-free water at 70 °C before elution and then incubating the elution column (containing preheated water) at 70 °C for 10 min increases the DNA yield. This is because DNA bound to spin columns dissolves better in warmer solution; hence, it is eluted in higher quantities. Nuclease-free water may serve as a better eluent than elution buffers provided in the kits because the latter may contain reagents that inhibit downstream processing such as amplification.

9. Heating the DNA eluates at 65 °C will concentrate the DNA allowing longer storage durations for a given yield, will not denature DNA, but can denature enzymes that may degrade DNA, and will cause the evaporation of volatile buffer contaminants, including ethanol, which may interfere with the 260/280 and 260/230 ratios and/or inhibit many subsequent reactions, including PCR. Shaking the eluates while heating minimizes, in a speed-dependent manner, the incubation time necessary to concentrate the samples. A volume of 50 μL used for elution will yield on average 13.4 ± 2.4 μL eluate after a heat incubation at 65 °C for 65 min (without shaking). Avoid complete dryness, which can lead to decreased DNA yields because dried DNA is difficult to dissolve.

10. DNA can be bisulfite-treated according to established protocols; using for example EZ DNA Methylation Kit (Zymo Research), according to manufacturer's instructions. Locus-specific and genome-wide methylation can be studied by various techniques outlined in this book, including Pyrosequencing® (*see also* Chapter 22) and Illumina Human Methylation 450 BeadChips (*see also* Chapter 16), respectively, which were our methods of choice. Briefly, Pyrosequencing and Illumina HM450 arrays were performed using 20 ng and 300 ng of bisulfite-converted DNA, respectively. For Pyrosequencing, the regions of interest (*Line1* and *AluYb8* transposable elements, which are markers of global methylation levels) were amplified with forward and reverse primers, one of which is biotinylated (btn), and then the methylation levels of the amplified region were analyzed using a sequencing primer. The forward, reverse and sequencing primers used for *Line1* were the following, respectively, and adopted from Daskalos et al. [20]: 5′-btn-TAGGGAGTGTTAGATAGTGG-3′, 5′-AACTCCCTAACCCCTTAC-3′ and 5′-CAAATAAAAC AATACCTC-3′. The forward, reverse and sequencing primers used for *AluYb8* were the following, respectively: 5′-AGATTATTTTGGTTAATAAG-3′, 5′-btn-AACTACRAA CTACAATAAC-3′ and 5′-GTTTGTAGTTTTAGTTATT-3′, as previously described [21]. Infinium HM450 arrays were processed according to manufacturer's instructions. GenomeStudio was used to analyze quality controls. Data was imported into R (3.0.0), using the *minfi* package version 1.2.0 (http://www.bioconductor.org), filtered from background noise, normalized [22], log-transformed [23] and analyzed for differential methylation using *minfi* and *lumi* R packages [24]. More details on the analysis of Infinium arrays can be found in Chapter 16.

Acknowledgments

This work was partially supported by the Association pour le Recherche Contre le Cancer (l'ARC). A.G. is supported by the IARC Postdoctoral Fellowship and Marie Curie Actions-People-COFUND. We thank the International Childhood Cancer Cohort Consortium for contributing to the coordination between participating cohorts. This manuscript is dedicated to the participating children and their families.

Disclaimer: The findings and conclusions in this report are those of the authors and do not necessarily represent the views of the National Children's Study, National Institutes of Health, or U.S. Department of Health and Human Services. Supported in part by NICHD Contracts HHSN275200503414C, HHSN275200503411C, HHSN275200603416C, HHSN275200503415C, HHSN275200503413C, HHSN275200503410C, HHSN275200503396C, HHSN275201000121U, and HHSN275200900010C.

References

1. Burczynski ME, Rockett JC (2006) Surrogate tissue analysis: genomic, proteomic and metabolomic approaches. CRC, Boca Raton
2. Guthrie R, Susi A (1963) A simple phenylalanine method for detecting phenylketonuria in large populations of newborn infants. Pediatrics 32:338–343
3. Mei JV, Alexander JR, Adam BW et al (2001) Use of filter paper for the collection and analysis of human whole blood specimens. J Nutr 131:1631S–1636S
4. He H, Argiro L, Dessein H et al (2007) Improved technique that allows the performance of large-scale SNP genotyping on DNA immobilized by FTA technology. Infect Genet Evol 7:128–132
5. Coyne SR, Craw PD, Norwood DA et al (2004) Comparative analysis of the Schleicher and Schuell IsoCode Stix DNA isolation device and the Qiagen QIAamp DNA mini kit. J Clin Microbiol 42:4859–4862
6. Luckwell J, Denniff P, Capper S et al (2013) Assessment of the within- and between-lot variability of Whatman FTA((R)) DMPK and 903((R)) DBS papers and their suitability for the quantitative bioanalysis of small molecules. Bioanalysis 5:2613–2630
7. Brown RC, Dwyer T, Kasten C et al (2007) Cohort profile: the International Childhood Cancer Cohort Consortium (I4C). Int J Epidemiol 36:724–730
8. Bundo M, Sunaga F, Ueda J et al (2012) A systematic evaluation of whole genome amplification of bisulfite-modified DNA. Clin Epigenetics 4:22
9. Reins J, Mossner M, Richter L et al (2011) Whole-genome amplification of sodium bisulfite-converted DNA can substantially impact quantitative methylation analysis using pyrosequencing. BioTechniques 50:161–164
10. Ghantous A, Saffery R, Cros MP et al (2014) Optimized DNA extraction from neonatal dried blood spots: application in methylome profiling. BMC Biotechnol 14:60
11. Wong N, Morley R, Saffery R et al (2008) Archived Guthrie blood spots as a novel source for quantitative DNA methylation analysis. Biotechniques 45:423–428
12. Hollegaard MV, Grauholm J, Norgaard-Pedersen B et al (2013) DNA methylome profiling using neonatal dried blood spot samples: a proof-of-principle study. Mol Genet Metab 108:225–231
13. Hollegaard MV, Thorsen P, Norgaard-Pedersen B et al (2009) Genotyping whole-genome-amplified DNA from 3- to 25-year-old neonatal dried blood spot samples with reference to fresh genomic DNA. Electrophoresis 30:2532–2535
14. Nguyen Thi Hue NDHC, Phong PT, Linh NTT, Giang NDT (2012) Extraction of human genomic DNA from dried blood spots

and hair roots. Int J Biosci Biochem Bioinform 2:21–26

15. Sjoholm MI, Dillner J, Carlson J (2007) Assessing quality and functionality of DNA from fresh and archival dried blood spots and recommendations for quality control guidelines. Clin Chem 53:1401–1407
16. Staunstrup NH, Starnawska A, Nyegaard M et al (2016) Genome-wide DNA methylation profiling with MeDIP-seq using archived dried blood spots. Clin Epigenetics 8:81
17. Murphy SC, Daza G, Chang M et al (2012) Laser cutting eliminates nucleic acid cross-contamination in dried-blood-spot processing. J Clin Microbiol 50:4128–4130
18. Dugué PA, English DR, MacInnis RJ et al (2016) Reliability of DNA methylation measures from dried blood spots and mononuclear cells using the HumanMethylation450k BeadArray. Sci Rep 6:30317
19. Bonne N, Clark P, Shearer P et al (2008) Elimination of false-positive polymerase chain reaction results resulting from hole punch carryover contamination. J Vet Diagn Investig 20:60–63
20. Daskalos A, Nikolaidis G, Xinarianos G et al (2009) Hypomethylation of retrotransposable elements correlates with genomic instability in non-small cell lung cancer. Int J Cancer 124:81–87
21. Choi SH, Worswick S, Byun HM et al (2009) Changes in DNA methylation of tandem DNA repeats are different from interspersed repeats in cancer. Int J Cancer 125:723–729
22. Maksimovic J, Gordon L, Oshlack A (2012) SWAN: subset-quantile within array normalization for illumina infinium HumanMethylation450 BeadChips. Genome Biol 13:R44
23. Du P, Zhang X, Huang CC et al (2010) Comparison of Beta-value and M-value methods for quantifying methylation levels by microarray analysis. BMC Bioinformatics 11:587
24. Du P, Kibbe WA, Lin SM (2008) lumi: a pipeline for processing Illumina microarray. Bioinformatics 24:1547–1548

Chapter 32

DNA Methylation Analysis of Free-Circulating DNA in Body Fluids

Maria Jung, Glen Kristiansen, and Dimo Dietrich

Abstract

Circulating cell-free DNA in body fluids is an analyte of great interest in basic and clinical research. The analyses of DNA methylation and hydroxymethylation patterns in body fluids might allow one to determine the certain state of a disease, in particular of cancer. DNA methylation biomarkers in liquid biopsies, i.e. blood plasma samples, may help optimizing personalized therapy for individual patients. DNA methylation analyses of specific loci usually require a bisulfite conversion of the DNA, which requires a sufficiently high amount of DNA at the appropriate concentration. However, free-circulating DNA is generally low concentrated. Therefore, high volumes of body fluids need to be analyzed. This high volume needs to be reduced in order to facilitate the bisulfite conversion. In addition, disease-related free-circulating DNA is even less abundant than normal DNA in the total amount of free-circulating DNA. Accordingly, analytical and pre-analytical methods are needed, which permit an accurate and sensitive quantification of single methylated DNA copies in the presence of unmethylated DNA in abundance.

This protocol describes two methods for DNA enrichment from body fluids: DNA extraction by means of magnetic beads and polymer-mediated enrichment of DNA. Subsequent bisulfite conversion is achieved by means of a high-speed conversion protocol. Adaptions of the workflow required for the analysis of hydroxymethylation via oxidation 5-hydroxymethylcytosines to 5-formylcytosines prior to the bisulfite conversion are introduced. A quantitative real-time PCR based on the methylation-specific and HeavyMethyl PCR methodologies is introduced. This qPCR assay allows for an accurate and sensitive quantification of single copies of the DNA methylation biomarkers *SHOX2* and *SEPT9* in blood plasma. Specific issues regarding the analysis of body fluids and respective trouble shooting approaches are discussed.

Key words DNA methylation, DNA hydroxymethylation, Polymer-mediated enrichment (PME), Magnetic bead extraction, Bisulfite conversion, Body fluid, Methylation-specific PCR (MSP), HeavyMethyl (HM) PCR, Quantitative real-time PCR

1 Introduction

Methylation and hydroxymethylation of cytosines within the CpG dinucleotide context are epigenetic mechanisms, which play an important role in biological processes, such as epigenetic reprogramming, and pluripotency [1]. In addition, DNA methylation

Jörg Tost (ed.), *DNA Methylation Protocols*, Methods in Molecular Biology, vol. 1708,
https://doi.org/10.1007/978-1-4939-7481-8_32,

and hydroxymethylation are involved in carcinogenesis and tumor progression [2].

In order to conduct clinically relevant research on DNA methylation and hydroxymethylation one must be able to analyze clinically relevant specimens. Body fluids, i.e. blood plasma, blood serum, urine, pleural effusion, ascites, aspirates, and lavages are clinically relevant and promising sample types. In particular, blood plasma and blood serum hold the potential to improve the management of malignant diseases in the future. These liquid biopsies can be obtained in minimally invasive procedures and might hold valuable information about the current presence or absence of cancer in patients, disease progression, and tumor response to certain therapies. Cancer-specific methylation of tumor-derived circulating cell-free DNA in body fluids has already been established as a biomarker for early cancer detection (screening), diagnostics and prognostics. An extensive observational prospective screening trial, including more than 7,000 asymptomatic subjects, showed that methylated copies of the *SEPT9* gene in plasma is a powerful biomarker for colorectal cancer screening [3, 4]. *SHOX2* is another blood plasma based DNA methylation biomarker used for the diagnosis of lung cancer [5].

DNA methylation analyses are based on determining the abundance of methylated, in contrast to unmethylated, cytosines. However, methylated cytosines and unmethylated cytosines exhibit the same base pairing behavior and therefore are difficult to distinguish from each other using conventional molecular biological techniques. In 1992, Frommer et al. [6] published a protocol illustrating how a deamination of cytosines to uracils using bisulfite was achieved leaving the methylated cytosines unaffected. The procedure allowed for a positive display of 5-methylcytosine since the epigenetic information of the DNA is transformed into sequence information, which can be studied via PCR and other hybridization based methods. Several technological advances have now led to protocols which are much more convenient and user-friendly as compared to the original 16-h protocol suggested by Frommer et al. [7–12]. Numerous commercial kits and published protocols are available giving even inexperienced researchers easy access to methylation analyses. However, these kits and protocols are mainly designed to enable the bisulfite conversion of extracted high molecular DNA. Only a few kits allow for the modification of DNA from challenging input sample material, i.e. body fluids. Currently, there is only one kit commercially available (innuCONVERT Bisulfite Body Fluids Kit, Analytik Jena, Germany) suitable for methylation analyses of large sample volumes up to 3 mL of pleural effusions, ascites, urine, blood plasma, and blood serum [13].

DNA methylation analysis of free-circulating DNA in body fluids is a complex multistep process. All comprised steps, i.e. sample preparation (Subheading 3.1 or 3.2), DNA enrichment

(Subheadings 3.3 and 3.4), bisulfite conversion and subsequent purification (Subheading 3.5), and downstream PCR analytics (Subheading 3.6), are described herein. Body fluids often only contain low concentrations of total DNA and an even lower abundance of tumor markers. The vast majority of free-circulating DNA in body fluids is usually not disease-related DNA. This background DNA complicates the analysis of the small fraction of disease-related free-circulating DNA. Accordingly, high volumes of body fluids must be analyzed to be above the limit of quantification (LOQ) or the limit of detection (LOD) of the specific biomarkers.

Due to the high volumes of body fluids, a volume reduction and enrichment of the DNA is required prior to the bisulfite conversion. Two established methods for this DNA concentration are suitable: Polymer-mediated enrichment (PME, Subheading 3.3) and magnetic bead extraction (Subheading 3.4). In the following paragraphs both methods are introduced.

During the method of polymer-mediated enrichment (PME, Subheading 3.3) the body fluid is incubated with polymer forming substrates. Subsequently, pellet formation is accomplished, whereby DNA molecules are captured in between polymerizing molecules. After centrifugation, the supernatant is removed from the DNA containing gel and the gel is washed to remove further impurities. Finally, the gel is dissolved and DNA is released in a low volume.

The extraction of DNA via magnetic beads (Subheading 3.4) is based on the binding of DNA to silica (SiO_2). The usage of magnetic particles with silica surface coatings enables an extraction without centrifugation steps and allows for the processing of high volume samples. Efficient binding of DNA to the silica surface strongly depends on pH and high concentration of chaotrophic salts. Sample impurities can be removed by washing with buffers containing at least 70% ethanol. Elution is performed in a low-salt buffer or water with a pH of 7.0–8.5.

Subheading 3.5 describes a high-speed bisulfite conversion protocol, using a high molar bisulfite solution and a DNA protecting denaturation buffer. This protocol enables the quick and efficient preparation of high integrity, purity and quality DNA. A high purity of the bisulfite converted DNA is mandatory since the low abundance of tumor DNA usually requires a high volume of the eluted DNA for input into the PCR reaction. This increases the risk of carrying over impurities. These impurities, i.e. bisulfite, wash buffers and ethanol, might lead to severe PCR inhibition.

DNA methylation analysis of circulating cell-free DNA in body fluids in clinically relevant questions usually requires the sensitive and accurate quantification of single or few methylated gene copies in a background of an abundance of unmethylated copies. Quantitative methylation-specific PCR (qMSP) (*see also* Chapter 24), quantitative HeavyMethyl (HM) PCR, and digital MethylLight

PCR technologies (*see also* Chapter 25) [5, 13–16], or combinations thereof, are methodologies which show a low limit of quantification and detection and are therefore well suited for the methylation analyses of body fluids. Bisulfite converted artificially methylated DNA is well suited for calculating relative methylation levels based on the ΔΔCT method [17, 18].

Subheading 3.6 provides an exemplary description of a quantitative real-time PCR triplex assay [18]. This assay allows for the accurate and sensitive quantification of single copies of the DNA methylation biomarkers *SHOX2* and *SEPT9*. It combines methylation-specific PCR (MSP) [19], HeavyMethyl (HM) technology [16] together with the TaqMan™ probe technology for real-time PCR detection. Quantification of total DNA content is based on the *ACTB* gene in a methylation independent manner. All assays (*SEPT9*, *ACTB*, and *SHOX2*) are performed in a single-tube triplex real-time PCR assay. The calculation of relative methylation levels is introduced in Subheading 3.7 and in Subheading 3.8 an exemplary evaluation of analytical performance is described.

Subheading 4 should help researchers avoid issues related to the analysis of these challenging sample types.

2 Materials

Specific plasticware for the sample preparation of blood plasma and serum is mentioned in Subheading 2.1. The reagents, kits, plasticware, and instruments used for DNA enrichment are listed in Subheading 2.2. Furthermore, Subheading 2.3 provides an overview of reagents, kits, and plasticware recommended for the bisulfite conversion of enriched DNA from body fluids. All components needed for polymer-mediated enrichment of DNA in body fluids, and the subsequent bisulfite conversion and cleanup, are contained in the innuCONVERT Bisulfite Body Fluids Kit (Analytik Jena, Jena, Germany). In Subheading 2.4 reagents and the instrument required for the quantitative real-time PCR are specified.

2.1 Sample Preparation

1. 9 mL (92 × 16 mm) S-Monovette® K3 EDTA.
2. 9 mL (92 × 16 mm) S-Monovette® Z-Gel.
3. 10 mL (16 × 100 mm) BD Vacutainer® K2E (EDTA).
4. 15 mL centrifuge tubes, polypropylene, conical bottom w/CentriStar™ cap, sterile (Corning Incorporated) or equivalent.

2.2 DNA Enrichment

1. SB3 Tube Rotator Variable Speed model (Bibby Scientific Limited) or equivalent.
2. DynaMag™-15 Magnet (Life Technologies) or equivalent.

3. innuCONVERT Bisulfite Body Fluids Kit (Analytik Jena) including Enrichment Reagent VCR-1 and -2, Lysis Solution GS, Carrier RNA, Proteinase K, Binding Solution VL, Spin Filter, Receiver Tubes, Washing Solution C, Washing Solution BS.
4. innuPREP MP Basic Kit A (Analytik Jena) including Lysis Solution RL, MAG suspension, Binding Solution VL, Washing Solution HS, Washing Solution LS.
5. Ethanol.

2.3 Bisulfite Conversion and Cleanup

1. innuCONVERT Bisulfite Body Fluids or Basic Kits (Analytik Jena) including Conversion Reagent, Conversion Buffer, Binding Buffer, Desulfonation Buffer, Elution Buffer, Silica Membrane Spin Column.

2.4 DNA Methylation Analysis: qPCR

1. 5(6)-Carboxy-X-Rhodamine (ROX).
2. Dimethyl Sulfoxide (DMSO).
3. *ACTB* forward primer (sequence: 5′-gtgatggaggaggtttagt aagtt-3′).
4. *ACTB* reverse primer (sequence: 5′-ccaataaaacctactcctcccttaa-3′).
5. *ACTB* probe (sequence: 5′-ATTO 647 N-accaccacccaacacacaa-taacaaacaca-BHQ-2-3′).
6. *SHOX2* forward primer (sequence: 5′-taacccgacttaaacgacga-3′).
7. *SHOX2* reverse primer (sequence: 5′-gtttttggatagttaggtaat-3′).
8. *SHOX2* blocker (sequence: 5′-taatttttgttttgtttgtttgattggggttgt atga-SpacerC3-3′).
9. *SHOX2* probe (sequence: 5′-6-FAM-ctcgtacgaccccgatcg-BBQ650-3′).
10. *SEPT9* forward primer (sequence: 5′-aaataatcccatccaacta-3′).
11. *SEPT9* reverse primer (sequence: 5′-gttgtttattagttattatgt-3′).
12. *SEPT9* blocker (sequence: 5′-gttattatgttggattttgtggttaatgtg tag-SpacerC3-3′).
13. *SEPT9* probe (sequence: 5′-JOE-ttaaccgcgaaatccgac-BHQ-1-3′).
14. Tris–HCl, pH 8.4.
15. $MgCl_2$.
16. KCl.
17. Glycerol.
18. FastStart Taq DNA Polymerase, dNTPack (Roche Applied Science) or equivalent.
19. 7500 Fast Real-Time PCR System (Life Technologies) or equivalent.

3 Methods

3.1 Sample Preparation: Plasma and Serum

Several systems are in use for the preparation of plasma from whole blood. These systems differ in the way the blood is collected (vacuum or aspiration) and, more importantly, the anticoagulant, i.e. EDTA (K_2EDTA, K_3EDTA), heparin, and citrate. The effect of the anticoagulant on the desired downstream analysis, i.e. qPCR needs to be evaluated carefully. EDTA plasma has previously been shown to be well suited for DNA methylation analysis [3–5, 13]. The preparation of serum is similar to the preparation of plasma with the exception of the usage of respective serum collection tubes (without anticoagulant). S-Monovette® 9 mL, 92 × 16 mm, Z-Gel is a suitable blood collection system for the preparation of serum for downstream methylation analyses.

1. Separate the cellular fraction of plasma or serum by centrifugation at 1,500 × *g* for 10 min.
2. Transfer the supernatant into a new 15 mL centrifugation tube and subsequently centrifuge at 1,500 × *g* for 10 min.
3. The supernatant can be stored at −20 °C (short-term storage up to 4 weeks) or at −80 °C (long-term storage) (*see* **Note 1**).

3.2 Sample Preparation: Urine, Pleural Effusion, Bronchial Aspirates

Ascites, pleural effusion, and bronchial aspirate samples can be used fixed (with formalin or Saccomanno's fixative) or naïve (*see* **Note 2**).

1. Separate the cellular fraction of the body fluid by centrifugation at 4,000 × *g* for 10 min.
2. The supernatant can be stored at −20°C (short-term storage up to 4 weeks) or at −80°C (long-term storage) (*see* **Note 3**).

3.3 DNA Enrichment: Polymer-Mediated Enrichment (PME)

The DNA enrichment step is challenging and requires the handling of large volumes. Two established methods for this DNA concentration are suitable: Polymer-mediated enrichment (PME) and magnetic bead extraction. Choose either the polymer-mediated enrichment, which is described in this section, or the method described in Subheading 3.4, and either way proceed with the steps of Subheading 3.5. All required reagents for the PME are included in the innuCONVERT Bisulfite Body Fluids Kit (*see* **Note 4**).

1. Prepare the proteinase K solution (20 mg/mL proteinase K in water).
2. Prepare the carrier RNA solution (200 ng/μL polyA in water).
3. Prepare the washing working solution by mixing nine parts of absolute ethanol with one part of washing solution BS. 1.3 mL washing working solution is required for each sample.

4. Vortex 3 mL of body fluid, 100 μL enrichment reagent VCR-1, and 10 μL carrier RNA solution in a 15 mL centrifugation tube for 10 s.
5. Add 600 μL enrichment reagent VCR-2 and vortex for 10 s.
6. Incubate for 10 min at room temperature and sediment the gel by centrifugation at 4500 × *g* for 10 min.
7. Discard the supernatant.
8. Wash the gel by adding 5 mL water without removing the pellet.
9. Centrifuge at 4,500 × *g* for 5 min and discard the supernatant.
10. Dissolve the pellet in 600 μL lysis solution GS.
11. Transfer the mixture to a 1.5 mL reaction tube, add 25 μL proteinase K solution, vortex for 10 s and incubate for 15 min at 70 °C in a water bath.
12. Centrifuge at 14,000 × *g* for 2 min.
13. Transfer the supernatant to a 1.5 mL reaction tube.
14. Add 300 μL binding solution VL, mix by pipetting up and down without introducing air bubbles.
15. Add spin filter to the receiver tube and transfer 600 μL of the sample to the spin filter.
16. Centrifuge at 11,000 × *g* for 1 min. Discard the receiver tube with the filtrate and place the spin filter into a new 2.0 mL receiver tube.
17. Add the residual sample to the spin filter and centrifuge at 11,000 × *g* for 1 min. Discard the receiver tube with the filtrate and place the spin filter into a new 2.0 mL receiver tube.
18. Add 500 μL washing solution C to the spin filter and centrifuge at 11,000 × *g* for 1 min. Discard the receiver tube with the filtrate and place the spin filter into a new 2.0 mL receiver tube.
19. Add 650 μL washing working solution (with ethanol) to the spin filter and centrifuge at 11,000 × *g* for 1 min. Discard the receiver tube with the filtrate and place the spin filter into a new 2.0 mL receiver tube.
20. Repeat the last wash step (**step 19**) one more time.
21. Centrifuge for 3 min at 14,000 × *g* to remove all traces of ethanol. Discard the 2 mL receiver tube and place the spin filter into a new 2 mL reaction tube.
22. Open the cap of the spin filter and dry the spin filter by incubation at 60 °C for 10 min in a thermal mixer. This step removes all traces of ethanol.

23. Elute the DNA by addition of 50 μL water to the spin filter, incubate at room temperature for 2 min and centrifuge for 1 min at 11,000 × *g*.
24. Discard the spin filter and store the DNA at 4 °C until bisulfite conversion.

3.4 DNA Enrichment: Magnetic Bead Extraction

All required reagents for the Magnetic Bead Extraction are included in the innuPREP MP Basic Kit A (*see* **Note 4**).

1. Prepare the proteinase K solution (20 mg/mL proteinase K in water).
2. Prepare the carrier RNA solution (200 ng/μL polyA in water).
3. Mix the following components properly in a 15 mL centrifugation tube: 3 mL of body fluid, 3 mL lysis solution RL, 50 μL proteinase K, and 10 μL carrier RNA solution.
4. Incubate for 30 min at 70 °C in a water bath or thermal mixer.
5. Extensively vortex the MAG suspension.
6. Add 100 μL MAG suspension and 6 mL binding solution VL to the 15 mL centrifugation tube containing the sample (*see* **Note 5**).
7. Incubate at room temperature for 20 min in a rotator at 20 rpm.
8. Perform separation of magnetic beads by placing the 15 mL centrifugation tube in a magnetic trap for 1 min.
9. Discard the supernatant while the 15 mL centrifugation tube remains in the magnetic trap.
10. Add 1.5 mL washing solution LS, mix thoroughly and transfer the magnetic beads to a 2 mL reaction tube.
11. Place the tube in a magnetic trap for 1 min and discard the supernatant while the tube remains in the magnetic trap. Make sure that the whole volume of washing solution LS is removed.
12. Place the open tube in a thermal mixer at 60 °C for 10 min for complete removal of ethanol.
13. Add 50 μL water, mix thoroughly, spin down briefly, and incubate for 10 min at 85 °C and 1000 rpm in a thermal mixer.
14. Place the tube in a magnetic trap for 1 min and transfer the eluate to a new 2 mL tube.

3.5 Bisulfite Conversion and Cleanup

The bisulfite conversion of DNA is a chemical reaction under harsh chemical conditions, i.e., low pH, high temperature, and elongated incubation times, causing significant DNA degradation [20–22]. Hayatsu summarized the principle of the bisulfite reaction [23]: Bisulfite (HSO_3^-) reacts with cytosines in a single stranded DNA molecule through the formation of a 5,6-dihydrocytosine-6-

sulfonate intermediate. This reaction would preferably take place at an acidic pH of around pH 5. Bisulfite induces the hydrolytic deamination of the intermediate [23, 24]. Subsequently, bisulfite is eliminated at alkaline pH [23]. The optimal reaction conditions result from a balanced control of all desired (conversion of cytosines) and undesired reactions (DNA degradation and inappropriate conversion of methylcytosines). In summary, high bisulfite concentrations and high temperature at prolonged incubation times will lead to a complete conversion of all cytosines to uracils on the one hand, but will cause DNA degradation and inappropriate conversion of methylated cytosines to thymines on the other hand [13, 25]. An incomplete conversion of cytosines and a high rate of inappropriate conversion of methylcytosines will lead to false positive and false negative methylation signals, respectively [13, 25]. DNA degradation is mainly caused by depurination and depyrimidation leading to abasic sites [20, 22], followed by DNA strand breaks due to N-glycoside bond cleavage. High-molarity, high-temperature protocols have been developed to allow for a rapid bisulfite conversion within less than 1 h [10–12, 23]. These high-molarity and high-temperature conditions are preferable because they yield greater homogeneity of bisulfite conversion, higher DNA integrity and thus yield more reliable data [13, 25]. In addition to optimized bisulfite reaction conditions, a highly efficient cleanup of the converted DNA is mandatory in order to allow for the input of high volumes of eluted DNA into the PCR without causing PCR inhibition.

The protocol described in the following paragraph is designed to prepare highly pure DNA with 99.0% converted cytosines and less than 1% inappropriately converted methylcytosines within 2 h including cleanup [13]. Conversion efficiency can be increased by adapting the reaction conditions (*see* **Note 6**). These adaptations are suited for the deamination of formylcytosine as required for oxidative bisulfite conversion for 5-hydroxymethylation analyses (*see* **Note** 7). All required reagents are included in the innuCONVERT Bisulfite kits.

1. Mix the following components in a 2 mL reaction tube: 50 μL enriched DNA, 30 μL conversion buffer (*see* **Note 8**) and 70 μL conversion reagent (*see* **Note 9**).
2. Mix properly and spin down briefly in order to remove drops from the tube caps.
3. Incubate 45 min at 85 °C in a thermal mixer or in a water bath (*see* **Note 10**).
4. Prepare the binding buffer working solution by mixing one part of absolute ethanol with one part of binding solution GS. 700 μL binding buffer working solution are required for each sample.

5. Prepare the desulfonation buffer working solution by mixing three parts of absolute ethanol with one part of desulfonation buffer. 700 μL desulfonation buffer working solution are required for each sample.
6. Prepare the washing buffer working solution by mixing nine parts of absolute ethanol with one part of washing solution BS. 1 mL washing buffer working solution is required for each sample.
7. Add 700 μL binding buffer working solution (including ethanol) to the bisulfite reaction mixture and mix thoroughly. Spin down briefly in order to remove drops from the tube caps.
8. Add the spin filter to the receiver tube and transfer the sample to the spin filter.
9. Spin 3 min at 14,000 × *g*. Discard the receiver tube with the flow-through and place the spin filter into a new 2.0 mL receiver tube.
10. Add 200 μL washing buffer working solution (including ethanol) to the membrane of the spin filter and centrifuge at 14,000 × *g* for 1 min. Discard the receiver tube with the flow-through and place the spin filter into a new 2 mL receiver tube.
11. Open the spin filter and add 700 μL desulfonation buffer working solution (prepared in **step 5**), and incubate at room temperature for 10 min. Centrifuge at 14,000 × *g* for 1 min. Discard the receiver tube with the flow-through and place the spin filter into a new 2 mL receiver tube.
12. Add 400 μL washing buffer working solution (prepared in **step 6**), close the cap and centrifuge at 14,000 × *g* for 1 min. Discard the flow-through and place the spin filter into the same 2 mL receiver tube.
13. Add 400 μL washing buffer working solution (prepared in **step 6**), close the cap and centrifuge at 14,000 × *g* for 1 min. Discard the receiver tube with the flow-through and place the spin filter into a new 2 mL receiver tube.
14. Centrifuge for 3 min at 14,000 × *g* to remove all traces of ethanol. Discard the 2 mL receiver tube and place the spin filter into a 1.5 mL reaction tube.
15. Open the cap of the spin filter and dry the spin filter by incubation at 60 °C for 10 min in a thermal mixer. This step removes all traces of ethanol.
16. Add 50 μL elution buffer to the spin filter and incubate at room temperature for 1 min (*see* **Note 11**). Centrifuge for 1 min at 8,000 × *g*.
17. The eluted DNA is stable for up to 4 weeks at −20 °C [13] (*see* **Note 12**).

3.6 DNA Methylation Analysis: qPCR

Due to the low concentration of bisulfite converted DNA in the eluted sample, an input of a high volume of the sample into the PCR reaction is needed. This might lead to the carryover of impurities, which might impair the PCR performance (*see* **Note 13**). Quality control criteria are described in **Note 14**. The choice of the PCR buffer conditions, an optimized $MgCl_2$ concentration and PCR cycling program, is mandatory for the performance of the assay. Once an assay is established, the PCR buffer conditions should not be changed. Control DNA, which mimics the conditions (DNA concentration, relative methylation) expected in the samples of interest should be prepared. DNA from human sperm is unmethylated at several CpG islands and can be used as a surrogate template DNA. Mixtures with artificially methylated DNA are well suited for assay optimization purposes. The *ACTB/SHOX2/SEPT9*-assay is described in the following:

1. Add 1.5 mL of prewarmed DMSO (40 °C) to 25 mg of 5(6)-Carboxy-X-rhodamine (ROX).
2. Vortex for 20 s, and incubate the solution at 99 °C for 4 h at 1,400 rpm in a thermal mixer.
3. Centrifuge the solution for 10 min at 20,000 × *g* and transfer the supernatant to a new 2 mL reaction tube without transferring residual ROX crystals.
4. Adjust the ROX solution to $OD_{580nm} = 1.74$.
5. Prepare the mix of oligonucleotides (4×) (*see* **Note 15**). The mix should include all required primers, blocker oligonucleotides, and probes for simultaneous quantification of a reference gene and the target genes. The mix should include each oligonucleotide in a four times higher concentration as required in the qPCR reaction. An exemplary oligonucleotide mix for simultaneous quantification of *ACTB*, *SHOX2*, and *SEPT9* should be composed as follows: 1.6 μM of each *SHOX2* and *SEPT9* primer, 3 μM of the *SHOX2* and *SEPT9* blocker oligonucleotide, respectively, 1.2 μM of the *SHOX2* and *SEPT9* probes, 0.3 μM of the *ACTB* primers, and 0.04 μM of *ACTB* probe. The analytical performance of the *ACTB/SHOX2/SEPT9*-Assay is described in Subheading 3.8.
6. Prepare the qPCR mastermix (2×) consisting of 70 mM Tris–HCl, pH 8.4, 12 mM $MgCl_2$, 100 mM KCl, 8% glycerol, 0.5 mM of each dNTP (*see* **Note 16**), 2 U FastStart Taq DNA polymerase, and 0.006 μL ROX solution in a volume of 10 μL.
7. Add 5 μL mix of oligonucleotides and 10 μL qPCR mastermix to 5 μL sample or calibrator.
8. Perform PCR with the following cycling protocol: 20 min at 95 °C and 50 cycles with 2 s at 62 °C, 45 s at 56 °C (each at 100% ramp rate) and 15 s, 95 °C (at 75% ramp rate). Data are collected during the elongation step (45 s at 56 °C).

3.7 Calculation of Relative Methylation Levels (ΔΔ CT Method)

The quantification of DNA methylation levels is based on the ΔΔCT method and is adapted for DNA methylation analyses as described earlier [26]. A calibrator sample is required which ideally consists of a 100% methylated and bisulfite converted DNA. The calibrator allows for the relative methylation quantification and for the assessment of inter-run variability. The quantification of a reference gene in a methylation independent manner is necessary for quantification of total DNA input and thereby enables the normalization of the threshold cycles (CTs) of the target genes.

1. Define baseline and cycle threshold settings for each amplification product in a way that the cycle threshold is defined in the exponential amplification phase of the respective target or reference gene. The baseline and threshold settings were previously described as cycle 3–24 (baseline) for all amplification products and thresholds for *ACTB*, *SHOX2*, and *SEPT9* were defined as 0.02, 0.015, and 0.01, respectively [18].
2. Define quality criteria for sufficient DNA yield by defining a CT cut-off which dichotomizes the samples with a CT higher than the cut-off as invalid and the samples with a CT equal or lower than the cut-off as valid. Quality criteria for *ACTB*, *SHOX2*, and *SEPT9* were previously described as $CT_{Sample/ACTB} = 31.5$ [18, 27], $CT_{Sample/SHOX2} = 35$ [18], and $CT_{Sample/SEPT9} = 40$ [18].
3. The methylation specific amplification of a target is normalized by calculating the difference of the threshold cycles (CT) of the target gene and the CT of the reference gene ($\Delta CT_{Sample} = CT_{Sample/Ref\_Gene} - CT_{Sample/Target\_Gene}$, or $\Delta CT_{Calibrator} = CT_{Calibrator/Ref\_Gene} - CT_{Calibrator/Target\_Gene}$), respectively.
4. After normalization of the CTs of the target gene both in the sample and the calibrator, the difference between the ΔCT of the sample and the calibrator has to be calculated ($\Delta\Delta CT_{Sample} = \Delta CT_{Sample} - \Delta CT_{Calibrator}$).
5. The methylation level [in %] is calculated as follows: $\text{Methylation level} = 100\% \cdot 2^{\Delta\Delta CT_{Sample}}$.

3.8 Analytical Performance Evaluation

The analytical performance of methylation assays can be evaluated using mixtures of bisulfite converted artificially methylated DNA and unmethylated DNA. In the following example, a background of unmethylated DNA as expected in clinical samples was simulated by using 3,000 copies of unmethylated DNA template molecules. Methylated copies of the human genome were spiked in different numbers (0–3,072) into this background DNA. These mixtures were loaded into six single PCR reactions. The assay showed a high analytical performance (high linearity, low variability) over a wide dynamic range of 6–3,072 copies of methylated DNA (Fig. 1). At high numbers of methylated gene copies (approximately >50

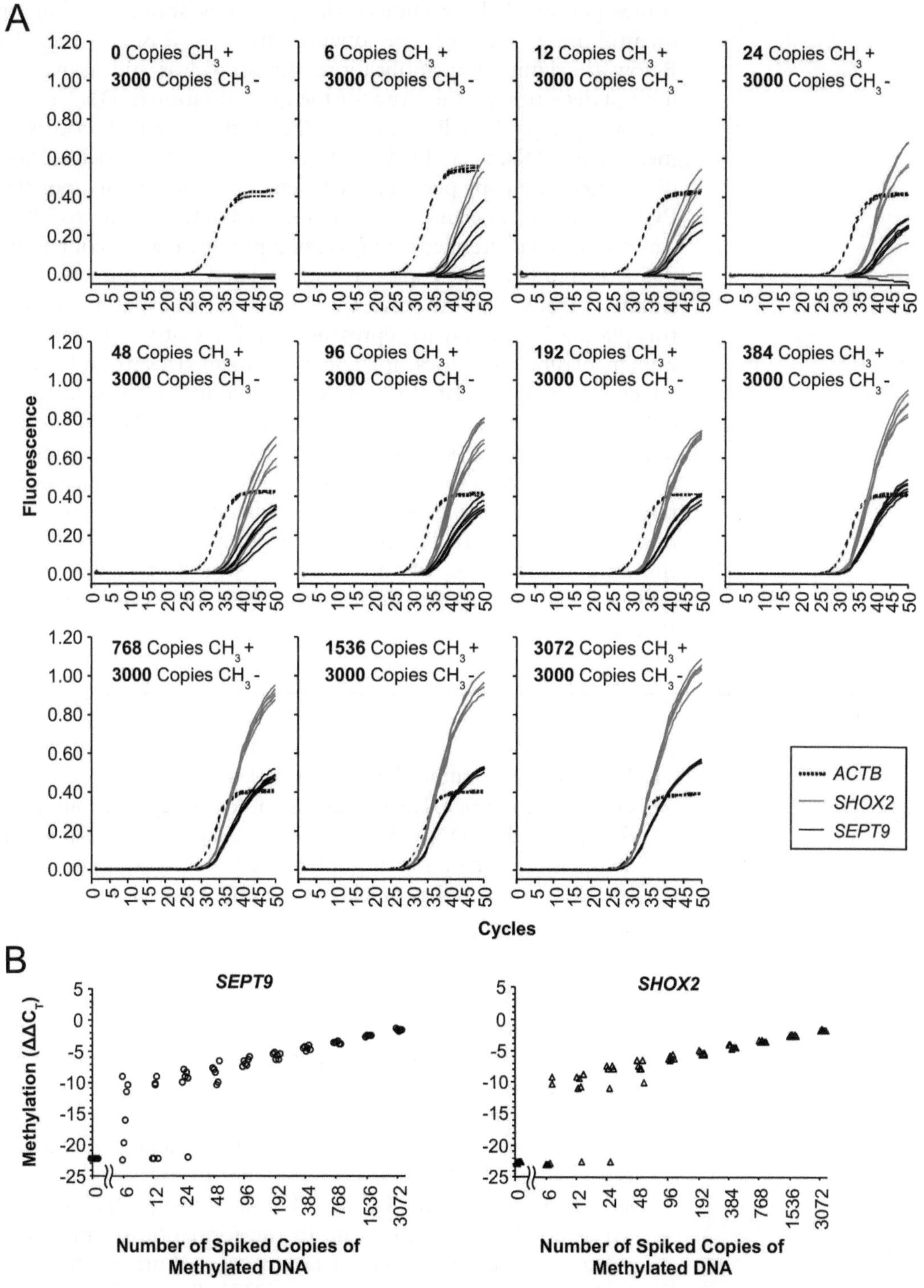

Fig. 1 Analytical performance of the quantitative real-time PCR for simultaneous quantification of *SEPT9* and *SHOX2* methylation levels. A CpG-free locus within *ACTB* is used as reference gene to quantify the total DNA amount. Different numbers of methylated (CH_3+) DNA (0–3,072 copies) and a constant number of 3,000 copies unmethylated (CH_3−) DNA were used as template DNA. Shown are real-time PCR amplification plots (**a**) and $\Delta\Delta$CT-values (**b**)

copies per six PCR reactions), the assay was above the limit of quantification and accurate quantification is feasible (Fig. 1). Below the limit of quantification, the assay was still above the limit of detection and allowed for the quantification of DNA methylation by digital PCR. Figure 1 shows that even six copies of methylated DNA in six PCR reactions (equivalent to stochastically distributed one copy per each PCR reaction) resulted in a positive PCR detection. The analytical performance of the whole workflow was further characterized by processing porcine plasma containing defined numbers of copies of methylated and unmethylated (human) DNA. Porcine plasma was used in this experiment since this plasma is believed to represent a similar matrix compared to human plasma, but does not contain human background DNA, which serves as template for the PCR. Accordingly, the spike-in of defined numbers of unmethylated and methylated DNA copies was possible. Figure 2 shows that approximately 100 copies of unmethylated DNA in 3 mL porcine plasma were quantified accurately. Around ten copies of spiked methylated DNA copies resulted in sporadic positive PCR reactions (Fig. 2), therefore representing the limit of detection of this multistep workflow comprised of polymer-mediated DNA enrichment, bisulfite conversion and PCR analysis.

4 Notes

1. Until now, no studies have described the stability of cell free-circulating tumor DNA in blood samples. A degradation of cell free-circulating DNA during the time lag between sampling and plasma preparation might be possible. Accordingly, plasma should be prepared as quickly as possible after sampling.
2. Application of formalin fixation should be avoided as it leads to a chemical modification of DNA.
3. The optimal storage of urine samples is highly controversial. Reported storage conditions range from storage at room temperature with 0.25% sodium azide for 30 days [28] to storage for 25 years without preservative at −20 °C [29]. Furthermore, these studies did not focus on cell-free DNA. Minor differences in urine sampling have been reported to influence the stability of the sample [30]. It has been frequently observed that, after storage of urine samples, no PCR-amplifiable DNA can be extracted. This might be due to the formation of inhibitory substances during the degradation process of urine. Thus, it is highly recommended to prepare DNA from urine samples immediately (within 2 days of sampling). The enriched DNA can be stored at −20 °C (short-term storage for up to 4 weeks) or at −80 °C (long-term storage).

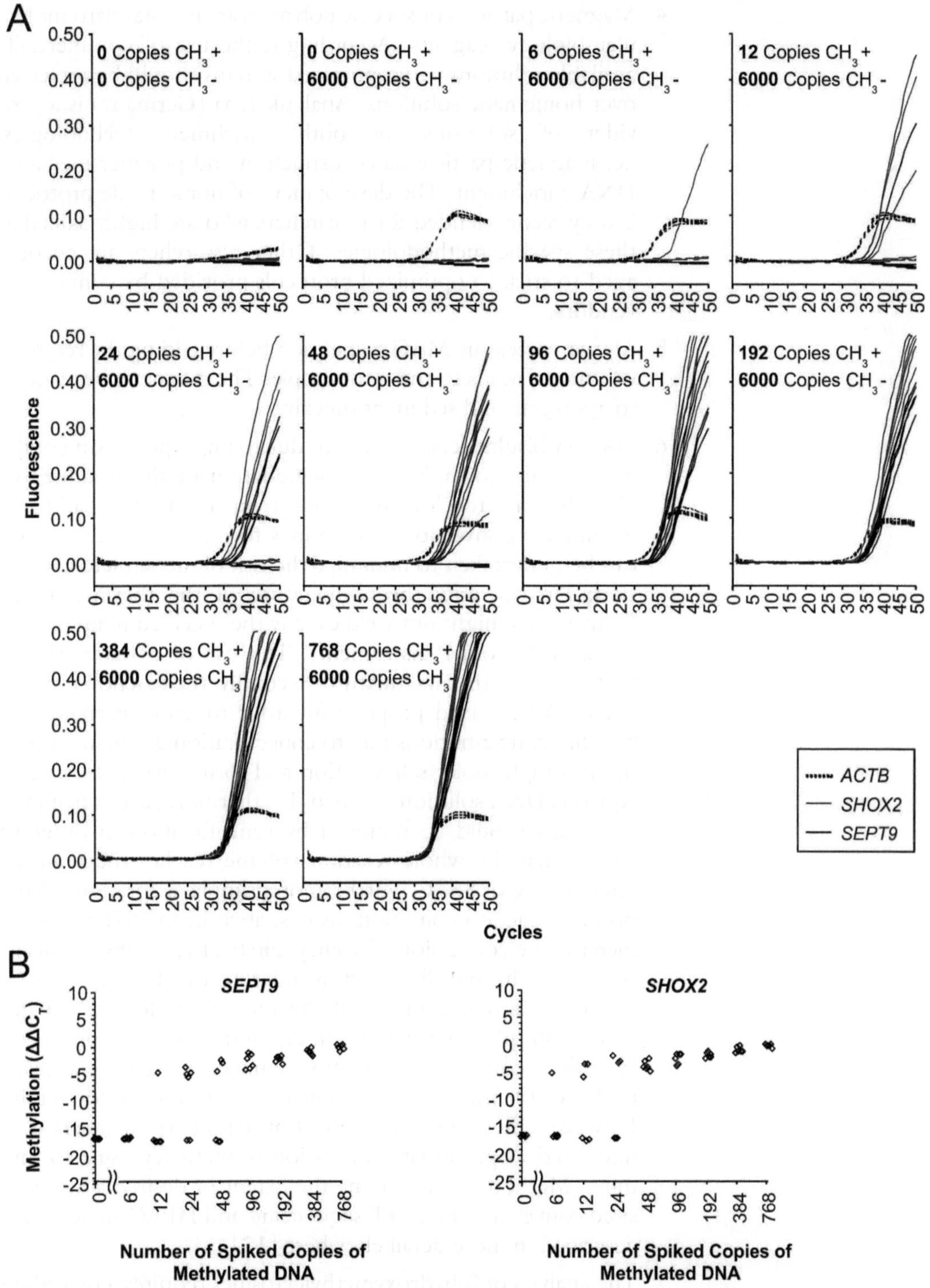

Fig. 2 Analytical performance of entire workflow including plasma preparation, polymer-mediated enrichment of the DNA, bisulfite conversion and cleanup, and PCR. Porcine plasma was spiked with different amounts of methylated (CH_3+) DNA (0–768 copies) and a constant amount of 6,000 copies unmethylated (CH_3−) DNA. Shown are real-time PCR amplification plots (a) and ΔΔCT-values (b)

4. Magnetic particles or specific polymers are not standard molecular biology reagents. Accordingly, the use of commercially available solutions with optimized systems should be preferred over homemade solutions. Analytik Jena (Germany) is a provider of solutions for both enrichment technologies, i.e. magnetic particle based extraction and polymer-mediated DNA enrichment. The development of home-made protocols is only recommended for researchers who are highly skilled at these specific methodologies. Other researchers are encouraged to stick to optimized protocols provided by commercial vendors.
5. Always vortex the MAG suspension before adding the required volume of beads to the next sample. The magnetic beads tend to aggregate and sediment quickly.
6. Too low bisulfite concentrations due to improper mixing of the bisulfite reaction mixture, a degradation of the bisulfite (*see* **Note 9**), or a too low incubation temperature will lead to an incomplete conversion of cytosines to uracils. Several actions can be undertaken to minimize the risk of incomplete conversion: Firstly, verify the temperature of the thermal mixer. Instruments might not yield exactly the specified temperature, in particular older instruments. The use of a water bath is preferred to a thermal mixture. Secondly, the reaction mixture needs to be mixed properly in order to avoid partially low bisulfite concentrations due to concentration gradients. Bisulfite is a high molar salt solution and forms two phases when aqueous DNA solution is added. Furthermore, drops from the tube caps should be removed by centrifugation in order to ensure that the whole reaction volume reaches the required temperature of 85 °C. Furthermore, simple adaptations of the bisulfite reaction conditions as described in the following can increase the conversion efficiency. Firstly, the bisulfite concentration in the bisulfite reaction mixture can be increased by adapting the composition of the mixture as follows: 40 μL DNA, 30 μL conversion buffer, and 100 μL conversion reagent. Secondly, the incubation temperature can be increased to 95 °C. Or the incubation time can be prolonged to 90 min. However, these adaptations will also lead to an increase of undesired inappropriate conversion of methylcytosine to thymine. Methods to determine the extent of desired and undesired conversion, i.e., CFP sequencing and HPLC analyses, are described in more detail elsewhere [13].
7. The analysis of 5-hydroxymethylcytosines requires a preceding oxidation of 5-hydroxymethylcytosines to 5-formylcytosines and subsequent conversion of the newly formed 5-formylcytosines to uracils by means of bisulfite in order to discriminate 5-hydroxymethylcytosines and 5-methylcytosines

[31]. Oxidation of 5-hydroxymethylcytosines to 5-formylcytosines is achieved by potassium perruthenate [32] (*see also* Chapter 34). This oxidation step can easily be incorporated between the polymer-mediated DNA enrichment step and the bisulfite conversion step. The TrueMethyl™ kit provided by Cambridge Epigenetix (CEGX, UK) contains all required reagents to oxidize 5-hydroxymethylcytosines to 5-formylcytosines. However, bisulfite-induced deamination of 5-formylcytosine is slower as compared to the deamination of unmethylated cytosine therefore requiring harsher reaction conditions. Adaptations of the bisulfite reaction as described above (*see* **Note 6**) are a suitable means for optimizing the reaction with regard to the analysis of 5-hydroxymethylcytosines.

8. The conversion buffer contains 1-methoxy-2-propanol and a radical scavenger in order to denature the DNA and to protect the DNA from degradation. This solution might change the color over time. This has no effect on the performance of the conversion buffer.

9. Dissolving sulfur dioxide in water leads to an acidic solution. This solution is called a sulfurous acid (H_2SO_3). However, there is no evidence that sulfurous acid exists in solution and the molecule has only been detected in the gas phase. The conjugate bases of this elusive acid are, however, common anions, i.e. sulfite (SO_3^{2-}) and hydrogen sulfite (aka bisulfite, HSO_3^{-}). While sulfurous acid does not exist in solution, the conjugate salts commonly used for DNA methylation analyses, i.e. sodium and ammonium bisulfite, are only stable in aqueous solution and not as solid salts. The concentration of bisulfite by evaporation leads to the formation of disulfite ($S_2O_5^{2-}$) after the elimination of water. The conversion reagent used within this protocol contains ammonium bisulfite. Several specific properties of bisulfite require a careful handling of this compound in the course of methylation analyses. Under acidic conditions, bisulfite liberates sulfur dioxide gas. Accordingly, the concentration of bisulfite decreases over time. More importantly, bisulfite and sulfite are reducing agents and oxygen in the air slowly oxidizes the solution to the more stable sulfuric acid and sulfate. The oxidation of ammonium bisulfite is indicated by a decreasing pH, a decreasing viscosity, ammonia odor, and lightly lucid yellow color. Once, the oxidation of bisulfite exceeds a critical level the bisulfite conversion of DNA is impaired due to the low pH and the low bisulfite concentration. Furthermore, sulfate might form solid crystals especially when using ethanol containing wash buffers. These sulfates will be dissolved when eluting the bisulfite converted DNA from the solid silica phase and will act as potent PCR inhibitors.

Accordingly, storage of bisulfite in the absence of oxygen is mandatory. Bisulfite solutions should not be used once expired. Preferably, bisulfite solution should be purchased from vendors that provide it in gas-tight vials without any oxygen. Analytik Jena (Jena, Germany) is a suitable supplier offering such single use vials. The vials should not be used for more than 1 month after they have been opened. If in doubt, the pH value of the conversion reagent should be tested with pH-paper. Do not use the conversion reagent if pH is lower than 5.1.

10. The cleanup of the bisulfite converted DNA should be performed immediately after the incubation time in order to avoid DNA degradation.
11. The bisulfite converted DNA should be eluted using small volumes of elution buffer in order to yield a high concentration of the bisulfite converted DNA. This allows for the input of high amounts of DNA into the PCR at a low volume, thereby limiting the carryover of impurities into the PCR. The design of the spin column is important for enabling elution with low volumes, and spin columns with small silica membranes are preferable.
12. Bisulfite converted DNA is stable and can be stored at −20 °C [13]. For long-term storage the converted DNA should be buffered in 10 mM Tris–HCl, pH 8.4, 10 ng/μL polyA.
13. Two distinct types of PCR inhibition may occur. Firstly, bisulfite salts and a low pH of the eluted DNA can lead to a bleaching of probe dyes, in particular of Cy5 and other Cy dyes. Lower background fluorescence and a decrease of this background over time is indicative of a bleaching of the dyes. Second, impurities can lead to an inhibition of the polymerase thereby impairing the PCR. The increase of the total PCR volume and/or the simultaneous decrease of the relative portion of eluted DNA volume in a PCR are a suitable means of decreasing both types of PCR inhibition. Usually, a 20 μL PCR reaction should not contain more than 10 μL eluted DNA. The input of 5 μL into a 20 μL PCR is preferred. The maximum volume of input DNA might be dependent on the used polymerase and the respective buffer conditions. Higher concentration of polymerase in the PCR might help reduce the PCR inhibition. An experiment with a dilution series of eluted DNA as input into the PCR should be conducted to find the maximum input volume of eluted DNA for a respective buffer system. Such inhibition experiments are described in more detail elsewhere [13].
14. The concentration of bisulfite converted DNA from body fluids is usually too low to allow for quantification by UV spectrophotometry. Furthermore, the carrier RNA used in

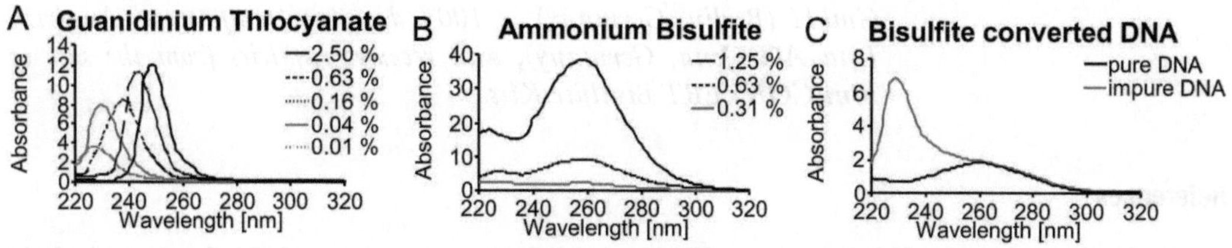

Fig. 3 UV spectrum of guanidinium thiocyanate (**a**), ammonium bisulfite (**b**), and bisulfite converted DNA extracted from ascites. The concentration of ammonium bisulfite and guanidinium thiocyanate is indicated, respectively. For pure and impure DNA one exemplary UV spectrum is shown, respectively (**c**)

the protocols interferes with the UV spectrum of bisulfite converted DNA. However, UV is a suitable means to determine the purity of the converted DNA. Guanidinium thiocyanate shows a strong absorption at around 230–250 nm. The exact absorption maximum is concentration dependent (Fig. 3a). Ammonium bisulfite shows a similar UV spectrum as compared with DNA (Fig. 3b). Guanidinium thiocyanate is contained in DNA binding buffers and represents a well-suited surrogate measure for the carryover of impurities. Accordingly, a bisulfite converted DNA from body fluids should show relatively low absorption at 230 nm (guanidinium thiocyanate) and a high absorption at 260 nm (bisulfite converted DNA and carrier RNA). A typical UV spectrum of a DNA well suited for PCR amplification is shown in Fig. 3c. The yield of bisulfite converted DNA should be determined using a conversion-specific real-time PCR assay, e.g., the *ACTB/SHOX2/SEPT9*-assay as described in Subheading 3.6.

15. It is crucial for an accurate analytical performance of the PCR that the oligonucleotides used are of high quality. HPLC purified oligonucleotides should be used in order to avoid truncated oligonucleotides. Furthermore, the concentration of the oligonucleotides should be verified by means of a UV measurement.
16. The quality of the dNTPs and the polymerase is a critical parameter for the performance of the PCR assay. dNTPs should be stored as single use aliquots and should not be thawed several times. The performance of dNTPs from different suppliers is highly variable and suppliers should not be changed once the assay is established.

Competing Interests: *Dimo Dietrich has been an employee and is a stockholder of Epigenomics AG, a company that aims to commercialize the DNA methylation biomarkers* SEPT9 *and* SHOX2. *Dimo Dietrich is coinventor and owns patents on methylation biomarkers and related technologies. These patents are commercially exploited by Epigenomics AG. Dimo Dietrich receives inventor's compensation from Epigenomics AG. Dimo Dietrich is a consultant for AJ Innuscreen*

GmbH (Berlin, Germany), a 100% daughter company of Analytik Jena AG (Jena, Germany), and receives royalties from the sale of innuCONVERT Bisulfite Kits.

References

1. Guibert S, Weber M (2013) Functions of DNA methylation and hydroxymethylation in mammalian development. Curr Top Dev Biol 104:47–83
2. Sarkar S, Horn G, Moulton K et al (2013) Cancer development, progression, and therapy: an epigenetic overview. Int J Mol Sci 14:21087–21113
3. Church TR, Wandell M, Lofton-Day C et al (2014) Prospective evaluation of methylated SEPT9 in plasma for detection of asymptomatic colorectal cancer. Gut 63:317–325
4. deVos T, Tetzner R, Model F et al (2009) Circulating methylated SEPT9 DNA in plasma is a biomarker for colorectal cancer. Clin Chem 55:1337–1346
5. Kneip C, Schmidt B, Seegebarth A et al (2011) SHOX2 DNA methylation is a biomarker for the diagnosis of lung cancer in plasma. J Thorac Oncol 6:1632–1638
6. Frommer M, McDonald LE, Millar DS et al (1992) A genomic sequencing protocol that yields a positive display of 5-methylcytosine residues in individual DNA strands. Proc Natl Acad Sci U S A 89:1827–1831
7. Darst RP, Pardo CE, Ai L et al (2010) Bisulfite sequencing of DNA. Curr Protoc Mol Biol 15 Chapter 7:Unit 7.9.1–17
8. Millar DS, Warnecke PM, Melki JR et al (2002) Methylation sequencing from limiting DNA: embryonic, fixed, and microdissected cells. Methods 27:108–113
9. Boyd VL, Zon G (2004) Bisulfite conversion of genomic DNA for methylation analysis: protocol simplification with higher recovery applicable to limited samples and increased throughput. Anal Biochem 326:278–280
10. Hayatsu H, Negishi K, Shiraishi M (2004) Accelerated bisulfite-deamination of cytosine in the genomic sequencing procedure for DNA methylation analysis. Nucleic Acids Symp Ser 48:261–262
11. Hayatsu H, Shiraishi M, Negishi K (2008) Bisulfite modification for analysis of DNA methylation. Curr Protoc Nucleic Acid Chem Chapter 6:Unit 6.10
12. Shiraishi M, Hayatsu H (2004) High-speed conversion of cytosine to uracil in bisulfite genomic sequencing analysis of DNA methylation. DNA Res 11:409–415
13. Holmes EE, Jung M, Meller S et al (2014) Performance evaluation of kits for bisulfite-conversion of DNA from tissues, cell lines, FFPE tissues, aspirates, lavages, effusions, plasma, serum, and urine. PLoS One 9:e93933
14. Weisenberger DJ, Trinh BN, Campan M et al (2008) DNA methylation analysis by digital bisulfite genomic sequencing and digital MethyLight. Nucleic Acids Res 36:4689–4698
15. Ostrow KL, Hoque MO, Loyo M et al (2010) Molecular analysis of plasma DNA for the early detection of lung cancer by quantitative methylation-specific PCR. Clin Cancer Res 16:3463–3472
16. Cottrell SE, Distler J, Goodman NS et al (2004) A real-time PCR assay for DNA-methylation using methylation-specific blockers. Nucleic Acids Res 32:e10
17. Dietrich D, Hasinger O, Bañez LL et al (2013) Development and clinical validation of a real--time PCR assay for PITX2 DNA methylation to predict prostate-specific antigen recurrence in prostate cancer patients following radical prostatectomy. J Mol Diagn 15:270–279
18. Dietrich D, Jung M, Puetzer S et al (2013) Diagnostic and prognostic value of SHOX2 and SEPT9 DNA methylation and cytology in benign, paramalignant and malignant pleural effusions. PLoS One 8:e84225
19. Herman JG, Graff JR, Myöhänen S et al (1996) Methylation-specific PCR: a novel PCR assay for methylation status of CpG islands. Proc Natl Acad Sci U S A 93:9821–9826
20. Raizis AM, Schmitt F, Jost JP (1995) A bisulfite method of 5-methylcytosine mapping that minimizes template degradation. Anal Biochem 226:161–166
21. Grunau C, Clark SJ, Rosenthal A (2001) Bisulfite genomic sequencing: systematic investigation of critical experimental parameters. Nucleic Acids Res 29:e65
22. Tanaka K, Okamoto A (2007) Degradation of DNA by bisulfite treatment. Bioorg Med Chem Lett 17:1912–1915
23. Hayatsu H (2008) The bisulfite genomic sequencing used in the analysis of epigenetic states, a technique in the emerging environmental genotoxicology research. Mutat Res 659:77–82

24. Jin L, Wang W, Hu D (2013) The conversion of protonated cytosine-SO3(−) to uracil-SO3 (−): insights into the novel induced hydrolytic deamination through bisulfite catalysis. Phys Chem Chem Phys 15:9034–9042

25. Genereux DP, Johnson WC, Burden AF et al (2008) Errors in the bisulfite conversion of DNA: modulating inappropriate- and failed-conversion frequencies. Nucleic Acids Res 36: e150

26. Dietrich D, Hasinger O, Liebenberg V et al (2012) DNA methylation of the homeobox genes PITX2 and SHOX2 predicts outcome in non-small-cell lung cancer patients. Diagn Mol Pathol 21:93–104

27. Dietrich D, Kneip C, Raji O et al (2012) Performance evaluation of the DNA methylation biomarker SHOX2 for the aid in diagnosis of lung cancer based on the analysis of bronchial aspirates. Int J Oncol 40:825–832

28. Vu NT, Chaturvedi AK, Canfield DV (1999) Genotyping for DQA1 and PM loci in urine using PCR-based amplification: effects of sample volume, storage temperature, preservatives, and aging on DNA extraction and typing. Forensic Sci Int 102:23–34

29. van der Hel OL, van der Luijt RB, Bueno de Mesquita HB (2002) Quality and quantity of DNA isolated from frozen urine in population-based research. Anal Biochem 304:206–211

30. Cannas A, Kalunga G, Green C et al (2009) Implications of storing urinary DNA from different populations for molecular analyses. PLoS One 4:e6985

31. Booth MJ, Ost TW, Beraldi D et al (2013) Oxidative bisulfite sequencing of 5-methylcytosine and 5-hydroxymethylcytosine. Nat Protoc 8:1841–1851

32. Booth MJ, Balasubramanian S (2014) Methods for detection of nucleotide modification. US Patent 14/235,707, 26 June 2014

Part VII

Hydroxymethylation

Chapter 33

Tet-Assisted Bisulfite Sequencing (TAB-seq)

Miao Yu, Dali Han, Gary C. Hon, and Chuan He

Abstract

5-Hydroxymethylcytosine (5hmC) is a modified form of cytosine, which has recently been found in mammalian cells and tissues. 5hmC is derived from 5-methylcytosine (5mC) by Ten-eleven translocation (TET) family protein-mediated oxidation and may regulate gene expression. Numerous affinity-based profiling methods have been developed to help understand the exact function of 5hmC in the genome. However, these methods have a relatively low resolution (~100 bp) without quantitative information of the modification percentage on each site. Here we demonstrated the detailed procedure of Tet-Assistant Bisulfite Sequencing (TAB-Seq), which can detect 5hmC at single-base resolution and quantify its abundance at each site. In this protocol, the genomic DNA is first treated with βGT and recombinant mTet1 consecutively to convert 5hmC to 5gmC and 5mC to 5caC, respectively. The treated genomic DNA can be directly applied to bisulfite treatment to detect 5hmC on specific loci or applied to whole-genome bisulfite sequencing as needed.

Key words TAB-Seq, 5-Hydroxymethylcytosine, Bisulfite sequencing, DNA methylation, Glycosylation

1 Introduction

The significance of 5-hydroxymethylcytosine (5hmC), which was first reported in 1972, did not receive much attention until its presence in the mammalian genome was recently confirmed in Purkinje neurons and mouse embryonic cells [1–3]. 5hmC is the direct oxidative product of 5-methylcytosine (5mC) through Ten-eleven translocation (TET) family protein-mediated oxidation in a Fe^{II}- and 2-oxoglutarate-dependent manner [3]. 5hmC has been suggested to be an intermediate in active and passive demethylation pathways, whereas the uneven distribution of 5hmC in different cell types and tissues indicates its potential roles in gene regulation [4–11]. Although the exact function of 5hmC is still elusive, studies on 5hmC and TET proteins demonstrated that 5hmC was closely related to embryonic stem cells maintenance, brain development, and zygote development [12–20]. 5hmC can be further oxidized to 5-formylcytosine (5fC) and

Jörg Tost (ed.), *DNA Methylation Protocols*, Methods in Molecular Biology, vol. 1708,
https://doi.org/10.1007/978-1-4939-7481-8_33,

5-carboxylcytosine (5caC), both can be excised by thymine DNA glycosylase (TDG) and converted back to unmodified cytosine through base excision repair (BER) [5, 21–24].

Antibodies specific against 5hmC have been developed and applied to profile 5hmC in genome (hMeDIP) shortly after the discovery of 5hmC [12, 25–27] (*see also* Chapter 35). Other affinity-based profiling methods were also reported, which take advantage of enzymes or chemicals to reduce the pull-down bias toward high-density hydroxymethylated regions [11, 28–31] (*see also* Chapter 35). However, profiling methods cannot give base resolution information and the quantitative information on the modification percentage is lost during enrichment. Bisulfite sequencing has been widely used to read out 5mC quantitatively by deaminating only unmodified cytosines to uracils [32–34]. Unfortunately, the same strategy is not feasible for 5hmC detection because both 5mC and 5hmC resist bisulfite-mediated deamination and are read as C, which also indicates that the previous identified 5mC sites are the sum of 5mC and 5hmC [35, 36] (Fig. 1a). To differentiate these two modifications, we modified the traditional bisulfite sequencing method and developed Tet-Assisted Bisulfite Sequencing (TAB-Seq), in which two enzymatic treatment steps were introduced before bisulfite conversion [37]. First, β-glucosyltransferase (βGT) is utilized to convert 5hmC in genomic DNA to β-glucosyl-5-hydroxymethylcytosine (5gmC) which cannot be further oxidized to 5fC or 5caC by the Tet proteins [38]. Next, the glucosylated genomic DNA is treated with an excess of highly active recombinant mouse Tet1 (mTet1) to oxidize most 5mC to 5caC, which can be readily deaminated upon

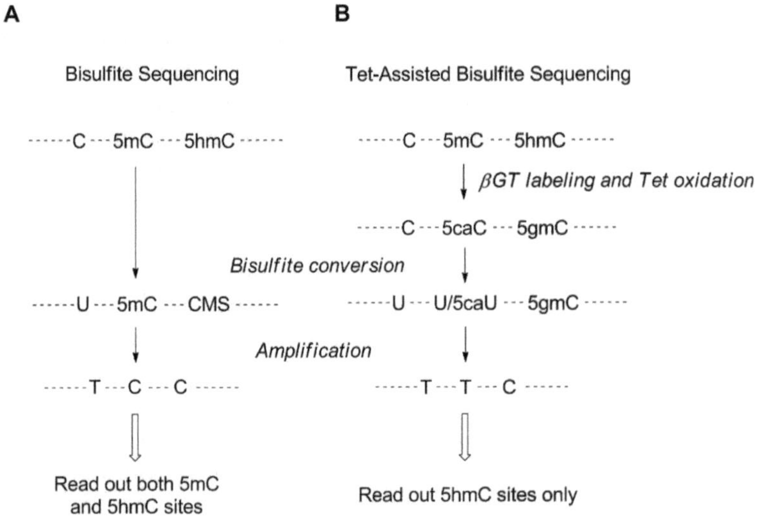

Fig. 1 Scheme of (**a**) traditional Bisulfite Sequencing and (**b**) Tet-Assisted Bisulfite Sequencing. *CMS* cytosine 5-methylenesulfonate, *5caC* 5-carboxylcytosine, *5gmC* β-glucosyl-5-hydroxymethylcytosine

standard bisulfite treatment [5]. After bisulfite treatment and amplification, 5hmC is sequenced as C, whereas both C and 5mC are sequenced as T, therefore distinguishing 5hmC sites from C and 5mC sites (Fig. 1b).

Besides providing the precise location of 5hmC, TAB-Seq can also provide information about the abundance of 5hmC at each site. To accurately estimate 5hmC abundance, three key parameters are required: (1) the conversion rate of C to T; (2) the conversion rate of 5mC to T; (3) the protection rate of 5hmC. To assess these three parameters in each TAB-Seq sample, proper controls bearing C, 5mC, or 5hmC need to be spiked into the genomic DNA prior to the enzymatic treatment steps. There are several rules to follow to determine the control sequence: (1) complex enough and have multiple modification sites (5mC or 5hmC); (2) at least 1 kb long, allowing for PCR duplicate removal during data analysis in order to accurately calculate the conversion and protection rate; (3) do not align to the target genome after bisulfite conversion. In this protocol, we use lambda DNA and a pUC19 vector as template to generate the spike-in controls, which are compatible with human and mouse genomes. If samples from other species are sequenced, proper controls with a different sequence may be needed and the PCR primers should be redesigned for the conversion/protection test described in Subheadings 3.2 and 3.5.

The sequencing depth required for each whole-genome TAB--Seq depends on both the abundance of 5hmC in the sequenced sample and the 5mC conversion rate. Less sequencing is required for samples with higher 5hmC abundance and a higher 5mC conversion rate. For example, if the median abundance of 5hmC at 5hmC sites in the genomic DNA is ~20%, to resolve base-resolution 5hmC at this level, a depth of ~25 times per cytosine, or ~50 times the haploid genome size is needed when the 5mC conversion rate is 97.8%. The number will increase to ~30 times per cytosine, or ~60 times the haploid genome size if the 5mC conversion rate drops to 96.5%. On the other hand, a depth of ~15 times per cytosine is sufficient to resolve base-resolution 5hmC with 30% abundance when 5mC conversion is 97.8%. With a 5mC conversion rate of 99%, a depth of ~16 times per cytosine is sufficient to resolve base-resolution 5hmC with 20% abundance.

2 Materials

2.1 Expression and Purification of Recombinant mTet1

1. Bac-to-Bac Baculovirus Expression System (Thermo Fisher Scientific).
2. pFastBac Dual-mTet1 plasmid (available upon request from the corresponding author).
3. Adherent High Five cells.

4. Grace's insect medium, supplemented.
5. Fetal Bovine Serum (FBS).
6. Penicillin/Streptomycin/L-Glutamine (Pen/Strep/L-Gln) mixture 25,000 U/25000 U/200 mM.
7. Insect cell culture medium: Add 55 mL FBS, 2.25 mL Pen/-Strep/L-Gln to 500 mL of Grace's insect medium, supplemented and filter (0.22 μm filter unit). Store at 4 °C.
8. Binding buffer: 20 mM Tris–HCl, pH 8.0, 500 mM NaCl, 1 mM Tris(2-carboxyethyl)phosphine hydrochloride (TCEP), 1 mM Phenylmethylsulfonyl fluoride (PMSF), 1 μg/mL Leupeptin, 1 μg/mL Pepstatin (*see* **Note 1**).
9. Econo-Column chromatography column.
10. Anti-flag M2 Affinity Gel (Sigma-Aldrich).
11. Econo-Column chromatography columns (Bio-Rad, 2.5 × 20 cm).
12. 3× Flag peptide (Sigma-Aldrich).
13. GF running buffer: 20 mM HEPES, pH 8.0, 150 mM NaCl, 1 mM DTT (*see* **Note 2**).
14. Elution buffer: Dissolve 3× Flag peptide in precooled GF running buffer at 0.2 mg/mL. Add 1 mM PMSF, 1 μg/mL Leupeptin and Pepstatin just before use.
15. Amicon Ultra centrifugal filters (4 mL, 50K cutoff).
16. Superdex-200 columns.
17. Glycerol.

2.2 Activity Test of Recombinant mTet1

1. Qubit 2.0 Fluorometer (Thermo Fisher Scientific).
2. Qubit dsDNA HS Assay Kit (Thermo Fisher Scientific).
3. Tet oxidation reagent 1: 1.5 mM $Fe(NH_4)_2(SO_4)_2$. Add 14.7 mg $Fe(NH_4)_2(SO_4)_2{\cdot}6H_2O$ to 1 mL of Milli-Q water, then make a 24-fold dilution (*see* **Note 3**).
4. Tet oxidation reagent 2: 333 mM NaCl, 167 mM HEPES (pH 8.0), 4 mM ATP, 8.3 mM DTT, 3.3 mM α-KG, 6.7 mM L-Ascorbic acid (*see* **Note 4**).
5. Proteinase K (20 mg/mL).
6. Micro Bio-Spin 30 Columns (Bio-Rad).
7. QIAquick PCR purification kit (Qiagen) or DNA purification kits from other suppliers.
8. Epitect Bisulfite kit (Qiagen) or MethylCode Bisulfite Conversion kit (Thermo Fisher Scientific) or similar.
9. P*fu*Turbo C_x Hotstart DNA polymerase.

10. 5mC test primer-F: 5′-TTTGGGTTATGTAAGTTGATTT-TATG.
11. 5mC test primer-R: 5′-CACCCTACTTACTAAAATTT ACACC.
12. dNTP mix.
13. Zero Blunt TOPO PCR Cloning Kit.
14. Agarose.

2.3 Generation of Spike-In Controls

1. *Sss*I CpG methyltransferase with associated buffer and S-adenosylmethionine (e.g., New England Biolabs).
2. Unmethylated lambda DNA from *E. coli* strain lacking both the *dam* and *dcm* methylase activities.
3. 5hmC dNTP mix.
4. pUC19 vector.
5. Zymo*Taq*™ DNA polymerase.
6. 5hmC control primer-F: 5′-GCAGATTGTACTGAGAGTGC.
7. 5hmC control primer-R: 5′-TGCTGATAAATCTGGAGCCG.
8. Agarose
9. Gel extraction kit.

2.4 Glycosylation and Oxidation of Genomic DNA

1. 10× βGT protection buffer: 500 mM HEPES, pH 8.0, 250 mM $MgCl_2$.
2. T4 Phage β-glucosyltransferase (T4-βGT, New England Biolabs)
3. 10 mM UDP-Glucose: Dissolve UDP-Glucose in Milli-Q water and store at −20 °C.
4. MinElute PCR purification kit (Qiagen) or equivalent.

All other required reagents can be found in Subheading 2.2.

2.5 Verification of Oxidized Genomic DNA

1. 5hmC test primer-F: 5′-GTAGATTGTATTGAGAGTGT.
2. 5hmC test primer-R: 5′-TACCCAACTTAATCGCCTTG.

Other required reagents can be found in Subheading 2.2.

2.6 Sequencing of Oxidized Genomic DNA

1. Self-designed primer sets for loci-specific TAB-Seq.
2. End-It™ DNA End-Repair Kit (Lucigen).
5. Klenow Fragment (3′ → 5′ exo-).
6. dATP.
7. Quick Ligation Kit (New England Biolabs).
8. Agarose (super fine resolution, APEX).
9. MinElute Gel Extraction Kit (Qiagen).

10. KAPA HiFi HotStart ReadyMix PCR Kit.
11. TruSeq® DNA Sample Prep Kit (Illumina).
12. Library primer-F: 5′-AATGATACGGCGACCACCGAGATC TACAC.
13. Library primer-R: 5′-CAAGCAGAAGACGGCATACGAGAT.
14. Illumina Hi-Seq 2000 or 2500.
15. Bismark software (http://www.bioinformatics.babraham.ac.uk/projects/bismark/) [39].

Other required reagents can be found in Subheading 2.2.

3 Methods

3.1 Expression and Purification of Recombinant mTet1

Recombinant mTet1 needs to be carefully expressed and purified to ensure high oxidation efficiency (over 97% conversion of 5mC to 5caC) for the TAB-Seq assay. Perform all the purification steps on ice or at 4 °C and try to finish in 1 day, if time allows. The TAB-Seq kit is also commercially available from Wisegene and it provides required enzymes and buffers for the glycosylation and oxidation of genomic DNA in Subheading 3.4.

1. Generate the bacmid and produce baculovirus P1, P2, and P3 with the pFastBac Dual-mTet1 plasmid using the Bac-to-Bac Baculovirus Expression System following the manufacturer's instructions.
2. Infect adherent High Five cells with 1 mL of P3 baculovirus for each 150 mm culture dish and incubate at 27 °C for 48 h.
3. Harvest 20–40 dishes of infected cells by centrifuging at 3800 × *g* for 10 min at 4 °C. Discard the medium and resuspend the cell pellets in 30–50 mL of binding buffer (*see* **Note 5**).
4. Break cells by passing though the homogenizer twice and collect the supernatant by centrifuging at 30,000 × *g* for 40 min. Clarify the cell lysate by passing through a 0.45 μM filter.
5. Add 6 mL of 50% anti-flag M2 affinity gel into the Econo-Column (2.5 × 20 cm) and allow the gel bed to drain. Add 50 mL of binding buffer to wash out the glycerol and drain the remaining buffer until it reaches the top of the gel bed.
6. Load the clarified cell lysate onto the equilibrated gel with the bottom closed and resuspend the gel using a pipette. Incubate for 10 min.
7. Let the cell lysate draw through the column under gravity at the rate of one drop per second and collect the flow-through.

8. Load the flow-through from **step** 7 back to the column and repeat **steps 6** and 7.
9. Wash the gel with 200–300 mL of binding buffer under gravity flow. Periodically resuspend the gel with a pipette during the washing.
10. Resuspend the gel in 3 mL of elution buffer and collect the eluate with the flow rate of one drop per second after incubating for 5 min (*see* **Note 6**).
11. Repeat **step 10** twice (*see* **Note** 7).
12. Combine the eluate and concentrate to less than 2 mL with an Amicon Ultra 50K filter. Load it onto a 120 mL Superdex-200 column equilibrated with precooled GF running buffer.
13. Pool fractions and concentrate purified mTet1 with an Amicon Ultra 4 mL 50K filter to about 7 mg/mL as quantified using a Bradford assay.
14. Add glycerol to the final concentration of 30% and aliquot the protein to a smaller volume as needed. Freeze with liquid nitrogen and store at −80 °C (*see* **Note 8**).

3.2 Activity Test of Recombinant mTet1

The amount of DNA is quantified with a Qubit Assay unless indicated otherwise.

1. Prepare the reaction as follows. Add each component in the order listed and mix well.

Component	Amount
Milli-Q water	To final volume of 50 μL
Sheared genomic DNA of interest (from Subheading 3.4)	300 ng
Sheared CpG methylated lambda DNA (from Subheading 3.3.1)	2.5 ng
Tet oxidation reagent 2	15 μL
Tet oxidation reagent 1	3.5 μL
mTet1 protein (5 mg/mL)	4 μL

2. Incubate at 37 °C for 80 min.
3. Add 1 μL Proteinase K (20 mg/mL) and incubate at 50 °C for 1 h.
4. Clean up with a Micro Bio-Spin 30 Column and purify with the QIAquick PCR purification kit following the manufacturer's instructions. Elute DNA in 30 μL of Milli-Q water (*see* **Note 9**).
5. Use 20 μL of the oxidized DNA from the previous step for the EpiTect or MethylCode Bisulfite Conversion kit following the manufacturer's instructions (*see* **Note 10**).

6. Prepare the following PCR in a final volume of 50 μL:

Component	Amount
10× *Pfu*Turbo C_x reaction buffer	5 μL
dNTP Mix (10 mM each dNTP)	1 μL
5mC test primer-F (10 μM)	1 μL
5mC test primer-R (10 μM)	1 μL
Bisulfite-treated DNA (from **step 5**)	2–4 μL
*Pfu*Turbo C_x Hotstart DNA polymerase (2.5 U/μL)	1 μL (*see* **Note 11**)
Milli-Q water	To a final volume of 50 μL

7. Run the PCR program as provided below.

Initial denaturation	95 °C	2 min
Denaturation	95 °C	30 s
Annealing	57 °C	30 s
Extension	72 °C	1 min
	40 cycles	
Final extension	72 °C	7 min
Hold	4 °C	

8. Verify the PCR product by running 5 μL of the reaction on a 2% agarose gel. There should be a single, clear band at ~300 bp.
9. TOPO clone 2 μL of the reaction using the Zero Blunt TOPO PCR Cloning Kit following the manufacturer's instruction and isolate plasmids from at least 20 clones for Sanger sequencing.
10. Calculate the 5mC conversion rate. The ideal conversion rate of 5mC is over 97% (*see* **Note 12**).
    Conversion rate of 5mC equals:
    number of 'T' reads at CpG cytosines/(number of CpGs on amplicon × number of clones sequenced)

3.3 Generation of Spike-In Controls

3.3.1 C/5mC Spike-In Control

1. Prepare an in vitro methylation reaction. Add each component in the order listed and mix well (*see* **Note 13**).

Component	Amount
Milli-Q water	44 μL
10× NEBuffer 2	10 μL
SAM (32 mM)	2 μL

Unmethylated lambda DNA (450 μg/ mL)	40 μL
*Sss*I methylase (20 U/μL)	4 μL
Final volume	100 μL

2. Incubate at 37 °C for 2 h followed by heating at 65 °C for 20 min to stop the reaction.
3. Clean-up DNA with phenol chloroform extraction and ethanol precipitation, or use proper commercial DNA purification kits (*see* **Note 14**).

3.3.2 5hmC Spike-In Control

1. Prepare the following PCR in a final volume of 50 μL:

Component	Amount
2× Reaction buffer	25 μL
5hmC dNTP Mix (10 mM each dNTP)	1 μL
5hmC control primer-F (10 μM)	2 μL
5hmC control primer-R (10 μM)	2 μL
pUC 19 vector (1 ng/μL)	1 μL
Zymo*Taq*™ DNA Polymerase (5 U/μL)	0.4 μL
Milli-Q water	18.6 μL
Final volume	50 μL

2. Run the PCR program as provided below.

Initial denaturation	95 °C	10 min
Denaturation	94 °C	30 s
Annealing	57 °C	30 s
Extension	72 °C	1.5 min
	40 cycles	
Final extension	72 °C	7 min
Hold	4 °C	

3. Verify the PCR product by running 5 μL of the reaction on a 1.2% agarose gel. There should be a single, clear band at ~1.6 kb.
4. Run the remaining reaction on a 1.2% agarose gel. Cut the gel and purify the PCR product with a commercial gel extraction kit. Elute DNA in Milli-Q water (*see* **Note 15**).

3.4 Glycosylation and Oxidation of Genomic DNA

1. Add 0.5% (w/w) CpG methylated lambda DNA (C/5mC spike-in control) and 0.25% (w/w) 5hmC control (generated from Subheading 3.3) to the genomic DNA of interest (*see* **Note 16**).
2. Shear the mixed genomic DNA to desired size range. For whole-genome bisulfite sequencing, the size range depends on the downstream sequencing parameters. We typically shear to an average of 400 bp. For loci-specific analysis, shear the genomic DNA to average 1 kb or longer (*see* **Note 17**).
3. Purify the sheared genomic DNA with MinElute PCR purification kit and elute in 10 μL of Milli-Q water (*see* **Note 18**).
4. Prepare the glucosylation reaction as follows:

Component	Amount
Sheared DNA from **step 3**	1–3 μg
10 mM UDP-Glucose	0.4 μL
10× βGT protection buffer	2 μL
T4-βGT (10 U/μL)	2 μL
Milli-Q water	To a final volume of 20 μL

5. Pipette up and down to mix and incubate at 37 °C for 1 h.
6. Clean up the DNA using a QIAquick PCR purification kit and elute in 30–50 μL of Milli-Q water. After purification, 60–70% of DNA can be recovered.
7. Prepare the oxidation reactions as follows. Add each component in the order listed and mix well (*see* **Note 19**).

Component	Amount
Milli-Q water	To a final volume of 50 μL
Glucosylated DNA from **step 6**	300 ng
Tet oxidation reagent 2	15 μL
Tet oxidation reagent 1	3.5 μL (*see* **Note 20**)
mTet1 protein (5 mg/ ml)	4 μL (*see* **Note 21**)

8. Incubate at 37 °C for 80 min.
9. Add 1 μL of Proteinase K (20 mg/mL) per reaction and incubate at 50 °C for 1 h.
10. Clean up the DNA with Micro Bio-Spin 30 Columns followed by the QIAquick PCR purification kit. Elute DNA in 50 μL of Milli-Q water. Normally, 40–60% of DNA can be recovered.

3.5 Verification of Oxidized Genomic DNA

1. Convert 10 ng treated of oxidized genomic DNA from Subheading 3.4. with the EpiTect or the MethylCode Bisulfite Conversion kit following the manufacturer's instructions.
2. Follow **steps 6–10** in Subheading 3.2 using bisulfite-treated DNA from **step 1** as the PCR template (*see* **Note 22**).

3.5.1 5mC Conversion Test

3.5.2 5hmC Protection Test

1. Prepare the following PCR in a final volume of 50 μL:

Component	Amount
10 × *Pfu*Turbo C_x reaction buffer	5 μL
dNTP Mix (10 mM each dNTP)	1 μL
5hmC test primer-F (10 μM)	1 μL
5hmC test primer-R (10 μM)	1 μL (*see* **Note 23**)
Bisulfite-treated DNA (from **step 1** in Subheading 3.5.1)	2–4 μL (*see* **Note 24**)
*Pfu*Turbo C_x Hotstart DNA polymerase (2.5 U/μL)	1 μL (*see* **Note 25**)
Milli-Q water	To a final volume of 50 μl

2. Run the PCR program as follows:

Initial denaturation	95 °C	2 min
Denaturation	95 °C	30 s
Annealing	51 °C	30 s
Extension	72 °C	1 min
	40 cycles	
Final extension	72 °C	7 min
Hold	4 °C	

3. Verify the PCR product by running 5 μL of the reaction on a 2% agarose gel. There should be a single, clear band at ~200 bp.
4. TOPO clone 2 μL of the reaction using a Zero Blunt TOPO PCR Cloning Kit following the manufacturer's instructions and isolate plasmids from at least 20 clones for Sanger sequencing.
5. Calculate the 5hmC protection rate. The ideal protection rate of 5hmC is over 80% (*see* **Note 26**).

   Protection rate of 5hmC equals:
   number of 'C' reads at cytosines/(number of cytosines × number of clones sequenced)

3.6 Sequencing of Oxidized Genomic DNA

3.6.1 Loci-Specific Sequencing

1. Convert 150 ng of treated genomic DNA from Subheading 3.4. using the EpiTect or MethylCode Bisulfite Conversion kit following the manufacturer's instructions.
2. Prepare the following PCR in a final volume of 50 μL:

Component	Amount
10× *Pfu*Turbo C_x reaction buffer	5 μL
dNTP Mix (10 mM each dNTP)	1 μL
Self-designed primer-F (10 μM)	1 μL (*see* **Note 27**)
Self-designed primer-R (10 μM)	1 μL
Bisulfite-treated DNA (from previous step)	1–2 μL
*Pfu*Turbo C_x Hotstart DNA polymerase (2.5 U/μL)	1 μL
Milli-Q water	To a final volume of 50 μL

3. Run the PCR program as follows:

Initial denaturation	95 °C	2 min
Denaturation	95 °C	30 s
Annealing	Primer dependent (*see* **Note 28**)	30 s
Extension	72 °C	1 min
	40 cycles	
Final extension	72 °C	7 min
Hold	4 °C	

4. Verify the PCR product by running 5 μL of the reaction on a 2% agarose gel. Purify the PCR product with a QIAquick PCR purification kit if a single band is observed. Otherwise, cut the band with the correct size and purify with a gel extraction kit.
5. Apply the purified PCR product to Sanger sequencing or perform colony picking with a Zero Blunt TOPO PCR Cloning Kit if quantitative information is needed. The signal of 'C' represents the position of 5hmC.

3.6.2 Whole-Genome Sequencing

1. Take 300–600 ng of oxidized genomic DNA (from Subheading 3.4) and prepare the end repair reaction as follows.

Component	Amount
Treated genomic DNA (300–600 ng)	Variable depending on its concentration
dNTP mix (10 mM each dNTP)	5 μL

10× End-Repair Buffer	5 μL
ATP (10 mM)	5 μL
End-Repair Enzyme Mix	1 μL
Milli-Q water	To a final volume of 50 μL

2. Mix well and incubate at room temperature for 45 min.
3. Purify DNA with the MinElute PCR purification kit and elute in 16 μL of EB buffer twice (32 μL in total).
4. Prepare 3′ end adenylation reaction as follows and mix well.

Component	Amount
DNA from **step 3**	32 μL
10× NEBuffer 2	5 μL
dATP (1 mM)	10 μL
Klenow fragment (3′ → 5′ exo-, 5 U/μL)	3 μL

5. Incubate at 37 °C for 30 min.
6. Purify DNA with the MinElute PCR purification kit and elute in 10 μL of EB buffer twice (20 μL in total), then speed vac to 7 μL.
7. Prepare ligation reaction as follows and mix well.

Component	Amount
DNA from **step 6**	7 μL
2× Quick ligase reaction buffer	10 μL
TruSeq adapter (Illumina)	2 μL
Quick ligase	1 μL

8. Incubate at room temperature for 15 min.
9. Purify the DNA with a MinElute PCR purification kit and elute in 10 μL of EB buffer twice (20 μL in total), then speed vac to 10 μL.
10. Run the ligation product on a 2% super fine resolution agarose gel and excise the band between 250 and 650 bp.
11. Purify DNA fragments from the gel using the MinElute gel extraction kit and elute in 10 μL of EB buffer twice (20 μL in total).
12. Perform bisulfite conversion using the EpiTect or Methyl-Code Bisulfite Conversion kit following the manufacturer's instructions.
13. Prepare the following PCR to amplify the bisulfite converted DNA.

Component	Amount
Bisulfite-converted DNA (from **step 12**)	7 μL
Library primer-F (10 μM)	3 μL
Library primer-R (10 μM)	3 μL
2× KAPA HiFi HotStart ReadyMix	25 μL
Milli-Q water	To a final volume of 50 μL

14. Run the PCR program as follows:

Initial denaturation	95 °C	1 min
Denaturation	98 °C	15 s
Annealing	60 °C	30 s
Extension	72 °C	30 s
	10 cycles	
Final extension	72 °C	1 min
Hold	4 °C	

15. Purify the DNA with a MinElute PCR purification kit and elute in 10 μL of EB buffer twice (20 μL in total), then speed vac to 10 μL.
16. Run the purified library on a 2% super fine resolution agarose gel and excise the band between 250 and 650 bp.
17. Purify the final library from the gel using the MinElute gel extraction kit and elute in 10 μL of EB buffer twice (20 μL in total).
18. The libraries generated are sequenced using an Illumina Hi-Seq 2000 or 2500 platform according to manufacturer's instruction.

3.6.3 Data Analysis for Whole-Genome Sequencing

TAB-Seq sequencing analysis is performed following a pipeline modified from a previously published paper [37].

1. Trimming: Trim sequencing reads for low-quality reads and adaptors sequences.
2. Mapping: Align sequencing reads with the bismark software. First, prepare a bismark genome index with "bismark_genome_preparation". Then, align the trimmed sequencing data with the following parameters: "bismark -n 1 -l 40 -chunkmbs 512 -q/yourpath/genome yourfastqdata".
3. Calling 5hmC: For each C site in the genome, count the number of 'C' reads as hydroxymethylated (denoted N_C) and the number of 'T' bases as not hydroxymethylated (denoted N_T). To calculate the probability of observing N_C or greater

cytosines by chance, use the binomial distribution with parameters N as the sequencing depth ($N_C + N_T$) and p as the 5mC nonconversion rate.

4 Notes

1. The binding buffer should be precooled on ice before use. Add fresh TCEP, PMSF, Leupeptin, and Pepstatin before use.
2. The GF running buffer should be precooled on ice before use. Add fresh DTT before use.
3. We recommended making small aliquots of Tet oxidation reagent 1 as needed. This buffer can be stored at −80 °C for up to 3 months without affecting the oxidation efficiency. Avoid freeze–thaw cycles.
4. Tet oxidation reagent 2 can be prepared by mixing 2 M NaCl, 1 M HEPES (pH 8.0), 24 mM ATP, 50 mM DTT, 20 mM α-KG, 40 mM L-Ascorbic acid stock solution at 1:1:1:1:1:1 ratio. We recommended making small aliquots of Tet oxidation reagent 2 as needed. This buffer can be stored at −80 °C for up to 3 months without affecting the oxidation efficiency. Avoid freeze-thaw cycles.
5. The cell pellets can be frozen with liquid nitrogen and stored at −80 °C for future purification. However, fresh cells are always recommended. Since recombinant mTet1 is prone to degrade in cell lysate, make sure TCEP, PMSF, Leupeptin, and Pepstatin are added to all binding buffers used during purification.
6. Make sure PMSF, Leupeptin, and Pepstatin are added to the elution buffer before use.
7. Or repeat **step 10** until no more protein is eluted (determined by protein gel) to recover as much protein as possible.
8. Storage of recombinant mTet1 above −80 °C or in less than 30% glycerol may lead to decreased oxidation efficiency. Avoid repeated freeze-thaw cycles. Each aliquot can be used up to three times.

   After adding glycerol, the concentration of mTet1 is ~5 mg/ mL.
9. The Micro Bio-Spin 30 Column is used to remove ATP in the reaction. This step may be skipped, but please note there will be ATP left in the purified DNA and will affect the concentration measure if a NanoDrop spectrophotometer is used.
10. Use the regular thermal cycling program described in the manual for the MethylCode Bisulfite Conversion kit. Alternative thermal cycling programs 1 and 2 may result in incomplete

5caC to 5caU/U conversion. If using milder bisulfite conversion conditions, an additional test of 5caC deamination efficiency is needed.

11. A different hot start polymerase may be used. Change the concentration of dNTPs, primers, polymerase, and thermal cycling program accordingly.
12. Increasing the amount of recombinant mTet1 used in **step 1** may enhance the conversion rate by up to ~1%. However, it is not recommended to add more than 7 μL of mTet1 per 50 μL reaction since the glycerol in the protein will affect the oxidation efficiency.

    There are 22 CpGs on the amplicon.
13. Use fresh SAM for the methylation reaction.
14. Generally, over 97% of CpG sites can be methylated. The methylation percentage on CpG sites can be determined with the same procedures described in Subheading 3.2. (**Steps 5–9**, apply the methylated lambda DNA to bisulfite conversion). If the methylation rate is not good, repeat the methylation reaction and purification steps.

    Lambda DNA is 48,502 base pairs in length.
15. Since there is always dCTP contamination in commercial 5hmdCTP, the percentage of 5hmC in the 5hmC control is not 100% (less than 95%). To estimate the real protection rate of 5hmC in the TAB-Seq sample, the precise percentage of 5hmC on this control needs to be determined by conventional bisulfite sequencing performed in parallel.
16. Use RNA-free genomic DNA for treatment. Include a RNA digestion step during extraction of the genomic DNA.

    For whole-genome bisulfite sequencing, starting from 3 μg is recommended but 1 μg is also acceptable. For loci-specific analysis, 1 μg is sufficient for screening 8–9 loci.
17. For loci-specific analysis, longer fragments can yield better results in PCR amplification after bisulfite treatment. However, unsonicated DNA is not recommended because of the tedious purification steps and low recovery rate.
18. Typically 10–20% of sheared DNA will be lost after purification.
19. If more than 300 ng of glucosylated DNA is applied, prepare several oxidation reactions in parallel.
20. After mixing Tet oxidation reagents 2 and 1, the solution should slowly change from transparent to light brown. For a large number of samples, a master solution of reagents 1 and 2 can be prepared. In that case, the color change will be much more obvious. If no color change is observed, the oxidation reagents need to be reprepared.

21. The amount of mTet1 used here is determined by the activity test results in Subheading 3.2 and may vary from batch to batch.
22. A 5mC conversion test is strongly recommended for each different genomic sample.
23. For the reverse primer design, assume those C's on the control are 100% of 5hmC and will not be deaminated (read as C) after bisulfite treatment.
24. Increase the amount of template if it is difficult to get the PCR product.
25. Other polymerases may have a bias toward protected 5hmC (5gmC) during PCR amplification and need a separate test before using.
26. Here 80% refers to the protection rate without normalization with the real percentage of 5hmC on the control.

    There are 37 cytosines on the amplicon (except for the primer part).
27. Follow the general rules for bisulfite sequencing based PCR to design the primers.

    Test whether self-designed primers works well on standard bisulfite-treated genomic DNA before applying to TAB-Seq samples to save the template.
28. The annealing temperature is primer dependent and we usually start screening from 3 °C below the T_m of the primers.

Acknowledgments

This work is supported by National Institutes of Health 1R01 HG006827 to C.H.

References

1. Penn NW, Suwalski R, O'Riley C et al (1972) The presence of 5-hydroxymethylcytosine in animal deoxyribonucleic acid. Biochem J 126:781–790
2. Kriaucionis S, Heintz N (2009) The nuclear DNA base 5-hydroxymethylcytosine is present in Purkinje neurons and the brain. Science 324:929–930
3. Tahiliani M, Koh KP, Shen Y et al (2009) Conversion of 5-methylcytosine to 5-hydroxymethylcytosine in mammalian DNA by MLL partner TET1. Science 324:930–935
4. Inoue A, Zhang Y (2011) Replication-dependent loss of 5-hydroxymethylcytosine in mouse preimplantation embryos. Science 334:194
5. He YF, Li BZ, Li Z et al (2011) Tet-mediated formation of 5-carboxylcytosine and its excision by TDG in mammalian DNA. Science 333:1303–1307
6. Cortellino S, Xu J, Sannai M et al (2011) Thymine DNA glycosylase is essential for active DNA demethylation by linked deamination-base excision repair. Cell 146:67–79
7. Guo JU, Su Y, Zhong C et al (2011) Hydroxylation of 5-methylcytosine by TET1 promotes active DNA demethylation in the adult brain. Cell 145:423–434

8. Globisch D, Munzel M, Muller M et al (2010) Tissue distribution of 5-hydroxymethylcytosine and search for active demethylation intermediates. PLoS One 5: e15367
9. Munzel M, Globisch D, Bruckl T et al (2010) Quantification of the sixth DNA base hydroxymethylcytosine in the brain. Angew Chem Int Ed Engl 49:5375–5377
10. Szwagierczak A, Bultmann S, Schmidt CS et al (2010) Sensitive enzymatic quantification of 5-hydroxymethylcytosine in genomic DNA. Nucleic Acids Res 38:e181
11. Song CX, Szulwach KE, Fu Y et al (2011) Selective chemical labeling reveals the genome-wide distribution of 5-hydroxymethylcytosine. Nat Biotechnol 29:68–72
12. Wu H, D'Alessio AC, Ito S et al (2011) Genome-wide analysis of 5-hydroxymethylcytosine distribution reveals its dual function in transcriptional regulation in mouse embryonic stem cells. Genes Dev 25:679–684
13. Ito S, D'Alessio AC, Taranova OV et al (2010) Role of Tet proteins in 5mC to 5hmC conversion, ES-cell self-renewal and inner cell mass specification. Nature 466:1129–1133
14. Dawlaty MM, Ganz K, Powell BE et al (2011) Tet1 is dispensable for maintaining pluripotency and its loss is compatible with embryonic and postnatal development. Cell Stem Cell 9:166–175
15. Gu TP, Guo F, Yang H et al (2011) The role of Tet3 DNA dioxygenase in epigenetic reprogramming by oocytes. Nature 477:606–610
16. Iqbal K, Jin SG, Pfeifer GP et al (2011) Reprogramming of the paternal genome upon fertilization involves genome-wide oxidation of 5-methylcytosine. Proc Natl Acad Sci U S A 108:3642–3647
17. Koh KP, Yabuuchi A, Rao S et al (2011) Tet1 and Tet2 regulate 5-hydroxymethylcytosine production and cell lineage specification in mouse embryonic stem cells. Cell Stem Cell 8:200–213
18. Szulwach KE, Li X, Li Y et al (2011) Integrating 5-hydroxymethylcytosine into the epigenomic landscape of human embryonic stem cells. PLoS Genet 7:e1002154
19. Szulwach KE, Li X, Li Y et al (2011) 5-hmC-mediated epigenetic dynamics during postnatal neurodevelopment and aging. Nat Neurosci 14:1607–1616
20. Kriukiene E, Liutkeviciute Z, Klimasauskas S (2012) 5-Hydroxymethylcytosine—the elusive epigenetic mark in mammalian DNA. Chem Soc Rev 41:6916–6930
21. Ito S, Shen L, Dai Q et al (2011) Tet proteins can convert 5-methylcytosine to 5-formylcytosine and 5-carboxylcytosine. Science 333:1300–1303
22. Pfaffeneder T, Hackner B, Truss M et al (2011) The discovery of 5-formylcytosine in embryonic stem cell DNA. Angew Chem Int Ed Engl 50:7008–7012
23. Maiti A, Drohat AC (2011) Thymine DNA glycosylase can rapidly excise 5-formylcytosine and 5-carboxylcytosine: potential implications for active demethylation of CpG sites. J Biol Chem 286:35334–35338
24. Zhang L, Lu X, Lu J et al (2012) Thymine DNA glycosylase specifically recognizes 5-carboxylcytosine-modified DNA. Nat Chem Biol 8:328–330
25. Ficz G, Branco MR, Seisenberger S et al (2011) Dynamic regulation of 5-hydroxymethylcytosine in mouse ES cells and during differentiation. Nature 473:398–402
26. Williams K, Christensen J, Pedersen MT et al (2011) TET1 and hydroxymethylcytosine in transcription and DNA methylation fidelity. Nature 473:343–348
27. Xu Y, Wu F, Tan L et al (2011) Genome-wide regulation of 5hmC, 5mC, and gene expression by Tet1 hydroxylase in mouse embryonic stem cells. Mol Cell 42:451–464
28. Pastor WA, Pape UJ, Huang Y et al (2011) Genome-wide mapping of 5-hydroxymethylcytosine in embryonic stem cells. Nature 473:394–397
29. Pastor WA, Huang Y, Henderson HR et al (2012) The GLIB technique for genome-wide mapping of 5-hydroxymethylcytosine. Nat Protoc 7:1909–1917
30. Huang Y, Pastor WA, Zepeda-Martinez JA et al (2012) The anti-CMS technique for genome-wide mapping of 5-hydroxymethylcytosine. Nat Protoc 7:1897–1908
31. Robertson AB, Dahl JA, Vagbo CB et al (2011) A novel method for the efficient and selective identification of 5-hydroxymethylcytosine in genomic DNA. Nucleic Acids Res 39:e55
32. Cokus SJ, Feng S, Zhang X et al (2008) Shotgun bisulphite sequencing of the Arabidopsis genome reveals DNA methylation patterning. Nature 452:215–219
33. Lister R, O'Malley RC, Tonti-Filippini J et al (2008) Highly integrated single-base resolution maps of the epigenome in Arabidopsis. Cell 133:523–536

34. Lister R, Pelizzola M, Dowen RH et al (2009) Human DNA methylomes at base resolution show widespread epigenomic differences. Nature 462:315–322
35. Jin SG, Kadam S, Pfeifer GP (2010) Examination of the specificity of DNA methylation profiling techniques towards 5-methylcytosine and 5-hydroxymethylcytosine. Nucleic Acids Res 38:e125
36. Huang Y, Pastor WA, Shen Y et al (2010) The behaviour of 5-hydroxymethylcytosine in bisulfite sequencing. PLoS One 5:e8888
37. Yu M, Hon GC, Szulwach KE et al (2012) Base-resolution analysis of 5-hydroxymethylcytosine in the mammalian genome. Cell 149:1368–1380
38. Josse J, Kornberg A (1962) Glucosylation of deoxyribonucleic acid. III. alpha- and beta-Glucosyl transferases from T4-infected Escherichia coli. J Biol Chem 237:1968–1976
39. Krueger F, Andrews SR (2011) Bismark: a flexible aligner and methylation caller for bisulfite-Seq applications. Bioinformatics 27:1571–1572

Chapter 34

Multiplexing for Oxidative Bisulfite Sequencing (oxBS-seq)

Kristina Kirschner, Felix Krueger, Anthony R. Green, and Tamir Chandra

Abstract

DNA modifications, especially methylation, are known to play a crucial part in many regulatory processes in the cell. Recently, 5-hydroxymethylcytosine (5hmC) was discovered, a DNA modification derived as an intermediate of 5-methylcytosine (5mC) oxidation. Efforts to gain insights into function of this DNA modification are underway and several methods were recently described to assess 5hmC levels using sequencing approaches. Here we integrate adaptation based multiplexing and high-efficiency library prep into the oxidative Bisulfite Sequencing (oxBS-seq) workflow reducing the starting amount and cost per sample to identify 5hmC levels genome-wide.

Key words Oxidative bisulfite sequencing, Multiplexing, 5-Hydroxymethylcytosine, Pooling, DNA, Methylation

1 Introduction

5-Hydroxmethylcytosine is a recently discovered DNA modification. The role of 5hmC has mainly been studied in brain tissue and embryonic stem (ES) cells where this DNA modification is very prominent and thought to be important in development and differentiation respectively [1]. Enrichment of 5hmC at promoters and other gene regulatory elements suggests a role in gene regulation, as does its appearance at poised chromatin regions in developmentally regulated ES cell genes [1]. Moreover, in ES cells the balance between pluripotency and lineage commitment seems to be linked to the balance of 5hmC and 5mC levels [1].

To distinguish 5-hydroxymethylcytosine and 5-methylcytosine, oxidative bisulfite sequencing has been developed [2]. Another technique for the detection of 5hmC, Tet- assisited Bisulfite Sequencing (TAB-seq), is described in Chapter 33 and immunoprecipitation and enrichment methods for 5hmC are described in Chapter 35. Conventional bisulfite conversion cannot distinguish between 5mC and 5hmC. The combination of oxBS-seq and Bisulfite Sequencing (BS-seq) can distinguish unmodified Cytosine (C), 5mC and 5hmC,

Jörg Tost (ed.), *DNA Methylation Protocols*, Methods in Molecular Biology, vol. 1708,
https://doi.org/10.1007/978-1-4939-7481-8_34, © Springer Science+Business Media, LLC 2018

because each state shows a unique pattern in BS-seq and oxBS-seq (Table 1), resulting in: 5hmC = Unconverted C in BS-seq − Converted C in oxBS-seq.

Workflow and reagents have been described in a recent protocol [3] and combined in a kit (TrueMethyl™ 6 kit (http://www.cambridge-epigenetix.com/). However, low starting amounts of DNA and cost per sample might limit the use of this method.

Here we report a method to multiplex low input DNA samples with TruSeq Illumina adapters. In short, samples are prepared for adapter ligation, pooled and subjected to oxBS treatment. This

Table 1
Conversion scheme

	C	5mC	5hmC
BS	U	C	C
oxBS	U	C	U

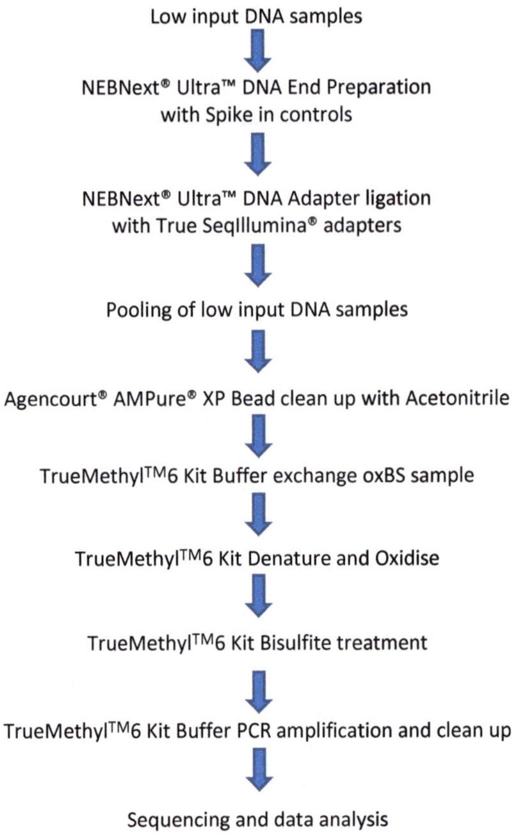

Fig. 1 Workflow for multiplexing of samples prior to bisulfite and oxidative bisulfite conversion. Protocol steps are summarized as text with *blue arrows* indicating the order of the workflow

approach will reduce costs per sample and allow for low input material to be treated using the TrueMethyl™ 6 kit (Fig. 1). Pooled samples being chemically converted in single tubes increases the comparability between samples.

However, the success of 5hmC detection genome wide depends on the abundance of the modification and the depth of sequencing. Brain tissue and embryonic stem cells contain high amounts of 5hmC, which can therefore be easily detected genome-wide whereas most somatic tissues contain low levels of 5hmC making it hard to detect difference in 5hmC abundance.

2 Materials

2.1 Preparation of the Next Generation Sequencing (NGS) Library

1. E220 focused ultrasonicator (Covaris).
2. Sonication tubes (microTUBE AFA Fiber Pre-Slit Snap-Cap 6 × 16 mm, Covaris).
3. NanoDrop (Thermo Fisher Scientific).
4. Quant-it™ Pico Green® dsDNA assay (Thermo Fisher Scientific).
5. TE buffer: 10 mM Tris–HCl, 1 mM EDTA, pH 8.
6. Spectrophotometer with plate reader option.
7. NEBNext® Ultra™ DNA library prep kit for Illumina (New England Biolabs).
8. Illumina® TruSeq adapters.
9. Spike in controls from the TruMethyl™ Kit (Cambridge Epigenetix).
10. DNA LoBind® tubes (Eppendorf) or similar.
11. TipOne®RPT (low retention) tips.
12. Acetonitrile, HPLC grade.
13. Magnetic stand for 2 mL tubes.
14. Agencourt® AMPure® XP Beads (Beckman Coulter).
15. Kapa Library Quantification kit for Illumina® platforms (KapaBiosystems).

2.2 oxBS-seq and BS-seq

1. TrueMethyl™ 6 Kit (Cambridge Epigenetix, supplied Bio-Rad® P6 Micro-Bio spin columns need to be stored at 4 °C).
2. Thermocycler.
3. PCR tubes or strips (0.2 mL).
4. Minicentrifuge for 0.2 mL PCR strips.
5. Primers for Illumina® TrueSeq indexed adapters.

6. Ethanol, HPLC grade.
7. Agencourt® AMPure® XP Beads (Beckman Coulter).
8. Magnetic stand for 2 mL tubes.
9. Quant-it™ Pico Green® dsDNA assay (Thermo Fisher Scientific).
10. 2100 Bioanalyzer Instrument (Agilent Technologies).
11. High sensitivity DNA Kit (Agilent Technologies).

2.3 Data Analysis

1. Seqmonk (http://www.bioinformatics.babraham.ac.uk/projects/seqmonk/).
2. Bismark (http://www.bioinformatics.babraham.ac.uk/projects/bismark/) [4].

3 Methods

3.1 Preparation of the Next Generation Sequencing (NGS) Library

The ideal input amount into the oxBS-seq workflow is 1 μg. To allow for smaller amounts of input DNA per sample, we pool samples after multiplexing to reach the desired amount (*see* **Note 1**). In this protocol we started with six samples of 150 ng each. Accurate and identical starting amounts are crucial in this workflow (*see* **Note 2**). We used the Covaris® E220 system to fragment our DNA and quantified the fragmented DNA using the Quant-it ™ Pico Green® dsDNA assay. For small amounts of DNA to be converted into a NGS library, pipetting steps should be minimalized to avoid loss of sample. Therefore, the end repair and adapter ligation steps are performed using the NEBNext® Ultra™ DNA library prep kit for Illumina® where all reactions take place in the same tube without a cleanup step before the adapter ligation. To allow for multiple samples to be sequenced together multiplexing adapters are used during the ligation step. We are using Illumina® TruSeq adapters for multiplexing in this protocol. For the oxidation reaction to work optimally, ethanol should be avoided during the cleanup procedure, as any residue of ethanol will affect the oxidation step. TrueMethyl™ 6 Kit Ultrapure water is used for elution and 80% acetonitrile is used for washing the beads as an alternative to ethanol since it can be tolerated in the oxidation step.

1. Fragment samples in 130 μL volume with Covaris® E220 (Settings: Intensity 4, cycles per burst 200, 80 s) to approximately 300–400 base pairs (bp) in size.
2. Follow the Quant-it ™ Pico Green® dsDNA assay to specify the input amount of DNA for each sample. Use 150 ng in 55 μL (Water or TE) per sample for input into the library generation protocol.
3. Prepare 150 ng sheared DNA in 55 μL per sample as input material in PCR tubes.

4. Add 3 μL End Prep Enzyme Mix and 6.5 μL 10× End repair Reaction Buffer per sample to a total volume of 65 μL.
5. Add 0.5 μL of spike in sequencing control (8 ng/μL) supplied with the TrueMethyl™ 6 Kit directly to the PCR tube (3% w/w; 3.8 μL/μg) (*see* **Note 3**).
6. Mix by pipetting and incubate on thermocycler with heated lid on for 30 min at 20 °C and 30 min at 65 °C. Hold at 4 °C.
7. Add 3.5 μL of TruSeq Illumina® adapters (1:10 dilution in water) for multiplexing. Be careful to not use NEB adapters at this stage.
8. Add 15 μL of blunt/TA Ligase Master Mix and 1 μL ligation enhancer. Mix and spin briefly to collect all sample material.
9. Incubate for 15 min at 20 °C on a thermal cycler. Heat-inactivate ligase for 10 min at 65 °C if pooling during cleanup.
10. The libraries can be pooled in a DNA LoBind® tube at this point. The six pooled libraries should be approximately 500 μL (83 × 6) (*see* **Note 4**).
11. Add 325 μL (0.65×) of Agencourt® AMPure® XP Beads to the pooled ligation reaction and mix well by pipetting.
12. Use a beads ratio of 0.65× to remove any adapter dimers here. However, size selection could also be done after library amplification if starting amounts are extremely low.
13. Incubate for 5 min at room temperature. Put tube on the magnetic stand to separate beads from supernatant until the solution is clear. Remove the supernatant without disturbing the beads.
14. Wash the beads with 1 mL 80% acetonitrile (*see* **Note 5**) per tube whilst on the magnetic stand. Discard supernatant. Perform wash step with 80% acetonitrile once more and discard the supernatant again. Let the beads air dry for 5 min until the beads look dry and/or start to crack (*see* **Note 5**).
15. Keep tube on the magnetic stand and add 40 μL ultrapure water to elute the DNA from the beads. Remove the tubes from the magnet and resuspend the beads by pipetting. Place the tubes onto the magnetic stand until solution has cleared.
16. Remove and keep the supernatant as input to the TrueMethyl™ 6 workflow. Purified DNA can be stored up for 24 h at 2–8 °C. For longer term storage, freeze at −20 °C.

3.2 DNA Oxidation

Oxidation of DNA is performed on half of each sample resulting in the 5hmC rate and location of the 5hmC mark for each sample after sequencing. In essence, site-specific oxidation and conversion of 5hmC bases to uracil is performed during the protocol using mild conditions to ensure maximum sample recovery. All centrifugation steps are done at room temperature. Only use the ultrapure water supplied with the kit.

1. Separate your sample into two parts for oxBS and BS conversion if pooled material is more than 1 μg. If pooled material is less than 1 μg then the sample does not need to be divided for the buffer exchange columns, but must be divided for the oxidation step.
2. Set the thermocycler to 37 °C before starting.
3. Resuspend the matrix of the Bio-Rad® P6 Micro-Bio spin column by flicking the column. Centrifuge quickly at 1000 × *g*. Remove the top cap and the bottom tip of the column and place into a new 2.0 mL tube. Centrifuge column at 1000 × *g* for 2 min to pack the column and remove any residual buffer. Discard the flow through.
4. Add 500 μL ultrapure water and centrifuge the column at 1000 × *g* for 1 min. Discard the flow through. Repeat the wash step three more times, always discarding the flow through. For the final wash, centrifuge the column for 2 min, discard the flow through and place column into a new collection tube.
5. Add the pooled sample DNA to the washed column. Centrifuge at 1000 × *g* for 2 min and keep the eluate, which should be approximately 22 μL (maximum 25 μL). Check by pipetting.
6. Start thawing the Oxidant Solution on ice in the dark. To denature dsDNA take a new 0.2 mL microcentrifuge tube or strips. Add the buffer exchanged dsDNA sample and add reagents according to Table 2, mix and quickly spin. Incubate the DNA denaturing solution at 37 °C for 30 min in a thermocycler. Place the tubes or strips on ice after denaturation and proceed immediately to next step as the oxidation reaction does not oxidize dsDNA and reannealing at a later time point will affect the 5hmC conversion rate.

Table 2
DNA denaturing master mix

Solution	Volume per reaction (μL) (oxBS sample)	Volume per reaction (μL) (BS sample)
DNA (buffer exchanged from **step 5**)	22.75	
dsDNA		22.75
Denaturing solution	1.25	1.25
Ultrapure water	Adjust to total volume	Adjust to total volume
Total volume	24	24

Table 3
Oxidation master mix

Solution	Volume per reaction (μL) (oxBS sample)	Volume per reaction (μL) (BS sample)
Denatured DNA	24	24
Ultrapure water	–	1
Oxidant solution	1	–
Total volume	25	25

7. Set thermocycler to 40 °C before pipetting. *Split the pooled DNA into two samples here, one for oxidation (oxBS) and one for mock oxidation (BS).* For the oxidation reaction add reagents to the PCR tubes or strips containing the denatured DNA from the previous step according to Table 3. The reaction mix should be orange. Mix the reaction and centrifuge. Incubate the samples at 40 °C for 30 min in a thermocycler. The solution should still be orange after the 30 min incubation. Compare the samples with Appendix 2 in the user manual of the TrueMethyl™ 6 conversion kit as the color of the oxidation reaction indicates a successful oxidation process.

8. Centrifuge samples in a bench-top mini centrifuge for PCR strips at 1177 × *g*, for 10 min to pellet any black precipitate and take the orange supernatant forward into PCR tubes or strips. Be careful not to carry any precipitate over as this can inhibit downstream steps. Proceed immediately to the next step.

3.3 Bisulfite Conversion

Bisulfite treatment is performed on both, the oxidized and non-treated sample. Dissolve the Bisulfite Reagent with the Bisulfite Diluent provided and dissolve fully before use (*see* **Note 6**).

1. Bring the oxidized and mock-oxidized samples from the previous step to room temperature. Add 700 μL Bisulfite Diluent to each aliquot of Bisulfite Reagent and incubate at 60 °C for 15 min, vortex until the bisulfite mix is completely dissolved.
2. For the bisulfite conversion add each component to PCR tubes or strips from the DNA oxidation step in the order listed in Table 4. Vortex thoroughly and centrifuge. The bisulfite DNA conversion is performed as shown in Table 5 in a thermocycler. Use the heated lid and set reaction volume to 200 μL or highest possible volume setting available on the thermocycler. The reaction should take approximately 8 h to complete and is best done overnight.

Table 4
Bisulfite conversion master mix

Solution	Volume per reaction (μL)
DNA	25
Bisulfite reagent solution	170
Bisulfite additive	5
Total volume	200

Table 5
Thermocycler conditions for Bisulfite conversion

Step	Time (min)	Temperature (°C)
Denaturing	5	95
Incubation	20	60
Denaturing	5	95
Incubation	40	60
Denaturing	5	95
Incubation	165	60
Denaturing	5	95
Incubation	20	60
Denaturing	5	95
Incubation	40	60
Denaturing	5	95
Incubation	165	60
Store o/n	Hold	20

3. Centrifuge to pellet precipitates once the bisulfite conversion is complete using a bench-top mini centrifuge for PCR strips at 1177 × *g* for 10 min. Carefully remove 190 μL of the supernatant and transfer to a new 1.5 mL microcentrifuge tube without disturbing the pellet.
4. Add 310 μL Desulfonation Buffer to each sample, mix and centrifuge. For each sample prepare an Amicon® filter provided with the kit as shown in Figure 3 on page 19 of the manual and transfer the sample to the Amicon® filter (approx. 500 μL). Avoid touching the filter as not to damage the filter membrane and affect the diafiltration step. Centrifuge filter for 5 min at 14,000 × *g* and discard the flow through.

5. Add 450 μL of Desulfonation Buffer to the filter, centrifuge for 5 min at 14,000 × *g* and discard the flow through.
6. Add 450 μL of Desulfonation Buffer to the filter and incubate for 5 min at room temperature before spinning for 5 min at 14,000 × *g*. Discard the flow through.
7. Add 450 μL of Wash Buffer to the filter, centrifuge for 5 min at 14,000 × *g* and discard the flow through. Repeat wash step twice more.
8. Elute the sample by inverting the Amicon® filter into a new 1.5 mL Amicon® collection tube as shown on page 20 Figure 4 of the manual provided with the kit and centrifuge at 2000 × *g* for 2 min. The bisulfite-converted sample should be around 10 μL total volume. Purified DNA can be stored up for 24 h at 2–8 °C. For longer term storage, freeze at −20 °C.

3.4 PCR Amplification

An amplification step must be added to enrich for NGS adapted, multiplexed and (oxidized) bisulfite-converted DNA prior to sequencing. The number of PCR cycles should be adapted to the input material used for NGS library making. The polymerase provided in the kit has been engineered to resist uracil stalling, therefore allowing accurate quantitation of bisulfite-converted libraries (*see* **Note 7**). The primers used are Illumina® TruSeq Primers for indexed adapters. For the PCR amplification, we only use half of the bisulfite-converted samples, keeping half of the sample as a back-up. Alternative strategies could be used to estimate optimal cycle numbers.

1. Thaw all reagents on ice prior to use, mix and centrifuge briefly. Prepare the PCR mix (outlined in Table 6) in 0.2 mL PCR tube or strip adding all reagents in order. If multiple samples are processed, prepare a master mix without the TrueMethyl™

Table 6
PCR reaction master mix

Solution	Volume per reaction (μL)
Nuclease-free water	31.5
5× TrueMethyl™ polymerase buffer	10
10 mM dNTP	1
PCR Illumina TruSeq primers	2
Bisulfite-converted DNA	5
TrueMethyl™ DNA polymerase (2 U/μL)	0.5
Total volume	50

Table 7
Thermocycler conditions for PCR amplification

Number of cycles	Time (s)	Temperature (°C)
1	120	95
	30	95
15× repeat	20	60
	45	72
1	Hold	4

DNA polymerase. Add the TrueMethyl™ DNA polymerase last and mix by pipetting. Amplify the DNA using the thermal cycling settings as outlined in Table 7.

2. Clean up the PCR products with Agencourt® AMPure® XP Beads. Use 1× beads to volume of PCR reaction. Mix beads and PCR reaction and incubate for 5 min at room temperature. Put the tubes on the magnetic stand to separate beads from supernatant until the solution is clear (~5 min). Remove the supernatant without disturbing the beads.
3. Wash the beads with 1 mL freshly prepared 80% ethanol per tube whilst on the magnetic stand. Discard supernatant. Perform wash step with 80% ethanol once more and discard the supernatant again. Let the beads air dry for 5 min until the beads are completely dry and/or look cracked.
4. Keep tube on the magnetic stand and add 20 μL ultrapure water supplied with TrueMethyl™ 6 Kit to elute the DNA from the beads. Remove the tubes from the magnet and resuspend the beads completely in water by pipetting. Place the tubes onto the magnetic stand until solution has cleared. Remove and keep the supernatant as input for quantification and quality control steps.
5. Run a High Sensitivity DNA Bioanalyzer ChIP and Quant-it ™ Pico Green® dsDNA assay before submission for sequencing (*see* **Note 8**). A representative result is shown in Fig. 2.
6. We recommend using 100 bp paired end sequencing on an Illumina® HiSeq platform.

3.5 Data Analysis

Sequencing costs can be a deterrent when performing whole genome bisulfite sequencing. The protocol described here resorts to the subtraction of the oxBS-seq from BS-seq dataset to obtain the levels of 5hmC. Plotting the values for BS-seq versus oxBS-seq in scatter plots allows a preliminary analysis of how many regions display higher methylation in BS-seq (as expected) or in oxBS

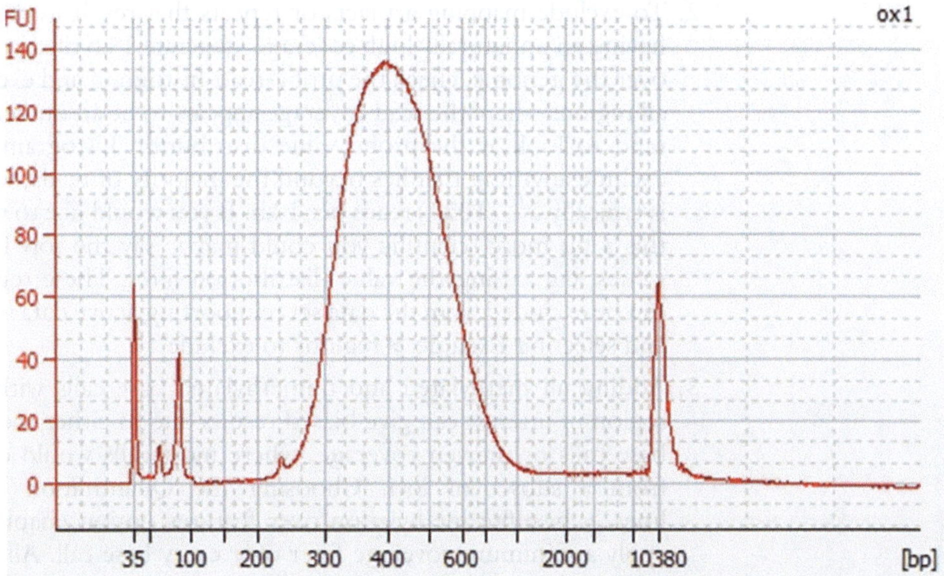

Fig. 2 Bioanalyzer electropherogram of an oxBS library. Representative profile of a library derived from a multiplexed sample

(which might be a measure for false positives and noise in the system). Higher overall sequencing depth can reduce this noise; however, this would imply increased sequencing costs. Combining the protocol described here with a downstream analysis that pools regions of biological interest, can be used to gain depth over groups of features without the need for further sequencing. Depending on the biological question/system being studied it is possible to analyse relevant categories of genes that are known to be modulated with the treatment/biological process under study and investigate if 5hmC plays a role in the known variation in the expression of these genes. Alternatively, it is possible to do an unfiltered analysis to understand if modulation of 5hmC is enriched in any specific biological category (*see* **Note 9**). This subheading is meant as a brief overview on how to do an initial exploration of the oxBS-seq data sets. For data exploration we use SeqMonk after the data has been mapped using Bismark [4]. A comprehensive protocol on how to map oxBS/BS data using Bismark is available online [5] and beyond the scope of this chapter.

1. Import methylation calls into SeqMonk using the generic text import; to keep the file size small import only CpG calls. SeqMonk will import methylated and unmethylated Cs individually as 1 base pair (bp) reads. The methylation state is encoded in the strandedness of the read (forward = methylated, reverse = unmethylated). Bisulfite datasets can be extensive. If you want to look at non-CG methylation import CpG and non-CG separately.

2. To exclude mapping artifacts or repeats that result in regions having an unnaturally high coverage, generate 1 kb tiled probes over the genome, check the probe-read-histogram and exclude all regions where the read coverage appears to be an outlier; we tend to look at the probe value distribution histogram and exclude anything that lies beyond the majority of reads, which is typically 2^{10} (1024) reads per 1 kb. If you would like to make this a bit more scientific you could maybe say the top 1% of values, e.g., using the value distribution filter. These regions can be excluded from the data set reimporting active data stores and selecting Exclude > regions to exclude.
3. Adding all methylated and unmethylated base calls within a region of interest can give biased results due to either too few base calls or uneven coverage, where most calls would come from a subset of loci. Choosing the Quantitation pipelines > bisulfite methylation over features option enables to apply a minimum coverage filter over every base call. All base calls meeting the coverage requirements are counted individually and weighed equally to calculate the average methylation level. In addition one can filter for a minimum number of cytosine positions (which also need to meet the coverage criteria) per feature before it is considered measured well enough. For 5mC levels compare the oxBS data sets. Here we run through an example to compare CpG Islands (CGIs) for a reasonably deep sequencing experiment (e.g., 10×).
4. Run the Quantitation pipelines > bisulfite methylation over features option with settings being: Features to quantitate > CGI, Minimum count to include position >4 (coverage over individual CpG), Minimum observations to include filter >5 (minimum number of CpGs above threshold for features to be counted), this has been replaced in the latest version of SeqMonk to produce NaN values (not a number), which means that they do no longer have to be filtered out as they do not count as numbers at all (similar to the R concept of NA values) > Run pipeline. This will yield a single value (percent methylation) for each CGI in the genome.
5. Create probes first. This can be done as unbiased tiled probes (e.g., 1 kb or 5 kb), choosing the tile length so that the tile is covered, which will depend on the overall methylation level. Alternatively, probes can be chosen covering certain number of CpG positions or covering certain features—such as, promoters, CGI, gene bodies or any other group of functional relevance, e.g., gene ontology (GO) terms. Lastly, by using windows (of variable length) spanning a fixed number of consecutive CpG positions, e.g., 50 CpG, probes can be generated across the genome.

6. To get a value for hydroxymethylation use the methylation over feature pipeline to determine average values of the feature of interest and then subtract oxBS-seq from the BS-seq data set using the Relative Quantitation method.
7. Doing trend plots of hydroxymethylation around features of interest (± a few kilobases (kb) up and downstream) shows if 5hmC is important in the modulation known to occur in the treatment/biological process of interest.
8. By generating probes over features, then generating running window probes within the features and quantitating the features as described above one can get CGIs of interest. Filtering using the windowed replicate test filter makes it possible to identify regions where there is a certain trend within the features. These can then be analysed to identify categories of interest using GO analysis.

4 Notes

1. At the moment Illumina® TruSeq adapters (which come methylated) allow for multiplexing of up to 24 samples.
2. Accurate library quantitation is essential, indeed is more important than in standard workflows. Any existing bias in the library pool going into the chemical conversion reactions will be carried through to the final libraries for sequencing, which will result in uneven coverage.
3. Spike-in controls can be added here since there is no clean-up step before adapter ligation as the spike in controls are too small to be recovered by SPRI beads if they are not adapter ligated. We found very consistent conversion rates between multiplexed samples. It could therefore be considered whether it is sufficient to add the spike-in control to only one or two of the pooled samples.
4. Before pooling we recommend to run a Kapa Library Quantification kit for Illumina® platforms to accurately quantify each library to achieve optimal pooling conditions.
5. When using 80% ethanol instead of acetonitrile make sure that beads are completely dry before elution step (cracks starting to show).
6. The Bisulfite Reagent comes in aliquots for four samples each. Once the Bisulfite Reagent is dissolved it cannot be stored. Considering the coverage two aliquots can be used for nine samples. When performing more than four reactions, mix the different aliquots of dissolved Bisulfite solution prior to use.

Table 8
PCR reaction master mix for Kapa HiFi Uracil+

Solution	Volume per reaction (μL)
Nuclease Free Water	18
PCR Illumina TruSeq primers	2
Bisulfite converted DNA	5
Kapa HiFi Uracil 2× Master Mix	25
Total volume	

Table 9
Thermocycler conditions for Kapa HiFi Uracil + PCR amplification

Number of cycles	Time (s)	Temperature (°C)
1	120	95
	20	95
15× repeat	30	65
	30	72
1	300	72
1	Hold	4

7. Kappa HiFi Uracil+ can be used as an alternative and works well with low amounts of DNA. *See* Table 8 for reaction mix and Table 9 for PCR conditions.
8. If sample has adapter dimer contamination clean with 0.65–0.8× Agencourt® AMPure® XP Beads as described in **steps 2–4** of Subheading 3.4.
9. It should be mentioned that while this chapter focuses on the combination of oxBS-seq and BS-seq for the identification of 5hmC levels, oxBS-seq by itself is superior to BS-seq to accurately map 5mC levels because BS-seq 5mC calls include 5hmC calls.

References

1. Fu Y, He C (2012) Nucleic acid modifications with epigenetic significance. Curr Opin Chem Biol 16:516–524
2. Booth MJ, Branco MR, Ficz G et al (2012) Quantitative sequencing of 5-methylcytosine and 5-hydroxymethylcytosine at single-base resolution. Science 336:934–937
3. Booth MJ, Ost TW, Beraldi D et al (2013) Oxidative bisulfite sequencing of 5-methylcytosine and 5-hydroxymethylcytosine. Nat Protoc 8:1841–1851
4. Krueger F, Andrews SR (2011) Bismark: a flexible aligner and methylation caller for Bisulfite-Seq applications. Bioinformatics 27:1571–1572
5. Krueger FA, Andrews SR. (2012) Quality control, Trimming and alignment of risulfite Seq data (Prot 57). http://www.epigenesys.eu/images/stories/protocols/pdf/20120720103700_p57.pdf

Chapter 35

Affinity-Based Enrichment Techniques for the Genome-Wide Analysis of 5-Hydroxymethylcytosine

John P. Thomson and Richard R. Meehan

Abstract

Since its initial characterization in 2009 there has been a great degree of interest in comparative profiling of 5-hydroxymethylcytosine (5hmC) nucleotides in vertebrate DNA. Through a host of genome-wide studies the distribution of 5hmC has been mapped in a range of cell lines, tissue types and organisms; the majority of which have been generated through affinity-based methods for 5hmC enrichment. Although recent advances in the field have resulted in the ability to investigate the levels of both methylated and hydroxymethylated cytosines at single base resolution, such studies are still relatively cost-prohibitive as well as technically challenging. As such affinity-based methods for the enrichment of 5hmC-modified DNA fragments represent a cost-effective and highly informative method for profiling 5hmC residency in genomic DNA. Here we will outline protocols for two independent affinity based methods to generate 5hmC enriched fractions for subsequent locus specific and genome-wide analysis; immunoprecipitation using highly specific 5hmC antibodies as well as a chemical capture based method.

Key words 5-Hydroxymethylcytosine, 5-Methylcytosine, hMeDIP, Affinity enrichment, qPCR, Sequencing, hMeSeal, ß-Glucosyltransferase, Glucosyl-hydroxymethylcytosine

1 Introduction

In mammals cytosine, usually in the context of the dinucleotide cytosine-phosphate-guanine (CpG), acts as a substrate for direct enzymatic modification to 5-methylcytosine (5mC) by DNA methyltransferases and subsequently to 5-hydroxymethylcytosine (5hmC), 5-formyl-cytosine (5fC) and 5-carboxyl-cytosine; (5caC) by the reiterative action of the Ten-eleven-translocation (TET) family of Fe(II), ascorbic acid and α-KG-dependent dioxygenases (Tet1, Tet2, and Tet3) [1]. The majority of these modified CpG dinucleotides occur as 5mC and this form of methylated DNA has been linked with silencing of repetitive elements, regulation of X-chromosome inactivation and imprinting; in addition to the maintenance of normal transcriptomic patterns in somatic tissues and stem cells [2–4]. 5hmC was initially reported in 1952 to be

Jörg Tost (ed.), *DNA Methylation Protocols*, Methods in Molecular Biology, vol. 1708,
https://doi.org/10.1007/978-1-4939-7481-8_35, © Springer Science+Business Media, LLC 2018

present in certain bacteriophage [5], many years later it was comprehensively identified in human and mouse genomes, especially in neural tissue [6, 7]. The enzymes that were responsible for the conversion of methylated cytosine residues into hydroxymethylated cytosines were also identified at the same time [7]. The Ten-eleven translocation 1 (TET1) gene has homology to the trypanosomal protein JBP1, which is known to oxidize thymine to 5-hydroxymethyl-uracil [8]. Two further 5mC oxidases were discovered (TET2 and TET3) along in concert with (albeit incredibly low in abundance) the 5fC and 5caC modifications [9]. The presence of 5hmC, 5fC, and 5caC in genomes has been linked with DNA demethylation programs in development in which 5hmC serves as an intermediate in this process [1, 10–14].

A wealth of research has focused on characterizing the role of these modified bases through mapping their presence throughout the genome in a host of cell and tissue types [10, 15–18]. Initial work demonstrated that traditional bisulfite based methods for distinguishing between C and 5mC were unable to discriminate between 5mC modified and 5hmC marked DNA, necessitating a reinterpretation of many data sets [19–21]. Subsequent work focused on the profiling of this modification across the genome through several affinity-based methods [10, 22–28]. These experiments made the important point that 5hmC is typically depleted over the TSSs and enriched in the bodies of highly expressed genes. Subsequent studies highlighted potential bias of anti-5hmC antibodies for recognizing modification dense regions as well as showing some bias towards CA repeat sequences [29, 30]. Overall the majority of 5hmC in the genome is found over the genic portions of the genome but is also present in a tissue specific manner at enhancer elements and promoter proximal regions [26]. The functional relationships between of the presence of 5hmC at regulatory regions and underlying transcriptional state are still poorly understood [27, 31]. To date there is a greater tendency to study the more abundant (and informative) 5hmC modifications in DNA rather that the less abundant 5fC and 5caC containing sequences [1].

There are now a range of methods available to study principally 5hmC and to a lesser extent the 5fC and 5caC modifications in DNA [11]. These include locus-specific methods of detection such as those which rely on the relative activity of sets of restriction enzymes for digestion of CpG containing recognition sequences that are modified by either 5mC and 5hmC [32], as well as those which can detect relative levels of 5hmC during qPCR amplification in the presence of boronic acid [33]. In addition to these locus specific assays, the global distribution of the modification can be studied following specific enrichment of 5hmC marked DNA fragments; most frequently by either antibody or affinity based enrichment methods [25]. More recently methods are now commercially distributed that allow researchers to study the distribution of the 5hmC modification at a single CpG resolution in a similar manner to the bisulfite based methods of 5mC analysis (*see also* Chapters 33 and 34). However, although such single CpG-based approaches have the ability to best map the true patterns of 5hmC distribution throughout the genome the costs of carrying out such an experiment due to the sequencing depth required (predicted to be at least 100-fold) and the bioinformatics support required, makes such approaches largely unaffordable as yet, for the majority of researchers [17, 28, 34]. In this chapter we will therefore describe the protocols for two widely used and accurate affinity based methods of 5hmC enrichment (antibody based hydroxymethyl DNA immunoprecipitation; hMeDIP and hydroxymethyl selective chemical labeling; hMeSeal) which enable the downstream analysis of 5hmC marked portions of the genome either for a specific candidate locus or through genome-wide sequencing/arraying approaches.

2 Materials

2.1 hMeDIP

1. Spectrophotometer: NanoDrop™ 8000 (Thermo Scientific).
2. Benchtop hot block: for example Thermomixer comfort (Eppendorf).
3. Sonicator: Bioruptor® Standard—chilled (Diagenode®) or similar.
4. Test tube rotator: LD-79 (Labinco BV) or similar.
5. Magnetic rack: DynaMag™-2 (Life Technologies).
6. TE: 10 mM Tris–HCl, 1 mM EDTA, pH 8.0. For 10 mL add 0.1 mL of 1 M Tris–HCl (pH 8.0) and 0.02 mL EDTA (0.5 M) and 9.88 mL H_2O.
7. 0.1% PBS-BSA: For 10 mL add 1 mL of 10 mg/mL Bovine Serum Albumin (BSA) and 9 mL Phosphate Buffered Saline (PBS).
8. 10× IP Buffer: 100 mM Na-Phosphate pH 7.0 (mono-dibasic), 1.4 M NaCl, 0.5% Triton X-100. For 10 mL add 2 mL 0.5 M Na-Phosphate pH 7.0 (mono-dibasic), 2.8 mL 5 M NaCl, 0.5 mL 10% Triton X-100, and 4.7 mL H_2O. Store at 4 °C.
9. 1× Digestion Buffer: 50 mM Tris–HCl pH 8.0, 10 mM EDTA, 0.5% SDS. For 10 mL add 0.5 mL 1 M Tris–HCl pH 8.0, 0.2 mL 0.5 M EDTA, 0.25 mL 20% SDS and 9.05 mL H_2O. Store at 4 °C. 0.5 M Potassium phosphate bibasic and monobasic: For 100 mL, add 21.1 mL 1 M

Potassium Phosphate Monobasic, 28.9 mL 1 M Potassium Phosphate Dibasic and 50 mL H_2O.

10. 5-Hydroxymethylcytosine (5hmC) antibody (pAb) (Active Motif).
11. [Optional] 5hmC-spike in control DNA (New England Biolabs or Zymo Research).
12. Dynabeads protein G (Life Technologies).
13. Proteinase K dissolved in water to 20 mg/mL.

2.2 hMeSeal

1. Spectrophotometer: NanoDrop™ 8000 (Thermo Scientific).
2. Bench-top hot-block: Thermomixer comfort (Eppendorf) or similar.
3. Sonicator: Bioruptor® Standard—chilled (Diagenode®) or similar.
4. Test tube rotator: LD-79 (Labinco BV) or similar.
5. Magnetic rack: DynaMag™-2 (Life Technologies).
6. Qubit fluorometer or similar.
7. TE: 10 mM Tris–HCl 1 mM EDTA, pH 8.0. For 10 mL add 0.1 mL of 1 M Tris–HCl (pH 8.0) and 0.02 mL EDTA (0.5 M) and 9.88 mL H_2O.
8. βGT Reaction Buffer: NEB buffer 4 (New England Biolabs).
9. βGT enzyme: T4 Phage β-glucosyltransferase (New England Biolabs) (*see* **Notes 1** and **2**).
10. Modified glucose moiety: 2 mM UDP-6-N_3-Glc in H_2O (*see* **Note 2**).
11. Disulfide biotin linker: 1 mM in DMSO (*see* **Note 2**).
12. DNA purification columns: (e.g., Micro Bio-Spin® 6, Bio-Rad; QIAquick® Nucleotide Removal Kit and MinElute® PCR Purification Kit, Qiagen).
13. Magnetic streptavidin beads for pull down: Dynabeads® MyOne™ Streptavidin C1 (Life Technologies)
14. 2× Binding and washing (B&W) Buffer: 10 mM Tris–HCl (pH 7.5), 1 mM EDTA, 2 M NaCl, and 0.01% Tween 20.
15. 100 mM Dithiothreitol (DTT)

3 Protocols

3.1 Considerations Prior to Starting

Ultimately both the antibody enrichment and chemical labeling/enrichment based strategies generate extremely similar 5hmC patterns across the genomes, with the majority of the modification found to reside in genic, promoter proximal and enhancer portions of the genome, with a small but significant number of

Table 1
Overview of some common advantages and disadvantages in using hmeDIP and hMeSeal based enrichment methods

	Advantages	Disadvantages
hMeDIP	Relatively simple and robust protocol Efficient enrichment of 5hmC marked DNA fragments vs background noise Good final yield of DNA	Enriched DNA single stranded: If genome-wide sequencing is desired, the protocol requires the addition of sequencing adapters to the sonicated DNA prior to the antibody enrichment Although low, there is some enrichment found at non CpG simple tandem repeat elements in the genome
hMeSeal	Can work with low levels of starting DNA (~5 ng) Efficient enrichment of 5hmC marked DNA fragments vs background noise	Low final yield of DNA Protocol contains steps requiring efficient enzymatic reactions to occur for the complete enrichment of 5hmC marked. Incomplete reactions/enzymatic issues will affect the overall enrichment efficiency dramatically

transcriptional start sites also marked strongly by 5hmC in a host of tissue and cell types (for a comprehensive review please *see* [25]). Each of the two techniques has their own advantages as is outlined in Table 1. Ultimately the choice of strategy will most likely depend on the amount of DNA available for the assay. When limited material is available the hMeSeal strategy is preferred as this technology is reported to enrich 5hmC from as little as 5–500 ng of starting DNA (depending on sample source) [35, 36].

It is also critical at this stage to plan out what form of analysis will be carried out on the enriched 5hmC fragments following purification. Normally this would lead to one of three outcomes, (1) quantitative PCR (qPCR) for locus-specific analysis, (2) microarray-based analysis or (3) genome-wide sequencing. For a summary of the downstream applications commonly used *see* Subheading 3.4 below. If genome-wide sequencing is desired the protocols for the addition of sequencing adapters vary greatly depending on the affinity based enrichment method selected. As the antibody-based method (hMeDIP) requires the DNA to be denatured in order to expose the antigen for subsequent enrichment, the enriched material is also single stranded. As such genome-wide sequencing adapters must be ligated immediately following sonication prior to the enrichment steps. In contrast the hMeSeal-based method maintains double stranded DNA fragments throughout the protocol allowing for a simpler addition of sequencing adapters post enrichment (similar to traditional ChIP-seq experiments). For the addition of sequencing adapters either before or after enrichment please refer to the suppliers' protocols.

Finally, It is important to remember that the levels of 5-hydroxy modified cytosines is far lower than that of 5-methylcytosine, which means that downstream analysis of the enriched DNA can be challenging. For example qPCR analysis is more prone to technical error due to the fact that a greater number of cycles are needed to reach the threshold cycle (often referred to as the Ct value), which is required to calculate the relative amount of PCR product. This is more of an issue for hMeSeal based enrichment strategies where the total yield of enriched material is typically lower than for hMeDIP based methods [25].

3.2 Hydroxymethyl DNA Immunoprecipitation (hMeDIP) Protocol

Antibody based enrichment techniques are frequently used to study a host of epigenetic modifications in DNA and histones due to the relative simplicity of the protocols with respect to other methodologies. The most critical point towards deciding whether or not to use such an approach is the overall quality of the antibody, i.e., how specific is the antibody for the epigenetic modification of interest and what is the level of background noise. In this regard the 5hmC antibody outlined here is excellent in specifically recognizing 5hmC marked DNA sequences with little to no cross talk observed for regions marked by methylated or nonmodified CpG dinucleotides [26]. Here we will detail our protocol for the enrichment of 5hmC marked DNA using antibody precipitation, which can be completed within a day and a half (Fig. 1).

3.2.1 Fragmentation of the DNA by Sonication

1. Dilute 20 μg genomic DNA in 400 μL of TE in a 1.5 mL microcentrifuge tube. Ensure that this step is carried out on ice (*see* **Note 3**).
2. Place samples in a prechilled (4 °C) sonicator and carry out sonication to achieve the desired fragment sizes. Optimal fragment sizes will range from 100 base pairs (bp) to ~1000 bp with the majority of sequences around 300–400 bp (*see* **Note 4**). Prior to sonication, it is vital to first optimize the settings for the instrument in order to achieve the desired fragmentation pattern for your sample(s).
3. Following fragmentation, determine the DNA concentration of the samples through multiple measurements (*see* **Note 5**). Run approximately 250 ng of sonicated DNA on a 1.5% agarose gel for 1.5 h at 100 V (*see* **Note 6**). The size range of the smear should be approximately 100–1000 bp, with a modal size of 500–700 bp (*see* **Note 7**).
4. If the 5hmC enriched material is to be run on a sequencer, specific adapter sequences should be added here. Please see suppliers' notes for adapter ligation protocols.

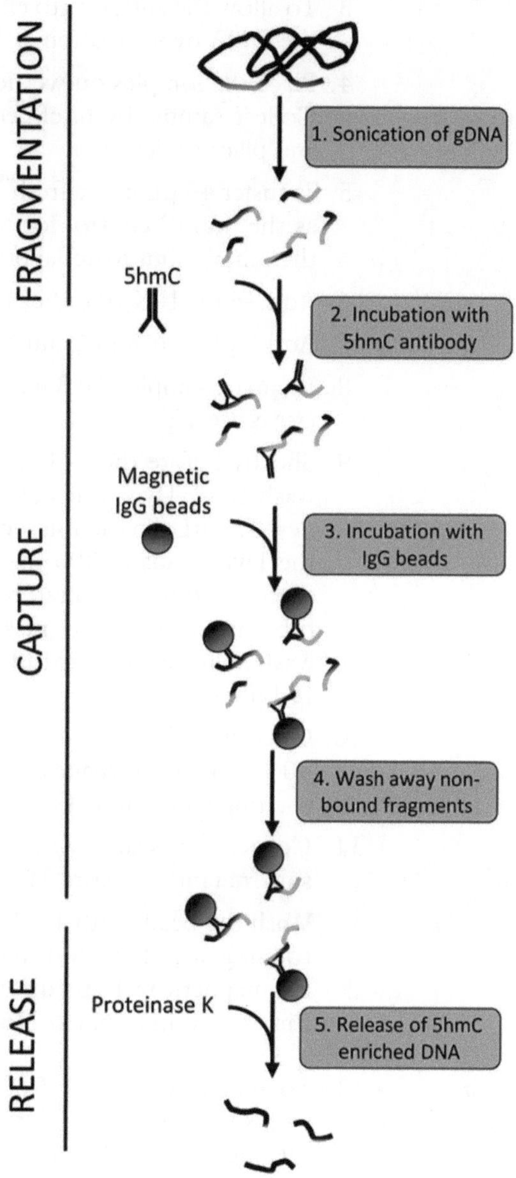

Fig. 1 Overview of the hMeDIP protocol. Genomic DNA (gDNA) is sheared by sonication prior to incubation with a specific 5hmC antibody. Following antibody incubation, magnetic IgG beads are added which bind to the antibody. Unbound DNA fragments are removed through a series or washes and the 5hmC marked DNA released from the antibody–bead conjugate through proteinase K treatment

3.2.2 Capture of 5hmC Enriched DNA Fragments

1. From the sonicated material, dilute 4 μg of sonicated DNA in 450 μL TE in a 1.5 mL screw-top tube.
2. [Optional] Spike in a 5hmC rich DNA template to act as a positive control for immunoprecipitation of 5hmC-containing DNA (*see* **Note 8**).

3. To allow the antibody to efficiently bind to the DNA, denature the DNA by incubation at 100 °C for 10 min in a hot-block.
4. Place the samples on wet-ice for 5 min (*see* **Notes 9** and **10**). Collect sample by brief centrifugation in a chilled centrifuge and place on ice.
5. Transfer 45 μL into a fresh 1.5 mL tube. This sample will serve as the 'input' control for the IP sample (representing 10% of the initial input material prior to enrichment).
6. Add 45 μL 10X IP buffer to remaining sample. Mix well.
7. Add 3 μL of α-5hmC antibody (*see* **Note 11**).
8. Incubate samples for 3 h at 4 °C on a rotating wheel at 25 rpm (*see* **Note 12**).
9. Shortly before the end of the incubation period, prepare and wash 40 μL Dynabeads Protein G (30 mg/mL) per sample in a new 1.5 mL flip cap microcentrifuge tube (*see* **Note 13**). Wash the Dynabeads in 800 mL PBS-BSA for 5 min on a rotating wheel at room temperature. Remove the liquid by placing the tube in a magnetic rack and carefully pipetting off the wash solution. Resuspend the washed beads in 40 μL of 1× IP buffer (4 °C)
10. Once the 3-h incubation with the antibody is complete, add 40 μL of the Dynabead mixture and incubate for 1 h on a rotating wheel at 4 °C (*see* **Note 14**).
11. Collect the beads with the magnetic rack and aspirate the supernatant (*see* **Note 15**).
12. Wash the beads with 1 mL 1X IP buffer (4 °C) for 5 min on a rotating wheel at room temperature. Collect the beads using the magnetic rack and discard the supernatant. Repeat this step three times in total (*see* **Note 16**).

3.2.3 Release and Cleanup of 5hmC Enriched DNA Fragments

1. To release the DNA from the beads, resuspend the beads in 250 μL digestion buffer, and add 10 μL Proteinase K (20 mg/mL stock). Incubate samples for a minimum of 3 h at 50 °C (or overnight) in a shaking heat block at 800 rpm (*see* **Note 17**).
2. Purify the immunoprecipitated DNA (IP sample) fragments *and* the INPUT sample using a PCR purification column or by phenol–chloroform extraction followed by ethanol precipitation (*see* **Note 18**). Resuspend or elute in TE (volume depends on subsequent analysis, *see* Subheading 3.4 below).

3.3 Hydroxymethyl Selective Chemical Labeling (hMeSeal) Protocol

A common alternative to antibody-based 5hmC enrichment methods is to use a chemical labeling based approach termed hydroxymethyl selective chemical labeling. Developed in the He lab at the University of Chicago, this approach relies on the enzymatic addition of a biotin conjugated glucose moiety specifically to 5hmC residues to form biotinylated glucosyl-hydroxymethylcytosine using β-glucosyltransferse [36]. Modified residues are then specifically enriched through streptavidin bead purification. This approach returns lower levels of background binding ("noise"), but at the cost of returning globally lower levels of DNA compared to the hMeDIP-based method. As this approach is reported to work on as little as 5–500 ng of starting DNA is particularly suited to projects where there is limited or precious material [35]. However, where available we suggest starting with up to 30 μg of DNA. Here we present our own protocol, which is based off of the initial protocols defined in the He lab (Fig. 2) [36]. In addition hMeSeal enrichment can also be carried out through the use of commercially available kits such as the Hydroxymethyl Collector™ kit marketed by Active Motif [35].

3.3.1 Fragmentation of the DNA by Sonication

1. Dilute 30 μg genomic DNA in 400 μL of TE in a 1.5 mL microcentrifuge tube. Ensure that this step is carried out on ice (*see* **Note 3**).
2. Place samples in a prechilled (4 °C) sonicator and carry out sonication to achieve the desired fragment sizes. Optimal fragment sizes will range from 100 bp to ~1000 bp with the majority of sequences around 300–00 bp (*see* **Note 4**). Prior to sonicating the sample it is vital to first optimize the settings for the instrument in order to achieve such a fragmentation pattern.
3. Following fragmentation, determine the DNA concentration of the samples through multiple measurements (*see* **Note 5**). Run approximately 250 ng of sonicated DNA on a 1.5% agarose gel for 1.5 h at 100 V (*see* **Note 6**). The size range of the smear should be approximately 100–1000 bp, with a modal size of 300–400 bp (*see* **Note 7**).

3.3.2 Glucosylation Reaction

4. In a clean 1.5 mL microcentrifuge tube add the desired amount of sonicated DNA (*see* **Note 19**). Remove 10% of this to act as an input control and store at −20 °C (*see* **Note 20**).
5. To check for the relative enrichment of 5hmC over background noise, set up a negative control reaction for each sample to be tested (*see* **Note 21**).
6. Add the following reagents in the order listed below to the microcentrifuge tube:

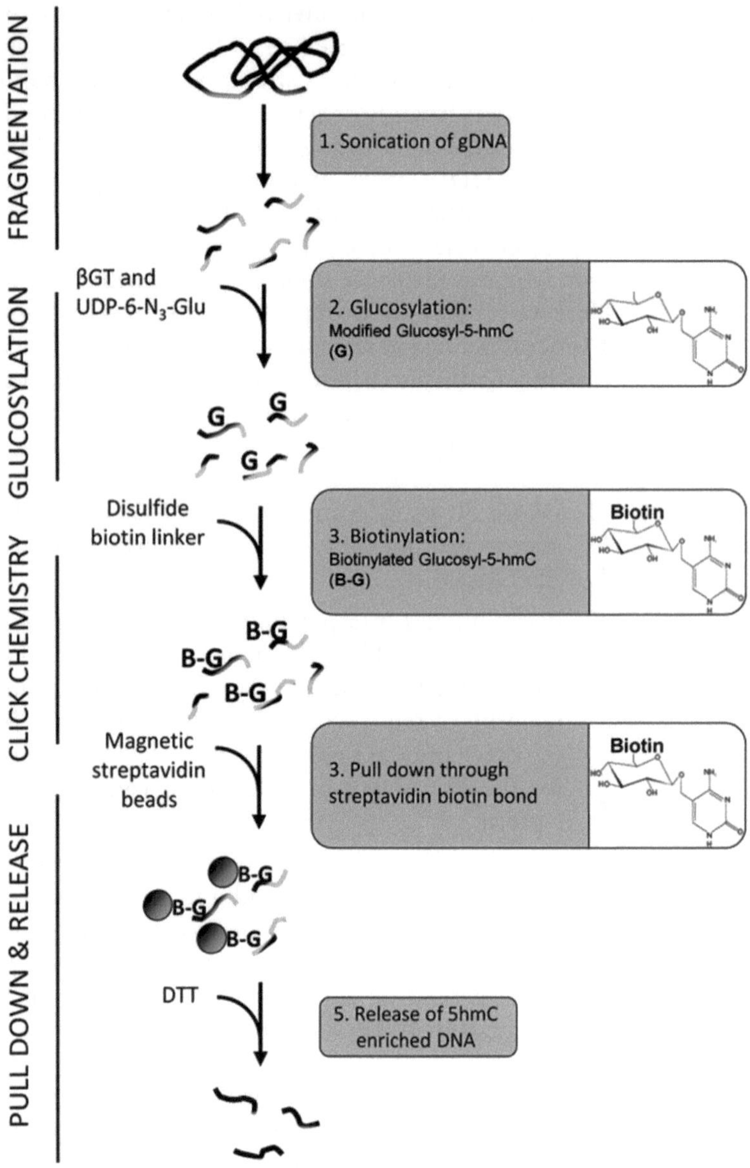

Fig. 2 Overview of the hMeSeal protocol. Genomic DNA (gDNA) is sheared by sonication prior to the conversion of 5hmC marks to modified glucosyl-5-hmC (G) following the addition of β-glucosyltransferase enzyme (βGT) and a modified UDP-6-N_3-Glu moiety. Disulfide biotin linkers are then added which bind to the modified glucosyl-5-hmC DNA fragments (B). Modified 5hmC fragments are then purified through the addition of magnetic streptavidin beads, which bind specifically to the biotin group. Following a series of washes to remove unbound DNA, the 5hmC marked DNA is released through the addition of 100 mM DTT

Component	Final concentration	Sample	Negative control
Sterile H_2O		_ μL	_ μL
10× βGT Reaction Buffer	1×	3 μL	3 μL
30 μg genomic DNA	1 μg/μL	_ μL	_ μL
UDP-6-N_3-Glu	100 μM	1.5 μL	n/a
βGT enzyme (10 U/μL)	50 units	5 μL	5 μL
Total volume		30 μL	30 μL

7. Gently mix by pipetting and incubate reactions for 2 h at 37 °C in a hot block or water bath.
8. Following the incubation, carry out a buffer exchange into H_2O through the use of Micro Bio-Spin columns. Prepare the columns with sterile water four times prior to the buffer exchange; following the manufacturer's protocols (*see* **Note 22**).

3.3.3 Click Chemistry

1. Following buffer exchange, briefly centrifuge the sample and add disulfide-biotin (1 mM) to a final concentration of 150 μM. For example, if the volume of eluted DNA is 40 μL, add 7 μL of 1 mM disulfide-biotin linker.
2. Gently mix several times by pipetting and incubate reactions for 2 h at 37 °C in a hot block or water bath.
3. Following completion of the incubation period, purify DNA fragments using the QIAquick Nucleotide Removal Kit (following manufacturer's instructions) to a final volume of 100 μL in H_2O. Depending on the starting amount of DNA used the number of purification tubes used must be adjusted (maximum of 10 μL per purification tube). For example, if 30 μg of starting DNA was used, the sample must be split into three separate tubes; each eluted into 30 μL H_2O and the combined elution volume taken to 100 μL.

3.3.4 Pull Down of 5hmC Marked DNA

1. Prepare 50 μL of Dynabeads® MyOne™ Streptavidin C1 following the manufacturer's instructions regarding the general washing procedure and immobilization of nucleic acids methods outlined in the manual.
2. Resuspend the washed beads in 100 μL of the 2X Binding and washing (B&W) Buffer.
3. Add the biotinylated sample (100 μL) to the washed beads and incubate at room temperature for 30 min with gentle agitation such as rotation (*see* **Note 23**).

4. To release the 5hmC marked DNA fragments, add 50 μL of 100 mM DTT in H_2O (*see* **Note 24**) and incubate at room temperature with gentle agitation for 2 h.
5. Capture the beads using the magnetic rack and transfer the 5hmC marked DNA in the unbound solution to a Micro Bio-Spin columns, this time without the need to exchange the buffer to H_2O.
6. Remove the residual DTT in the solution by spinning the samples through the Micro Bio-Spin columns, following the manufacturer's protocols (*see* **Note 25**).
7. Purify the eluted DNA from the previous step into 30 μL EB buffer through the use of the MinElute PCR Purification Kit, following the manufacturer's instructions.
8. Measure the DNA concentration using a Qubit flurometer. As the DNA is maintained in a double stranded manner it is ready for deep sequencing library preparation (e.g., End-Repair).

3.4 Downstream Applications

As stated above there are three main avenues of interrogation following the enrichment of 5hmC fragments (1) quantitative PCR (qPCR), (2) microarray analysis and (3) genome-wide sequencing [25, 28]. Here we will outline some important considerations, which must be taken if one is to pursue the particular analytical technique.

3.4.1 Preparing for qPCR Analysis

To prepare for qPCR analysis it is recommended that the enriched DNA be eluted into ~80 μL of TE. Subsequently, 3 μL of the INPUT and IP samples are used as templates in quantitative PCR (thus when used in triplicate there is enough material for at least 8 different primer pairs). qPCR parameters will vary with machine type, assay type, and PCR primer efficiency.

Typically, the enrichment values are displayed as IP copy number/INPUT copy number as determined by qPCR. For human brain and liver tissues, *H19* and *Tex19.1* promoter loci can be used as positive endogenous control loci, whilst the *GAPDH* promoter can be used as a negative endogenous control locus) (*see* **Notes 26** and **27** for more information and for quality control human and mouse primer sequences). Note, these same primer sequences can be used independent of the enrichment method selected.

3.4.2 Preparing for Microarray Analysis

If the enriched material is to be hybridized onto a microarray we suggest eluting the DNA into 20 μL of TE. Remove 10 μL and amplify the DNA using the Whole Genome Amplification Kit (WGA, Sigma) following the manufacturer's instructions. Keep the remaining 10 μL and dilute to 50 μL in TE for qPCR validation (see above and hMeDIP **Note 27** for quality control regions).

Amplified material can then be labeled and hybridized onto microarrays following the selected manufacturer's protocols. For examples of hMeDIP and hMeSeal derived mouse microarrays please refer to the GEO series GSE51577, which contains datasets for both mouse brain and liver 5hmC [25] (http://www.ncbi.nlm.nih.gov/geo).

3.4.3 Preparing for Genome-Wide Sequencing Analysis

As stated earlier the methods for adding sequencing adapters will differ depending on the affinity enrichment method selected (see Subheading 3.1 for more information). As with the microarray preparations it is important to keep a portion of the eluted DNA for quality control qPCR analysis. For an example of human and mouse genome-wide datasets please refer to GEO accession numbers GSM611199 (hMeDIP-seq on mouse ES cells), GSM936817 (hMeDIP-seq on human ES cells) or GSM909322 (hMeSeal-seq on human ES cells) [28, 37–39].

4 Notes

1. Make sure you do NOT use the uridine diphosphoglucose (UDP-Glc) supplied with the enzyme as this glucose moiety has not been modified for use in hMeSeal purification.
2. Reagents available on request from the He lab (chuanhe@uchicago.edu). Alternatively several similar components are available from Active Motif as part of the Hydroxymethyl Collector™ Kit (β-glucosyltransferase enzyme, UDP-Azide-Glucose, Disulfide biotin linker solution, Streptavidin Beads, reaction buffers etc.)
3. It is important to ensure that the genomic DNA used is free from contaminating agents and does not show signs of degradation. We have optimized this protocol for use with 20 μg DNA, which allows for several hMeDIPs from the same sonicated material. If the DNA is limiting, as little as 1 μg can be used; however, sonication conditions will have to be optimized for altered DNA concentrations. Always keep ~250 ng of the sonicated sample to run on an agarose gel to check for fragmentation.
4. It is critical to optimize this step to the machine of use/amount of starting DNA. We typically find that 12 cycles, 30 s on/off on a high setting works well on our Diagenode Bioruptor®. Before starting this protocol it is imperative that sonication conditions are established to obtain DNA fragments ranging from 200 bp to 1000 bp, with the majority fragmented to ~500 bp in size. It is possible to reduce the fragment size to a mean of ~250 bp, however, one must ensure that downstream PCR primers will operate efficiently within this size range.

5. Although each sample should have a similar DNA concentration, due to the viscosity of genomic DNA the actual amount of DNA added to each sample may vary. It is therefore important to determine sample concentrations after sonication.
6. Gel running conditions may vary depending upon the apparatus used. Optimize gel electrophoresis conditions to obtain good resolution in the range of 100–1500 bp.
7. It is not uncommon for the fragmentation profiles to differ slightly between samples. This is fine so long as so long as the general fragment size range is the same for all samples (i.e., the majority of DNA between 200 and 1000 bp peaking at 500 bp). It is not advised to perform hMeDIP on samples exhibiting significantly different fragment size ranges. If required, additional cycles of sonication can be performed on samples requiring further fragmentation. We suggest that samples to be compared by hMeDIP be sonicated simultaneously to reduce inter-sample variation in fragment size.
8. Our knowledge of locus-specific quantities of 5hmC in mammalian genomes is still rather limited, as is our knowledge of how these levels vary between tissues. Thus, in the absence of endogenous positive control regions, the addition of an exogenous 5hmC-DNA positive control is highly recommended. 5hmC control templates are made by amplification of DNA in which canonical dCTP has been replaced by hydroxymethyl dCTP (dhmCTP) in the polymerase chain reaction (PCR). Any short (200–1000 bp), CpG rich DNA template may be used to generate the 5hmC control. Note that all cytosines in the amplified product will be hydroxymethylcytosine, not just those in the context of a CpG. To accurately mimic a hydroxymethylated single-copy locus in the genome, the number of control molecules added should be equal to the number of haploid genomes in the hMeDIP [19]. 5hmC-containing control DNA is also available commercially (New England Biolabs or Zymo Research). We recently reported endogenous regions at the *IGF2/H19* imprinted locus that were relatively enriched for 5hmC in eight normal human tissues and several cell lines [40]. These loci may be used as endogenous positive control regions. However, as the region is imprinted, 5hmC content may vary with developmental stage. The unmethylated CpG islands of housekeeping genes such as *GAPDH* and *ACTB* may be used as negative control regions.
9. Denaturation of the DNA is not required as the 5hmC antibody employed here also recognizes dsDNA, however its sensitivity is significantly higher for ssDNA. Thus, using the same conditions for denaturation for every sample is important as varying amounts of ssDNA to dsDNA between samples may affect antibody binding affinity and consequently enrichment

values. Snap-chilling on wet-ice is important, given the large sample volume; thin-walled tubes should be used if possible.

10. It is important at this stage to determine if one requires double or single stranded DNA for downstream applications. Although it is possible to purify double stranded DNA for genome-wide sequencing without denaturation, we do not recommend this as the final yield will be low and thus hard to prepare libraries from. Instead we suggest that any genome-wide sequencing adapters are ligated on prior to this stage and denaturation and immunoprecipitation carried out on these adapter mediated DNA fragments.
11. Although several 5hmC antibodies are currently available we and others have found the polyclonal 5hmC antibody from ActiveMotif to be the most sensitive and specific of those available [22, 24, 26, 41]. Be aware that the whole serum or purified IgG versions of the antibody have very different titers. We used the IgG purified version in this method. Each batch of antibody should be tested, as different batches of a polyclonal antibody can have very different binding affinities for the epitope.
12. Overnight incubation can be used; however this may increase nonspecific binding.
13. Using Dynabeads Protein A (Thermo Fisher Scientific) will give similar results.
14. Longer incubations times can result in increased nonspecific binding of DNA to the beads.
15. The supernatant, or 'flow-through', contains the DNA not bound by the antibody, and thus depleted for 5hmC. This can be retained to assay regions depleted for 5hmC.
16. The wash steps serve to remove any contaminating DNA not bound to the antibody, and thus reduce the signal to noise ratio in the results.
17. The samples are shaken to prevent the beads from sedimenting at the bottom of the tube and consequently reducing the efficiency of the proteinase reaction. It is recommended to perform this step overnight, however, a 3 h incubation is also sufficient.
18. We use the QIAquick PCR kit (Qiagen) to purify the INPUT sample and IP sample DNA, eluting in 47 μL of TE, to yield a final volume of 45 μL. We found this approach gave best reproducibility between experiments.
19. It has been reported that if using the Hydroxymethyl Collector™ Kit (Active Motif) the reaction can be carried out on as little as 5 ng of starting material [35]. However in the protocol outlined here we typically start by using far more DNA. Ideally we using between 5 and 30 μg of sonicated DNA.

20. A separate 10% input is required for *each* sample being processed.
21. Negative control reactions do not contain any of the modified UDP-Azide-Glucose base and should therefore not result in the purification following streptavidin bead capture.
22. Clean up through a buffer exchange into H_2O is essential as residual βGT Reaction Buffer will interfere with the subsequent biotinylation reaction.
23. Gentle agitation is essential to ensure that the beads to not accumulate at the bottom of the microcentrifuge tube greatly reducing the pull down efficiency.
24. It is important that the DTT is freshly made fresh prior to the release of the DNA fragments. Do not store the DTT for use a second time.
25. Removal of residual DTT in the solution is essential for the subsequent purification steps.
26. Unlike global 5mC levels, which are broadly similar between mammalian tissues, we and others have shown that global 5hmC levels vary markedly between tissue types [40, 42]. For example, normal human breast has a lower global 5hmC content than normal human brain and consequently shows lower enrichment values at all endogenous loci tested. Thus, inter-tissue differences in global 5hmC content should be considered when designing hMeDIP experiments.

    As with 5mC antibodies, 5hmC antibodies exhibit slight sequence-biases in their binding affinities [25]. Given the lack of high quality genome-wide, quantitative data for 5hmC, the extent of such effects, if any, is difficult to determine at present. A meta-analysis of genome-wide 5hmC profiles determined by hMeDIP and other 5hmC profiling techniques suggested that the antibody used here may bind unmethylated CA- and CT-, but this cross-reactivity has yet to be confirmed [25, 30].
27. Quality control qPCR primers (for use following either the hMeDIP or the hMeSeal method):

Target name	Forward primer (5′ → 3′)	Reverse primer (5′ → 3′)
Human H19 (+ve control)	CTCAGCTCTGGGATGATGTG	AGCCCAACATCAAAGACACC
Human Gapdh (−ve control)	CGGCTACTAGCGGTTTTACG	AAGAAGATGCGGCTGACTGT
Mouse Tex19.1 (+ve control)	GGGAGATATGTAAATGAGCTGG	CATCCTTACCTCCCTGACTGAG
Mouse Gapdh (−ve control)	CCACTCCCCTTCCCAGTTTC	CCTATAAATACGGACTGCAGC

Acknowledgments

We thank Dr. Jamie Hackett and Dr. Colm Nestor for their roles in developing the hMeDIP protocol presented here. Also we wish to acknowledge the He lab at the University of Chicago for their development of the hMeSeal protocol outlined in this chapter. The work outlined here is funded in part by the Innovative Medicine Initiative Joint Undertaking (IMI JU) under grant agreement number [115001] (MARCAR project). J.T. is a recipient of IMI-MARCAR funded career development fellowships at the MRC HGU. R.M. is supported by Medical Research Council. Work in R.M.'s lab is supported by IMI-MARCAR, the BBSRC, and the MRC (Ref: MC_PC_U127574433).

References

1. Wu H, Zhang Y (2014) Reversing DNA methylation: mechanisms, genomics, and biological functions. Cell 156:45–68
2. Bird A (2002) DNA methylation patterns and epigenetic memory. Genes Dev 16:6–21
3. Crichton JH, Dunican DS, Maclennan M et al (2014) Defending the genome from the enemy within: mechanisms of retrotransposon suppression in the mouse germline. Cell Mol Life Sci 71:1581–1605
4. Marks H, Stunnenberg HG (2014) Transcription regulation and chromatin structure in the pluripotent ground state. Biochim Biophys Acta 1839:129–137
5. Wyatt GR, Cohen SS (1952) A new pyrimidine base from bacteriophage nucleic acids. Nature 170:1072–1073
6. Kriaucionis S, Heintz N (2009) The nuclear DNA base 5-hydroxymethylcytosine is present in Purkinje neurons and the brain. Science 324:929–930
7. Tahiliani M, Koh KP, Shen Y et al (2009) Conversion of 5-methylcytosine to 5-hydroxymethylcytosine in mammalian DNA by MLL partner TET1. Science 324:930–935
8. Yu Z, Genest PA, ter Riet B et al (2007) The protein that binds to DNA base J in trypanosomatids has features of a thymidine hydroxylase. Nucleic Acids Res 35:2107–2115
9. Koh KP, Rao A (2013) DNA methylation and methylcytosine oxidation in cell fate decisions. Curr Opin Cell Biol 25:152–161
10. Song CX, Yi C, He C (2012) Mapping recently identified nucleotide variants in the genome and transcriptome. Nat Biotechnol 30:1107–1116
11. Laird A, Thomson JP, Harrison DJ et al (2013) 5-hydroxymethylcytosine profiling as an indicator of cellular state. Epigenomics 5:655–669
12. Thomson JP, Moggs JG, Wolf CR et al (2013) Epigenetic profiles as defined signatures of xenobiotic exposure. Mutat Res 764–765:3–9
13. Ito S, Shen L, Dai Q et al (2011) Tet proteins can convert 5-methylcytosine to 5-formylcytosine and 5-carboxylcytosine. Science 333:1300–1303
14. Hackett JA, Sengupta R, Zylicz JJ et al (2012) Germline DNA demethylation dynamics and imprint erasure through 5-hydroxymethylcytosine. Science 339:448–452
15. Shen L, Wu H, Diep D et al (2013) Genome-wide analysis reveals TET- and TDG-dependent 5-methylcytosine oxidation dynamics. Cell 153:692–706
16. Song CX, Szulwach KE, Dai Q et al (2013) Genome-wide profiling of 5-formylcytosine reveals its roles in epigenetic priming. Cell 153:678–691
17. Booth MJ, Branco MR, Ficz G et al (2012) Quantitative sequencing of 5-methylcytosine and 5-hydroxymethylcytosine at single-base resolution. Science 336:934–937
18. Raiber EA, Beraldi D, Ficz G et al (2012) Genome-wide distribution of 5-formylcytosine in embryonic stem cells is associated with transcription and depends on thymine DNA glycosylase. Genome Biol 13: R69
19. Nestor C, Ruzov A, Meehan R et al (2010) Enzymatic approaches and bisulfite sequencing cannot distinguish between 5-methylcytosine and 5-hydroxymethylcytosine in DNA. Biotechniques 48:317–319
20. Jin SG, Kadam S, Pfeifer GP (2010) Examination of the specificity of DNA methylation profiling techniques towards 5-methylcytosine

and 5-hydroxymethylcytosine. Nucleic Acids Res 38:e125
21. Huang Y, Pastor WA, Shen Y et al (2010) The behaviour of 5-hydroxymethylcytosine in bisulfite sequencing. PLoS One 5:e8888
22. Wu H, D'Alessio AC, Ito S et al (2011) Genome-wide analysis of 5-hydroxymethylcytosine distribution reveals its dual function in transcriptional regulation in mouse embryonic stem cells. Genes Dev 25:679–684
23. Pastor WA, Huang Y, Henderson HR et al (2012) The GLIB technique for genome-wide mapping of 5-hydroxymethylcytosine. Nat Protoc 7:1909–1917
24. Ficz G, Branco MR, Seisenberger S et al (2011) Dynamic regulation of 5-hydroxymethylcytosine in mouse ES cells and during differentiation. Nature 473:398–402
25. Thomson JP, Hunter JM, Nestor CE et al (2013) Comparative analysis of affinity-based 5-hydroxymethylation enrichment techniques. Nucleic Acids Res 41:e206
26. Thomson JP, Lempiainen H, Hackett JA et al (2012) Non-genotoxic carcinogen exposure induces defined changes in the 5-hydroxymethylome. Genome Biol 13:R93
27. Serandour AA, Avner S, Oger F et al (2012) Dynamic hydroxymethylation of deoxyribonucleic acid marks differentiation-associated enhancers. Nucleic Acids Res 40:8255–8265
28. Thomson JP, Fawkes A, Ottaviano R et al (2015) DNA immunoprecipitation semiconductor sequencing (DIP-SC-seq) as a rapid method to generate genome wide epigenetic signatures. Sci Rep 5:9778
29. Pastor WA, Pape UJ, Huang Y et al (2011) Genome-wide mapping of 5-hydroxymethylcytosine in embryonic stem cells. Nature 473:394–397
30. Matarese F, Carrillo-de Santa Pau E, Stunnenberg HG (2011) 5-Hydroxymethylcytosine: a new kid on the epigenetic block? Mol Syst Biol 7:562
31. Nestor CE, Ottaviano R, Reinhardt D et al (2015) Rapid reprogramming of epigenetic and transcriptional profiles in mammalian culture systems. Genome Biol 16:11
32. Davis T, Vaisvila R (2011) High sensitivity 5-hydroxymethylcytosine detection in Balb/C brain tissue. J Vis Exp (48):pii:2661
33. Zhao C, Wang H, Zhao B et al (2014) Boronic acid-mediated polymerase chain reaction for gene- and fragment-specific detection of 5-hydroxymethylcytosine. Nucleic Acids Res 42:e81
34. Booth MJ, Ost TW, Beraldi D et al (2013) Oxidative bisulfite sequencing of 5-methylcytosine and 5-hydroxymethylcytosine. Nat Protoc 8:1841–1851
35. Anonymous (2016) Hydroxymethyl collector. http://www.activemotif.com/catalog/775/hydroxymethyl-collector
36. Song CX, Szulwach KE, Fu Y et al (2011) Selective chemical labeling reveals the genome-wide distribution of 5-hydroxymethylcytosine. Nat Biotechnol 29:68–72
37. Wang T, Wu H, Li Y et al (2013) Subtelomeric hotspots of aberrant 5-hydroxymethylcytosine-mediated epigenetic modifications during reprogramming to pluripotency. Nat Cell Biol 15:700–711
38. Williams K, Christensen J, Pedersen MT et al (2011) TET1 and hydroxymethylcytosine in transcription and DNA methylation fidelity. Nature 473:343–348
39. Kim M, Park YK, Kang TW et al (2014) Dynamic changes in DNA methylation and hydroxymethylation when hES cells undergo differentiation toward a neuronal lineage. Hum Mol Genet 23:657–667
40. Nestor CE, Ottaviano R, Reddington J et al (2012) Tissue type is a major modifier of the 5-hydroxymethylcytosine content of human genes. Genome Res 22:467–477
41. Thomson JP, Hunter JM, Lempiainen H et al (2013) Dynamic changes in 5-hydroxymethylation signatures underpin early and late events in drug exposed liver. Nucleic Acids Res 41:5639–5654
42. Szwagierczak A, Bultmann S, Schmidt CS et al (2010) Sensitive enzymatic quantification of 5-hydroxymethylcytosine in genomic DNA. Nucleic Acids Res 38:e181

Index

A

B

Jörg Tost (ed.), *DNA Methylation Protocols*, Methods in Molecular Biology, vol. 1708,
https://doi.org/10.1007/978-1-4939-7481-8, © Springer Science+Business Media, LLC 2018

C

D

K

L

M

N

O

P

Q

T

U

W

X

Z

MIX
Papier aus verantwortungsvollen Quellen
Paper from responsible sources
FSC® C105338

If you have any concerns about our products,
you can contact us on
ProductSafety@springernature.com

In case Publisher is established outside the EU,
the EU authorized representative is:
Springer Nature Customer Service Center GmbH
Europaplatz 3, 69115 Heidelberg, Germany

Printed by Libri Plureos GmbH
in Hamburg, Germany